Advances in Computational Electrodynamics

The Finite-Difference Time-Domain Method

For a complete listing of the *Artech House Antenna Library*,
turn to the back of this book.

Advances in Computational Electrodynamics

The Finite-Difference Time-Domain Method

Allen Taflove

Artech House
Boston • London

Library of Congress Cataloging-in-Publication Data
Taflove, Allen
Advances in computational electrodynamics: the finite-difference time-domain method / Allen Taflove.
p. cm. — (Artech House antenna library)
Includes bibliographical references and index.
ISBN 0-89006-834-8 (alk. paper)
1. Electromagnetism—Mathematics. 2. Electromagnetism—Industrial applications.
3. Finite differences. 4. Time-domain analysis. I. Title. II. Series.
QC760.T3 1998
537.6'0151—dc21 98-19035
CIP

British Library Cataloguing in Publication Data
Taflove, Allen
Advances in computational electrodynamics: the finite-difference time-domain method—(Artech House antenna library)
1. Electrodynamics 2. Time-domain analysis
I. Title
537.6

ISBN 0-89006-834-8

Cover design by Lynda Fishbourne

685 Canton Street
Norwood, MA 02062

International Standard Book Number: 0-89006-834-8
Library of Congress Catalog Card Number: 98-19035

10 9 8 7 6 5 4 3 2 1

To my family, Sylvia, Mike, and Nate,
and to my graduate students over the years.

Contents

Preface

The predecessor of this book, *Computational Electrodynamics: The Finite Difference Time Domain Method*, appeared in summer, 1995. I am gratified with its high level of use by both the university and industrial research communities. It is frequently cited in refereed journal papers as the primary background reference for FDTD methods and applications.

Since the publication of the 1995 book, there has been much progress in this field. When the publisher contacted me regarding a possible updating and expansion of the original work, I responded with a proposal for an entirely new book that would instead complement the original tutorial material. The basic ideas behind this book are:

- Provide a convenient single-source reference on the state of the FDTD art for university graduate students and faculty and professional engineers and scientists involved in electromagnetics technology;
- Build upon the strong base in FDTD theory presented in the 1995 book, while minimizing the duplication of content;
- Select the most important advances made in FDTD theory and applications since 1995;
- Present the absolute latest research results possible, organized into major themes that are mutually reinforcing;
- Engage experts in each topic area to write invited chapters that provide in-depth coverage;
- Encourage the chapter authors to provide ample and well-explained tutorial material in sufficient depth to permit the readers to replicate their results;
- Conduct a very active editorial role so that the final chapter manuscripts have a uniform style, read with a common "voice," and have common symbols and notation. That is, the book should read like a *book*, and not a disjointed collection of material cobbled together.

Chapter 1 is intended to provide an annotated FDTD literature review that is so comprehensive that it alone merits constant usage of the book. In this chapter, Dr. Shlager and Prof. Schneider substantially expand and update the review that they recently published in *IEEE Antennas and Propagation Magazine*. While their new review lays out the literature background for all of the material in the chapters that follow, it is noteworthy that these authors had no knowledge of the identities of the other contributors to this book. This was a strategy that Shlager and Schneider proposed to ensure complete impartiality in their review.

The theme of Chapters 2 and 3 involves new techniques aimed at reducing numerical dispersion and computer resources in FDTD approaches. In Chapter 2, Prof. Turkel reports recent progress in compact fourth-order spatial-differencing methods. The Ty(2,4) and Ty(4,4) algorithms investigated in this chapter permit 8:1 reductions in gridding density in each Cartesian direction relative to the classic Yee algorithm for comparable solution accuracy. Ty(2,4) is especially intriguing since it can be directly and easily overlaid on existing Yee grids, and uses Yee's leapfrog time stepping. Prof. Turkel further demonstrates how abrupt material boundaries can be treated in a manner that preserves the fourth-order accuracy of the basic numerical method. This approach has significant promise for incorporation in many existing FDTD codes.

In Chapter 3, Prof. Katehi, Dr. Harvey, and Prof. Tentzeris review their recent pioneering work in applying wavelet technology to FDTD methods. Their multiresolution time-domain technique permits an unprecedented control of the distance and time scales of the electromagnetic phenomena to be modeled, and yields significant reductions in numerical dispersion and computer burdens. Unique to this method is its ability to automatically adapt to changing needs for local space-time resolution as the various wave species being modeled propagate through the mesh.

The theme of Chapters 4, 5, and 6 involves recent advances in fundamental aspects of grid-based time-domain Maxwell's equations algorithms: (1) generalized meshes; (2) perfectly matched layer absorbing boundary conditions; and (3) periodic structures. In Chapter 4, a multi-disciplinary group of authors having university, national laboratory, and aerospace industry backgrounds (Prof. Gedney, Dr. Roden, Dr. Madsen, Dr. Mohammadian, Dr. Hall, and Dr. Shankar, and Mr. Rowell) combine forces to provide a comprehensive discussion of finite-difference and finite-volume time-domain techniques implemented on generalized meshes, including those that are unstructured. Their discussion extends over 100 pages and is intended to be both broad and deep. The authors' careful and detailed exposition of the theory is supported by computed examples for electromagnetic wave interactions with a variety of three-dimensional structures.

In Chapter 5, Prof. Gedney thoroughly discusses recent advances in PML absorbing boundary conditions, including means to terminate lossy and dispersive media. He details the most recent understanding of the nature and mutual relationships of the Berenger, stretched-coordinate, and uniaxial PML formulations. Comprehensive parametric studies are reported to allow the reader to optimally choose the PML parameters for best efficiency under a variety of practical modeling circumstances.

In Chapter 6, Dr. Maloney and Dr. Kesler review in detail their promising new work in modeling periodic structures within the confines of finite-sized FDTD space lattices. Several potential approaches are examined with regard to numerical stability, dispersion, accuracy, and efficiency. Practical examples involving a microwave photonic bandgap structure, a frequency selective surface, and an antenna array are discussed. This work opens new possibilities for FDTD modeling of an entire class of structures that previously could be investigated only by applying frequency-domain mode-matching or moment-method techniques.

The theme of Chapters 7 through 11 involves recent progress in FDTD modeling of specific, high-interest engineering applications in microwaves, millimeter waves, and optics technology. In Chapter 7, Dr. Maloney and Prof. Smith provide a detailed tutorial discussion of the key aspects of FDTD modeling of antennas. They then report a detailed benchmark FDTD study of a standard-gain microwave horn antenna used for transmitting and receiving, and demonstrate that superb correspondence can be obtained with high-quality measurements when the FDTD modeling is done with care and proper attention to details. The chapter concludes with examples of emerging FDTD modeling applications involving microwave photonic bandgap structures, ground-penetrating radar, and antenna-radome interaction.

In Chapter 8, Dr. Houshmand, Prof. Itoh, and Prof. Piket-May show how FDTD methods can be efficiently and systematically applied to model high-speed electronic circuits of all types, whether microwave amplifiers or digital circuits. On the microwave side, a key theoretical advance involves the development of simple (yet robust) Norton's and Thevenin's equivalent circuits "looking into" the FDTD space lattice. These equivalent-circuit connections into the FDTD solver permits its direct interfacing with the popular SPICE circuit-analysis software or with special-purpose state-variable circuit algorithms. On the digital-circuit side, specific FDTD techniques are presented to calculate the effective inductance of complicated, multiplane, power-distribution systems for multichip modules. In addition, an engineering case history is reviewed involving the use of FDTD in designing an ultrahigh-speed, 2.5-GHz clock-distribution scheme for a multichip module employing low-impedance Josephson-junction logic. Design trades-off are discussed.

In Chapter 9, Prof. El-Ghazaly reviews his pioneering work in combining FDTD electromagnetic wave modeling with the physics of charge transport within millimeter-wave MESFETs. Such a first-principles model is especially useful at frequencies in the many tens or even hundreds of gigahertz, where the lumped-circuit transistor models used in Chapter 8 (to link with Norton's or Thevenin's equivalent circuits of the FDTD grid) may lose accuracy. In this elevated-frequency regime, the simultaneous motion of semiconductor charges and electromagnetic fields across the transistor leads to mutual charge-field interactions that require a self-consistent model. After discussing the algorithmic aspects of the combined charge and electromagnetic wave simulator, the author goes into considerable detail regarding examples of full-physics MESFET design. A global design process is detailed wherein the MESFET physics is modeled while accounting for the input and output matching networks using a convolutional approach.

In Chapter 10, Prof. Hagness discusses recent applications of FDTD modeling to micron-scale optical resonators useful for low-threshold, high-speed lasers and practical optical signal processors. A variety of such structures are reviewed, including vertical-cavity surface-emitting lasers, photonic-bandgap structures, and strongly confined, waveguide-coupled, microrings and microdisks. The microrings and microdisks are particularly challenging for FDTD simulations since: the waveguide coupling is via the evanescent field (causing the sensitivity to gap dimensions to be high); resonator Q factors range up to 10,000; multimoding in the disk occurs due to various radial whispering-gallery modes; and resonance splitting can occur due to nuances of surface roughness. The key aspects of these modeling challenges are thoroughly discussed, and engineering tradeoff studies of important design parameters are conducted.

In Chapter 11, Prof. Gandhi reviews his pioneering work in using FDTD to evaluate the dosimetry and engineering aspects of human exposure to a wide variety of electromagnetic fields including those generated by 60 Hz power lines, VHF plane waves, ultrawideband electromagnetic pulses, and handheld wireless personal communications devices. Rigorous analytical, numerical, and experimental validations of the FDTD modeling approach are provided for canonical antennas adjacent to brain-equivalent phantoms. Experimental validations are also provided for FDTD predictions of far-field radiation patterns and near-field absorption within skull and brain tissues when actual cellular telephones are positioned adjacent to the human head. Here, highly detailed anatomical models derived from MRI scans are used in the FDTD investigations.

The twelfth and final chapter (by Prof. Chew) has its own theme: to provide an excellent, highly detailed tutorial discussion of imaging and inverse problems in electromagnetics. In addition to the widely used frequency-domain forward-scattering techniques used for this purpose, this chapter shows how FDTD modeling can be advantageously applied. Numerous examples of two- and three-dimensional reconstructions are provided to indicate both the potential and limitations of existing inverse-scattering methods and algorithms.

In assembling this book, I gratefully acknowledge the chapter contributors. Each one devoted considerable time wrenched from their busy schedules to create detailed and expert scholarly works, always under deadline pressure. Their biographical sketches appear at the end of the book. Also acknowledged are the helpful contributions of my graduate students, especially Milica Popovic, who assisted mightily in dealing with a variety of electronic downloads. I bid a fond farewell to Susan Hagness who now goes on to have graduate students of her own as a professor.

Finally, I acknowledge my wife, Sylvia, and younger son, Nate, now completing his sophomore year at New Trier High School. Quoting in part from the preface of the 1995 book, they "somehow were able to keep their composure while sharing a home with a very driven person..." In this case, it meant having to deal with a husband/dad who slaved on the computer for months on end until the wee hours every night, and generally wasn't much fun at all. (My older son, Mike, missed his Dad's isolation and grumpiness. He has been away from home as a freshman at the University of Iowa during most of the editing of this book. Lucky!)

The FDTD story is continuing. As indicated by the scope of the experts' contributions in this book, the electromagnetics engineering community is indeed rapidly moving "to develop detailed FDTD models of microchips, microlasers, and microcells, and bring the power of Maxwell's equations to bear upon society's needs in ultrahigh-speed communications technology." More so than ever before, I believe that electromagnetics engineers have a special responsibility to utilize their technical knowledge to enable people to freely communicate. We can best understand each other when we talk to each other. This is what our human society is all about.

Allen Taflove
Wilmette, Illinois
June 1998

Chapter 1

A Survey of the Finite-Difference Time-Domain Literature

Kurt L. Shlager and John B. Schneider

1.1 INTRODUCTION

The finite-difference time-domain (FDTD) method is arguably the most popular numerical method for the solution of problems in electromagnetics. Although the FDTD method has existed for over 30 years, its popularity continues to grow as computing costs continue to decline. In addition, extensions and enhancements to the method are continually being published, which further broaden its appeal.

Because of the tremendous amount of FDTD-related research activity, tracking the literature in this area can be a daunting task. In this chapter, we present a selective survey of 496 FDTD publications. This survey presents some of the significant works that made the FDTD method so popular, and tracks its development up to the present-day state-of-the-art in several areas. In addition, we provide at the end of the chapter (without discussion) an additional 68 citations in a selected bibliography of novel and emerging FDTD areas.

The FDTD method, as first proposed by Yee in 1966 [1], is a simple and elegant way to discretize the differential form of Maxwell's equations. Yee used an electric-field (E) grid offset both spatially and temporally from a magnetic-field (H) grid to obtain update equations that yield the present fields throughout the computational domain in terms of the past fields. The update equations are used in a leapfrog scheme to incrementally march the E and H fields forward in time. Despite the simplicity and elegance of Yee's algorithm, it did not receive much interest immediately after its publication. One could attribute the lack of attention to the high computational cost of the day as well as to some of the limitations inherent in the original publication (such as the inability to model an "open" problem for any significant period of time). However, as the shortcomings of the original FDTD implementation were alleviated and the cost of computing fell, interest in FDTD soared. Based on the information we have gathered, by 1996 the yearly number of publications related to FDTD was 430, a sixty-fold increase relative to the 1985 number of seven publications. The history of the remarkable growth of FDTD publications since Yee's 1966 paper is shown in Fig. 1.1.

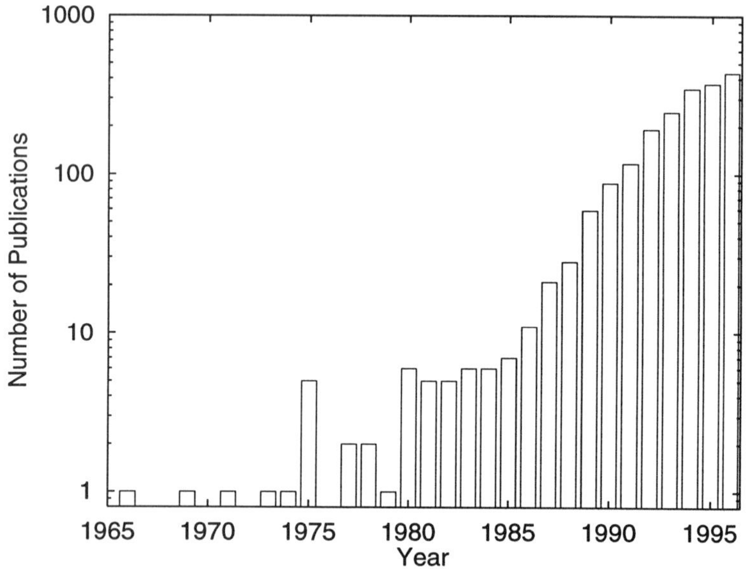

Fig. 1.1 Yearly number of publications related to FDTD per year. The last year for which reliable data are available is 1996.

Since the body of FDTD literature is so large (we have cataloged over 2300 FDTD-related publications as of December 1997), the survey presented here is necessarily incomplete. Therefore, our goal is to highlight some of the more successful extensions to and applications of the FDTD method.

1.2 FUNDAMENTAL ISSUES

The original Yee FDTD algorithm is second-order accurate in both space and time. Numerical dispersion and grid anisotropy errors are kept small by having a sufficient number of grid spaces per wavelength. Taflove was the first to present the correct stability criterion for the original orthogonal-grid Yee algorithm [2]. Taflove was also among the first to rigorously analyze the numerical dispersion and anisotropy errors [3].

The FDTD method is used to calculate either scattered fields or total fields. When calculating only scattered fields, the source of the fields is a function of the known incident field and the difference in material parameters from those of the background medium [4, 5]. When using total fields, the total fields are often calculated only over an interior subsection of the computational domain [6 – 8], while scattered fields are calculated in the remaining (exterior) portion of the grid. By using scattered fields in this way, the field incident on the absorbing boundary condition (Section 1.3) is more readily absorbed. To obtain this division of the computational domain into scattered-field and total-field regions, the incident field is specified over the boundary between these two regions. Holland and Williams presented a comparison of scattered-field formulations (wherein only the scattered fields are computed throughout the computational domain) and total-field formulations (wherein total fields are computed in a subdomain that contains the object under study) [8]. They determined that, due to numerical dispersion, the total-field FDTD approach is superior to the scattered-field approach. Further, the scattered-field approach has the disadvantage that it does not easily accommodate nonlinear media. However, for problems involving only linear media and not containing shielded cavities, the scattered-field formulation is the more desired approach [9]. The relative merits of the total-field and scattered-field formulations were also explored by Fang [10].

1.3 ABSORBING BOUNDARY CONDITIONS

To model open-region problems, an absorbing boundary condition (ABC) is often used to truncate the computational domain, since the tangential components of the electric field along the outer boundary of the computational domain cannot be updated using the Yee algorithm. The quest for an ABC that produces negligible reflections has been, and continues to be, one of the most active areas of FDTD research. Most of the popular ABCs are grouped into two categories: ABCs derived from differential equations or ABCs employing a material absorber. The perfectly matched layer (PML) technique originated by Berenger in 1994 can be considered to employ a material absorber. However, its formulation is a radical departure from that of all previously proposed grid termination techniques. In application after application, the PML approach has been shown to provide significantly better accuracy than most other ABCs. Thus, the PML technique has become the "gold standard" to which other techniques are compared. Because of the quality afforded by the PML scheme, much of the sizable previous literature on ABCs was rendered obsolete. Hence, the following section covering non-PML ABCs is somewhat terse relative to the number of publications on this topic.

1.3.1 Differential-Equation-Based and Other Nonmaterial ABCs

Differential-equation-based ABCs, or "traditional" ABCs, are obtained by factoring the wave equation and allowing a solution that permits only outgoing waves.

Another approach to the construction of these ABCs uses the discretized form of a one-way boundary operator, or the product of two or more one-way boundary operators. Traditional ABCs are local in the sense that the field at any point on the boundary of the computational domain only depends on the fields local to that point. Traditional ABCs were reported by, among others, Merewether [11], Engquist and Majda [12], Lindman [13], and Mur [14]. These early techniques were improved in the mid-1980s by formulations presented by Keys [15], Liao et al. [16], and Higdon [17, 18].

In theory, one can construct a traditional ABC of arbitrary order to obtain arbitrarily high accuracy. In practice, however, stability has limited the construction of higher-order ABCs. To help circumvent this shortcoming, Moghaddam and Chew [19] presented a method to stabilize Liao's ABC, although the method was found to be machine-dependent. In his 1987 paper, Higdon introduced a dissipative term to help control the stability of his ABC [18]. The stability of the Higdon ABC was further explored by Fang [20, 21], among others. Having obtained a stable fifth-order Higdon ABC, Wu and Fang [22] used this ABC to terminate a waveguide and showed that this ABC gives similar performance to that of a 16-layer PML (although they concluded the PML technique is more robust in general).

It has often been reported that, in practice, the performance of high-order ABCs does not live up to theory. Prescott and Shuley [23] attributed this discrepancy to the fact that previous analyses used the continuous wavenumber rather than the discrete numeric wavenumber, and presented an appropriately extended analysis.

In situations where the speed of propagation is a function of the frequency, such as the termination of a waveguide or a microstrip, traditional ABCs (which assume a constant speed of propagation) do not work well. Several researchers developed dispersive boundary conditions to overcome this deficiency, including Moglie [24], Railton et al. [25], Fang [21], Werthen et al. [26, 27], Mrozowski et al. [28], and Alimenti et al. [29]. Both Mrozowski et al. and Alimenti et al. provided a comparison of results using their techniques to results obtained using the PML technique.

In 1992, Mei and Fang [30] proposed a "superabsorption" technique applicable to many differential-equation-based ABCs to improve their performance. In this technique, the ABC is applied both to the nodes along the termination of the computational domain and to interior nodes adjacent to the termination. From these two applications of the ABC, an error term is estimated and used to correct the values at the termination of the computational domain. More recently, Ramahi [31] presented a grid-termination scheme that uses two complementary boundary operators whose reflection coefficients are identically opposite. Thus, the errors associated with reflections from the ABCs are canceled either by performing two separate simulations, as originally proposed, or by employing the two complementary operators concurrently in a single simulation [32]. Schneider and Ramahi [33] have shown that the performance of this scheme is comparable with, or superior to, that of the PML technique.

Several researchers have presented comparisons of the accuracy of various ABCs. With the introduction of the PML technique, many of these papers are now only of minor interest. However, recent papers by Andrew et al. [34] and Kantartzis and Tsiboukis [35] do present a comparison of traditional ABCs with the PML technique.

Finally, we note that exact grid-termination techniques have been presented that yield accurate results, but since they are nonlocal, they are computationally expensive. Such approaches have been investigated by Ziolkowski et al. [36], Olivier [37], De Moerloose and De Zutter [38], and Tromp and Olivier [39].

1.3.2 Material ABCs

Material-based ABCs are realized by surrounding the computational domain with a lossy material that dampens the outgoing fields. The idea of using an absorbing-material boundary condition has existed for some time [8]. However, early material ABCs did not provide a sufficiently low level of boundary reflection because the characteristic impedance of the material boundary was matched to the impedance of free space only at normal incidence.

The use of material ABCs was significantly advanced in 1994 with the work of Berenger [40], who introduced the PML ABC. Berenger postulated a "split-field" extension of Maxwell's equations that ensures that the electromagnetic-wave phase velocity in the PML region is the same as in free space for all frequencies and for all angles of incidence. Large amounts of loss can be present in the PML region without affecting the phase velocity. Thus, a PML region that surrounds the computational domain can be terminated with a perfect electric conductor (PEC), and yet the reflections from this PEC that reenter the interior computational space are often insignificant. Berenger's original paper introduced the PML technique in two dimensions, but it was extended to three dimension by, among others, Chew and Weedon [41], Katz et al. [42], and Berenger [43].

The field-splitting method originally used to construct PML regions is considered merely a mathematical convenience that has no correspondence to the physical world. However, Chew and Weedon [41] obtained a perfectly matched layer by introducing complex coordinate stretching into Maxwell's equations. In their development of an ABC for finite-element problems, Sacks et al. [44] formulated a PML as an anisotropic active material. This work led to the development of PML schemes that do not require the splitting of fields. Researchers presenting these types of formulations include Veihl and Mittra [45], Gedney [46], Zhao and Cangellaris [47], and Sullivan [48]. Alternatively, Ziolkowski [49] provided an interpretation of the Berenger PML medium showing that, in contrast to previous work, it is a passive lossy electric and magnetic medium with particular conductivity and Debye dispersion characteristics. Although the original split-field Berenger PML medium is physically unrealizable, Ziolkowski developed a physically realizable medium that has characteristics similar to that of Berenger's PML [49, 50].

A perfectly matched layer is actually only "perfect" in the continuous world. When implemented in discrete form, it is not perfect. These numerical errors have been explored by several researchers, including Wittwer and Ziolkowski [51], Fang and Wu [52], Chew and Jin [53], Prescott and Shuley [54], and Berenger [55, 56].

Berenger's original paper employed exponential differencing for the spatial derivatives. Several researchers, including Wu and Fang [57] and Zhao and Cangellaris [47], questioned the need for exponential differencing. Petropoulos provided a rigorous comparison of exponential differencing versus central differencing [58]. The stability of the PML technique has been examined by, among others, Nehrbass et al. [59], Naishadham [60], and Abarbanel and Gottlieb [61].

The PML technique has been modified in a number of ways so that it can be applied to many grids with planar interfaces beyond the standard Yee grid. For example, Navarro et al. [62], Wu et al. [63], and Roden and Gedney [64] used the PML technique in a nonorthogonal grid. Maloney et al. [65] presented a PML scheme for cylindrical geometries, while Teixeira and Chew [66] presented a scheme in both cylindrical and spherical coordinates. Roberts and Joubert [67] used a PML technique together with high-order differencing. Zhao et al. [68] and Perez et al. [69] presented a PML formulation that permits the truncation of spaces filled with anisotropic dielectrics, while Fang and Wu [70], among others, demonstrated one suitable for truncating regions filled with lossy media. Finally, Gedney [71] developed a formulation that can match a second-order Lorentz dispersive material.

The original implementation of the perfectly matched layer only absorbs propagating waves. Modified versions of the PML technique that can also absorb evanescent energy have been presented by Fang and Wu [70], Gedney [71], Chen et al. [72], and Gribbons et al. [73].

1.4 GRIDDING

1.4.1 Orthogonal Grids

As originally formulated, the Cartesian grids used in the FDTD method dictate that a smoothly varying surface must be approximated by one that is "staircased." This approximation may lead to significant errors in certain problems. Such errors have been analyzed rigorously by Cangellaris and Wright [74] for the staircased representation of a planar surface tilted 45° to the grid, while Holland [75] analyzed errors associated with a slightly tilted planar scatterer. More recently, Schneider and Shlager [76] analyzed staircasing errors using narrow-angled horns as a test bed, and developed a guideline for the discretization necessary to avoid general staircasing errors.

In addition to problems caused by staircasing, if an object under consideration has small-scale structure such as a narrow slot, the original method would have to use an excessively fine grid to accurately model the associated fields. Several solutions, as discussed below, have been proposed to address these shortcomings.

If the object under consideration is more naturally described in an orthogonal coordinate system other than Cartesian, it is rather simple to develop update equations appropriate for that coordinate system, as was done by Merewether [11] and by Holland [77]. Alternatively, a grid with varying spatial increments along the different coordinate directions can be used. In general, for a Cartesian grid, this results in rectangular cells and permits finer discretization in areas of rapid field fluctuation. Kunz and Lee [78, 79] used this approach to calculate the external response of an aircraft to an electromagnetic pulse. Monk and Suli showed that these grids preserve the second-order accuracy of the original constant cell-size algorithm [80, 81]. Svigelj and Mittra [82] also presented a theoretical and numerical analysis of the dispersion error of these grids.

1.4.2 Subgridding

Subdomains are gridded more finely than the rest of the problem space. These "subgridding" techniques have been put forward by a number of researchers, including Yee [83], Kasher and Yee [84], Kim and Hoefer [85], and Zivanovic et al. [86].

The key issue in these techniques is the coupling of the fine and coarse grids. (Away from the boundary between these two grids, the standard FDTD update equations are used.) Yee, Kasher and Yee, and Kim and Hoefer used linear interpolation in both time and space to advance the fields on the boundary between the fine and coarse grids. Monk [87] presented an error analysis of this subgridding scheme as well as an analysis of a simple subgridding scheme based on piecewise-constant interpolation. The technique proposed by Zivanovic et al. employs a discretized form of the wave equation to obtain the fields on the boundary between the grids. Prescott and Shuley [88] presented a modification of the technique of Zivanovic et al. that improved its efficiency. In 1997, White et al. [89] provided a detailed description of the application of this technique to three-dimensional problems. That same year, Okoniewski et al. [90] presented a subgridding technique that uses spatial and temporal interpolation as well as temporal extrapolation to obtain the boundary fields. The fine grid is offset from the coarse grid, and the fine-grid step size must be half that of the coarse grid. Recursive grid refinement is used to obtain even finer discretizations.

Chevalier et al. [91] presented a subgridding technique that permits nonmagnetic or perfectly conducting materials to traverse the boundary between the coarse and fine grids. In this scheme, the coarse-grid step size is assumed to be an odd multiple of the fine-grid step size. White and Iskander [92] later presented a scheme that also permits material traverse of the boundary, but their approach does not require that the coarse-grid step size be an odd multiple of the fine-grid step size.

An alternative subgridding scheme was proposed by Kunz and Simpson [93]. Their formulation requires two runs. The first is done for a coarse grid that spans the entire computational domain, while the second is done for the finely gridded subdomain and takes its boundary values from the stored values calculated during the coarse simulation. More recently, a similar two-pass approach was used by Holland [94] to compute the fields behind a cavity-backed aperture.

1.4.3 Subcellular Techniques

In 1980 and 1981, Holland and Simpson [95, 96] presented the implementation and analysis of an FDTD thin-strut formalism. This permits the modeling of arbitrarily thin wires in the FDTD lattice without the need to reduce the size of the space cells to the thickness of the wire. Results were obtained for linear dipole and loop antennas. Soon after, Gilbert and Holland [97] presented an FDTD thin-slot formulation (TSF) that permits the modeling of conducting plates having narrow slots. Their approach assumes that the slot is narrow compared with the cells in which it lies, and that the slot is long in the direction of the slot. The approximation involves increasing the effective permittivity of the cells containing the slot, while correspondingly reducing the effective permeability of these cells.

Following the work of Yee [98], Umashankar et al. [99] and Taflove et al. [100] derived field-update equations suitable for modeling subcellular structures such as wires, narrow slots, and lapped joints in conducting screens. These equations were obtained from the integral form of Faraday's law rather than the differential form, and result in modified updates only for space cells where the subcellular structure is present.

Demarest [101] presented an FDTD technique utilizing Babinet's principle to model narrow apertures in conducting scatterers. Subsequently, Turner and Bacon [102] evaluated the accuracy of Gilbert and Holland's thin-slot formalism. Realizing the inherent limitation of the assumptions in Gilbert and Holland's approach, Riley and Turner [103] developed a hybrid thin-slot algorithm (HTSA) based on an integro-differential equation which accurately models slots that are narrow in both width and depth with respect to an FDTD cell. They soon extended their work to include lossy slots characterized by a surface-impedance boundary condition [104]. Later, Wang [105] presented an enhanced thin-slot formalism that includes the singularities of the field variation near the slot. In this work, Wang showed that this approach is more accurate than the original Gilbert and Holland formulation. Wang [106] also presented a formalism for small-hole coupling based on Bethe's small-hole coupling theory. A similar approach was taken by Oates and Shin [107], who developed a model for scattering by a small aperture in a conducting screen. Here, the aperture is replaced with equivalent, oppositely directed electric and magnetic dipoles on either side of the short-circuited screen.

Recent work on subcellular structures has involved comparing the accuracy of several of the above approaches. Ma et al. [108] compared the Gilbert and Holland TSF approach with the Riley and Turner HTSA approach for thin slots in planes and loaded cavities. They concluded that the HTSA approach is more susceptible to late-time instabilities. Li et al. [109] compared results obtained using the TSF approach of Gilbert and Holland with those obtained using the method of moments. They showed that this method, although originally formulated for slots in infinite planes, also works well for slots near corners.

It is possible for the FDTD grid to be aligned with the material boundaries in a problem and still not yield acceptable results. This happens when the space cells are large relative to the local variation of the fields; for example, when singular fields such as those found at microstrip discontinuities are present. To improve the fidelity of the FDTD method in these situations, Shorthouse and Railton [110] proposed a technique that incorporates correction factors based on local static-field behavior in the vicinity of a PEC into the update equations. Shortly thereafter, Railton et al. [111] demonstrated the application of this technique to narrow microstrips (i.e., microstrips that have a width less than a unit cell). Railton also presented extensions to handle curved metallic boundaries [112, 113], and therefore, this method could also be considered a conformal technique. Unfortunately, these schemes do not always produce numerically stable results, although a method to ensure a stable solution was presented by Craddock and Railton [114]. More recently, Esselle et al. [115] presented an extension to handle situations where sharp metal edges are not aligned with the grid.

Finally, Anderson et al. [116] presented a simple technique to improve the accuracy of the staircased FDTD method when a flat metal plane or metallic corner is not aligned with the grid. Their technique also incorporates a correction factor based on the local electrostatic behavior of the fields.

1.4.4 Conformal Grids

In 1983, Holland published the first FDTD algorithm for generalized nonorthogonal coordinates [117]. In this approach, the fields are expressed in terms of their covariant components (flow along a coordinate direction) and contravariant components (flow through a constant-coordinate surface), and an integral formulation is used to obtain the update equations. In an orthogonal coordinate system, the covariant and contravariant components are collinear. However, in a nonorthogonal system, these components are not collinear and auxiliary equations must be obtained to express one form in terms of the other.

Holland's formulation was revisited by Fusco [118, 119], Lee et al. [120, 121], and Harms et al. [122, 123]. Fusco applied the algorithm to cylindrical scatterers, while Lee et al. applied it to waveguide problems and derived the stability equations. Harms et al. computed the resonances of dielectric-filled cavities and compared results to those obtained using the staircased FDTD method. Harms et al. [124] also used the nonorthogonal FDTD method to study vias. Kashiwa et al. [125] used nonorthogonal grids in their study of microstrip antennas, while Kitamura et al. [126] used nonorthogonal grids to study microstrip bends and the crosstalk between microstrip lines that do not cross orthogonally. Navarro et al. [62, 127] and Wu et al. [63] have studied fundamental aspects of nonorthogonal grids, including numerical dispersion and nonzero divergence, and adapted the Berenger PML ABC for use in such grids. Prior to this work, Ray [128] studied the numerical dispersion and stability of nonorthogonal two-dimensional grids having uniform skew. Recently, Janaswamy and Liu [129] presented a similar scheme but used collocated fields and forward and backward finite-differences for the spatial derivatives. Results were presented only for two-dimensional problems, although the pertinent equations for three-dimensional structures were also given.

Recently, Yang and Railton presented a two-dimensional method to couple a nonorthogonal grid with the standard Cartesian FDTD grid. In their scheme, similar to the contour-path approach described below, the nonorthogonal grid is used only in the immediate vicinity of material interfaces and can be used for either PEC structures [130] or dielectrics [131]. Finally, note that Holland [75] cautioned that his original nonorthogonal grid scheme may be of limited utility compared to other more versatile conformal techniques.

Mei et al. [132] presented a conformal technique that employs second-order polynomials and six grid points (ten grid points for three dimensions). Subsequently, Gwarek [133] presented a scheme to incorporate geometric information based on the assumption of a quasistatic field distribution into the update equations for two-dimensional simulations. This work was extended to three dimensions by Celuch-Marcysiak and Gwarek [134]. In the same paper, the authors presented an approach derived by assuming linear variation of continuous field components between nodes that maintains second-order accuracy at material interfaces. Celuch-Marcysiak and Gwarek [135] also introduced a generalized transmission line matrix (TLM) scheme that permits curved boundaries. They showed the formal equivalence of this scheme to an FDTD approach employing a modified grid.

In order to take advantage of the work that was done using standard Cartesian grids and yet gain the benefit of a conformal approach, Yee et al. [136] developed an overlapping-grid technique for FDTD that uses a conformal grid in the neighborhood of material boundaries and a Cartesian grid elsewhere. Unlike this overlapping-grid scheme, which uses spatial interpolation to connect the grids, Riley and Turner presented a method wherein the standard Cartesian grid and the conformal grid are directly connected [137, 138].

Liao et al. [139] presented a technique where a transformation is first used to map a conformal grid into a Cartesian grid, and then standard, second-order centered differences are used to advance the fields. This method, which borrows techniques developed for hydrodynamics, has only been demonstrated in two-dimensions using a first-order absorbing boundary condition.

For some PEC objects, a simple quasi-conformal scheme is realized by constructing the grid so that the PEC boundary is either aligned with the edge of a cell or passes through the diagonal of a cell. A slightly modified update equation is then used for cells that have the PEC along the diagonal. This scheme was described by, among others, Mezzanotte et al. [140]. (See also Weiland [141] and Mezzanotte [142] for alternative descriptions of this approach.)

Jurgens et al. [143] and Jurgens and Taflove [144] extended the technique previously used to model subcellular structures so that it is suitable for modeling PEC and dielectric curved surfaces. This "contour path" (CP) FDTD method uses a Cartesian grid everywhere except in the vicinity of the material boundaries where a distorted grid is used to conform to the boundary. The original CP-FDTD technique employs a nonreciprocal borrowing of neighboring fields that may lead to instability. A modified version of the CP-FDTD method that uses reciprocal borrowing, and hence is stable, was presented by Railton et al. [145] for two-dimensional PEC problems and by Railton and Craddock [146] for three-dimensional PEC problems. The CP-FDTD method is based on the standard integral forms of Ampere's law and Faraday's law (i.e., line and surface integrals). In 1996, Garcia et al. [147] presented a similar conformal technique for PEC structures, but based it upon the volume-integral form of Faraday's and Ampere's laws. Their method, like the original CP-FDTD method, also used a nonreciprocal borrowing of fields for certain cells. Anderson et al. [148] presented a simple algorithm for incorporating geometric information about PEC surfaces into the FDTD update equations. Their algorithm results in a solution similar to that provided by the CP-FDTD method, but an additional factor based on electrostatics is used to modify the borrowed fields.

An alternative locally conformal FDTD scheme for PEC structures was presented by Fang and Ren [149]. Similar to the CP-FDTD method, the standard FDTD update equations are used away from the PEC boundary. For cells at the boundary, local corrections are made to the update equations to account for the presence of the metal. However, unlike the CP-FDTD method, this method does not employ nonreciprocal borrowing of fields.

A new conformal technique for PEC surfaces was recently put forward by Dey and Mittra [150] and Dey et al. [151]. It does not involve any borrowing and is arguably the simplest conformal technique to implement. All electric field update equations are the same as those used in the conventional FDTD scheme, while the magnetic field update equations for cells adjacent to a metallic boundary are only slightly modified to include geometric factors. The "cost" of this simple scheme is a reduction of the Courant number in order to obtain a stable solution. The Dey / Mittra method, together with many of the aforementioned methods, have been studied in the context of narrow-angled transverse electromagnetic (TEM) horns by Schneider and Shlager [76].

Conformal modeling continues to be an area of active research and several techniques have been proposed. Many of these conformal methods, such as the finite-volume time-domain (FVTD) method [152, 153], the discrete surface integral (DSI) method [154], and the generalized Yee algorithm [155], are related to the original FDTD method. However, in the interest of brevity, we do not discuss these methods here. Instead, some of the papers that employ these conformal techniques are cited in later sections as they pertain to modeling of particular applications.

1.5 MATERIAL MODELING

FDTD modeling of "complicated" materials has recently gained much attention. One of the major advantages of the FDTD method relative to other numerical techniques is the ability to obtain wideband results using transient excitation. To obtain accurate results over a broad spectrum, it is often necessary to include the frequency-dependent properties of the material. That is, it may not be physically meaningful to treat the permittivity, conductivity, or permeability as constants over the entire spectrum. Several techniques have been proposed to account for this frequency dependence. Additionally, the use of a surface-impedance boundary condition or a thin-material-sheet model was shown to provide significant computational savings over a full FDTD model. Further, the FDTD algorithm has been extended to account for materials that are anisotropic and nonlinear.

1.5.1 Frequency Dispersive Material

In 1990, Luebbers et al. published the first frequency-dependent FDTD formulation [156] by using a recursive-convolution (RC) scheme to model Debye media. They did this by relating the electric flux density to the electric field through a convolution integral, discretizing the integral as a running sum, and assuming that the susceptibility function is described by a decaying exponential. Independently, in 1991 Bui et al. [157] also developed an RC FDTD model for modeling Debye media.

Subsequently, Luebbers et al. [158] modified their RC formulation to study wave propagation in a Drude material, and generalized the approach to treat *M*th-order dispersive media [159]. Their approach requires storage of $M/2$ complex variables per electric field location, while the original scheme for a first-order Debye medium requires storage of M real numbers. Hunsberger et al. [160] extended the RC FDTD approach to model an anisotropic magnetoactive plasma, while Luebbers et al. [161] formulated a scattered-field frequency-dispersive method and applied it to three-dimensional spheres. In 1994, Pontalti et al. [162] independently proposed an RC approach that could model Debye media of *M*th order. Melon et al. [163] extended the frequency-dispersive RC FDTD approach to ferrites, while Hawkins and Kallman [164] simplified the RC approach for Debye media by using a causality argument in their formulation.

In 1996, Kelley et al. [165] improved the accuracy of the RC approach by using piecewise-linear integration. Independently, Siushansian and LoVetri [166] achieved the same improved accuracy but with a discrete trapezoidal integration. Their approach requires less storage than the method of Kelley et al.

Concurrent with the development of the frequency-dispersive RC FDTD technique was the formulation of an alternate approach termed the auxiliary differential equation (ADE) method. The first papers utilizing this approach were by Kashiwa and coworkers [167–169] in 1990 for Debye media, Lorentz media, and media obeying the Cole-Cole circular-arc law, respectively. While this research was progressing, Joseph et al. [170] independently developed a similar ADE model for Debye and Lorentz media. Goorjian and Taflove [171] soon extended this model to include effects for nonlinear dispersive media. Independently, a third research group headed by Gandhi proposed the ADE method for treating *M*th-order dispersive media [172, 173]. An approach similar to the ADE method for modeling the dispersive ionosphere was developed by Nickisch and Franke [174] in 1992. Additional FDTD modeling of wave propagation in a plasma using the joint equations of Euler and Maxwell was developed by Young [175] in 1994. Young [176] also proposed a higher-order FDTD dispersive method for wave propagation in an isotropic cold plasma. Recently, Torres and Jecko [177] presented an alternate FDTD formulation for a Cole-Cole dispersive medium.

Originally, the disadvantage of the ADE approach was that it required storage on the order of $2M$-1 additional real variables, nearly twice that of the original RC method. However, further research has shown how to reduce this storage requirement. For example, state-space approaches by Pereda et al. [178] and Young [179] have reduced the storage of the frequency-dispersive ADE method to a level similar to that of the RC approach. Recently, Okoniewski et al. [180] also reformulated the ADE method for Debye and Lorentz media so that it has storage similar to the piecewise-linear RC method.

An alternative to the more common RC and ADE methods was presented in 1992 by Sullivan [181], who proposed a dispersive formulation based on Z transforms. Sullivan [182, 183] has since extended this approach to treat nonlinear optical phenomena. Recently, Weedon and Rappaport [184] presented a general formulation for FDTD modeling of frequency-dispersive media using Padé approximations of the complex permittivity or conductivity in conjunction with Z transforms.

Applications of the FDTD RC method include Sullivan's modeling of three-dimensional biological problems [185] and Hum et al.'s [186] experimentally verified model of a cylindrical cavity composed of Debye media. Recent work by Joseph and Taflove [187] gives a detailed description of the extension of the ADE method to nonlinear dispersive media, especially with regard to the full-wave modeling of temporal and spatial optical solitons in two dimensions. Recently, Torres and Jecko [188] have used the ADE method in conjunction with a discretized form of the heat-transport equation to evaluate the heating process in a microwave oven.

Finally, comparisons of the dispersion error of several frequency-dispersive FDTD methods have been provided by Petropoulos [189], Young et al. [190], Siushansian and LoVetri [191], and Cummer [192]. In his paper, Cummer also presented two new FDTD methods for modeling isotropic cold plasmas, one based on exponential fitting. Meanwhile, Hulse and Knoesen [193] used digital signal processing techniques to unify the various dispersive FDTD approaches and evaluate their strengths and weaknesses.

1.5.2 Surface-Impedance Boundary Conditions

In 1992, several independent surface-impedance boundary condition (SIBC) techniques were proposed for the FDTD method. Maloney and Smith [194] presented a frequency-dispersive formulation that is used over the spectrum of the incident pulse. In this work, E and H are related by a convolution sum subsequently modified using Prony's method. Beggs et al. [195] also presented a frequency-dispersive SIBC formulation, although their SIBC is limited to the case of high conductivities. Yee et al. [196] presented two SIBC algorithms, one for an inductive SIBC and one for a capacitive SIBC, that is used for monochromatic excitation. Kashiwa et al. [197] also presented an SIBC algorithm, although their algorithm is limited to problems expressed by an equivalent circuit.

Subsequently, Kellali et al. [198, 199] presented a frequency-dispersive SIBC formulation that is valid for any single angle of incidence, unlike previous formulations that assume normal incidence. Jansson et al. [200] reported an SIBC technique where a system of differential equations describing the behavior of the SIBC is solved concurrently with normal Yee updating. This approach, like most of the previous ones, assumes normal incidence and, like the Beggs et al. approach, is limited to the high-conductivity case.

Another SIBC approach was proposed by Lee et al. [201], wherein the tangential fields are related to their normal derivatives by a partial differential equation. Wang [202, 203] extended this approach to both parallel and perpendicularly polarized electromagnetic waves in two dimensions. Wang [204] also used a similar SIBC to model superconducting interconnects.

Additionally, Oh and Schutt-Aine [205] presented an efficient implementation of an SIBC that eliminates the preprocessing time of the Maloney and Smith approach, and yet is not restricted to high conductivities like the SIBC of Beggs et al. Recently, Penney et al. [206] presented a wideband SIBC for use in radar cross-section calculations. This is of a form similar to a Debye medium and uses Z-transform theory to derive suitable FDTD update equations. Meanwhile, Kellali et al. [207] recently presented an SIBC for lossless Debye media. Finally, Yee and Chen [208] recently extended their single-frequency SIBC for use in conjunction with their FDTD/FVTD hybrid approach to model the surface impedance of nonstaircased scatterers.

1.5.3 Thin Material Sheets

In 1990, Railton and McGeehan [209] presented the first FDTD model for thin sheets of conducting or dielectric material. A year later, Tirkas and Demarest [210] presented an FDTD model for thin dielectric sheets that treated not only the tangential component of the electric field at the air/sheet interface, but also the normal component of the field in the sheet. Independently, in 1992 Maloney and Smith [211] presented a similar formulation, but also allowed for the modeling of thin conducting sheets. Luebbers and Kunz [212] and Boonzaaier and Pistorius [213] also presented methods for treating thin material sheets, all of which treat only the tangential component of the electric field at the sheet/air interface. Finally, a comparison of the FDTD methods for modeling thin dielectric and conducting sheets was given by Maloney and Smith [214].

1.5.4 Anisotropic Material

The original Yee algorithm can be used without modification to model anisotropic materials provided that the permittivity, permeability, and conductivity tensors can be diagonalized. In 1986, Choi and Hoefer [215] exploited this capability to analyze a microstrip cavity on an anisotropic substrate. Later, Beker et al. [216] presented results for a discrete scatterer with a diagonal anisotropic permittivity tensor. Nevertheless, it was recognized fairly early (see [216] and the citations therein) that it would be possible to use a modified Yee scheme to model materials with full tensors.

In 1993, Schneider and Hudson [217] described the application of the FDTD method to materials with full permittivity and conductivity tensors. The same year, Xiao et al. [218] developed a compact two-dimensional algorithm to model arbitrarily shaped waveguides that permit full permittivity tensors. Garcia et al. [219] recently compared several approaches that can be used to treat dielectrics with full permittivity tensors. Recognizing that traditional ABCs based on one-way operators cannot be applied to general anisotropic materials, Zhao et al. [68] and Perez et al. [69] extended Berenger's PML to permit matching of arbitrary waves propagating in anisotropic material.

Hunsberger et al. [160] used a recursive-convolution FDTD scheme to analyze magnetized plasmas (which are both dispersive and anisotropic). The analysis was conducted in one dimension so that the difficulties associated with the lack of collocated fields in higher dimensions could be avoided.

A great deal of effort has gone into applying the FDTD method to the analysis of magnetized ferrites having a frequency-dispersive permeability tensor with off-diagonal terms. Early work was done by, among others, Reineix et al. [220], Okoniewski and Okoniewska [221], and Pereda et al. [222]. These papers employed a discretized form of auxiliary differential equation that describes in the time domain the relationship between the magnetic field and the magnetization current. Here, the ferrite is assumed to be saturated with the biasing field aligned with one of the coordinate axes. On the other hand, Schuster and Luebbers [223] analyzed magnetized ferrites using a recursive-convolution technique that permits arbitrary orientation of the biasing field. Finally, Pereda et al. [224] developed a method for modeling partially magnetized ferrites, although the biasing field was again assumed to be aligned with one of the axes.

1.5.5 Nonlinear Material

In addition to modeling nonlinear lumped elements (see Section 1.6), the FDTD method is used to model propagation in nonlinear media. Merewether and Radasky [225] were the first to use FDTD in this way. In their study of the electromagnetic fields within a partially evacuated cylindrical cavity, the conductivity was assumed to be a function of the electric field. Apparently unaware of the Merewether and Radasky work, Wang et al. [226] reported a similar application of FDTD in 1996. Here, the authors considered coupling processes that occur when a high-power microwave pulse is incident on a cylindrical cavity with a cylindrical aperture.

The quest to model optical devices led to the development of several schemes to incorporate the nonlinear behavior of materials directly into FDTD simulations. In 1992, Zheng and Chen [227] presented a technique to model media having a Kerr-type (instantaneous) nonlinearity. Goorjian et al. [228], Joseph et al. [229], and Joseph and Taflove [187, 230] incorporated both Kerr and dispersive nonlinear effects in their FDTD models, and succeeded in obtaining the first temporal and spatial optical solitons from simulations of the full-wave time-domain Maxwell's equations. In the approach used by this group, the nonlinear relationship between the electric field and the electric flux density is discretized and used to advance the electric field.

An alternative approach put forward by Ziolkowski and Judkins [231, 232] and Ziolkowski [233] relies on the simultaneous time-stepping of quantities such as the bound polarization current and the polarization vector in addition to the electric and magnetic fields. Flesch et al. [234] used an ADE approach to study the formation of shocks in materials having Lorentz dispersion and a Kerr nonlinearity. Forysiak and coworkers [235, 236] used an FDTD scheme that coupled the Bloch equations with Maxwell's equations to study the self-reflection of optical pulses. Prior to this work, Ziolkowski et al. [237] also introduced a combined Maxwell-Bloch/FDTD scheme (see also [233]).

Sullivan presented a Z-transform-based FDTD scheme [182, 183] that is used to model nonlinear dispersive material. Others reporting the use of the FDTD method to model nonlinear phenomena include Jamid and Al-Bader [238], who considered an instantaneous nonlinearity in the permittivity, and Basinger and Brady [239], who modeled dispersive nonlinear Fabry-Perot cavities. Reineix et al. [240] also used the FDTD method to model the propagation of femtosecond pulses in dispersive and nondispersive nonlinear media. Shiozawa and Yoshitake [241] used the method to study the efficiency of a Cherenkov laser. Tran [242, 243] developed an alternative FDTD scheme for Kerr nonlinearities where the electric field is obtained from the flux density via the solution of a cubic equation.

In addition to work done for optical media, Zheng and Chen [244] presented a scheme to accommodate the nonlinear effects of a ferrite. Additionally, Luebbers et al. [245] and Holland [246, 247] presented techniques to model nonlinear magnetic materials.

1.6 ACTIVE AND PASSIVE DEVICE MODELING

Another area of FDTD research that has gained recent attention is active and passive device modeling. Two basic approaches are used. In the first, analytical device models are used directly with the FDTD method. In the second, lumped-element subgrid models are used in which the device behavior is determined by other software. The second approach may be preferable in the modeling of active devices with complicated equivalent circuit models.

In 1992, Sui et al. [248] reported a two-dimensional FDTD model of lumped circuit elements, including nonlinear elements such as diodes and transistors. Independently, Tsuei et al. [249] and Piket-May et al. [250] extended the two-dimensional analysis of Sui et al. to model full-wave propagation in three-dimensional circuits containing both active and passive loads. Toland and coworkers [251, 252] also used FDTD to model nonlinear active circuit elements, including a cavity oscillator and a two-element active antenna. Luebbers et al. [253] investigated antennas with both linear and nonlinear loads. Recently, Kuo et al. [254] presented a large-signal analysis of packaged, nonlinear active microwave circuits. Meanwhile, Ciampolini et al. [255] incorporated a variable time step that is adaptively adjusted during the simulation to further enhance the lumped-element approach for modeling nonlinear circuit elements. Ciampolini et al. [256] also used numerical device models to model single-pole, double-throw switches.

Finally, Alsunaidi et al. [257] developed an active device model that couples the Yee update equations with the solution of the current-continuity equation, the energy-conservation equation, and the momentum-conservation equations.

In 1992, Wolff [258] used the FDTD method to model planar microwave circuits containing nonlinear active components. Their approach was to use FDTD in conjunction with harmonic-balance method (HBM) software to calculate S parameters of an in-phase, line-unified field-effect transistor (FET) combiner. In 1994, Thomas et al. [259] developed an approach for coupling SPICE lumped elements into the FDTD method. Subsequently, Thomas et al. [260] used this method to model the same active antenna originally investigated by Toland. Thomas et al. [261] later coupled the FDTD method with the Los Alamos PISCES device modeler to simulate a diode connected to the end of a transmission line. Following Thomas, Kuo and coworkers [262, 263] presented a small-signal analysis of a microwave amplifier containing matching and dc-bias circuits and a three-terminal active device.

Recognizing that the original passive-element formulation of Sui et al. is valid only across one FDTD cell, Durney et al. [264] reported an extension that permits modeling of lumped elements that span many FDTD cells. This extension was found to be useful in the analysis of active devices connected across a transmission line containing more than one dielectric. Chen and Fusco [265] used the extension of Durney et al. to model an FET connected across a unilaterally gapped slotline. Recently, Xu et al. [266] determined that the algorithm of Durney et al. is stable only for structures having large permittivity, and presented a stable algorithm for both small and large permittivity.

1.7 TRANSFORMATIONS

In the first FDTD calculations of the radar cross section of discrete scatterers, time-harmonic illumination was used to determine the equivalent electric and magnetic currents on a virtual surface that completely encloses the scatterer [7, 267, 268]. These currents are then transformed to the far field. An alternative far-field transformation for harmonic illumination was presented in 1991 by Lee et al. [269]. However, by using harmonic illumination in this way, a different simulation has to be run for each frequency of interest. Alternatively, as shown by Furse et al. [270], results were obtained efficiently at several frequencies by using pulsed illumination and a discrete Fourier transform (DFT) of the equivalent surface currents.

These approaches are not well-suited for the determination of the temporal far fields. In 1989, Britt [271] reported temporal far-field results, but did not fully describe the means by which he obtained these results. Therefore, it was not until the independent work of Yee et al. [272], Luebbers et al. [273], Barth et al. [274], and Shlager and Smith [275, 276] that efficient three-dimensional, time-domain, near-to-far-field transformations were developed and described in detail. Luebbers et al. [277] also proposed a two-dimensional, time-domain, near-to-far-field transformation.

A two-dimensional time-domain technique suitable for both near-to-near-field and near-to-far-field transformations was recently presented by Kragalott et al. [278]. Finally, Demarest et al. [279] described an FDTD frequency-domain near-to-far-field transformation for scatterers buried in stratified grounds.

1.8 DIGITAL SIGNAL-PROCESSING TECHNIQUES

By using digital signal-processing techniques in conjunction with the FDTD method, relevant data such as frequency-domain scattering parameters can be extracted from FDTD simulations of shorter duration than would otherwise be possible. Ko and Mittra [280], Pereda et al. [281], and Naishadham and Lin [282] all used Prony's method in this manner, while Houshmand et al. [283], Huang et al. [284], Craddock et al. [285], and Kuempel and Wolff [286] used the system-identification method. Bi et al. [287] used digital filtering and spectral-estimation techniques to improve the FDTD method for eigenvalue problems. Jandhyala et al. [288, 289] and Chen et al. [290] employed autoregressive methods to reduce FDTD computation time.

More recently, Wu et al. [291] used the finite-impulse-response neural-network technique in conjunction with FDTD to shorten the run time needed to analyze a microwave filter. Meanwhile, Goswami and Mittra [292] employed the generalized pencil-of-function approach with FDTD to model coplanar waveguides and slotlines. Finally, Ritter and Arndt [293] used the matrix-pencil technique with FDTD to analyze waveguide structures more efficiently. Like Prony's method, the matrix-pencil technique approximates a transient waveform by a sum of damped complex exponentials. However, this technique requires less numerical effort than Prony's method.

1.9 TECHNIQUES TO REDUCE NUMERICAL DISPERSION ERRORS

While the second-order-accurate Yee algorithm has been the primary FDTD modality to date, several related methods have been introduced with the goal of reducing the numerical dispersion error. These methods have typically relied upon higher-order approximations to the space or time derivatives. However, some recently proposed techniques, such as those employing wavelets, make a significant departure from traditional methods in order to reduce dispersion errors.

In 1989, Fang [10] proposed a (2, 4) FDTD algorithm that is second-order accurate in time and fourth-order accurate in space, and a (4, 4) algorithm that is fourth-order accurate in both time and space. Independently, Deveze et al. [294, 295] published a similar (2, 4) FDTD method and also developed a matching absorbing boundary condition. Because of the problems associated with the enlarged spatial stencils of these algorithms, they have not enjoyed widespread use. Recently, Hadi and Piket-May [296] presented a two-dimensional, modified (2, 4) FDTD algorithm for reducing dispersion errors and used it to model the propagation of electromagnetic waves in a building. An alternative (4, 4) scheme was presented by Young et al. [297] that uses an implicit, compact operator for the space derivatives and a four-stage Runge-Kutta scheme to advance the fields in time.

Another FDTD method for reducing numerical dispersion was recently put forward by Cole [298]. In this method, nonstandard finite-difference operators are used to substantially reduce the numerical dispersion over that of the Yee scheme at a single frequency. Cole stated that the scheme remains superior to the Yee algorithm at other frequencies, although it does not offer the improvement seen at the design frequency.

Another new technique is the pseudospectral time-domain (PSTD) method put forward by Liu [299]. Like previous researchers, Liu used fast Fourier transforms (FFTs) to calculate the space derivatives. In the past, the utility of this approach was limited by the inherent periodicity of the FFT; fields that propagate out of one side of the computational domain reappear at the other side, giving rise to a "wraparound" effect. However, unlike previous researchers, Liu was able to eliminate wraparound with an appropriate implementation of Berenger's PML. The PSTD method has the appealing property that the numerical dispersion error is made arbitrarily small simply by reducing the Courant number.

New techniques developed from the theory of wavelets applied to Maxwell's equations appear to be another promising area. One such method is the multiresolution time-domain (MRTD) scheme reported by Krumpholz and coworkers [300, 301]. In this approach, the unknown field components are represented by scaling functions and wavelets in space and pulse functions in time. Battle-Lemarie scaling and wavelet functions may be chosen, resulting in an algorithm with an enlarged spatial stencil similar to those seen in high-order FDTD techniques. A somewhat different approach recently put forward by Werthen and Wolff [302] uses Daubechies wavelets. Unlike the MRTD method, these wavelets do not excite spurious modes. A further discussion of the MRTD scheme and a comparison with higher-order FDTD methods is found in Tam [303].

A comparison of the dispersion error of several FDTD methods was performed by Shlager and coworkers [304]. Later, Liu [305] presented a further comparison of the dispersion and dissipation errors of FDTD algorithms, and showed that centered methods are more accurate and efficient than upwind schemes such as the modified FVTD approach of Shankar et al. [306]. Recently, Shlager and Schneider [307] analyzed the dispersion of the MRTD approach that uses Battle-Lemarie scaling functions. They showed that its dispersion errors are much better than those of the Yee algorithm for a monochromatic wave, but may be worse for wideband pulses.

1.10 RADIATING STRUCTURES

The FDTD method was initially used primarily to model the scattering from objects. However, by including the feeding source within the grid, it is possible to model radiating structures. This has led to the use of FDTD to model antennas ranging from simple monopoles and dipoles to more complex practical antennas radiating in the presence of material structures.

1.10.1 Simple Antennas

In 1990, Maloney et al. [308] presented accurate results for the radiation from rotationally symmetric simple antennas such as cylindrical and conical monopoles. Boonzaaier and Pistorius [309, 310] presented results for the radiation from thin-wire dipoles and thin-wire Yagi antennas. Yagi antennas were also studied by Kashiwa et al. [311]. In 1992, Tirkas and Balanis [312] presented results for a monopole over a ground plane, while Luebbers and coworkers [313, 314] presented mutual coupling and gain computations for a pair of wire dipoles. Subsequently, Maloney et al. [315] presented a simple one-dimensional TEM feed model for the FDTD method, and used it to analyze a monopole backed by a plane reflector.

1.10.2 Horn Antennas

In 1991, Katz et al. [316] used the FDTD method to analyze both two- and three-dimensional horn antennas, while a year later, Tirkas and Balanis [312] also modeled three-dimensional horn antennas. In 1994, Tirkas and Balanis [317] extended the contour-path FDTD method and used it to analyze pyramidal horns with composite inner E-plane walls. Reig et al. [318] recently analyzed E-plane sectoral horn antennas using the cell-splitting technique described by Mezzanotte et al. [140]. A broadband choke horn used as a feed in a Gregorian dual-reflector system was also recently analyzed by Ying and Kildal [319].

1.10.3 Antennas for Pulse Radiation

In 1992, Maloney and Smith [320] optimized a two-dimensional parallel-plate structure for pulse radiation using the FDTD method. A year later, they studied the radiation from a Wu-King resistive monopole [321] and used FDTD to optimize a conical antenna for pulse radiation [322]. Meanwhile, Shlager and coworkers [323–325] analyzed and optimized bowtie and TEM horn antennas for pulse radiation. Hagness et al. [326] recently reported FDTD optimization of an impulsively excited, resistively loaded bowtie antenna situated at the skin surface for the ultralow reverberation needed for biomedical sensing applications. Hagness et al. [327] also discussed FDTD analysis of impulsively excited fixed-focus and array antennas at the skin surface for confocal detection of early-stage breast cancers. Last, Montoya and Smith [328] used FDTD to compare the performance of several broadband loaded monopoles.

1.10.4 Microstrip Antennas

Reineix and Jecko [329] were the first to apply the FDTD method to the analysis of microstrip antennas. The following year, Sheen et al. [330] also used FDTD to analyze a rectangular patch antenna. In 1992, Leveque et al. [331] modeled a dispersive microstrip antenna by incorporating the RC approach of Luebbers et al. [156] for Debye media.

Their Debye model showed substantially better agreement with measured results than did a constant permittivity model of the substrate. Meanwhile, Wu et al. [332] used FDTD to accurately calculate the reflection coefficient of various microstrip patch configurations. Uehara and Kagoshima [333] presented an analysis of the mutual coupling between two microstrip antennas, while Oonishi et al. [334] and Kashiwa et al. [125] used a curvilinear FDTD approach to analyze microstrip antennas on a curved surface. In 1994, Qian et al. [335] used the FDTD method to design twin-slot antennas. Reineix and coworkers [336–338] expanded their FDTD analysis to include the input impedance of microstrips with slots to obtain the radar cross section of microstrip patch antennas, and to model the radiation from microstrip patches with a ferrite substrate.

1.10.5 Dielectric Resonator Antennas

Shum and Luk [339–343] conducted FDTD analyses of aperture-coupled dielectric resonator antennas. In [339], results are presented for a dielectric ring resonator operating in the endfire mode and fed by an air gap, while [342] shows results for an antenna operating in the fundamental broadside mode. In [341] and [343], results are given for more complex geometries, including dielectric resonators having a dielectric coating, and stacked dielectric resonators. Dielectric resonator antennas have also been investigated by Esselle [344, 345].

1.10.6 Handheld Antennas

In 1992, Luebbers et al. [346] and Chen et al. [347] analyzed handheld antennas using an FDTD model of a monopole antenna on a conducting or dielectric box. Toftgard et al. [348] calculated the effects on the radiation from such an antenna due to the presence of a person. Jensen and Rahmat-Samii [349] presented results for the input impedance and gain of monopoles, planar inverted-*F* antennas (PIFAs), and loop antennas on handheld transceivers. Coupled patch antennas for personal communications have also been recently studied by Navarro et al. [350]. Recently, PIFAs have also been studied by Liu et al. [351] and Rowell and Murch [352]. The interaction of a handheld antenna and the human head has been studied by many researchers, including Dimbylow and Mann [353], Chen and Wang [354], Jensen and Rahmat-Samii [355], Martens et al. [356], Excell et al. [357], Okoniewski and Stuchly [358], Watanabe et al. [359], and Gandhi and coworkers [360, 361]. (Note that the use of FDTD to model electromagnetic wave absorption by the human body can be traced back to Taflove's and Brodwin's second paper of 1975 [362].)

1.10.7 Antenna Arrays

Cherry and Iskander [363, 364] applied the FDTD method to model an array of interstitial antennas for hyperthermia. In 1994, Ren et al. [365] analyzed two-dimensional phased-array antennas with Floquet boundary conditions. Also in 1994, Thiele and Taflove

[366] presented experimental and three-dimensional FDTD modeling results for the radiation from a single Vivaldi flared horn antenna (realized in microstrip), a coplanar pair of Vivaldi horns, a quad array element comprised of perpendicular-bisecting coplanar pairs, and phased arrays containing up to eight quad elements. Naito et al. [367] reported results for a single radiator of a circularly polarized printed array composed of strips and slots. Uehara and Kagoshima [368] analyzed microstrip phased-array antennas. Recently, Crouch [369] utilized symmetry to reduce the computational resources needed to analyze finite phased arrays.

1.10.8 Other Radiating Structures

In 1993, Beggs et al. [370] used the FDTD method to analyze the radiation from shaped-end radiators, while Maeshima et al. [371] investigated a two-dimensional cavity-backed antenna. Toland et al. [252] and Thomas et al. [260] investigated active antennas using FDTD. In 1994, Penney and Luebbers [372, 373] analyzed both square and semicircular spiral antennas using FDTD.

Tsai and York [374] analyzed coplanar waveguide slot antennas, while Simons et al. [375] analyzed linearly tapered slot antennas. Recently, Rayner et al. [376] used FDTD to model short backfire antennas used to feed reflectors. The analysis of metal-strip-loaded, dielectric leaky wave antennas was recently performed by Chen et al. [377]. Finally, Toland et al. [378] and Ruoss et al. [379] analyzed practical rotationally symmetric antennas. Toland et al. modeled a dielectric rod and a compact Cassegrain antenna, while Ruoss et al. investigated lens antennas.

1.11 MICROWAVE DEVICES AND GUIDING STRUCTURES

The FDTD method has also found widespread applications in modeling microwave devices and guiding structures such as waveguides, resonators, junctions, microstrips, vias, interconnects, and transmission lines. In this area, several techniques to improve the efficiency of FDTD have been developed. Such techniques include the compact two-dimensional FDTD method for guided-wave structures, absorbing boundary conditions specifically developed for guided-wave problems, and the previously discussed advanced schemes for gridding and reducing numerical dispersion.

1.11.1 Waveguides, Feeds, Junctions, and Resonators

In 1985, DePourcq [380] used the FDTD method to analyze various three-dimensional waveguide devices. A year later, Choi and Hoefer [215] applied FDTD to model a finline cavity and an anisotropic microstrip. Olivier and McNamara applied FDTD to study edge slots, an *H*-plane *T*-junction, waveguide apertures, and discontinuities in homogeneous waveguides [381–385]. Bi et al. [386] and Navarro et al. [387, 388] also applied FDTD to *H*-plane waveguide discontinuities. Navarro and coworkers [389–391] also investigated rectangular, circular, and T junctions in square coaxial waveguides and narrow-wall multiple-slot couplers.

In 1990, Chu [392] and Chu and Chaudhuri [393] initiated FDTD modeling of dielectric waveguides for integrated microwave and optical circuits. Later, Jarem [394] presented results for a probe-sleeve-fed rectangular waveguide. Alinikula and Kunz [395] investigated waveguide aperture coupling using FDTD. In 1992, Van Hese and De Zutter [396] modeled coaxial waveguide structures with discontinuities using FDTD in both Cartesian and cylindrical coordinates. That same year, Ko [397] used FDTD to model various microwave integrated circuit components including branch-line couplers and filters. Feng and Junmei [398] analyzed a dielectric post in a rectangular waveguide, while Dib and Katehi [399] modeled the transition from a rectangular waveguide to a shielded dielectric image guide. Kraut et al. [400] analyzed edge slots in waveguides. Ferrite-loaded waveguides were investigated by Pereda and coworkers [222, 224, 401, 402]. Okoniewski and Okoniewska [221] proposed an alternative FDTD algorithm for ferrites and successfully applied it to ferrite-loaded waveguides.

In the early 1990s, Navarro et al. [403, 404] used the FDTD method to model cylindrical homogeneous and dielectric-loaded resonators. Shen et al. [405] and Pereda et al. [281] extended FDTD analysis to open cylindrical dielectric resonators. Wang et al. [406] also studied the Q factors of resonators using the FDTD method. Recently, Hagness et al. [407] reported two-dimensional FDTD models of dielectric-waveguide-coupled ring and disk resonators having Q factors between 5,000 and 10,000. Resonant frequencies, Q factors, the effects of varying the gap between the waveguide and resonator, and the evolution of higher-order whispering-gallery modes in the disk resonators were explored in this study. While this work was primarily aimed at designing very-large-scale-integration (VLSI) microoptical devices, the principles explored apply equally well to microwave circuits employing dimensionally scaled-up versions of the devices.

1.11.2 Microstrips

There has been tremendous interest in applying the FDTD method to microstrip circuits. Fang, Zhang, Mei, and Liu [408–410] were among the first to investigate the dispersive characteristics of microstrips using FDTD. In about the same time period, Liang et al. [411] used FDTD to analyze coplanar waveguides and slotlines. In 1990, Sheen et al. [330] presented FDTD results for various microstrip structures including a rectangular patch antenna, a low-pass filter, and a branch-line coupler. Independently, Moore and Ling [412] and Feix et al. [413] analyzed 90° microstrip bends using FDTD. Meanwhile, Railton and McGeehan [209] used a nonuniform, staircased FDTD mesh to model microstrips having both rectangular and trapezoidal cross sections, with and without a thin passivation layer. Similar to the work of [411], Shibata and Sano [414, 415] presented results for coplanar waveguides and striplines including T junctions, while Wu and Chang [416] investigated a waveguide-shielded microstrip line. Other early-1990s FDTD microstrip work was reported by Zheng and Chen, who investigated microstrips having a variety of substrate materials, including ferrite [417], nonlinear dielectric [227], nonlinear ferrite [244], and anisotropic dielectric with a permittivity tensor including off-diagonal elements [418].

In 1992, Shorthouse and Railton [110] incorporated static-field solutions into the FDTD method for microstrip discontinuities. Railton et al. [111] also introduced an approach to model narrow microstrip lines. Fang and Ren [149] presented a locally conformal FDTD algorithm for arbitrary planar strips. In 1994, Qian et al. [419] used FDTD to predict the insertion loss of a probe head in microwave measurements. Finally, both Zheng and Chen [420] and Kitamura et al. [421] analyzed cylindrical microstrip lines using FDTD in cylindrical coordinates.

A dispersive ABC to terminate microstrip problems was presented in 1992 by Bi et al. [422]. Their ABC was designed so that it would perfectly absorb waves incident at two different velocities. These velocities were determined from the effective dielectric constant of the microstrip over the range of frequencies of interest. That same year, Betz and Mittra [423] presented a modified version of the ABC of Bi et al., and compared its performance with that of several other ABCs designed for the grid truncation plane being perpendicular to the direction of propagation. Over the next two years, Betz and Mittra [424] and Fang [21] independently presented ABCs suitable for terminating grid truncations tangential to the direction of propagation. In these papers, an additional term in Higdon's absorbing boundary condition was used to absorb evanescent waves. Such an additional term (which is not required to absorb traveling waves) was originally put forward by Higdon as a means of controlling the stability of his ABC [18]. However, the generalized PML technique [70] has arguably become the most popular ABC for terminating dispersive microstrip structures, owing to its excellent absorption properties.

1.11.3 Vias, Interconnects, and Transmission Lines

In 1991, Maeda et al. [425] used a nonuniform FDTD mesh to compute the S-parameters of a microstrip-to-via-to-microstrip connection. The following year, Becker et al. [426] presented FDTD results for both rectangular and cylindrical vias, applying a nonorthogonal FDTD mesh for the latter. In a similar manner, Harms et al. [124] presented further results for cylindrical vias. Meanwhile, Mezzanotte et al. [427] analyzed packaged and unpackaged, single and coupled microstrip vias. Yook et al. [428] modeled a coplanar-waveguide-to-via-to-microstrip transition. Further results for microstrip lines with vias were presented in 1995 by Cherry and Iskander [429]. Recently, Pillai and coworkers [430, 431] studied several via configurations including stripline-to-stripline and microstrip-to-stripline coax vias. The latter paper reported a design that reduced crosstalk and improved impedance matching relative to that of other vias.

The characterization of interconnect transitions in multichip and microwave circuit modules has been pursued via FDTD modeling by several groups. Lam et al. [432] used a nonuniform mesh to model microstrip-to-via-to-stripline connections both where the microstrip lines are parallel and fed oppositely from one another, and where they are perpendicular. Gribbons et al. [433] studied interconnects having perforated reference planes. Visan et al. [434] characterized interconnects between microwave-integrated-circuit and monolithic-microwave-integrated-circuit modules. Piket-May et al. [250] studied pulse propagation and crosstalk in a computer module consisting of a stack of four multilayer printed circuit boards, each with more than ten metal-dielectric-metal layers, and each containing many vias. Later, Becker et al. [435] presented further results of FDTD modeling of high-speed electronic packages. Recently, Ghouz and El-Sharawy [436, 437] analyzed two flip-chip and via interconnect packages, one a stripline-to-stripline transition and the other a coplanar-waveguide-to-coplanar-waveguide transition. Finally, Wang [204] analyzed superconducting interconnects by implementing a surface-impedance boundary condition.

The problem of analyzing and reducing crosstalk between microstrip transmission lines has also been investigated using FDTD by several researchers. In 1991, Maeda et al. [438] used a nonuniform grid to model the crosstalk between two parallel microstrips. The same year, Pothecary and Railton [439] investigated the coupling between three parallel microstrip lines. Meanwhile, Cerri et al. [440] modeled two crossed, perpendicular, noncoplanar bundles of parallel microstrip lines. Kitamura et al. [126] used a nonorthogonal mesh to investigate the crosstalk between a microstrip line and a microstrip bend, and also between nonperpendicular microstrip crossovers. Recently, Rizvi and LoVetri [441] analyzed the crosstalk on microstrip structures used in multichip modules where the lines are terminated in lumped resistors. Last, Paul [442] studied lumped terminal conditions in multiconductor transmission lines.

1.11.4 Algorithm Improvements

There has been active research in reducing the computations needed to analyze three-dimensional guided-wave devices. The approaches utilized include the compact two-dimensional FDTD method, the MRTD method, and conformal FDTD methods.

In 1992, Xiao et al. [443] presented a new compact FDTD method for analyzing guided-wave structures. By introducing a phase shift along the direction of propagation, the authors reduced the three-dimensional mesh to two dimensions. Independently, Asi and Shafai [444, 445] and Brankovic et al. [446] introduced similar methods. In 1993, Cangellaris [447] presented a stability and dispersion analysis of the method of Asi and Shafai. Noting that these algorithms introduce complex numbers into an otherwise real process, Mroczkowski and Gwarek [448], Okoniewski [449], and Krupeuzevic et al. [450] presented alternate schemes using the vector wave equation that permit retention of only real variables. Celuch-Marcysiak and Gwarek [451] presented both real-valued and complex-valued algorithms for modeling waveguiding structures having periodic or uniform cross sections. Xiao and Vahldieck [218] introduced a variable transformation to their original compact FDTD method so that it contains only real variables. This approach was used by Hofschen and Wolff [452] and Hong and coworkers [453, 454] to model coplanar waveguides and unilateral finlines. The compact two-dimensional FDTD method was also utilized in conjunction with digital signal processing techniques by Fujii and Kobayashi [455] and Hofschen and Wolff [456] for further benefit.

The amount of computation needed to analyze waveguidelike structures has also been reduced by using the MRTD technique and conformal FDTD approaches. Krumpholz and Katehi [300] applied MRTD to analyze a partially filled dielectric cavity. The nonorthogonal FDTD method has been used by Lee et al. [120, 121] to analyze an H-plane waveguide junction and a circular waveguide with irises. It has also been applied to dielectric-filled cavities by Harms et al. [122, 123]. Influenced by Madsen's DSI work [154], Gedney et al. [155] implemented a generalized Yee algorithm on unstructured grids and used this technique to study Wilkinson and Gysel power dividers. Recently, Shi and Drewniak [457] used the DSI method to model waveguide couplers.

In 1992, a modal termination condition was proposed by Moglie et al. [24] to implement an ABC in waveguides. Improved modal ABCs were presented by Tentzeris et al. [458], Jung et al. [459], Park and Nam [460], Alimenti et al. [461], Mrozowski et al. [28, 462], and Werthen et al. [26, 27]. Recently, Alimenti et al. [29] presented a review of the modal ABCs and their application to waveguide problems. In this paper, the authors showed that the original impedance formulation of [24] is less accurate than the improved methods in [26, 27, 462]. These improved methods are capable of providing spurious reflections below -100 dB. The generalized PML technique [70, 72, 73] for absorbing evanescent waves is another strong candidate for terminating waveguide structures due to its high absorption properties. This technique was reported to yield reflections below -80 dB [70]. However, Alimenti et al. [29] claimed that the modal ABC is more efficient than PML. Similarly, Mrozowski et al. [28] reported that their improved modal approach is also capable of outperforming PML.

1.12 DISCRETE SCATTERERS

Since its inception, a primary application area for FDTD has been the study of plane-wave electromagnetic wave scattering from discrete objects. Here, the object fits completely within the computational domain, and there are no localized sources within the grid. In addition to Yee's original paper [1], other early applications of FDTD to such problems were by Taylor et al. [463], Merewether and coworkers [6, 11], Taflove and Umashankar and coworkers [2, 7, 267, 268], Holland [4], and Kunz et al. [78, 79].

In 1989, Harfoush et al. [464] used FDTD to analyze scattering from relativistically moving surfaces in one and two dimensions. That same year, Britt [271] was among the first to use a pulse excitation in conjunction with a near-to-far-field transformation to compute the radar cross section (RCS) of both two- and three-dimensional perfectly conducting and dielectric scatterers. Britt also presented transformations and governing equations used to analyze scatterers with axial symmetry (i.e., body-of-revolution or BOR problems). Furse et al. [270] described how a DFT is used with pulsed excitation to obtain results efficiently at several frequencies in a single simulation, and how to reduce memory requirements by eliminating the storage of lattice cells interior to the scatterer.

In addition to modeling perfectly conducting or dielectric scatterers, the FDTD method is used to model the RCS of scatterers that are described by either an SIBC, thin resistive sheets, or by frequency-dispersive media. Yee et al. [196] presented such results using a monochromatic SIBC, while more recently Penney et al. [206] presented a wideband SIBC. Wu and Han [465, 466] used a thin-resistive-sheet formulation to analyze the RCS of thin, resistively coated dielectric strips and a rectangular conducting slot filled with a thin resistive dielectric strip. Finally, Luebbers et al. [161] presented RCS calculations for spheres composed of Debye, Lorentz, and plasma material. Other papers related to frequency-dispersive media, SIBCs, and thin resistive sheets are reviewed in Sections 1.5.1–1.5.3.

While the above research used the standard staircasing FDTD method, several researchers in the 1990s developed and applied conformal FDTD approaches to model scattering problems. Some of this research includes the work by Fusco [118, 119], Lee [467], Holland et al. [152], Jurgens et al. [143, 144], and Yee and coworkers [136, 153]. In addition, Yee and Chen [208] recently presented RCS results using an FDTD/FVTD hybrid that employs a surface-impedance boundary condition.

1.13 INFINITE AND PERIODIC STRUCTURES

In 1991, Chan et al. [468] were the first to use the FDTD method in conjunction with periodic boundary conditions. In that work, the scattering cross sections of two-dimensional, randomly rough surfaces under normal incidence were investigated. Later, Fung et al. [469] modeled the scattering from three-dimensional rough surfaces. The incident field was spatially tapered so that only a small portion of the surface need be modeled. In 1995, Hastings et al. [470] reported scattering cross sections of two-dimensional, randomly rough surfaces under oblique incidence. As in the work of [468], a Monte Carlo technique was used to obtain the cross sections of infinite surfaces by averaging the cross sections of a set of finite surfaces.

In 1994, Luebbers and Penney [471] presented a scattered-field formulation for arbitrary apertures in infinite ground planes. The following year, Demarest et al. [472] reported a total-field/scattered-field formulation for a plane wave incident on stratified media. In a separate paper [279], these authors discussed a near-to-far-field transformation for scatterers buried in stratified media. At about the same time, Wong et al. [473] presented a total-field/scattered-field formulation for modeling the scattering from objects located near the boundary of an unbounded half-space. Their work included a near-to-far-field transformation, but was limited to the two-dimensional half-space case.

Recently, Stratis et al. [474] reported FDTD calculation of the diffraction coefficients of two-dimensional, infinite 90° wedges comprised of either a perfect conductor, lossless dielectric, or lossy dielectric. Results were obtained by time-windowing the impulse response of the corresponding finite rectangular scatterer to eliminate unwanted edge diffractions, and then performing a DFT. Comparison with uniform theory of diffraction (UTD) results for the infinite PEC wedge case showed very good agreement for both transverse magnetic (TM) and transverse electric (TE) polarizations, thereby showing promise of the procedure for accurately obtaining the diffraction coefficients of infinite wedges of arbitrary composition.

Employing periodic boundary conditions similar to those used by Chan et al. [468], Navarro et al. [475] used the FDTD method to model scattering from a two-dimensional metallic strip grating. Tsay and Pozar [476] investigated a similar problem using a periodic boundary condition. Both these studies presented results only for normal incidence. Ko and Mittra [477] were the first to permit oblique incidence on an infinite periodic structure. Their work was based upon a generalization of the compact FDTD method presented by Xiao et al. [443]. Extensions and improvements were presented shortly after by Prescott and Shuley [478]. Independently, Veysoglu et al. [479] presented a two-dimensional technique for the oblique-incidence case where the electric and magnetic fields are transformed to periodic functions of space. These periodic functions are then time-stepped using the FDTD algorithm and subsequently inverse-transformed to obtain the electric and magnetic fields.

Ren et al. [365] presented a two-dimensional FDTD analysis for periodic phased-array antennas where the fields at the boundaries are determined by time-delayed or time-advanced copies of the boundary fields. This requires modeling three periods of structures having one-dimensional periodicity, or a three-period by three-period portion of structures having two-dimensional periodicity. Shortly afterward, Harms et al. [480] extended the phase-based FDTD analysis of periodic structures to three dimensions.

In 1993, Cangellaris et al. [481] used a hybrid spectral-FDTD method to investigate the propagation of waves in a periodic guided-wave structure. More recently, Celuch-Marcysiak and Gwarek [451] presented and compared different algorithms that can be used to investigate periodic waveguiding structures.

1.14 GROUND-PENETRATING RADAR

As indicated in the previous section, the FDTD method is used to study the scattering from a discrete object or the radiation from a source in the presence of a planar boundary. This capability has prompted the use of FDTD to study problems related to subsurface radar and ground-penetrating radar (GPR).

In 1991, Moghaddam et al. [482, 483] applied FDTD to model a subsurface radar excited by a canonical point source or line source. Maeshima et al. [484] modeled a two-dimensional subsurface radar using a more complex cavity-backed antenna. He et al. [485] also analyzed two-dimensional scattering from the air-ground interface. Recently, Bourgeois and Smith [486] conducted three-dimensional simulations of the detection of buried pipes. Their model incorporated transmitting and receiving bowtie antennas fed by parallel-wire transmission lines, and included the frequency-dispersive characteristics of the soil. Bourgeois and Smith [487] also presented results for a GPR system employing a separated aperture sensor. Zook [488] performed three-dimensional FDTD studies of antennas used in GPR applications and compared results with those obtained using a method-of-moments code. Rappaport and Weedon [489] developed a scheme to model the dispersive properties of soils typically observed at frequencies used in GPR applications. Finally, the work of Demarest et al. [279, 472] and Wong et al. [473], described in the previous section, is applicable to GPR systems where the radar unit is far from the air-ground interface (such as with airborne radar).

1.15 HYBRID TECHNIQUES

The FDTD method has been used in conjunction with other techniques to attack problems where it would be difficult or infeasible to model the complete geometry in a single space lattice. In 1982, Taflove and Umashankar [490] used a hybrid FDTD / method-of-moments approach based upon Schelkunoff's field-equivalence theorems to investigate coupling problems involving loaded cavities. More recently, other hybrid FDTD techniques were introduced. Aoyagi et al. [491] used the Yee algorithm in conjunction with the scalar wave equation to reduce the computations needed to model a Vivaldi antenna. Cangellaris et al. [481] reported a hybrid spectral / FDTD method to model propagation in anisotropic, inhomogeneous, periodic structures. Lee and Chia [492] introduced a hybrid ray / FDTD method and used it to investigate scattering from a cavity with a complex termination. In 1994, Mrozowski [493] introduced a hybrid FDTD / partial-eigenfunction-expansion method to speed up the Cartesian-grid FDTD method when solving shielded-structure problems. Subsequently, Mrozowski et al. [494] extended this hybrid approach for use in cylindrical coordinates. Finally, Beillard et al. [495] described a hybrid FDTD / asymptotic method that utilizes the uniform theory of diffraction in conjunction with FDTD.

1.16 CONCLUSIONS

Over the past 30 years, the FDTD method has been shown to be a remarkably robust and accurate tool for the analysis of problems in electromagnetics. Indeed, FDTD or one of its variant forms has been successfully applied in a number of disciplines outside of "traditional" electromagnetics, including acoustics, elastodynamics, optics, and biology.

Since the FDTD method has attracted attention from so many researchers, the number of relevant publications is quite large and keeping track of this growing body of literature can be challenging. For those interested in exploring a broad range of FDTD-related issues from a single source, Kunz and Luebbers [9] and Taflove [496] have written books that are devoted to the FDTD method. Further, an on-line database listing publications related to FDTD is available via anonymous ftp and the World Wide Web (WWW). This database currently contains over 2,300 entries and grows almost daily. Entries include journal publications, Ph.D. theses, technical reports, conference presentations, and so forth. Conference presentations are listed only if the associated material was also published in conference proceedings (i.e., conferences that merely publish abstracts are not included). To find out more, use anonymous ftp to connect to the host `ftp.eecs.wsu.edu` and peruse the material in the subdirectory `/pub/FDTD`. Alternatively, use a WWW browser and connect to the URL address `http://www.eecs.wsu.edu/~schneidj/fdtd-bib.html`.

REFERENCES

[1] Yee, K. S., "Numerical solution of initial boundary value problems involving Maxwell's equations in isotropic media," *IEEE Trans. Antennas and Propagation*, Vol. 14, 1966, pp. 302–307.

[2] Taflove, A. and M. E. Brodwin, "Numerical solution of steady-state electromagnetic scattering problems using the time-dependent Maxwell's equations," *IEEE Trans. Microwave Theory and Techniques*, Vol. 23, 1975, pp. 623–630.

[3] Taflove, A., "Review of the formulation and applications of the finite-difference time-domain method for numerical modeling of electromagnetic wave interactions with arbitrary structures," *Wave Motion*, Vol. 10, 1988, pp. 547–582.

[4] Holland, R., "THREDE: A free-field EMP coupling and scattering code," *IEEE Trans. Nuclear Science*, Vol. 24, 1977, pp. 2416–2421.

[5] Holland, R., L. Simpson, and K. Kunz, "Finite-difference analysis of EMP coupling to lossy dielectric structures," *IEEE Trans. Electromagnetic Compatibility*, Vol. 22, 1980, pp. 203–209.

[6] Merewether, D. E., R. Fisher, and F. W. Smith, "On implementing a numeric Huygen's source scheme in a finite-difference program to illuminate scattering bodies," *IEEE Trans. Nuclear Science*, Vol. 27, 1980, pp. 1829–1833.

[7] Taflove, A., and K. Umashankar, "Radar cross section of general three-dimensional scatterers," *IEEE Trans. Electromagnetic Compatibility*, Vol. 25, 1983, pp. 433–440.

[8] Holland, R., and J. W. Williams, "Total-field versus scattered-field finite-difference codes: A comparative assessment," *IEEE Trans. Nuclear Science*, Vol. 30, 1983, pp. 4583–4588.

[9] Kunz, K. S., and R. J. Luebbers, *The Finite Difference Time Domain Method for Electromagnetics*, Boca Raton, FL: CRC Press, 1993.

[10] Fang, J., *Time Domain Finite Difference Computation for Maxwell's Equations*, Ph.D. dissertation, University of California at Berkeley, Berkeley, CA, 1989.

[11] Merewether, D. E., "Transient currents on a body of revolution by an electromagnetic pulse," *IEEE Trans. Electromagnetic Compatibility*, Vol. 13, 1971, pp. 41–44.

[12] Engquist, B., and A. Majda, "Absorbing boundary conditions for the numerical simulation of waves," *Mathematics of Computation*, Vol. 31, 1977, pp. 629–651.

[13] Lindman, E. L., "Free-space boundary conditions for the time dependent wave equation," *J. Computational Physics*, Vol. 18, 1975, pp. 66–78.

[14] Mur, G., "Absorbing boundary conditions for the finite-difference approximation of the time-domain electromagnetic field equations," *IEEE Trans. Electromagnetic Compatibility*, Vol. 23, 1981, pp. 377–382.

[15] Keys, R. G., "Absorbing boundary conditions for acoustic media," *Geophysics*, Vol. 50, 1985, pp. 892–902.

[16] Liao, Z. P., H. L. Wong, B.-P. Yang, and Y.-F. Yuan, "A transmitting boundary for transient wave analysis," *Scientia Sinica, Series A*, Vol. 27, 1984, pp. 1063–1076.

[17] Higdon, R. L., "Absorbing boundary conditions for difference approximations to the multi-dimensional wave equation," *Mathematics of Computation*, Vol. 47, 1986, pp. 437–459.

[18] Higdon, R. L., "Numerical absorbing boundary conditions for the wave equation," *Mathematics of Computation*, Vol. 49, 1987, pp. 65–90.

[19] Moghaddam, M., and W. C. Chew, "Stabilizing Liao's absorbing boundary conditions using single precision arithmetic," *Proc. IEEE Antennas and Propagat. Soc. Int. Symp.*, London, Ontario, 1991, pp. 430–433.

[20] Fang, J., "Investigation on the stability of absorbing boundary conditions for the time-domain finite-difference method," *Proc. IEEE Antennas and Propagat. Soc. Int. Symp.*, Chicago, IL, 1992, pp. 548–551.

[21] Fang, J., "Absorbing boundary conditions applied to model wave propagation in microwave integrated circuits," *IEEE Trans. Microwave Theory and Techniques*, Vol. 42, 1994, pp. 1506–1513.

[22] Wu, Z., and J. Fang, "Numerical implementation and performance of perfectly matched layer boundary condition for waveguide structures," *IEEE Trans. Microwave Theory and Techniques*, Vol. 43. 1995, pp. 2676–2683.

[23] Prescott, D. T., and N. V. Shuley, "Reflection analysis of FDTD boundary conditions– Part I: time-space absorbing boundaries," *IEEE Trans. Microwave Theory and Techniques*, Vol. 45, 1997, pp. 1162–1170.

[24] Moglie, F., T. Rozzi, P. Marcozzi, and A. Schiavoni, "A new termination condition for the application of FDTD techniques to discontinuity problems in close homogeneous waveguide," *IEEE Microwave and Guided Wave Lett.*, Vol. 2, 1992, pp. 475–477.

[25] Railton, C. J., E. M. Daniel, D. L. Paul, and J. P. McGeehan, "Optimized absorbing boundary conditions for the analysis of planar circuits using the finite-difference time-domain method," *IEEE Trans. Microwave Theory and Techniques*, Vol. 41, 1993, pp. 290–297.

[26] Werthen, M., M. Rittweger, and I. Wolff, "FDTD simulation of waveguide junctions using a new boundary condition for rectangular waveguides," *Proc. 24th European Microwave Conf.*, Cannes, France, 1994, pp. 1715–1719.

[27] Werthen, M., M. Rittweger, and I. Wolff, "Multi-mode simulation of homogeneous waveguide components using a combination of the FDTD and FDFD method," *Proc. 25th European Microwave Conf.*, Bologna, Italy, 1995, pp. 234–237.

[28] Mrozowski, M., M. Niedzwiecki, and P. Suchomski, "Improved wideband highly dispersive absorbing boundary condition," *Electronics Lett.*, Vol. 32, 1996, pp. 1109–1111.

[29] Alimenti, F., P. Mezzanotte, L. Roselli, and R. Sorrentino, "Modal absorption in the FDTD method: A critical review," *Int. J. Numerical Modeling of Electronic Networks, Devices, and Fields*, Vol. 10, 1997, pp. 245–264.

[30] Mei, K. K., and J. Fang, "Superabsorption– A method to improve absorbing boundary conditions," *IEEE Trans. Antennas and Propagation*, Vol. 40, 1992, pp. 1001–1010.

[31] Ramahi, O. M., "Complementary boundary operators for wave propagation problems," *J. Computational Physics*, Vol. 133, 1997, pp. 113–128.

[32] Ramahi, O. M., "Concurrent implementation of the complementary operators method in 2-D space," *IEEE Microwave and Guided Wave Lett.*, Vol. 7, 1997, pp. 165–167.

[33] Schneider, J., and O. M. Ramahi, "A comparison and evaluation of the PML and COM mesh truncation techniques for FDTD simulation," *Proc. IEEE Antennas and Propagat. Soc. Int. Symp.*, Montreal, Canada, 1997, pp. 1904–1907.

[34] Andrew, W. V., C. A. Balanis, and P. A. Tirkas, "A comparison of the Berenger perfectly matched layer and the Lindman higher-order ABC's for the FDTD method," *IEEE Microwave and Guided Wave Lett.*, Vol. 5, 1995, pp. 192–194.

[35] Kantartzis, N. V., and T. D. Tsiboukis, "A comparative study of the Berenger perfectly matched layer, the superabsorption technique, and several higher-order ABCs for the FDTD algorithm in two and three-dimensional problems," *IEEE Trans. Magnetics*, Vol. 33, 1997, pp. 1460–1463.

[36] Ziolkowski, R. W., N. K. Madsen, and R. C. Carpenter, "Three-dimensional computer modeling of electromagnetic fields: A global lookback lattice truncation scheme," *J. Computational Physics*, Vol. 50, 1983, pp. 360–408.

[37] Olivier, J. C., "On the synthesis of exact free space absorbing boundary conditions for the finite-difference time-domain method," *IEEE Trans. Antennas and Propagation*, Vol. 40, 1992, pp. 456–459.

[38] De Moerloose, J., and D. De Zutter, "Surface integral representation radiation boundary condition for the FDTD method," *IEEE Trans. Antennas and Propagation*, Vol. 41, 1993, pp. 890–896.

[39] Tromp, E. N., and J. C. Olivier, "Synthesis of absorbing boundary conditions for the FDTD method: Numerical results," *IEEE Trans. Antennas and Propagation*, Vol. 43, 1995, pp. 213–215.

[40] Berenger, J.-P., "A perfectly matched layer for the absorption of electromagnetic waves," *J. Computational Physics*, Vol. 114, 1994, pp. 185–200.

[41] Chew, W. C., and W. H. Weedon, "A 3D perfectly matched medium from modified Maxwell's equations with stretched coordinates," *Microwave and Optical Technology Lett.*, Vol. 7, 1994, pp. 599–604.

[42] Katz, D. S., E. T. Thiele, and A. Taflove, "Validation and extension to three dimensions of the Berenger PML absorbing boundary condition for FDTD meshes," *IEEE Microwave and Guided Wave Lett.*, Vol. 4, 1994, pp. 268–270.

[43] Berenger, J.-P., "Three-dimensional perfectly matched layer for the absorption of electromagnetic waves," *J. Computational Physics*, Vol. 127, 1996, pp. 363–379.

[44] Sacks, Z. S., D. M. Kingsland, R. Lee, and J.-F. Lee, "A perfectly matched anisotropic absorber for use as an absorbing boundary condition," *IEEE Trans. Antennas and Propagation*, Vol. 43, 1995, pp. 1460–1463.

[45] Veihl, J. C. and R. Mittra, "An efficient implementation of Berenger's perfectly matched layer PML for finite-difference time-domain mesh truncation," *IEEE Microwave and Guided Wave Lett.*, Vol. 6, 1996, pp. 94–96.

[46] Gedney, S. D., "An anisotropic perfectly matched layer absorbing medium for the truncation of FDTD lattices," *IEEE Trans. Antennas and Propagation*, Vol. 44, 1996, pp. 1630–1639.

[47] Zhao, L., and A. C. Cangellaris, "GT-PML: Generalized theory of perfectly matched layers and its application to the reflectionless truncation of finite-difference time-domain grids," *IEEE Trans. Microwave Theory and Techniques*, Vol. 44, 1996, pp. 2555–2563.

[48] Sullivan, D. M., "An unsplit step 3-D PML for use with the FDTD method," *IEEE Microwave and Guided Wave Lett.*, Vol. 7, 1997, pp. 184–186.

[49] Ziolkowski, R. W., "The design of Maxwellian absorbers for numerical boundary conditions and for practical applications using engineered artificial materials," *IEEE Trans. Antennas and Propagation*, Vol. 45, 1997, pp. 656–671.

[50] Ziolkowski, R. W., "Time-derivative Lorentz material model-based absorbing boundary condition," *IEEE Trans. Antennas and Propagation*, Vol. 45, 1997, pp. 1530–1535.

[51] Wittwer, D. C., and R. W. Ziolkowski, "How to design the imperfect Berenger PML," *Electromagnetics*, Vol. 16, 1996, pp. 465–485.

[52] Fang, J., and Z. Wu, "Closed-form expression of numerical reflection coefficient at PML interfaces and optimization of PML performance," *IEEE Microwave and Guided Wave Lett.*, Vol. 6, 1996, pp. 332–334.

[53] Chew, W. C., and J. M. Jin, "Perfectly matched layers in the discretized space: An analysis and optimization," *Electromagnetics*, Vol. 16, 1996, pp. 325–340.

[54] Prescott, D. T., and N. V. Shuley, "Reflection analysis of FDTD boundary conditions– Part II: Berenger's PML absorbing layers," *IEEE Trans. Microwave Theory and Techniques*, Vol. 45, 1997, pp. 1171–1178.

[55] Berenger, J.-P., "Perfectly matched layer for the FDTD solution of wave-structure interaction problems," *IEEE Trans. Antennas and Propagation*, Vol. 44, 1996, pp. 110–117.

[56] Berenger, J.-P., "Improved PML for the FDTD solution of wave-structure interaction problems," *IEEE Trans. Antennas and Propagation*, Vol. 45, 1997, pp. 466–473.

[57] Wu, Z., and J. Fang, "Performance of the perfectly matched layer in modeling wave propagation in microwave and digital circuit interconnects," *11th Annual Review of Progress in Applied Computational Electromagnetics*, Monterey, CA, 1995, pp. 504–511.

[58] Petropoulos, P. G., "Analysis of exponential time-differencing for FDTD in lossy dielectrics," *IEEE Trans. Antennas and Propagation*, Vol. 45, 1997, pp. 1054–1057.

[59] Nehrbass, J. W., J.-F. Lee, and R. Lee, "Stability analysis for perfectly matched layered absorbers," *Electromagnetics*, Vol. 16, 1996, pp. 385–397.

[60] Naishadham, K., "Stability considerations in the application of PML absorbing boundary condition to FDTD simulation of microwave circuits," *Proc. IEEE Microwave Theory and Techniques Soc. Int. Symp.*, San Francisco, CA, 1996, pp. 581–584.

[61] Abarbanel, S., and D. Gottlieb, "A mathematical analysis of the PML method, *J. Computational Physics*, Vol. 134, 1997, pp. 357–363.

[62] Navarro, E. A., C. Wu, P. Y. Chung, and J. Litva, "Application of PML superabsorbing boundary condition to non-orthogonal FDTD method," *Electronics Lett.*, Vol. 30, 1994, pp. 1654–1656.

[63] Wu, C., E. A. Navarro, P. Y. Chung, and J. Litva, "Modeling of waveguide structures using the nonorthogonal FDTD method with a PML absorbing boundary," *Microwave and Optical Technology Lett.*, Vol. 8, 1995, pp. 226–228.

[64] Roden, J. A., and S. D. Gedney, "Efficient implementation of the uniaxial-based PML media in three-dimensional nonorthogonal coordinates with the use of the FDTD technique," *Microwave and Optical Technology Lett.*, Vol. 14, 1997, pp. 71–75.

[65] Maloney, J., M. Kesler, and G. Smith, "Generalization of PML to cylindrical geometries," *13th Annual Review of Progress in Applied Computational Electromagnetics*, Monterey, CA, 1997, pp. 900–908.

[66] Teixeira, F. L., and W. C. Chew, "Systematic derivation of anisotropic PML absorbing media in cylindrical and spherical coordinates," *IEEE Microwave and Guided Wave Lett.*, Vol. 7, 1997, pp. 371–373.

[67] Roberts, A. R., and J. Joubert, "PML absorbing boundary condition for higher-order FDTD schemes," *Electronics Lett.*, Vol. 33, 1997, pp. 32–34.

[68] Zhao, A. P., J. Juntunen, and A. V. Raisanen, "Material independent PML absorbers for arbitrary anisotropic dielectric media," *Electronics Lett.*, Vol. 33, 1997, pp. 1535–1536.

[69] Perez, I. V., S. G. Garcia, R. G. Martin, and B. G. Olmedo, "Extension of Berenger's absorbing boundary conditions to match dielectric anisotropic media," *IEEE Microwave and Guided Wave Lett.*, Vol. 7, 1997, pp. 302–304.

[70] Fang, J., and Z. Wu, "Generalized perfectly matched layer for the absorption of propagating and evanescent waves in lossless and lossy media," *IEEE Trans. Microwave Theory and Techniques*, Vol. 44, 1996, pp. 2216–2222.

[71] Gedney, S. D., "An anisotropic PML absorbing media for the FDTD simulation of fields in lossy and dispersive media," *Electromagnetics*, Vol. 16, 1996, pp. 399–415.

[72] Chen, B., D. G. Fang, and B. H. Zhou, "Modified Berenger PML absorbing boundary condition for FDTD meshes," *IEEE Microwave and Guided Wave Lett.*, Vol. 5, 1995, pp. 399–401.

[73] Gribbons, M. A., W. P. Pinello, and A. C. Cangellaris, "A stretched coordinate technique for numerical absorption of evanescent and propagating waves in planar waveguiding structures," *IEEE Trans. Microwave Theory and Techniques*, Vol. 43, 1995, pp. 2883–2889.

[74] Cangellaris, A. C., and D. B. Wright, "Analysis of the numerical error caused by the stair–stepped approximation of a conducting boundary in FDTD simulations of electromagnetic phenomena," *IEEE Trans. Antennas and Propagation*, Vol. 39, 1991, pp. 1518–1525.

[75] Holland, R., "Pitfalls of staircase meshing," *IEEE Trans. Electromagnetic Compatibility*, Vol. 35, 1993, pp. 434–439.

[76] Schneider, J. B., and K. L. Shlager, "FDTD simulations of TEM horns and the implications for staircased representations," *IEEE Trans. Antennas and Propagation*, Vol. 45, 1997, pp. 1830–1838.

[77] Holland, R., "THREDS: a finite-difference time-domain EMP code in 3D spherical coordinates," *IEEE Trans. Nuclear Science*, Vol. 30, 1983, pp. 4592–4595.

[78] Kunz, K. S., and K. M. Lee, "A three-dimensional finite-difference solution of the external response of an aircraft to a complex transient EM environment: I – The method and its implementation," *IEEE Trans. Electromagnetic Compatibility*, Vol. 20, 1978, pp. 328–333.

[79] Kunz, K. S., and K. M. Lee, "A three-dimensional finite-difference solution of the external response of an aircraft to a complex transient EM environment: II – Comparison of predictions and measurements," *IEEE Trans. Electromagnetic Compatibility*, Vol. 20, 1978, pp. 333–341.

[80] Monk, P., and E. Suli, "A convergence analysis of Yee's scheme on non-uniform grids," *SIAM J. Numerical Analysis*, Vol. 31, 1994, pp. 393–412.

[81] Monk, P., and E. Suli, "Error estimates for Yee's method on non-uniform grids," *IEEE Trans. Magnetics*, Vol. 30, 1994, pp. 3200–3203.

[82] Svigelj, J., and R. Mittra, "Grid dispersion error using the nonuniform orthogonal finite-difference time-domain method," *Microwave and Optical Technology Lett.*, Vol. 10, 1995, pp. 199–201.

[83] Yee, K. S., *A Subgridding Method for the Finite Difference Time Domain Algorithm to Solve Maxwell's Equations*, Tech. Report UCRL-96772, Lawrence Livermore National Laboratory, 1987.

[84] Kasher, J. C., and K. S. Yee, "A numerical example of a 2-D scattering problem using a subgrid," *Applied Comp. Electromag. Soc. J. and Newsletter*, Vol. 2, 1987, pp. 75–102.

[85] Kim, I. S., and W. J. R. Hoefer, "A local mesh refinement algorithm for the time domain finite difference method using Maxwell's curl equations," *IEEE Trans. Microwave Theory and Techniques*, Vol. 38, 1990, pp. 812–815.

[86] Zivanovic, S. S., K. S. Yee, and K. K. Mei, "A subgridding method for the time-domain finite-difference method to solve Maxwell's equations," *IEEE Trans. Microwave Theory and Techniques*, Vol. 39, 1991, pp. 471–479.

[87] Monk, P., "Sub-gridding FDTD schemes," *J. Applied Computational Electromagnetics Soc.*, Vol. 11, 1996, pp. 37–46.

[88] Prescott, D. T., and N. V. Shuley, "A method for incorporating different sized cells into the finite-difference time-domain analysis technique," *IEEE Microwave and Guided Wave Lett.*, Vol. 2, 1992, pp. 434–436.

[89] White, M. J., and M. F. Iskander, "Development of a multigrid FDTD code for three-dimensional applications," *IEEE Trans. Antennas and Propagation*, Vol. 45, 1997, pp. 1512–1517.

[90] Okoniewski, M., E. Okoniewska, and M. A. Stuchly, "Three-dimensional subgridding algorithm for FDTD," *IEEE Trans. Antennas and Propagation*, Vol. 45, 1997, pp. 422–429.

[91] Chevalier, M. W., R. J. Luebbers, and V. P. Cable, "FDTD local grid with material traverse," *IEEE Trans. Antennas and Propagation*, Vol. 45, 1997, pp. 411–421.

[92] White, M. J, and M. F. Iskander, "A new 3D FDTD multi-grid technique with dielectric-traverse capabilities," *Proc. IEEE Antennas and Propagat. Soc. Int. Symp.*, Montreal, Canada, 1997, pp. 2160–2163.

[93] Kunz, K. S., and L. Simpson, "A technique for increasing the resolution of finite-difference solutions of the Maxwell's equations," *IEEE Trans. Electromagnetic Compatibility*, Vol. 23, 1981, pp. 419–422.

[94] Holland, R., "Two-pass finite-difference time-domain (FDTD) calculations on a fighter aircraft," *IEEE Trans. Antennas and Propagation*, Vol. 44, 1996, pp. 659–664.

[95] Holland, R., and L. Simpson, "Implementation and optimization of the thin-strut formalism in THREDE," *IEEE Trans. Nuclear Science*, Vol. 27, 1980, pp. 1625–1630.

[96] Holland, R., and L. Simpson, "Finite-difference analysis of EMP coupling to thin struts and wires," *IEEE Trans. Electromagnetic Compatibility*, Vol. 23, 1981, pp. 88–97.

[97] Gilbert, J., and R. Holland, "Implementation of the thin-slot formalism in the finite-difference EMP code THREDII," *IEEE Trans. Nuclear Science*, Vol. 28, 1981, pp. 4269–4274.

[98] Yee, K. S., *Numerical Solution to Maxwell's Equations with Non-Orthogonal Grids*, Tech. Report UCRL–93268, Lawrence Livermore National Laboratory, 1987.

[99] Umashankar, K. R., A. Taflove, and B. Beker, "Calculation and experimental validation of induced currents on coupled wires in an arbitrary shaped cavity," *IEEE Trans. Antennas and Propagation*, Vol. 35, 1987, pp. 1248–1257.

[100] Taflove, A., K. R. Umashankar, B. Beker, F. Harfoush, and K. S. Yee, "Detailed FDTD analysis of electromagnetic fields penetrating narrow slots and lapped joints in thick conducting screens," *IEEE Trans. Antennas and Propagation*, Vol. 36, 1988, pp. 247–257.

[101] Demarest, K. R., "A finite difference time domain technique for modeling narrow apertures in conducting scatterers," *IEEE Trans. Antennas and Propagation*, Vol. 35, 1987, pp. 826–831.

[102] Turner, C. D., and L. D. Bacon, "Evaluation of a thin-slot formalism for finite-difference time-domain electromagnetics codes," *IEEE Trans. Electromagnetic Compatibility*, Vol. 30, 1988, pp. 523–528.

[103] Riley, D. J., and C. D. Turner, "Hybrid thin-slot algorithm for the analysis of narrow apertures in finite-difference time-domain calculations," *IEEE Trans. Antennas and Propagation*, Vol. 38, 1990, pp. 1943–1950.

[104] Riley, D. J., and C. D. Turner, "The inclusion of wall loss in finite-difference time-domain thin-slot algorithms," *IEEE Trans. Electromagnetic Compatibility*, Vol. 33, 1991, pp. 304–311.

[105] Wang, B.-Z., "Enhanced thin-slot formalism for the FDTD analysis of thin-slot penetration," *IEEE Microwave and Guided Wave Lett.*, Vol. 5, 1995, pp. 142–143.

[106] Wang, B.-Z., "Small-hole formalism for the FDTD simulation of small-hole coupling," *IEEE Microwave and Guided Wave Lett.*, Vol. 5, 1995, pp. 15–17.

[107] Oates, J. J., and R. T. Shin, "Small aperture modeling for EMI applications using the finite-difference time-domain technique," *J. Electromagnetic Waves and Applications*, Vol. 9, 1995, pp. 37–69.

[108] Ma, K.-P., M. Li, J. L. Drewniak, T. H. Hubing, and T. P. Van Doren, "Comparison of FDTD algorithms for subcellular modeling of slots in shielding enclosures," *IEEE Trans. Electromagnetic Compatibility*, Vol. 39, 1997, pp. 147–155.

[109] Li, M., K.-P. Ma, D. M. Hockanson, J. L. Drewniak, T. H. Hubing, and T. P. Van Doren, "Numerical and experimental corroboration of an FDTD thin-slot model for slots near corners of shielding enclosures," *IEEE Trans. Electromagnetic Compatibility*, Vol. 39, 1997, pp. 225–232.

[110] Shorthouse, D. B., and C. J. Railton, "The incorporation of static field solutions into the finite difference time domain algorithm," *IEEE Trans. Microwave Theory and Techniques*, Vol. 40, 1992, pp. 986–994.

[111] Railton, C. J., D. B. Shorthouse, and J. P. McGeehan, "Modeling of narrow microstrip lines using finite difference time domain method," *Electronics Lett.*, Vol. 28, 1992, pp. 1168–1169.

[112] Railton, C. J., "Use of static field solutions in the FDTD method for the efficient treatment of curved metal surfaces," *Electronics Lett.*, Vol. 29, 1993, pp. 1466–1467.

[113] Railton, C. J., "An algorithm for the treatment of curved metallic laminas in the finite difference time domain method," *IEEE Trans. Microwave Theory and Techniques*, Vol. 41, 1993, pp. 1429–1438.

[114] Craddock, I. J., and C. J. Railton, "Stable inclusion of a-priori knowledge of field behavior in the FDTD algorithm: Application to the analysis of microstrip lines," *Proc. IEEE Antennas and Propagat. Soc. Int. Symp.*, Baltimore, MD, 1996, pp. 1300–1303.

[115] Esselle, K. P., M. Okoniewski, and M. A. Stuchly, "Accurate modeling of field singularities at metal edges diagonal to the FDTD grid," *Proc. IEEE Antennas and Propagat. Soc. Int. Symp.*, Montreal, Canada, 1997, pp. 2176–2179.

[116] Anderson, J., M. Okoniewski, and S. S. Stuchly, "Subcell treatment of 90° metal corners in FDTD," *Electronics Lett.*, Vol. 31, 1995, pp. 2159–2160.

[117] Holland, R., "Finite difference solutions of Maxwell's equations in generalized nonorthogonal coordinates," *IEEE Trans. Nuclear Science*, Vol. 30, 1983, pp. 4589–4591.

[118] Fusco, M., "FDTD algorithm in curvilinear coordinates," *IEEE Trans. Antennas and Propagation*, Vol. 38, 1990, pp. 76–89.

[119] Fusco, M. A., M. V. Smith, and L. W. Gordon, "A three-dimensional FDTD algorithm in curvilinear coordinates," *IEEE Trans. Antennas and Propagation*, Vol. 39, 1991, pp. 1463–1471.

[120] Lee, J.-F., R. Palandech, and R. Mittra, "Modeling three-dimensional discontinuities in waveguides using non-orthogonal FDTD algorithm," *IEEE Trans. Microwave Theory and Techniques*, Vol. 40, 1992, pp. 346–352.

[121] Lee, J.-F., R. Palandech, and R. Mittra, "Corrections to 'Modeling three-dimensional discontinuities in waveguides using non-orthogonal FDTD algorithm'," *IEEE Trans. Microwave Theory and Techniques*, Vol. 40, 1992, p. 1736.

[122] Harms, P. H., J.-F. Lee, and R. Mittra, "A study of the nonorthogonal FDTD method versus the conventional FDTD technique for computing resonant frequencies of cylindrical cavities," *IEEE Trans. Microwave Theory and Techniques*, Vol. 40, 1992, pp. 741–746.

[123] Harms, P. H., J.-F. Lee, and R. Mittra, "Corrections to 'A study of the nonorthogonal FDTD method versus the conventional FDTD technique for computing resonant frequencies of cylindrical cavities'," *IEEE Trans. Microwave Theory and Techniques*, Vol. 40, 1992, pp. 2115–2116.

[124] Harms, P. H., J.-F. Lee, and R. Mittra, "Characterizing the cylindrical via discontinuity," *IEEE Trans. Microwave Theory and Techniques*, Vol. 41, 1993, pp. 153–156.

[125] Kashiwa, T., T. Onishi, and I. Fukai, "Analysis of microstrip antennas on a curved surface using the conformal grids FDTD method," *IEEE Trans. Antennas and Propagation*, Vol. 42, 1994, pp. 423–427.

[126] Kitamura, T., S. Nakamura, M. Hira, and S. Kurazono, "Analysis of microstrip crossovers using the nonorthogonal finite-difference time-domain method," *Electronics and Communications in Japan, Part 2 – Electronics*, Vol. 77, 1994, pp. 26–34.

[127] Navarro, E. A., C. Wu, P. Y. Chung, and J. Litva, "Some considerations about the finite difference time domain method in general curvilinear coordinates," *IEEE Microwave and Guided Wave Lett.*, Vol. 4, 1994, pp. 396–398.

[128] Ray, S. L., "Numerical dispersion and stability characteristics of time-domain methods on nonorthogonal meshes," *IEEE Trans. Antennas and Propagation*, Vol. 41, 1993, pp. 233–235.

[129] Janaswamy, R., and Y. Liu, "An unstaggered collocated finite-difference scheme for solving time-domain Maxwell's equations in curvilinear coordinates," *IEEE Trans. Antennas and Propagation*, Vol. 45, 1997, pp. 1584–1591.

[130] Yang, H., and C. J. Railton, "A new FDTD model in the study of hollow conducting elliptical waveguides and cylindrical cavity resonator," *Proc. IEEE Antennas and Propagat. Soc. Int. Symp.*, Montreal, Canada, 1997, pp. 994–997.

[131] Yang, H., and C. J. Railton, "Efficient and accurate FDTD algorithm for the treatment of curved material boundaries," *IEE Proc. H: Microwaves, Antennas, and Propagation*, Vol. 144, 1997, pp. 382–388.

[132] Mei, K. K., A. Cangellaris, and D. J. Angelakos, "Conformal time domain finite difference method," *Radio Science*, Vol. 19, 1984, pp. 1145–1147.

[133] Gwarek, W. K., "Analysis of an arbitrarily shaped planar circuit– A time-domain approach," *IEEE Trans. Microwave Theory and Techniques*, Vol. 33, 1985, pp. 1067–1072.

[134] Celuch-Marcysiak, M., and W. K. Gwarek, "Higher order modeling of media interfaces for enhanced FDTD analysis of microwave circuits," *Proc. 24th European Microwave Conf.*, Cannes, France, 1994, pp. 1530–1535.

[135] Celuch-Marcysiak, M., and W. K. Gwarek, "Generalized TLM algorithms with controlled stability margin and their equivalence with finite-difference formulations for modified grids," *IEEE Trans. Microwave Theory and Techniques*, Vol. 43, 1995, pp. 2081–2089.

[136] Yee, K. S., J. S. Chen, and A. H. Chang, "Conformal finite-difference time-domain (FDTD) with overlapping grids," *IEEE Trans. Antennas and Propagation*, Vol. 40, 1992, pp. 1068–1075.

[137] Riley, D. J., and C. D. Turner, "Interfacing unstructured tetrahedron grids to structured-grid FDTD," *IEEE Microwave and Guided Wave Lett.*, Vol. 5, 1995, pp. 284–286.

[138] Riley, D. J., and C. D. Turner, "VOLMAX: A solid-model-based, transient volumetric Maxwell solver using hybrid grids," *IEEE Antennas and Propagation Magazine*, Vol. 39, no. 1, 1997, pp. 20–33.

[139] Liao, C., Y.-S. Zhao, and W.-G. Lin, "A new method for calculating electromagnetic scattering of complex objects," *Microwave and Optical Technology Lett.*, Vol. 9, 1995, pp. 144–147.

[140] Mezzanotte, P., L. Roselli, and R. Sorrentino, "A simple way to model curved metal boundaries in FDTD algorithm avoiding staircase approximation," *IEEE Microwave and Guided Wave Lett.*, Vol. 5, 1995, pp. 267–269.

[141] Weiland, T., "Comment on 'A simple way to model curved metal boundaries in FDTD algorithm avoiding staircase approximation'," *IEEE Microwave and Guided Wave Lett.*, Vol. 6, 1996, pp. 183–184.

[142] Mezzanotte, P., and L. Roselli, "Reply to Comment on 'A simple way to model curved metal boundaries in FDTD algorithm avoiding staircase approximation'," *IEEE Microwave and Guided Wave Lett.*, Vol. 6, 1996, p. 184.

[143] Jurgens, T. G., A. Taflove, K. Umashankar, and T. G. Moore, "Finite-difference time-domain modeling of curved surfaces," *IEEE Trans. Antennas and Propagation*, Vol. 40, 1992, pp. 357–366.

[144] Jurgens, T. G., and A. Taflove, "Three-dimensional contour FDTD modeling of scattering from single and multiple bodies," *IEEE Trans. Antennas and Propagation*, Vol. 41, 1993, pp. 1703–1708.

[145] Railton, C. J., I. J. Craddock, and J. B. Schneider, "The analysis of general two-dimensional PEC structures using a modified CPFDTD algorithm," *IEEE Trans. Microwave Theory and Techniques*, Vol. 44, 1996, pp. 1728–1733.

[146] Railton, C. J., and I. J. Craddock, "Stabilised CPFDTD algorithm for the analysis of arbitrary 3D PEC structures," *IEE Proc. H: Microwaves, Antennas, Propagation*, Vol. 143, 1996, pp. 367–372.

[147] Garcia, S. G., T. M. Hung-Bao, B. G. Olmedo, and R. G. Martin, "Volume-conformation method to study scattering by PEC objects with FDTD," *IEE Proc. H: Microwaves, Antennas, and Propagation*, Vol. 143, 1996, pp. 131–136.

[148] Anderson, J., M. Okoniewski, and S. S. Stuchly, "Practical 3-D contour / staircase treatment of metals in FDTD," *IEEE Microwave and Guided Wave Lett.*, Vol. 6, 1996, pp. 146–148.

[149] Fang, J., and J. Ren, "A locally conformed finite-difference time-domain algorithm of modeling arbitrary shape planar metal strips," *IEEE Trans. Microwave Theory and Techniques*, Vol. 41, 1993, pp. 830–838.

[150] Dey, S., and R. Mittra, "A locally conformal finite-difference time-domain (FDTD) algorithm for modeling three-dimensional perfectly conducting objects," *IEEE Microwave and Guided Wave Lett.*, Vol. 7, 1997, pp. 273–275.

[151] Dey, S., R. Mittra, and S. Chebolu, "A technique for implementing the FDTD algorithm on a nonorthogonal grid," *Microwave and Optical Technology Lett.*, Vol. 14, 1997, pp. 213–215.

[152] Holland, R., V. P. Cable, and L. C. Wilson, "Finite-volume time-domain (FVTD) techniques for EM scattering," *IEEE Trans. Electromagnetic Compatibility*, Vol. 33, 1991, pp. 281–294.

[153] Yee, K. S., and J. S. Chen, "Conformal hybrid finite difference time domain and finite volume time domain," *IEEE Trans. Antennas and Propagation*, Vol. 42, 1994, pp. 1450–1455.

[154] Madsen, N. K., "Divergence preserving discrete surface integral methods for Maxwell's curl equations using non-orthogonal unstructured grids," *J. Computational Physics*, Vol. 119, 1995, pp. 34–45.

[155] Gedney, S. D., F. S. Lansing, and D. L. Rascoe, "Full wave analysis of microwave monolithic circuit devices using a generalized Yee algorithm based on an unstructured grid," *IEEE Trans. Microwave Theory and Techniques*, Vol. 44, 1996, pp. 1393–1400.

[156] Luebbers, R. J., F. Hunsberger, K. S. Kunz, R. B. Standler, and M. Schneider, "A frequency-dependent finite-difference time-domain formulation for dispersive materials," *IEEE Trans. Electromagnetic Compatibility*, Vol. 32, 1990, pp. 222–227.

[157] Bui, M. D., S. S. Stuchly, and G. I. Costache, "Propagation of transients in dispersive dielectric media," *IEEE Trans. Microwave Theory and Techniques*, Vol. 39, 1991, pp. 1165–1172.

[158] Luebbers, R. J., F. Hunsberger, and K. S. Kunz, "A frequency-dependent finite-difference time-domain formulation for transient propagation in plasma," *IEEE Trans. Antennas and Propagation*, Vol. 39, 1991, pp. 29–34.

[159] Luebbers, R. J., and F. Hunsberger, "FDTD for Nth-order dispersive media," *IEEE Trans. Antennas and Propagation*, Vol. 40, 1992, pp. 1297–1301.

[160] Hunsberger, F., R. J. Luebbers, and K. S. Kunz, "Finite-difference time-domain analysis of gyrotropic media. I: magnetized plasma," *IEEE Trans. Antennas and Propagation*, Vol. 40, 1992, pp. 1489–1495.

[161] Luebbers, R., D. Steich, and K. Kunz, "FDTD calculation of scattering from frequency-dependent materials," *IEEE Trans. Antennas and Propagation*, Vol. 41, 1993, pp. 1249–1257.

[162] Pontalti, R., L. Cristoforetti, R. Antolini, and L. Cescatti, "A multi-relaxation $(FD)^2$-TD method for modeling dispersion in biological tissues," *IEEE Trans. Microwave Theory and Techniques*, Vol. 42, 1994, pp. 526–527.

[163] Melon, C., Ph. Leveque, T. Monediere, A. Reineix, and F. Jecko, "Frequency dependent finite-difference time-domain formulation applied to ferrite material," *Microwave and Optical Technology Lett.*, Vol. 7, 1994, pp. 577–579.

[164] Hawkins, R. J., and J. S. Kallman, "Linear electronic dispersion and finite-difference time-domain calculations: a simple approach," *J. Lightwave Technology*, Vol. 11, 1993, pp. 1872–1874.

[165] Kelley, D. F., and R. J. Luebbers, "Piecewise linear recursive convolution for dispersive media using FDTD," *IEEE Trans. Antennas and Propagation*, Vol. 44, 1996, pp. 792–797.

[166] Siushansian, R., and J. LoVetri, "A comparison of numerical techniques for modeling electromagnetic dispersive media," *IEEE Microwave and Guided Wave Lett.*, Vol. 5, 1995, pp. 426–428.

[167] Kashiwa, T., N. Yoshida, and I. Fukai, "A treatment by the finite-difference time-domain method of the dispersive characteristics associated with orientation polarization," *Trans. IEICE*, Vol. E73, 1990, pp. 1326–1328.

[168] Kashiwa, T., and I. Fukai, "A treatment by the FDTD method of the dispersive characteristics associated with electronic polarization," *Microwave and Optical Technology Lett.*, Vol. 3, 1990, pp. 203–205.

[169] Kashiwa, T., Y. Ohtomo, and I. Fukai, "A finite-difference time-domain formulation for transient propagation in dispersive media associated with Cole-Cole's circular arc law," *Microwave and Optical Technology Lett.*, Vol. 3, 1990, pp. 416–419.

[170] Joseph, R. M., S. C. Hagness, and A. Taflove, "Direct time integration of Maxwell's equations in linear dispersive media with absorption for scattering and propagation of femtosecond electromagnetic pulses," *Optics Lett.*, Vol. 16, 1991, pp. 1412–1414.

[171] Goorjian, P. M., and A. Taflove, "Direct time integration of Maxwell's equations in nonlinear dispersive media for propagation and scattering of femtosecond electromagnetic solitons," *Optics Lett.*, Vol. 17, 1992, pp. 180–182.

[172] Gandhi, O. P., B. Q. Gao, and J. Y. Chen, "A frequency-dependent finite-difference time-domain formulation for induced current calculations in human beings," *Bioelectromagnetics*, Vol. 13, 1992, pp. 543–556.

[173] Gandhi, O. P., B. Q. Gao, and J. Y. Chen, "A frequency-dependent finite-difference time-domain formulation for general dispersive media," *IEEE Trans. Microwave Theory and Techniques*, Vol. 41, 1993, pp. 658–665.

[174] Nickisch, L. J., and P. M. Franke, "Finite-difference time-domain solution of Maxwell's equations for the dispersive ionosphere," *IEEE Antennas and Propagation Magazine*, Vol. 34, no. 5, 1992, pp. 33–39.

[175] Young, J. L., "A full finite difference time domain implementation for radio wave propagation in a plasma," *Radio Science*, Vol. 29, 1994, pp. 1513–1522.

[176] Young, J. L., "A higher order FDTD method for EM propagation in a collisionless cold plasma," *IEEE Trans. Antennas and Propagation*, Vol. 44, 1996, pp. 1283–1289.

[177] Torres, F., P. Vaudon, and B. Jecko, "Application of fractional derivatives to the FDTD modeling of pulse propagation in a Cole-Cole dispersive medium," *Microwave and Optical Technology Lett.*, Vol. 13, 1996, pp. 300–304.

[178] Pereda, J. A., L. A. Vielva, A. Vegas, and A. Prieto, "State-space approach for the FDTD formulation for dispersive media," *IEEE Trans. Magnetics*, Vol. 31, 1995, pp. 1602–1605.

[179] Young, J. L., "Propagation in linear dispersive media: finite-difference time-domain methodologies," *IEEE Trans. Antennas and Propagation*, Vol. 43, 1995, pp. 422–426.

[180] Okoniewski, M., M. Mrozowski, and M. A. Stuchly, "Simple treatment of multi-term dispersion in FDTD," *IEEE Microwave and Guided Wave Lett.*, Vol. 7, 1997, pp. 121–123.

[181] Sullivan, D. M., "Frequency-dependent FDTD methods using *Z* transforms," *IEEE Trans. Antennas and Propagation*, Vol. 40, 1992, pp. 1223–1230.

[182] Sullivan, D. M., "Nonlinear FDTD formulations using *Z* transforms," *IEEE Trans. Microwave Theory and Techniques*, Vol. 43, 1995, pp. 676–682.

[183] Sullivan, D. M., "Z-transform theory and the FDTD method," *IEEE Trans. Antennas and Propagation*, Vol. 44, 1996, pp. 28–34.

[184] Weedon, W. H., and C. M. Rappaport, "General method for FDTD modeling of wave propagation in arbitrary frequency-dispersive media," *IEEE Trans. Antennas and Propagation*, Vol. 45, 1997, pp. 401–410.

[185] Sullivan, D. M., "A frequency-dependent FDTD method for biological applications," *IEEE Trans. Microwave Theory and Techniques*, Vol. 40, 1992, pp. 532–539.

[186] Hum, P. J., M. S. Leong, P. S. Kooi, and T. S. Yeo, "$(FD)^2$-TD and experimental comparison for a cylindrical cavity with lossy dielectric," *Microwave and Optical Technology Lett.*, Vol. 6, 1993, pp. 869–871.

[187] Joseph, R. M., and A. Taflove, "FDTD Maxwell's equations models for nonlinear electrodynamics and optics," *IEEE Trans. Antennas and Propagation*, Vol. 45, 1997, pp. 364–374.

[188] Torres, F., and B. Jecko, "Complete FDTD analysis of microwave heating processes in frequency-dependent and temperature-dependent media," *IEEE Trans. Microwave Theory and Techniques*, Vol. 45, 1997, pp. 108–117.

[189] Petropoulos, P. G., "Stability and phase error analysis of FDTD in dispersive dielectrics," *IEEE Trans. Antennas and Propagation*, Vol. 42, 1994, pp. 62–69.

[190] Young, J. L., A. Kittichartphayak, Y. M. Kwok, and D. Sullivan, "On the dispersion errors related to $(FD)^2$-TD type schemes," *IEEE Trans. Microwave Theory and Techniques*, Vol. 43, 1995, pp. 1902–1910.

[191] Siushansian, R., and J. LoVetri, "Efficient evaluation of convolution integrals arising in FDTD formulations of electromagnetic dispersive media," *J. Electromagnetic Waves and Applications*, Vol. 11, 1997, pp. 101–117.

[192] Cummer, S. A., "An analysis of new and existing FDTD methods for isotropic cold plasma and a method for improving their accuracy," *IEEE Trans. Antennas and Propagation*, Vol. 45, 1997, pp. 392–400.

[193] Hulse, C., and A. Knoesen, "Dispersive models for the finite-difference time-domain method: design, analysis, and implementation," *J. Optical Society of America A: Optics, Image Science, and Vision*, Vol. 11, 1994, pp. 1802–1811.

[194] Maloney, J. G., and G. S. Smith, "The use of surface impedance concepts in the finite-difference time-domain method," *IEEE Trans. Antennas and Propagation*, Vol. 40, 1992, pp. 38–48.

[195] Beggs, J. H., R. J. Luebbers, K. S. Yee, and K. S. Kunz, "Finite-difference time-domain implementation of surface impedance boundary conditions," *IEEE Trans. Antennas and Propagation*, Vol. 40, 1992, pp. 49–56.

[196] Yee, K. S., K. Shlager, and A. H. Chang, "An algorithm to implement a surface impedance boundary condition for FDTD," *IEEE Trans. Antennas and Propagation*, Vol. 40, 1992, pp. 833–837.

[197] Kashiwa, T., O. Chiba, and I. Fukai, "A formulation for surface impedance boundary conditions using the finite-difference time-domain method," *Microwave and Optical Technology Lett.*, Vol. 5, 1992, pp. 486–490.

[198] Kellali, S., B. Jecko, and A. Reineix, "Implementation of a surface impedance formalism at oblique incidence in FDTD method," *IEEE Trans. Electromagnetic Compatibility*, Vol. 35, 1993, pp. 347–356.

[199] Kellali, S., B. Jecko, and A. Reineix, "Surface impedance boundary conditions at oblique incidence in FDTD," *Annales des Telecommunications*, Vol. 48, 1993, pp. 268–276.

[200] Jansson, M. C. O., S. P. O. Ljung, M. G. Backstrom, and B. I. Wahlgren, "Efficient implementation of a submodel for composite materials to be combined with the FDTD algorithm," *IEEE Trans. Magnetics*, Vol. 30, 1994, pp. 3188–3191.

[201] Lee, C. F., R. T. Shin, and J. A. Kong, "Time domain modeling of impedance boundary condition," *IEEE Trans. Microwave Theory and Techniques*, Vol. 40, 1992, pp. 1847–1850.

[202] Wang, B. Z., "Time-domain modeling of the impedance boundary condition for an oblique incident parallel-polarization plane wave," *Microwave Optical Technology Lett.*, Vol. 7, 1994, pp. 19–22.

[203] Wang, B. Z., "Time-domain modeling of the impedance boundary condition for an oblique incident perpendicular-polarization plane wave," *Microwave and Optical Technology Lett.*, Vol. 7, 1994, pp. 355–359.

[204] Wang, B. Z., "Analysis of superconductive interconnects by the finite-difference-time-domain method," *Microwave and Optical Technology Lett.*, Vol. 7, 1994, pp. 837–840.

[205] Oh, K. S., and J. E. Schutt-Aine, "An efficient implementation of surface impedance boundary conditions for the finite-difference time-domain method," *IEEE Trans. Antennas and Propagation*, Vol. 43, 1995, pp. 660–666.

[206] Penney, C., R. J. Luebbers, and J. W. Schuster, "Scattering from coated targets using a frequency-dependent, surface impedance boundary condition in FDTD," *IEEE Trans. Antennas and Propagation*, Vol. 44, 1996, pp. 434–443.

[207] Kellali, S., A. Reineix, P. Leveque, and B. Jecko, "Time-domain surface impedances of lossless Debye media," *Annales des Telecommunications*, Vol. 50, 1995, pp. 705–710.

[208] Yee, K. S., and J. S. Chen, "Impedance boundary condition simulation in the FDTD / FVTD hybrid," *IEEE Trans. Antennas and Propagation*, Vol. 45, 1997, pp. 921–925.

[209] Railton, C. J., and J. P. McGeehan, "An analysis of microstrip with rectangular and trapezoidal conductor cross sections," *IEEE Trans. Microwave Theory and Techniques*, Vol. 38, 1990, pp. 1017–1022.

[210] Tirkas, P. A., and K. R. Demarest, "Modeling of thin dielectric structures using finite-difference time-domain technique," *IEEE Trans. Antennas and Propagation*, Vol. 39, 1991, pp. 1338–1344.

[211] Maloney, J. G., and G. S. Smith, "The efficient modeling of thin material sheets in the finite-difference time-domain (FDTD) method," *IEEE Trans. Antennas and Propagation*, Vol. 40, 1992, pp. 323–330.

[212] Luebbers, R. J., and K. Kunz, "FDTD modeling of thin impedance sheets," *IEEE Trans. Antennas and Propagation*, Vol. 40, 1992, pp. 349–350.

[213] Boonzaaier, J. J., and C. W. Pistorius, "Scattering by thin lossy dielectric plates– A finite-difference time-domain approach," *Microwave and Optical Technology Lett.*, Vol. 6, 1993, pp. 326–332.

[214] Maloney, J. G., and G. S. Smith, "A comparison of methods for modeling electrically thin dielectric and conducting sheets in the finite-difference time-domain (FDTD) method," *IEEE Trans. Antennas and Propagation*, Vol. 41, 1993, pp. 690–694.

[215] Choi, D. H., and W. J. R. Hoefer, "The finite-difference time-domain method and its application to eigenvalue problems," *IEEE Trans. Microwave Theory and Techniques*, Vol. 34, 1986, pp. 1464–1470.

[216] Beker, B., K. R. Umashankar, and A. Taflove, "Numerical analysis and validation of the combined field surface integral equations for electromagnetic scattering by arbitrary shaped two-dimensional anisotropic objects," *IEEE Trans. Antennas and Propagation*, Vol. 37, 1989, pp. 1573–1581.

[217] Schneider, J. B., and S. Hudson, "The finite-difference time-domain method applied to anisotropic material," *IEEE Trans. Antennas and Propagation*, Vol. 41, 1993, pp. 994–999.

[218] Xiao, S., and R. Vahldieck, "An efficient 2-D FDTD algorithm using real variables," *IEEE Microwave and Guided Wave Lett.*, Vol. 3, 1993, pp. 127–129.

[219] Garcia, S. G., T. M. Hung-Bao, R. G. Martin, and B. G. Olmedo, "On the application of finite methods in time domain to anisotropic dielectric waveguides," *IEEE Trans. Microwave Theory and Techniques*, Vol. 44, 1996, pp. 2195–2206.

[220] Reineix, A., T. Monediere, and B. Jecko, "Ferrite analysis using the finite-difference time-domain (FDTD) method," *Microwave and Optical Technology Lett.*, Vol. 5, 1992, pp. 685–686.

[221] Okoniewski, M., and E. Okoniewska, "FDTD analysis of magnetized ferrites: A more efficient algorithm," *IEEE Microwave and Guided Wave Lett.*, Vol. 4, 1994, pp. 169–171.

[222] Pereda, J. A., L. A. Vielva, M. A. Solano, A. Vegas, and A. Prieto, "FDTD analysis of magnetized ferrites: Application to the calculation of dispersion characteristics of ferrite-loaded waveguides," *IEEE Trans. Microwave Theory and Techniques*, Vol. 43, 1995, pp. 350–357.

[223] Schuster, J. W., and R. J. Luebbers, "Finite difference time domain analysis of arbitrarily biased magnetized ferrites," *Radio Science*, Vol. 31, 1996, pp. 923–929.

[224] Pereda, J. A., L. A. Vielva, A. Vegas, and A. Prieto, "An extended FDTD method for the treatment of partially magnetized ferrites," *IEEE Trans. Magnetics*, Vol. 31, 1995, pp. 1666–1669.

[225] Merewether, D. E., and W. A. Radasky, "Nonlinear electromagnetic fields within a cylindrical cavity excited by ionizing radiation," *IEEE Trans. Nuclear Science*, Vol. 21, 1974, pp. 998–1005.

[226] Wang, J., Y. Chen, R. Fan, H. Yu, and D. Ge, "Numerical studies on nonlinear coupling of high-power microwave pulses into a cylindrical cavity," *IEEE Trans. Plasma Science*, Vol. 24, 1996, pp. 193–197.

[227] Zheng, G. W., and K. S. Chen, "Transient analysis of dielectric step discontinuity of microstrip lines containing a nonlinear layer," *Int. J. Infrared and Millimeter Waves*, Vol. 13, 1992, pp. 1127–1137.

[228] Goorjian, P. M., A. Taflove, R. M. Joseph, and S. C. Hagness, "Computational modeling of femtosecond optical solitons from Maxwell's equations," *IEEE J. Quantum Electronics*, Vol. 28, 1992, pp. 2416–2422.

[229] Joseph, R. M., P. M. Goorjian, and A. Taflove, "Direct time integration of Maxwell's equations in two-dimensional dielectric waveguides for propagation and scattering of femtosecond electromagnetic solitons," *Optics Lett.*, Vol. 18, 1993, pp. 491–493.

[230] Joseph, R. M., and A. Taflove, "Spatial soliton deflection mechanism indicated by FDTD Maxwell's equations modeling," *IEEE Photonics Technology Lett.*, Vol. 6, 1994, pp. 1251–1254.

[231] Ziolkowski, R. W., and J. B. Judkins, "Applications of the nonlinear finite difference time domain (NL-FDTD) method to pulse propagation in nonlinear media: Self-focusing and linear-nonlinear interfaces," *Radio Science*, Vol. 28, 1993, pp. 901–911.

[232] Ziolkowski, R. W., and J. B. Judkins, "Full-wave vector Maxwell equations modeling of the self-focusing of ultrashort optical pulses in a nonlinear Kerr medium exhibiting a finite response time," *J. Optical Society of America B: Optical Physics*, Vol. 10, 1993, pp. 186–198.

[233] Ziolkowski, R. W., "The incorporation of microscopic material models into the FDTD approach for ultrafast optical pulse simulations," *IEEE Trans. Antennas and Propagation*, Vol. 45, 1997, pp. 375–391.

[234] Flesch, R. G., A. Pushkarev, and J. V. Moloney, "Carrier wave shocking of femtosecond optical pulses," *Physical Review Lett.*, Vol. 76, 1996, pp. 2488–2491.

[235] Forysiak, W., R. G. Flesch, J. V. Moloney, and E. M. Wright, "Doppler shift of self-reflected optical pulses at an interface: Dynamic nonlinear optical skin effect," *Physical Review Lett.*, Vol. 76, 1996, pp. 3695–3698.

[236] Forysiak, W., J. V. Moloney, and E. M. Wright, "Nonlinear focusing of femtosecond pulses as a result of self-reflection from a saturable absorber," *Optics Lett.*, Vol. 22, 1997, pp. 239–241.

[237] Ziolkowski, R. W., J. M. Arnold, and D. M. Gogny, "Ultrafast pulse interactions with two-level atoms," *Physical Review A*, Vol. 52, 1995, pp. 3082–3094.

[238] Jamid, H. A., and S. J. Al-Bader, "Finite-difference time-domain approach to nonlinear guided waves," *Electronics Lett.*, Vol. 29, 1993, pp. 83–84.

[239] Basinger, S. A., and D. J. Brady, "Finite-difference time-domain modeling of dispersive nonlinear Fabry-Perot cavities," *J. Optical Society of America B: Optical Physics*, Vol. 11, 1994, pp. 1504–1511.

[240] Reineix, A., M. Ariaudo, B. Jecko, and A. Barthelemy, "Adaptation of the FDTD method to the study of femtosecond electromagnetic soliton," *Microwave and Optical Technology Lett.*, Vol. 7, 1994, pp. 507–511.

[241] Shiozawa, T., and T. Yoshitake, "Efficiency enhancement in a Cherenkov laser loaded with a Kerr-like medium," *IEEE J. Quantum Electronics*, Vol. 31, 1995, pp. 539–545.

[242] Tran, P., "Optical switching with a nonlinear photonic crystal: A numerical study," *Optics Lett.*, Vol. 21, 1996, pp. 1138–1140.

[243] Tran, P., "Optical limiting and switching of short pulses by use of a nonlinear photonic bandgap structure with a defect," *J. Optical Society of America B: Optical Physics*, Vol. 14, 1997, pp. 2589–2595.

[244] Zheng, G. W., and K. S. Chen, "The nonlinear study of microstrip lines containing ferrite dielectric layers," *Int. J. Infrared and Millimeter Waves*, Vol. 13, 1992, pp. 1599–1608.

[245] Luebbers, R., K. Kumagai, S. Adachi, and T. Uno, "FDTD calculation of transient pulse propagation through a nonlinear magnetic sheet," *IEEE Trans. Electromagnetic Compatibility*, Vol. 35, 1993, pp. 90–94.

[246] Holland, R., "Finite-difference time-domain (FDTD) analysis of magnetic diffusion," *IEEE Trans. Electromagnetic Compatibility*, Vol. 36, 1994, pp. 32–39.

[247] Holland, R., "FDTD analysis of nonlinear magnetic diffusion by reduced *c*," *IEEE Trans. Antennas and Propagation*, Vol. 43, 1995, pp. 653–659.

[248] Sui, W., D. A. Christensen, and C. H. Durney, "Extending the two-dimensional FDTD method to hybrid electromagnetic systems with active and passive lumped elements," *IEEE Trans. Microwave Theory and Techniques*, Vol. 40, 1992, pp. 724–730.

[249] Tsuei, Y.-S., A. C. Cangellaris, and J. L. Prince, "Rigorous electromagnetic modeling of chip-to-package (first-level) interconnections," *IEEE Trans. Components, Hybrids, and Manufacturing Technology*, Vol. 16, 1993, pp. 876–883.

[250] Piket-May, M., A. Taflove, and J. Baron, "FDTD modeling of digital signal propagation in 3-D circuits with passive and active loads," *IEEE Trans. Microwave Theory and Techniques*, Vol. 42, 1994, pp. 1514–1523.

[251] Toland, B., B. Houshmand, and T. Itoh, "Modeling of nonlinear active regions with the FDTD method," *IEEE Microwave and Guided Wave Lett.*, Vol. 3, 1993, pp. 333–335.

[252] Toland, B., J. Lin, B. Houshmand, and T. Itoh, "FDTD analysis of an active antenna," *IEEE Microwave and Guided Wave Lett.*, Vol. 3, 1993, pp. 423–425.

[253] Luebbers, R., J. Beggs, and K. Chamberlin, "Finite-difference time-domain calculations of transients in antennas with nonlinear loads," *IEEE Trans. Antennas and Propagation*, Vol. 41, 1993, pp. 566–573.

[254] Kuo, C.-N., B. Houshmand, and T. Itoh, "Full-wave analysis of packaged microwave circuits with active and nonlinear devices: An FDTD approach," *IEEE Trans. Microwave Theory and Techniques*, Vol. 45, 1997, pp. 819–826.

[255] Ciampolini, P., P. Mezzanotte, L. Roselli, and R. Sorrentino, "Accurate and efficient circuit simulation with lumped-element FDTD technique," *IEEE Trans. Microwave Theory and Techniques*, Vol. 44, 1996, pp. 2207–2215.

[256] Ciampolini, P., L. Roselli, and G. Stopponi, "Mixed-mode circuit simulation with full-wave analysis of interconnections," *IEEE Trans. Electron Devices*, Vol. 44, 1997, pp. 2098–2105.

[257] Alsunaidi, M. A., S. M. Sohel Imtiaz, and S. M. El-Ghazaly, "Electromagnetic wave effects on microwave transistors using a full-wave time-domain model," *IEEE Trans. Microwave Theory and Techniques*, Vol. 44, 1996, pp. 799–808.

[258] Wolff, I., "Finite difference time-domain simulation of electromagnetic fields and microwave circuits," *Int. J. Numerical Modeling*, Vol. 5, 1992, pp. 163–182.

[259] Thomas, V. A., M. E. Jones, M. Piket-May, A. Taflove, and E. Harrigan, "The use of SPICE lumped circuits as sub-grid models for FDTD analysis," *IEEE Microwave and Guided Wave Lett.*, Vol. 4, 1994, pp. 141–143.

[260] Thomas, V. A., K. M. Ling, M. E. Jones, B. Toland, J. Lin, and T. Itoh, "FDTD analysis of an active antenna," *IEEE Microwave and Guided Wave Lett.*, Vol. 4, 1994, pp. 296–298.

[261] Thomas, V. A., M. E. Jones, and R. J. Mason, "Coupling of the PISCES device modeler to a 3-D Maxwell FDTD solver," *IEEE Trans. Microwave Theory and Techniques*, Vol. 43, 1995, pp. 2170–2172.

[262] Kuo, C.-N., V. A. Thomas, S. T. Chew, B. Houshmand, and T. Itoh, "Small signal analysis of active circuits using FDTD algorithm," *IEEE Microwave and Guided Wave Lett.*, Vol. 5, 1995, pp. 216–218.

[263] Kuo, C.-N., R.-B. Wu, B. Houshmand, and T. Itoh, "Modeling of microwave active devices using the FDTD analysis based on the voltage-source approach," *IEEE Microwave and Guided Wave Lett.*, Vol. 6, 1996, pp. 199–201.

[264] Durney, C. H., W. Sui, D. A. Christensen, and J. Zhu, "A general formulation for connecting sources and passive lumped-circuit elements across multiple 3-D FDTD cells," *IEEE Microwave and Guided Wave Lett.*, Vol. 6, 1996, pp. 85–87.

[265] Chen, Q., and V. F. Fusco, "Hybrid FDTD large-signal modeling of three-terminal active devices," *IEEE Trans. Microwave Theory and Techniques*, Vol. 45, 1997, pp. 1267–1270.

[266] Xu, J., A. P. Zhao, and A. V. Raisanen, "A stable algorithm for modeling lumped circuit source across multiple FDTD cells," *IEEE Microwave and Guided Wave Lett.*, Vol. 7, 1997, pp. 308–310.

[267] Umashankar, K. R., and A. Taflove, "A novel method to analyze electromagnetic scattering of complex objects," *IEEE Trans. Electromagnetic Compatibility*, Vol. 24, 1982, pp. 397–405.

[268] Taflove, A., K. R. Umashankar, and T. G. Jurgens, "Validation of FDTD modeling of the radar cross section of three-dimensional structures spanning up to nine wavelengths," *IEEE Trans. Antennas and Propagation*, Vol. 33, 1985, pp. 662–666.

[269] Lee, S. M., W. C. Chew, M. Moghaddam, M. A. Nasir, S.-L. Chuang, R. W. Herrick, and C. L. Balestra, "Modeling of rough-surface effects in an optical turning mirror using the finite-difference time-domain method," *J. Lightwave Technology*, Vol. 9, 1991, pp. 1471–1480.

[270] Furse, C. M., S. P. Mathur, and O. P. Gandhi, "Improvements to the finite-difference time-domain method for calculating the radar cross section of a perfectly conducting target," *IEEE Trans. Microwave Theory and Techniques*, Vol. 38, 1990, pp. 919–927.

[271] Britt, C. L., "Solution of electromagnetic scattering problems using time domain techniques," *IEEE Trans. Antennas and Propagation*, Vol. 37, 1989, pp. 1181–1192.

[272] Yee, K. S., D. Ingham, and K. Shlager, "Time-domain extrapolation to the far field based on FDTD calculations," *IEEE Trans. Antennas and Propagation*, Vol. 39, 1991, pp. 410–413.

[273] Luebbers, R. J., K. S. Kunz, M. Schneider, and F. Hunsberger, "A finite-difference time-domain near zone to far zone transformation," *IEEE Trans. Antennas and Propagation*, Vol. 39, 1991, pp. 429–433.

[274] Barth, M. J., R. R. McLeod, and R. W. Ziolkowski, "A near and far-field projection algorithm for finite-difference time-domain codes," *J. Electromagnetic Waves and Applications*, Vol. 6, 1992, pp. 5–18.

[275] Shlager, K. L., and G. S. Smith, "Near-field to near-field transformation for use with FDTD method and its application to pulsed antenna problems," *Electronics Lett.*, Vol. 30, 1994, pp. 1262–1264.

[276] Shlager, K. L., and G. S. Smith, "Comparison of two FDTD near-field to near-field transformations applied to pulsed antenna problems," *Electronics Lett.*, Vol. 31, 1995, pp. 936–938.

[277] Luebbers, R. J., D. Ryan, and J. Beggs, "A two-dimensional time-domain near-zone to far-zone transformation," *IEEE Trans. Antennas and Propagation*, Vol. 40, 1992, pp. 848–851.

[278] Kragalott, M., M. S. Kluskens, and W. P. Pala, "Time-domain fields exterior to a two-dimensional FDTD space," *IEEE Trans. Antennas and Propagation*, Vol. 45, 1997, pp. 1655–1663.

[279] Demarest, K., Z. Huang, and R. Plumb, "An FDTD near- to far-zone transformation for scatterers buried in stratified grounds," *IEEE Trans. Antennas and Propagation*, Vol. 44, 1996, pp. 1150–1157.

[280] Ko, W. L., and R. Mittra, "A combination of FDTD and Prony's methods for analyzing microwave integrated circuits," *IEEE Trans. Microwave Theory and Techniques*, Vol. 39, 1991, pp. 2176–2181.

[281] Pereda, J. A., L. A. Vielva, A. Vegas, and A. Prieto, "Computation of resonant frequencies and quality factors of open dielectric resonators by a combination of the finite-difference time-domain (FDTD) and Prony's methods," *IEEE Microwave and Guided Wave Lett.*, Vol. 2, 1992, pp. 431–433.

[282] Naishadham, K., and X. P. Lin, "Application of spectral domain Prony's method to the FDTD analysis of planar microstrip circuits," *IEEE Trans. Microwave Theory and Techniques*, Vol. 42, 1994, pp. 2391–2398.

[283] Houshmand, B., T. W. Huang, and T. Itoh, "Microwave structure characterization by a combination of FDTD and system identification methods," *IEEE Microwave and Guided Wave Lett.*, Vol. 3, 1995, pp. 262–264.

[284] Huang, T. W., B. Houshmand, and T. Itoh, "Fast sequential FDTD diakoptics method using the system identification technique," *IEEE Microwave and Guided Wave Lett.*, Vol. 3, 1993, pp. 378–380.

[285] Craddock, I. J., P. G. Turner, and C. J. Railton, "Reducing the computational overhead of the near-field transform through system identification," *Electronics Lett.*, Vol. 30, 1994, pp. 1609–1610.

[286] Kuempel, W., and I. Wolff, "Digital signal processing of time domain field simulation results using the system identification method," *IEEE Trans. Microwave Theory and Techniques*, Vol. 42, 1994, pp. 667–671.

[287] Bi, Z., Y. Shen, K. Wu, and J. Litva, "Fast finite-difference time-domain analysis of resonators using digital filtering and spectrum estimation techniques," *IEEE Trans. Microwave Theory and Techniques*, Vol. 40, 1992, pp. 1611–1619.

[288] Jandhyala, V., E. Michielssen, and R. Mittra, "FDTD signal extrapolation using the forward-backward autoregressive (AR) model," *IEEE Microwave and Guided Wave Lett.*, Vol. 4, 1994, pp. 163–165.

[289] Jandhyala, V., E. Michielssen, and R. Mittra, "On the performance of different AR methods in the spectral estimation of FDTD waveforms," *Microwave and Optical Technology Lett.*, Vol. 7, 1994, pp. 690–692.

[290] Chen, J., C. Wu, T. K. Lo, K.-L. Wu, and J. Litva, "Using linear and nonlinear predictors to improve the computational efficiency of the FDTD algorithm," *IEEE Trans. Microwave Theory and Techniques*, Vol. 42, 1994, pp. 1992–1997.

[291] Wu, C., A. Navarro, and J. Litva, "Combination of finite impulse response neural network technique with FDTD method for simulation of electromagnetic problems," *Electronics Lett.*, Vol. 32, 1996, pp. 1112–1113.

[292] Goswami, J. C., and R. Mittra, "An application of FDTD in studying the end effects of slotline and coplanar waveguide with anisotropic substrates," *IEEE Trans. Microwave Theory and Techniques*, Vol. 45, 1997, pp. 1653–1657.

[293] Ritter, J., and F. Arndt, "Efficient FDTD / matrix-pencil method for the full-wave scattering parameter analysis of waveguiding structures," *IEEE Trans. Microwave Theory and Techniques*, Vol. 44, 1996, pp. 2450–2456.

[294] Deveze, T., L. Beaulie, and W. Tabbara, "An absorbing boundary condition for the fourth order FDTD scheme," *Proc. IEEE Antennas and Propagat. Soc. Int. Symp.*, Chicago, IL, 1992, pp. 342–345.

[295] Deveze, T., L. Beaulie, and W. Tabbara, "A fourth order scheme for the FDTD algorithm applied to Maxwell equations," *Proc. IEEE Antennas and Propagat. Soc. Int. Symp.*, Chicago, IL, 1992, pp. 346–349.

[296] Hadi, M. F., and M. Piket-May, "A modified FDTD (2,4) scheme for modeling electrically large structures with high phase accuracy," *IEEE Trans. Antennas and Propagation*, Vol. 45, 1997, pp. 254–264.

[297] Young, J. L., D. Gaitonde, and J. S. Shang, "Toward the construction of a fourth-order difference scheme for transient EM wave simulation: staggered grid approach," *IEEE Trans. Antennas and Propagation*, Vol. 45, 1997, pp. 1573–1580.

[298] Cole, J. B., "A high-accuracy realization of the Yee algorithm using non-standard finite differences," *IEEE Trans. Microwave Theory and Techniques*, Vol. 45, 1997, pp. 991–996.

[299] Liu, Q. H., "The PSTD algorithm: A time-domain method requiring only two cells per wavelength," *Microwave and Optical Technology Lett.*, Vol. 15, 1997, pp. 158–165.

[300] Krumpholz, M., and L. P. B. Katehi, "MRTD: New time-domain schemes based on multi-resolution analysis," *IEEE Trans. Microwave Theory and Techniques*, Vol. 44, 1996, pp. 555–571.

[301] Krumpholz, M., H. G. Winful, and L. P. B. Katehi, "Nonlinear time-domain modeling by multiresolution time domain (MRTD)," *IEEE Trans. Microwave Theory and Techniques*, Vol. 45, 1997, pp. 385–393.

[302] Werthen, M., and I. Wolff, "A novel wavelet based time domain simulation approach," *IEEE Microwave and Guided Wave Lett.*, Vol. 6, 1996, pp. 438–440.

[303] Tam, W. Y., "Comments on "New prospects for time domain analysis"," *IEEE Microwave and Guided Wave Lett.*, Vol. 6, 1996, p. 422.

[304] Shlager, K. L., J. G. Maloney, S. L. Ray, and A. F. Peterson, "Relative accuracy of several finite-difference time-domain methods in two and three dimensions," *IEEE Trans. Antennas and Propagation*, Vol. 41, 1993, pp. 1732–1737.

[305] Liu, Y., "Fourier analysis of numerical algorithms for Maxwell's equations," *J. Computational Physics*, Vol. 124, 1996, pp. 396–416.

[306] Shankar, V., W. F. Hall, and A. Mohammadian, "A time-domain differential solver for electromagnetic scattering problems," *Proc. IEEE*, Vol. 77, 1989, pp. 709–721.

[307] Shlager, K. L., and J. B. Schneider, "Analysis of the dispersion properties of the multiresolution time-domain (MRTD) scheme," *Proc. IEEE Antennas and Propagat. Soc. Int. Symp.*, Montreal, Canada, 1997, pp. 2144–2147.

[308] Maloney, J. G., G. S. Smith, and W. R. Scott, Jr., "Accurate computation of the radiation from simple antennas using the finite-difference time-domain method," *IEEE Trans. Antennas and Propagation*, Vol. 38, 1990, pp. 1059–1068.

[309] Boonzaaier, J. J., and C. W. Pistorius, "Thin wire dipoles — a finite-difference time-domain approach," *Electronics Lett.*, Vol. 26, 1990, pp. 1891–1892.

[310] Boonzaaier, J. J., and C. W. Pistorius, "Thin-wire Yagi antenna radiation patterns using the finite-difference time-domain method," *Microwave and Optical Technology Lett.*, Vol. 4, 1991, pp. 311–313.

[311] Kashiwa, T., S. Tanaka, and I. Fukai, "Time-domain analysis of Yagi-Uda antennas using the FDTD method," *Electronics and Communications in Japan, Part 1: Communications*, Vol. 77, 1994, pp. 96–105.

[312] Tirkas, P. A., and C. A. Balanis, "Finite-difference time-domain method for antenna radiation," *IEEE Trans. Antennas and Propagation*, Vol. 40, 1992, pp. 334–340.

[313] Luebbers, R., and K. Kunz, "Finite difference time domain calculations of antenna mutual coupling," *IEEE Trans. Electromagnetic Compatibility*, Vol. 34, 1992, pp. 357–359.

[314] Luebbers, R. J., and J. Beggs, "FDTD calculation of wide-band antenna gain and efficiency," *IEEE Trans. Antennas and Propagation*, Vol. 40, 1992, pp. 1403–1407.

[315] Maloney, J. G., K. L. Shlager, and G. S. Smith, "A simple FDTD model for transient excitation of antennas by transmission lines," *IEEE Trans. Antennas and Propagation*, Vol. 42, 1994, pp. 289–292.

[316] Katz, D. S., M. J. Piket-May, A. Taflove, and K. R. Umashankar, "FDTD analysis of electromagnetic wave radiation from systems containing horn antennas," *IEEE Trans. Antennas and Propagation*, Vol. 39, 1991, pp. 1203–1212.

[317] Tirkas, P. A., and C. A. Balanis, "Contour path FDTD methods for analysis of pyramidal horns with composite inner *E*-plane walls," *IEEE Trans. Antennas and Propagation*, Vol. 42, 1994, pp. 1476–1483.

[318] Reig, C., E. A. Navarro, and V. Such, "FDTD analysis of *E*-sectoral horn antennas for broad-band applications," *IEEE Trans. Antennas and Propagation*, Vol. 45, 1997, pp. 1484–1487.

[319] Ying, Z., and P.-S. Kildal, "FDTD analysis of a broadband feed with chokes for the radio telescope in Arecibo," *Proc. IEICE Int. Symp. on Antennas and Propagat.*, Chiba, Japan, 1996, pp. 305–308.

[320] Maloney, J. G., and G. S. Smith, "Optimization of pulse radiation from a simple antenna using resistive loading," *Microwave and Optical Technology Lett.*, Vol. 5, 1992, pp. 299–303.

[321] Maloney, J. G., and G. S. Smith, "A study of transient radiation from the Wu-King resistive monopole - FDTD analysis and experimental measurements," *IEEE Trans. Antennas and Propagation*, Vol. 41, 1993, pp. 668–676.

[322] Maloney, J. G., and G. S. Smith, "Optimization of a conical antenna for pulse radiation: an efficient design using resistive loading," *IEEE Trans. Antennas and Propagation*, Vol. 41, 1993, pp. 940–947.

[323] Shlager, K. L., G. S. Smith, and J. G. Maloney, "Optimization of bow-tie antennas for pulse radiation," *IEEE Trans. Antennas and Propagation*, Vol. 42, 1994, pp. 975–982.

[324] Shlager, K. L., G. S. Smith, and J. G. Maloney, "TEM horn antenna for pulse radiation: an improved design," *Microwave and Optical Technology Lett.*, Vol. 12, 1996, pp. 86–90.

[325] Shlager, K. L., G. S. Smith, and J. G. Maloney, "Accurate analysis of TEM horn antennas for pulse radiation," *IEEE Trans. Electromagnetic Compatibility*, Vol. 38, 1996, pp. 414–423.

[326] Hagness, S. C., A. Taflove, and J. E. Bridges, "Wideband ultralow reverberation antenna for biological sensing," *Electronics Lett.*, Vol. 33, 1997, pp. 1594–1595.

[327] Hagness, S. C., A. Taflove, and J. E. Bridges, "FDTD analysis of a pulsed microwave confocal system for breast cancer detection," *Proc. Int. Conf. IEEE Engineering in Medicine and Biology Soc.*, Chicago, IL, 1997, pp. 2506–2508.

[328] Montoya, T. P., and G. S. Smith, "A study of pulse radiation from several broad-band loaded monopoles," *IEEE Trans. Antennas and Propagation*, Vol. 44, 1996, pp. 1172–1182.

[329] Reineix, A., and B. Jecko, "Analysis of microstrip patch antennas using finite difference time domain method," *IEEE Trans. Antennas and Propagation*, Vol. 37, 1989, pp. 1361–1369.

[330] Sheen, D. M., S. M. Ali, M. D. Abouzahra, and J. A. Kong, "Application of the three-dimensional finite-difference time-domain method to the analysis of planar microstrip circuits," *IEEE Trans. Microwave Theory and Techniques*, Vol. 38, 1990, pp. 849–857.

[331] Leveque, P., A. Reineix, and B. Jecko, "Modeling dielectric losses in microstrip patch antennas: Application of FDTD method," *Electronics Lett.*, Vol. 28, 1992, pp. 539–540.

[332] Wu, C., K.-L. Wu, Z.-Q. Bi, and J. Litva, "Accurate characterization of planar printed antennas using finite-difference time-domain method," *IEEE Trans. Antennas and Propagation*, Vol. 40, 1992, pp. 526–533.

[333] Uehara, K., and K. Kagoshima, "FDTD method analysis of mutual coupling between microstrip antennas," *IEICE Trans. Communications*, Vol. E76-B, 1993, pp. 762–764.

[334] Oonishi, T., T. Kashiwa, and I. Fukai, "Analysis of microstrip antennas on a curved surface using the conformal grids FDTD method," *Electronics and Communications in Japan, Part 1: Communications*, Vol. 76, 1993, pp. 73–81.

[335] Qian, Y., S. Iwata, and E. Yamashita, "Optimal design of an offset-fed, twin-slot antenna element for millimeter-wave imaging arrays," *IEEE Microwave and Guided Wave Lett.*, Vol. 4, 1994, pp. 232–234.

[336] Reineix, A., and B. Jecko, "A time domain theoretical method for the analysis of microstrip antennas composed by slots," *Annales des Telecommunications*, Vol. 48, 1993, pp. 29–34.

[337] Reineix, A., J. Paillol, and B. Jecko, "FDTD method applied to the study of radar cross section of microstrip patch antennas," *Annales des Telecommunications*, Vol. 48, 1993, pp. 589–593.

[338] Reineix, A., C. Melon, T. Monediere, and F. Jecko, "The FDTD method applied to the study of microstrip patch antennas with a biased ferrite substrate," *Annales des Telecommunications*, Vol. 49, 1994, pp. 137–142.

[339] Shum, S. M., and K. M. Luk, "Characteristics of dielectric ring resonator antenna with an air gap," *Electronics Lett.*, Vol. 30, 1994, pp. 277–278.

[340] Shum, S. M., and K. M. Luk, "Analysis of aperture coupled rectangular dielectric resonator antenna," *Electronics Lett.*, Vol. 30, 1994, pp. 1726–1727.

[341] Shum, S. M., and K. M. Luk, "Numerical study of a cylindrical dielectric-resonator antenna coated with a dielectric layer," *IEE Proc. H: Microwaves, Antennas, and Propagation*, Vol. 142, 1995, pp. 189–191.

[342] Shum, S. M., and K. M. Luk, "FDTD analysis of probe-fed cylindrical dielectric resonator antenna operating in fundamental broadside mode," *Electronics Lett.*, Vol. 31, 1995, pp. 1210–1212.

[343] Shum, S. M., and K. M. Luk, "Stacked annular ring dielectric resonator antenna excited by axi-symmetric coaxial probe," *IEEE Trans. Antennas and Propagation*, Vol. 43, 1995, pp. 889–892.

[344] Esselle, K. P., "Finite-difference time-domain analysis of a rectangular dielectric resonator antenna," *J. Electrical and Electronics Engineering, Australia*, Vol. 15, 1995, pp. 63–69.

[345] Esselle, K. P., "A low-profile rectangular dielectric-resonator antenna," *IEEE Trans. Antennas and Propagation*, Vol. 44, 1996, pp. 1296–1297.

[346] Luebbers, R., L. Chen, T. Uno, and S. Adachi, "FDTD calculation of radiation patterns, impedance, and gain for a monopole antenna on a conducting box," *IEEE Trans. Antennas and Propagation*, Vol. 40, 1992, pp. 1577–1583.

[347] Chen, L., T. Uno, S. Adachi, and R. J. Luebbers, "FDTD analysis of a monopole antenna mounted on a conducting box covered with a layer of dielectric," *IEICE Trans. Communications*, Vol. E76-B, 1993, pp. 1583–1586.

[348] Toftgard, J., S. N. Hornsleth, and J. B. Andersen, "Effects on portable antennas of the presence of a person," *IEEE Trans. Antennas and Propagation*, Vol. 41, 1993, pp. 739–746.

[349] Jensen, M. A., and Y. Rahmat-Samii, "Performance analysis of antennas for hand-held transceivers using FDTD," *IEEE Trans. Antennas and Propagation*, Vol. 42, 1994, pp. 1106–1113.

[350] Navarro, E. A., N. T. Sangary, C. Wu, and J. Litva, "Analysis of a coupled patch antenna with application in personal communications," *IEE Proc. H: Microwaves, Antennas, and Propagation*, Vol. 142, 1995, pp. 495–497.

[351] Liu, Z. D., P. S. Hall, and D. Wake, "Dual-frequency planar inverted-F antenna," *IEEE Trans. Antennas and Propagation*, Vol. 45, 1997, pp. 1451–1458.

[352] Rowell, C. R., and R. D. Murch, "A capacitively loaded PIFA for compact mobile telephone handsets," *IEEE Trans. Antennas and Propagation*, Vol. 45, 1997, pp. 837–842.

[353] Dimbylow, P. J., and S. M. Mann, "SAR calculations in an anatomically realistic model of the head for mobile communication transceivers at 900 MHz and 1.8 GHz," *Physics in Medicine and Biology*, Vol. 39, 1994, pp. 1537–1553.

[354] Chen, H. Y., and H. H. Wang, "Current and SAR induced in a human head model by electromagnetic fields irradiated from a cellular phone," *IEEE Trans. Microwave Theory and Techniques*, Vol. 42, 1994, pp. 2249–2254.

[355] Jensen, M. A., and Y. Rahmat-Samii, "EM interaction of handset antennas and a human in personal communications," *Proc. IEEE*, Vol. 83, 1995, pp. 7–17.

[356] Martens, L., J. De Moerloose, D. De Zutter, J. De Poorter, and C. De Wagter, "Calculation of the electromagnetic fields induced in the head of an operator of a cordless telephone," *Radio Science*, Vol. 30, 1995, pp. 283–290.

[357] Excell, P. S., P. Olley, and N. N. Jackson, "Modeling of an arbitrarily-oriented mobile telephone handset in the finite-difference time-domain field computation method," *J. Applied Computational Electromagnetics Soc.*, Vol. 11, 1996, pp. 55–65.

[358] Okoniewski, M., and M. A. Stuchly, "A study of the handset antenna and human body interaction," *IEEE Trans. Microwave Theory and Techniques*, Vol. 44, 1996, pp. 1855–1864.

[359] Watanabe, S., M. Taki, T. Nojima, and O. Fujiwara, "Characteristics of the SAR distributions in a head exposed to electromagnetic fields radiated by a hand-held portable radio," *IEEE Trans. Microwave Theory and Techniques*, Vol. 44, 1996, pp. 1874–1883.

[360] Gandhi, O. P., and J.-Y. Chen, "Electromagnetic absorption in the human head from experimental 6-GHz handheld transceivers," *IEEE Trans. Electromagnetic Compatibility*, Vol. 37, 1995, pp. 547–558.

[361] Gandhi, O. P., G. Lazzi, and C. M. Furse, "Electromagnetic absorption in the human head and neck for mobile telephones at 835 and 1900 MHz," *IEEE Trans. Microwave Theory and Techniques*, Vol. 44, 1996, pp. 1884–1897.

[362] Taflove, A., and M. E. Brodwin, "Computation of the electromagnetic fields and induced temperatures within a model of the microwave irradiated human eye," *IEEE Trans. Microwave Theory and Techniques*, Vol. 23, 1975, pp. 888–896.

[363] Cherry, P. C., and M. F. Iskander, "FDTD analysis of power deposition patterns of an array of interstitial antennas for use in microwave hyperthermia," *IEEE Trans. Microwave Theory and Techniques*, Vol. 40, 1992, pp. 1692–1700.

[364] Cherry, P. C., and M. F. Iskander, "Calculations of heating patterns of an array of microwave interstitial antennas," *IEEE Trans. Biomedical Engineering*, Vol. 40, 1993, pp. 771–779.

[365] Ren, J. R., O. P. Gandhi, L. R. Walker, J. Fraschilla, and C. R. Boerman, "Floquet-based FDTD analysis of two-dimensional phased array antennas," *IEEE Microwave and Guided Wave Lett.*, Vol. 4, 1994, pp. 109–111.

[366] Thiele, E., and A. Taflove, "FDTD analysis of Vivaldi flared horn antennas and arrays," *IEEE Trans. Antennas and Propagation*, Vol. 42, 1994, pp. 633–641.

[367] Naito, M., S.-I. Matsuzawa, and K. Ito, "FDTD analysis of unit radiator for a circularly polarized printed array antenna composed of strips and slots," *IEICE Trans. Communications*, Vol. 77, 1994, pp. 1621–1627.

[368] Uehara, K., and K. Kagoshima, "Rigorous analysis of microstrip phased array antennas using a new FDTD method," *Electronics Lett.*, Vol. 30, 1994, pp. 100–101.

[369] Crouch, D., "On the use of symmetry to reduce the computational requirements for FDTD analyses of finite phased arrays," *Microwave and Optical Technology Lett.*, Vol. 13, 1996, pp. 123–128.

[370] Beggs, J. J., R. J. Luebbers, and B. G. Ruth, "Analysis of electromagnetic radiation from shaped-end radiators using the finite difference time domain method," *IEEE Trans. Antennas and Propagation*, Vol. 41, 1993, pp. 1324–1327.

[371] Maeshima, O., T. Uno, Y. He, and S. Adachi, "FDTD analysis of two-dimensional cavity-backed antenna for subsurface radar," *IEICE Trans. Electronics*, Vol. E-76-C, 1993, pp. 1468–1473.

[372] Penney, C. W., and R. J. Luebbers, "Input impedance, radiation pattern, and radar cross-section of spiral antennas using FDTD," *IEEE Trans. Antennas and Propagation*, Vol. 42, 1994, pp. 1328–1332.

[373] Penney, C. W., and R. J. Luebbers, "Radiation and scattering of a square Archimedean spiral antenna using FDTD," *Electromagnetics*, Vol. 14, 1994, pp. 87–98.

[374] Tsai, H.-S., and R. A. York, "FDTD analysis of CPW-fed folded-slot and multiple-slot antennas on thin substrates," *IEEE Trans. Antennas and Propagation*, Vol. 44, 1996, pp. 217–226.

[375] Simons, R. N., N. I. Dib, R. Q. Lee, and L. P. Katehi, "Integrated uniplanar transition for linearly tapered slot antenna," *IEEE Trans. Antennas and Propagation*, Vol. 43, 1995, pp. 998–1002.

[376] Rayner, M., A. D. Olver, and A. D. Monk, "FDTD design of short backfire antennas," *IEE Proc. H: Microwaves, Antennas, and Propagation*, Vol. 44, 1997, pp. 1–6.

[377] Chen, M., B. Houshmand, T. Itoh, "FDTD analysis of a metal-strip-loaded dielectric leaky-wave antenna," *IEEE Trans. Antennas and Propagation*, Vol. 45, 1997, pp. 1294–1301.

[378] Toland, B., C. C. Liu, and P. G. Ingerson, "Design and analysis of arbitrarily shaped dielectric antennas," *Microwave J.*, Vol. 40, 1997, pp. 278–286.

[379] Ruoss, H.-O., P. M. Maier, F. M. Landstorfer, and G. Villino, "Lens antennas for mobile broadband systems in the millimetre-wave range," *Electronics Lett.*, Vol. 33, 1997, pp. 1510–1512.

[380] DePourcq, M., "Field and power-density calculations in closed microwave systems by three-dimensional finite differences," *IEE Proc. H: Microwaves, Antennas, and Propagation*, Vol. 132, 1985, pp. 360–368.

[381] Olivier, J. C., and D. A. McNamara, "Finite-difference time-domain (FDTD) analysis of discontinuities in homogeneous, dispersive waveguides," *Electronics Lett.*, Vol. 25, 1989, pp. 1006–1007.

[382] Olivier, J. C., and D. A. McNamara, "Analysis of edge slots in rectangular waveguide using finite-difference time-domain method," *Electronics Lett.*, Vol. 26, 1990, pp. 1135–1136.

[383] Olivier, J. C., and D. A. McNamara, "Analysis of multiport rectangular waveguide devices using pulsed finite-difference time-domain (FDTD)," *Electronics Lett.*, Vol. 28, 1992, pp. 129–130.

[384] Olivier, J. C., "Mutual coupling between waveguide apertures mounted on a common conducting surface using a time- and Fourier-gated pulsed FDTD method," *IEEE Microwave and Guided Wave Lett.*, Vol. 3, 1993, pp. 177–179.

[385] Olivier, J. C., and D. A. McNamara, "Analysis of multiport discontinuities in waveguide using a pulsed FDTD approach," *IEEE Trans. Microwave Theory and Techniques*, Vol. 42, 1994, pp. 2229–2238.

[386] Bi, Z., K. Wu, and J. Litva, "Application of the FDTD method to the analysis of *H*-plane waveguide discontinuities," *Electronics Lett.*, Vol. 26, 1990, pp. 1897–1898.

[387] Navarro, E. A., V. Such, B. Gimeno, and J. L. Cruz, "Analysis of *H*-plane waveguide discontinuities with an improved finite-difference time domain algorithm," *IEE Proc. H: Microwaves, Antennas, and Propagation*, Vol. 139, 1992, pp. 183–185.

[388] Navarro, E. A., L. Gallart, J. L. Cruz, B. Gimeno, and V. Such, "Accurate absorbing boundary conditions for the FDTD analysis of *H*-plane waveguide discontinuities," *IEE Proc. H: Microwaves, Antennas, and Propagation*, Vol. 141, 1994, pp. 59–61.

[389] Navarro, E. A., V. Such, B. Gimeno, and J. L. Cruz, "*T*-junctions in square coaxial waveguide: an FDTD approach," *IEEE Trans. Microwave Theory and Techniques*, Vol. 42, 1994, pp. 347–350.

[390] Navarro, E. A., and V. Such, "Study of TE and TM modes in waveguides of arbitrary cross-section using an FDTD formulation," *IEE Proc. H: Microwaves, Antennas, and Propagation*, Vol. 139, 1992, pp. 491–494.

[391] Navarro, E. A., "Numerical analysis of rectangular waveguide multiple-slot narrow-wall couplers using a 2D FDTD algorithm," *Microwave and Optical Technology Lett.*, Vol. 7, 1994, pp. 696–699.

[392] Chu, S. T., *Modeling of Guided-Wave Optical Structures by the Finite-Difference Time-Domain Method*, Ph.D. dissertation, University of Waterloo, Waterloo, Canada, 1990.

[393] Chu, S. T., and S. K. Chaudhuri, "Combining modal analysis and the finite-difference time-domain method in the study of dielectric waveguide problems," *IEEE Trans. Microwave Theory and Techniques*, Vol. 38, 1990, pp. 1755–1760.

[394] Jarem, J. M., "A method of moments analysis and a finite-difference time-domain analysis of a probe-sleeve fed rectangular waveguide cavity," *IEEE Trans. Microwave Theory and Techniques*, Vol. 39, 1991, pp. 444–451.

[395] Alinikula, P., and K. S. Kunz, "Analysis of waveguide aperture coupling using the finite-difference time-domain method," *IEEE Microwave and Guided Wave Lett.*, Vol. 1, 1991, pp. 189–191.

[396] Van Hese, J., and D. De Zutter, "Modeling of discontinuities in general coaxial waveguide structures by the FDTD method," *IEEE Trans. Microwave Theory and Techniques*, Vol. 40, 1992, pp. 547–556.

[397] Ko, W., "Time domain solution of electromagnetic problems," *Electromagnetics*, Vol. 12, 1992, pp. 403–433.

[398] Feng, Z., and F. Junmei, "Efficient analysis of a dielectric post in a rectangular waveguide based on the two-dimensional FDTD method," *Microwave and Optical Technology Lett.*, Vol. 6, 1993, pp. 407–410.

[399] Dib, N. I., and L. P. B. Katehi, "Analysis of the transition from rectangular waveguide to shielded dielectric image guide using the finite-difference time-domain method," *IEEE Microwave and Guided Wave Lett.*, Vol. 3, 1993, pp. 327–329.

[400] Kraut, E. A., J. C. Olivier, and J. B. West, "FDTD solution of Maxwell's equations for an edge slot penetrating adjacent broadwalls of a finite wall thickness waveguide," *IEEE Trans. Antennas and Propagation*, Vol. 42, 1994, pp. 1646–1648.

[401] Pereda, J. A., L. A. Vielva, A. Vegas, A. Prieto, "FDTD analysis of magnetized ferrites: An approach based on the rotated Richtmyer difference scheme," *IEEE Microwave and Guided Wave Lett.*, Vol. 3, 1993, pp. 322–324.

[402] Pereda, J. A., L. A. Vielva, A. Vegas, and A. Prieto, "A treatment of magnetized ferrites using the FDTD method," *IEEE Microwave and Guided Wave Lett.*, Vol. 3, 1993, pp. 136–138.

[403] Navarro, A., M. J. Nunez, and E. Martin, "Finite difference time domain FFT method applied to axially symmetrical electromagnetic resonant devices," *IEE Proc. H: Microwaves, Antennas, and Propagation*, Vol. 137, 1990, pp. 193–196.

[404] Navarro, A., M. J. Nunez, and E. Martin, "Study of TE_0 and TM_0 modes in dielectric resonators by a finite difference time-domain method coupled with the discrete Fourier transform," *IEEE Trans. Microwave Theory and Techniques*, Vol. 39, 1991, pp. 14–17.

[405] Shen, Y., Z. Bi, K. Wu, and J. Litva, "FDTD analysis of open cylindrical dielectric resonators," *Microwave and Optical Technology Lett.*, Vol. 5, 1992, pp. 261–265.

[406] Wang, C., B.-Q. Gao, and C.-P. Ding, "Q factor of a resonator by the finite-difference time-domain method incorporating perturbation techniques," *Electronics Lett.*, Vol. 29, 1993, pp. 1866–1867.

[407] Hagness, S. C., D. Rafizadeh, S. T. Ho, and A. Taflove, "FDTD microcavity simulations: Design and experimental realization of waveguide-coupled single-mode ring and whispering-gallery-mode disk resonators," *J. Lightwave Technology*, Vol. 15, 1997, pp. 2154–2165.

[408] Fang, J., X. Zhang, and K. K. Mei, "Dispersion characteristics of microstrip lines in the vicinity of a coplanar ground," *Electronics Lett.*, Vol. 23, 1987, pp. 1142–1143.

[409] Zhang, X., J. Fang, K. K. Mei, and Y. Liu, "Calculations of the dispersive characteristics of microstrips by the time-domain finite difference method," *IEEE Trans. Microwave Theory and Techniques*, Vol. 36, 1988, pp. 263–267.

[410] Zhang, X., and K. K. Mei, "Time-domain finite difference approach to the calculation of the frequency-dependent characteristics of microstrip discontinuities," *IEEE Trans. Microwave Theory and Techniques*, Vol. 36, 1988, pp. 1775–1787.

[411] Liang, G.-C., Y.-W. Liu, and K. K. Mei, "Full-wave analysis of coplanar waveguide and slotline using the time-domain finite-difference method," *IEEE Trans. Microwave Theory and Techniques*, Vol. 37, 1989, pp. 1949–1957.

[412] Moore, J., and H. Ling, "Characterization of a 90° microstrip bend with an arbitrary miter via time-domain finite difference method," *IEEE Trans. Microwave Theory and Techniques*, Vol. 38, 1990, pp. 405–410.

[413] Feix, N., M. Lalande, and B. Jecko, "Harmonical characterization of a microstrip bend via the finite difference time domain method," *IEEE Trans. Microwave Theory and Techniques*, Vol. 40, 1992, pp. 955–961.

[414] Shibata, T., and E. Sano, "Characterization of MIS structure coplanar transmission lines for investigation of signal propagation in integrated circuits," *IEEE Trans. Microwave Theory and Techniques*, Vol. 38, 1990, pp. 881–890.

[415] Shibata, T., and E. Sano, "Analysis of coplanar lines utilizing the finite difference time-domain technique," *Electronics and Communications in Japan: Part 2 Electronics*, Vol. 73, no. 11, 1990, pp. 61–70.

[416] Wu, L. K., and Y. C. Chang, "Characterization of the shielding effects on the frequency-dependent effective dielectric constant of a waveguide-shielded microstrip using the finite-difference time-domain method," *IEEE Trans. Microwave Theory and Techniques*, Vol. 39, 1991, pp. 1688–1693.

[417] Zheng, G. W., and K. S. Chen, "Transient analysis of microstrip lines with ferrite substrate by extended FDTD method," *Int. J. Infrared and Millimeter Waves*, Vol. 13, 1992, pp. 1115–1125.

[418] Zheng, G. W., and K. S. Chen, "Transient response of microstrip step discontinuities on anisotropic substrate," *Int. J. Infrared and Millimeter Waves*, Vol. 13, 1992, pp. 1609–1617.

[419] Qian, Y., S. Iwata, and E. Yamashita, "Characterization of the perturbation effect of a probe head using FDTD method," *IEEE Microwave and Guided Wave Lett.*, Vol. 4, 1994, pp. 317–319.

[420] Zheng, G. W., and K. S. Chen, "The studies of cylindrical microstrip line with the FDTD method in cylindrical coordinate system," *Int. J. Infrared and Millimeter Waves*, Vol. 13, 1992, pp. 1421–1431.

[421] Kitamura, T., T. Koshimae, M. Hira, and S. Kurazono, "Analysis of cylindrical microstrip lines utilizing the finite-difference time-domain method," *IEEE Trans. Microwave Theory and Techniques*, Vol. 42, 1994, pp. 1279–1282.

[422] Bi, Z., K. Wu, C. Wu, and J. Litva, "A dispersive boundary condition for microstrip component analysis using the FDTD method," *IEEE Trans. Microwave Theory and Techniques*, Vol. 40, 1992, pp. 774–776.

[423] Betz, V., and R. Mittra, "Comparison and evaluation of boundary conditions for the absorption of guided waves in an FDTD simulation," *IEEE Microwave and Guided Wave Lett.*, Vol. 2, 1992, pp. 499–501.

[424] Betz, V., and R. Mittra, "A boundary condition to absorb both propagating and evanescent waves in a finite-difference time-domain simulation," *IEEE Microwave and Guided Wave Lett.*, Vol. 3, 1993, pp. 182–184.

[425] Maeda, S., T. Kashiwa, and I. Fukai, "Full wave analysis of propagation characteristics of a through hole using the finite-difference time-domain method," *IEEE Trans. Microwave Theory and Techniques*, Vol. 39, 1991, pp. 2154–2159.

[426] Becker, W. D., P. H. Harms, and R. Mittra, "Time-domain electromagnetic analysis of interconnects in a computer chip package," *IEEE Trans. Microwave Theory and Techniques*, Vol. 40, 1992, pp. 430–451.

[427] Mezzanotte, P., M. Mongiardo, L. Roselli, R. Sorrentino, and W. Heinrich, "Analysis of packaged microwave integrated circuits by FDTD," *IEEE Trans. Microwave Theory and Techniques*, Vol. 42, 1994, pp. 1796–1801.

[428] Yook, J.-G., N. I. Dib, and L. P. B. Katehi, "Characterization of high frequency interconnects using finite difference time domain and finite element methods," *IEEE Trans. Microwave Theory and Techniques*, Vol. 42, 1994, pp. 1727–1736.

[429] Cherry, P. C., and M. F. Iskander, "FDTD analysis of high frequency electronic interconnection effects," *IEEE Trans. Microwave Theory and Techniques*, Vol. 43, 1995, pp. 2445–2451.

[430] Pillai, E., and W. Wiesbeck, "FDTD analysis of wideband aperture coupled interconnect," *Electronics Lett.*, Vol. 31, 1995, pp. 982–983.

[431] Pillai, E. R., "Coax via — A technique to reduce crosstalk and enhance impedance match at vias in high-frequency multilayer packages verified by FDTD and MoM modeling," *IEEE Trans. Microwave Theory and Techniques*, Vol. 45, 1997, pp. 1981–1985.

[432] Lam, C.-W., S. M. Ali, and P. Nuytkens, "Three-dimensional modeling of multichip module interconnects," *IEEE Trans. Components, Hybrids, and Manufacturing Technology*, Vol. 16, 1993, pp. 699–704.

[433] Gribbons, M., A. C. Cangellaris, and J. L. Prince, "Finite-difference time-domain analysis of pulse propagation in multichip module interconnects," *IEEE Trans. Components, Hybrids, and Manufacturing Technology*, Vol. 16, 1993, pp. 490–498.

[434] Visan, S., O. Picon, and V. Fouad Hanna, "3D FDTD characterization of interconnections between MMIC and MIC modules," *Microwave and Optical Technology Lett.*, Vol. 7, 1994, pp. 279–281.

[435] Becker, W. D., S. Chebolu, and R. Mittra, "Finite difference time domain modeling of high speed electronic packages," *Microwave J.*, Vol. 38, 1995, no. 12, pp. 62–74.

[436] Ghouz, H. H., and E.-B. El-Sharawy, "Finite-difference time-domain analysis of flip-chip interconnects with staggered bumps," *IEEE Trans. Microwave Theory and Techniques*, Vol. 44, 1996, pp. 960–963.

[437] Ghouz, H. H., and E.-B. El-Sharawy, "An accurate equivalent circuit model of flip chip and via interconnects," *IEEE Trans. Microwave Theory and Techniques*, Vol. 44, 1996, pp. 2543–2554.

[438] Maeda, S., T. Kashiwa, and I. Fukai, "Analysis of crosstalk between parallel microstrips using finite-difference time-domain method," *Electronics and Communications in Japan, Part 2: Electronics*," Vol. 74, no. 11, 1991, pp. 23–31.

[439] Pothecary, N. M., and C. J. Railton, "Analysis of cross-talk on high-speed digital circuits using the finite difference time-domain method," *Int. J. Numerical Modeling*, Vol. 4, 1991, pp. 225–240.

[440] Cerri, G., R. De Leo, V. M. Primiani, and A. Schiavoni, "FDTD analysis of crosstalk in multilayer printed circuits," *Int. J. Numerical Modeling*, Vol. 7, 1994, pp. 209–221.

[441] Rizvi, M., and J. LoVetri, "Modeling and reduction of crosstalk on coupled microstrip line structures and multichip modules: An FDTD approach," *Int. J. Microwave and Millimeter-Wave Computer-Aided Engineering*, Vol. 6, 1996, pp. 58–68.

[442] Paul, C. R., "Incorporation of terminal constraints in the FDTD analysis of transmission lines," *IEEE Trans. Electromagnetic Compatibility*, Vol. 36, 1994, pp. 85–91.

[443] Xiao, S., R. Vahldieck, and H. Jin, "Full-wave analysis of guided wave structures using a novel 2-D FDTD," *IEEE Microwave and Guided Wave Lett.*, Vol. 2, 1992, pp. 165–167.

[444] Asi, A., and L. Shafai, "Dispersion analysis of anisotropic inhomogeneous waveguides using compact 2D FDTD," *Electronics Lett.*, Vol. 28, 1992, pp. 1451–1452.

[445] Asi, A., and L. Shafai, "Correction to 'Dispersion analysis of anisotropic inhomogeneous waveguides using compact 2D FDTD'," *Electronics Lett.*, Vol. 29, 1993, p. 423.

[446] Brankovic, V. J., D. V. Krupeuzevic, and F. Arndt, "An efficient two-dimensional graded mesh finite-difference time-domain algorithm for shielded or open waveguide structures," *IEEE Trans. Microwave Theory and Techniques*, Vol. 40, 1992, pp. 2272–2277.

[447] Cangellaris, A. C., "Numerical stability and numerical dispersion of a compact 2D FDTD method used for the dispersion analysis of waveguides," *IEEE Microwave and Guided Wave Lett.*, Vol. 3, 1993, pp. 3–5.

[448] Mroczkowski, C., and W. K. Gwarek, "Microwave circuits described by two-dimensional vector wave equation and their analysis by FDTD method," *Proc. 21st European Microwave Conf.*, Stuttgart, Germany, 1991, pp. 866–871.

[449] Okoniewski, M., "Vector wave equation 2-D FDTD method for guided wave problems," *IEEE Microwave and Guided Wave Lett.*, Vol. 3, 1993, pp. 307–309.

[450] Krupeuzevic, D. V., V. J. Brankovic, and F. Arndt, "The wave-equation FDTD method for the efficient eigenvalue analysis and *S*-matrix computation of waveguide structures," *IEEE Trans. Microwave Theory and Techniques*, Vol. 41, 1993, pp. 2109–2115.

[451] Celuch-Marcysiak, M., and W. K. Gwarek, "Spatially looped algorithms for time-domain analysis of periodic structures," *IEEE Trans. Microwave Theory and Techniques*, Vol. 43, 1995, pp. 860–865.

[452] Hofschen, S., and I. Wolff, "Simulation of an elevated coplanar waveguide using 2-D FDTD," *IEEE Microwave and Guided Wave Lett.*, Vol. 6, 1996, pp. 28–30.

[453] Hong, I. P., and H. K. Park, "Dispersion characteristics of a unilateral fin-line using 2D FDTD," *Electronics Lett.*, Vol. 32, 1996, pp. 1992–1994.

[454] Hong, I. P., Y. K. Lee, and H. K. Park, "Dispersion characteristics of a broadside-coupled coplanar waveguide," *Electronics Lett.*, Vol. 33, 1997, pp. 965–966.

[455] Fujii, M., and S. Kobayashi, "Accurate analysis of losses in waveguide structures by compact two-dimensional FDTD method combined with autoregressive signal analysis," *IEEE Trans. Microwave Theory and Techniques*, Vol. 44, 1996, pp. 970–975.

[456] Hofschen, S., and I. Wolff, "Improvements of the two-dimensional FDTD method for the simulation of normal and superconducting planar waveguides using time series analysis," *IEEE Trans. Microwave Theory and Techniques*, Vol. 44, 1996, pp. 1487–1490.

[457] Shi, H., and J. L. Drewniak, "Modeling waveguide couplers with an explicit FDTD discrete surface integral method," *Microwave and Optical Technology Lett.*, Vol. 15, 1997, pp. 208–213.

[458] Tentzeris, E. M., M. Krumpholz, N. Dib, and L. P. B. Katehi, "A waveguide absorber based on analytic Green's functions," *Proc. 25th European Microwave Conf.*, Bologna, Italy, 1995, pp. 251–254.

[459] Jung, K.-Y., H. Kim, and K.-C. Ko, "An improved unimodal absorbing boundary condition for waveguide problems," *IEEE Microwave and Guided Wave Lett.*, Vol. 7, 1997, pp. 368–370.

[460] Park, M.-J., and S. Nam, "FDTD ABC for waveguides with a new discrete Green's function and its effective implementation," *Electronics Lett.*, Vol. 32, 1996, pp. 2128–2130.

[461] Alimenti, F., P. Mezzanotte, L. Roselli, and R. Sorrentino, "On the analysis of multiple waveguide discontinuities by combined 1D - 3D FDTD method," *Proc. 25th European Microwave Conf.*, Bologna, Italy, 1995, pp. 449–452.

[462] Mrozowski, M., M. Niedzwiecki, and P. Suchomski, "A fast recursive highly dispersive absorbing boundary condition using time domain diakoptics and Laguerre polynomials," *IEEE Microwave and Guided Wave Lett.*, Vol. 5, 1995, pp. 183–185.

[463] Taylor, C. D., D. H. Lam, and T. H. Shumpert, "Electromagnetic pulse scattering in time-varying inhomogeneous media," *IEEE Trans. Antennas and Propagation*, Vol. 17, 1969, pp. 585–589.

[464] Harfoush, F., A. Taflove, and G. A. Kriegsmann, "A numerical technique for analyzing electromagnetic wave scattering from moving surfaces in one and two dimensions," *IEEE Trans. Antennas and Propagation*, Vol. 37, 1989, pp. 55–63.

[465] Wu, L.-K., and L.-T. Han, "Implementation and application of resistive sheet boundary condition in the finite-difference time-domain method," *IEEE Trans. Antennas and Propagation*, Vol. 40, 1992, pp. 628–633.

[466] Wu, L.-K., and L.-T. Han, "*E*-polarized scattering from a conducting rectangular cylinder with an infinite axial slot filled by a resistively coated dielectric strip," *IEEE Trans. Antennas and Propagation*, Vol. 40, 1992, pp. 731–733.

[467] Lee, J.-F., "Numerical solutions of TM scattering using an obliquely Cartesian finite difference time domain algorithm," *IEE Proc. H: Microwaves, Antennas, Prop.*, Vol. 140, 1993, pp. 23–28.

[468] Chan, C. H., S. H. Lou, L. Tsang, and J. A. Kong, "Electromagnetic scattering of waves by rough surfaces: a finite-difference time-domain approach," *Microwave and Optical Technology Lett.*, Vol. 4, 1991, pp. 355–359.

[469] Fung, A. K., M. R. Shah, and S. Tjuatja, "Numerical simulation of scattering from three-dimensional randomly rough surfaces," *IEEE Trans. Geoscience and Remote Sensing*, Vol. 32, 1884, pp. 986–994.

[470] Hastings, F. D., J. B. Schneider, and S. L. Broschat, "A Monte-Carlo FDTD technique for rough surface scattering," *IEEE Trans. Antennas and Propagation*, Vol. 43, 1995, pp. 1183–1191.

[471] Luebbers, R. J., and C. Penney, "Scattering from apertures in infinite ground planes using FDTD," *IEEE Trans. Antennas and Propagation*, Vol. 42, 1994, pp. 731–736.

[472] Demarest, K., R. Plumb, and Z. Huang, "FDTD modeling of scatterers in stratified media," *IEEE Trans. Antennas and Propagation*, Vol. 43, 1995, pp. 1164–1168.

[473] Wong, P. B., G. L. Tyler, J. E. Baron, E. M. Gurrola, and R. A. Simpson, "A three-wave FDTD approach to surface scattering with applications to remote sensing of geophysical surfaces," *IEEE Trans. Antennas and Propagation*, Vol. 44, 1996, pp. 504–513.

[474] Stratis, G., V. Anantha, and A. Taflove, "Numerical calculation of diffraction coefficients of generic conducting and dielectric wedges using FDTD," *IEEE Trans. Antennas and Propagation*, Vol. 45, 1997, pp. 1525–1529.

[475] Navarro, E. A., B. Gimeno, and J. L. Cruz, "Modeling of periodic structures using the finite-difference time-domain method combined with the Floquet theorem," *Electronics Lett.*, Vol. 29, 1993, pp. 446–447.

[476] Tsay, W.-J., and D. M. Pozar, "Application of the FDTD technique to periodic problems in scattering and radiation," *IEEE Microwave and Guided Wave Lett.*, Vol. 3, 1993, pp. 250–252.

[477] Ko, W. L., and R. Mittra, "Implementation of Floquet boundary condition in FDTD for FSS analysis," *Proc. IEEE Antennas and Propagat. Soc. Int. Symp.*, Ann Arbor, MI, 1993, pp. 14–17.

[478] Prescott, D. T., and N. V. Shuley, "Extensions to the FDTD method for the analysis of infinitely periodic arrays," *IEEE Microwave and Guided Wave Lett.*, Vol. 4, 1994, pp. 352–354.

[479] Veysoglu, M. E., R. T. Shin, and J. A. Kong, "A finite-difference time-domain analysis of wave scattering from periodic surfaces: Oblique incidence case," *J. Electromagnetic Waves and Applications*, Vol. 7, 1993, pp. 1595–1608.

[480] Harms, P. H., R. Mittra, and W. Ko, "Implementation of the periodic boundary condition in the finite-difference time-domain algorithm for FSS structures," *IEEE Trans. Antennas and Propagation*, Vol. 42, 1994, pp. 1317–1324.

[481] Cangellaris, A. C., M. Gribbons, and G. Sohos, "A hybrid spectral / FDTD method for the electromagnetic analysis of guided waves in periodic structures," *IEEE Microwave and Guided Wave Lett.*, Vol. 3, 1993, pp. 375–377.

[482] Moghaddam, M., W. C. Chew, B. Anderson, E. Yannakakis, and Q. H. Liu, "Computation of transient electromagnetic waves in inhomogeneous media," *Radio Science*, Vol. 26, 1991, pp. 265–273.

[483] Moghaddam, M., E. J. Yannakakis, W. C. Chew, and C. Randall, "Modeling of the subsurface interface radar," *J. Electromagnetic Waves and Applications*, Vol. 5, 1991, pp. 17–39.

[484] Maeshima, O., T. Uno, Y. He, and S. Adachi, "FDTD analysis of transient cylindrical electromagnetic wave scattered by objects buried in a two-dimensional inhomogeneous medium," *Trans. IEICE B-II*, Vol. J76B-II, 1993, pp. 138–141.

[485] He, Y., T. Uno, and S. Adachi, "FDTD analysis of two-dimensional transient scattering of cylindrical wave via buried conducting objects," *Electronics and Communications in Japan, Part 1: Communications*, Vol. 77, 1994, pp. 93–102.

[486] Bourgeois, J. M., and G. S. Smith, "A fully three-dimensional simulation of a ground-penetrating radar: FDTD theory compared with experiment," *IEEE Trans. Geoscience and Remote Sensing*, Vol. 34, 1996, pp. 36–44.

[487] Bourgeois, J. M., and G. S. Smith, "A complete electromagnetic simulation of a ground penetrating radar for mine detection: Theory and experiment," *Proc. IEEE Antennas and Propagat. Soc. Int. Symp.*, Montreal, Canada, 1997, pp. 986–989.

[488] Zook, B. J., "Validation of FDTD modeling of ground-penetrating radar antennas," *Proc. 12th Annual Review of Progress in Applied Computational Electromagnetics*, Monterey, CA, 1996, pp. 508–515.

[489] Rappaport, C. M., and W. H. Weedon, "Efficient modeling of electromagnetic characteristics of soil for FDTD ground penetrating radar simulation," *Proc. IEEE Antennas and Propagat. Soc. Int. Symp.*, Baltimore, MD, 1996, pp. 620–623.

[490] Taflove, A., and K. Umashankar, "A hybrid moment method / finite-difference time-domain approach to electromagnetic coupling and aperture penetration into complex geometries," *IEEE Trans. Antennas and Propagation*, Vol. 30, 1982, pp. 617–627.

[491] Aoyagi, P. H., J.-F. Lee, and R. Mittra, "A hybrid Yee algorithm / scalar-wave equation approach," *IEEE Trans. Microwave Theory and Techniques*, Vol. 41, 1993, pp. 1593–1600.

[492] Lee, R., and T. T. Chia, "Analysis of electromagnetic scattering from a cavity with a complex termination by means of a hybrid-ray FDTD method," *IEEE Trans. Antennas and Propagation*, Vol. 41, 1993, pp. 1560–1569.

[493] Mrozowski, M., "A hybrid PEE-FDTD algorithm for accelerated time domain analysis of electromagnetic waves in shielded structures," *IEEE Microwave and Guided Wave Lett.*, Vol. 4, 1994, pp. 323–325.

[494] Mrozowski, M., M. Okoniewski, and M. A. Stuchly, "Hybrid PEE-FDTD method for efficient field modeling in cylindrical coordinates," *Electronics Lett.*, Vol. 32, 1996, pp. 194–195.

[495] Beillard, B., J. Andrieu, and B. Jecko, "Coupling of methods: Finite difference time domain and asymptotic methods," *Electronics Lett.*, Vol. 32, 1996, pp. 308–309.

[496] Taflove, A., *Computational Electrodynamics: The Finite-Difference Time-Domain Method*, Norwood, MA: Artech House, 1995.

SELECTED BIBLIOGRAPHY: NOVEL AND EMERGING FDTD AREAS

Bellanca, G., R. Semprini, and P. Bassi, "FDTD modeling of spatial soliton propagation," *Opt. Quantum Electronics*, Vol. 29, 1997, pp. 233–241.

Bian, R. X., R. C. Dunn, X. S. Xie, and P. T. Leung, "Single molecule emission characteristics in near-field microscopy," *Phys. Review Lett.*, Vol. 75, 1995, pp. 4772–4775.

Botteldooren, D., "Acoustical finite-difference time-domain simulation in a quasi-Cartesian grid," *J. Acoustical Soc. of America*, Vol. 95, 1994, pp. 2313–2319.

Bromage, J., S. Radic, G. P. Agrawal, C. R. Stroud Jr., P. M. Fauchet, and R. Sobolewski, "Spatiotemporal shaping of terahertz pulses," *Optics Lett.*, Vol. 22, 1997, pp. 627–629.

Chen, H.-Y., I.-Y. Tarn, and Y.-J. Hwang, "Infrared extinction of the powder of brass 70Cu/30Zn calculated by the FDTD and turning bands methods," *IEEE Trans. Geoscience and Remote Sensing*, Vol. 33, 1995, pp. 1321–1324.

Chen, J. C., P. R. Villenueve, H. A. Haus, and J. D. Joannopoulos, "Narrow-band transmission filters from highly resonant gratings," *Proc. Integrated Photonics*, Boston, MA, 1996, pp. 406–409.

Chen, J. C., H. A. Haus, S. Fan, P. Villeneuve, and J. D. Joannopoulis, "Optical filters from photonic band gap air bridges," *J. Lightwave Technology*, Vol. 14, 1996, pp. 2575–2580.

Chen, J. C., S. Fan, A. Mekis, I. Kurkland, P. R. Villeneuve, K. Li, H. A. Haus, and J. D. Joannopoulos, "Finite differencing of periodic structures," *Proc. SPIE - Int. Soc. Opt. Eng.*, San Jose, CA, 1997, pp. 275–286.

Chew, W. C., and Q. H. Liu, "Using perfectly matched layers for elastodynamics," *Proc. IEEE Antennas and Propagat. Soc. Int. Symp.*, Baltimore, MD, 1996, pp. 366–369.

Christensen, D. A., "Analysis of near field tip patterns including object interaction using finite-difference time-domain calculations," *Ultramicroscopy*, Vol. 57, 1995, pp. 189–195.

Christensen, D. A., and D. Fowers, "Modeling SPR sensors with the finite-difference time-domain method," *Biosensors and Bioelectronics*, Vol. 11, 1996, pp. 677–684.

Conn, D., H. X. Wu, and M. Zhang, "Full wave electromagnetic simulation of electrooptic high-speed probes," *Optical and Quantum Electronics*, Vol. 28, 1996, pp. 765–782.

Cummings, K. E., J. G. Maloney, and M. A. Clements, "Modeling speech production using Yee's finite difference method," *Proc. 1995 Int. Conf. on Acoustics, Speech, and Signal Processing*, Detroit, MI, 1995, pp. 672–675.

Debroux, P. S., "3D modeling of the electromagnetic response of geophysical targets using the FDTD method," *Geophysical Prospecting*, Vol. 44, 1996, pp. 457–468.

DeFord, J. F., G. Kamin, G. D. Craig, and L. Walling, "Dispersive soft ferrite models for time-domain simulation and their application to accelerator component modeling," *Particle Accelerators*, Vol. 45, 1994, pp. 135–148.

Doi, T., A. Iwata, and M. Hirose, "Analysis and design of low loss and low mode-shift integrated optical waveguides using finite-difference time-domain method," *IEICE Trans. Electronics*, Vol. E80-C, 1997, pp. 625–631.

Dunn, A., C. Smithpeter, A. J. Twelch, and R. Richards-Korturn, "Finite-difference time-domain simulation of light scattering from single cells," *J. Biomedical Optics*, Vol. 2, 1997, pp. 262–266.

Furukawa, H., and S. Kawata, "Analysis of image formation in a near-field scanning optical microscope: effects of multiple scattering," *Optics Communications*, Vol. 132, 1996, pp. 170–178.

Greenfield, R. J., and S. T. Wu, "Electromagnetic wave propagation in disrupted coal seams," *Geophysics*, Vol. 56, 1991, pp. 1571–1577.

Grotjohn, T. A., W. Tan, V. Gopinath, and A. K. Srivastava, "Modeling the electromagnetic excitation of a compact ECR ion/free radical source," *Review of Scientific Instruments*, Vol. 65, 1994, pp. 1761–1765.

Hagness, S. C., R. M. Joseph, and A. Taflove, "Subpicosecond electrodynamics of distributed Bragg reflector microlasers: results from finite difference time domain simulations," Radio Science, Vol. 31, 1996, pp. 931–941.

Hastings, F. D., *A Study of Rough Surface Scattering in Electromagnetics, Acoustics, and Elastodynamics Applications Using Finite-Difference Time-Domain Simulations*, Ph.D. dissertation, Washington State University, Pullman, WA, 1997.

Hawkins, R. J., and J. S. Kallman, "Lasing in tilted-waveguide semiconductor laser amplifiers," *Opt. Quantum Electron.*, Vol. 26, 1994, pp. S207–S217.

Hawkins, R. J., N. K. Madsen, J. S. Kallman, M. D. Feit, C. C. Shang, B. W. Shore, and J. F. DeFord, "Full-wave simulation of the thumbtack laser," *Integrated Photonics Research Technical Digest*, Palm Springs, CA, 1993, pp. 116–119.

He, Y., and T. Kojima, "Three-dimensional analysis of light-beam scattering from magneto-optical disk structure by FDTD method," *Proc. IEEE Antennas and Propagat. Soc. Int. Symp.*, Montreal, Canada, 1997, pp. 2148–2151.

Hinrikus, H., H. Krasavin, K. Beilenhoff, and H. L. Hartnagel, "Calculation of microwave radiometric signal in multilayered biological tissue," *Proc. IEEE Engineering in Medicine and Biology 17th Annual Conf. and 21st Canadian Medical and Biological Engineering Conf.*, Montreal, Canada, 1995, pp. 1583–1584.

Hong, K. C., S.-K. Park, and J.-W. Ra, "Reconstruction of a high-contrast penetrable object in pulsed time domain by using the genetic algorithm," *Proc. IGARSS '97. Int. Geoscience and Remote Sensing Symp.*, Singapore, 1997, pp. 136–138.

Iskander, M. F., R. L. Smith, A. O. Andrade, H. Kimrey, Jr., and L. M. Walsh, "FDTD simulation of microwave sintering of ceramics in multimode cavities," *IEEE Trans. Microwave Theory and Techniques*, Vol. 42, 1994, pp. 793–800.

Iwabuchi, K., T. Kubota, and T. Kashiwa, "Analysis of electromagnetic fields in a mass-produced microwave oven using the finite-difference time-domain method," *J. Microwave Power*, Vol. 31, 1996, pp. 188–196.

Jambunathan, R., Y. Yuan, J. Singh, and P. Bhattacharya, "Theoretical and experimental studies on large-bandwidth 1.55 μm integrated InP-based strained MQW laser-modulators," *Proc. SPIE - Int. Soc. Opt. Eng.*, San Jose, CA, 1997, pp. 222–232.

Jambunathan, R., and J. Singh, "Design studies for distributed Bragg reflectors for short-cavity edge-emitting lasers," *IEEE J. Quantum Electronics*, Vol. 33, 1997, pp. 1180–1189.

Judkins, J. B., C. W. Haggans, and R. W. Ziolkowski, "Two-dimensional finite-difference time-domain simulation for rewritable optical disk surface structure design," *Applied Optics*, Vol. 35, 1996, pp. 2477–2487.

Kann, J. L., T. D. Milster, F. F. Foehlich, R. W. Ziolkowski, and J. B. Judkins, "Linear behavior of a near-field optical scanning system," *J. Optical Society of America A: Optics, Image Science, and Vision*, Vol. 12, 1995, pp. 1677–1682.

Kann, J. L., T. D. Milster, F. Foehlich, R. W. Ziolkowski, and J. Judkins, "Numerical analysis of a two-dimensional near-field probe," *Ultramicroscopy*, Vol. 57, 1995, pp. 251–256.

Kelly, P. K., J. G. Maloney, B. L. Shirley, and R. L. Moore, "Photonic band structures of finite thickness: theory and experiment," *Proc. IEEE Antennas and Propagat. Soc. Int. Symp.*, Seattle, WA, 1994, pp. 718–721.

Kessler, M. P., J. G. Maloney, B. L. Shirley, and G. S. Smith, "Antenna design with the use of photonic band-gap materials as all-dielectric planar reflectors," *Microwave and Optical Technology Lett.*, Vol. 11, 1996, pp. 169–174.

Li, B.-J., and P.-L. Liu, "Numerical analysis of microdisk lasers with rough boundaries," *IEEE J. Quantum Electronics*, Vol. 33, 1997, pp. 791–795.

Li, B.-J., and P.-L. Liu, "Analysis of far-field patterns of microdisk resonators by the finite-difference time-domain method," *IEEE J. Quantum Electronics*, Vol. 33, 1997, pp. 1489–1491.

Liang, T., and R. W. Ziolkowski, "Mode conversion of ultrafast pulses by grating structures in layered dielectric waveguides," *J. Lightwave Technology*, Vol. 15, 1997, pp. 1966–1973.

Maloney, J. G., and K. E. Cummings, "Adaptation of FDTD techniques to acoustic modeling," *Proc. 11th Annual Review of Progress in Applied Computational Electromagnetics*, Monterey, CA, 1995, pp. 724–731.

Maloney, J. G., M. P. Kessler, B. L. Shirley, G. S. Smith, "A simple description for waveguiding in photonic bandgap materials," *Microwave and Optical Technology Lett.*, Vol. 14, 1997, pp. 261–266.

Manry, C. W., and S. L. Broschat, "FDTD simulations for ultrasound propagation in a 2-D breast model," *Ultrasonic Imaging*, Vol. 18, 1996, pp. 25–34.

Marshall, T., and M. Piket-May, "Finite-difference time-domain modeling of light-trapping in solar cells," *J. Applied Computational Electromagnetics Soc.*, Vol. 13, no. 3, 1997, pp. 31–42.

Muta, H., Y. Ueda, and Y. Kawai, "Three-dimensional simulation of microwave propagation in an electron cyclotron resonance plasma," *Japanese J. Applied Physics, Part 1*, Vol. 36, 1997, pp. 4773–4776.

Piket-May, M., and A. Taflove, "Electrodynamics of visible-light interactions with the vertebrate retinal rod," *Optics Lett.*, Vol. 18, 1993, pp. 568–570.

Radic, S., and N. George, "Ultrafast pulse propagation in periodic optical media: A generalized finite-difference time-domain approach," *Optics Lett.*, Vol. 19, 1994, pp. 1064–1066.

Reineix, A., and B. Jecko, "A new photonic band gap equivalent model using finite difference time domain method," *Annales des Telecommunications*, Vol. 51, 1996, pp. 656–662.

Riipuld, J., H. Hinrikus, K. Beilenhoff, and H. Hartnagel, "Simulation of breast tissues temperature measurement using Dicke radiometer by FDTD method," *Proc. 1st Int. Conf. on Bioelectromagnetism*, Tampere, Finland, 1996, Vol. 34, pp. 135–136.

Schroter, U., S. Seider, S. Tode, and D. Heitman, "Surface plasmon reflection at edges and resonance effects in metal bars," *Ultramicroscopy*, Vol. 68, 1997, pp. 223–230.

Schuster, J. W., and R. J. Luebbers, "Comparison of GTD and FDTD predictions for UHF radio wave propagation in a simple outdoor urban environment," *Proc. IEEE Antennas and Propagat. Soc. Int. Symp.*, Montreal, Canada, 1997, pp. 2022–2025.

Sohel Imtiaz, S. M., and S. M. El-Ghazaly, "Physical simulation of complete millimeter-wave amplifiers using full-wave FDTD technique," *Proc. IEEE MTT-S Int. Microwave Symp.*, Denver, CO, 1997, pp. 79–82.

Sterkenburgh, T., R. M. Michels, P. Dress, and H. Franke, "Explicit finite-difference simulation of optical integrated devices on massive parallel computers," *Applied Optics*, Vol. 36, 1997, pp. 1191–1197.

Strickel, M., A. Taflove, and K. Umashankar, "Finite-difference time-domain formulation of an inverse scattering scheme for remote sensing of conducting and dielectric targets: Part II– Two dimensional case," *J. Electromagnetic Waves and Applications*, Vol. 8, 1994, pp. 509–530.

Sundberg, R., P. O. Risman, P.-S. Kildal, and T. Ohlsson, "Analysis and design of industrial microwave ovens using the finite difference time domain method," *J. Microwave Power and Electromagnetic Energy*, Vol. 31, 1996, pp. 142–157.

Tan, W., and T. A. Grotjohn, "Modeling the electromagnetic field and plasma discharge in a microwave plasma diamond deposition reactor," *Diamond and Related Materials*, Vol. 4, 1995, pp. 1145–1154.

Tang, C., and K. Aydin, "Scattering from ice crystals at 94 and 220 GHz millimeter wave frequencies," *IEEE Trans. Geoscience and Remote Sensing*, Vol. 33, 1995, pp. 93–99.

Taylor, H. C., M. Burl, and J. W. Hand, "Experimental verification of numerically predicted electric field distributions produced by a radiofrequency coil," *Physics of Medicine and Biology*, Vol. 42, 1997, pp. 1395–1402.

Thode, L., G. Csanak, R. Hotchkis, C. Snell, and M. Campbell, "Time-dependent model for vertical-cavity surface-emitting laser," *Proc. SPIE - Int. Soc. Opt. Eng.*, San Jose, CA, 1995, pp. 348–359.

Villeneuve, P. R., S. Fan, and J. D. Joannopoulos, "Microcavities in photonic crystals: Mode symmetry, tunability and coupling efficiency," *Phys. Review B*, Vol. 54, 1996, pp. 7837–7842.

Villeneuve, P. R., D. S. Abrams, S. Fan, and J. D. Joannopoulos, "Single-mode waveguide microcavity for fast optical switching," *Optics Lett.* Vol. 21, 1996, pp. 2017–2019.

Vurgaftman, I., and J. R. Meyer, "Generalized two-dimensional mesoscopic quantum transport: application to disordered quantum wires," *Phys. Review B*, Vol. 55, 1997, pp. 4494–4502.

Wang, X., and R. J. Hwu, "FDTD simulation of ultrashort pulse generation via optoelectronic switches," *Microwave and Optical Technology Lett.*, Vol. 14, 1997, pp. 204–207.

Wu, X., S. Zhou, and D. Xu, "The photoinduced carrier distribution in a semiconductor waveguide and the FDTD analysis for millimeter-wave propagation," *Microwave and Optical Technology Lett.*, Vol. 13, 1996, pp. 35–38.

Yamauchi, J., M. Mita, S. Aoki, and H. Nakano, "Analysis of antireflection coatings using the FDTD method with the PML absorbing boundary condition," *IEEE Photonics Technol. Lett.*, Vol. 8, 1996, pp. 239–241.

Yamauchi, J., K. Nishio, and H. Nakano, "Hybrid numerical technique combining the finite-difference beam-propagation method and the finite-difference time-domain method," *Optics Lett.*, Vol. 22, 1997, pp. 259–261.

Yang, P., and K. N. Liou, "Finite-difference time domain method for light scattering by small ice crystals in three-dimensional space," *J. Optical Society of America A: Optics, Image Science, and Vision*, Vol. 13, 1996, pp. 2072–2085.

Zhao, A. P., J. Juntunen, and A. V. Raisanen, "Analysis of hybrid modes in channel multilayer optical waveguides with the compact 2-D FDTD method," *Microwave and Optical Technology Lett.*, Vol. 15, 1997, pp. 398–403.

Ziolkowski, R. W., and T. Liang, "Design and characterization of a grating-assisted coupler enhanced by a photonic-band-gap structure for effective wavelength-division multiplexing," *Optics Lett.*, Vol. 22, 1997, pp. 1033–1035.

Chapter 2

High-Order Methods

Eli Turkel

2.1 INTRODUCTION

In this chapter we concentrate on fourth-order methods with applications to the time-dependent Maxwell equations and the Helmholtz equation. Before beginning, we define what we mean by order of accuracy.

We consider a general first-order system of equations for u, which we symbolically denote by

$$\frac{\partial u}{\partial t} = \mathcal{L} u \tag{2.1}$$

where $\mathcal{L}$ is a general linear operator, for example, the spatial-derivative portion of Maxwell's equations. We have a one-step approximation, v, given by

$$\frac{v^{n+1} - v^n}{\Delta t} = \mathcal{L}_h v \tag{2.2}$$

Assume that u is a sufficiently smooth solution of (2.1). Let Δt denote the time step and h the mesh size in all spatial directions. The approximation $\mathcal{L}_h$ has order of accuracy p in time and q in space if the truncation error, τ, satisfies

$$\tau = \frac{u^{n+1} - u^n}{\Delta t} - \mathcal{L}_h u = O\left[(\Delta t)^p + h^q\right] \tag{2.3}$$

We usually calculate the order of accuracy (p, q) by a Taylor series. Note that if the solution is not sufficiently smooth, then the order of accuracy is reduced. This can happen if the coefficients are not smooth, for example, due to the permittivity jump across the interface of two dielectric media.

2.1.1 Lax-Richtmyer Equivalence Theorem

The local definition of accuracy is important because of the Lax-Richtmyer equivalence theorem. This theorem states that if a scheme has a truncation error of order (p, q) and the scheme is stable, then the difference between the analytic solution and the numerical solution in an appropriate norm is of the order $(\Delta t)^p + h^q$ for all finite time. This equivalence theorem can be extended to variable coefficients and even smooth nonlinear problems. Further, Gustafsson has shown that if the numerical boundary treatment is one order less accurate than the interior accuracy, then the order of the global accuracy is preserved [1].

We stress that the equivalence theorem deals with errors measured in some norm, for example, least squares. Hence, we deal with the total error. At times, we may be mainly interested in either the phase (velocity) of the solution or its amplitude rather than the entire information. It is possible to construct schemes that have higher-order-accurate phase error but are only lower order when measured in the least-square norm. We present some examples in the next section. For a system with variable coefficients, we need to consider the group velocity rather than the phase velocity.

For hyperbolic, wavelike equations, it is usually desirable that the accuracy in time and space are equal. This is because hyperbolic systems behave in a similar manner in both space and time. In fact for the simplest equation, $\partial u / \partial t = \partial u / \partial x$, one cannot distinguish between the time and space directions. Nevertheless, we also consider schemes that are fourth-order-accurate in space but only second-order-accurate in time. This is a reasonable option since the temporal accuracy is improved by choosing a smaller time step. This increases the work only linearly and does not increase the storage. A fourth-order scheme in space enables us to choose a coarser mesh. This decreases the work in each space dimension and also decreases the storage. Hence, the accuracy in space is of greater importance then the accuracy in time. As a consequence, one should choose a time step based on accuracy conditions and not on stability grounds [2]. In computational electromagnetics, we are frequently interested in long-term time integrations, and so we want the dissipation to be kept to a minimum. For nonlinear problems with discontinuous solutions, it is necessary to include a strong dissipation mechanism [1, 3, 4].

Since the order of accuracy is found by a Taylor series expansion, it only gives information in the limit of sufficiently fine meshes. Equivalently, it stresses the lower frequencies. For high frequencies not resolved by the mesh, the order of accuracy has no relevance. We can consider methods that are formally lower-order-accurate but have better behavior for higher frequencies. For example, one can consider a fourth-order scheme that contains a free parameter. Instead of choosing the free parameter to increase the formal order of accuracy, the parameter is chosen to optimize some property for higher frequencies (for example, see [5, 6]). One difficulty with this approach is that it relies entirely on a Fourier transform. Hence, it is valid only for constant coefficients and uniform meshes in an infinite or periodic domain. There is no equivalent of the Lax-Richtmyer theorem that extends these results to variable coefficients and includes the boundary conditions and forcing functions. Similarly, Taflove [7] constructs a scheme with no dispersion error at specific angles. We do not discuss these schemes further. See [6, 8] for further details.

2.1.2 Basic Numerical Approaches

There are several basic numerical approaches that have been used. The first is the finite-difference method. In this approach, the unknowns are located at certain mesh locations. A Taylor series expansion is then constructed about a particular location to develop a solution scheme with a specified order of accuracy. Another approach is the finite-volume scheme (see for example [8–10]). Here, we begin with the equations in integral rather than differential form. The basic unknown is now an average of the variables over a volume (not necessarily the same volume for all the variables). Other approaches appear naturally in a finite-element setting. Here, the unknowns are expanded in a series of basis functions (local for finite elements or global for spectral methods). The order of accuracy is then determined by the smoothness of these basis functions and the accuracy at the nodes. This is known as a collocation scheme. An alternative is to demand that the error be orthogonal to a certain space (Galerkin approach). The order of accuracy is then determined by the size of this space.

2.1.3 Accuracy-Definition Considerations

It is important to stress that orders of accuracy between the different approaches are not equivalent. Usually a second-order scheme is second-order accurate with all approaches. However, a given approach can be fourth-order accurate when interpreted as a finite-volume scheme and only second-order accurate as a finite-difference scheme, or vice-versa. This occurs because the cell average is fourth-order accurate, but replacing the cell average by the point value at the center is only second-order accurate. In effect, an integral is equal to the size of the cell times the value at the cell center (the midpoint rule for integration) only with second-order accuracy. When plotting solutions, point values are commonly used since it is not easy to plot cell averages. Hence, for a finite-volume or Galerkin scheme, it is important either to use an appropriate plotting package or else construct a method of deriving point values from the appropriate high-order information. We describe here a simple example of this phenomenon [11, 12]. Consider the equation

$$\frac{\partial u}{\partial t} + \frac{\partial F(u)}{\partial x} = 0 \qquad u = u(x, t) \tag{2.4a}$$

with the approximation

$$U_i^{n+1} = U_i^n - \frac{\Delta t}{\Delta x}(F_{i+1/2} - F_{i-1/2}) \tag{2.4b}$$

In order for this formula to be fourth-order accurate in space as a finite-difference scheme, we must choose

$$F_{i+1/2} = \frac{7}{12}(F_{i+1} + F_i) - \frac{1}{12}(F_{i+2} + F_{i-1}) \tag{2.5}$$

so that

$$F_{i+1/2} - F_{i-1/2} = \frac{2}{3}(F_{i+1} - F_{i-1}) - \frac{1}{12}(F_{i+2} - F_{i-2}) \tag{2.6}$$

We note that (2.5) is not fourth-order accurate by itself but the difference given in (2.6) is fourth-order accurate.

We next rewrite (2.4a) in integral form. This yields

$$\int_{x_{i-1/2}}^{x_{i+1/2}} \frac{\partial u}{\partial t} dx + F\left[u(x_{i+1/2})\right] - F\left[u(x_{i-1/2})\right] = 0 \tag{2.7a}$$

or equivalently

$$\frac{\partial U}{\partial t} + \frac{F\left[u(x_{i+1/2})\right] - F\left[u(x_{i-1/2})\right]}{x_{i+1/2} - x_{i-1/2}} = 0 \tag{2.7b}$$

$$U(x, t) = \frac{1}{x_{i+1/2} - x_{i-1/2}} \frac{\partial}{\partial t} \int_{x_{i-1/2}}^{x_{i+1/2}} u(x, t)\, dx \tag{2.7c}$$

This formula is exact. However, we need to calculate u from U and then to evaluate F at the cell center. To calculate U with fourth-order accuracy in space, we interpolate $F[u(x_{i+1/2})]$ to fourth-order accuracy. This yields

$$F_{i+1/2} = \frac{9}{16}(F_{i+1} + F_i) - \frac{1}{16}(F_{i+2} + F_{i-1}) + O(\Delta x)^4 \tag{2.8}$$

Hence, any high-order method needs to specify whether it is higher order for point values, cell averages, or some finite-element meaning.

2.2 EXAMPLE OF THE CONCEPT OF ORDER OF ACCURACY

We next examine the concept of order of accuracy for the simplest one-dimensional wave-propagation system

$$\frac{\partial u}{\partial t} = \frac{\partial v}{\partial x}, \qquad \frac{\partial v}{\partial t} = \frac{\partial u}{\partial x} \tag{2.9}$$

which is equivalent to

$$\frac{\partial^2 u}{\partial t^2} = \frac{\partial^2 u}{\partial x^2}, \qquad \frac{\partial^2 v}{\partial t^2} = \frac{\partial^2 v}{\partial x^2} \tag{2.10}$$

Define $\delta_t u = u^{n+1} - u^n$ and $\delta_t v = v^{n+3/2} - v^{n+1/2}$. Then we consider scheme I:

$$\alpha(\delta_t u_{i-1} + \delta_t u_{i+1}) + (1 - 2\alpha)\delta_t u_i = \frac{\Delta t}{\Delta x}\left(v_{i+1/2}^{n+1/2} - v_{i-1/2}^{n+1/2}\right) \tag{2.11a}$$

$$\alpha(\delta_t v_{i-1/2} + \delta_t v_{i+3/2}) + (1 - 2\alpha)\delta_t v_{i+1/2} = \frac{\Delta t}{\Delta x}\left(u_{i+1}^{n+1} - u_i^{n+1}\right) \tag{2.11b}$$

This is fourth-order accurate if $\alpha = 1/24$. Reducing this to a second-order equation and letting $\delta_t^2 u = u^{n+1} - 2u^n + u^{n-1}$, we obtain

$$\begin{aligned} &\alpha^2 \delta_t^2 (u_{i+2} + u_{i-2}) + 2\alpha(1 - 2\alpha)\delta_t^2 (u_{i+1} + u_{i-1}) \\ &\quad + (1 - 4\alpha + 6\alpha^2)\delta_t^2 u_i = \left(\frac{\Delta t}{\Delta x}\right)^2 \left(u_{i+1}^n - 2u_i^n + u_{i-1}^n\right) \end{aligned} \tag{2.12}$$

Though this is fourth-order accurate, it uses the points u_{i+2} and u_{i-2}. We get a more compact implicit approximation to the wave equation by

$$B\delta_t^2 (u_{i+1} + u_{i-1}) + (1 - 2B)\delta_t^2 u_i = \left(\frac{\Delta t}{\Delta x}\right)^2 \left(u_{i+1}^n - 2u_i^n + u_{i-1}^n\right) \tag{2.13}$$

with $B = [1 - (\Delta t / \Delta x)^2]/12$. To achieve the equivalent of this with a system of equations, we consider scheme II:

$$A\delta_t u_{i+1} + (1 - A)\delta_t u_i = \frac{\Delta t}{\Delta x}\left(v_{i+1/2}^{n+1/2} - v_{i-1/2}^{n+1/2}\right) \tag{2.14a}$$

$$A\delta_t v_{i-1/2} + (1 - A)\delta_t v_{i+1/2} = \frac{\Delta t}{\Delta x}\left(u_{i+1}^{n+1} - u_i^{n+1}\right) \tag{2.14b}$$

Each equation in scheme II is first-order, while the system is equivalent to (2.13), and hence fourth-order if $B = A(1 - A)$, or equivalently $A = (1 \pm \sqrt{1 - 4B})/2$.

To resolve this dilemma, it is necessary to consider the semidiscrete equations and Fourier transform the equations in space. We denote the Fourier transforms by $\hat{u}$ and $\hat{v}$ and consider the generic formulas

$$\frac{\partial \hat{u}}{\partial t} = \lambda_1(\theta)\,\hat{v}\,, \qquad \frac{\partial \hat{v}}{\partial t} = \lambda_2(\theta)\,\hat{u} \tag{2.15a}$$

The characteristic variables are

$$\sqrt{\lambda_2}\,\hat{u} + \sqrt{\lambda_1}\,\hat{v} \tag{2.15b}$$

We are interested in the case where $\lambda_1 \neq \lambda_2$. For general λ_1 and λ_2, there is no way to locally diagonalize the equations in physical space. Reducing this to a second-order equation, we get

$$\frac{\partial^2 \hat{u}}{\partial t^2} = \lambda^2(\theta)\,\hat{u}\,, \qquad \frac{\partial^2 \hat{v}}{\partial t^2} = \lambda^2(\theta)\,\hat{v} \tag{2.16a}$$

where

$$\lambda^2 = \lambda_1 \lambda_2 \tag{2.16b}$$

In summary, we have for scheme I

$$\lambda_1 = \lambda_2 = \frac{2\,j\sin(\theta/2)}{\Delta x\left[1 - 4\alpha\sin^2(\theta/2)\right]} \tag{2.17a}$$

$$\lambda^2 = -\frac{4\sin^2(\theta/2)}{(\Delta x)^2\left[1 - 2B\sin^2(\theta/2)\right]} \tag{2.17b}$$

and for scheme II

$$\lambda_1 = \frac{2\,j\sin(\theta/2)}{\Delta x(Ae^{j\theta} + 1 - A)} \tag{2.18a}$$

$$\lambda_2 = \frac{2\,j\sin(\theta/2)}{\Delta x(Ae^{-j\theta} + 1 - A)} \tag{2.18b}$$

$$\lambda^2 = -\frac{4\sin^2(\theta/2)}{(\Delta x)^2\left[1 - 2B\sin^2(\theta/2)\right]} \tag{2.18c}$$

λ_1 and λ_2 are first-order approximations to λ_{exact}, but λ is a fourth-order approximation to λ_{exact}. λ^2 is negative, and so λ is purely imaginary. Gottlieb, Turkel, and Abarbanel [13] find the general solution to this equation as a function of the initial data. They show that scheme II is first-order in the L_2 norm, but is fourth-order-accurate for the phase velocity. Further, they show how to preprocess the initial data and postprocess the final solution so that true fourth-order accuracy is recovered. This can also be extended to the two-dimensional TM and TE equations of electromagnetic wave propagation. The conclusions are:

1. Without preprocessing or postprocessing, $\lambda \sim \lambda_{exact}$ to the order of the scheme, and the phase velocity is equal to that of the second-degree equation. In particular, the phase of scheme II is fourth-order accurate.
2. For the TM electromagnetic system, we preprocess the initial conditions for only H_x and H_y and not E_z. For linear problems, we postprocess only at the conclusion of the computation, not after each time step.
3. For general λ_1 and λ_2, we cannot entirely eliminate the amplitude error in physical space. Instead, we minimize it to the order of accuracy of the scheme.
4. λ is purely imaginary so the schemes are conservative (i.e., $A^2 + B^2 = 1$).
5. The schemes are first-order unless we preprocess and postprocess. If we choose $u_i^0 = u(i\Delta x, 0)$, the scheme has a higher-order phase error but the amplitude error is first-order. To make the scheme higher-order, we must preprocess and postprocess the numerical solution.

Hence, we conclude that the Taylor series expansion indeed gives the formal accuracy of the scheme. However, the phase error of the system may be of higher order than the total error. Further, it may be possible to preprocess and postprocess the solution to extract the higher-order information that is inherent within the scheme. Conversely, it is possible to construct schemes that are higher-order accurate for the amplitude but not for the phase [14]. This is all in addition to the fact, previously discussed, that fourth-order accuracy depends on the nature of the discretization whether it is finite-difference, finite-volume, finite-element, or spectral.

2.3 FOURTH-ORDER SPACE DERIVATIVES

Fourth-order methods have been investigated in acoustics for many years [4, 6, 15–25]. We first consider how to approximate the space derivative, $(Du)_i = (\partial u / \partial x)_i$ to fourth-order accuracy for a finite-difference scheme. To establish the notation and the nomenclature, the second-order-accurate Yee space-derivative operator is written as

$$(D_0 u)_i = \frac{u_{i+1/2} - u_{i-1/2}}{\Delta x} \tag{2.19}$$

The Yee operator is said to be explicit and working on a staggered grid where the derivative is evaluated at a point i one-half grid cell between adjacent values of u_i. Note that there is no value of u located at point i where the derivative is evaluated.

2.3.1 Explicit Fourth-Order Schemes

In the following, fourth-order-accurate finite-difference expressions are provided for the space derivatives that work directly on the local values of u centered about point i. We consider two possibilities, either the collocated grid where u and its derivative are located at i, or the staggered grid similar to that utilized by the Yee operator, D_0:

Explicit collocated scheme

$$(D_1 u)_i = \frac{8(u_{i+1} - u_{i-1}) - (u_{i+2} - u_{i-2})}{12\,\Delta x} \tag{2.20}$$

Explicit staggered scheme

$$(D_2 u)_i = \frac{27(u_{i+1/2} - u_{i-1/2}) - (u_{i+3/2} - u_{i-3/2})}{24\,\Delta x} \tag{2.21}$$

2.3.2 Implicit Fourth-Order Schemes

We can also construct implicit schemes that require the inversion of a tridiagonal matrix to obtain in one step the complete set of derivatives along a line through the mesh. These are given by the following expressions:

Implicit collocated scheme

$$\frac{(D_3 u)_{i+1} + (D_3 u)_{i-1}}{6} + \frac{2}{3}(D_3 u)_i = \frac{u_{i+1} - u_{i-1}}{2\,\Delta x} \tag{2.22}$$

Implicit staggered scheme

$$\frac{(D_4 u)_{i+1} + (D_4 u)_{i-1}}{24} + \frac{11}{12}(D_4 u)_i = \frac{u_{i+1/2} - u_{i-1/2}}{\Delta x} \tag{2.23}$$

The implicit staggered scheme of (2.23) is of particular interest and usefulness in this chapter since it employs a compact stencil compatible with the Yee grid. We designate this as the *Ty* space-derivative operator of order four, and subsequent FDTD algorithms based on this as *Ty*(p, 4), where p is the order of accuracy of the time stepping.

2.3.3 Nonuniform Mesh

We can extend the above implicit staggered scheme to nonuniform meshes. For simplicity, we consider the one-dimensional system

$$\frac{\partial u}{\partial t} = \frac{\partial v}{\partial x}, \qquad \frac{\partial v}{\partial t} = \frac{\partial u}{\partial x} \tag{2.24}$$

The variable u is at location x_i with $\Delta x_i = x_{i+1} - x_i$. We place the variable $v_{i+1/2}$ half way between x_i and x_{i+1}. Hence, $h_{i-1} = [(\Delta x)_i + (\Delta x)_{i-1}]/2$. We then have the semi-discrete formula

$$A_1 \left.\frac{\partial u}{\partial t}\right|_{i+1} + (1 - B_1)\left.\frac{\partial u}{\partial t}\right|_{i} + C_1 \left.\frac{\partial u}{\partial t}\right|_{i-1} = \frac{v_{i+1/2} - v_{i-1/2}}{h_{i-1}} \tag{2.25}$$

$$A_2 \left.\frac{\partial v}{\partial t}\right|_{i+3/2} + B_2 \left.\frac{\partial v}{\partial t}\right|_{i+1/2} + C_2 \left.\frac{\partial v}{\partial t}\right|_{i-1/2} = \frac{u_{i+1} - u_i}{(\Delta x)_i} \tag{2.26}$$

This is third-order accurate for arbitrary meshes and fourth-order accurate for quasi-uniform meshes [26] if we choose

$$A_1 = \frac{-2(\Delta x)_i^2 + (\Delta x)_{i-1}^2 + 2(\Delta x)_i(\Delta x)_{i-1}}{12(\Delta x)_{i-1}\left[(\Delta x)_i + (\Delta x)_{i-1}\right]} \tag{2.27a}$$

$$C_1 = \frac{(\Delta x)_i^2 - 2(\Delta x)_{i-1}^2 + 2(\Delta x)_i(\Delta x)_{i-1}}{12(\Delta x)_i\left[(\Delta x)_i + (\Delta x)_{i-1}\right]} \tag{2.27b}$$

$$B_1 = 1 - A_1 - C_1 \tag{2.27c}$$

$$A_2 = \frac{(\Delta x)_i^2}{12 h_{i-1}(h_i + h_{i-1})} \tag{2.27d}$$

$$C_2 = \frac{h_{i-1}}{h_i} A_2 \tag{2.27e}$$

$$B_2 = 1 - A_2 - C_2 \tag{2.27f}$$

If we choose a uniform grid except near the boundary, then we have fourth-order local accuracy in the interior and third-order accuracy near the boundary. This gives a global accuracy of fourth order.

2.3.4 Accuracy

We define the truncation error of the spatial derivative operators of (2.20) – (2.23) as $T = Du - (du/dx) = \tau h^4$. By a Taylor series expansion, we obtain for these operators the following truncation errors, respectively:

$$\tau_1 = \frac{1}{30} \cong 0.0333, \qquad \tau_2 = \frac{3}{640} \cong 0.00469$$
$$\tau_3 = \frac{1}{180} \cong 0.00555, \qquad \tau_4 = \frac{17}{5760} \cong 0.00295 \tag{2.28}$$

The *Ty* operator has the smallest truncation error of this group.

We next consider the semidiscrete case where the time derivative is continuous and the space derivative is discretized. Kreiss and Oliger [27] present an analysis to calculate the number of points per wavelength, M, needed to resolve a propagating wave, $u(x, t) = \exp[j(kx - \omega t)]$, over K periods with the desired accuracy, $\mathrm{e} = |\pm \Delta u / u|$. This was later rederived for Maxwell's equations by Petropoulos [28]. The number of points per wavelength is given by

$$M = 2\pi (2\pi\tau)^{1/q} \left(\frac{K}{\mathrm{e}}\right)^{1/q} \tag{2.29}$$

where q is the order of the finite-difference space derivative. If we choose $\mathrm{e} = 0.01$ (1% error), we get for Yee ($q = 2$) and for the four $q = 4$ cases:

$$M_0 \sim 32.2\, K^{1/2}, \qquad M_1 \sim 13.4\, K^{1/4}$$
$$M_2 \sim 8.23\, K^{1/4}, \qquad M_3 \sim 8.59\, K^{1/4}, \qquad M_4 \sim 7.33\, K^{1/4} \tag{2.30}$$

We conclude that using either staggering or the implicit methods gives substantial improvement over the simplest fourth-order-accurate method. Combining staggering with an implicit method (the *Ty* approach) gives a little more improvement (see also [29]). Staggering also helps in the imposition of the boundary conditions and simplifies the conversion of an existing Yee-algorithm FDTD code to fourth-order accuracy.

We note from (2.30) that the number of points per wavelength needed for a given accuracy increases with the length of the integration in time, K. The higher the order of accuracy, q, the slower the increase as a function of time. The benefits of a fourth-order method compared with a second-order method improve with demands for higher accuracy (for example, 0.1% vs. 1%) and with longer times of integration. Sixth-order methods are more efficient only for very long time integrations. Since sixth-order finite-difference methods complicate the boundary treatment, we choose to emphasize fourth-order schemes in this chapter.

2.3.5 Boundary Treatment

In all cases, we must add special treatment for the boundaries. We now describe one possibility for the Ty (implicit staggered) scheme.

At the first and last points, we use a fourth-order-accurate one-sided approximation to the derivative. Note that this is used only to globally approximate the derivative. No physical boundary conditions are included at this stage. Define

$$A = \frac{1}{24}\begin{pmatrix} 26 & -5 & 4 & -1 & . & . & 0 \\ 1 & 22 & 1 & 0 & . & . & 0 \\ 0 & 1 & 22 & 1 & 0 & . & 0 \\ . & . & . & . & . & . & . \\ 0 & . & . & 0 & 1 & 22 & 1 \\ 0 & . & . & -1 & 4 & -5 & 26 \end{pmatrix} = \begin{pmatrix} a_{11} & a_{12} & \cdots & a_{1n} \\ . & . & . & . \\ . & . & . & . \\ a_{n1} & a_{n2} & \cdots & a_{nn} \end{pmatrix} \tag{2.31}$$

The matrix form of these approximations is

$$A\frac{\partial}{\partial x}\begin{bmatrix} U_{1/2} \\ U_{3/2} \\ \vdots \\ U_{(2p-1)/2} \end{bmatrix} = \frac{1}{\Delta x}\left(\begin{bmatrix} U_1 \\ U_2 \\ \vdots \\ U_p \end{bmatrix} - \begin{bmatrix} U_0 \\ U_1 \\ \vdots \\ U_{p-1} \end{bmatrix}\right) \tag{2.32}$$

If we eliminate the nontridiagonal elements of the first row by Gaussian elimination, we find that the first row is equivalent to

$$\left(\frac{\partial U}{\partial x}\right)_{3/2} = \frac{27(U_2 - U_1) - (U_3 - U_0)}{24\Delta x} \tag{2.33}$$

Thus, instead of an approximation to the derivative at $i = 0.5$, we have two different fourth-order-accurate approximations to the derivative at $i = 1.5$. We therefore replace A by $\hat{A}$:

$$\hat{A} = \begin{pmatrix} 0 & 1 & . & . & . & . & 0 \\ 1 & 22 & 1 & 0 & . & . & 0 \\ 0 & 1 & 22 & 1 & 0 & . & 0 \\ . & . & . & . & . & . & . \\ 0 & . & . & 0 & 1 & 22 & 1 \\ 0 & . & . & . & . & 1 & 0 \end{pmatrix} \tag{2.34}$$

and the right-hand side of the first equation by the right-hand side of (2.33). This is done in a similar manner for variables shifted by one-half cellwidth. The matrix is now tridiagonal and so requires only about 5N multiplications to invert it (or more properly to solve the system). A^{-1} is never evaluated, but rather the decomposition $A = LU$ is calculated once in each direction. Further, the L and U decomposition can be stored. These matrices depend only on the number of mesh points in the line. Hence, we store these only once in each direction independent of the number of variables. Explicitly, given the matrix A of (2.31), we have

$$L = \begin{pmatrix} 1 & 0 & 0 & 0 & \cdots \\ \ell_{21} & 1 & 0 & 0 & \cdots \\ 0 & \ell_{32} & 1 & 0 & \cdots \\ 0 & 0 & \ell_{43} & 1 & 0 \\ \vdots & \vdots & \vdots & \vdots & \cdots \\ 0 & \ell_{n,n-3} & \ell_{n,n-2} & \ell_{n,n-1} & 1 \end{pmatrix} \tag{2.35}$$

$$U = \begin{pmatrix} a_{11} & a_{12} & a_{13} & a_{14} & 0 \\ 0 & u_{22} & u_{23} & u_{24} & 0 \\ 0 & 0 & u_{33} & u_{34} & 0 \\ 0 & 0 & 0 & u_{44} & u_{45} \\ \vdots & \vdots & \vdots & \vdots & \cdots \\ 0 & 0 & 0 & 0 & u_{n,n} \end{pmatrix} \tag{2.36}$$

$$\begin{aligned} \ell_{21} &= \frac{a_{21}}{a_{11}}, & u_{22} &= a_{22} - \ell_{21} a_{12} \\ u_{23} &= a_{23} - \ell_{21} a_{13}, & u_{24} &= a_{24} - \ell_{21} a_{14} \\ \ell_{32} &= \frac{a_{32}}{u_{22}}, & u_{33} &= a_{33} - \ell_{32} u_{23} \\ u_{34} &= a_{34} - \ell_{32} u_{24} \end{aligned} \tag{2.37}$$

As usual, for a tridiagonal matrix we then have

$$\begin{aligned} \ell_{j,j-1} &= \frac{a_{j,j-1}}{u_{j-1,j-1}}, & u_{j,j} &= a_{j,j} - \ell_{j,j-1} u_{j-1,j} \\ u_{j,j+1} &= a_{j,j+1} = 1/24 & j &= 4, \ldots, n-1 \end{aligned} \tag{2.38}$$

and finally

$$
\begin{aligned}
\ell_{n,n-3} &= \frac{a_{n,n-3}}{u_{n-3,n-3}}, & \ell_{n,n-2} &= \frac{a_{n,n-2} - \ell_{n,n-3}\, u_{n-3,n-2}}{u_{n-2,n-2}} \\
\ell_{n,n-1} &= \frac{a_{n,n-1} - \ell_{n,n-2}\, u_{n-2,n-1}}{u_{n-1,n-1}}, & u_{n,n} &= a_{n,n} - \ell_{n,n-1}\, u_{n-1,n}
\end{aligned}
\tag{2.39}
$$

Having calculated the matrices L and U, we now solve the system $Ax = b$ by implementing $Ly = b$ followed by $Ux = y$. Again, writing this explicitly, we have

$$
\begin{aligned}
y_1 &= b_1 \\
y_j &= b_j - \ell_{j,j-1}\, y_{j-1} \qquad j = 2,\ \dots,\ n-1 \\
y_n &= b_n - \ell_{n,n-1}\, y_{n-1} - \ell_{n,n-2}\, y_{n-2} - \ell_{n,n-3}\, y_{n-3}
\end{aligned}
\tag{2.40}
$$

followed by

$$
\begin{aligned}
x_n &= \frac{y_n}{u_{n,n}} \\
x_j &= \frac{y_j - u_{j,j+1}\, y_{j+1}}{u_{j,j}} \qquad j = n-1,\ n-2,\ \dots,\ 3 \\
x_2 &= \frac{y_2 - u_{23}\, y_3 - u_{24}\, y_4}{u_{22}} \\
x_1 &= \frac{y_1 - a_{12}\, y_2 - a_{13}\, y_3 - a_{14}\, y_4}{u_{22}}
\end{aligned}
\tag{2.41}
$$

Of course, we store $1/u_{jj}$ rather than u_{jj} so that we can perform a multiply rather than a divide. Thus, except for the edge elements, we need three multiplications and two additions per point to solve the system. We need to invert D such systems for each equation where D is the number of spatial dimensions. In the computations section, we present results to demonstrate the speedup in computer time using fourth-order methods to achieve the same level of accuracy as the Yee algorithm.

2.4 TIME-STEPPING SCHEMES

Consider the simple one-dimensional hyperbolic system:

$$
\frac{\partial u}{\partial t} = c\frac{\partial v}{\partial x} \tag{2.42a}
$$

$$
\frac{\partial v}{\partial t} = c\frac{\partial u}{\partial x} \tag{2.42b}
$$

2.4.1 Second-Order-Accurate Leapfrog

As described in Section 2.3, u and v are located either at the same space point or else staggered with u located at a grid-cell vertex, j, and v located at the grid-cell center, $j+1/2$. We choose to approximate the space derivative by one of the fourth-order-accurate operators, D_k. Finally, we can approximate the time derivative with a second-order-accurate leapfrog scheme. Hence, for a staggered-grid, our approximation to (2.42) is

$$\frac{u_i^{n+1} - u_i^n}{\Delta t} = c\,(D_k\, v^{n+1/2})_i \tag{2.43a}$$

$$\frac{v_{i+1/2}^{n+3/2} - v_{i+1/2}^{n+1/2}}{\Delta t} = c\,(D_k\, u^{n+1})_{i+1/2} \tag{2.43b}$$

For D_2 and D_4 described in Section 2.3, this scheme is second-order-accurate in time and fourth-order-accurate in space.

2.4.2 Fourth-Order Time Stepping, Explicit Space Derivatives

We now convert the explicit scheme of (2.43) to fourth-order accuracy in time [30]:

$$\frac{u_i^{n+1} - u_i^n}{\Delta t} = \left(\frac{\partial u}{\partial t}\right)^{n+1/2} + \frac{(\Delta t)^2}{24}\left(\frac{\partial^3 u}{\partial t^3}\right)^{n+1/2} + O(\Delta t)^4 \tag{2.44}$$

Hence, we can replace (2.43a) by

$$\left(\frac{\partial u}{\partial t}\right)^{n+1/2} = \frac{u_i^{n+1} - u_i^n}{\Delta t} - \frac{(\Delta t)^2}{24}\left(\frac{\partial^3 u}{\partial t^3}\right)^{n+1/2} + O(\Delta t)^4 = c\,(D_2\, v^{n+1/2})_i \tag{2.45}$$

Calculating $\partial^3 u/\partial t^3$ directly would require additional time levels. Instead, we use the basic system differential equations (2.42) to replace time derivatives by space derivatives; for example, $\partial^3 u/\partial t^3 = c^3\,(\partial^3 v/\partial x^3)$. Substituting in (2.45) yields

$$\frac{u_i^{n+1} - u_i^n}{\Delta t} = c\,(D_2\, v^{n+1/2})_i + \frac{c^3(\Delta t)^2}{24}\left(\frac{\partial^3 v}{\partial x^3}\right)^{n+1/2} + O(\Delta x)^4 + O(\Delta t)^4 \tag{2.46}$$

Since $D_2\, v$ and $\partial^3 v/\partial x^3$ both require a five-point stencil for this case, there is no additional work required to get the fourth-order accuracy in time. Combining (2.46) with the explicit staggered scheme we get

$$\frac{u_i^{n+1} - u_i^n}{\Delta t} = \frac{c}{24\,\Delta x}\left\{\begin{array}{l}\left[27 - 3\left(\frac{c\Delta t}{\Delta x}\right)^2\right]\left(v_{i+1/2}^{n+1/2} - v_{i-1/2}^{n+1/2}\right) - \\ \left[1 - \left(\frac{c\Delta t}{\Delta x}\right)^2\right]\left(v_{i+3/2}^{n+1/2} - v_{i-3/2}^{n+1/2}\right)\end{array}\right\} \tag{2.47}$$

For a nonstaggered mesh this is replaced by

$$\frac{u_i^{n+1} - u_i^{n-1}}{2\,\Delta t} = \frac{c}{6\,\Delta x}\left\{\begin{array}{l}\left[8 - 2\left(\frac{c\Delta t}{\Delta x}\right)^2\right]\left(v_{i+1}^{n} - v_{i-1}^{n}\right) - \\ \left[1 - \left(\frac{c\Delta t}{\Delta x}\right)^2\right]\left(v_{i+2}^{n} - v_{i-2}^{n}\right)\end{array}\right\} \tag{2.48}$$

For small CFL factors, the correction term is negligible.

For Maxwell's equations with $\vec{J} = \sigma \vec{E}$ and ε and μ independent of time, it can be shown that

$$\mu \frac{\partial^3 \vec{H}}{\partial t^3} = \nabla \times \left[\frac{1}{\varepsilon}\nabla \times \left(\frac{1}{\mu}\nabla \times \vec{E}\right) + \frac{\sigma}{\varepsilon^2}\left(\nabla \times \vec{H} - \vec{J}\right)\right] \tag{2.49}$$

$$\begin{aligned}\varepsilon \frac{\partial^3 \vec{E}}{\partial t^3} &= -\nabla \times \left\{\frac{1}{\mu}\nabla \times \left[\frac{1}{\varepsilon}\left(\nabla \times \vec{H} - \vec{J}\right)\right]\right\} \\ &\quad + \frac{\sigma}{\varepsilon}\nabla \times \left(\frac{1}{\mu}\nabla \times \vec{E}\right) + \frac{\sigma^2}{\varepsilon^2}\left(\nabla \times \vec{H} - \vec{J}\right)\end{aligned} \tag{2.50}$$

For the special case where ε and μ are constants and $\vec{J} = 0$, these expressions can be simplified. Under these assumptions, the (4, 4) scheme becomes

$$\vec{E}^{n+1/2} = \vec{E}^{n-1/2} + \frac{\Delta t}{\varepsilon}\nabla \times \left[\vec{H}^n + \left(\frac{c\Delta t}{24}\right)^2 \Delta \vec{H}^n\right] \tag{2.51}$$

$$\vec{H}^{n+1} = \vec{H}^n - \frac{\Delta t}{\mu}\nabla \times \left[\vec{E}^{n+1/2} + \left(\frac{c\Delta t}{24}\right)^2 \Delta \vec{E}^{n+1/2}\right] \tag{2.52}$$

Here, the spatial differences including the Laplacians, $\Delta \vec{E}$ and $\Delta \vec{H}$, are implemented by the usual staggered central differences.

2.4.3 Fourth-Order Time Stepping, Implicit Space Derivatives

The use of a fourth-order-accurate implicit space-derivative scheme requires only a three-point stencil. This does not match the five-point stencil needed to implement $\partial^3 v/\partial x^3$. Instead, we use the relation

$$\frac{\partial^3 u}{\partial t^3} = c^2 \frac{\partial^3 u}{\partial t\,\partial x^2} \tag{2.53}$$

Denoting

$$(\delta_t u)_i = \frac{u_i^{n+1} - u_i^n}{\Delta t} \tag{2.54a}$$

$$D_0 u_i = \frac{u_{i+1/2} - u_{i-1/2}}{\Delta x} \tag{2.54b}$$

then (2.46) is replaced by

$$\frac{1}{24}\left[(\delta_t u)_{i+1} + 22(\delta_t u)_i + (\delta_t u)_{i-1} - c(\Delta t)^2 \frac{\partial^3 u}{\partial t\,\partial x^2}\right] = c(D_0 v)_i + O(\Delta x)^4 + O(\Delta t)^4 \tag{2.55}$$

We also write

$$\frac{\partial^3 u}{\partial t\,\partial x^2} \cong \frac{(\delta_t u)_{i+1} - 2(\delta_t u)_i + (\delta_t u)_{i-1}}{(\Delta x)^2} \tag{2.56}$$

so on a staggered mesh (2.56) becomes

$$\frac{1}{24}\left\{\begin{array}{l}\left[1-\left(\dfrac{c\Delta t}{\Delta x}\right)^2\right](\delta_t u)_{i+1} + \left[22 + 2\left(\dfrac{c\Delta t}{\Delta x}\right)^2\right](\delta_t u)_i + \\ \left[1-\left(\dfrac{c\Delta t}{\Delta x}\right)^2\right](\delta_t u)_{i-1}\end{array}\right\} = c(D_0 v)_i + O(\Delta x)^4 + O(\Delta t)^4 \tag{2.57}$$

For a nonstaggered mesh we have

$$\frac{1}{6}\left\{\begin{array}{l}\left[1-\left(\frac{c\Delta t}{\Delta x}\right)^2\right](\delta_t u)_{i+1} + \left[4+2\left(\frac{c\Delta t}{\Delta x}\right)^2\right](\delta_t u)_i + \\ \left[1-\left(\frac{c\Delta t}{\Delta x}\right)^2\right](\delta_t u)_{i-1}\end{array}\right\} = \tag{2.58}$$

$$c\left(\frac{v_{i+1}-v_{i-1}}{2\Delta x}\right) + O(\Delta x)^4 + O(\Delta t)^4$$

This requires the inversion of a tridiagonal matrix as before. Thus, for the model system of (2.42), we also achieve fourth-order accuracy in time for the implicit scheme with essentially no additional work.

However, when we extend this scheme to the multidimensional Maxwell's equations, some difficulties appear. With the explicit schemes, the extension to fourth-order accuracy in time is straightforward as long as ε and μ are constant. In this case, we again replace the third time derivative of any electric or magnetic field component by appropriate third-order spatial derivatives. Since we are already using a five-point stencil in each coordinate direction, no extra points are required. When ε and μ are not constant or a current is present, the expressions become more complicated. When ε and μ are not smooth (for example, across material interfaces), we expect difficulties with this approach since it introduces higher-order spatial derivatives.

2.4.4 Fourth-Order Time Stepping, Implicit Space Derivatives in Two Dimensions

We next consider the implicit space-derivative schemes in two space dimensions. We begin with the scalar equation

$$\frac{\partial u}{\partial t} = \frac{\partial f}{\partial x} + \frac{\partial g}{\partial y} \tag{2.59}$$

where f and g are arbitrary fluxes. As in the one-dimensional case, we replace each space derivative by a compact implicit approximation. Denoting the central difference by δ, we write (2.59) symbolically as

$$\frac{\partial u}{\partial t} = \frac{\delta_x}{I+\alpha\delta_x^2}f + \frac{\delta_y}{I+\alpha\delta_y^2}g \tag{2.60}$$

This scheme is fourth-order accurate in space for a given α depending on whether staggering is introduced. Clearing fractions, we obtain

$$\left(I + \alpha\delta_x^2\right)\left(I + \alpha\delta_y^2\right)\frac{\partial u}{\partial t} = \left(I + \alpha\delta_y^2\right)\frac{\partial f}{\partial x} + \left(I + \alpha\delta_x^2\right)\frac{\partial g}{\partial y} \tag{2.61}$$

These two forms are identical except for boundary conditions. Form (2.60) requires slightly fewer operations and the boundary treatment is done on the fluxes without regard for the quantities which are physically specified. Form (2.61) has additional averaging on the right-hand side and also requires boundary conditions on the change in u, which is harder to specify. In order to obtain fourth-order accuracy in time, we replace (2.60) by

$$\frac{\partial u}{\partial t} = \frac{u_i^{n+1} - u_i^n}{\Delta t} - \frac{(\Delta t)^2}{24}\frac{\partial^3 u}{\partial t^3} + O(\Delta t)^4 = \frac{\delta_x}{I + \alpha\delta_x^2} f + \frac{\delta_y}{I + \alpha\delta_y^2} g \tag{2.62}$$

For Maxwell's equations without any forcing functions and constant ε and μ, each E and H component satisfies a wave equation. Hence, we replace (2.61) by

$$(\delta_t u)_i - \frac{(c\Delta t)^2}{24}\left[\frac{\partial^2(\delta_t u)}{\partial x^2} + \frac{\partial^2(\delta_t u)}{\partial y^2}\right]_i + O(\Delta t)^4 = \frac{\delta_x}{I + \alpha\delta_x^2} f + \frac{\delta_y}{I + \alpha\delta_y^2} g \tag{2.63}$$

Multiplying by the denominators and ignoring lower-order terms, we get

$$\left(I + \alpha\delta_x^2\right)\left(I + \alpha\delta_y^2\right)(\delta_t u)_i - \frac{(c\Delta t)^2}{24}\left[\frac{\partial^2(\delta_t u)}{\partial x^2} + \frac{\partial^2(\delta_t u)}{\partial y^2}\right]_i = \left(I + \alpha\delta_y^2\right)\delta_x f + \left(I + \alpha\delta_x^2\right)\delta_y g \tag{2.64}$$

This requires the very expensive inversion of a full two-dimensional operator. Instead, we replace this by the usual alternating-direction-implicit (ADI) approximation:

$$\left\{I + \left[\alpha - \frac{1}{24}\left(\frac{c\Delta t}{\Delta x}\right)^2\right]\delta_x^2\right\}\left\{I + \left[\alpha - \frac{1}{24}\left(\frac{c\Delta t}{\Delta y}\right)^2\right]\delta_y^2\right\}(\delta_t u)_i = \left(I + \alpha\delta_y^2\right)\delta_x f + \left(I + \alpha\delta_x^2\right)\delta_y g \tag{2.65}$$

This requires one tridiagonal matrix inversion per direction per equation. However, the inversions are now operating on $(\delta_t u)_i$, which as stated, is less desirable. As previously discussed, we store one set of factors in each direction (independent of the number of equations). Then, each tridiagonal inversion requires three multiplications and two additions per point. As will be seen later, the time step is chosen based on accuracy requirements, and therefore is usually much smaller than the stability limit. Hence, the stability limit is usually not of great importance for these schemes.

2.4.5 Fourth-Order Time Stepping Using the Runge-Kutta Method

We now describe a robust alternative to achieving fourth-order accuracy in time—the Runge-Kutta (R-K) method. This method has the significant advantage that it is straightforward to program and does not depend on the use of a wave-equation relation to eliminate the third time derivative. Here, the variables are staggered in space but not in time. This formulation works well in the presence of forcing functions, inhomogeneities, currents, and even nonlinearities. However, it has the disadvantage that it requires four stages for fourth-order accuracy and hence four times the work.

There is a choice of several versions of the fourth-order R-K technique, including the classical approach and the higher-resolution scheme proposed by Hu et al. [31]. For a linear set of equations without a forcing function, we simplify the classical R-K system for $\partial U / \partial t = f(U)$ to

$$\begin{aligned} U^{(1)} &= U^n + \frac{\Delta t}{4} f(U^n) \\ U^{(2)} &= U^n + \frac{\Delta t}{3} f\left[U^{(1)}\right] \\ U^{(3)} &= U^n + \frac{\Delta t}{2} f\left[U^{(2)}\right] \\ U^{n+1} &= U^n + \Delta t\, f\left[U^{(3)}\right] \end{aligned} \tag{2.66}$$

Although this scheme is only second-order accurate for the general function, f, it is fourth-order accurate for the linear functions occurring in Maxwell's equations. Storage is needed for U^n and the last iterate of U throughout the grid. We store f throughout the grid, or else calculate it along each grid line and then update so that f is stored only along a single direction. Thus, effectively only two storage locations are required for the electric and magnetic fields. There are also other low-storage algorithms for nonlinear equations [32–34].

Zingg [35] shows that (2.66) is fourth-order accurate only for homogenous linear problems. He suggests some other low-storage schemes for the inhomogeneous case. However, for problems with forcing functions, we have not seen any difference in the numerical solutions between (2.66) and the schemes suggested by Zingg.

2.4.6 Time Step

For each of the methods described above, a time step is chosen for the numerical integration. This time step is based on two considerations: stability and accuracy. Since all the schemes have stability limits, this places an upper bound on the time step.

Numerical stability is governed by the amplification factor, a, given by

$$a = 1 + \frac{1}{2}(\lambda \Delta t)^2 \pm \sqrt{\left[1 + \frac{1}{2}(\lambda \Delta t)^2\right]^2 - 1} \qquad \text{leapfrog} \tag{2.67a}$$

$$a = 1 + (\lambda \Delta t) + \frac{1}{2}(\lambda \Delta t)^2 + \frac{1}{6}(\lambda \Delta t)^3 + \frac{1}{24}(\lambda \Delta t)^4 \qquad \text{4-stage R-K} \tag{2.67b}$$

where λ is an eigenvalue of the Fourier transform of the spatial finite-difference operator. Defining

$$D = \left[\frac{1}{(\Delta x)^2} + \frac{1}{(\Delta y)^2}\right]^{1/2} \tag{2.68}$$

then we have

$$|\lambda_{\text{Yee}}| = 2\left[\frac{\sin^2(\theta/2)}{(\Delta x)^2} + \frac{\sin^2(\theta/2)}{(\Delta y)^2}\right]^{1/2} \leq 2D \tag{2.69a}$$

$$|\lambda_{\text{Ty}}| = 24\left[\frac{\sin^2(\theta/2)}{(\Delta x)^2(11+\cos\theta)^2} + \frac{\sin^2(\theta/2)}{(\Delta y)^2(11+\cos\theta)^2}\right]^{1/2} \leq 2.4D \tag{2.69b}$$

Consider the following comparison of the four-stage R-K method versus leapfrog time stepping:

1. *Time step.* Without staggering in time, the four-stage R-K method has a time step that is potentially 2.8 times larger than leapfrog. Since the Yee algorithm is staggered in time, the R-K scheme loses a factor of two.
2. *Dissipation.* For an imaginary eigenvalue, λ, the leapfrog method is not dissipative, but the four-stage R-K method is dissipative.
3. *Phase lead or lag.* The leapfrog scheme has a phase lead for time steps within the stability limit. The four-stage R-K method has a phase lag for CFL factors less than 1.88, and a phase lead for CFL factors between 1.88 and $\sqrt{8}$.

Numerical accuracy requires limiting the time step. A basic premise is that the overall accuracy of the numerical technique should not be determined by the time integration. Hence, the temporal error should be equal to or less than the spatial error. For the Yee scheme, the time step is chosen close to the limit set by numerical stability. For the fourth-order space-derivative approaches, the optimal time step may well be much less than that required for stability. The issues of optimal time step and algorithm efficiency are discussed in Section 2.6 in the context of numerical experiments involving two-dimensional electromagnetic wave propagation.

2.5 COMBINED DISCRETIZATION

Thus far, we have discussed spatial and time errors independently. We now investigate the numerical errors from the combined discretization in the context of the variation of the normalized numerical wave phase velocity $\tilde{v}_p/c$.

Assuming a uniform grid-cell size, $\Delta x = \Delta y = h$, and a grid resolution of $M = 2\pi/kh$ points per wavelength, then $\tilde{v}_p/c$ for the Yee algorithm is given by

$$\frac{\tilde{v}_p}{c} = \frac{M}{\pi}\left[\sin^2\left(\frac{\pi\cos\theta}{M}\right) + \sin^2\left(\frac{\pi\sin\theta}{M}\right)\right]^{1/2} \tag{2.70a}$$

Under the same grid conditions, $\tilde{v}_p/c$ for the Ty(2,4) algorithm is

$$\frac{\tilde{v}_p}{c} = \frac{12M}{\pi}\left\{\left[\frac{\sin\left(\dfrac{\pi\cos\theta}{M}\right)}{11 + \cos\left(\dfrac{2\pi\cos\theta}{M}\right)}\right]^2 + \left[\frac{\sin\left(\dfrac{\pi\sin\theta}{M}\right)}{11 + \cos\left(\dfrac{2\pi\sin\theta}{M}\right)}\right]^2\right\}^{1/2} \tag{2.70b}$$

Fig. 2.1 is a polar diagram of $\tilde{v}_p/c$ versus wave propagation angle in the grid for $M = 1, 2, 4, 8$, and 16 for Yee and Ty(2,4). Fig. 2.2 uses a line graph to provide the same comparison for $M = 20$. For gridding densities $M > 4$, it is clear that the total velocity error, $[1 - (\tilde{v}_p/c)]$, for Ty is much less than that of Yee. The error is largest along the grid axes and smallest along the grid diagonals. At $M = 20$, the velocity anisotropy error is 0.2% for Yee but only 0.002% for Ty.

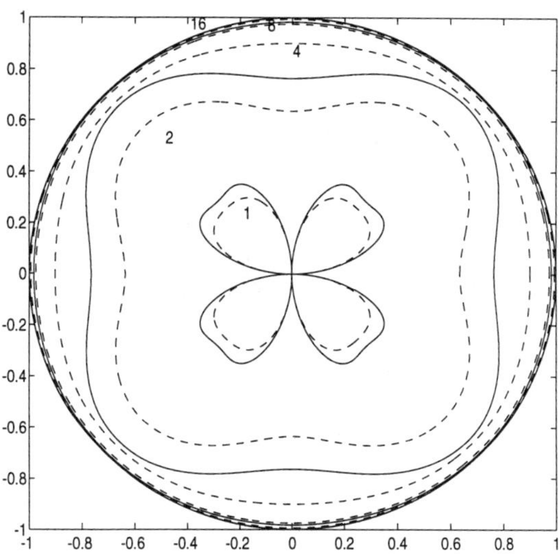

Fig. 2.1 Polar diagram of the numerical phase velocity. Yee: - - - -, Ty (2,4): ——.

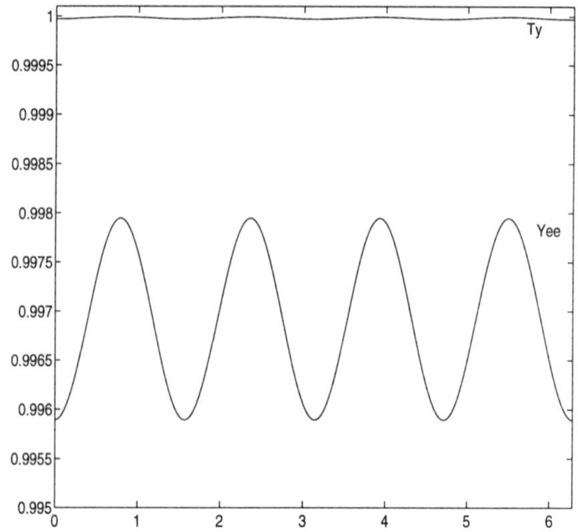

Fig. 2.2 Comparison of the numerical phase velocity versus wave propagation angle in the grid for M = 20 points per wavelength.

Additional insight is provided by expanding $\tilde{v}_p / c$ to obtain

$$1 - \frac{\tilde{v}_p}{c} = \frac{\pi^2}{M^2}\left(\frac{1}{8} + \frac{1}{24}\cos 4\theta\right) \qquad \text{Yee} \tag{2.71a}$$

$$1 - \frac{\tilde{v}_p}{c} = \frac{17\pi^4}{2880\,M^4}(5 + 3\cos 4\theta) \qquad \text{Ty} \tag{2.71b}$$

This shows that the leading dispersive error is proportional to M^{-2} for the Yee algorithm, but drops to M^{-4} for the Ty(2,4) algorithm. This is a reflection that Yee's scheme is second-order accurate while the Ty scheme is fourth-order accurate.

2.6 COMPUTATIONAL RESULTS

In this section, we compare the accuracy and efficiency of three different FDTD algorithms: standard Yee; Ty(2,4) (second-order leapfrog time stepping coupled with fourth-order implicit space derivatives); and Ty(4,4) (fourth-order R-K time stepping coupled with fourth-order implicit space derivatives). All of the test cases involve a two-dimensional TM grid with field components E_z, H_x, and H_y. In each case, the Berenger PML absorbing outer boundary is used.

2.6.1 Relation Between the Yee, Ty(2,4), and Ty(4,4) Methods

The computational results presented in this section are obtained using a common staggered space grid. The relation between the three methods implemented on this grid is now summarized. We first present Yee's finite-difference equations in a form that enables us to generalize them easily:

$$E_z\Big|_{i,j}^{n+1} = E_z\Big|_{i,j}^{n} + \frac{\Delta t}{\varepsilon_{i,j}}\delta_x H_y\Big|_{i,j}^{n+1/2} - \frac{\Delta t}{\varepsilon_{i,j}}\delta_y H_x\Big|_{i,j}^{n+1/2} \tag{2.72a}$$

$$H_x\Big|_{i,j-1/2}^{n+1/2} = H_x\Big|_{i,j-1/2}^{n-1/2} - \frac{\Delta t}{\mu_{i,j-1/2}}\delta_y E_z\Big|_{i,j-1/2}^{n} \tag{2.72b}$$

$$H_y\Big|_{i-1/2,j}^{n+1/2} = H_y\Big|_{i-1/2,j}^{n-1/2} + \frac{\Delta t}{\mu_{i-1/2,j}}\delta_x E_z\Big|_{i-1/2,j}^{n} \tag{2.72c}$$

where for any field variable, U, on the staggered mesh we have

$$\begin{aligned} \delta_x U_{i,j} &= (U_{i+1/2,j} - U_{i-1/2,j}) / \Delta x \\ \delta_y U_{i,j} &= (U_{i,j+1/2} - U_{i,j-1/2}) / \Delta y \end{aligned} \tag{2.73}$$

As described in Section 2.3, we directly replace (2.73) by the fourth-order-accurate Ty implicit-staggered scheme to produce a much more accurate solution. For example, in the y-direction, the Ty approximation to the spatial derivatives is given by

$$\delta_y U\text{:}\ \frac{U_{j+1} - U_j}{\Delta y} = \left(\frac{11}{12}\frac{\partial U}{\partial y}\right)_{j+1/2} + \frac{1}{24}\left[\left(\frac{\partial U}{\partial y}\right)_{j+3/2} + \left(\frac{\partial U}{\partial y}\right)_{j-1/2}\right] \tag{2.74}$$

In fact, the only difference between the Yee algorithm and the Ty(2,4) algorithm is the replacement of (2.73) by (2.74), since the same staggered space grid and leapfrog time stepping is used for both techniques. Implementation of Ty(4,4) can also be accomplished on the same staggered space grid as Yee, but with the replacement of the leapfrog time stepping with a four-step R-K method. Finally, the Berenger PML ABC can be implemented with equal success with any of these approaches.

Maxwell's equations have the property that, in the absence of charge, $\nabla\cdot\vec{D} = 0$ and $\nabla\cdot\vec{B} = 0$. It is important that the spatial finite-difference approximations and the boundary conditions also have this property. Otherwise, fictitious (nonphysical) sources can be introduced into the numerical solution. Both the Yee and Ty schemes can be shown to have the required divergence properties. Applying δ_y to (2.72b) and δ_x to (2.72c) and adding, we get for both schemes

$$\delta_t\left[\delta_x(\mu H_x) + \delta_y(\mu H_y)\right] = 0 \tag{2.75}$$

which satisfies the required divergence property for H assuming zero initial conditions.

2.6.2 Example 1: Sinusoidal Line Source in Free Space

We first consider an isotropic, sinusoidal line source of wavelength $\lambda = 1/4$ that is switched on at $t = 0$ in the domain $0 \leq x, y \leq 0.9$. The grid parameters are:

Yee	$h = 1/160 = \lambda/40$	$\Delta t = 2h/3 = 1/240$
Ty(2,4)	$h = 1/20 = \lambda/5$	$\Delta t = h/18 = 1/360$
Ty(4,4)	$h = 1/20 = \lambda/5$	$\Delta t = h/4 = 1/80$

The line source is modeled by the current $I_z(t) = 0.01 \sin(8\pi t)\, H(t)$ located at the point $\bar{x}_s = (x, y) = (0.2, 0.2)$, where $H(t)$ is the Heaviside unit-step function. This radiates a rotationally symmetric outgoing wave having the exact solution

$$E_z(|\bar{r}|, t) = -\frac{Z}{2\pi}\int_0^\infty \frac{\frac{\partial}{\partial t} I_z\left(t - \sqrt{|\bar{r}|^2 + \xi^2}\right)}{\sqrt{|\bar{r}|^2 + \xi^2}}\, d\xi \quad \Bigg\} \quad |\bar{r}| = |\bar{x} - \bar{x}_s| \tag{2.76}$$

Fig. 2.3 plots as a function of the simulated time the errors in the L_2 norm (relative to the exact solution) along a diagonal grid line running through the source for the three FDTD methods. We see that the errors are comparable, but recall that the grid for the Yee algorithm is a factor of 8 times as dense in each coordinate direction.

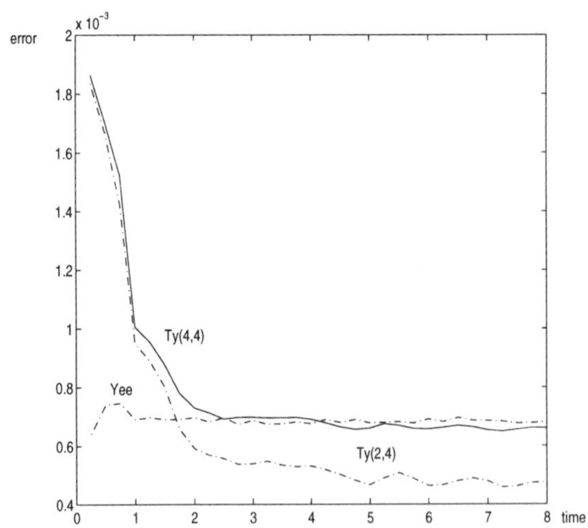

Fig. 2.3 Errors in the L_2 norm for the computed radially propagating wave as a function of the simulated time. Low resolution ($h = 1/20$) used for the Ty(2,4) and Ty(4,4) grids; high resolution ($h = 1/160$) used for the Yee grid.

Table 2.1 presents the computational time required for each solution. (Each program was written in MATLAB and runs on an SGI Challenge computer.) We see that, for this example, either Ty(2,4) or Ty(4,4) runs about 23 times faster than does the Yee algorithm to achieve similar accuracy. Note that this is a consequence of the Ty grids being very coarse relative to the Yee grid (requiring only 1/64th of the cells). This permissible coarseness is decisive in accelerating the performance of the Ty simulations, more than compensating for the extra time required by the scalar tridiagonal inversions.

Table 2.1

Computer Time for Radial Pulse Propagation Using Various FDTD Schemes

Scheme	h	Δt	CPU Time
Yee	1/160	1/240	1771 sec
Ty (2, 4)	1/20	1/360	76.5
Ty (4, 4)	1/20	1/80	74.3

2.6.3 Example 2: Parallel-Plate Waveguide

We next consider a parallel-plate waveguide with perfectly conducting walls located at $y = 0,\ 1$ and propagating the TE_3 mode. We choose the same computation grids and parameters as for the line source in the previous example. Fig. 2.4 plots as a function of the simulated time the errors in the L_2 norm (relative to the exact solution) along the longitudinal axis of the waveguide for the Yee and both Ty schemes. We see that the error behavior is very similar for all three methods, but again recall that the grid for the Yee algorithm is a factor of 8 times as dense in each coordinate direction.

2.6.4 Example 3: Obliquely Propagating Wave in Free Space— Parametric Studies of Solution Accuracy

This section summarizes the results of parametric studies of the accuracy of the Yee, Ty(2,4), and Ty(4,4) algorithms. The goal is to pinpoint the optimal time step for the Ty approaches and the grid-resolution advantage of these schemes relative to the Yee algorithm. For this purpose, the grid is excited with boundary conditions that simulate a wave of wavelength $\lambda = 1/2$ obliquely propagating in free space.

For the first study, $h = \Delta x = \Delta y = 1/40$ and $\Delta t = 1/80,\ 1/160,\ \ldots,\ 1/1200$. Hence, the CFL number, $\Delta t/h$, ranges from 1/2 to 1/30. Fig. 2.5 shows the error for Ty(2,4) as a function of $\Delta t/h$ in an expanded range from 1/30 to 1/10. There is an error minimum at $\Delta t/h = 1/18$, indicating the optimum choice of Δt for this algorithm.

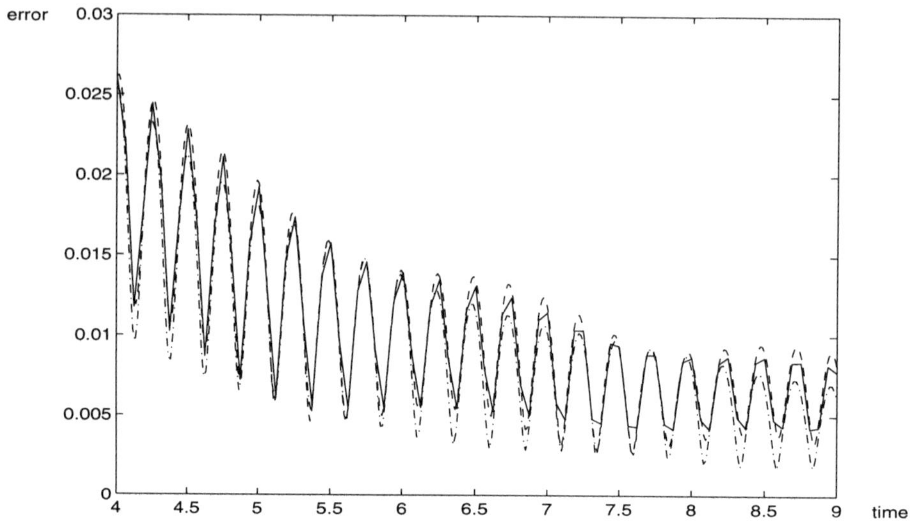

Fig. 2.4 Errors in the L_2 norm for the computed waveguide modal propagation as a function of the simulated time. Low-resolution ($h = 1/20$) Ty(2,4) and Ty(4,4) coincide and are drawn as the solid line. High-resolution ($h = 1/160$) Yee results are drawn as the dashed line.

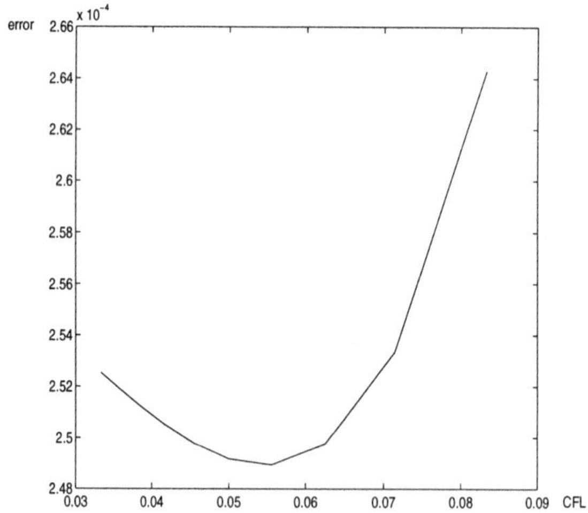

Fig. 2.5 Computed error generated by the Ty(2,4) algorithm as a function of $\text{CFL} = \Delta t / h$ for oblique wave propagation. There is a distinct error minimum at 1/18.

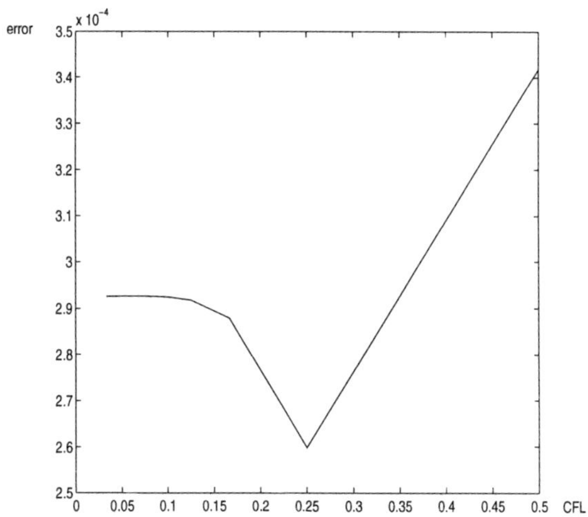

Fig. 2.6 Computed error generated by the Ty(4,4) algorithm as a function of CFL $= \Delta t / h$ for oblique wave propagation. There is a distinct error minimum at 1/4.

Fig. 2.6 repeats this error study for the Ty(4,4) algorithm. From this figure, we see that there is a distinct error minimum at $\Delta t / h = 1/4$.

To compare the Yee and Ty schemes, we choose for the Yee algorithm $h = \Delta x = \Delta y$ and $\Delta t = (2/3)h$, where h varies between 1/20 and 1/360. As shown in Fig. 2.7, the error for the Yee algorithm when $h = 1/360$ is comparable to the error for Ty(2,4) with $h = 1/40$ and $\Delta t = 1/720$; and comparable to the error for Ty(4,4) with $h = 1/40$ and $\Delta t = 1/160$. Fig. 2.8 plots the logarithm of each error type as a function of $\log(h)$.

We conclude that both the Ty(2,4) and Ty(4,4) methods are much more accurate and efficient than the Yee algorithm, which requires a mesh about 8 times finer in each direction to achieve similar accuracy. The accuracy of Ty(2,4) and Ty(4,4) is made comparable by choosing an appropriate time step. For Ty(4,4), $\Delta t = h/4$ seems to be optimal. Since this is a four-stage method, this is equivalent to $\Delta t = h/16$ per stage. For Ty(2,4), the time step must be smaller to achieve optimal accuracy. Here, $\Delta t = h/18$ is found to be best. Overall, Ty(2,4) is only slightly less efficient than Ty (4,4). The major advantage of Ty(2,4) is that it requires less programming if there is an existing code based on the Yee algorithm. Also, Ty(2,4) is nondissipative, while Ty(4,4), an R-K scheme, introduces a slight dissipation.

2.7 DISCONTINUOUS COEFFICIENTS

The usual procedure to determine the order of accuracy of a numerical method is to expand the solution in a Taylor series. However, when the coefficients of the equation are not smooth, this approach may not be justified. Similar remarks apply when the

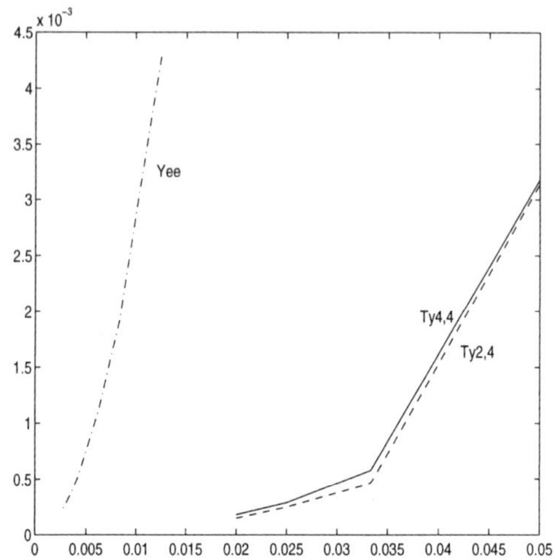

Fig. 2.7 Computed error as a function of grid-cell size, h, achieved by the Yee, Ty(2,4), and Ty(4,4) schemes for oblique wave propagation. h ranges from 1/360 to 1/20.

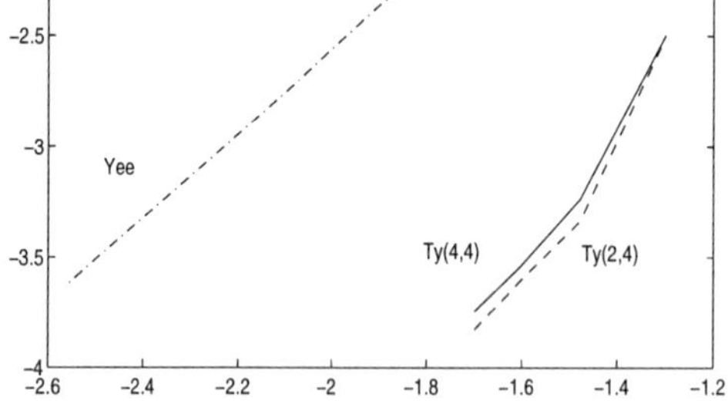

Fig. 2.8 Logarithm of the computed error as a function of $\log(h)$ achieved by Yee, Ty(2,4), and Ty(4,4) for oblique wave propagation.

domain of integration is not smooth. In particular, this applies to scattering from a smoothly curved body if the body surface is replaced by a series of polygons, in effect staircasing the surface. For both the discontinuous-coefficients case and the nonsmooth-domain case, the full accuracy of the numerical method may not be realized.

2.7.1 Initial Examination

To see some of the effects of nonsmooth coefficients, we Fourier-transform Maxwell's equations in time to obtain the Helmholtz equation. To further simply the matter, we consider the ordinary differential equation version of the Laplace equation

$$\frac{d}{dx}\left[a(x)\frac{d}{dx}\right]u(x) = f \tag{2.77}$$

We approximate this by

$$a_{i+1/2}\left(u_{i+1} - u_i\right) - a_{i-1/2}\left(u_i - u_{i-1}\right) = (\Delta x)^2 f_i \tag{2.78}$$

Assuming that $a(x)$ is only defined at the nodes x_i, we decide on how to calculate $a_{i+1/2}$. One simple choice is

$$a_{i+1/2} = \frac{a_i + a_{i+1}}{2} \tag{2.79}$$

Wesseling [36] shows that this leads to a first-order method. He further shows that if the arithmetic average is replaced by a geometric average

$$a_{j+1/2} = \sqrt{a_j\, a_{j+1}} \tag{2.80}$$

then second-order accuracy is achieved. Similarly, for Maxwell's equations, ε and μ at the interface are defined as an appropriate average of their values in the adjacent media.

2.7.2 Fourth-Order Smooth Approximation

Here, we consider a different approach found to markedly reduce error. This approach replaces the discontinuous function, ε, by a fourth-order smooth implicit approximation. (A similar strategy can be applied to jumps in μ.) Relative to the use of a polynomial approximation to ε, this strategy avoids the well-known overshoot artifact. We note that, with an implicit interpolation, ε varies in the entire domain and not just near the interface. However, far from the interface, the variation is quite small.

A second-order interpolation is achieved using

$$\begin{bmatrix} \varepsilon_1 \\ \varepsilon_2 \\ \cdot \\ \cdot \\ \varepsilon_{p-1} \end{bmatrix} = \frac{1}{2}\left(\begin{bmatrix} \varepsilon_{3/2} \\ \varepsilon_{5/2} \\ \cdot \\ \varepsilon_{p-3/2} \\ \varepsilon_{p-1/2} \end{bmatrix} + \begin{bmatrix} \varepsilon_{1/2} \\ \varepsilon_{3/2} \\ \cdot \\ \varepsilon_{p-5/2} \\ \varepsilon_{p-3/2} \end{bmatrix} \right) \tag{2.81}$$

A fourth order interpolation is achieved using

$$\frac{1}{8}\begin{bmatrix} 10 & -5 & 4 & -1 & \cdot & \cdot & 0 \\ 1 & 6 & 1 & 0 & \cdot & \cdot & 0 \\ 0 & 1 & 6 & 1 & 0 & \cdot & 0 \\ \cdot & \cdot & \cdot & \cdot & \cdot & \cdot & \cdot \\ 0 & \cdot & \cdot & 0 & 1 & 6 & 1 \\ 0 & \cdot & \cdot & -1 & 4 & -5 & 10 \end{bmatrix} \begin{bmatrix} \varepsilon_1 \\ \varepsilon_2 \\ \cdot \\ \cdot \\ \varepsilon_{p-1} \end{bmatrix} = \frac{1}{2}\left(\begin{bmatrix} \varepsilon_{3/2} \\ \varepsilon_{5/2} \\ \cdot \\ \varepsilon_{p-3/2} \\ \varepsilon_{p-1/2} \end{bmatrix} + \begin{bmatrix} \varepsilon_{1/2} \\ \varepsilon_{3/2} \\ \cdot \\ \varepsilon_{p-5/2} \\ \varepsilon_{p-3/2} \end{bmatrix} \right) \tag{2.82}$$

In two dimensions, this is replaced by

$$\left(I + \frac{h^2}{8}\Delta \right)\varepsilon_{i,j} = \frac{\varepsilon_{i+1/2,j+1/2} + \varepsilon_{i-1/2,j+1/2} + \varepsilon_{i+1/2,j-1/2} + \varepsilon_{i-1/2,j-1/2}}{4} \tag{2.83}$$

where Δ is the standard approximation to the Laplacian. This is solved once by either multigrid or approximate factorization.

2.7.3 Computed Example: Dielectric Slab

We now consider a rectangular domain $[-1, 1] \times [0, 1]$ which contains air and lossless dielectric with $\varepsilon_r = 4$ as shown in Fig. 2.9(a). The exact solution for the electric field in this case is given by

$$E_z = \begin{cases} 2\cos(2\pi x/3)\sin(K_y y)\cos\omega t & |x| \le 0.5, \quad 0 \le y \le 1 \\ \exp\left[(\pi\sqrt{3}/3)\cdot(1 - 2|x|)\right]\sin(K_y y)\cos\omega t & |x| > 0.5, \quad 0 \le y \le 1 \end{cases} \tag{2.84}$$

where $K_y = (2\pi/3)\sqrt{13/3}$ and $\omega = 4\pi/(3\sqrt{3})$. We use the exact solutions for the electric and magnetic fields for the boundary and initial conditions, and develop L_2-normed errors of the calculated Yee and Ty fields relative to the exact solution. Initially, we choose $\Delta x = \Delta y = 1/40$ for the Yee, Ty(2,4), and Ty(4,4) grids with time-steps $\Delta t_{\text{Yee}} = 1/60$, $\Delta t_{\text{Ty(2,4)}} = 1/720$, and $\Delta t_{\text{Ty(4,4)}} = 1/160$.

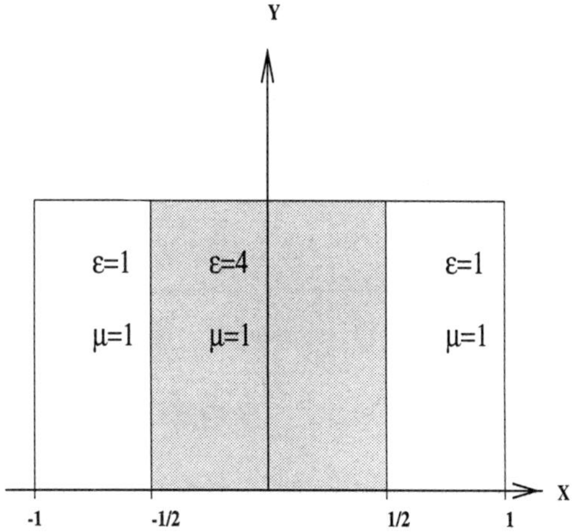

(a)

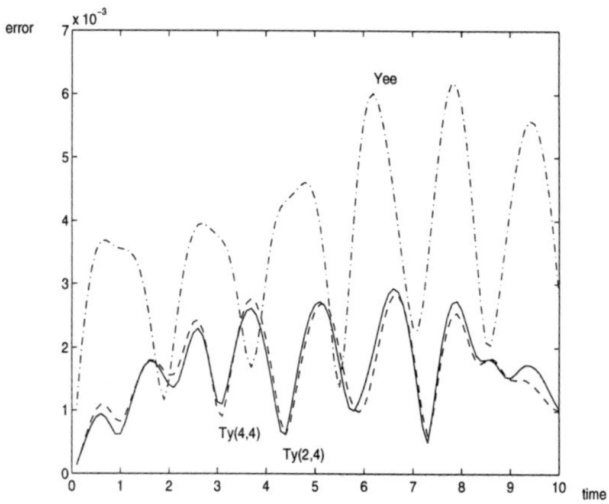

(b)

Fig. 2.9 (a) Computational domain for the discontinuous-coefficient modeling example; (b) L_2 errors for the case of the arithmetic average of the permittivity used at the dielectric interfaces of (a) for the Yee, Ty(2,4) and Ty(4,4) schemes. The same grid resolution $\Delta x = \Delta y = 1 / 40$ is used for each approach with an optimal choice of time step used for the Ty schemes.

Fig. 2.9(b) shows the numerical errors versus the simulation time for the three methods for the case where ε at the interface is set to the arithmetic average of the values on both sides. While the two fourth-order methods have less error than Yee, their error performance is clearly hurt by the treatment of the interface, which gives only second-order accuracy.

Fig. 2.10(a) shows the corresponding numerical errors for the case where the dielectric coefficients are smoothed, as described in Section 2.7.2. While there is a modest improvement in the accuracy of the Yee scheme, there is a much greater improvement in the accuracy of Ty(2,4) and Ty(4,4). Fig. 2.10(b) shows how the error provided by the Yee algorithm can be reduced to that of the Ty schemes by having the Yee grid resolution eight times finer than that of Ty.

(a)

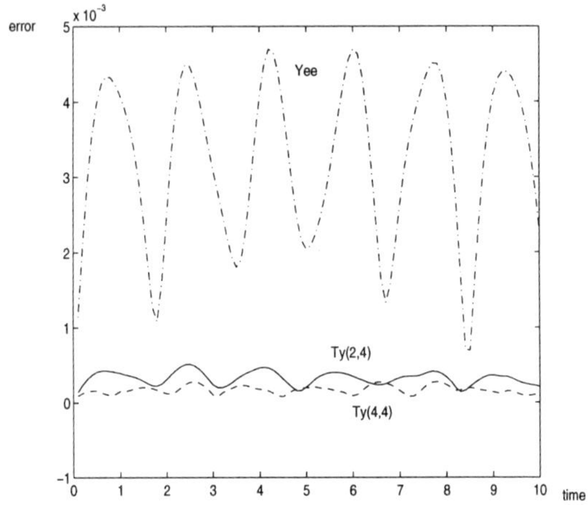

(b)

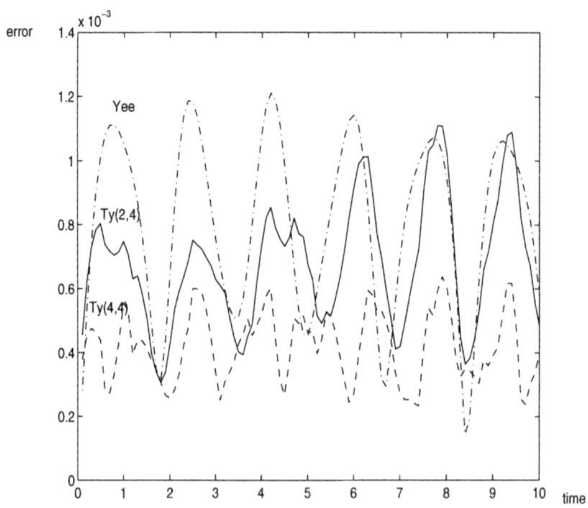

Fig. 2.10 (a) Same as Fig. 2.9(b), but here fourth-order implicit interpolation of the dielectric coefficients is used for all three schemes with $\Delta x = \Delta y = 1/40$; (b) same as (a), but here the Yee grid resolution is $\Delta x = \Delta y = 1/160$, eight-times finer than that of Ty ($\Delta x = \Delta y = 1/20$) to give comparable errors.

In Fig. 2.10(b), the Yee grid is refined such that $\Delta x = \Delta y = 1/160$ and $\Delta t = 1/240$. To equalize the error for the Ty(2,4) and Ty(4,4) grids, we choose an eightfold-coarser cell size of $\Delta x = \Delta y = 1/20$ along with optimal time steps of $\Delta t_{\mathrm{Ty(2,4)}} = 1/360$ and $\Delta t_{\mathrm{Ty(4,4)}} = 1/80$. Overall, we see that the coefficient smoothing prescribed in Section 2.7.2 for abrupt permittivity jumps succeeds in preserving the basic fourth-order-accuracy advantage of Ty noted previously for the homogeneous case (Section 2.6).

2.7.4 Computed Example: Dielectric-Coated Conductor

We next consider a perfect electric conductor (PEC) coated with a dielectric layer of thickness 0.5, as shown in Fig. 2.11(a). As in the previous example, we use implicit interpolation for ε. Since $\varepsilon < 1$ for this case, the time step is adjusted for stability. We choose for the Yee algorithm $\Delta x = \Delta y = 1/160$ and $\Delta t = 1/320$, and for the Ty(2,4) and Ty(4,4) methods $\Delta x = \Delta y = 1/20$, $\Delta t_{\mathrm{Ty(2,4)}} = 1/700$, and $\Delta t_{\mathrm{Ty(4,4)}} = 1/200$. The exact solution for the electric field in this case is

$$E_z = \begin{cases} -\sin(\pi x)\sin\left(\sqrt{5}\,\pi y - 3\pi t\right) & 0 \le x \le 0.5 \\ \cos(2\pi x)\sin\left(\sqrt{5}\,\pi y - 3\pi t\right) & 0.5 < x \le 1.25 \end{cases} \tag{2.85}$$

with the boundary conditions

$$\begin{aligned} E_z(x=0,\, y,\, t) \;&=\; E_z(x=1.25,\, y,\, t) \;=\; 0 \\ E_z(0 \le x \le 0.5,\; y=0,\; t) \;&=\; \sin(\pi x)\sin(3\pi t) \\ E_z(0.5 < x \le 1.25,\; y=0,\; t) \;&=\; -\cos(2\pi x)\sin(3\pi t) \end{aligned} \tag{2.86}$$

Fig. 2.11(b) shows that the errors for all three schemes are about the same even though Ty(2,4) and Ty(4,4) use an eightfold-coarser mesh in each coordinate direction than Yee. This supports the results and conclusion of the previous example of Section 2.7.3.

2.8 PML ABSORBING BOUNDARY

Berenger's PML absorbing boundary and related formulations (see [37, 38] and Chapter 5) have become preferred means to truncate FDTD domains for open-region modeling problems. An important feature of all of these approaches is that the only spatially differentiated terms are the physical field variables. Hence, we can calculate all of the required space derivatives by the implicit method used in Ty(2,4) and Ty(4,4) across the entire FDTD domain, including both the interior working volume and the PML or PML-like absorber. Once the derivatives are approximated, we time-advance each system independently in the interior and in the PML. Thus, the Ty approaches are completely compatible with the PML techniques proposed to date.

(a)

(b)

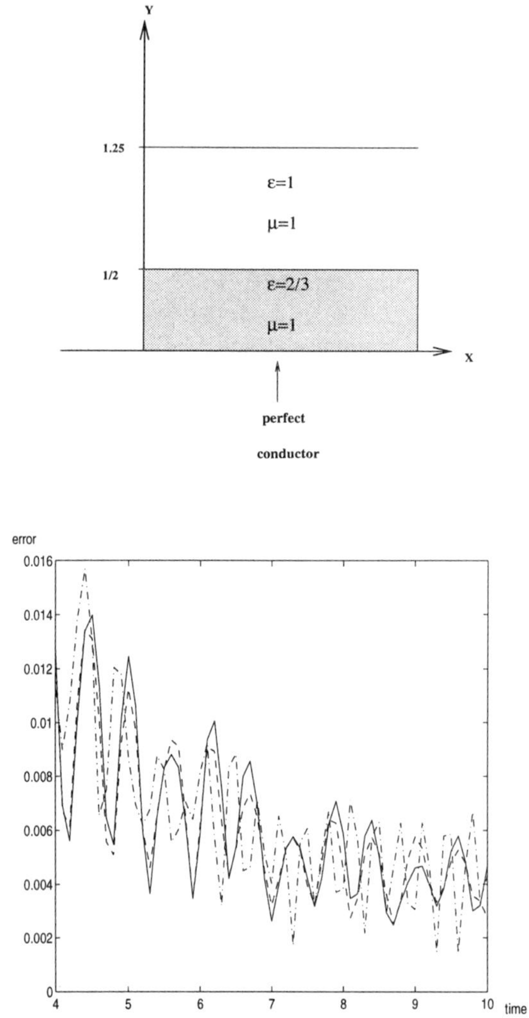

Fig. 2.11 (a) Computational domain for the dielectric-coated conductor example; (b) L_2 errors for the case of implicit interpolation of the dielectric coefficients of (a).

2.9 SIMULTANEOUS APPROXIMATION TERM SOLID BOUNDARY

The imposition of the electromagnetic field boundary condition at the surface of a rectangular-shaped perfect conductor is straightforward using a Cartesian-coordinate Yee grid. Curved surfaces are clearly more difficult to model in a rectangular grid. The simplest possibility is a staircasing approximation of the actual surface contour.

However, this can introduce significant errors [39] due to the introduction of artificial corners in the surface. This may degrade the overall solution to only first-order accuracy.

One way to overcome this difficulty is to use a finite-volume method [9, 40]. A different approach proposed by Abarbanel and Ditkowski known as the simultaneous approximation term (SAT) method [41 – 43] is to add a penalty term that accounts for the shape of the body. In this section, we present results for solving the scalar wave equation within the circular annular region $1/6 < r < 1/2$ using the SAT method. (Refer to [41 – 43] for the details of the theory of the SAT technique.) The exact solution that we consider is

$$U = \left[J_1(\omega r) + a Y_1(\omega r)\right] \cos(\omega t + \theta) \tag{2.87}$$

where $\omega = 9.8137$ and $a = 1.7637$. Fig. 2.12 shows the grids for the staircasing and SAT methods, respectively. For staircasing, note that the grid cells at the bounding contours can have portions outside of the domain of integration, whereas for SAT all of the grid cells are entirely within the computational domain.

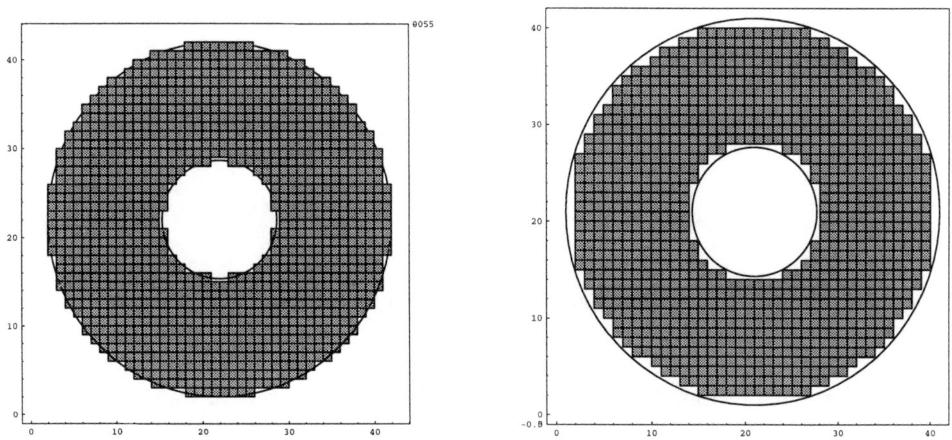

(a) (b)

Fig. 2.12 Circular annular domain for solution of the scalar wave equation: (a) grid for the standard staircased method; (b) grid for the SAT method.

Table 2.2 presents the L_2 errors for the standard staircased and SAT methods. As the mesh is refined, the staircasing technique yields first-order accuracy as expected. However, the SAT approach seems to be almost third-order accurate. It is not clear why third-order accuracy is achieved, since a second-order-accurate central-difference scheme is used to discretize the wave equation in the interior region in both cases. In all cases, the L_2 error of the SAT scheme is about 1/100 (or less) of the error due to staircasing.

TABLE 2.2

Convergence of the Staircased and SAT Boundary Schemes for the Wave Equation Solved in a Circular Annular Region

Scheme	h	Time	L_2 Error	Ratio
Staircase	1/40	1	.102	
Staircase	1/80	1	.050	2.04
SAT	1/40	1	.0012	
SAT	1/80	1	.00017	7.05
Staircase	1/40	10	.793	
Staircase	1/80	10	.473	1.67
SAT	1/40	10	.0084	
SAT	1/80	10	.00083	10.12

2.10 HELMHOLTZ EQUATION

In this section, we consider fourth-order methods for the Helmholtz equation

$$\frac{\partial^2 u}{\partial x^2} + \frac{\partial^2 u}{\partial y^2} + k^2 u = F(x, y) \tag{2.88}$$

We only consider the constant-coefficient case, but the extension to variable $k(x, y)$ is straightforward [44].

2.10.1 Interior Region

Let $\phi_{i,j}$ be the numerical approximation to $u(x_i, y_j)$, and let $F_{i,j} = F(x_i, y_j)$ be a known function. We wish to have symmetric stencils in both the x and y directions, using at most nine points. A scheme with these properties has the form

$$A_0 \phi_{i,j} + A_s \sigma_s + A_c \sigma_c = B_0 F_{i,j} + B_s \Gamma_s + B_c \Gamma_c \tag{2.89}$$

where

$$\sigma_s = \phi_{i,j+1} + \phi_{i+1,j} + \phi_{i,j-1} + \phi_{i-1,j} \tag{2.90a}$$

$$\Gamma_s = F_{i,j+1/2} + F_{i+1/2,j} + F_{i,j-1/2} + F_{i-1/2,j} \tag{2.90b}$$

$$\sigma_c = \phi_{i+1,j+1} + \phi_{i+1,j-1} + \phi_{i-1,j-1} + \phi_{i-1,j+1} \quad (2.90c)$$

$$\Gamma_c = F_{i+1/2,j+1/2} + F_{i+1/2,j-1/2} + F_{i-1/2,j-1/2} + F_{i-1/2,j+1/2} \quad (2.90d)$$

We first describe the standard second-order approximation we call the pointwise representation:

$$D_{xx}\phi_{i,j} + D_{yy}\phi_{i,j} + k^2\phi_{i,j} = F_{i,j} \quad (2.91a)$$

or

$$\phi_{i,j+1} + \phi_{i+1,j} + \phi_{i,j-1} + \phi_{i-1,j} - 4\phi_{i,j} + k^2h^2\phi_{i,j} = h^2 F_{i,j} \quad (2.91b)$$

This is a five-point representation with the coefficients

$$A_0 = -4 + k^2h^2 \qquad A_s = 1 \qquad B_0 = h^2 \qquad A_c = B_s = B_c = 0 \quad (2.91c)$$

We next consider higher-order representations, beginning with a Padé approximation:

$$\frac{D_{xx}}{1 + (h^2 D_{xx}/12)}\phi_{i,j} + \frac{D_{yy}}{1 + (h^2 D_{yy}/12)}\phi_{i,j} + k^2\phi_{i,j} = F_{i,j} \quad (2.92)$$

This may be generalized to

$$\left(D_{xx} + D_{yy} + \frac{h^2}{6}D_{xx}D_{yy}\right)\phi_{i,j} + k^2\left[1 + \frac{h^2}{12}\left(D_{xx} + D_{yy}\right) + \gamma\frac{h^4}{144}D_{xx}D_{yy}\right]\phi_{i,j}$$

$$= \left[1 + \frac{h^2}{12}\left(D_{xx} + D_{yy}\right) + \delta\frac{h^4}{144}D_{xx}D_{yy}\right]F_{i,j} \quad (2.93)$$

Here, γ and δ are free parameters selected to improve the properties in directions other than along the grid lines [43]. These parameters do not affect the accuracy of the scheme because they multiply an $O(h^4)$ factor. $D_{xx}D_{yy}$ in (2.93) is replaced by

$$D_{xx}D_{yy}\phi_{i,j} = D_{yy}D_{xx}\phi_{i,j} = \frac{\sigma_c - 2\sigma_s + 4\phi_{i,j}}{h^4} \quad (2.94a)$$

$$D_{xx}D_{yy}F_{i,j} = D_{yy}D_{xx}F_{i,j} = \frac{16\Gamma_c - 32\Gamma_s + 64F_{i,j}}{h^4} \quad (2.94b)$$

Expressing the scheme in the form of (2.89), we get

$$A_0 = -\frac{10}{3} + k^2h^2\left(\frac{2}{3} + \frac{\gamma}{36}\right), \quad A_s = \frac{2}{3} + \frac{k^2h^2}{12}\left(1 - \frac{\gamma}{6}\right)$$

$$A_c = \frac{1}{6} + \frac{\gamma k^2 h^2}{144} \tag{2.95a}$$

$$B_0 = h^2\left(-\frac{1}{3} + \frac{4\delta}{9}\right), \quad B_s = h^2\left(\frac{1}{3} - \frac{2\delta}{9}\right), \quad B_c = \frac{h^2\delta}{9} \tag{2.95b}$$

A second approach to achieve a fourth-order algorithm is referred to as the equation-based approach. As in the time-dependent case, we use the differential equation itself to find higher derivatives. Thus:

$$\frac{\partial^4 u}{\partial x^4} = \frac{\partial^2 F}{\partial x^2} - \frac{\partial^4 u}{\partial y^2 \partial x^2} - k^2\frac{\partial^2 u}{\partial x^2} \tag{2.96a}$$

$$\frac{\partial^4 u}{\partial y^4} = \frac{\partial^2 F}{\partial y^2} - \frac{\partial^4 u}{\partial x^2 \partial y^2} - k^2\frac{\partial^2 u}{\partial y^2} \tag{2.96b}$$

Using this in the Taylor series expansion of the Laplacian, we get

$$\begin{aligned}(D_{xx} + D_{yy})u &= \frac{\partial^2 u}{\partial x^2} + \frac{\partial^2 u}{\partial y^2} \\ &+ \frac{h^2}{12}\left[\frac{\partial^2 F}{\partial x^2} + \frac{\partial^2 F}{\partial y^2} - 2\frac{\partial^4 u}{\partial x^2 \partial y^2} - k^2\left(\frac{\partial^2 u}{\partial x^2} + \frac{\partial^2 u}{\partial y^2}\right)\right] + O(h^4) \\ &= \left(1 - \frac{k^2h^2}{12}\right)\left(\frac{\partial^2 u}{\partial x^2} + \frac{\partial^2 u}{\partial y^2}\right) \\ &+ \frac{h^2}{12}\left[\left(D_{xx} + D_{yy}\right)F - 2D_{xx}D_{yy}u\right] + O(h^4)\end{aligned} \tag{2.97}$$

and the $O(h^4)$ approximation

$$\frac{\partial^2 u}{\partial x^2} + \frac{\partial^2 u}{\partial y^2} \cong \frac{\left(D_{xx} + D_{yy} + \frac{h^2}{6}D_{xx}D_{yy}\right)u - \frac{h^2}{12}\left(D_{xx} + D_{yy}\right)F}{1 - \frac{k^2h^2}{12}} \tag{2.98}$$

The discrete solution to (2.88) now satisfies

$$\left(D_{xx} + D_{yy} + \frac{h^2}{6} D_{xx} D_{yy}\right) \phi_{i,j} + k^2\left(1 - \frac{k^2h^2}{12}\right)\phi_{i,j} = \left[1 - \frac{k^2h^2}{12} + \frac{h^2}{12}\left(D_{xx} + D_{yy}\right)\right] F_{i,j} \tag{2.99}$$

In the format of (2.89), we have

$$A_0 = -\frac{10}{3} + k^2h^2\left(1 - \frac{k^2h^2}{12}\right), \quad A_s = \frac{2}{3}, \quad A_c = \frac{1}{6} \tag{2.100a}$$

$$B_0 = h^2\left(-\frac{1}{3} - \frac{k^2h^2}{12}\right), \quad B_s = \frac{h^2}{3}, \quad B_c = 0 \tag{2.100b}$$

As in the higher-order approximation of (2.93), we can add free parameters that multiply factors of $O(h^4)$ and do not affect the accuracy of the scheme. We omit these parameters for simplicity.

2.10.2 Boundary Conditions

We wish to approximate the boundary conditions for the Helmholtz equation with the same accuracy as the interior. For the pointwise stencil, we can use $O(h^2)$ approximations, but for the higher-order and the equation-based stencils, we need $O(h^4)$ approximations. The boundary approximations should be symmetric and use only nearby points. For a Dirichlet boundary condition, the function, $f(x)$, is known and specified along the boundary and thus is suitable for all the schemes:

$$\phi_{i,j} = f(x_i, y_j) \tag{2.101}$$

We now consider a rectangular domain and assume that a Neumann boundary condition is given on the line $x = 0$:

$$\left.\frac{\partial u}{\partial x}\right|_{x=0} = g(y) \tag{2.102}$$

We introduce the artificial line $i = -1$, which we will eventually eliminate. The essential idea is to get a fourth-order central approximation to the boundary condition that includes this line. We also impose the differential equation at the boundary. Thus, there are two relations along the boundary that are used to eliminate the artificial line. Some flexibility is available to keep the final formula symmetric. After all of these steps, the final formula for the fourth-order approximation to (2.102) is given by

$$
\begin{aligned}
A_s \phi_{1,j} + A_c(\phi_{1,j+1} + \phi_{1,j-1}) + \frac{1}{2} A_s(\phi_{0,j+1} + \phi_{0,j-1}) + \frac{1}{2} A_0 \phi_{0,j} \\
= \frac{1}{2} B_0 F_0 + \frac{1}{2} B_s \Gamma_s + \frac{1}{2} B_c \Gamma_c + \alpha\left(h g_j + \frac{h^3}{6} \hat{F}_x \right)
\end{aligned}
\tag{2.103}
$$

where A and B are given from the interior scheme (2.89) (see [45] for details). Note that if g in (2.102) is not a given function but rather a function of the unknown, as occurs in a Robin or Sommerfeld boundary condition, it presents no additional complications.

2.10.3 Computed Example

We consider a simple example given by

$$
\Delta u + k^2 u = (k^2 - 5) \sin x \sin 2y \quad \Big\} \quad 0 \le x \le \pi\,, \quad 0 \le y \le \pi \tag{2.104}
$$

and the boundary conditions

$$
u(\pi, y) = u(x, 0) = u(x, \pi) = 0\,, \qquad \left.\frac{\partial u}{\partial x}\right|_{(0,y)} = \sin 2y \tag{2.105}
$$

The exact solution is

$$
u(x, y) = \sin x \sin 2y \tag{2.106}
$$

Fig. 2.13 graphs the accuracy of the pointwise, higher-order, and equation-based methods as a function of $\log_2(N)$, where N is the number of grid cells spanning each side of the domain. In this figure, the vertical scale is the negative logarithm of the ℓ_∞-normed error of each approach for $k = 6.4$. Thus, an increasing vertical coordinate denotes an increasingly negative logarithm of the error, and hence, a diminishing error. It is clear from this figure that the higher-order and equation-based methods have superior accuracy and a superior trend-line slope relative to the pointwise method for all grids having more than $2^4 = 16$ cells spanning a side of the domain. We also see the effect of the free parameter, γ, in the higher-order scheme (here denoted as p).

2.11 SPECTRAL METHODS

Until now, we have described methods that are fourth-order accurate in space and (possibly) time. It is straightforward to extend these methods to sixth-order accuracy or higher in space. It is more difficult to achieve higher-order accuracy in time.

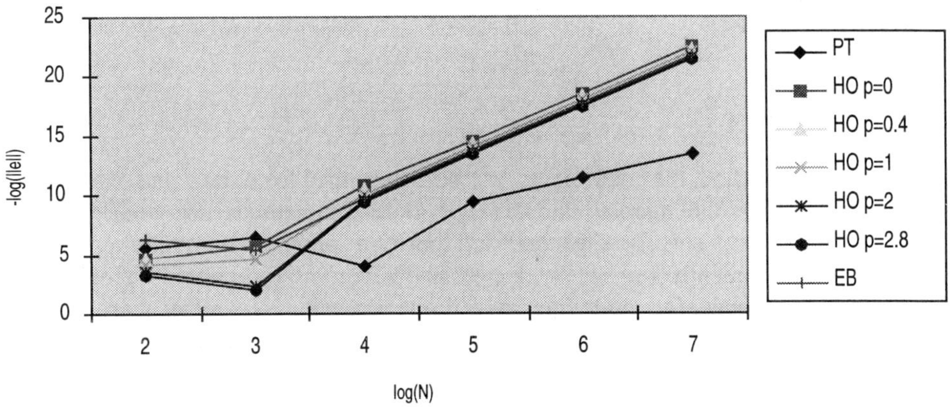

Fig. 2.13 Accuracy of the pointwise (PT), higher-order (HO), and equation-based (EB) solution methods for the Helmholtz equation as a function of the number of grid cells spanning each side of the domain of the sample problem of (2.104) and (2.105) for $k = 6.4$.

Higher-order spatial schemes require a larger stencil and usually require a smaller time step for stability. Since a much smaller time step is required by considerations of accuracy, this is not an important limitation. However, a larger stencil creates difficulties near boundaries. Special treatment is now required not only for points on a boundary but also for nearby mesh points, and a higher accuracy is needed for these neighboring points. These difficulties becomes more severe near media interfaces where ε and μ are discontinuous. Hence, a method beyond sixth-order accuracy in space is almost never used. Even sixth-order accuracy is hard to achieve with a stable, accurate, and robust boundary treatment [46].

One alternative is to use essentially non-oscillatory (ENO) methods [47]. Here the stencil is chosen to increase the smoothness of the solution. These methods are most suited for discontinuous solutions and so are not pursued further in computational electromagnetics.

Another alternative to finite-difference methods is to use spectral methods (Galerkin, tau, or collocation) [48]. We now describe the collocation algorithm. We assume that we have a set of basis functions $\phi_i(x)$. In contrast to finite-element methods, this basis consists of global functions. Therefore, all the points in the mesh are connected. The basis functions usually arise from a singular Sturm-Liouville equation so that expansions in terms of these variables converge with "infinite" order of accuracy (i.e., the accuracy is higher than any polynomial order). In fact, we can show that for analytic functions, the convergence rate is exponential.

Consider

$$\frac{\partial u}{\partial t} + a\frac{\partial u}{\partial x} = 0 \tag{2.107}$$

$$u(x) = \sum_{n=0}^{N} a_n \phi_n(x) \tag{2.108}$$

For a collocation method, the coefficients, a_n, are determined from the value of $u(x)$ at N distinct points within the interval (the tau and Galerkin methods use other ways of determining the a_n). This is a general interpolation problem that requires $O(N^2)$ operations. We frequently use either sine/cosine functions for periodic problems or Chebyshev polynomials for problems in a finite domain. In either case, the a_n coefficients are found by an FFT. Given (2.107) and (2.108), we have

$$\frac{du}{dx} = \sum_{n=0}^{N} a_n \phi_n'(x) \tag{2.109}$$

For a Fourier series, $\phi_n'(x)$ is given trivially in terms of $\phi_n(x)$. For a Chebyshev polynomial, we construct a recursion formula that evaluates $\phi_n'(x)$ in terms of $\phi_i(x)$, $i = 1, 2, ..., n$. We then calculate the sum in (2.109), usually by an FFT. We can also construct a matrix that gives du/dx at the N collocation points in terms of $u(x)$ at these N points. This requires $O(N^2)$ operations rather than $O(N \log N)$ points that the FFT requires. Nevertheless, calculating the derivative through matrix multiplication is more efficient when N is small enough (approximately 64 to 128, depending on the computer). Optimized software for evaluating the spectral derivatives is available through the Internet (`http://www.cfm.brown.edu/people/wsdon/home.html`). See [49–51] for further details.

Given the derivative of u at each point, we update the solution similar to the finite-difference approach. A fourth order Runge-Kutta formula is frequently used to advance the solution in time. This method automatically updates the solution even along the boundaries. Boundary conditions must be superimposed on the calculated solution; otherwise the spectral method will be unstable. As previously noted, PML-like layers at the outer grid truncations require only the spatial derivatives of physical quantities. Hence, we can evaluate the spectral derivative throughout the domain, including the PML absorber. For multidimensional problems, the derivatives are calculated independently in each direction.

Expansions in Chebyshev polynomials have the disadvantage that they require the use of very small time steps, $\Delta t \sim O(N^2)$, rather than the usual $\Delta t \sim O(\Delta x)$. Because of the time-step restriction and also in order to deal with complex geometries, it is frequently advisable to combine spectral methods with domain decomposition or spectral elements [51–53]. Computations using these techniques are presented in [29, 54–56].

2.12 SUMMARY AND CONCLUSIONS

This chapter focused on fourth-order FDTD methods with applications to the time-dependent Maxwell's equations and to the Helmholtz equation. After a discussion examining the concept of the order of accuracy, explicit and implicit fourth-order realizations of the space derivatives were presented. Of particular interest and emphasis here was a fourth-order implicit staggered scheme employing a compact stencil that is compatible with the Yee grid. This was designated the Ty space-derivative operator. Subsequently, the accuracy and boundary treatment of this operator was studied. Then, a variety of time-stepping schemes were discussed, including second-order leapfrog, fourth-order time stepping in the context of explicit and implicit space derivatives, and Runge-Kutta approaches. The theoretical numerical dispersion error was obtained for Ty space differencing coupled with second-order leapfrog time stepping, and evidenced substantial improvement relative to the classic Yee algorithm.

With the theory established, the chapter provided a series of computed examples that demonstrated the optimum operation of the Ty scheme with respect to Courant number. Both second-order leapfrog and fourth-order Runge-Kutta time stepping coupled to the Ty operator were studied in this context. Further, these examples showed that, when run at their respective optimum Courant numbers, both the Ty / leapfrog and Ty / Runge-Kutta FDTD algorithms permit reduction of the meshing density in each Cartesian direction by about 8:1 relative to the classic Yee algorithm while achieving comparable accuracy. In two dimensions, the resulting 64:1 drop in computer storage relative to Yee's approach is complemented by an approximate 23:1 reduction in computer running time.

The discussion then shifted to the impact of abrupt dielectric boundaries upon the accuracy of the Yee and Ty approaches. The replacement of the discontinuous permittivity function by a fourth-order-smooth, implicit approximation was proposed. Two computed canonical problems involving dielectric slab and dielectric-coated conductor geometries showed that this technique permits the Ty approaches to retain their fourth-order accuracy and 8:1 grid-coarseness advantage (in each Cartesian direction) relative to the Yee algorithm.

Additional discussions were then presented with regard to the compatibility of the Ty FDTD algorithms with PML ABCs, the treatment of curved surfaces using the SAT method, the application of fourth-order methods for the Helmholtz equation, and (briefly) spectral methods.

A logical conclusion of the discussions presented in this chapter is that fourth-order compact implicit schemes, especially the Ty approaches, have substantial promise for future FDTD modeling applications. The combination of Ty space-differencing with second-order leapfrog time stepping is especially attractive due to its simplicity, high accuracy, computational efficiency, and lack of dissipation. Further, this Ty algorithm has the significant advantage of direct-overlay compatibility with existing FDTD codes based on the classic Yee grid and leapfrog time stepping, as well as PML ABCs.

REFERENCES

[1] Gustafsson, B., H.-O. Kreiss and J. Oliger, *Time Dependent Problems and Difference Methods*, New York: Wiley-Interscience, 1995.

[2] Gottlieb, D., and E. Turkel, "Dissipative two-four methods for time dependent problems," *Mathematics of Computations*, Vol. 30, 1976, pp. 703–723.

[3] Gustafsson, B., and P. Olsson, "Fourth-order difference methods for hyperbolic initial boundary-value problems," *J. Computational Physics*, Vol. 117, 1995, pp. 300–317.

[4] Hixon, R., and E. Turkel, "High-accuracy compact McCormack-type schemes for computational acoustics," AIAA paper 98-0365, 1998.

[5] Tam, C. K. W., and J. C. Webb, "Dispersion-relation-preserving finite-difference schemes for computational acoustics," *J. Computational Physics*, Vol. 107, 1993, pp. 262–281.

[6] Zingg, D. W., "A review of high-order and optimized finite-difference methods for simulating linear wave phenomena," AIAA paper 97–2088, 1997.

[7] Taflove, A., *Computational Electrodynamics: The Finite-Difference Time-Domain Method*, Norwood, MA: Artech House, 1995.

[8] Gaitonde, D., J. S. Shang, and J. L. Young, "Practical aspects of high-order-accurate finite-volume schemes for electromagnetics," AIAA paper 97–0363, 1997.

[9] Madsen, N. K., and R. W. Ziolkowski, "A 3-D modified finite-volume technique for Maxwell's equations," *Electromagnetics*, Vol. 10, 1990, pp. 147–161.

[10] Yee, K. S., and J. S. Chen, "The finite-difference time-domain (FDTD) and the finite-volume time-domain (FVTD) methods in solving Maxwell's equations," *IEEE Trans. Antennas and Propagation*, Vol. 45, 1997, pp. 354–363.

[11] Colella, P., and P. R. Woodward, "The piecewise-parabolic method (PPM) for gas-dynamical simulations," *J. Computational Physics*, Vol. 54, 1984, pp. 174–184.

[12] Zalesak, S. T., "A physical interpretation of the Richtmyer two-step Lax-Wendroff scheme, and its generalization to higher spatial order," in *Advances in Computer Methods for Partial Differential Equations*, Vichnevetsky, V. R., and R. S. Stepleman, eds., IMACS, 1984.

[13] Gottlieb, D., E. Turkel, and S. Abarbanel, "Phase and amplitude errors for systems of equations," to appear.

[14] Turkel, E., and G. Zwas, "Explicit large time step schemes for the shallow-water equations," *Advances in Computer Methods for Partial Differential Equations III*, 1979, pp. 65–69.

[15] Alford, R. M., K. R. Kelly, and D. M. Boore, "Accuracy of finite-difference modeling of the acoustic wave equation," *Geophysics*, Vol. 39, 1974, pp. 834–842.

[16] Bayliss, A., K. E. Jordan, B. J. LeMesurier, and E. Turkel, "A fourth-order-accurate finite-difference scheme for the computation of elastic waves," *Bulletin Seismological Society*, Vol. 76, 1986, pp. 1115–1132.

[17] Cohen, G., and P. Joly, "Fourth-order schemes for the heterogeneous acoustics equation," *Computational Methods in Applied Mechanics and Engineering*, Vol. 80, 1990, pp. 397–407.

[18] Dablain, M. A., "The application of high-order differencing to the scalar wave equation," *Geophysics*, Vol. 51, 1986, pp. 54–66.

[19] Fung, K.-Y., R. S. Man, and S. Davis, "Implicit high-order compact algorithm for computational acoustics," *AIAA J.*, Vol. 34, 1996, pp. 2029–2038.

[20] Gaitonde. D., and J. S. Shang, "High-order finite-volume schemes in wave propagation phenomena," AIAA 96–2335, 1996.

[21] Goodrich, J. W., and J. Hardin, "Accurate finite-difference algorithms for computational aeroacoustics," *CFD Review*, 1997.

[22] Kim, J. W., and D. J. Lee, "Optimized compact implicit finite-difference schemes with maximum resolution," *AIAA J.*, Vol. 34, 1996, pp. 887–893.

[23] Marfurt, K. J., "Accuracy of finite-difference and finite-element modeling of the scalar and elastic wave equations," *Geophysics*, Vol. 49, 1984, pp. 533–549.

[24] Roberts, K. V., and N. O. Weiss, "Convective difference schemes," *Mathematics of Computations*, Vol. 22, 1966, pp. 272–299.

[25] Shubin, G. R., and J. B. Bell, "A modified equation approach to constructing fourth-order methods for acoustic wave propagation," *SIAM J. Scientific and Statistical Computing*, Vol. 8, 1987.

[26] Monk, P., and E. Süli, "Error estimates for Yee's method on nonuniform grids," *IEEE Trans. Magnetics*, 1994, pp. 3200–3203.

[27] Kreiss, H.-O., and J. Oliger, "Comparison of accurate methods for the integration of hyperbolic equations," *Tellus*, Vol. 24, 1972, pp. 199–215.

[28] Petropoulos, P. G., "Phase-error control for FDTD methods of second- and fourth-order accuracy," *IEEE Trans. Antennas and Propagation*, Vol. 42, 1994, pp. 859–864.

[29] Gottlieb, D., and B. Yang, "Comparisons of staggered and non-staggered schemes for Maxwell's equations," *12th Annual Review of Progress in Applied Computational Electromagnetics*, 1996, pp. 1122–1131

[30] Fang, J., *Time-Domain Finite-Difference Computations for Maxwell's Equations*, Ph.D. dissertation, EECS Dept., University of California, Berkeley, CA, 1989.

[31] Hu, F. Q., M. Y. Hussaini, and J. Manthey, "Low-dissipation and low-dispersion Runge-Kutta schemes for computational acoustics," *J. Computational Physics*, Vol. 124, 1996, pp. 177–191.

[32] Carpenter, M. H., and C. A. Kennedy, "Fourth-order 2N-storage Runge-Kutta schemes," *NASA TM-109112*, 1994.

[33] Fyfe, D. J., "Economical evaluation of Runge-Kutta formulae," *Mathematics of Computations*, Vol. 20, 1966, pp. 392-398.

[34] Williamson, J. H., "Low-storage Runge-Kutta schemes," *J. Computational Physics*, Vol. 35, 1980, pp. 44–56.

[35] Zingg, D. W., and T. T. Chisholm, "Runge-Kutta methods for linear problems," 12th AIAA CFD Conference, 1995, AIAA paper 95–1756–CP.

[36] Wesseling, P., *An Introduction to Multigrid Methods*, New York: Wiley, 1992.

[37] Berenger, J.-P., "A perfectly matched layer for the absorption of electromagnetic waves,'' *J. Computational Physics*, Vol. 114, 1994, pp. 185–200.

[38] Turkel, E., and A. Yefet, "Absorbing PML boundary layers for wave-like equations," to appear in *Applied Numerical Mathematics*.

[39] Holland, R., "The case against staircasing," *Proc. 6th Ann. Rev. Progress Applied Computational Electromagnetics*, 1990, pp. 89–95.

[40] Holland, R., V. P. Cable, and L.C. Wilson, "Finite-volume time-domain (FVTD) techniques for EM Scattering," *IEEE Trans. Electromagnetic Compatibility*, Vol. 33, 1991, pp. 281–294.

[41] Abarbanel, S., and A. Ditkowski, "Asymptotically stable fourth-order-accurate schemes for the diffusion equation on complex shapes," *J. Computational Physics*, Vol. 133, 1997.

[42] Abarbanel, S., and A. Ditkowski, "Multi-dimensional asymptotically stable schemes for advection-diffusion equations," to appear in *Computers and Fluids.*

[43] Abarbanel, S., A. Ditkowski, and A. Yefet, "Bounded-error finite-difference schemes for the wave equation on complex domains," in preparation.

[44] Harari, I., and E. Turkel, "Accurate finite-difference methods for time-harmonic wave propagation," *J. Computational Physics*, Vol. 119, 1995, pp. 252–270.

[45] Singer, I., and E. Turkel, "High-order finite-difference methods for the Helmholtz equation," to appear in *Computer Methods for Applied Mechanics & Engineering.*

[46] Carpenter, M. H., D. Gottlieb, and A. Abarbanel, "The stability of numerical boundary treatments for compact high-order finite-difference schemes," *J. Computational Physics*, Vol. 108, 1993, pp. 272–295.

[47] Harten, A., B. Engquist, S. Osher, and S. R. Chakravarthy, "Uniformly high-order-accurate essentially non-oscillatory schemes, III," *J. Computational Physics*, Vol. 71, 1987, pp. 231–303.

[48] Gottlieb, D., and S. Orszag, *Numerical Analysis of Spectral Methods: Theory and Applications*, Philadelphia, PA: SIAM-CBMS, 1977.

[49] Canuto, C., M. Y. Husseini, A. Quarteroni, and T. A. Zang, *Spectral Methods in Fluid Dynamics*, New York: Springer-Verlag, 1986.

[50] Fornberg, B., *A Practical Guide to Pseudospectral Methods,* Cambridge University Press, 1996.

[51] Funaro, D., *Spectral Elements for Transport-Dominated Equations*, Lecture Notes in Computational Science and Engineering, Vol. 1, Heidelberg, Germany: Springer-Verlag, 1997.

[52] Kopriva, D. A., "Multidomain spectral solutions of high-speed flows over blunt cones," *AIAA J.*, Vol. 31, 1993, pp. 2227–2231.

[53] Hesthaven, J. S., "A stable penalty method for the compressible Navier-Stokes equations III: multi-dimensional domain-decomposition schemes," to appear in *SIAM J. Scientific Computations.*

[54] Kabakian, A. V., "A spectral algorithm for electromagnetic wave scattering in the time domain: application to RCS computation," AIAA paper 96–2334, 1996.

[55] Yang, B., D. Gottlieb, and J. S. Hesthaven, "Spectral simulations of electromagnetic scattering," *J. Computational Physics*, Vol. 134, 1997, pp. 216–230.

[56] Yang, B., and J. S. Hesthaven, "A pseudospectral method for time-domain computation of electromagnetic scattering by bodies of revolution," *IEEE Trans. Antennas and Propagation*, submitted.

SELECTED BIBLIOGRAPHY

Bayliss, A., C .I. Goldstein, and E. Turkel, "On accuracy conditions for the numerical computation of waves," *J. Computational Physics*, Vol. 59, 1985, pp. 396–404.

Cangellaris, A. C., and R. Lee, "On the accuracy of numerical wave simulations using finite methods," *J. Electromagnetic Waves and Applications*, Vol. 6, 1992, pp. 1635–1653.

Carpenter, M. H., D. Gottlieb, and S. Abarbanel, "Time-stable boundary conditions for finite-difference schemes solving hyperbolic systems: methodology and application to high-order compact schemes," *J. Computational Physics*, Vol. 111, 1994, pp. 220–236.

Gottlieb, D., and E. Turkel, "Topics in spectral methods for time-dependent problems," in *Numerical Methods in Fluid Dynamics*, Brezzi, F. (ed.), Berlin, Germany: Springer-Verlag, 1985, pp. 115–155.

Gustafsson, B., H.-O. Kreiss, and A. Sundström, "Stability theory of difference approximations for mixed initial boundary value problems, II," *Mathematics of Computations*, Vol. 26, 1972, pp. 649–686.

Gustafsson, B., "The convergence rate for difference approximations to mixed initial boundary value problems," *Mathematics of Computations*, Vol. 29, 1975, pp. 396–406.

Hadi, M. F. and M. Piket-May, "A modified FDTD(2,4) scheme for modeling electrically large structures with high-phase accuracy," *IEEE Trans. Antennas and Propagation*, Vol. 45, 1997, pp. 254–264.

Lee, R., and A. C. Cangellaris, "A study of discretization error in finite-element approximation of wave solutions," *IEEE Trans. Antennas and Propagation*, Vol. 40, 1992, pp. 542–549.

Lele, S. K., "Compact finite-difference schemes with spectral-like resolution," *J. Computational Physics*, Vol. 103, 1992, pp. 16–42.

Shang, J., "A perspective of computational electromagnetics in the time domain," AIAA paper 97–2356, 1997.

Shlager, K. L., J. G. Maloney, S. L. Ray, and A. F. Peterson, "Relative accuracy of several finite-difference time-domain methods in two and three dimensions," *IEEE Trans. Antennas and Propagation*, Vol. 41, 1993, pp. 1732–1737.

Yang, B., D. Gottlieb, and J. S. Hesthaven, "On the use of PML ABCs in spectral time-domain simulations of electromagnetic scattering," *Proc. ACES 13th Annual Review Progress in Applied Computational Electromagnetics*, 1997, pp. 926–934.

Yang, B., and J. S. Hesthaven, "A pseudospectral method for time-domain computation of electromagnetic scattering by bodies of revolution," *IEEE Trans. Antennas and Propagation*, submitted.

Yee, K. S., "Numerical solution of initial boundary value problems involving Maxwell's equations in isotropic media," *IEEE Trans. Antennas and Propagation*, Vol. AP-14, 1966, pp. 302–307.

Young, J. L., D. Gaitonde, and J. S. Shang, "Toward the construction of a fourth-order difference scheme for transient EM wave simulation: staggered grid approach," *IEEE Trans. Antennas and Propagation*, Vol. 45, 1997, pp. 1573–1580.

Chapter 3

Time-Domain Analysis Using Multiresolution Expansions

Linda P. B. Katehi, James F. Harvey, and Emmanouil Tentzeris

3.1 INTRODUCTION

The FDTD technique has many advantages that makes it the method of choice for several classes of electromagnetic problems. It is relatively straightforward to implement; it easily accommodates complex geometrical features and nonuniform materials; it efficiently handles high-bandwidth waveforms; and it is easy to incorporate nonlinear device elements. Recent advances in ABCs have allowed a wide-dynamic-range truncation of FDTD space lattices very close to the structures being modeled, thereby enhancing the ability of FDTD to address complicated problems of practical significance.

Problems appropriate for FDTD modeling include, for example, the analysis of multiple nonplanar antennas on a helicopter or vehicular body that require accurate electromagnetic characterization of the feed structures and the complex body geometry consisting of metal and composite materials. Another example is the near-field-radar mine detection problem that requires the accurate representation of complicated-shaped targets of varying composition, highly nonuniformly shaped clutter of different materials, and a highly nonuniform air-ground interface and ground-material composition. This problem also involves pulsed or other complex waveforms. Yet a third example is the electromagnetic analysis of high-speed electronic circuits. This problem also requires the treatment of complex geometries of varied material composition, as well as the self-consistent modeling of the nonlinear circuit elements.

The main limitation to the FDTD method is its required dense volumetric discretization, which significantly limits the size of the problem that can be addressed. While evolving computer technology has enabled application of the original FDTD technique to ever larger problems, it remains a key goal to increase the computational efficiency of the method in a manner that retains its many desirable features.

This chapter demonstrates that multiresolution time-domain (MRTD) techniques involving the application of a wavelet basis to FDTD further increases the size and scope of the problems that can be addressed. MRTD techniques have two major potential impacts:

1. MRTD provides a unifying theoretical structure that relates various FDTD algorithms to each other and to other techniques of electromagnetic analysis.
2. MRTD can reduce computer running time and memory requirements by more than one order of magnitude for complex problems.

Wavelets provide a flexible basis set in which the differential equations governing an electromagnetic problem are discretized using the method of moments (MoM). This is essentially the same theoretical procedure used for decades to discretize electromagnetic integral equations. For example, discretization of the differential form of Maxwell's equations in the Haar basis using the Galerkin technique yields the classic Yee algorithm. Discretization in other bases produces higher-order FDTD algorithms having varying degrees of localization, depending on the nature of the basis functions. Better knowledge of the physical properties of the explicit basis functions can result in better understanding of the higher-order techniques and the ability to optimize them for particular classes of problems.

A wavelet basis for multiresolution analysis is generated from a scaling function basis by well-established mathematical operations. The use of wavelet functions in addition to the underlying scaling functions results in a very natural basis for describing different levels of resolution. For this reason, multiresolution analysis is used extensively in signal processing and image analysis. The same kind of advantages can be applied to electromagnetic boundary value problems. Namely, FDTD algorithms can be developed that provide a rigorous and natural approach to multilevel grids for describing different degrees of resolution at specified locations in the problem grid. Effectively, this permits a fine grid resolution to be applied where the electromagnetic field is changing rapidly, and a coarse grid resolution in other regions to conserve computational resources. In addition, the time dimension can be discretized with a wavelet basis, resulting in temporal multiresolution. Here, the resolution in time is correlated with the spatial resolution to preserve numerical stability. Finally, the gridding can be adaptive in both space and time, thereby allowing the tracking of a propagating pulse with a locally refined grid.

The savings in computer run time and memory requirements of MRTD compared with the traditional Yee FDTD algorithm [1] result because: (1) a coarser grid is used for the same level of accuracy; (2) the adaptive MRTD grid provides high resolution only where and when needed; and (3) basis functions with extended support produce higher-order FDTD algorithms even at the scaling-function level. For either the Yee FDTD or MRTD techniques, a grid size is chosen to adequately resolve the critical features in the volume of interest, generally 1/10 to 1/20 of a wavelength. If some limited region in the volume has finer features, the Yee FDTD technique must either use a finer mesh for the entire volume or attempt to further subdivide the grid in the limited region. The problem with using a finer grid in a limited region is that the Yee algorithm calculates the field value at a point based on the field values at the neighboring points. However, the points on the boundary of the finer-gridded region may not have calculated field values at all of the neighboring points. Therefore, some of these values are estimated or interpolated, introducing numerical error.

On the other hand, MRTD techniques do not require subdivision of the original grid to achieve higher resolution in a localized region. Instead, the field values between grid points are expanded in a wavelet basis and the coefficients of the expansion are calculated by testing functions at each grid point. Higher resolution is obtained *simply by including more basis functions in the expansion* (the scaling function plus higher resolution wavelet functions) at each point in the original grid. The expansion coefficients at any grid point certainly depend on the coefficients at the neighboring grid points. However, in a region where the higher-resolution wavelets are not required to describe the fields, their coefficients are small enough to be neglected and do not affect the calculation of the coefficients on the boundary of, or interior to, the high-resolution region. Thus, the MRTD technique provides a rigorous method for terminating a localized higher-resolution region. The wavelet coefficients themselves are used as test elements to determine when a higher resolution is needed. As they become larger, additional wavelet coefficients are calculated for a higher level of resolution. As they become smaller, they are neglected and their calculation discontinued. In this manner, the region of higher resolution adaptively changes with time during the calculation. The wavelet coefficients are regarded as providing information about the spatial or temporal rate of change of the fields at the next lowest level of resolution.

It should be emphasized that FDTD and MRTD achieve higher resolution in a different manner. For FDTD, the space-lattice resolution is determined by the finest structural feature to be resolved. In the MRTD technique, the scaling function is thought of as a low-pass filter, with the corresponding wavelet functions being bandpass filters at higher spatial frequencies. Here, the scaling function is optimally sized at a lowest order of resolution. Higher resolution is provided by the wavelet functions. Attempting to size the scaling functions based on the highest required resolution results in a nonoptimal calculation. Likewise, mixing different resolutions of scaling functions results in numerical problems. These issues are discussed in more detail below.

3.2 INTRODUCTION TO WAVELET EXPANSIONS

Wavelet theory is very close to the theory of orthogonal expansions wherein a function is represented by a sum of orthogonal functions [2]. Historically, orthogonal expansions have been closely associated with physical applications. The classic example is Fourier's invention of trigonometric series in 1822 to solve the partial differential equation associated with heat conduction. Other 19th-century orthogonal series involved polynomials in similar attempts to solve boundary-value problems characterized by partial differential equations.

The development of the Lebesque integral in the early 20th century allowed the pursuit of a general theory of orthogonal systems. While not specifically oriented to physical applications, the Lebesque integral allowed the subsequent introduction of new orthogonal systems, such as those of Haar and Shannon, which have proved very useful in the important area of signal processing.

While orthogonal sequences of wavelets are generalizations of the Haar and Shannon systems, it is recognized that the primary developments in the systematic theory of wavelets, representing a seminal advance in applied mathematics, occurred much later in the 20th century, in the late 1980s. Since then, wavelet theory has evolved very rapidly, and specific portions of the theory have reached a certain maturity. This development now provides novel analytical techniques to systematically formulate and accurately solve difficult problems, and is contributing significantly to the computational sciences in general.

Wavelets have a number of unique properties that make them useful in data compression, signal processing, and numerical analysis of vector and scalar fields. They are particularly useful in digitizing continuous functions because of their decomposition and reconstruction algorithms, and in solving boundary-value problems because their convergence and stability properties are superior to those of conventional orthogonal systems. There are noteworthy advantages in using wavelets instead of conventional orthogonal expansions:

1. Wavelet series converge uniformly and pointwise while Fourier series do not [3];
2. Wavelet series are more localized and pick up sharp discontinuities much better than Fourier series;
3. Wavelets use fewer coefficients to represent certain signals, functions or images, and provide much better stability in the numerical solution of boundary value problems than Fourier series.

To visualize the capability of wavelet expansions to provide accurate solutions while using many fewer coefficients than conventional expansions, we provide a simplified example of sampling a signal. Consider a signal/field in the form of a stepwise continuous function as shown in Fig. 3.1(a). We follow two different approaches for sampling this function. In the first, as shown in Fig. 3.1(b), we digitize the signal by taking the arithmetic average of the values of the signal at the points x_n and $x_n + \Delta x$: $\{ f_n = f(x_n) = [f(x_n + \Delta x) + f(x_n)]/2, \quad n = 0, 1, 2, 3, \dots \}$. For the reconstruction of this signal, all of the discretized values are very important, and the continuous function, $f(x)$, is recreated by the following formula:

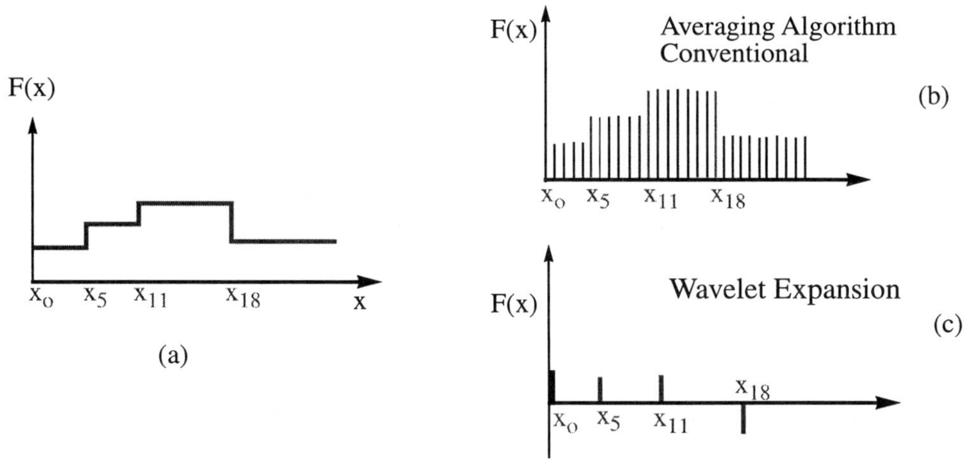

Fig. 3.1 Expansions of a stepwise continuous signal: (a) the original function; (b) a conventional expansion with pulse functions; and (c) wavelet expansions with the Haar system.

$$f(x) = \sum_{n=0}^{\infty} f_n \, p(x) \tag{3.1}$$

where $p(x)$ is a pulse function given by

$$p(x) = \begin{cases} 1 & x_n \le x \le x_{n+1} \\ 0 & \text{otherwise} \end{cases} \tag{3.2}$$

In the second approach, illustrated in Fig. 3.1(c), the function is discretized by taking the arithmetic differences between the values of the function at points x_n and x_{n+1}: $\{ w_n = f(x_{n+1}) - f(x_n), \quad n = 0, 1, 2, 3, \ldots \}$. For the reconstruction of this signal, only the value of the signal at x_0 and the nonzero differences, w_n, are needed, as shown below:

$$f(x) = f(x_0) + \sum_n w_n \, u(x - x_n) \tag{3.3}$$

where $u(x - x_n)$ is a step function given by

$$u(x - x_n) = \begin{cases} 1 & x \ge x_n \\ 0 & x < x_n \end{cases} \tag{3.4}$$

The major advantage of the second approach is the small number of sampling points needed compared with the first. It is the first approach that simulates a conventional expansion using orthogonal series, while the second approach simulates very closely a wavelet expansion. The second approach can be called a multiresolution expansion.

In this chapter, we first review the Haar and Shannon systems, introduce the concept of multiresolution analysis, discuss properties of scaling and wavelet functions, and summarize the popular Franklin and Battle-Lemarie wavelet systems. Then, we discuss the development and numerical properties of a two-dimensional MRTD analysis for electromagnetic waves based upon the Battle-Lemarie system. The chapter concludes with the application of the MRTD technique to a two-dimensional stripline problem.

3.3 THE HAAR SYSTEM—A PREDECESSOR TO MULTIRESOLUTION ANALYSIS

The Haar orthogonal system was developed in 1906 by A. Haar for the solution of eigenvalue problems [4]. The development of this system begins with $\phi(x)$, the characteristic function of the unit-interval pulse function:

$$\phi(t) = \chi_{[0,1)}(t) \tag{3.5}$$

and generates the orthogonal sequence $\{\phi_n(x)\}$ shown in Fig. 3.2:

$$\phi_n(x) = \phi(x-n) = \chi_{[n,n+1)}(x) \qquad (n \neq 0),\ (n \in Z) \tag{3.6}$$

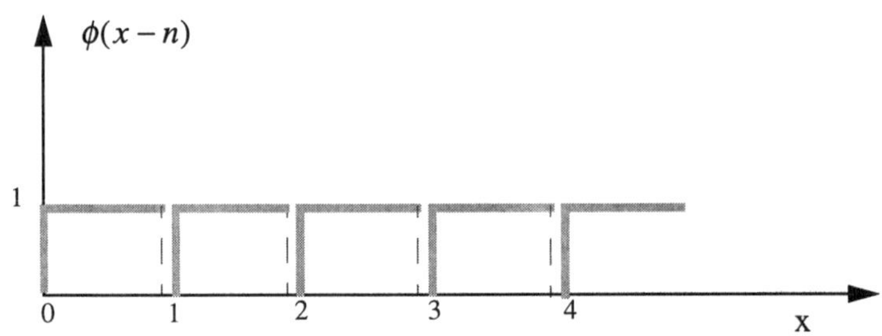

Fig. 3.2 The orthogonal sequence $\{f_n\}$.

However, while orthogonal, this sequence is not complete in $L^2(R)$, as explained previously. The same is true for all piecewise functions whose derivatives have jumps at discrete points on the real axis. In order to include more functions and make the system complete, we consider the dilated versions of $\phi(x)$, $\{\phi_m(x): \phi_m(x) = \phi(2^m x)\}$ shown in Fig. 3.3.

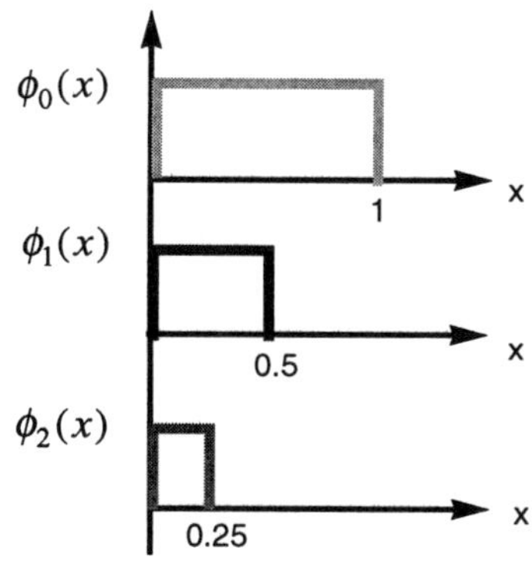

Fig. 3.3 Sequences of scaling functions.

Further inclusion of functions is achieved by use of *dilated* and *translated* versions of $\phi(x)$ given by the sequence $\{\phi_{nm}(x)\}$ (see Fig. 3.4):

$$\phi_{mn}(x) = 2^{m/2}\phi(2^m x - n) \tag{3.7}$$

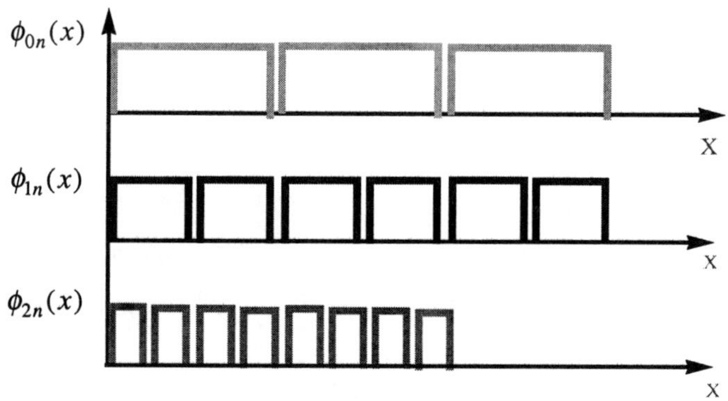

Fig. 3.4 Complete set of scaling functions.

The system defined by (3.7) is complete but not orthogonal in any sense. To keep the completeness and also provide some form of orthogonality, we introduce a new set of functions coming as dilations and translations of a "new" generating function $\psi(x)$:

$$\psi(x) = \phi(2x) - \phi(2x - 1) \tag{3.8}$$

This function is shown in Fig. 3.5 and leads to a group of new orthogonal sequences.

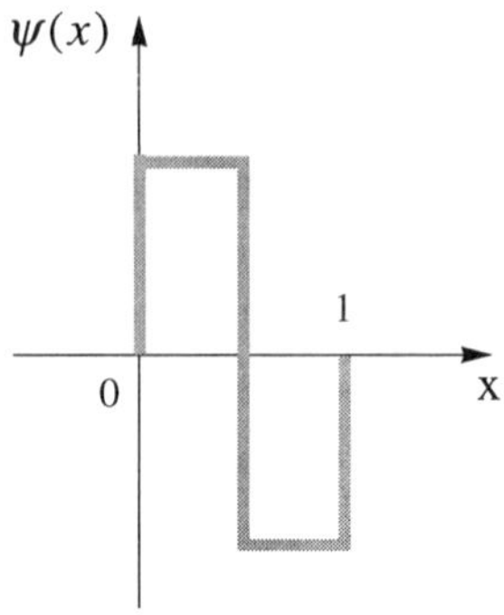

Fig. 3.5 Generating function $\Psi(x)$.

The new sequence of functions $\{\Psi_{mn} = \psi_{mn}(x), \quad m, n \in \aleph\}$ is given by

$$\psi_{mn}(x) \;=\; 2^{m/2}\psi(2^m x - n) \qquad m, n \in Z \tag{3.9}$$

and its members are plotted in Fig. 3.6.

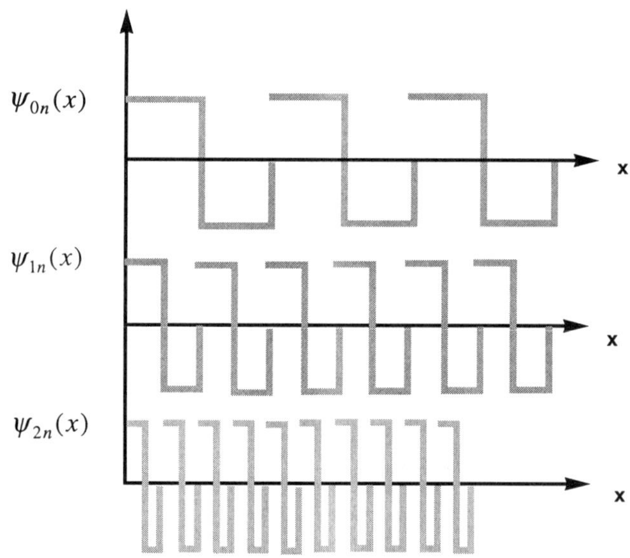

Fig. 3.6 Orthonormal sequence $\psi_{mn}(x)$.

This sequence constitutes a new orthonormal, complete system known as the *Haar system*. In this system, any doubly differentiable function can be expanded as the series:

$$f(x) = \sum_{m=-\infty}^{\infty} \sum_{n=-\infty}^{\infty} \langle f | \psi_{mn} \rangle \psi_{mn}(x) \tag{3.10}$$

If we truncate this series as shown here,

$$f_m(x) = \sum_{k=0}^{m} \sum_{n=-\infty}^{\infty} \langle f | \psi_{kn} \rangle \psi_{kn}(x) \tag{3.11}$$

then it can be proven that f_m converges to f uniformly. This is a very important property not shared by the trigonometric system [2, 5]. The lack of this uniform convergence capability in conventional expansions has been the source of many problems in computational electromagnetics. The two generating functions, $\phi(x)$ and $\psi(x)$, for this system are called, respectively, the *scaling function* and the *mother wavelet* for this system, and are shown in Fig. 3.7.

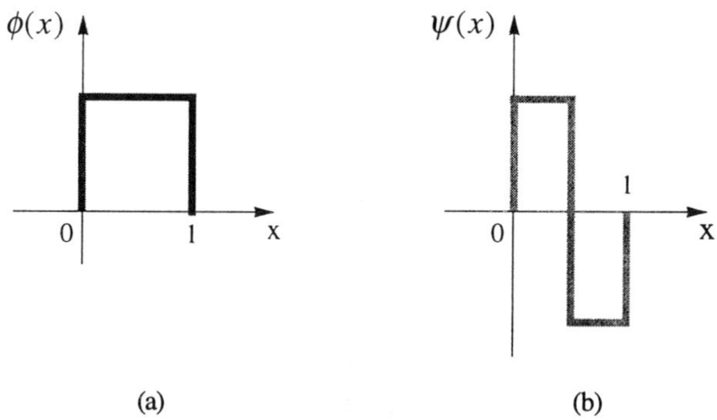

Fig. 3.7 Generating functions in the Haar system: (a) scaling function; (b) mother wavelet.

3.4 THE SHANNON SYSTEM — THE SECOND PROTOTYPE

The second prototype and predecessor of the wavelet systems is the *Shannon system*, developed in 1948 by Shannon for signal sampling and digitization. This prototype begins with the pulse function that is the characteristic function of an interval. Now, however, it is the Fourier transform of the scaling function that is taken to be equal to this pulse function:

$$\hat{\phi}(k_x) = \begin{cases} 1 & -\pi \le k_x \le \pi \\ 0 & \text{otherwise} \end{cases} \tag{3.12}$$

This is shown in Fig. 3.8.

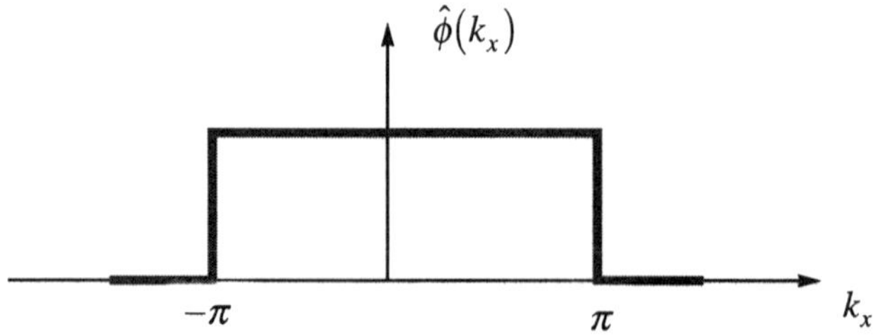

Fig. 3.8 The Fourier transform of the scaling function in the Shannon system.

As a result of the definition of (3.12), the scaling function for this system becomes the well-known sinc function:

$$\phi(x) = \frac{1}{2\pi}\int_{-\infty}^{\infty}\hat{\phi}(k_x)\, e^{jk_x x}\, dk_x = \frac{1}{2\pi}\int_{-\pi}^{\pi} e^{jk_x x}\, dk_x = \frac{\sin(\pi x)}{\pi x} \tag{3.13}$$

and is shown in Fig. 3.9.

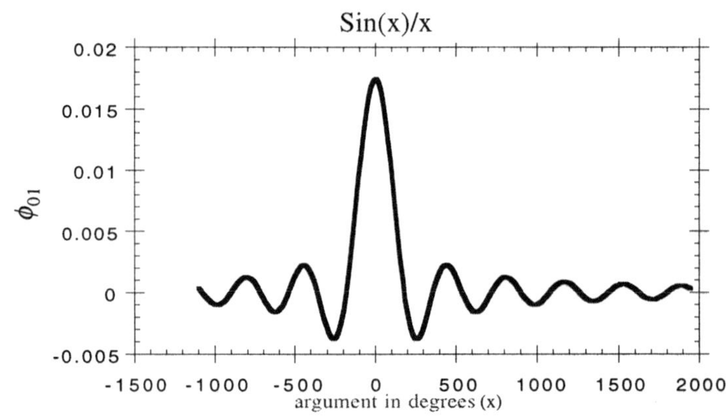

Fig. 3.9 The scaling function of the Shannon system.

Using the translations of the scaling function $\phi_n(x) = \sin[\pi(x-n)]/[\pi(x-n)]$ we can expand any function in L^2 in the following well-known series:

$$f(x) = \sum_n f(-n)\,\frac{\sin[\pi(x+n)]}{[\pi(x+n)]} \tag{3.14}$$

Equation (3.14) represents the *sampling theorem*, and enables recovery of a bandlimited function from its sampled values. As with the Haar system, we define a mother wavelet function

$$\psi(x) = 2\phi(2x) - \phi(x) \tag{3.15}$$

from its Fourier transform given by

$$\hat{\psi}(k_x) = \hat{\phi}\left(\frac{k_x}{2}\right) - \hat{\phi}(k_x) \tag{3.16}$$

This is shown in Fig. 3.10.

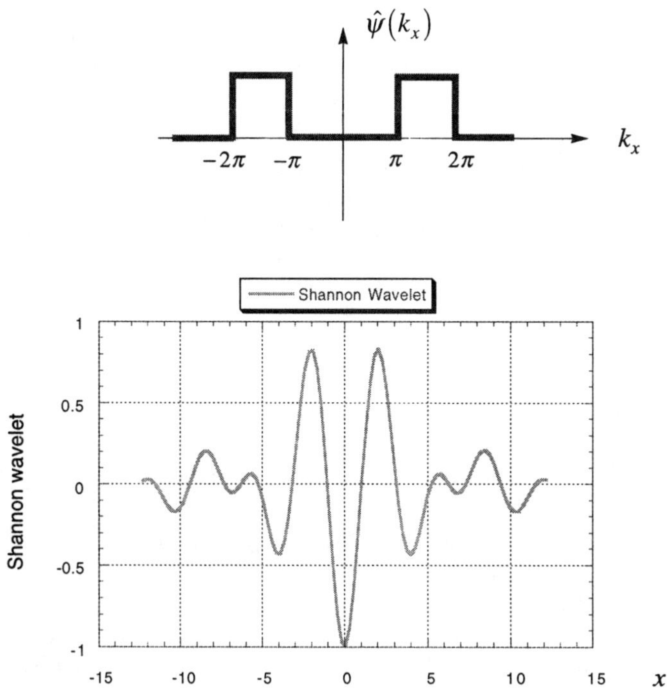

Fig. 3.10 The mother wavelet of the Shannon system.

It is very interesting to observe that the Shannon system includes functions with infinite support, while the Haar system includes functions with finite support. This property can impact the use of multiresolution analysis in various applications. In fact, this property divides wavelet systems into two main categories: (1) systems with infinite support, equivalent to entire-domain conventional expansions; and (2) systems with finite support (*intervalic wavelets*), equivalent to conventional subsectional basis expansion systems.

3.5 MULTIRESOLUTION ANALYSIS

The newly developed systems of orthogonal wavelets have a structure similar to that of the two prototypes, Haar and Shannon. However, all of wavelet systems and their Fourier transforms have a certain degree of smoothness that the prototypes do not have. It was the discovery of smooth orthogonal wavelets by Meyer and Daubechies and others that led to an explosion of activity on the subject. Wavelets began not as an orthogonal system but rather as an integral transform involving dilations and translations of a fixed function. This transform was found to give a better representation for certain seismic signals than Fourier methods [6]. The original theory was soon put into a rigorous framework by a number of mathematicians, and finally the orthogonal theory was discovered in the late 1980s by Battle [7], Daubechies [8], Coifman [9], Mallat [10], and others. The orthogonal wavelet expansion theory is based on the concept of multiresolution analysis.

In simple terms, multiresolution analysis is the generation of orthogonal sequences $\{\Phi_m = \phi_{mn}(x), \quad (n \in \aleph)\}$ and $\{\Psi_m = \psi_{mn}(x), \quad (n \in \aleph)\}$ where each corresponds to one resolution level and generates one resolution subspace, V_m, W_m, respectively:

$$\{\Phi_m\} \in V_m \tag{3.17a}$$

$$\{\Psi_m\} \in W_m \tag{3.17b}$$

From the previous definition of Φ_m and Ψ_m, we write the following:

$$\Phi_m \rightarrow \phi_{mn}(x) = 2^{m/2}\phi(2^m x - n) \qquad m, n \in Z \tag{3.18a}$$

$$\Psi_m \rightarrow \psi_{mn}(x) = 2^{m/2}\psi(2^m x - n) \qquad m, n \in Z \tag{3.18b}$$

From these equations we see that as m varies from negative to positive values, the resolution becomes finer and finer, thus leading into more detail. To have a multiresolution analysis, the closed subspaces, V_m, must satisfy the following properties:

$$\ldots \subseteq V_{-m} \subseteq \ldots \subseteq V_{-1} \subseteq V_0 \subseteq V_1 \subseteq \ldots \subseteq V_m \subseteq \ldots \tag{3.19a}$$

$$\bigcap_m V_m = \varnothing \tag{3.19b}$$

$$\bigcup_m V_m = L^2(\Re) \tag{3.19c}$$

$$V_{j+1} = V_j \oplus W_j \tag{3.19d}$$

$$W_j \cap W_k = \{\varnothing\} \tag{3.19e}$$

$$V_m = \ldots \oplus W_{m-3} \oplus W_{m-2} \oplus W_{m-1} \tag{3.19f}$$

$$f(x) \in V_m \Leftrightarrow f(2x) \in V_{m+1} \tag{3.19g}$$

To understand what amount of information is contained in each subspace, let us consider a function discretized by Haar scaling functions alone. The reconstruction of this function in different subspaces, shown in Fig. 3.11, illustrates the fact that as m increases, the corresponding subspace increases also and includes all the subspaces of smaller values of m.

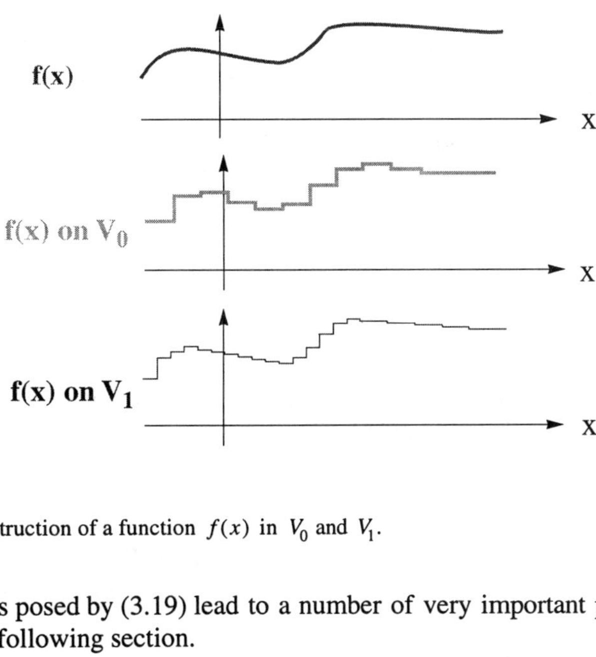

Fig. 3.11 Reconstruction of a function $f(x)$ in V_0 and V_1.

The requirements posed by (3.19) lead to a number of very important properties that are discussed in the following section.

3.6 PROPERTIES OF SCALING AND WAVELET FUNCTIONS

The most important properties of the scaling and wavelet functions defining a multiresolution analysis are now summarized.

1. All scaling and wavelet functions are localized both in the frequency and the space / time domains.

2. The Fourier transform of a scaling function is a low-pass function, while the Fourier transform of the corresponding mother wavelet is a bandpass function. This implies that, in a multiresolution expansion, scaling functions can be utilized to sample the low-frequency content of the field / signal, while wavelets sample the high-frequency content.
3. The scaling functions of one resolution level, m, are orthonormal to the scaling functions of the same resolution level, but not the scaling functions of other resolution levels:

$$\left\langle \phi_{mn}, \phi_{m\nu} \right\rangle = \delta_{n\nu} \tag{3.20}$$

This implies that, when using scaling functions in an expansion, we cannot mix different resolution levels. Expansions based only on scaling functions of the same resolution level are the basis of the conventional MoM technique in electromagnetics. Due to limiting the expansion to one resolution level, conventional orthogonal expansion techniques seek convergence by increasing the order of resolution, m, in addition to the number of translations, n, in the expansions. This requires a dense uniform discretization. This property is the source of all of the convergence problems associated with nonuniform discretization schemes observed in the integral-equation, mode-matching, finite-element, and TLM / FDTD techniques.
4. The scaling functions are orthogonal to all wavelets of any higher-order resolution level:

$$\left\langle \phi_{mn}, \psi_{\mu\nu} \right\rangle = 0 \qquad m \leq \mu \tag{3.21}$$

while the wavelet functions of any resolution level are orthonormal to all other wavelets at any resolution level:

$$\left\langle \psi_{mn}, \psi_{\mu\nu} \right\rangle = \delta_{m\mu}\delta_{n\nu} \tag{3.22}$$

The above two properties allow us to mix scaling functions of one resolution level with wavelets of one or more resolution levels of the same or higher order, m.
5. Any scaling function of m resolution can be expanded in the form of wavelets of all lower resolution levels:

$$\phi_{mn}(x) = \sum_{j=-\infty}^{m-1} \sum_{i=-\infty}^{\infty} d_{ji}^{mn} \psi_{ji}(x) \tag{3.23}$$

6. Any doubly differentiable function, $f(x)$, can be expressed as a uniformly converging series of the form

$$f(x) = \sum_{n=-\infty}^{\infty} c_n \phi_{m_0 n}(x) + \sum_{m=m_0+1}^{\infty} \sum_{n=-\infty}^{\infty} d_{mn} \psi_{m_0 n}(x) \tag{3.24}$$

The above property leads to a technique we call the multiresolution expansion, an approach that has the potential to be more general and more accurate than the conventional MoM. The use of multiresolution expansions may lead to the development of a variety of new frequency-domain and time-domain techniques having an inherent capability to provide adaptive discretizations in space and / or time.

7. If the mother wavelet function, $\psi(x)$, has r continuous derivatives, then it is orthogonal to all polynomials of degree $k \leq r$:

$$\int_{-\infty}^{\infty} x^k \psi_{mn}(x) dx = 0 \qquad k = 0, 1, 2, \ldots, r \tag{3.25}$$

This property is responsible for the capability of wavelet expansions to effectively sample abrupt variations in the electromagnetic fields and store the information in the wavelet coefficients. This observation implies that, in field expansions such as (3.24) where both scaling and wavelet functions are employed, the scaling-function coefficients are much stronger than the wavelet coefficients unless the function experiences a very fast variation. Therefore, the resulting matrices in MoM solutions have many very small terms that can be neglected without losing valuable information.

Applying thresholding techniques therefore permits generation of very sparse matrices without compromising accuracy. In other words, multiresolution expansions generate matrices having condition numbers orders of magnitude better than those achieved using the conventional MoM. As demonstrated in [11], a matrix sparsity above 90% is always observable with any of the most-known wavelet systems. Also, as shown by (3.25), the greater the number of zero moments, the greater the possible sparsity. Some of the wavelet systems can provide sparsities of the order of 99.9%.

3.7 POPULAR WAVELET SYSTEMS

In addition to the two prototypes, a variety of wavelet systems have been developed. Their properties as well as their application to electromagnetic field expansions are discussed in the following sections. Two of these are now summarized: the Franklin system and the Battle-Lemarie system.

3.7.1 Franklin Wavelet System

The Franklin wavelet system is an example of a system where the generation of the scaling function and the mother wavelet is performed in the frequency domain. The generating function for this system is triangular with a Fourier transform that varies as sinc^2. The scaling function is defined so that its Fourier transform has the form

$$\hat{\phi}(k_x) = \left[\frac{\sin\left(k_x/2\right)}{\left(k_x/2\right)}\right]^2 \left[1 - \frac{2}{3}\sin^2\left(\frac{k_x}{2}\right)\right]^{-1/2} \tag{3.26}$$

This equation gives the scaling function shown in Fig. 3.12. From this figure, we see that the Fourier transform of the scaling function has a low-pass characteristic with steep skirts and substantial rejection outside the passband.

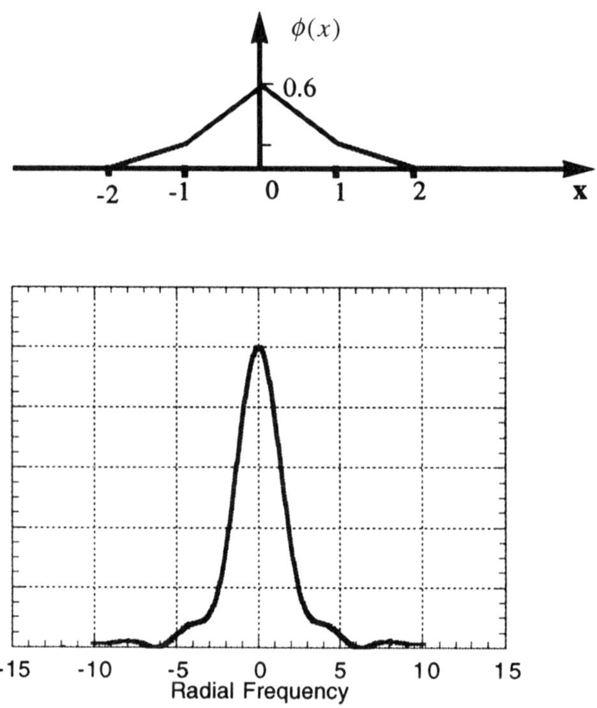

Fig. 3.12 Franklin scaling function (top) and its Fourier transform (bottom).

In a similar way, the mother wavelet is defined by its Fourier transform and takes the form shown in Fig. 3.13 [2]. From this figure, we see that the Fourier transform of the mother wavelet function has a bandpass characteristic.

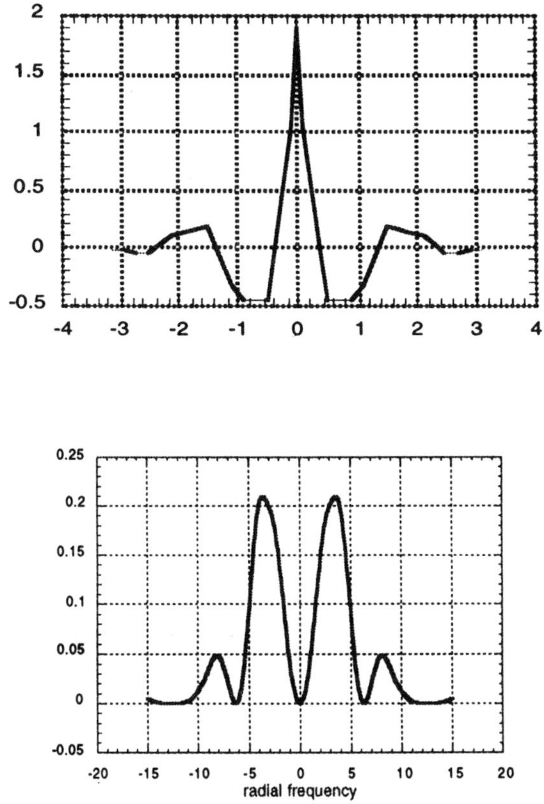

Fig. 3.13 Franklin mother wavelet (top) and its Fourier transform (bottom).

3.7.2 Battle-Lemarie System

The Battle-Lemarie system is generated from B-spline functions and provides scaling functions and wavelets of infinite support. Fig. 3.14 graphs the scaling function and its Fourier transform. We see that the scaling-function spectrum again has a low-pass characteristic with steep rejection outside the passband. Based on the reciprocal-spreading property of the Fourier transform, we conclude that scaling functions belonging to higher-resolution levels have broader passbands. By increasing the resolution level, we can sample more of the high-frequency content of the signal. This is why in the conventional MoM, dense discretization is necessary to ensure small numerical errors in the expansions.

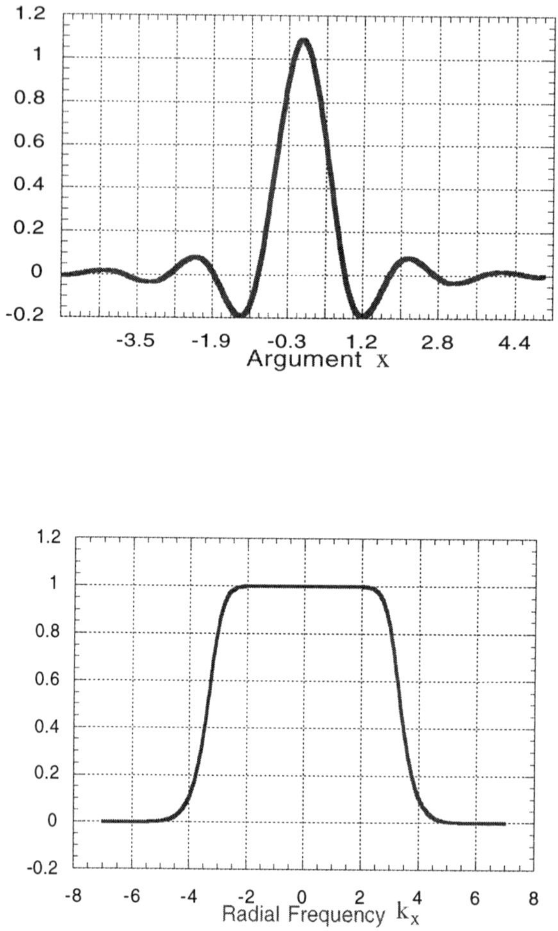

Fig. 3.14 Battle-Lemarie scaling function (top) and its Fourier transform (bottom).

Fig. 3.15 graphs the Battle-Lemarie mother wavelet function and its Fourier transform. We see that the spectrum of the wavelet function again has a bandpass characteristic. This allows accurate sampling of the high-frequency content of the signal without increasing the resolution level of the scaling functions. This property leads into very sparse matrices since the dense part of the matrix is due mostly to the scaling functions.

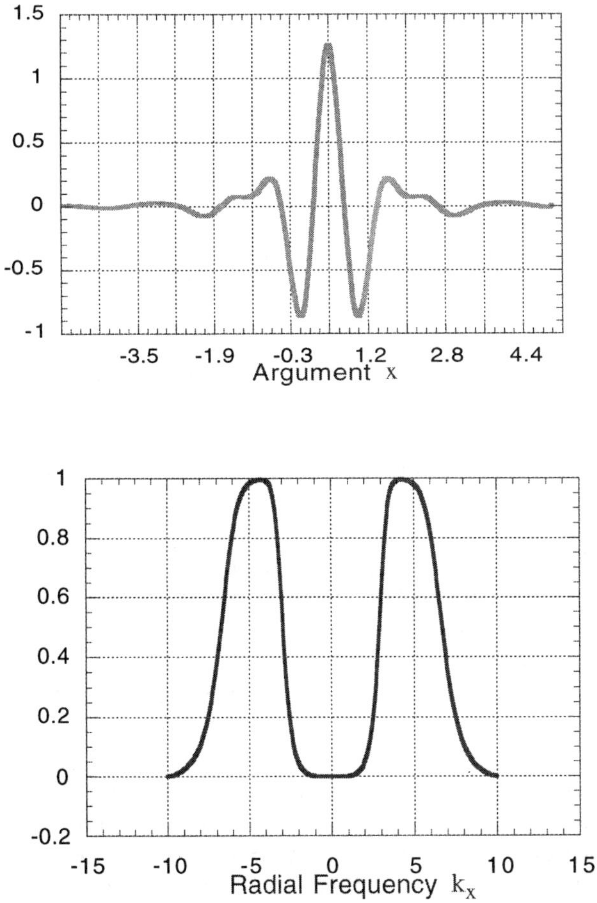

Fig. 3.15 Battle-Lemarie mother wavelet (top) and its Fourier transform (bottom).

Other wavelet systems in common use include the Daubechies system [8], the Coifman system [9], and intervalic wavelet systems [2].

3.8 TIME-DOMAIN SCHEMES BASED ON MULTIRESOLUTION ANALYSIS

Significant attention is currently being devoted to the design of high-speed circuits, antennas, and related equipment of importance in commercial and defense electronics. To understand high-frequency effects and incorporate them into the design process, there is a compelling need to implement full-wave electromagnetic models. There exists a variety of simulation techniques developed for this purpose, with many computer software packages available commercially. Despite this fact, many problems have been left untreated. This is due to the inability of existing techniques to calculate large-scale models with sufficiently fine resolution to pick up the electrically important, fine-grained geometrical details. The straightforward application of existing tools can require computer memory sizes and speeds that are beyond the capability of readily available machines.

Recently, a new technique was successfully applied [11–20] to a variety of microwave problems, and demonstrated novel and very promising properties. This technique, called the MRTD method, is derived by applying a multiresolution analysis based on Battle-Lemarie functions to the discretization of the time-domain Maxwell's equations. MRTD has been applied to linear as well as nonlinear propagation problems, and has demonstrated savings in computer time and memory by one and two orders of magnitude, respectively. In addition, this new technique provides space-adaptive and time-adaptive meshing without the problems encountered by the conventional FDTD method [17]. The capability to provide adaptive meshing is connected with the use of two separate sets of basis functions, the scaling and wavelet functions, and the capability to threshold the field coefficients due to the excellent conditioning of the formulated mathematical problem.

The use of multiresolution analysis in the time domain has shown that Yee's classic FDTD scheme [1] can be derived by applying the MoM for the discretization of Maxwell's equations using pulse basis functions for the expansion of the unknown fields. MRTD schemes are generalizations of Yee's FDTD, and can greatly extend its capabilities by improving the computational efficiency and substantially reducing the required computer resources.

For the derivation of the MRTD scheme, the electromagnetic fields are represented by an expansion in cubic spline Battle-Lemarie scaling functions [7, 8] with respect to space. For this type of scaling functions, the evaluation of the moment-method integrals is simplified due to the existence of closed-form expressions in the spectral domain and simple representations in terms of cubic spline functions in the space domain. The use of nonlocalized basis functions cannot accommodate localized boundary conditions. To overcome this difficulty, the image principle is used to model perfect electric and magnetic boundary conditions. Pulse functions are used as expansion and test functions in the time domain.

In this section, the Battle-Lemarie system is used to develop a two-dimensional time-domain electromagnetic scheme. The intent of this discussion is to familiarize the reader with the application of multiresolution analysis.

3.8.1 The Two-Dimensional MRTD Scheme

The choice of the scaling function for the development of this new time-domain scheme is the cubic cardinal spline ($m = 3$). After orthonormalization, the spectral expressions of the scaling and mother wavelet functions are:

$$\hat{\phi}(k_x) = \frac{1}{\sqrt{2\pi}}\left[\frac{\sin(k_x/2)}{(k_x/2)}\right]^4\left[1 - \frac{4}{3}\sin^2\left(\frac{k_x}{2}\right) + \frac{2}{5}\sin^4\left(\frac{k_x}{2}\right) - \frac{4}{315}\sin^6\left(\frac{k_x}{2}\right)\right]^{-1/2} \quad (3.27a)$$

$$\hat{\psi}(k_x) = e^{jk_x/2}\frac{\hat{\phi}(k_x + 2\pi)}{\hat{\phi}(k_x/2 + 2\pi)}\hat{\phi}(k_x/2) \quad (3.27b)$$

For simplicity, the two-dimensional MRTD scheme is analyzed for a homogeneous lossless medium with the permittivity, ε, and the permeability, μ. Maxwell's equations in two dimensions are written as

$$\frac{\partial H_x}{\partial t} = -\frac{1}{\mu}\frac{\partial E_z}{\partial y} \quad (3.28a)$$

$$\frac{\partial H_y}{\partial t} = \frac{1}{\mu}\frac{\partial E_z}{\partial x} \quad (3.28b)$$

$$\frac{\partial E_z}{\partial t} = \frac{1}{\varepsilon}\left(\frac{\partial H_y}{\partial x} - \frac{\partial H_x}{\partial y}\right) \quad (3.28c)$$

To derive the two-dimensional scheme, the E- and H-field components incorporated in these equations are expanded in a series of Battle-Lemarie scaling and wavelet functions in both the x and y directions and as pulse functions in time:

$$H_y(x, y, t) = \sum_{i,j,n}^{+\infty} \hat{H}_y\Big|_{i-1/2,j}^{n} h|^n(t) \quad (3.29a)$$

where

$$\hat{H}_y\Big|_{i-1/2,j}^{n} = \left[H_{y,\phi}\Big|_{i-1/2}^{n}\phi_{i-1/2}(x) + \sum_l H_{y,\psi_l}\Big|_{i-1/2}^{n}\psi_{l,i-1/2}(x)\right]\cdot \left[H_{y,\phi}\Big|_j^n \phi_j(y) + \sum_l H_{y,\psi_l}\Big|_j^n \psi_{l,j}(y)\right] \quad (3.29b)$$

Here, $\phi_i(x) = \phi(x/\Delta x - i)$ and $\psi_{l,i}(x) = \psi_l(x/\Delta x - i)$ represent the Battle-Lemarie scaling and lth-order-resolution wavelet functions, respectively, whose spectral transforms are given by (3.27a) and (3.27b). $H_{y,\phi}\big|_{i-1/2}^{n}$, $H_{y,\psi_l}\big|_{i-1/2}^{n}$, $H_{y,\phi}\big|_j^n$, and $H_{y,\psi_l}\big|_j^n$ are the coefficients for the field expansions in terms of scaling and wavelet functions. The indices i, j, and n are the discrete space (x and y) and time indices related to the space and time coordinates via $x = i\Delta x$, $y = j\Delta y$, and $t = n\Delta t$, where Δx and Δy are the space discretization intervals in the x and y directions and Δt is the time-discretization interval. For an accuracy of 0.1%, the above summations are truncated to a finite number of terms determined by the dispersion and stability requirements (typically between 22–26). The time-domain expansion function $h|^n(t)$ is defined as

$$h|^n(t) = h\left(\frac{t}{\Delta t} - n\right) \tag{3.30a}$$

with the rectangular pulse function

$$h(t) = \begin{cases} 1 & |t| \leq 0.5 \\ 0.5 & |t| = 0.5 \\ 0 & |t| > 0.5 \end{cases} \tag{3.30b}$$

In a similar way, we expand the other components of the E and H fields to give

$$H_x(x, y, t) = \sum_{i,j,n}^{+\infty} \hat{H}_x\Big|_{i,j-1/2}^n h|^n(t) \tag{3.31a}$$

where

$$\hat{H}_x\Big|_{i,j-1/2}^n = \left[H_{x,\phi}\Big|_i^n \phi_i(x) + \sum_l H_{x,\psi_l}\Big|_i^n \psi_{l,i}(x)\right] \cdot \left[H_{x,\phi}\Big|_{j-1/2}^n \phi_{j-1/2}(y) + \sum_l H_{x,\psi_l}\Big|_{j-1/2}^n \psi_{l,j-1/2}(y)\right] \tag{3.31b}$$

and

$$E_z(x, y, t) = \sum_{i,j,n}^{+\infty} \hat{E}_z\Big|_{i,j}^n h|^n(t) \tag{3.32a}$$

where

$$\hat{E}_z\Big|_{i,j}^n = \left[E_{z,\phi}\Big|_i^n \phi_i(x) + \sum_l E_{z,\psi_l}\Big|_i^n \psi_{l,i}(x)\right] \cdot \left[E_{z,\phi}\Big|_j^n \phi_j(y) + \sum_l E_{z,\psi_l}\Big|_j^n \psi_{l,j}(y)\right] \tag{3.32b}$$

The H-field components are shifted by half a discretization interval in space and time with respect to the E-field components (leapfrog). Upon inserting the field expansions, Maxwell's equations are sampled using pulse functions as time-domain test functions and scaling / wavelet functions as space-domain test functions.

For the sampling in time domain, the following integrals are utilized:

$$\int_{-\infty}^{\infty} h|^{n}(t)\; h|^{n'}(t)\, dt \;=\; \delta_{n,n'}\,\Delta t \tag{3.33a}$$

$$\int_{-\infty}^{\infty} h|^{n}(t)\; \frac{\partial h|^{n'+1/2}(t)}{\partial t}\, dt \;=\; \delta_{n,n'} - \delta_{n,n'+1} \tag{3.33b}$$

where $\delta_{n,n'}$ is the Kroenecker delta:

$$\delta_{n,n'} \;=\; \begin{cases} 1 & n = n' \\ 0 & n \neq n' \end{cases} \tag{3.34}$$

Sampling in the space domain is obtained by use of the orthogonality relationships for the scaling and wavelet functions [17]:

$$\int_{-\infty}^{\infty} \phi_i(x)\, \phi_{i'}(x)\, dx \;=\; \delta_{i,i'}\,\Delta x \tag{3.35a}$$

$$\int_{-\infty}^{\infty} \psi_{l,i}(x)\, \psi_{l',i'}(x)\, dx \;=\; \delta_{l,l'}\,\delta_{i,i'}\,\Delta x \tag{3.35b}$$

For the rest of the description of the MRTD technique, an expansion only in a series of scaling and 0-resolution wavelet functions is considered. Hints for the enhancement of additional wavelet resolutions are presented where needed. The integrals containing derivatives are approximated by the following expressions:

$$\int_{-\infty}^{\infty} \phi_i(x)\, \frac{\partial \phi_{i'+1/2}(x)}{\partial x}\, dx \;\cong\; \sum_{\nu=n_a}^{n_a-1} \alpha(\nu)\, \delta_{i+\nu,i'} \tag{3.36a}$$

where

$$\alpha(\nu) \;=\; \int_{0}^{\infty} \left|\hat{\phi}(k_x)\right|^2 k_x \,\sin\!\left[k_x(\nu + 0.5)\right] dk_x \tag{3.36b}$$

and

$$\int_{-\infty}^{\infty} \phi_i(x)\, \frac{\partial \psi_{l,i'+1/2}(x)}{\partial x}\, dx \;\cong\; \sum_{\nu=-n_c}^{n_c} c_l(\nu)\, \delta_{i+\nu,i'} \tag{3.37a}$$

where

$$c_l(\nu) = \int_0^\infty \hat{\phi}_i(k_x) \left| \hat{\psi}_l(k_x) \right| k_x \sin(k_x \nu)\, dk_x \tag{3.37b}$$

and

$$\int_{-\infty}^{\infty} \psi_i(x) \frac{\partial \psi_{l,i'+1/2}(x)}{\partial x}\, dx \cong \sum_{\nu=-n_b}^{n_b-1} b_l(\nu)\, \delta_{i+\nu,i'} \tag{3.38a}$$

where

$$b_l(\nu) = \int_0^\infty \left| \hat{\psi}_l(k_x) \right|^2 k_x \sin[k_x(\nu + 0.5)]\, dk_x \tag{3.38b}$$

The coefficients $\alpha(\nu)$, $b_0(\nu)$, and $c_0(\nu)$ are given in Table 3.1 [14]. The Battle-Lemarie scaling function has exponential decay; thus, the coefficients $\alpha(\nu)$ for $\nu \geq 12$ are not zero, but their value is negligible ($\leq 10^{-4}$).

TABLE 3.1

MRTD Coefficients

ν	$\alpha(\nu)$	$b_0(\nu)$	$c_0(\nu)$
0	1.29161	2.47254	0.
1	-0.15598	0.95623	-0.04660
2	0.05961	0.16606	0.05454
3	-0.02929	0.09392	-0.03700
4	0.01536	0.00314	0.02057
5	-0.00818	0.01349	-0.01115
6	0.00438	-0.00286	0.00598
7	-0.00234	0.00278	-0.00320
8	0.00125	-0.00113	0.00171
9	-0.00067	0.00071	-0.00092
10	0.00036	-0.00034	0.00049
11	-0.00019	0.00020	-0.00026
12	0.00010	-0.00010	0.00014
13	-0.00006	0.00005	-0.00008
14	0.00003	0.00003	0.00004
15	-0.00002	0.00002	-0.00002

Source: Tentzeris et al., *Proc. IEEE Int. Symp. on Microwave Theory and Techniques*, 1996, pp. 573-576, © 1996 IEEE.

For sake of simplicity, let us demonstrate the sampling of Maxwell's equations by using only scaling functions as expansion and testing functions. By applying Galerkin's method, the system of (3.28) takes the form:

$$\frac{1}{\Delta t}\left(H_{x,\phi\phi}\Big|_{i,j-1/2}^{n+1/2} - H_{x,\phi\phi}\Big|_{i,j-1/2}^{n-1/2} \right) = -\frac{1}{\mu\,\Delta y}\sum_{j'=j-n_a}^{j+n_a-1} \alpha(j')\, E_{z,\phi\phi}\Big|_{i,j'}^{n} \tag{3.39a}$$

$$\frac{1}{\Delta t}\left(H_{y,\phi\phi}\Big|_{i-1/2,j}^{n+1/2} - H_{y,\phi\phi}\Big|_{i-1/2,j}^{n-1/2} \right) = \frac{1}{\mu\,\Delta x}\sum_{i'=i-n_a}^{i+n_a-1} \alpha(i')\, E_{z,\phi\phi}\Big|_{i',j}^{n} \tag{3.39b}$$

$$\begin{aligned}\frac{1}{\Delta t}\left(E_{z,\phi\phi}\Big|_{i,j}^{n+1} - E_{z,\phi\phi}\Big|_{i,j}^{n} \right) &= \frac{1}{\varepsilon\,\Delta x}\sum_{i'=i-n_a}^{i+n_a-1} \alpha(i')\, H_{y,\phi\phi}\Big|_{i'-1/2,j}^{n+1/2} \\ &\quad - \frac{1}{\varepsilon\,\Delta y}\sum_{j'=j-n_a}^{j+n_a-1} \alpha(j')\, H_{x,\phi\phi}\Big|_{i,j'-1/2}^{n+1/2}\end{aligned} \tag{3.39c}$$

where i, j, and n are the discrete space and time indices. The selection of the stencil lengths n_a, n_b, and n_c depends on the accuracy and dispersion requirements.

3.8.2 Numerical Stability

For the analysis of numerical stability, the system of MRTD equations of (3.39) is decomposed into separate time- and space-eigenvalue problems. Assuming an expansion only in scaling functions, the left-hand-side time-differentiation parts are written as an eigenvalue problem:

$$\frac{1}{\Delta t}\left(H_x\Big|_{i,j-1/2}^{n+1/2} - H_x\Big|_{i,j-1/2}^{n-1/2} \right) = \Lambda_n\, H_x\Big|_{i,j-1/2} \tag{3.40a}$$

$$\frac{1}{\Delta t}\left(H_y\Big|_{i-1/2,j}^{n+1/2} - H_y\Big|_{i-1/2,j}^{n-1/2} \right) = \Lambda_n\, H_y\Big|_{i-1/2,j} \tag{3.40b}$$

$$\frac{1}{\Delta t}\left(E_z\Big|_{i,j}^{n+1} - E_z\Big|_{i,j}^{n} \right) = \Lambda_{n+1/2}\, E_z\Big|_{i,j} \tag{3.40c}$$

where Λ_n is the time eigenvalue. In order to avoid having any spatial mode increase without limit during normal stepping, the imaginary part of Λ_n must satisfy the equation

$$-\frac{2}{\Delta t} \le \mathrm{Im}(\Lambda_n) \le \frac{2}{\Delta t} \tag{3.41}$$

For each time step, n, the instantaneous values of the E and H fields distributed in space across the grid are Fourier-transformed with respect to the i and j coordinates to provide a spectrum of sinusoidal modes (plane-wave eigenmodes of the grid). Assuming an eigenmode of the spatial-frequency domain with $\tilde{k}_x$ and $\tilde{k}_y$ being the x and y components of the numerical wavevector, the field components are written as

$$E_z\big|_{I,J} = E_{z_0} \exp\left[j\left(\tilde{k}_x I\Delta x + \tilde{k}_y J\Delta y\right)\right] \tag{3.42a}$$

$$H_x\big|_{I,J-1/2} = H_{x_0} \exp\left\{j\left[\tilde{k}_x I\Delta x + \tilde{k}_y (J-1/2)\Delta y\right]\right\} \tag{3.42b}$$

$$H_y\big|_{I-1/2,J} = H_{y_0} \exp\left\{j\left[\tilde{k}_x (I-1/2)\Delta x + \tilde{k}_y J\Delta y\right]\right\} \tag{3.42c}$$

Substituting these expressions into (3.39) and applying Euler's identity, we obtain an expression for Λ. When used in conjunction with (3.41), this results in the inequality

$$\Delta t \le \frac{1}{c\left(\sum_{i'=0}^{n_a-1} |\alpha(i')|\right)\sqrt{\frac{1}{(\Delta x)^2} + \frac{1}{(\Delta y)^2}}} \tag{3.43}$$

For $\Delta x = \Delta y = \Delta$, this gives the stability criterion

$$\Delta t_{\text{MRTD}} \le \frac{\Delta}{c\sqrt{2}\sum_{i'=0}^{n_a-1} |\alpha(i')|} = s\,\frac{\Delta}{c\sqrt{2}} \tag{3.44}$$

From (3.44), we observe that the upper bound of the time step depends on the stencil size n_a. This dependence is expressed through the coefficient s, which decreases as the stencil size increases. Fig. 3.16 shows that s converges to the value 0.6371 after $n_a \ge 15$. Since usually $n_a \ge 8$, the value $s = 0.6371$ is used throughout the MRTD simulations to ensure numerical stability.

Considering that the numerical stability criterion for FDTD is given by

$$\Delta t_{\text{FDTD}} \le \frac{1}{c\sqrt{\frac{1}{(\Delta x)^2} + \frac{1}{(\Delta y)^2}}} \tag{3.45}$$

or

$$\Delta t_{\text{FDTD}} \le \frac{\Delta}{c\sqrt{2}} \tag{3.46}$$

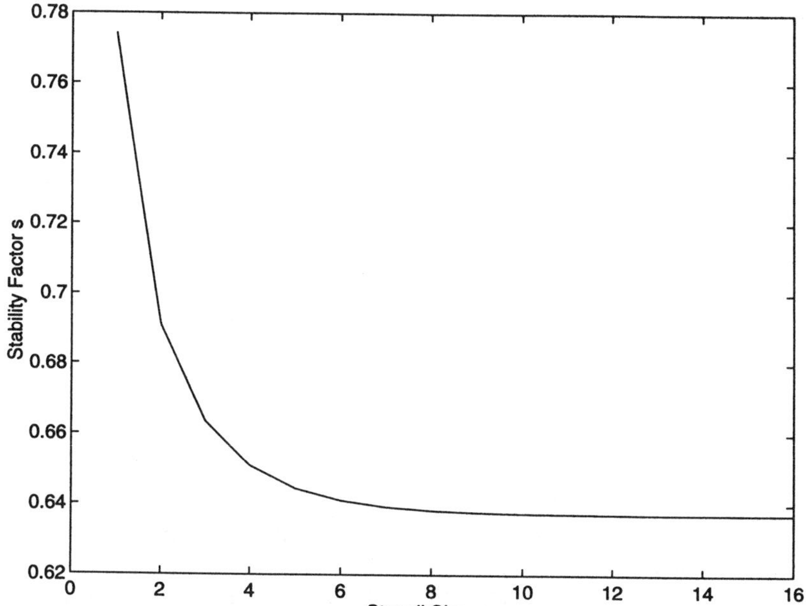

Fig. 3.16 Stability parameter s for MRTD employing scaling functions only.

we conclude that, for the same space discretization, the upper bounds of the time steps of FDTD and MRTD are comparable and related through the factor s. The numerical stability analysis is easily generalized to two- or three-dimensional MRTD schemes including wavelets [19].

3.8.3 Numerical Dispersion

To calculate the numerical dispersion of the two-dimensional MRTD scheme discussed above, plane monochromatic traveling-wave modes are substituted into the discretized Maxwell's equations. For example, the E_z component in (3.39c) has the form

$$E_z\big|_{I,J}^{n} = E_{z_0} \exp\left[j\left(\tilde{k}_x I\Delta x + \tilde{k}_y J\Delta y - \omega n\Delta t\right)\right] \tag{3.47}$$

where $\tilde{k}_x$ and $\tilde{k}_y$ are the x and y components of the numerical wavevector and ω is the wave angular frequency. Substituting the traveling-wave modes into (3.39), we obtain the following numerical dispersion relation for the two-dimensional MRTD scheme employing scaling functions only:

$$\left[\frac{1}{c\Delta t}\sin\left(\frac{\omega\Delta t}{2}\right)\right]^2 = \left\{\frac{1}{\Delta x}\sum_{i'=0}^{n_a-1}\alpha(i')\sin\left[\tilde{k}_x(i'+0.5)\Delta x\right]\right\}^2 + \left\{\frac{1}{\Delta y}\sum_{j'=0}^{n_a-1}\alpha(j')\sin\left[\tilde{k}_y(j'+0.5)\Delta y\right]\right\}^2 \tag{3.48}$$

For square grid cells ($\Delta x = \Delta y = \Delta$) and the wave propagating at angle θ with respect to the x axis ($\tilde{k}_x = \tilde{k}\cos\theta$ and $\tilde{k}_y = \tilde{k}\sin\theta$), (3.48) reduces to

$$\left[\frac{\Delta}{c\Delta t}\sin\left(\frac{\omega\Delta t}{2}\right)\right]^2 = \left\{\sum_{i'=0}^{n_a-1}\alpha(i')\sin\left[\tilde{k}(i'+0.5)\Delta\cos\theta\right]\right\}^2 + \left\{\sum_{j'=0}^{n_a-1}\alpha(j')\sin\left[\tilde{k}(j'+0.5)\Delta\sin\theta\right]\right\}^2 \tag{3.49}$$

This equation relates the numerical wavevector to the frequency, cell size, and time step, and is solved numerically for different propagation angles to quantify the numerical dispersion characteristics. Defining the Courant number, $C = (c\Delta t)/\Delta$, and the grid resolution, $R = \lambda_0/\Delta$ cells per wavelength, (3.49) is rewritten as a function of the ratio of the free-space wavelength λ_0 to the numerical wavelength, $\tilde{\lambda}$:

$$\left[\frac{1}{C}\sin\left(\frac{\pi C}{R}\right)\right]^2 = \left\{\sum_{i'=0}^{n_a-1}\alpha(i')\sin\left[\left(\frac{2\pi}{R}\right)\left(\frac{\lambda_0}{\tilde{\lambda}}\right)(i'+0.5)\cos\theta\right]\right\}^2 + \left\{\sum_{j'=0}^{n_a-1}\alpha(j')\sin\left[\left(\frac{2\pi}{R}\right)\left(\frac{\lambda_0}{\tilde{\lambda}}\right)(j'+0.5)\sin\theta\right]\right\}^2 \tag{3.50}$$

Upon numerical solution of (3.50) by applying the bisection-Newton-Raphson hybrid technique for different values of n_a, θ, and C, we obtain the phase error introduced by the MRTD algorithm. We note that $C < 0.45 = 0.6371/\sqrt{2}$ to satisfy the numerical stability requirements for the two-dimensional case.

Fig. 3.17(a) graphs the variation of the numerical phase error in degrees per wavelength as a function of the reciprocal of the stability parameter, s, for the case $n_a = 10$. Three grid resolutions are used: $R = 10$ (coarse), $R = 20$ (normal), and $R = 40$ (dense), and the results are compared with those of conventional FDTD. We see that the phase error of MRTD employing scaling functions only is significantly lower than that of FDTD for $R = 10$ and $R = 20$, with the best advantage for MRTD occurring at $R = 10$ for $1/s > 5$, equivalently $s < 0.2$. At $R = 40$, however, FDTD reclaims the advantage. Figs. 3.17(b) and 3.17(c) repeat the study of Fig. 3.17(a), but for the cases

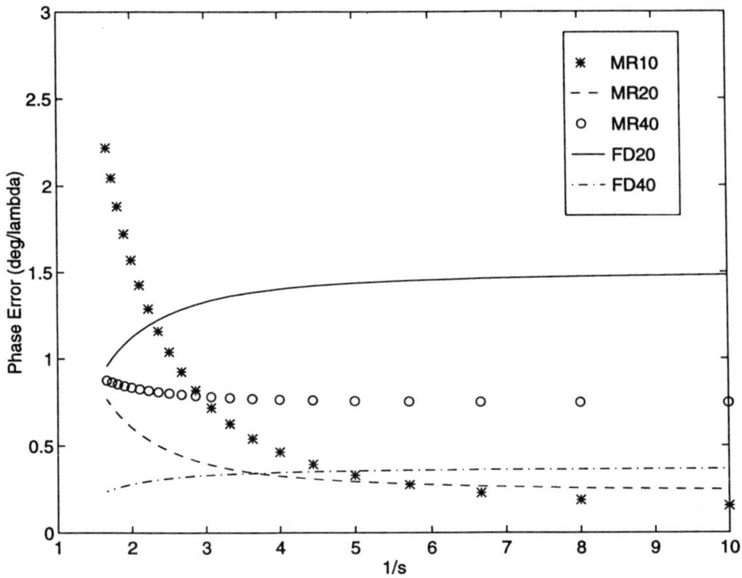

(a)

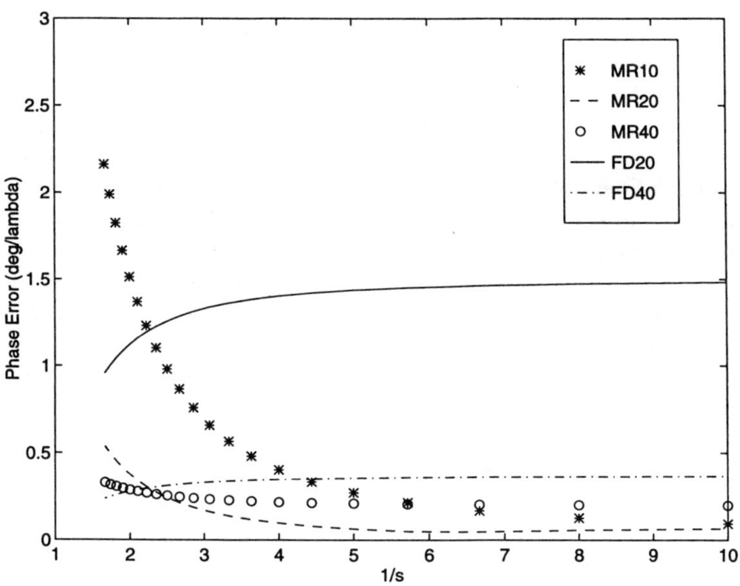

(b)

Fig. 3.17 Phase error of MRTD employing scaling functions only: (a) $n_a = 10$; (b) $n_a = 12$.

(c)

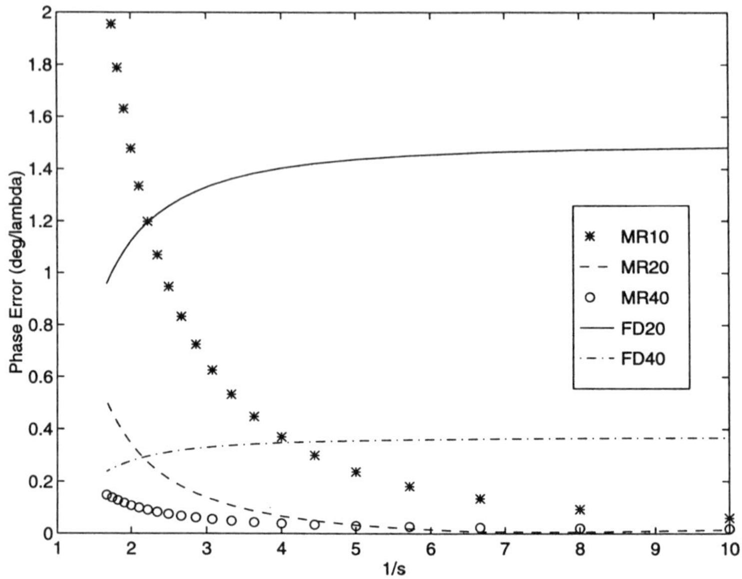

Fig. 3.17 *(cont.)* (c) $n_a = 14$.

$n_a = 12$ and $n_a = 14$, respectively. As R increases to 20 to 40 cells per wavelength, we see that the phase error of MRTD worsens unless the stencil increases. Increasing n_a improves the accuracy of the MRTD solution at all resolutions, with particular benefit at $R = 40$. For $n_a = 14$, there is no advantage for FDTD at any resolution for $1/s > 2$, equivalently $s < 0.5$. Note that the potential use of low resolutions for MRTD is very attractive because of the favorable implications of a coarse grid for reduced computer memory and running time, and increased electrical size of the modeled structure.

Consider the following argument for the favorable dispersion characteristics of MRTD versus conventional FDTD. FDTD is derived by expanding the fields in a pulse basis. As is well known, the Fourier transform of a pulse is a highly oscillating sinc function. On the other hand, as shown in Fig. 3.14, the Fourier transform of the Battle-Lemarie scaling function has a steeply skirted, monotonically decreasing, low-pass filter characteristic. This smooth spectral behavior offers a lower phase error even for coarse discretizations approaching $R = 3$.

It can be shown that the use of wavelets as well as scaling functions preserves the low-dispersion characteristic of MRTD for coarse discretizations close to $R = 2.5$. (The general relation for dispersion, including scaling and wavelet functions, is reported in [19].) Fig. 3.18 graphs the phase error over the grid-resolution range $2 \le R \le 10$ for conventional FDTD and for $n_a = 8$ MRTD with and without the use of wavelets. We see that MRTD offers low dispersion relative to FDTD even for sparse grids very close to the Nyquist limit.

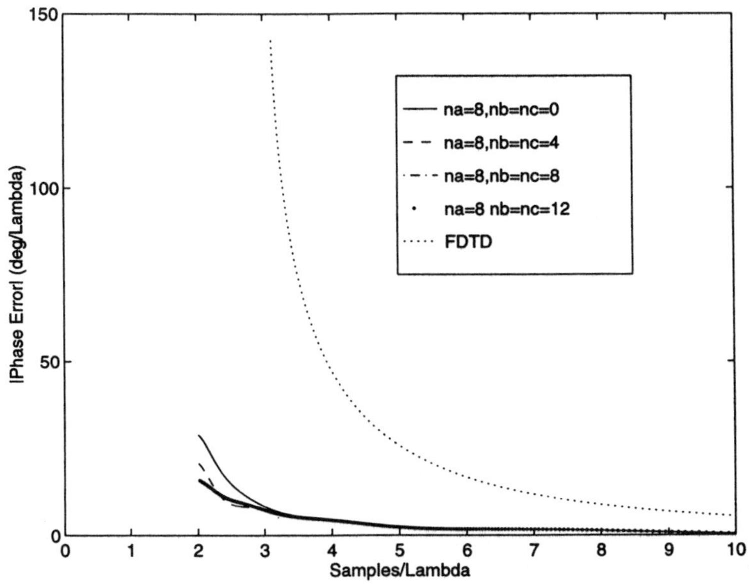

Fig. 3.18 Phase error of MRTD employing scaling functions and wavelets for $n_a = 8$. The coarse-resolution regime $2 \le R \le 10$ is considered.

3.9 TREATMENT OF BOUNDARY CONDITIONS IN MRTD

Following the development of a time-domain scheme, the next most challenging step is the numerical discretization of the boundary conditions in a self-consistent way. In this section, we discuss the treatment of the excitation, absorbing boundary conditions, hard boundaries such as perfect electric and perfect magnetic conductors, and dielectric interfaces.

3.9.1 Source Excitation

Let us consider a wave propagating along the positive x direction. In order to implement a single-pulse excitation $E_S(t)$ at $x = i\Delta x$, the space pulse is decomposed into scaling and wavelet functions:

$$E_{\text{exc}}\Big|_i^n \approx E_S(n\Delta t)\left[\sum_{l=-\zeta_\phi}^{\zeta_\phi} c_\phi(l)\phi_{i+l} + \sum_{r=0}^{r_{\max}}\sum_{\ell=-\zeta_\psi}^{\zeta_\psi} c_{\psi_{r,\ell}}(\ell)\psi_{r,i+\ell}\right] \tag{3.51}$$

where i indicates the position of the excitation along the x direction, l is the order of the scaling function on the 0-resolution level, r indicates the resolution level of the wavelets, and ℓ is the order of the wavelet on that resolution level. In this case, l and ℓ denote the translation of the scaling and wavelet functions around i. The total field in the MRTD scheme is then given by

$$E_{\phi,\text{total}}\Big|_{i+l}^{n} = E_{\phi}\Big|_{i+l}^{n} + E_{\text{exc}}\Big|_{i}^{n} \tag{3.52}$$

According to (3.51) and (3.52), it is possible to assume a simpler excitation that is represented by one single scaling function:

$$E_{\text{exc}}\Big|_{i}^{n} = E_{S}(n\Delta t)\,\phi_{i} \tag{3.53}$$

As a result of this assumption, the wavelet coefficients are excited by the coupling of the MRTD equations. This simplifying assumption is used extensively in the problems presented here due to its small computational overhead compared with other excitations.

3.9.2 Treatment of Hard Boundaries

Implicit in the application of boundary conditions in the FDTD method is its consistency with image theory. However, for MRTD schemes based on entire-domain functions, image theory must be applied *explicitly* at the locations of PECs or perfect magnetic conductors (PMCs). In MRTD, the total value of a field component at a specific cell is affected by a theoretically infinite — but practically finite — number of neighboring cells due to the fact that the basis functions extend from $-\infty$ to ∞. Some of these neighbors may be located on the other side of a conductor. This effect can be taken into consideration by applying image theory such that the physical boundary condition of zeroing-out the E field tangential to the PEC is automatically satisfied.

For example, even symmetry is applied for the E components that are normal to the PEC surface, and odd symmetry is applied for those E components parallel to the PEC. Image theory can be implemented automatically for an arbitrary number of hard boundaries. The use of wavelets in MRTD requires a special treatment of the wavelet components of the E field that are normal to the PEC, since the electrical domains on each side of the PEC surface are isolated from each other.

Consider a vertical PEC surface as shown in Fig. 3.19. At the surface, one wavelet component value of the normal (horizontal) E field would create a nonphysical electrical coupling. However, two wavelet components located infinitesimally close to the left and to the right of the PEC surface can be defined to satisfy the electrical isolation condition. The H field component parallel to the PEC has to be treated in a similar way. The rest of the E and H components have to be zeroed-out on both sides of the PEC, so one value is sufficient. In FDTD, the interleaved positioning of the field components in a Yee cell (which are the same as with the scaling-function components in the MRTD cell) requires that the normal E component is located one-half cell away from the conductor. In this way, the definition of only one field component per cell is sufficient.

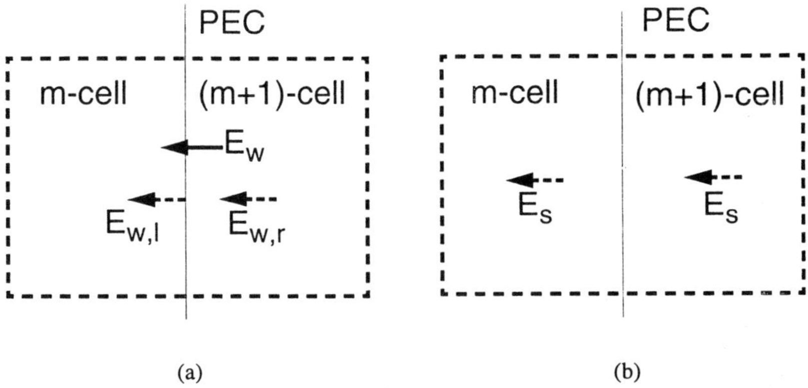

Fig. 3.19 Treatment of wavelet components of the normal-to-PEC electric field: (a) MRTD-wavelet; (b) MRTD-scaling or FDTD.

3.9.3 Application of the PML ABC

The PML ABC [21] is one of the most popular means to truncate FDTD lattices, routinely providing wave reflection coefficients below -90 dB. In this section, both split-field and nonsplit-field PML ABCs for the Battle-Lemarie MRTD are discussed, and their performance is validated [20].

For simplicity and without loss of generality, we consider a two-dimensional field mode having E_x, E_z, and H_y field components. (An analogous approach is utilized to provide the appropriate equations for the dual two-dimensional mode.) For this example, Maxwell's equations are given by

$$\varepsilon_0 \frac{\partial E_x}{\partial t} + \sigma_E E_x = -\frac{\partial H_y}{\partial z} \tag{3.54a}$$

$$\varepsilon_0 \frac{\partial E_z}{\partial t} + \sigma_E E_z = \frac{\partial H_y}{\partial x} \tag{3.54b}$$

$$\mu_0 \frac{\partial H_y}{\partial t} + \sigma_H H_y = \frac{\partial E_z}{\partial x} - \frac{\partial E_x}{\partial z} \tag{3.54c}$$

where we assume a free-space permittivity, ε_0, and permeability, μ_0, and the electric and magnetic conductivities (σ_E, σ_H).

Split-Field Formulation

Following the approach of [21], H_y is split into two subcomponents, H_{yx} and H_{yz}. Then, (3.54) is rewritten as a system of four equations:

$$\varepsilon_0 \frac{\partial E_x}{\partial t} + \sigma_E(z) E_x = -\frac{\partial}{\partial z}\left(H_{yx} + H_{yz}\right) \tag{3.55a}$$

$$\varepsilon_0 \frac{\partial E_z}{\partial t} = \frac{\partial}{\partial x}\left(H_{yx} + H_{yz}\right) \tag{3.55b}$$

$$\mu_0 \frac{\partial H_{yx}}{\partial t} = \frac{\partial E_z}{\partial x} \tag{3.55c}$$

$$\mu_0 \frac{\partial H_{yz}}{\partial t} + \sigma_H(z) H_{yz} = -\frac{\partial E_x}{\partial z} \tag{3.55d}$$

Here, without loss of generality, PML cells absorbing waves only in the z direction are considered. (The extension to the x and y directions is straightforward.) At each constant-z plane of the PML region, σ_E and σ_H are chosen to satisfy the relation

$$\frac{\sigma_E(z)}{\varepsilon_0} = \frac{\sigma_H(z)}{\mu_0} \tag{3.56}$$

for perfect absorption of the outgoing waves. Parabolic spatial distributions of σ_E and σ_H are used in the simulations:

$$\sigma_{E,H}(z) = \sigma_{E,H_{\max}}\left(1 - \frac{z}{\delta}\right)^p \tag{3.57}$$

where $p = 2$ for $0 \le z \le \delta$, although higher-order distributions such as $p = 3$ may be advantageously used in certain problems. The PML area is terminated with a PEC and usually has a thickness varying between 4 to 16 cells. The maximum value $\sigma_{E,\max}$ is determined by the designated reflection coefficient at normal incidence, given by

$$R(0) = \exp\left[-\frac{2}{\varepsilon_0 c}\int_0^\delta \sigma_E(z)\,dz\right] = \exp\left[-\frac{2\sigma_{E,\max}\delta}{\varepsilon_0 c(p+1)}\right] \tag{3.58}$$

For simplicity in this example, we expand E_x, E_z, H_{yx}, and H_{yz} in space in terms of scaling functions only, and use pulse functions in time domain. Upon applying Galerkin's technique, the following system is obtained:

$$\begin{aligned} E_{x,\phi\phi}\Big|_{i+1/2,k}^{n+1} &= e^{-\sigma_E(k)\Delta t/\varepsilon_0}\, E_{x,\phi\phi}\Big|_{i+1/2,k}^{n} \\ &+ \frac{e^{-\sigma_E(k)\Delta t/\varepsilon_0} - 1}{\sigma_E(k)\Delta z} \cdot \sum_{l=k-n_a}^{k+n_a-1} \alpha(l)\left(H_{yx,\phi\phi}\Big|_{i+1/2,l+1/2}^{n+1/2} + H_{yz,\phi\phi}\Big|_{i+1/2,l+1/2}^{n+1/2}\right) \end{aligned} \tag{3.59a}$$

$$E_{z,\phi\phi}\Big|_{i,k+1/2}^{n+1} = E_{z,\phi\phi}\Big|_{i,k+1/2}^{n} + \frac{\Delta t}{\varepsilon_0\,\Delta x}\sum_{l=i-n_a}^{i+n_a-1}\alpha(l)\left(H_{yx,\phi\phi}\Big|_{l+1/2,k+1/2}^{n+1/2} + H_{yz,\phi\phi}\Big|_{l+1/2,k+1/2}^{n+1/2}\right) \tag{3.59b}$$

$$H_{yx,\phi\phi}\Big|_{i+1/2,k+1/2}^{n+1/2} = H_{yx,\phi\phi}\Big|_{i+1/2,k+1/2}^{n-1/2} + \frac{\Delta t}{\mu_0\,\Delta x}\sum_{l=i-n_a}^{i+n_a-1}\alpha(l)\,E_{z,\phi\phi}\Big|_{l,k+1/2}^{n} \tag{3.59c}$$

$$H_{yz,\phi\phi}\Big|_{i+1/2,k+1/2}^{n+1/2} = e^{-\sigma_H(k+1/2)\Delta t/\mu_0}\,H_{yz,\phi\phi}\Big|_{i+1/2,k+1/2}^{n-1/2} + \frac{e^{-\sigma_H(k+1/2)\Delta t/\mu_0}-1}{\sigma_H(k+1/2)\,\Delta z}\cdot\sum_{l=k-n_a}^{k+n_a-1}\alpha(l)\,E_{x,\phi\phi}\Big|_{i+1/2,l}^{n} \tag{3.59d}$$

Exponential time-stepping is used in the system of (3.59) for the field components attenuated by the PML.

Due to the entire-domain nature of the Battle-Lemarie scaling functions, the PML conductivity must be sampled by these functions over at least 12 cells (6 cells per side). This implies that

$$\sigma_{E,H}(k) = \int_{k-6}^{k+6}\sigma_{E,H}(z)\,\phi_k(z)\,dz \tag{3.60}$$

Image theory is applied to extend the conductivity layer outside the terminating PECs.

Nonsplit-Field Formulation

An alternative to Berenger's original split-field concept [21], achieving an effective nonsplit-field implementation of the PML, is desirable since it would not require extra memory to store the two H_y subcomponents per cell. In line with this goal, consider making the following substitutions in the system of (3.54):

$$E_{x,z}(x, z, t) = \breve{E}_{x,z}(x, z, t)\exp\left[-\sigma_E(z)\,t/\varepsilon_0\right] \tag{3.61a}$$

$$H_y(x, z, t) = \breve{H}_y(x, z, t)\exp\left[-\sigma_H(z)\,t/\mu_0\right] \tag{3.61b}$$

This leads to the following system of equations:

$$\varepsilon_0 \frac{\partial \breve{E}_x}{\partial t} = -\frac{\partial \breve{H}_y}{\partial z} \tag{3.62a}$$

$$\varepsilon_0 \frac{\partial \breve{E}_z}{\partial t} = \frac{\partial \breve{H}_y}{\partial x} \tag{3.62b}$$

$$\mu_0 \frac{\partial \breve{H}_y}{\partial t} = \frac{\partial \breve{E}_z}{\partial x} - \frac{\partial \breve{E}_x}{\partial z} \tag{3.62c}$$

The system of (3.62) can be discretized in a manner similar to that of (3.55).

Validations

The two PML ABC formulations discussed above are now applied to terminate a simple, 48-mm-wide parallel-plate waveguide impulsively excited at one end with a Gabor time variation. A benchmark MRTD solution having zero reflections during the observation window is obtained by modeling a much longer waveguide of the same width. Quadratic variation of the PML conductivity is assumed for all cases, with a maximum theoretical reflection coefficient $R = 10^{-5}$ at normal incidence.

Figs. 3.20(a), 3.20(b), and 3.21(a) graph the MRTD-computed frequency variation of the nonsplit-field PML reflection coefficient, S_{11}, for the TEM mode over the range $0 \le f \le 0.9 f_c|_{\mathrm{TM}_1}$, where $f_c|_{\mathrm{TM}_1} = c/2d = 3.125$ GHz. The graphs are parameterized according to R. The cell size is equal to $\lambda_{\min}/10$, where $\lambda_{\min}$ is the wavelength at the highest frequency of excitation. For 4 PML cells, S_{11} is smaller than -40 dB; for 8 PML cells, S_{11} is smaller than -78 dB; and for 16 PML cells, S_{11} is smaller than -100 dB. Different values of R ranging from 10^{-5} to 10^{-8} do not significantly change the numerical performance of the absorber (causing a variation of only 4 to 5 dB). Fig. 3.21(b) shows that the performances of the split-field and nonsplit-field formulations are almost identical for the case of 8 PML cells with $R = 10^{-6}$.

3.9.4 Modeling of Dielectric Interfaces

Dielectric interfaces are modeled within MRTD by discretizing the constitutive relations

$$\vec{D} = [\varepsilon]\vec{E} \tag{3.63a}$$

$$\vec{B} = [\mu]\vec{H} \tag{3.63b}$$

where $\vec{D}$ and $\vec{B}$ are the electric and magnetic displacement vectors, $\vec{E}$ and $\vec{H}$ are the electric and magnetic field intensity vectors, and $[\varepsilon]$ and $[\mu]$ are the electric permittivity and magnetic permeability tensors. Equations (3.63a) and (3.63b) are discretized by expanding the fields into scaling and wavelet functions and applying Galerkin's method

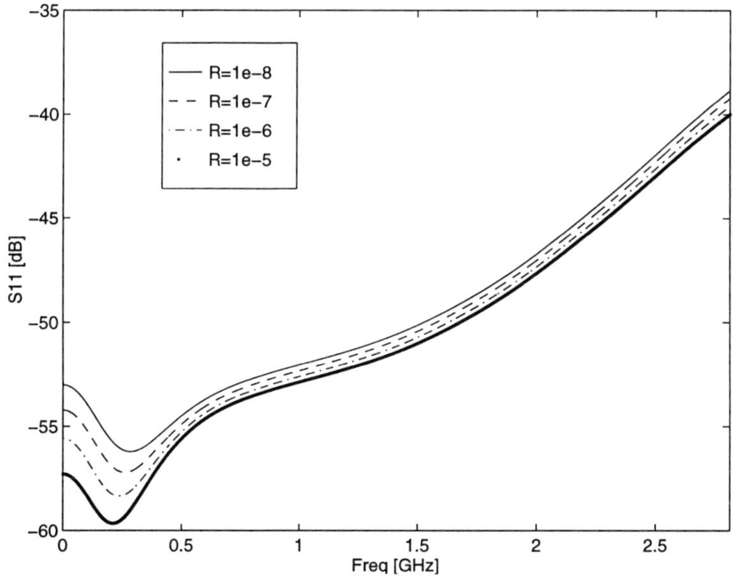

(a)

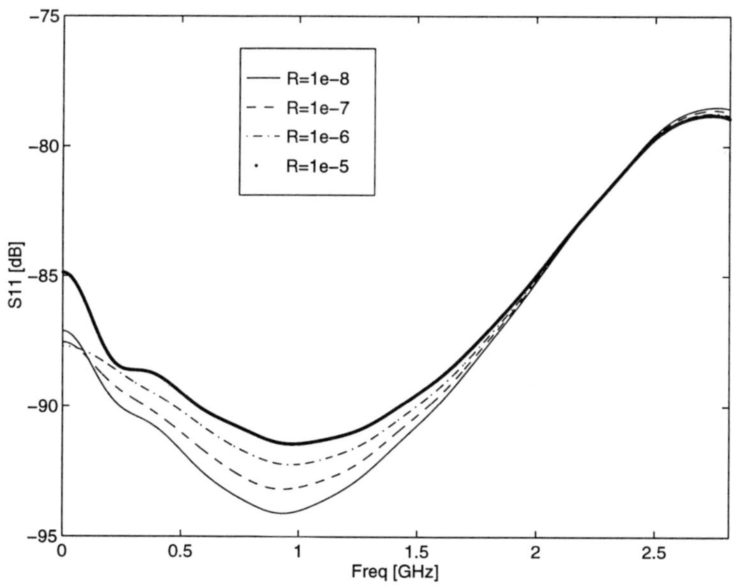

(b)

Fig. 3.20 MRTD-calculated nonsplit-field PML reflection coefficient S_{11} for a TEM mode propagating in a parallel-plate waveguide: (a) 4 PML cells; (b) 8 PML cells.

(a)

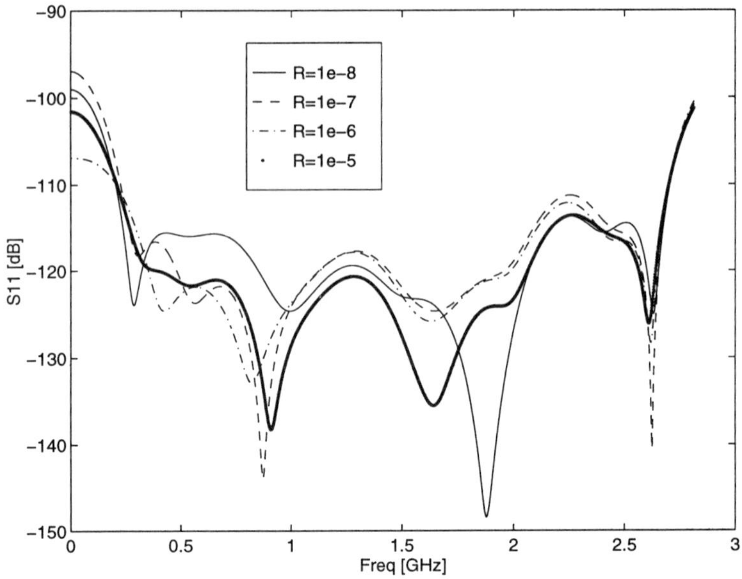

(b)

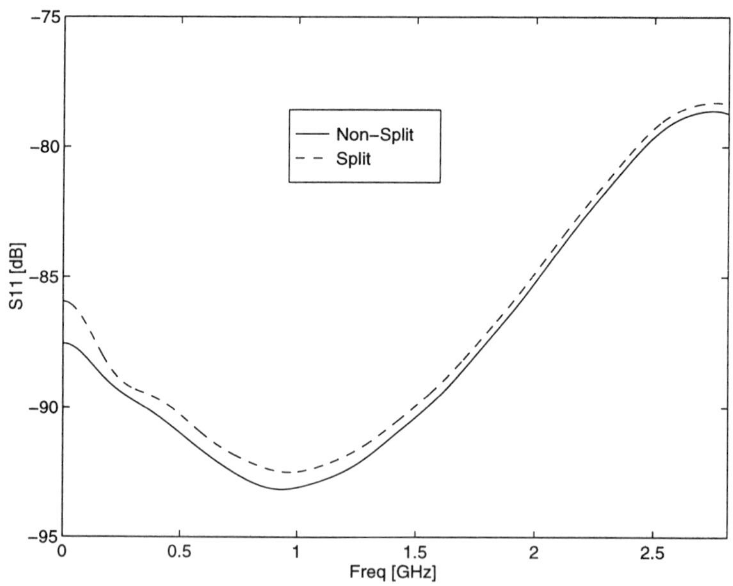

Fig. 3.21 MRTD-calculated PML reflection S_{11} for a TEM mode propagating in a parallel-plate waveguide: (a) 16 nonsplit PML cells; (b) comparison of 8 split and 8 nonsplit PML cells.

for testing. In the case of nonmagnetic materials, only (3.63a) is discretized. It reduces to the following two coupled linear equations:

$$D_\phi\Big|_{i,j}^{n} = \sum_{l=-l_\phi}^{l_\phi} c_{\phi,\phi}^{n}(l)\, E_\phi\Big|_{i,j+l}^{n} + \sum_{r=0}^{r_{\max}} \sum_{p=0}^{2^r-1} \sum_{l=-l_{\psi_{r,p}}}^{l_{\psi_{r,p}}} c_{\psi_{r,p},\phi}^{n}(l)\, E_{\psi_{r,p}}\Big|_{i,j+l}^{n} \tag{3.64a}$$

$$D_{\psi_{r',p'}}\Big|_{i,j}^{n} = \sum_{l=-\xi_\phi}^{\xi_\phi} c_{\phi,\psi_{r',p'}}^{n}(l)\, E_\phi\Big|_{i,j+l}^{n} + \sum_{r=0}^{r_{\max}} \sum_{p=0}^{2^r-1} \sum_{l=-\xi_{\psi_{r,p}}}^{\xi_{\psi_{r,p}}} c_{\psi_{r,p},\psi_{r',p'}}^{n}(l)\, E_{\psi_{r,p}}\Big|_{i,j+l}^{n} \tag{3.64b}$$

where

$$c_{\phi,\phi}^{n}(l) = \int_{-\infty}^{\infty} \varepsilon\, \phi_n(u)\, \phi_{n+l}(u)\, du \tag{3.65a}$$

$$c_{\phi,\psi_{r,p}}^{n}(l) = \int_{-\infty}^{\infty} \varepsilon\, \phi_n(u)\, \psi_{r,p}\Big|_{n+l}(u)\, du \tag{3.65b}$$

$$c_{\psi_{r,p},\phi}^{n}(l) = \int_{-\infty}^{\infty} \varepsilon\, \psi_{r,p}\Big|_{n}(u)\, \phi_{n+l}(u)\, du \tag{3.65c}$$

$$c_{\psi_{r,p},\psi_{r',p'}}^{n}(l) = \int_{-\infty}^{\infty} \varepsilon\, \psi_{r,p}\Big|_{n}(u)\, \psi_{r',p'}\Big|_{n+l}(u)\, du \tag{3.65d}$$

The system of (3.65) is the actual implementation of (3.63a). For geometries with dielectrics varying from air ($\varepsilon_r = 1$) to silicon ($\varepsilon_r = 11.5$), we observe that the above summations can be truncated to $l_\phi = l_{\psi_{r,p}} = \xi_\phi = \xi_{\psi_{r,p}} = 6$ when only one resolution of wavelets is used. Also, the integrals can be approximated by finite summations of 10 cells on each side of the central (k) cell. Due to the orthogonality relationship between the scaling and wavelet functions, for uniform dielectrics these integrals simplify to

$$c_{\phi,\phi}(l) = \varepsilon\, \delta_{l,0} \tag{3.66a}$$

$$c_{\phi,\psi_{r,p}}(l) = c_{\psi_{r,p},\phi}(l) = 0 \tag{3.66b}$$

$$c_{\psi_{r,p},\psi_{r',p'}}(l) = \varepsilon\, \delta_{l,0}\, \delta_{r,r'}\, \delta_{p,p'} \tag{3.66c}$$

and $[\varepsilon]$ becomes a diagonal matrix [13].

For structures containing dielectric discontinuities, none of these integrals has a zero value. In this case, the whole geometry is preprocessed before the initialization of the time loop. Coefficients $c_{\phi,\phi}(l)$, $c_{\phi,\psi_{r,p}}(l)$, $c_{\psi_{r,p},\phi}(l)$, and $c_{\psi_{r,p},\psi_{r',p'}}(l)$ are assigned to each cell (i, j) and included in the matrix. For each cell, the amplitude of these coefficients is compared with the amplitude of the self-term, $c_{\phi,\phi}(l)$. If all coefficients are below a threshold (usually $\leq 0.1\%$), they are set to zero and this cell is exempted and removed from the matrix. In this way, the initial matrix $[\varepsilon]$ is reduced to a submatrix of a smaller size (usually about 10% of the original size) and contains only cells close to the dielectric interface. The inverse of this submatrix is used for the calculation of the $\vec{E}$ components from the $\vec{D}$ components at each time step. Since the inversion takes place only once, it adds only negligible computational overhead to the algorithm.

3.10 SPACE- AND TIME-ADAPTIVE GRIDDING

Multiresolution analysis is based on the fact that the wavelets increase the local resolution of the expansion. Each added wavelet resolution is virtually equivalent to the use of a denser grid with half-cell size. In addition, wavelets have significant values close to discontinuities or near regions of fast field variation since they contain only high-spatial-frequency components.

There are a number of ways to exploit these wavelet characteristics to create a space- and time-adaptive gridding algorithm [11]. In digital signal processing, thresholding of the wavelet coefficients over a specific space and time window (5–10 points) contributes significant memory economy (a factor between 4 and 8 in comparison with scaling-only expansions), but increases the implementation complexity and the execution time. Sometimes the added computational overhead is greater than the previous execution time. The simplest way to create a dynamically changing grid is to threshold the wavelet components to a fraction (usually 0.1%) of the scaling function at the same cell (*space adaptiveness*) and/or to an absolute threshold (usually 0.0001 or a number smaller than 1/10,000 of the peak of the excitation time-domain function). This comparison is repeated for each time step (*time adaptiveness*). All components below the threshold are eliminated from the subsequent calculations. This is the simplest thresholding algorithm, one that has little impact on the execution time (usually $\leq 10\%$) and allows dynamic memory allocation in its software implementation by using the appropriate programming languages (e.g., C). However, it offers only a moderate savings in memory (a factor close to 2).

The principles of the space- and time-adaptive grid are illustrated in Fig. 3.22. In this figure, a pulse is propagating from left to right in a partially filled parallel-plate waveguide. At $t = 0$, the wavelets are localized at the excitation area. Subsequently, the wavelets follow the propagating pulse, thereby creating a moving dense subgrid. After the incident pulse is split into reflected and transmitted pulses, the wavelets increase the grid resolution only around these pulses. Elsewhere, the wavelet components have negligible values and are ignored.

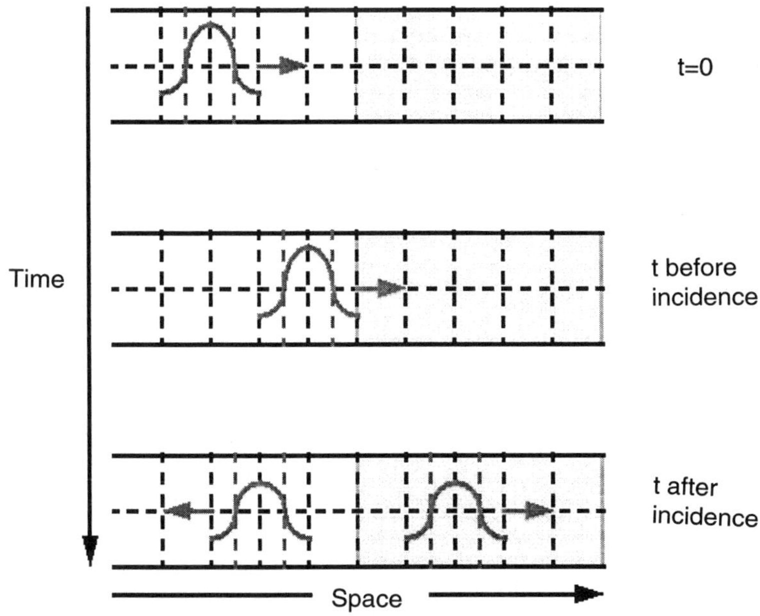

Fig. 3.22 Space- and time-adaptive grid.

3.11 APPLICATIONS OF MRTD TO MICROWAVE PROBLEMS

In this section, we describe a number of applications of MRTD to microwave problems and discuss the advantages of using this technique for these specific geometries.

3.11.1 Parallel-Plate Waveguide

The two-dimensional MRTD scheme is first applied to the analysis of a 2000-mm-long, air-filled parallel-plate waveguide having a plate spacing of 4.8 mm. The waveguide is terminated at each open end with a PML region of 22 grid cells and $\sigma_{E_{\max}} = 0.4$ S/m. Impulsive excitation is provided along a plane adjacent to one of the PML regions with a Gabor function of bandwidth 0 to 30 GHz.

For the analysis based on Yee's FDTD scheme, a uniform $4\text{-}\times 1120\text{-cell}$ mesh is used, whereas a $2\text{-}\times 160\text{-cell}$ mesh is chosen for MRTD. This size is based on the number of the scaling functions, since wavelets are used only when and where necessary. The time step is selected to be the same for both schemes and equal to one-fourth of the numerical stability limit for the MRTD algorithm. Three-thousand time steps are iterated for both cases. The longitudinal distance is chosen such that no reflections appear before the impulsive excitation function is complete. This permits the dispersion performance of the two schemes to be evaluated.

Fig. 3.23 graphs the time waveforms of the MRTD-computed E_x field at three grid locations along the waveguide: 70, 100, and 130 cells from the left grid boundary. Only minimal dispersion is observed.

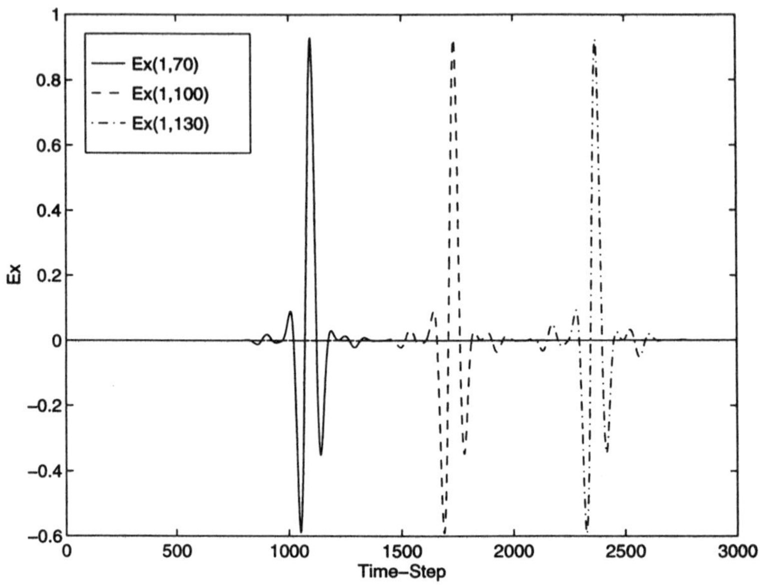

Fig. 3.23 Time waveforms of the MRTD-computed E_x field at three grid locations along the waveguide: 70, 100, and 130 cells from the left grid boundary.

Next, the capability of MRTD to provide space- and time-adaptive gridding is verified by thresholding the wavelet components at each grid cell at each time step at the larger of two possible values:

(1) 0.01% of the value of the scaling function at the same cell and time step;

(2) An absolute threshold of 10^{-5}.

The use of the absolute threshold enhances the efficiency of the algorithm for very small field values. We note that the accuracy obtained by using only a small number of wavelets is equal to that achieved if wavelets are used everywhere. Fig. 3.24 graphs the variation of the number of nonzero wavelets with time-step number, indicating a maximum value of 122 out of a total of 320 in the z direction. Relative to the Yee FDTD simulation, this results in a reduction in memory by a factor of 5.1 in the z direction and a reduction in program execution time by a factor of 3 to 4.

Fig. 3.25 illustrates the principle of the adaptive grid corresponding to the example of Figs. 3.23 and 3.24. In the top portion of this figure, the computed total E_x field at

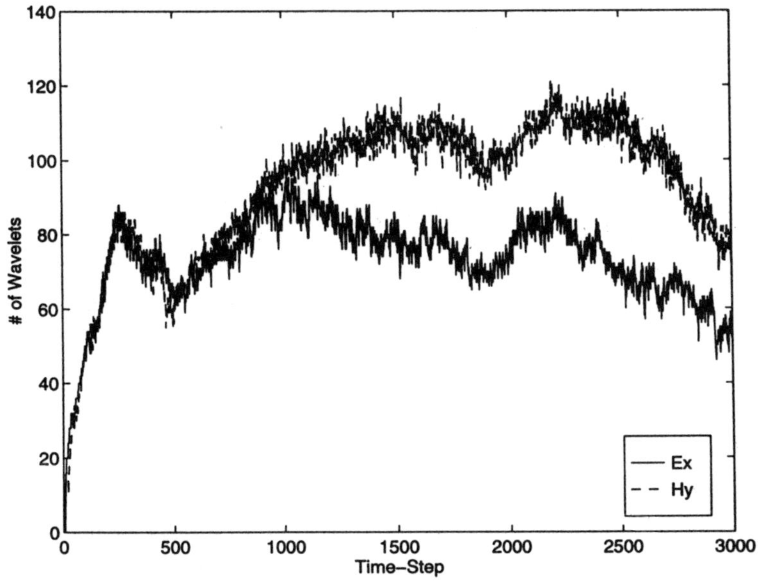

Fig. 3.24 Variation of the number of nonzero wavelets with time-step number in the model of Fig. 3.23, indicating a maximum value of 122 out of a total of 320 to the z direction.

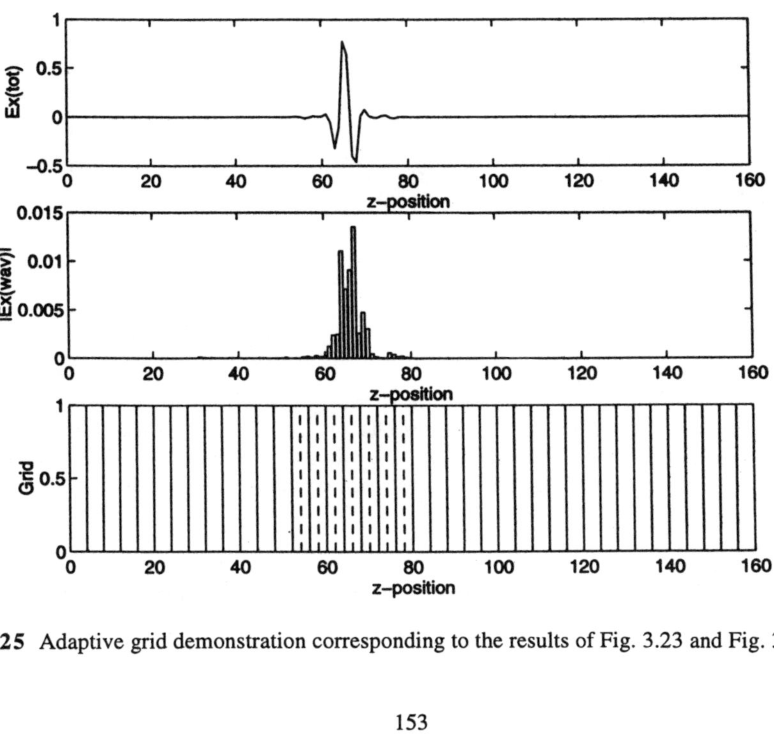

Fig. 3.25 Adaptive grid demonstration corresponding to the results of Fig. 3.23 and Fig. 3.24.

1000 time steps is graphed versus z position. The middle portion shows the contribution to E_x provided by the wavelets at this observation time. We see that the wavelets have a significant value only in the region of the propagating pulse, thus creating the locally dense grid shown in the bottom portion of the figure.

We now assume that the parallel-plate waveguide is filled with a dielectric of relative permittivity $\varepsilon_r = 2.56$ to the right of a planar interface. Sixteen-cell PML regions with $\sigma_{E_{\max}} = 0.8$ S/m terminate the grid, and wavelets of 0 resolution are used in the longitudinal z direction. The wideband impulsive excitation is the same as in the previous completely air-filled geometry, now originating in the air adjacent to the left PML region and impinging on the dielectric interface to the right. Fig. 3.26 compares the MRTD and FDTD calculations of the reflection coefficient, S_{11}, versus frequency, obtained by taking the ratio of the Fourier transforms of the incident and the reflected field pulses. We see that similar results are obtained for a $4\text{-}\times 640\text{-}$cell FDTD grid and a $2\text{-}\times 80\text{-}$cell MRTD grid having a relative threshold of 0.01% and an absolute threshold of 10^{-3}.

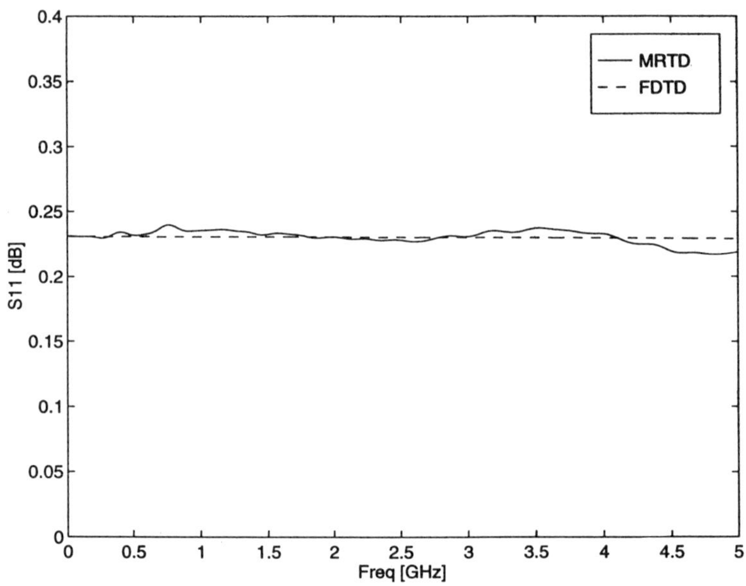

Fig. 3.26 Comparison of MRTD and FDTD results for S_{11} versus frequency of a parallel-plate waveguide filled with dielectric ($\varepsilon_r = 2.56$) to the right of an interface. Grids: FDTD, 4×640 cells; MRTD, 2×80 cells, with adaptive usage of wavelets.

Fig. 3.27 shows that the maximum number of wavelets used is 68, offering a $6.53:1$ reduction in memory in the z direction in comparison with the FDTD simulations. Fig. 3.28 illustrates the principle of the adaptive grid corresponding to this example.

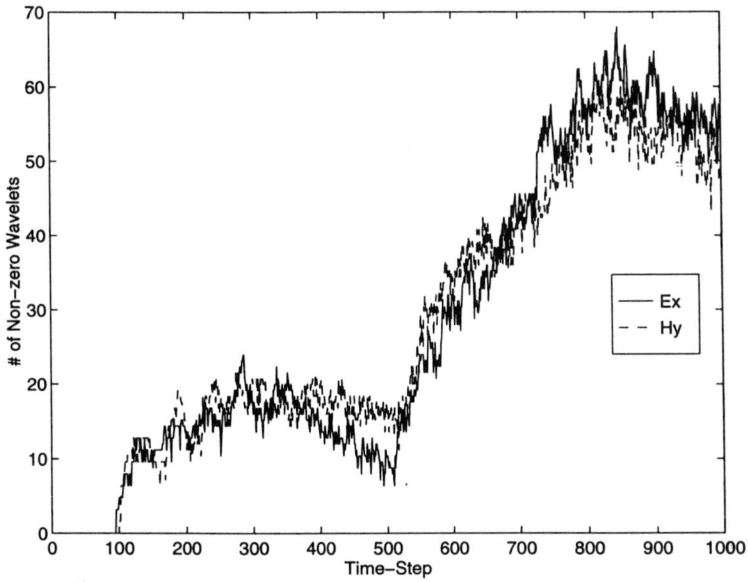

Fig. 3.27 Variation of the number of nonzero wavelets with time-step number in the model of Fig. 3.26, indicating a maximum of 68.

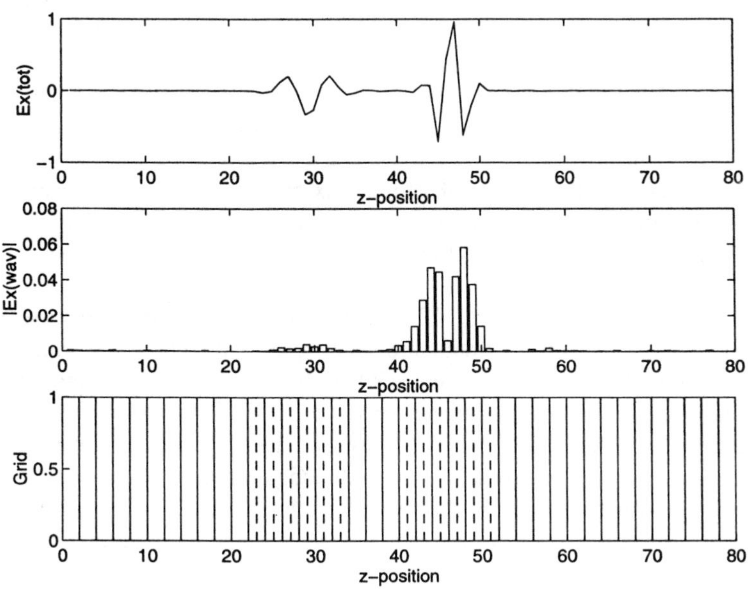

Fig. 3.28 Adaptive grid demonstration corresponding to the results of Fig. 3.26 and Fig. 3.27.

In the top portion of Fig. 3.28, the computed total E_x field is graphed versus z position at a time after the incident pulse reaches the air-dielectric interface. This shows a rightward-propagating transmitted pulse in the dielectric region and a leftward-propagating reflected pulse in the air region. The middle portion shows the contribution to E_x provided by the wavelets at this observation time. We see that the wavelets have significant values only in the regions of the transmitted and reflected pulses, thus creating the locally dense grid zones shown in the bottom portion. In effect, the wavelets follow the reflected and the transmitted pulses and have negligible values elsewhere. The location and the number of the wavelet coefficients with significant values are different for each time step, thereby creating a dense mesh in regions of strong field variations while maintaining a coarser mesh for the other cells.

3.11.2 Open Single and Coupled Striplines

In this section, MRTD is applied to model open single and coupled striplines. Only wavelets of 0 resolution are used since the value of the higher-resolution fields is negligible (smaller than 1%). The nonsplit-field PML algorithm presented in Section 3.9.3 is easily extended to this 2.5-dimensional case incorporating scaling and wavelet functions, and maintains the same performance. For each resolution added to the scheme, the conductivity is sampled with the additional wavelet functions. It is observed that S_{11} changes only by 1 – 1.6 dB after the enhancement of multiple resolutions.

We first consider the application of the 2.5-dimensional FDTD and MRTD schemes to model an open, air-filled stripline for the first (quasi-TEM) propagating mode. The thin strip conductor is assumed to be 23.8 mm wide and located 5.5 mm above the bottom ground plane and 16.5 mm below the top ground plane. A 4-cell PML absorber is applied to the left and the right sides of the structure with $\sigma_{E_{\max}} = 0.1$ S/m. For the FDTD model, a $42\text{-}\times 28$-cell mesh is used, while for MRTD a $12\text{-}\times 4$-cell and a $12\text{-}\times 8$-cell mesh is used. The time step is assumed to be the same for both FDTD and MRTD, set equal to one-tenth of the maximum value for numerical stability of MRTD. Twenty-thousand time steps are run using the factor $\beta = 30$. Overall, the MRTD execution time is reduced from that of FDTD by a factor of 4 to 5 when using the $12\text{-}\times 4$-cell grid.

Table 3.2 lists the FDTD- and MRTD-calculated frequencies of the dominant propagating mode. We see that all results are very close to the theoretical values, with the largest error less than 0.035%. Of course, MRTD obtains this high accuracy using mesh sizes much smaller than that needed for the conventional FDTD simulation.

Fig. 3.29 compares the FDTD- and MRTD-calculated distributions of E_y just below the strip conductor. Since the singularity due to the edge effect is prominent, a $12\text{-}\times 8$-cell grid with scaling functions and 0-resolution wavelets is used for the MRTD model.

The characteristic impedance for the quasi-TEM mode of the stripline is taken as

$$Z_0 = \frac{V}{I} = \int_{C_V} \vec{E}\cdot d\vec{l} \;/\; \oint_{C_I} \vec{H}\cdot d\vec{l} \tag{3.67}$$

TABLE 3.2

Dominant-Mode Frequency of the Open Stripline

Model	Frequency (GHz)	Relative Error (%)
Analytic	1.4324	0.
12 × 4 MRTD	1.4329	0.035
12 × 8 MRTD	1.4325	0.007
42 × 28 FDTD	1.4321	- 0.021

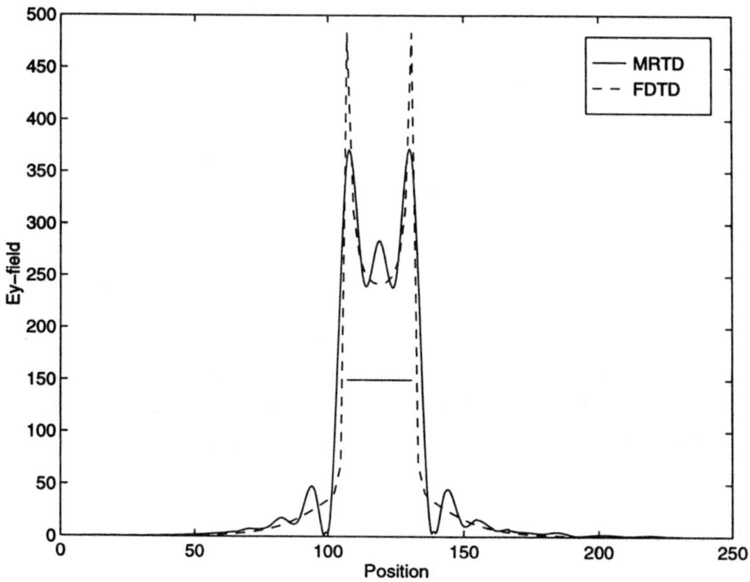

Fig. 3.29 Comparison of FDTD- and MRTD-computed distributions of the E_y field just below the strip conductor of the open single stripline corresponding to Table 3.2.

Table 3.3 lists the values of the stripline Z_0 obtained by implementing the contour integrals of (3.67) as sums of field values computed using FDTD and MRTD.

TABLE 3.3

Stripline Characteristic Impedance Z_0 for Different Mesh Sizes

Model	Z_0 (ohms)	Relative Error (%)
Analytic	56.83	0.
12 × 4 MRTD	57.24	0.72
12 × 8 MRTD	57.09	0.46
42 × 28 FDTD	54.96	- 3.29

We note that only one field value per cell is needed for the FDTD sums because FDTD uses pulse expansion functions that are constant for each cell. On the other hand, for the MRTD sums, the field values for a number of subpoints along the integration path have to be calculated since the expansion functions are not constant for each cell. Here, the accuracy is improved by increasing the number of subpoints per cell at which the field values are calculated. An accuracy of better than 1% is achieved if the field values are computed for more than nine subpoints per cell along the integration path.

A similar procedure is used to model an open, air-filled, coupled-stripline geometry for the first even and odd modes. Here, the two strips are parallel and coplanar and located 5.5 mm above the bottom ground plane and 16.5 mm below the top ground plane. Each strip is 23.8 mm wide, and the strips are separated by 23.8 mm. The MRTD PML is 4 cells thick (23.8 mm) with $\sigma_{E_{\max}} = 0.1$ S/m, and starts exactly at the outer edges of the strips. For the FDTD model, a 65 - × 20 - cell mesh is used. The same accuracy is achieved using a 20 - × 4 - cell MRTD mesh, resulting in a reduction of memory by a factor of 16.25. Figs. 3.30(a) and 3.30(b) visualize the spatial distribution of the electric field tangential to the strips (using a logarithmic gray scale) for the even and odd modes.

3.12 SUMMARY AND CONCLUSIONS

This chapter discussed the theory and numerical application of wavelet basis functions in the context of FDTD solutions of Maxwell's equations. Such a basis provides the possibility of a multiresolution analysis that is computationally much more efficient than the classic Yee algorithm.

The chapter began with a tutorial introduction to wavelet expansions. It then introduced two prototypes and predecessors of wavelet systems and multiresolution analysis: the Haar orthogonal system and the Shannon system. The fundamentals of multiresolution analysis were next presented, including the properties of scaling and

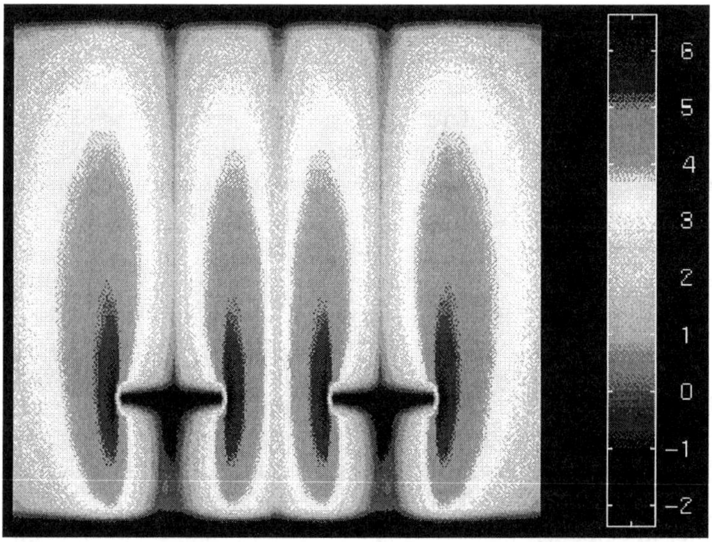

(a)

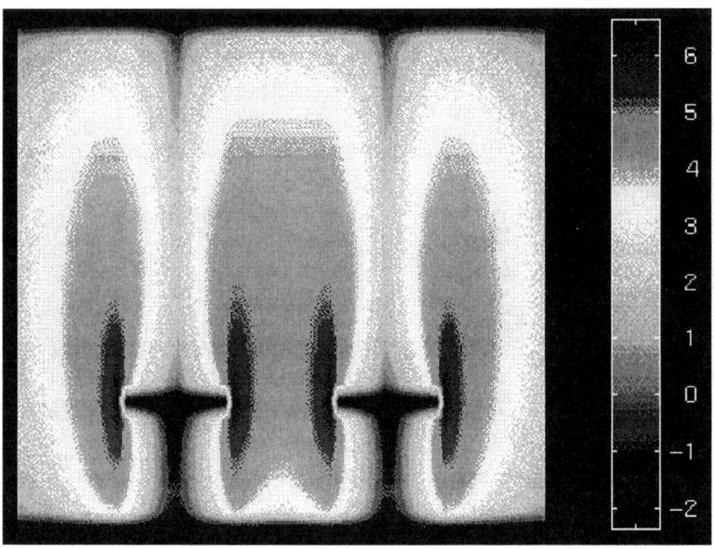

(b)

Fig. 3.30 Logarithmic gray-scale visualization of the spatial distribution of the MRTD-computed electric field tangential to the strips of the coupled stripline geometry: (a) even mode; (b) odd mode.

wavelet functions. Next, two popular wavelet systems were reviewed: the Franklin system and the Battle-Lemarie system.

Having established the basic theory of wavelets and multiresolution analysis, the chapter then applied the Battle-Lemarie wavelet system to develop a two-dimensional MRTD electromagnetic wave scheme. This discussion included the details of the MRTD algorithm and its numerical stability and dispersion properties. Next, the treatment of material boundary conditions in MRTD was discussed including the source excitation, hard boundaries such as PECs, dielectric interfaces, and the PML ABC. The latter was developed in both its original split-field formulation as well as a nonsplit-field formulation. This part of the chapter closed with a discussion of the potential of MRTD for space- and time-adaptive gridding.

The chapter concluded with examples of microwave applications of MRTD. First, MRTD was used to model a two-dimensional, air-filled, parallel-plate waveguide and one filled with dielectric beyond an interface plane. Space- and time-adaptive gridding was used to illustrate how MRTD can automatically track the motion of propagating and reflected pulses in the waveguide with locally dense grid zones moving along with the pulses. Relative to the classic Yee algorithm, MRTD provided a 5:1 to 6.5:1 reduction in computer memory in the direction of pulse propagation, and a 3:1 to 4:1 reduction in computer running time while maintaining a high level of accuracy comparable to the FDTD simulation. Second, MRTD was used to model 2.5-dimensional problems involving air-filled open and coupled striplines. For the latter problem, MRTD achieved the same accuracy as FDTD with a much coarser mesh, yielding an overall reduction of computer memory by a factor of 16:1.

We conclude that MRTD techniques represent a significant generalization of the original FDTD method. MRTD holds much promise in providing accurate and computationally efficient solutions to important microwave engineering problems. Its ability to automatically provide space- and time-adaptive gridding is unique relative to existing FDTD approaches, and may allow the solution of linear and nonlinear pulse propagation problems beyond the scope of current FDTD methods.

REFERENCES

[1] Yee, K. S., "Numerical solution of initial boundary value problems involving Maxwell's equations in isotropic media," *IEEE Trans. Antennas and Propagation*, Vol. 14, 1966, pp. 302–307.

[2] Walter, G. G., *Wavelets and Other Orthogonal Systems with Applications*, Boca Raton, FL: CRC Press, 1994.

[3] Alexits, G., *Convergence Problems of Orthogonal Series*, New York: Oxford, 1961.

[4] Haar, A., "Zur Theorie der Orthogonalen Funktionensysteme," *Math. Ann.*, Vol. 69, 1910, pp. 331–371.

[5] Chui, C. H., *An Introduction to Wavelets*, San Diego, CA: Academic Press, 1992.

[6] Morlet, J., G. Arens, I. Fourgeau, and D. Giard, "Wave propagation and sampling theory, Part II," *Geophysics*, Vol. 47, 1982, pp. 203–236.

[7] Battle, G., "A block spin construction of ondelettes, Part I: Lemarie functions," *Comm. Math. Phys.*, Vol. 110, 1987, pp. 601–615.

[8] Daubechies, I., "Orthonormal bases of compactly supported wavelets," *Comm. Pure Appl. Math.*, Vol. 41, 1988, pp. 909–996.

[9] Coifman, R. R., and Y. Meyer, "Remarques sur l'analyse de Fourier a fenetre, serie I," *C. R. Acad. Sci. Paris*, Vol. 312, 1991, pp. 259–261.

[10] Mallat, S., "Multiresolution approximation and wavelet orthonormal bases of L2(R)," *Trans. American Math. Soc.*, Vol. 315, 1989, pp. 69–88.

[11] E. Tentzeris, L. P. B. Katehi, J. F. Harvey, and A. Cangellaris, "A space- and time-adaptive MRTD gridding algorithm used for the analysis of 2D microwave geometries," *IEEE Trans. Microwave Theory and Techniques*, submitted.

[12] Krumpholz, M., and L. P .B. Katehi, "New prospects for time domain analysis," *IEEE Microwave and Guided Wave Lett.*, Vol. 5, 1995, pp. 382–384.

[13] Krumpholz, M., and L. P. B. Katehi, "MRTD: new time domain schemes based on multiresolution analysis," *IEEE Trans. Microwave Theory and Techniques*, Vol. 44, 1996, pp. 555–572.

[14] Tentzeris, E., M. Krumpholz, and L. P. B. Katehi, "Application of MRTD to printed transmission lines," *Proc. IEEE Int. Symp. on Microwave Theory and Techniques*, 1996, pp. 573–576.

[15] Tentzeris, E., R. Robertson, M. Krumpholz, and L. P. B. Katehi, "Application of the PML absorber to the MRTD technique," *Proc. IEEE Int. Symp. on Antennas and Propagation*, 1996, pp. 634–637.

[16] Goverdhanam, K., E. Tentzeris, M. Krumpholz, and L. P .B. Katehi, "An FDTD multigrid based on multiresolution analysis," *Proc. IEEE Int. Symp. on Antennas and Propagation*, 1996, pp. 352–355.

[17] Tentzeris, E., R. Robertson, A. C. Cangellaris, and L. P. B. Katehi, "Space- and time-adaptive gridding using MRTD," Proc. IEEE Int. Symp. on Microwave Theory and Techniques, 1997, pp. 337–340.

[18] Krumpholz, M., and L. P. B. Katehi, "MRTD modeling of nonlinear pulse propagation," *IEEE Trans. Microwave Theory and Techniques*, in press.

[19] Emmanouil, M., E. Tentzeris, R. L. Robertson, J. Harvey, and L. P. B. Katehi, "Stability and dispersion analysis of Battle-Lemarie-based MRTD schemes," *IEEE Trans. Microwave Theory and Techniques*, submitted.

[20] E. Tentzeris, R. L. Robertson, J. F. Harvey, and L. P. B. Katehi, "PML absorbing boundary conditions for the characterization of open microwave circuit components using multiresolution time-domain techniques (MRTD)," *IEEE Trans. Antennas and Propagation*, submitted.

[21] Berenger, J.-P., "A perfectly matched layer for the absorption of electromagnetic waves," *J. Comput. Physics*, Vol. 114, 1994, pp. 185–200.

SELECTED BIBLIOGRAPHY

Battle, G., "A block spin construction of ondelettes," *Comm. Math. Phys.*, Vol. 110, 1987, pp. 601–615.

Daubechies, I., *Ten Lectures on Wavelets*, Philadelphia, PA: Society for Industrial and Applied Math., 1992.

Krumpholz, M., and P. Russer, "Two-dimensional FDTD and TLM," *Int. Journal of Num. Modeling*, Vol. 7, 1993, pp. 141–153.

Lemarie, P. G., "Ondelettes a localisation exponentielle," *J. Math. Pures Appl.*, Vol. 67, 1988, pp. 227–236.

Perrier, V., and C. Basdevant, "La decomposition en ondelettes periodiques: un outil pour l'analyse des champs inhomogenes. Theorie et algorithmes," *La Recherche Aerospatiale*, 1989, pp. 53–67.

Pozar, D. M., *Microwave Engineering*, New York: Addison-Wesley, 1990.

Wadell, B. C., *Transmission Line Design Handbook*, Boston: Artech House, 1991, pp. 136–137.

Xiao, S., and R. Vahldieck, "An improved 2D FDTD algorithm for hybrid mode analysis of quasi-planar transmission lines," *Proc. IEEE Int. Symp. on Microwave Theory and Techniques*, 1993, Vol. 1, pp. 421–424.

Chapter 4

Explicit Time-Domain Solutions of Maxwell's Equations via Generalized Grids

Stephen D. Gedney, J. Alan Roden, Niel K. Madsen, Alireza H. Mohammadian, William F. Hall, Vijaya Shankar, and Chris Rowell

4.1 INTRODUCTION

The FDTD method has been a highly successful solution to a wide variety of electromagnetic wave interaction problems. However, a principal limitation of the classical FDTD method is its restriction to orthogonal grids, dating back to the initial paper by Kane Yee [1]. As a result, material boundaries that do not conform to an orthogonal coordinate system must be approximated using staircasing. Unfortunately, staircasing can lead to large solution errors. To rectify this, a number of variations of FDTD using conformal meshing have been proposed. Some of the earlier schemes implemented a local deformation of the uniform orthogonal grid near an irregular boundary. Such schemes could be labeled generically as subcell models. For example, contour-path subcell models implementing Faraday's law and Ampere's law in integral form were successfully applied to thin wires [2] and thin slots [3]. Contour-path techniques were later applied to arbitrarily shaped surfaces using what is known as the CP-FDTD [4, 5] scheme. An excellent summary of such techniques is provided in [6].

Variants of FDTD have pursued conformal gridding through the use of nonorthogonal and irregular structured grids. The earliest attempt at this was by Holland, who proposed a nonorthogonal FDTD (N-FDTD) scheme based on a global curvilinear coordinate system [7]. This was generalized by Lee to a nonorthogonal staggered grid scheme that assumes a local curvilinear coordinate system for each cell face [8–10].

Problems involving complex boundaries are often difficult to accurately mesh using structured grids. Unstructured meshing offers the versatility to easily and accurately model highly complex geometries. Several commercial computer-aided design (CAD) software packages are also available to perform automatic mesh generation. Madsen originally generalized Yee's approach to an unstructured grid [11, 12] using the discrete surface integral (DSI) method. This technique utilizes an edge-based discretization on a dual staggered grid, with Maxwell's curl equations enforced in their integral form over each face of the primary and secondary grids. The fields are also staggered in time, leading to an explicit leapfrog solution. Second-order accuracy is maintained via a projection of the face normals onto the edges using a local interpolation scheme. Similar schemes were also used by Gedney and Lansing [13–15], and are referred to here as the generalized Yee (GY) algorithm.

One problem encountered with the CP-FDTD, N-FDTD, and DSI algorithms as well as other schemes is late-time instability. It was demonstrated by Craddock et al. [5] that the source of instability of the original CP-FDTD algorithm [4] is due to its nonreciprocal nature. By analyzing the FDTD method as a passive circuit, an alternative stable and accurate CP-FDTD solution was proposed [5]. However, a similar approach to stabilize the N-FDTD and DSI methods is not likely because of the complexity of the projection operations used in these methods.

The schemes summarized above are all based on a dual grid with edge-based discretization and a staggering of the E and H fields in space and time, and can all be described as generalizations of the original Yee algorithm. An alternate class of finite-volume time-domain (FVTD) methods has been introduced [16–19] that uses a volumetric form of Maxwell's curl equations based on a single grid and collocating the E and H fields at the cell centers. While FVTD methods have only been recently introduced to computational electromagnetics (CEM) relative to FDTD, such methods have a long history of development and applications in computational fluid dynamics (CFD). The FVTD methods proposed by Shankar and Mohammadian et al. [17, 18] and Shang [19] are based on a discretization of the conservation equations and the use of a Lax-Wendroff time-integration scheme. These methods were originally based on irregular, nonorthogonal structured grids, but have been recently generalized to unstructured grids with a small grid stencil [20]. For completeness, we note that hybrid FVTD / FDTD methods have also been introduced [21, 22] that utilize unstructured grids and classic leapfrog time marching.

This chapter has two primary areas of concentration, and is divided accordingly:

1. *Nonorthogonal staggered-grid generalizations of FDTD* — the N-FDTD and DSI methods (Sections 4.2 – 4.7). Principal contributors: S. D. Gedney, J. A. Roden, and N. K. Madsen.
2. *FVTD techniques* (Sections 4.8 – 4.13). Principal contributors: A. H. Mohammadian, W. F. Hall, V. Shankar, and C. Rowell.

Sections 4.2 and 4.3 present the N-FDTD and DSI algorithms and their efficient implementation. Sections 4.4 and 4.5 discuss aspects of their numerical stability and well-posedness. In fact, Section 4.5 demonstrates that the late-time instabilities in the N-FDTD and DSI methods arise because their explicit operators are not well posed. This section further presents means to determine the degree of well-posedness and to achieve this desirable condition. Section 4.6 discusses the error mechanisms present in N-FDTD and DSI approaches and provides a systematic analysis. Finally, Section 4.7 applies N-FDTD and DSI techniques to a number of problems to demonstrate the accuracy and utility of these approaches relative to traditional FDTD.

The focus then shifts to FVTD methods in Section 4.8. Section 4.9 presents the conservation form of Maxwell's equations and then extends this to the time-dependent representation of dispersive media. Section 4.10 presents the FVTD solution of Maxwell's equations based on structured grids using a time-integration scheme based on the Lax-Wendroff method. Important issues such as numerical stability and boundary conditions are detailed. Generalizations of the boundary conditions to impedance and resistive boundaries are also discussed. Section 4.11 presents the extension of FVTD methods to unstructured grids and discusses issues relating to stability, multizoning, and boundary conditions. Section 4.12 provides numerical results and validation studies relating to radar cross-section (RCS) calculations. Finally, Section 4.13 discusses the performance of an unstructured FVTD code on a high-performance parallel computer.

4.2 NONORTHOGONAL FDTD METHOD

Holland's early work [7], based upon analyses by Stratton [23], demonstrated that if a geometry can conform to a global curvilinear coordinate system, then an FDTD Maxwell's equations algorithm can be constructed that is second-order accurate in both time and space. This is accomplished by expressing and solving the curl equations in the general curvilinear space. Specifically, a global set of edge vectors is defined analytically from which the curvilinear-coordinate constructs are defined. This is referred to as the globally nonorthogonal finite-difference time-domain (GN-FDTD) method.

Later, it was found that the nonorthogonal FDTD technique can be applied on a local basis [10] in cases where a global curvilinear coordinate system is not practical. Here, curvilinear coordinates are constructed on a local basis from the local cell faces in the irregular structured lattice. This is called the locally nonorthogonal finite-difference time-domain (LN-FDTD) method.

Recently it was shown that the LN-FDTD technique can be formulated in a more mathematically consistent manner by defining local curvilinear coordinates at each corner of a structured irregular lattice [24]. This technique was shown to be equivalent to the DSI/GY methods and, like these methods, can be applied on unstructured grids. This technique is called the divergenceless nonorthogonal (DN-FDTD) method.

Each of these techniques is formulated based on general curvilinear coordinate concepts, to be reviewed next. Subsequently, the analytic basis for these methods is presented and an accurate, efficient method of implementation is demonstrated. The implementation of the GN-FDTD method is detailed in [7] and is not presented here.

4.2.1 General Curvilinear Coordinate Concepts

A general curvilinear coordinate system is defined from a set of three independent vectors commonly called unitary vectors. From these vectors, a second set of three basis vectors called the reciprocal vectors are defined that are individually orthogonal to consecutive pairs of the unitary vectors in a right-handed sense.

For some problem geometries, we can analytically define a global curvilinear coordinate system that conforms to the problem constructs. Examples of such a coordinate system include the globally skewed coordinate system or the twisted coordinate system [25]. In such cases, a set of global curvilinear coordinates denoted by (u^1, u^2, u^3) is defined such that each position in the orthogonal space is given by a position vector of the form $\bar{r}(x, y, z) = \bar{r}(u^1, u^2, u^3)$. For a given point $P(u^1, u^2, u^3)$, a set of unitary vectors is defined by

$$\vec{A}_1 = \frac{\partial \bar{r}}{\partial u^1} \qquad \vec{A}_2 = \frac{\partial \bar{r}}{\partial u^2} \qquad \vec{A}_3 = \frac{\partial \bar{r}}{\partial u^3} \tag{4.1}$$

Note that these vectors are typically not of unit length.

For many geometries, the definition of a global curvilinear coordinate system is not practical. Here, discretization of the problem domain can be accomplished using an automated gridding tool. First, a primary grid is constructed that conforms to the material boundaries of the problem space and represents the unknown E-field vectors. Then, a secondary grid is constructed by connecting the barycenters of the primary grid. Two unique sets of unitary vectors are defined directly from the cell edges of the primary and secondary grids. A typical FDTD lattice construction of this type is shown in Fig. 4.1.

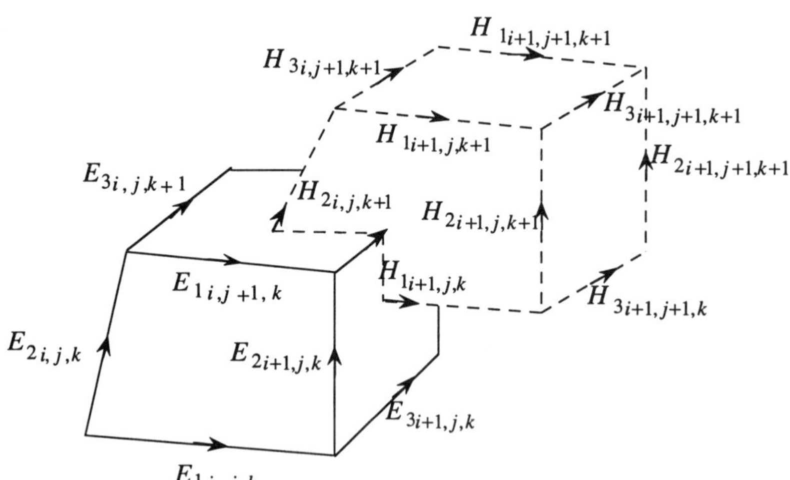

Fig. 4.1 Primary and secondary grids of an irregular structured lattice. The covariant (unitary) field vectors are indicated by arrows.

We refer to the unitary vectors as the covariant vectors of the FDTD lattice; these are denoted here by subscripts. Using these vectors, a set of reciprocal (contravariant) vectors is then defined; these are denoted by superscripts. The contravariant vectors are related to the covariant vectors through

$$\vec{A}^1 = \frac{\vec{A}_2 \times \vec{A}_3}{V}, \qquad \vec{A}^2 = \frac{\vec{A}_3 \times \vec{A}_1}{V}, \qquad \vec{A}^3 = \frac{\vec{A}_1 \times \vec{A}_2}{V} \tag{4.2}$$

where the volume V is determined from the relation

$$V = \vec{A}_i \cdot \left[\vec{A}_{(i+1)\,\mathrm{mod}_3} \times \vec{A}_{(i+2)\,\mathrm{mod}_3} \right] \tag{4.3}$$

The contravariant vectors are normal to each cell face in both the primary and secondary grids. Due to their reciprocal relation, it can be shown that

$$\vec{A}_i \cdot \vec{A}^j = \delta_{i,j} \tag{4.4}$$

As shown by Stratton [23], the vector fields in the orthogonal space are related to the vector fields of the general curvilinear space by

$$\vec{F} = \sum_{i=1}^{3} \vec{A}_i F^i, \qquad \vec{F} = \sum_{i=1}^{3} \vec{A}^i F_i \tag{4.5}$$

and the curvilinear field components are determined from the orthogonal fields from

$$F^i = \vec{A}^i \cdot \vec{F}, \qquad F_i = \vec{A}_i \cdot \vec{F} \tag{4.6}$$

Further, the contravariant and covariant field values are related by

$$F_i = \sum_{j=1}^{3} g_{i,j} F^j, \qquad F^i = \sum_{j=1}^{3} g^{i,j} F_j \tag{4.7}$$

where

$$g_{i,j} = \vec{A}_i \cdot \vec{A}_j, \qquad g^{i,j} = \vec{A}^i \cdot \vec{A}^j \tag{4.8}$$

The fields in the general curvilinear space are in geometry-dependent units. To obtain these values in geometry-independent units, the following normalization is applied:

$$\vec{F}_i = \frac{\vec{F}_i}{\sqrt{g_{i,i}}}, \qquad \vec{F}^i = \frac{\vec{F}^i}{\sqrt{g^{i,i}}} \tag{4.9}$$

where $\vec{F}_i$ and $\vec{F}^i$ are the physical fields.

4.2.2 Implementation of Maxwell's Equations in a General Curvilinear Coordinate System

The general curvilinear coordinate concepts defined in the previous section are applicable to either a global or local curvilinear space by proper definition of the unitary vectors. Then, the application of Maxwell's equations is straightforward, as demonstrated in [7]. For a globally defined curvilinear space, Maxwell's equations are easily implemented in their differential form. Ampere's law becomes

$$\frac{\partial D^1}{\partial t} + \frac{\sigma}{\varepsilon} D^1 = \frac{1}{V}\left(\frac{\partial H_3}{\partial u^2} - \frac{\partial H_2}{\partial u^3}\right) \tag{4.10a}$$

$$\frac{\partial D^2}{\partial t} + \frac{\sigma}{\varepsilon} D^2 = \frac{1}{V}\left(\frac{\partial H_1}{\partial u^3} - \frac{\partial H_3}{\partial u^1}\right) \tag{4.10b}$$

$$\frac{\partial D^3}{\partial t} + \frac{\sigma}{\varepsilon} D^3 = \frac{1}{V}\left(\frac{\partial H_2}{\partial u^1} - \frac{\partial H_1}{\partial u^2}\right) \tag{4.10c}$$

and the three components of Faraday's law are written as

$$\frac{\partial B^1}{\partial t} = -\frac{1}{V}\left(\frac{\partial E_3}{\partial u^2} - \frac{\partial E_2}{\partial u^3}\right) \tag{4.11a}$$

$$\frac{\partial B^2}{\partial t} = -\frac{1}{V}\left(\frac{\partial E_1}{\partial u^3} - \frac{\partial E_3}{\partial u^1}\right) \tag{4.11b}$$

$$\frac{\partial B^3}{\partial t} = -\frac{1}{V}\left(\frac{\partial E_2}{\partial u^1} - \frac{\partial E_1}{\partial u^2}\right) \tag{4.11c}$$

The implementation of these expressions in the globally nonorthogonal discrete space is straightforward and is not demonstrated here. The interested reader will find a thorough treatment of this implementation in [7, 24].

In an irregular nonorthogonal space, Maxwell's equations are most naturally applied in integral form. It is demonstrated in [7, 10] that these equations may be written in general curvilinear coordinates as

$$\frac{\partial}{\partial t}\iint_{F^i} \vec{D}\cdot d\vec{s}^{\,i} + \iint_{F^i} \frac{\sigma}{\varepsilon}\vec{D}\cdot d\vec{s}^{\,i} = \frac{1}{V}\oint_{C_i} \vec{H}\cdot d\vec{\ell}_i \tag{4.12a}$$

$$\frac{\partial}{\partial t}\iint_{F^i} \vec{B}\cdot d\vec{s}^{\,i} = -\frac{1}{V}\oint_{C_i} \vec{E}\cdot d\vec{\ell}_i \tag{4.12b}$$

The unitary vectors for the irregular nonorthogonal method are defined directly from the irregular lattice and therefore are scaled by the edge length. This allows the efficient numerical implementation of the line integral that constitutes the right-hand side of Ampere's law, (4.12a), which is expressed by a summation around each local cell face to yield

$$\begin{aligned} D^1\Big|_{i,j,k}^{n+\frac{1}{2}} &= \left(\frac{1/\Delta t - \sigma/2\varepsilon}{1/\Delta t + \sigma/2\varepsilon}\right) D^1\Big|_{i,j,k}^{n-\frac{1}{2}} \\ &+ \left(\frac{1/V_1^e\big|_{i,j,k}}{1/\Delta t + \sigma/2\varepsilon}\right)\left[\left(H_3\Big|_{i,j,k}^{n} - H_3\Big|_{i,j-1,k}^{n}\right) - \left(H_2\Big|_{i,j,k}^{n} - H_2\Big|_{i,j,k-1}^{n}\right)\right] \end{aligned} \tag{4.13a}$$

$$\begin{aligned} D^2\Big|_{i,j,k}^{n+\frac{1}{2}} &= \left(\frac{1/\Delta t - \sigma/2\varepsilon}{1/\Delta t + \sigma/2\varepsilon}\right) D^2\Big|_{i,j,k}^{n-\frac{1}{2}} \\ &+ \left(\frac{1/V_2^e\big|_{i,j,k}}{1/\Delta t + \sigma/2\varepsilon}\right)\left[\left(H_1\Big|_{i,j,k}^{n} - H_1\Big|_{i,j,k-1}^{n}\right) - \left(H_2\Big|_{i,j,k}^{n} - H_2\Big|_{i-1,j,k}^{n}\right)\right] \end{aligned} \tag{4.13b}$$

$$\begin{aligned} D^3\Big|_{i,j,k}^{n+\frac{1}{2}} &= \left(\frac{1/\Delta t - \sigma/2\varepsilon}{1/\Delta t + \sigma/2\varepsilon}\right) D^3\Big|_{i,j,k}^{n-\frac{1}{2}} \\ &+ \left(\frac{1/V_3^e\big|_{i,j,k}}{1/\Delta t + \sigma/2\varepsilon}\right)\left[\left(H_2\Big|_{i,j,k}^{n} - H_2\Big|_{i-1,j,k}^{n}\right) - \left(H_1\Big|_{i,j,k}^{n} - H_1\Big|_{i,j-1,k}^{n}\right)\right] \end{aligned} \tag{4.13c}$$

Likewise, Faraday's law, (4.12b), is discretized as

$$B^1\Big|_{i,j,k}^{n+1} = B^1\Big|_{i,j,k}^{n} - \frac{\Delta t}{\mu V_1^h\big|_{i,j,k}}\left[\left(E_3\Big|_{i,j+1,k}^{n+\frac{1}{2}} - E_3\Big|_{i,j,k}^{n+\frac{1}{2}}\right) - \left(E_2\Big|_{i,j,k+1}^{n+\frac{1}{2}} - E_2\Big|_{i,j,k}^{n+\frac{1}{2}}\right)\right] \tag{4.14a}$$

$$B^2\Big|_{i,j,k}^{n+1} = B^2\Big|_{i,j,k}^{n} - \frac{\Delta t}{\mu V_2^h\big|_{i,j,k}}\left[\left(E_1\Big|_{i,j,k+1}^{n+\frac{1}{2}} - E_1\Big|_{i,j,k}^{n+\frac{1}{2}}\right) - \left(E_3\Big|_{i+1,j,k}^{n+\frac{1}{2}} - E_3\Big|_{i,j,k}^{n+\frac{1}{2}}\right)\right] \tag{4.14b}$$

$$B^3\Big|_{i,j,k}^{n+1} = B^3\Big|_{i,j,k}^{n} - \frac{\Delta t}{\mu V_3^h\big|_{i,j,k}}\left[\left(E_2\Big|_{i+1,j,k}^{n+\frac{1}{2}} - E_2\Big|_{i,j,k}^{n+\frac{1}{2}}\right) - \left(E_1\Big|_{i,j+1,k}^{n+\frac{1}{2}} - E_1\Big|_{i,j,k}^{n+\frac{1}{2}}\right)\right] \tag{4.14c}$$

where the indices i, j, k correspond to the curvilinear coordinates, (u^1, u^2, u^3), respectively, and n is the discrete time step. The volume V is defined uniquely for each cell face in both the primary and secondary grids from the covariant edge (which is collocated with F^i) and the four dual-grid covariant vectors (which define the face through which this field vector passes). For example, from Fig. 4.1, the volume $V_3^h\big|_{i,j,k}$ of the face penetrated by the magnetic field vector $H_3\big|_{i,j,k}$ is defined by

$$V_3^h\Big|_{i,j,k} = \vec{A}_3^h\Big|_{i,j,k} \cdot \left\{ \left(\frac{\vec{A}_1^e\Big|_{i,j,k} + \vec{A}_1^e\Big|_{i,j+1,k}}{2} \right) \times \left(\frac{\vec{A}_2^e\Big|_{i,j,k} + \vec{A}_2^e\Big|_{i+1,j,k}}{2} \right) \right\} \tag{4.15}$$

The superscripts e and h that appear in (4.13a) to (4.15) designate electric field and magnetic field grid quantities, respectively (i.e., the primary and secondary grid edges, respectively).

4.2.3 Projection Strategies for Irregular Grids

The contravariant field updates of (4.13) and (4.14) are executed in a time-marching manner with each field value on the right-hand side known from previous computations. For nonorthogonal grids, the contravariant vectors and the covariant vectors are not aligned. Therefore, the computed contravariant field values must be projected onto each covariant edge using relations given in (4.7). Since all three E- or H-field components are not collocated in space, a simple field averaging scheme must be employed.

Facial-Average Method

The first projection strategy considered is facial averaging. This involves constructing a set of local curvilinear coordinates at each face in the lattice by averaging both the field values and the grid edge vectors for the local face. Each projected field component has contributions from nine distinct local coordinate systems. With this approach, the projections for the covariant electric fields become

$$E_1\Big|_{i,j,k}^{n+\frac{1}{2}} = \frac{1}{\varepsilon_{i,j,k}} \left[\begin{aligned} & g_{11} D^1\Big|_{i,j,k}^{n+\frac{1}{2}} + \frac{g_{12}}{4}\left(D^2\Big|_{i,j,k}^{n+\frac{1}{2}} + D^2\Big|_{i+1,j,k}^{n+\frac{1}{2}} + D^2\Big|_{i,j-1,k}^{n+\frac{1}{2}} + D^2\Big|_{i+1,j-1,k}^{n+\frac{1}{2}} \right) \\ & + \frac{g_{13}}{4}\left(D^3\Big|_{i,j,k}^{n+\frac{1}{2}} + D^3\Big|_{i+1,j,k}^{n+\frac{1}{2}} + D^3\Big|_{i,j,k-1}^{n+\frac{1}{2}} + D^3\Big|_{i+1,j,k-1}^{n+\frac{1}{2}} \right) \end{aligned} \right] \tag{4.16a}$$

$$E_2\Big|_{i,j,k}^{n+\frac{1}{2}} = \frac{1}{\varepsilon_{i,j,k}}\left[\begin{aligned} & g_{22} D^2\Big|_{i,j,k}^{n+\frac{1}{2}} + \frac{g_{21}}{4}\left(D^1\Big|_{i,j,k}^{n+\frac{1}{2}} + D^1\Big|_{i,j+1,k}^{n+\frac{1}{2}} + D^1\Big|_{i-1,j,k}^{n+\frac{1}{2}} + D^1\Big|_{i-1,j+1,k}^{n+\frac{1}{2}}\right) \\ & + \frac{g_{23}}{4}\left(D^3\Big|_{i,j,k}^{n+\frac{1}{2}} + D^3\Big|_{i,j+1,k}^{n+\frac{1}{2}} + D^3\Big|_{i,j,k-1}^{n+\frac{1}{2}} + D^3\Big|_{i,j+1,k-1}^{n+\frac{1}{2}}\right)\end{aligned}\right] \tag{4.16b}$$

$$E_3\Big|_{i,j,k}^{n+\frac{1}{2}} = \frac{1}{\varepsilon_{i,j,k}}\left[\begin{aligned} & g_{33} D^3\Big|_{i,j,k}^{n+\frac{1}{2}} + \frac{g_{31}}{4}\left(D^1\Big|_{i,j,k}^{n+\frac{1}{2}} + D^1\Big|_{i,j,k+1}^{n+\frac{1}{2}} + D^1\Big|_{i-1,j,k}^{n+\frac{1}{2}} + D^1\Big|_{i-1,j,k+1}^{n+\frac{1}{2}}\right) \\ & + \frac{g_{32}}{4}\left(D^2\Big|_{i,j,k}^{n+\frac{1}{2}} + D^2\Big|_{i,j,k+1}^{n+\frac{1}{2}} + D^2\Big|_{i,j-1,k}^{n+\frac{1}{2}} + D^2\Big|_{i,j-1,k+1}^{n+\frac{1}{2}}\right)\end{aligned}\right] \tag{4.16c}$$

The magnetic field projections are accomplished in an analogous manner.

For a regular lattice, the elements of the projection matrix, $g_{i,j}$, are easily defined. In the presence of an irregular lattice, the proper definition of the vectors that comprise (4.8) is ambiguous due to both the irregularity and the staggered nature of the lattice.

To illustrate the proper computation of the projection metrics for an irregular grid, consider the computation of $H^3|_{i,j,k}$ in Fig. 4.1. This contravariant magnetic field component is normal to the plane defined by each of the four edges that define the face through which $H_3|_{i,j,k}$ passes. Therefore, a local curvilinear coordinate system that includes this field component can be accurately defined from the covariant field vector collocated with $H_3|_{i,j,k}$ and the two vectors defined by averaging the opposing edges of the face defined by $\vec{A}^3|_{i,j,k}$. Specifically, the three vectors defining the local coordinate system for this primary grid face are

$$\vec{A}_1\Big|_{i,j,k} = \frac{1}{2}\left(\vec{A}_1^e\Big|_{i,j,k} + \vec{A}_1^e\Big|_{i,j-1,k}\right) \tag{4.17a}$$

$$\vec{A}_2\Big|_{i,j,k} = \frac{1}{2}\left(\vec{A}_2^e\Big|_{i,j,k} + \vec{A}_2^e\Big|_{i-1,j,k}\right) \tag{4.17b}$$

$$\vec{A}_3\Big|_{i,j,k} = \vec{A}_3^h\Big|_{i,j,k} \tag{4.17c}$$

The vectors composing the local curvilinear coordinate system of each face of the primary and secondary grids are computed in a similar manner. Subsequently, the projection metrics are uniquely defined for each face of the primary and secondary grids.

The facial-average projection strategy defined above is quite accurate, as shown in subsequent sections. However, with this approach, the vectors aligned with the projected fields and the projection metrics do not necessarily constitute a reciprocal set. Hence, the resultant projections are not necessarily divergence-free, which can lead to error. Further, creating a structured grid for many typical geometries can be quite difficult. It has been found that poorly constructed geometries can lead to inaccuracy and instability. Therefore, it is beneficial to formulate a projection technique that is applicable to either unstructured or structured grids and results in a reciprocal set of local basis. A recently developed approach that accomplishes these goals is the DN-FDTD technique.

The DN-FDTD technique is formulated by introducing local curvilinear coordinate systems at the vertex of each cell in both the primary and secondary grids. To this end, consider the two dual-grid cells, $i-1$ and i, that share the common face, i, defined by $\vec{A}^1$ and illustrated in Fig. 4.2.

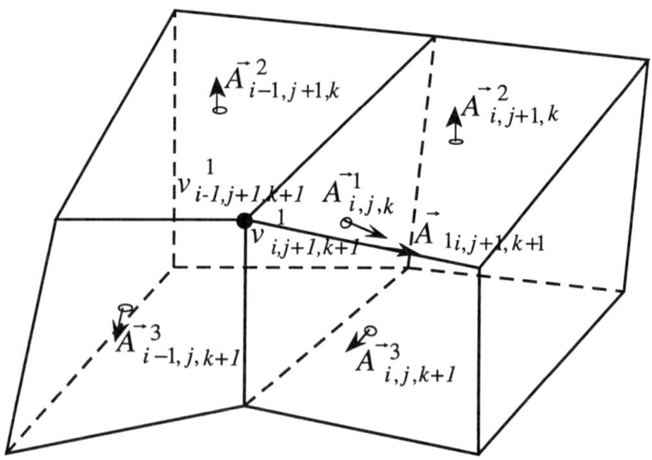

Fig. 4.2 Geometry constructs of local curvilinear coordinate systems constructed at the cell vertices of each primary and dual grid cell.

In Fig. 4.2, there are four vertices bounding the face, and each vertex is shared by two corners.

For example, consider the vertex $v^1|_{i,j+1,k+1}$. This vertex is shared by two corners composed of secondary grid faces. Hence, it is split into two vertices: $v^1|_{i-1,j+1,k+1}$ defined by contravariant vectors ($\vec{A}^1|_{i,j,k}$, $\vec{A}^2|_{i-1,j+1,k+1}$, $\vec{A}^3|_{i-1,j+1,k+1}$), and $v^1|_{i,j+1,k+1}$ defined by contravariant vectors ($\vec{A}^1\big|_{i,j,k}$, $\vec{A}^2\big|_{i,j+1,k+1}$, $\vec{A}^3\big|_{i,j+1,k+1}$). We note that each of these contravariant vectors is defined using the averaged covariant vectors that surround the face. Thus, the set of vectors is consistent with the field-update expressions.

The three contravariant vectors associated with each corner are now used to explicitly define a set of reciprocal unitary vectors at the vertices $v^1\big|_{i-1,j+1,k+1}$ and $v^1\big|_{i,j+1,k+1}$. The unitary vectors are mutually orthogonal to consecutive pairs of these contravariant vectors and form a reciprocal set. These vectors are uniquely given by the relations

$$\vec{C}_1 = \frac{\vec{A}^2 \times \vec{A}^3}{V}, \qquad \vec{C}_2 = \frac{\vec{A}^3 \times \vec{A}^1}{V}, \qquad \vec{C}_3 = \frac{\vec{A}^1 \times \vec{A}^2}{V} \tag{4.18}$$

where the volume is given by

$$V = \vec{A}^1 \cdot (\vec{A}^2 \times \vec{A}^3) \tag{4.19}$$

and the relation $\vec{A}^i \cdot \vec{C}_j = \delta_{i,j}$ is satisfied. Thus, for example, at vertex $v^1\big|_{i,j+1,k+1}$:

$$\vec{C}_1\Big|_{i,j+1,k+1} = \frac{\vec{A}^2\Big|_{i,j+1,k+1} \times \vec{A}^3\Big|_{i,j,k+1}}{V_c\Big|_{i,j+1,k+1}} \tag{4.20a}$$

$$\vec{C}_2\Big|_{i,j+1,k+1} = \frac{\vec{A}^3\Big|_{i,j,k+1} \times \vec{A}^1\Big|_{i,j,k}}{V_c\Big|_{i,j+1,k+1}} \tag{4.20b}$$

$$\vec{C}_3\Big|_{i,j+1,k+1} = \frac{\vec{A}^1\Big|_{i,j,k} \times \vec{A}^2\Big|_{i,j+1,k+1}}{V_c\Big|_{i,j+1,k+1}} \tag{4.20c}$$

where

$$V_c\Big|_{i,j+1,k+1} = \vec{A}^1\Big|_{i,j,k} \cdot \left(\vec{A}^2\Big|_{i,j+1,k+1} \times \vec{A}^3\Big|_{i,j,k+1} \right) \tag{4.20d}$$

The electric field associated with the vertex $v^1\big|_{i,j+1,k+1}$ is then expressed as an expansion of the three contravariant fields:

$$\vec{B}\Big|_{i,j+1,k+1} = \vec{C}_1\Big|_{i,j+1,k+1} B^1\Big|_{i,j,k} + \vec{C}_2\Big|_{i,j+1,k+1} B^2\Big|_{i,j+1,k+1} + \vec{C}_3\Big|_{i,j+1,k+1} B^3\Big|_{i,j,k+1} \tag{4.21}$$

Finally, the field is projected onto the dual-grid edge passing through the face through the dot product of (4.21) with $\vec{A}_1\big|_{i,j,k}$. This leads to

$$B_1\big|_{i,j+1,k+1} = \tilde{g}_{1,1}\big|_{i,j+1,k+1} B^1\big|_{i,j,k} + \tilde{g}_{1,2}\big|_{i,j+1,k+1} B^2\big|_{i,j+1,k+1} + \tilde{g}_{1,3}\big|_{i,j+1,k+1} B^3\big|_{i,j,k+1} \quad (4.22)$$

where

$$\tilde{g}_{1,m}\big|_{i,j+1,k+1} = \vec{A}_1\big|_{i,j,k} \cdot \vec{C}_m\big|_{i,j+1,k+1} \quad (4.23)$$

The application of (4.22) at each vertex pair leads to eight unique sets of projected fields for each face. These projected fields are then averaged to give the total interpolated field. For example, for the face shared by cells $(i-1, j, k)$ and (i, j, k) as illustrated in Fig. 4.2, this leads to

$$B_1\big|_{i,j,k} = \frac{\sum_{l=i-1}^{i} \sum_{m=j}^{j+1} \sum_{n=k}^{k+1} \tilde{B}_1\big|_{l,m,n} \left| w_1\big|_{l,m,n} \right|}{\sum_{l=i-1}^{i} \sum_{m=j}^{j+1} \sum_{n=k}^{k+1} \left| w_1\big|_{l,m,n} \right|} \quad (4.24)$$

It is shown in Section 4.3 that this projection scheme is equivalent to the DSI / GY methods if the weighting functions are chosen to be consistent with those defined by Madsen [12]. The weighting functions are defined by

$$w_1\big|_{l,m,n} = \left[\vec{A}^1\big|_{i,j,k} \cdot \left(\vec{A}^2\big|_{l,m,k} \times \vec{A}^3\big|_{l,j,n}\right)\right] \cdot \left(V_1\big|_{i,j,k}\, V_2\big|_{l,m,k}\, V_3\big|_{l,j,n}\right) \quad (4.25)$$

where the volume V is that associated with each face as computed similarly to (4.15).

Following the notation of (4.16) and using the projection in (4.24), the projected magnetic field is finally expressed as

$$H_1\big|_{i,j,k}^{n+1} = \frac{1}{\varepsilon_{i,j,k}} \begin{bmatrix} g_{11}\big|_{i,j,k} B^1\big|_{i,j,k}^{n+1} & + \; g_{12}\big|_{i,j,k} B^2\big|_{i,j,k}^{n+1} & + \; g_{12}\big|_{i+1,j,k} B^2\big|_{i+1,j,k}^{n+1} & + \\ g_{12}\big|_{i,j-1,k} B^2\big|_{i,j-1,k}^{n+1} & + \; g_{12}\big|_{i+1,j-1,k} B^2\big|_{i+1,j-1,k}^{n+1} & + \; g_{13}\big|_{i,j,k} B^3\big|_{i,j,k}^{n+1} & + \\ g_{13}\big|_{i+1,j,k} B^3\big|_{i+1,j,k}^{n+1} & + \; g_{13}\big|_{i,j,k-1} B^3\big|_{i,j,k-1}^{n+1} & + \; g_{13}\big|_{i+1,j,k-1} B^3\big|_{i+1,j,k-1}^{n+1} & \end{bmatrix}$$

(4.26)

where

$$g_{11}\Big|_{i,j,k} = \frac{\sum_{l=i-1}^{i} \sum_{m=j}^{j+1} \sum_{n=k}^{k+1} \tilde{g}_{11}\Big|_{l,m,n} \left| w_1\Big|_{l,m,n} \right|}{\sum_{l=i-1}^{i} \sum_{m=j}^{j+1} \sum_{n=k}^{k+1} \left| w_1\Big|_{l,m,n} \right|} \tag{4.27}$$

and the $\tilde{g}_{11}\big|_{l,m,n}$ are obtained from (4.23). The off-diagonal terms are composed of the superposition of only two terms from (4.24). For example, we have

$$g_{12}\Big|_{i,j,k} = \frac{\tilde{g}_{12}\Big|_{i,j,k} \left| w_1\Big|_{i,j,k} \right| + \tilde{g}_{12}\Big|_{i,j,k+1} \left| w_1\Big|_{i,j,k+1} \right|}{\sum_{l=i-1}^{i} \sum_{m=j}^{j+1} \sum_{n=k}^{k+1} \left| w_1\Big|_{l,m,n} \right|} \tag{4.28}$$

$$g_{13}\Big|_{i,j,k} = \frac{\tilde{g}_{13}\Big|_{i,j,k} \left| w_1\Big|_{i,j,k} \right| + \tilde{g}_{13}\Big|_{i,j+1,k} \left| w_1\Big|_{i,j+1,k} \right|}{\sum_{l=i-1}^{i} \sum_{m=j}^{j+1} \sum_{n=k}^{k+1} \left| w_1\Big|_{l,m,n} \right|} \tag{4.29}$$

Again, $\tilde{g}_{12}$ and $\tilde{g}_{13}$ are defined in (4.24). Finally, the projections for the electric field are determined in an analogous manner.

4.2.4 Boundary Conditions

The use of idealized boundaries is typical in electromagnetic field problems. These boundaries may represent electric or magnetic conductors that are present in the problem geometries, or they may be introduced to impose an even or odd symmetry condition on the fields, thereby leading to a reduced computational space. The proper implementation of these boundary conditions in general curvilinear coordinates is discussed next.

Perfect Electric Conductors

A PEC boundary is defined by the relation $E_{\tan} = 0$ at a surface. Further, since a PEC is incapable of supporting magnetic charge, the normal magnetic field is also zero. The implementation of the PEC boundary in an orthogonal FDTD grid is accomplished in a straightforward manner by setting the electric field edges that exist at the surface of the conductor to zero explicitly.

In a nonorthogonal lattice, two sets of fields exist at each grid location: the contravariant and covariant fields. These field relations are demonstrated in Fig. 4.3. By choosing the primary grid edges to lie along a PEC boundary, it is apparent that the covariant fields must be set to zero to properly implement the field relations of the PEC.

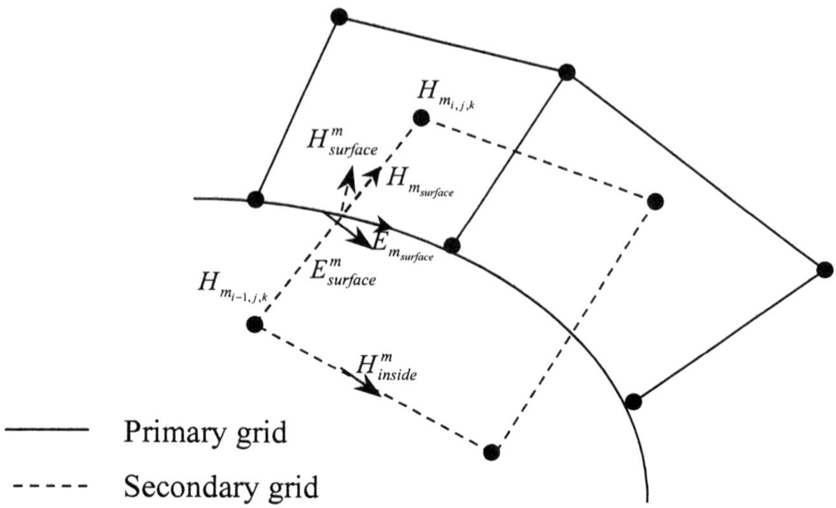

Fig. 4.3 Relationship between contravariant and covariant fields at the surface of a curved conductor.

Consider an irregular grid existing at the surface of the conductor. Specifically, assume that a secondary grid vector, $\vec{A}_m$, that passes through the conductor surface is not perpendicular to that surface. Then, the contravariant electric field vectors, E^m_{surface}, that surround $\vec{A}_m$ and lie on the surface of the conductor, are not zero. The nonzero values of E^m_{surface} are a direct result of the nonzero covariant magnetic field vectors, $H_{m_{\text{surface}}}$, that penetrate the conductor surface since these components contribute directly to the electric field updates. Therefore, it is essential that $H_{m_{\text{surface}}}$ be updated correctly.

The contravariant magnetic field at the surface, H^m_{surface}, is implicitly set to zero due to the enforcement of $E_{\tan} = 0$ at the conductor surface. However, H^m_{surface} is also related to the fields outside and inside the conductor through the off diagonals of $G_{m,n}$. Therefore, it is necessary to set the contributing contravariant magnetic fields inside the conductor to their proper value. To be consistent with the physics of the electric conductor, the contravariant magnetic fields within a conductor must be set to zero.

The line integral that appears in the field update for E^m_{surface} at the surface of the conductor also requires modification. Specifically, the two magnetic field components contributing to this field update that do not pass through the conductor surface should not contribute to the computation of E^m_{surface}. This is easily and efficiently enforced by explicitly setting the covariant magnetic field just inside the conductor to that just outside the conductor. For example, for a conductor whose upward normal is defined by $\vec{A}^1|_{i,j,k}$, the condition $H_m|_{i-1,j,k} = H_m|_{i,j,k}$ is enforced, effectively canceling these contributions. The other internal tangential magnetic field components are explicitly set in a similar manner. Enforcing this equality implicitly assumes that the secondary grid is imaged across the PEC surface. Where the enforcement of a symmetric grid is not possible, some error may be introduced. It has been found that this error is much smaller than that which results from setting E^m_{surface} explicitly to zero [24].

Perfect Magnetic Conductors

A PMC boundary is defined by the relation $H_{\tan} = 0$ at a surface. Typically, it is desirable to enforce this boundary condition on a primary grid edge. It has been demonstrated [26] that this is easily accomplished by setting the tangential values of the magnetic field below the conductor equal to the negative of the value directly above the conductor. If one assumes a linearly varying magnetic field across each secondary grid cell, this results in enforcing the appropriate boundary condition at the primary grid edge that passes through the center of the secondary grid cell. In a general curvilinear coordinate system with a u^1 normal again assumed, this is written as $H_m|_{i-1,j,k} = H_m|_{i,j,k}$. These field components are illustrated in Fig 4.3. Each other tangential field is imaged similarly.

The tangential and normal components of the magnetic field image differently across a PMC boundary. Therefore, it is also necessary to image the grid across the boundary. Since the primary and dual grids are irregular and nonorthogonal in the region of the conductor, image theory results in only an approximately correct solution for some geometries. For planar surfaces, however, the resulting boundary condition is exact. It is easily demonstrated that an oddly imaged grid coupled with oddly imaged covariant field components results in a proper set of imaged fields.

4.3 NONORTHOGONAL FDTD METHODS BASED ON UNSTRUCTURED GRIDS

Unstructured grids are so termed because they are composed of unstructured arrays of general fitted polyhedral cells. While every cell shares a common face and edge, the vertices of the grid can no longer be listed in a regular sense. These grids have the advantage that they can conform to highly complex shapes and surfaces while maintaining good cell aspect ratios and global uniformity. Unfortunately, the methods discussed in the previous section are not applicable to unstructured grids. More general methods must be used.

A general method to formulate an explicit time-marching solution of Maxwell's equations using unstructured grids was first introduced by Madsen [11, 12]. The method involves deriving an explicit operator based on the integral form of Maxwell's equations (4.12), and evaluating the discrete surface and line integrals. A primary issue involves the development of a scheme similar to (4.16) or (4.26) that projects the face-normal components of the fields onto the dual-grid edges. Due to the unstructured nature of the lattice, these schemes cannot be implemented directly. With Madsen's DSI method, the normal fields are projected onto the grid edges in a numerically stable and second-order accurate manner. This approach was later incorporated into the GY algorithms for microwave circuit analysis by Gedney and Lansing [14, 15, 27]. Overall, the DSI / GY algorithms presented in this section provide an efficient, second-order-accurate means for treating three-dimensional electromagnetic wave interaction problems based on unstructured grids.

The DSI algorithm is based on a direct solution of the time-dependent Maxwell's equations in their integral form (Ampere's and Faraday's laws) within a closed three-dimensional volume:

$$\frac{\partial}{\partial t}\iint_S \vec{D}\cdot d\vec{s} + \iint_S \sigma\vec{E}\cdot d\vec{s} + \iint_S \vec{J}\cdot d\vec{s} = \oint_C \vec{H}\cdot d\vec{\ell} \tag{4.30a}$$

$$\frac{\partial}{\partial t}\iint_S \vec{B}\cdot d\vec{s} = -\oint_C \vec{E}\cdot d\vec{\ell} \tag{4.30b}$$

The electric and magnetic fields are discretized over a dual lattice formed by a primary and secondary mesh, as illustrated in Fig. 4.4 for hexahedral lattice cells.

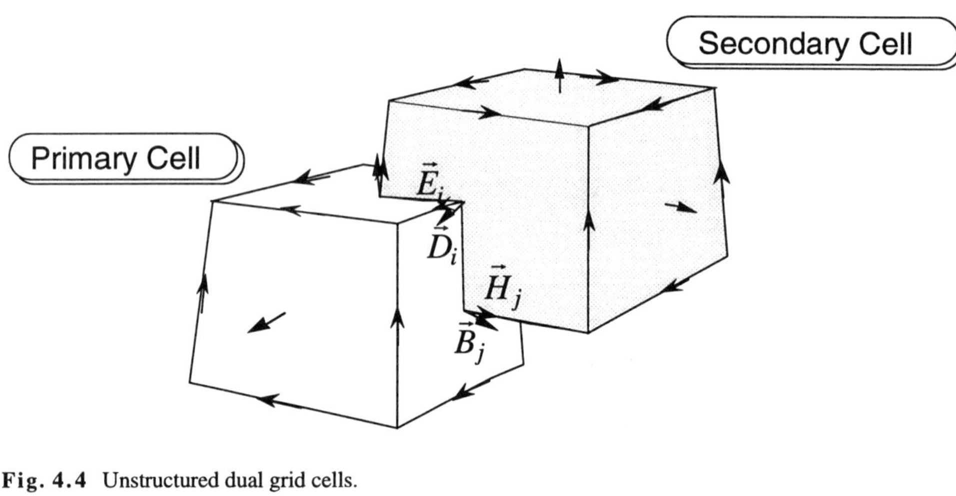

Fig. 4.4 Unstructured dual grid cells.

In general, the primary lattice is composed of arbitrary, convex, fitted polyhedra distributed throughout the volume. The secondary (or dual) lattice is built up of the closed polyhedra whose edges connect the centroids of adjacent primary cells, penetrating shared faces. Along each edge of the primary cells, an electric field vector is defined to be parallel to the edge and constant along its length. Similarly, a magnetic field vector is assumed to be parallel to each secondary-cell edge and constant along its length. Further, a magnetic flux density is assumed to be normal to each primary-cell face and constant over the face, and an electric flux density is similarly associated with each secondary-cell face.

A normalization is introduced for the electric and magnetic fields such that

$$\vec{e}_i = \frac{\vec{E}_i}{\ell_i \sqrt{\eta_0}}, \qquad \vec{h}_j = \frac{\vec{H}_j \sqrt{\eta_0}}{\ell_j} \tag{4.31}$$

where ℓ_i and ℓ_j are the edge lengths of the ith and jth primary and secondary lattice edges and η_0 is the free-space wave impedance. The flux densities are similarly normalized. Based on this discretization of the fields, Ampere's law (4.30a) is approximated by expressing the surface and line integrals over each primary cell face. The time derivative is then approximated using a central difference, leading to

$$d_i^{n+\frac{1}{2}} = \left(\frac{1 - \dfrac{c\Delta t \sigma_i \eta_0}{2\varepsilon_{r_i}}}{1 + \dfrac{c\Delta t \sigma_i \eta_0}{2\varepsilon_{r_i}}} \right) d_i^{n-\frac{1}{2}} + \left(\frac{\dfrac{c\Delta t \ell_i}{\varepsilon_{r_i} A_i}}{1 + \dfrac{c\Delta t \sigma_i \eta_0}{2\varepsilon_{r_i}}} \right) \sum_{j=1}^{N_{e_i}} v_j h_j^n \tag{4.32}$$

where $v_j = \pm 1$ (depending on the direction of h_j relative to the direction of the line integral); N_{e_i} is the number of edges bounding the ith face; d_i is the normalized flux density; c is the speed of light in vacuum; ℓ_i is the length of the ith primary edge passing through the face; A_i is the face area; ε_{r_i} and σ_i are the averaged relative permittivity and conductivity associated with the ith edge; and Δt is the time step resulting from the central-difference approximation of the time derivative. We note that the electric flux and magnetic field are staggered in time such that the time derivative is second-order accurate. Similarly, trapezoidal time integration is used to integrate the loss term. Implementing Faraday's law in a similar manner leads to

$$b_j^n = b_j^{n-1} - \frac{c\Delta t \ell_j}{A_j} \sum_{i=1}^{N_{e_j}} v_i e_i^{n-\frac{1}{2}} \tag{4.33}$$

Before (4.32) and (4.33) can be applied, the flux densities that are normal to their associated faces must be mapped onto the corresponding dual edges passing through their faces. Consequently, a projection operator similar to the contravariant-to-unique-covariant projection described in Section 4.2 must be introduced. However, because the grid is unstructured, a curvilinear coordinate system cannot be uniquely defined for each face. Madsen's interpolation scheme [11] addresses this problem. Referring to Fig. 4.5, consider the jth face of the primary grid with unit normal vector $\hat{n}_j$.

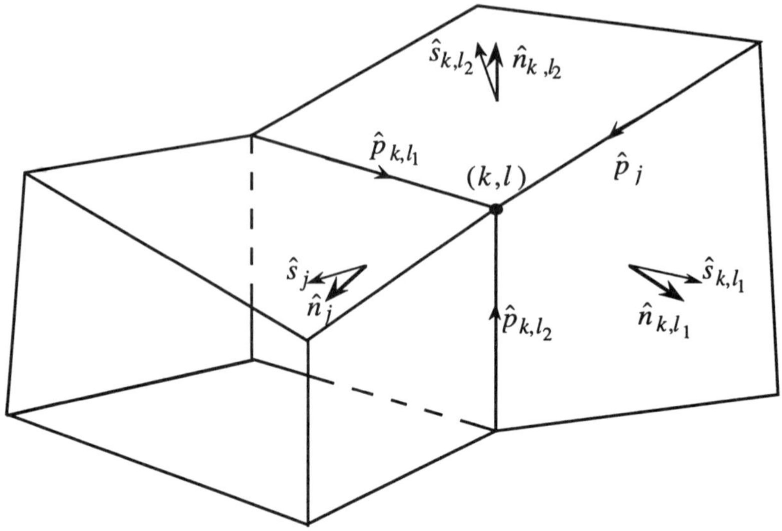

Fig. 4.5 Adjacent primary lattice cells sharing a common face with the magnetic fields normal to each face known.

The jth face is assumed to be bounded by N_{e_j} edges that connect N_{e_j} vertices (in Fig. 4.5, $N_{e_j} = 4$). Further, the face is uniquely shared by two cells unless it is on a Dirichlet or Neumann boundary.

We observe in Fig. 4.5 that the kth vertex $(k = 1, N_{e_j})$ of the jth face is shared by three faces of the lth cell ($l = 1$ or 2). It is assumed that (4.33) has been used to update the discrete normal magnetic flux densities. Then, the magnetic flux density associated with the kth vertex and the lth cell is computed by solving the following 3×3 system of equations:

$$\begin{aligned} \vec{B}_{k,l} \cdot \hat{n}_j &= \vec{B} \cdot \hat{n}_j = b_j \\ \vec{B}_{k,l} \cdot \hat{n}_{k,l_1} &= \vec{B} \cdot \hat{n}_{k,l_1} \\ \vec{B}_{k,l} \cdot \hat{n}_{k,l_2} &= \vec{B} \cdot \hat{n}_{k,l_2} \end{aligned} \tag{4.34}$$

where $\hat{n}_{k,l_1}$ and $\hat{n}_{k,l_2}$ are the normal vectors to each of the adjacent primary-cell faces. Since the right-hand sides are all known from (4.33), we use (4.34) to solve for the three orthogonal components of $\vec{B}_{k,l}$. Subsequently, this is performed for each of the vertices of the face ($k = 1, N_{e_j}$) and for each cell ($l = 1, 2$) shared by the face. At this point $\vec{B}_{k,l}$ is not an interpolation for the total field, but rather a local value associated with the

(k, l)th corner shared by the face. The magnetic flux density vector over the face is then expressed by the interpolation of the local field values as

$$\vec{B} = \frac{\sum_{l=1}^{2} \sum_{k=1}^{N_{ej}} \left| w_{k,l} \right| \vec{B}_{k,l}}{\sum_{l=1}^{2} \sum_{k=1}^{N_{ej}} \left| w_{k,l} \right|} \tag{4.35}$$

where $w_{k,l}$ are weighting factors to be determined. Finally, $\vec{B}$ is projected onto the dual edge with unit vector $\hat{s}_j$ along its length through a dot product (i.e., $\vec{B} \cdot \hat{s}_j$). Then, h_j is derived from (4.35) through the constitutive relationship. This leads to

$$\vec{h}_j = \frac{\vec{B} \cdot \hat{s}_j}{\mu_{r_j}} = \frac{\sum_{l=1}^{2} \sum_{k=1}^{N_{ej}} \left| w_{k,l} \right| \vec{B}_{k,l} \cdot \hat{s}_j}{\mu_{r_j} \sum_{l=1}^{2} \sum_{k=1}^{N_{ej}} \left| w_{k,l} \right|} \tag{4.36}$$

Employing simple algebra, it can be shown from (4.34) and (4.36) that

$$\begin{aligned} \vec{B}_{k,l} \cdot \hat{s}_j &= \frac{\hat{s}_j \cdot (\hat{n}_{k,l_1} \times \hat{n}_{k,l_2})}{\hat{n}_j \cdot (\hat{n}_{k,l_1} \times \hat{n}_{k,l_2})} \vec{B} \cdot \hat{n}_j + \frac{\hat{s}_j \cdot (\hat{n}_{k,l_2} \times \hat{n}_j)}{\hat{n}_j \cdot (\hat{n}_{k,l_1} \times \hat{n}_{k,l_2})} \vec{B} \cdot \hat{n}_{k,l_1} \\ &+ \frac{\hat{s}_j \cdot (\hat{n}_j \times \hat{n}_{k,l_1})}{\hat{n}_j \cdot (\hat{n}_{k,l_1} \times \hat{n}_{k,l_2})} \vec{B} \cdot \hat{n}_{k,l_2} \end{aligned} \tag{4.37}$$

This allows for a simple computation of the projection coefficients.

It was originally proposed by Madsen [11] to equate the weights $w_{k,l}$ to the triple scalar product of the normal area vectors of the (k, l)th corner. Specifically:

$$w_{k,l} = \vec{N}_j \cdot \left(\vec{N}_{k,l_1} \times \vec{N}_{k,l_2} \right) \tag{4.38}$$

where $\vec{N}_j = A_j \hat{n}_j$. This was also used by Gedney and Lansing [15]. This weight provides a geometric average when the local mesh is irregular. An alternate choice for $w_{k,l}$ advocated by Gedney and Roden [28] is to simply choose $w_{k,l} = 1$. This is discussed further in Section 4.5. A similar projection operator is derived for $\vec{e}_i$ based on a dual expression for (4.36).

At this point, it is instructive to compare the DSI method with the DN-FDTD method presented in Section 4.2.3. Specifically, let us compare (4.37) with (4.22). The $\vec{C}_i$ in

(4.23) are defined in (4.20) and are expressed by the cross products of the normal face vectors. Further, $\vec{A}_i$ passing through the face is aligned along the lattice edge. Thus, the two equations are directly analogous. The principal difference is that the unitary vectors $\vec{A}_i$ and $\vec{C}_i$ have the units of length. If the lattice is structured and nonorthogonal, the projections derived from (4.37) and (4.22) provide the same results for the computation of the discrete fields.

Interestingly, if the lattice is structured and regular but still nonorthogonal, the projections of the DSI and DN-FDTD methods are identical to those of the N-FDTD algorithm defined in (4.16a) through (4.16c). This occurs since $\vec{C}_i = \vec{A}_i$. Hence, from (4.23), $\tilde{g}_{i,j} = g_{i,j}$. Further, the weights defined in (4.25) and (4.38) are all equal. This property was also indirectly recognized by Xiao and Yabe, who realized that the DSI and N-FDTD methods have the same dispersion properties on a uniform skewed grid [29]. Based on these analogies, we see that the DSI method is simply a generalization of N-FDTD, in fact an extension of the DN-FDTD algorithm to unstructured meshes.

4.4 NUMERICAL STABILITY

In general, the N-FDTD and DSI methods are posed as a coupled set of difference equations expressed as a discrete linear operator. Specifically, the explicit update scheme derived using either of these techniques can be expressed as

$$d^{n+\frac{1}{2}} = d^{n-\frac{1}{2}} + \Delta t \, C_h \, A_b \, b^n \tag{4.39a}$$

$$b^n = b^{n-1} - \Delta t \, C_e \, D_\varepsilon \, A_d \, d^{n-\frac{1}{2}} \tag{4.39b}$$

where

- d and b are the discrete vector flux densities associated with the primary and secondary grid faces, respectively;
- C_e and C_h represent the discrete line integrals of the electric and magnetic fields about the primary and secondary cell faces, respectfully, weighted by the coefficients observed in (4.13) and (4.14) for the N-FDTD and DN-FDTD algorithms, or in (4.32) and (4.33) for the DSI algorithm;
- A_b and A_d represent the contravariant-to-covariant projections as derived for the magnetic and electric fields, respectively, from (4.16) for the N-FDTD algorithm, from (4.22) for the DN-FDTD algorithm, and from (4.36) for the DSI algorithm;
- D_ε is a diagonal matrix with entries representing the inverse of the relative permittivity assigned to the primary grid edge it is associated with (a lossless space is assumed here for simplicity).

The coupled difference equations in (4.39) are explicit in nature and only conditionally stable. To derive the numerical stability relation, these equations are re-posed as a single first-order difference equation. To this end, first substitute (4.39b) into (4.39a). Then, introducing the vector

$$w^n = \begin{bmatrix} b^n \\ d^{n+\frac{1}{2}} \end{bmatrix}, \qquad w^{n-1} = \begin{bmatrix} b^{n-1} \\ d^{n-\frac{1}{2}} \end{bmatrix} \tag{4.40}$$

(4.39) is re-posed as a first-order difference equation

$$w^n = G w^{n-1} \tag{4.41}$$

where

$$G = \begin{bmatrix} I & -\Delta t\, C_e\, D_\varepsilon\, A_d \\ \Delta t\, C_h\, A_b & I - (\Delta t)^2\, C_h\, A_b\, C_e\, D_\varepsilon\, A_d \end{bmatrix} \tag{4.42}$$

By definition, difference equation (4.41) is stable if it produces a bounded solution given any bounded input, and it is unstable if it produces an unbounded solution given a bounded input. Let w^0 be the initial condition or input into our system. Then, from (4.41), we have

$$w^M = [G]^M\, w^0 \tag{4.43}$$

where $[G]^M$ is G raised to the Mth power. To ensure stability for a passive linear system, the condition

$$\lim_{M\to\infty} \left\| w^M \right\| \le \left\| w^0 \right\| \tag{4.44a}$$

must be satisfied. This is true if

$$\left\| G \right\| \le 1 \tag{4.44b}$$

For this condition to hold, we must have $|\lambda(G)| \le 1$, or that all of the eigenvalues of G must lie within or on the unit circle. It can be shown that $|\lambda(G)| \le 1$ if [30]

$$\left| \lambda \left[(\Delta t)^2\, C_h\, A_b\, C_e\, D_\varepsilon\, A_d \right] \right| \le 4 \tag{4.45}$$

and G is a normal matrix. Strict stability of the time-dependent update expressions in (4.39) requires

$$\Delta t < \frac{2}{\sqrt{\left| \lambda_{\max}(C_h\, A_b\, C_e\, D_\varepsilon\, A_d) \right|}} \tag{4.46}$$

(Note that the square root is neglected in [30].) To evaluate this stability criterion, we can efficiently compute the maximum eigenvalue of the matrix product using an iterative technique such as an Arnoldi method or the power method.

4.5 WELL-POSED NONORTHOGONAL FDTD METHODS

Stability relation (4.46) is a necessary but not a sufficient condition to ensure the strict stability of the coupled difference equations. The formulation must also be well posed [31]. If the formulation is not well posed, then it can be unconditionally unstable.

4.5.1 The Nature of a Well-Posed FDTD Formulation

To discuss the possibility that (4.41) is well posed, the matrix

$$M = C_h A_b C_e D_\varepsilon A_d \tag{4.47}$$

is defined. Here, we assume that the number of independent variables defining the magnetic field exceeds the number of degrees of freedom defining the electric field.[1] The following definitions regarding the coupled system of (4.39) are then stated. The system is:

(a) Symmetric hyperbolic if M is a Hermitian matrix;
(b) Strictly hyperbolic if the eigenvalues are real and distinct;
(c) Strongly hyperbolic if the eigenvalues are real and there exists a complete system of eigenvectors;
(d) Weakly hyperbolic if the eigenvalues are real.

Note that the system is ill posed if not at least strongly hyperbolic [31].

Consider the case where the computation domain and its boundaries are lossless. If the eigenvalues of M are real and distinct and the time step satisfies (4.46), the eigenvalues of G are all distributed on the unit circle and located in the right-half plane. This is expected since, without dissipation, the total energy in the system is unchanged with time. If (4.46) is not satisfied, then a finite number of eigenvalues of G are outside of the unit circle and on the real axis. This leads to rapid instability, as expected.

If M has any complex eigenvalues, they occur as conjugate eigenpairs since M is a real matrix. Here, one of the eigenvalues appears in quadrant I of the complex plane, and the other in quadrant IV. Each complex-conjugate eigenvalue pair maps into a conjugate pair of eigenvalues of G that are outside of the unit circle, and a conjugate pair of eigenvalues that are inside of the unit circle. This is true for all Δt satisfying (4.46), providing that $\Delta t > 0$. Reducing Δt draws the eigenvalues closer to the unit circle but never onto it. The system is thus ill posed and unconditionally unstable, and exhibits an instability that typically arises in the late time with spectral content near the Nyquist rate.

---

[1] If this is not satisfied, one can easily re-pose (4.41) and define $M = C_e D_\varepsilon A_d C_h A_b$.

If the system is weakly hyperbolic, then it is ill conditioned and can exhibit instabilities. In such a case, the eigenvalue analysis in Section 4.4 is not sufficient to determine stability. Upon performing a pseudospectral analysis, it can be shown that an ill-conditioned system can be unstable despite its eigenvalues being on the unit circle. This is due to the sensitivity of the system to small perturbations [32, 33].

The questions that remain are:

Under what conditions does M have complex eigenvalues?

When does M have distinct real eigenvalues and a complete set of eigenvectors?

Initially, consider the simplest case of a dual, orthogonal, cubic lattice ($\Delta x = \Delta y = \Delta z$). Here, M is a real symmetric matrix. Consequently, the system is symmetric hyperbolic and well posed. For a noncubic grid, M is not symmetric. However, $M = C_h C_e$ can be made symmetric through a diagonal transformation, $P^{-1/2} M P^{1/2}$. Hence, we see that the classic FDTD method is strongly hyperbolic.

Next, assume that the mesh is unstructured and nonorthogonal. For the sake of illustration, neglect the projections, A_b and A_d, such that $M = C_h C_e D_\varepsilon$. Due to the complementary nature of the line integrals over the dual grid, the system is strongly hyperbolic, even for inhomogeneous media. While this formulation is only first-order accurate for general grids, it provides useful insight to the more general formulation.

Finally, for the most general case where the mesh is irregular and nonorthogonal and the projection operators are included as in (4.47), M is nonsymmetric. Here, because M is the product of matrices, it may not be symmetric even if $C_h C_e D_\varepsilon$ is symmetrizable. If matrices A_b and A_d are nonsymmetric, M likely has complex eigenvalues. As proposed in Sections 4.2 and 4.3, the N-FDTD and DSI projection operators, A_d and A_b, are nonsymmetric if the mesh is irregular and nonorthogonal. Consequently, both formulations potentially suffer from late-time instabilities, as demonstrated by Roden [24].

If we can formulate a technique based on DSI / GY or N-FDTD algorithms that leads to M having real and distinct eigenvalues, then the problem can be recast in a well-posed manner. We observe that if A_d and A_b are symmetric and well conditioned, and if ε_r is constant such that $D_\varepsilon = \varepsilon_r I$, then M has real and distinct eigenvalues. (What defines the ill-conditioned nature of A_d and A_b is described in the next section.) Further, if A_d and A_b are symmetric, the reactions between neighboring fields due to the projection operators are equivalent.

Unfortunately, if ε_r is inhomogeneous, this may introduce complex eigenvalues in M even if A_d and A_b are symmetric. This is principally due to the nonreciprocal nature that results from the projections (e.g., $D_\varepsilon A_d \neq A_d D_\varepsilon$). An effective way to force symmetry in the projection operator is to approximate (4.39b) as

$$b^n = b^{n-1} - \Delta t\, C_e D_\varepsilon^{1/2} A_d D_\varepsilon^{1/2} d^{n-1/2} \tag{4.48}$$

Then, G is rewritten as

$$G = \begin{bmatrix} I & -\Delta t\, C_e\, D_\varepsilon^{1/2}\, A_d\, D_\varepsilon^{1/2} \\ \Delta t\, C_h\, A_b & I - (\Delta t)^2\, C_h\, A_b\, C_e\, D_\varepsilon^{1/2}\, A_d\, D_\varepsilon^{1/2} \end{bmatrix} \tag{4.49}$$

and

$$M \approx M_s = C_h\, A_b\, C_e\, D_\varepsilon^{1/2}\, A_d\, D_\varepsilon^{1/2} \tag{4.50}$$

We realize that M_s is an approximation. To understand the impact of this approximation, let us look at its mathematical interpretation. The projection operation $e = D_\varepsilon^{1/2}\, A_d\, D_\varepsilon^{1/2}\, d$ maps the normal flux density to the field intensity projected on the edge passing through the face. Examining the ith entry of e, we observe

$$e_i = \frac{a_{d_{i,i}}\, d_i}{\varepsilon_{r_i}} + \sum_{j=1}^{M} \frac{a_{d_{i,j}}\, d_j}{\sqrt{\varepsilon_{r_i}\, \varepsilon_{r_j}}} \tag{4.51}$$

where $a_{d_{i,j}}$ is the entry of matrix A_d in the ith row and jth column. The diagonal term of the projection operation is unchanged by the use of M_s, whereas the off-diagonal terms are normalized by an effective permittivity expressed as the geometric mean of the permittivities of the two adjacent edges. It is demonstrated in Section 4.5.4 that this approximation does not degrade the accuracy of the N-FDTD or DSI/GY algorithms, yet it leads to a well-posed formulation.

4.5.2 Well-Posed Nonorthogonal FDTD Algorithm

The N-FDTD algorithm was presented in Section 4.2 with two projection strategies. In the first G^{avg} strategy, metrics are defined based on the vectors that constitute the faces of the primary and secondary grids. In the second strategy, the projections constituting the DN-FDTD technique are formulated from corner-based curvilinear-coordinate constructs. Due to differences in projection techniques, the analysis and improvement of the well-posed nature of these two approaches also differs. The well-posed nature of the DN-FDTD method, previously shown to be equivalent to the DSI family of techniques, is investigated in Section 4.5.3. In this section, the well-posed nature of the G^{avg} strategy is investigated, and a well-posed curvilinear-coordinate-based projection scheme is subsequently developed.

As discussed in Section 4.5.1, a necessary condition for a well-posed N-FDTD algorithm is a symmetric projection matrix, $G_{i,j}$. Specifically, each field involved in the projection must have a reciprocal relationship. In the G^{avg} method, the projections between two curvilinear coordinate components, i and j, are computed independently using two different sets of dual lattice vectors. Therefore, these projections are not symmetric for an irregular lattice. Further, the symmetry conversion of G^{avg} in the form previously presented is not easily accomplished because each off-diagonal projection operates on four separate contravariant field values. Consequently, an alternative projection strategy is desired in order to easily achieve a well-posed problem.

It was also demonstrated in Section 4.5.1 that the well-posed nature of any projection method is destroyed by improper implementation of inhomogeneous regions. Subsequently, it was demonstrated that this ill-posed condition can be corrected by applying a geometric-averaging scheme to the off-diagonal tensor elements. Roden recently introduced a simple, accurate, and well-posed projection scheme for this purpose [24], referred to here as the G^{sym} method. This method assumes that the primary and dual meshes are locally regular. Under this assumption, these meshes are locally equivalent. Consequently, the projections for these lattices can be defined independently.

The covariant field components for the primary mesh are computed by projecting each field component that appears in (4.9) individually. These projections are accomplished entirely from the edge vectors of the primary mesh. Likewise, the covariant fields for the secondary mesh are computed using only its edge vectors. Using this technique, the projections that appear in (4.16) are compactly stated as

$$E_m = \frac{g_{m,m}D^m}{\varepsilon_{r_m}} + \sum_{p=1}^{8} \frac{G_{m,p}D^p}{\sqrt{\varepsilon_{r_m}\varepsilon_{r_p}}} \tag{4.52a}$$

where

$$G_{m,p} = \frac{1}{4}\vec{A}_m \cdot \vec{A}_p \tag{4.52b}$$

and the index p scans through the eight adjacent faces having the contravariant flux's D^p. For example, the projection of the electric field onto the primary-mesh edge $\vec{A}_1$ is computed from (4.52) as

$$E_1\Big|_{i,j,k}^{n+\frac{1}{2}} = \vec{A}_1\Big|_{i,j,k} \cdot \vec{A}_1\Big|_{i,j,k} \frac{D^1\Big|_{i,j,k}^{n+\frac{1}{2}}}{\varepsilon_{r_{i,j,k}}} + \frac{\vec{A}_1\Big|_{i,j,k}}{4\sqrt{\varepsilon_{r1_{i,j,k}}}} \times$$
$$\left(\frac{\vec{A}_2\Big|_{i,j,k} D^2\Big|_{i,j,k}^{n+\frac{1}{2}}}{\sqrt{\varepsilon_{r2_{i,j,k}}}} + \frac{\vec{A}_2\Big|_{i+1,j,k} D^2\Big|_{i+1,j,k}^{n+\frac{1}{2}}}{\sqrt{\varepsilon_{r2_{i+1,j,k}}}} + \frac{\vec{A}_2\Big|_{i,j-1,k} D^2\Big|_{i,j-1,k}^{n+\frac{1}{2}}}{\sqrt{\varepsilon_{r2_{i,j-1,k}}}} + \frac{\vec{A}_2\Big|_{i+1,j-1,k} D^2\Big|_{i+1,j-1,k}^{n+\frac{1}{2}}}{\sqrt{\varepsilon_{r2_{i+1,j-1,k}}}} + \right.$$
$$\left. \frac{\vec{A}_3\Big|_{i,j,k} D^3\Big|_{i,j,k}^{n+\frac{1}{2}}}{\sqrt{\varepsilon_{r3_{i,j,k}}}} + \frac{\vec{A}_3\Big|_{i-1,j,k} D^3\Big|_{i-1,j,k}^{n+\frac{1}{2}}}{\sqrt{\varepsilon_{r3_{i-1,j,k}}}} + \frac{\vec{A}_3\Big|_{i,j,k-1} D^3\Big|_{i,j,k-1}^{n+\frac{1}{2}}}{\sqrt{\varepsilon_{r3_{i,j,k-1}}}} + \frac{\vec{A}_3\Big|_{i+1,j,k-1} D^3\Big|_{i+1,j,k-1}^{n+\frac{1}{2}}}{\sqrt{\varepsilon_{r3_{i+1,j,k-1}}}} \right) \tag{4.53}$$

Each of the other projections is computed in this same straightforward manner. We note that the computational burden of this method is somewhat higher than the original formulation of Lee [10] since $G_{m,p}$ is unique for each of the nine field components appearing on the right-hand side of (4.52a), and individual multiplications are required.

4.5.3 Well-Posed Discrete Surface Integral Algorithm

The projection matrices resulting from the DSI and DN-FDTD algorithms are made symmetric in a similar manner as the N-FDTD algorithm. The normal-to-edge projections of the fields are provided in (4.36) for the DSI algorithm. Recall that the contribution of each local field component is founded on a local interpolation based on the corners. Referring to Fig. 4.5, assume that we wish to compute the contribution to the projection onto the edge defined by the unit vector $\hat{s}_j$ (passing through a face having the normal $\hat{n}_j$) from the corner involving the faces having the normals $\hat{n}_j$, $\hat{n}_{k,l_1}$, and $\hat{n}_{k,l_2}$. This contribution is given by (4.37) as

$$\begin{aligned}\vec{B}_{k,l}\cdot\hat{s}_j &= \frac{\hat{s}_j\cdot(\hat{n}_{k,l_1}\times\hat{n}_{k,l_2})}{\hat{n}_j\cdot(\hat{n}_{k,l_1}\times\hat{n}_{k,l_2})}\vec{B}\cdot\hat{n}_j + \frac{\hat{s}_j\cdot(\hat{n}_{k,l_2}\times\hat{n}_j)}{\hat{n}_j\cdot(\hat{n}_{k,l_1}\times\hat{n}_{k,l_2})}\vec{B}\cdot\hat{n}_{k,l_1} \\ &+ \frac{\hat{s}_j\cdot(\hat{n}_j\times\hat{n}_{k,l_1})}{\hat{n}_j\cdot(\hat{n}_{k,l_1}\times\hat{n}_{k,l_2})}\vec{B}\cdot\hat{n}_{k,l_2}\end{aligned} \tag{4.54}$$

which is reprinted here for convenience. In conjunction with (4.36), this provides a partial contribution to the row entries of A_b. The first term contributes to a diagonal entry of A_b and the second two terms contribute to off-diagonal entries. Suppose that one projects onto the edge $\hat{s}_{k,l_1}$ passing through the face having the normal $\hat{n}_{k,l_1}$. When interpolating using (4.37), the same corner would be involved. The off-diagonal term contributing to $\vec{B}\cdot\hat{s}_{k,l_1}$ due to the flux $\vec{B}\cdot\hat{n}_j$ is

$$\frac{\hat{s}_{k,l_1}\cdot(\hat{n}_{k,l_1}\times\hat{n}_{k,l_2})}{\hat{n}_j\cdot(\hat{n}_{k,l_1}\times\hat{n}_{k,l_2})}\vec{B}\cdot\hat{n}_j. \tag{4.55}$$

This is the reciprocal term of the second term on the right-hand side of (4.54). As expected, the two terms share the same denominator, which is the volume of the parallelepiped bound by the three unit normal vectors. The numerators are the scalar triple products of the edge vector with the two adjacent normal vectors. An interesting observation is that

$$(\hat{n}_{k,l_1}\times\hat{n}_{k,l_2}) = \vec{p}_j, \quad (\hat{n}_{k,l_2}\times\hat{n}_j) = \vec{p}_{l,k_1}, \quad (\hat{n}_j\times\hat{n}_{k,l_1}) = \vec{p}_{l,k_2} \tag{4.56}$$

where the $\vec{p}$ are vectors parallel to the edges of the primary grid, as illustrated in Fig. 4.5. If the mesh is truly reciprocal, then the primary and secondary grids share the same curvilinear coordinates (i.e., $\hat{p}_j = \hat{s}_j$). This leads to a symmetric system. However, for an unstructured and irregular grid, this is only true for specialized discretizations.

The issue now is how to enforce symmetry in the projections without sacrificing accuracy. An effective means found is to average the off-diagonal terms of A_b and A_d:

$$a'_{b_{i,j}} = \frac{a_{b_{i,j}} + a_{b_{j,i}}}{2} \tag{4.57}$$

To investigate the mathematical significance of this averaging scheme, we see that the second term in (4.54) is modified as

$$\frac{1}{2}\left[\frac{\hat{s}_j \cdot (\hat{n}_{k,l_2} \times \hat{n}_j)}{\hat{n}_j \cdot (\hat{n}_{k,l_1} \times \hat{n}_{k,l_2})} + \frac{\hat{s}_{k,l_1} \cdot (\hat{n}_{k,l_1} \times \hat{n}_{k,l_2})}{\hat{n}_j \cdot (\hat{n}_{k,l_1} \times \hat{n}_{k,l_2})}\right] =$$

$$\frac{1}{2\hat{n}_j \cdot (\hat{n}_{k,l_1} \times \hat{n}_{k,l_2})}\left(\hat{s}_j \cdot \vec{p}_{k,l_1} + \hat{s}_{k,l_1} \cdot \vec{p}_j\right) \tag{4.58}$$

The third term in (4.54) is modified similarly. In essence, this averages the projections between the two dual coordinate systems. If the grid is regular and nonorthogonal, the averaged formulation reduces to the original algorithm in (4.54) since $\hat{p}_i = \hat{s}_i$.

Finally, consider the weights $w_{k,l}$ in (4.36). In the original formulation, Madsen proposed to use the triple scalar product $w_{k,l} = \vec{N}_j \cdot (\vec{N}_{k,l_1} \times \vec{N}_{k,l_2})$, where the $\vec{N}$ are the normal area vectors (i.e., $\vec{N}_j = A_j \hat{n}_j$ and A_j is the primary cell face area). The $w_{k,l}$ are identical for the corner shared by the two projections. However, the sum of the weights is not the same for an irregular grid, leading to asymmetry. Through experimentation, it has been found that choosing $w_{k,l} = 1$ is a sufficient and accurate weighting.

The projection matrix A_d can be made symmetric in an identical manner based on the secondary cell. Once the symmetric matrices A_b and A_d are constructed, the update process is carried out using (4.39a) and (4.48). This leads to a well-posed formulation.

4.5.4 Validations

Homogeneous Cavity Model

To demonstrate the above theory, an eigenvalue analysis of a simple problem is presented wherein an irregular unstructured mesh is generated to model a homogeneous cavity. Fig. 4.6 illustrates a cross section of the mesh. Because only TM_Z modes are excited by a vertical electric current source, it is sufficient to simulate the transverse magnetic fields and the vertical electric field. This is done to reduce the order of G, which undergoes a full eigenvalue analysis.

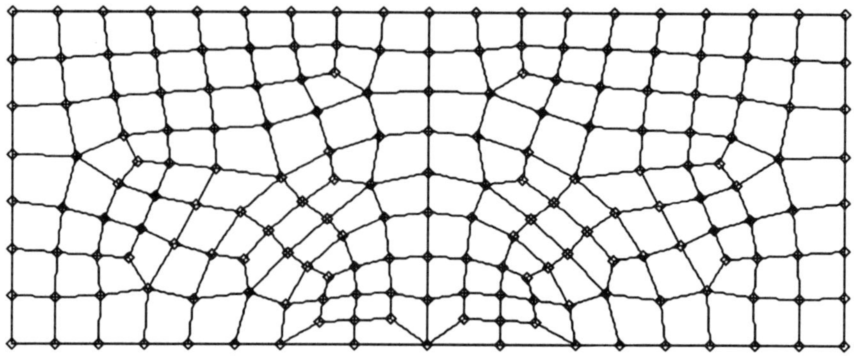

Fig. 4.6 Cross section of the small hexahedral mesh (primary grid) used for the eigenvalue analysis.

Matrices C_e and C_h are constructed according to (4.32) and (4.33), respectively. Projection matrix A_b is constructed using (4.36) to (4.38), leading to an asymmetric sparse matrix. A_b is also constructed using (4.57) and (4.58), leading to a symmetric matrix referred to as A_b^s. Note that for this example, A_d is an identity matrix since the transverse electric fields are assumed to be zero. Matrices $M = C_h A_b C_e$ and $M^s = C_h A_b^s C_e$ are explicitly constructed and their eigenvalues are computed using a numerical eigenvalue routine. In this case, the eigenvalues of M^s are found to be real numbers, and the set of eigenvalues of M contains three complex-conjugate pairs. This implies that this system is not well posed.

Matrix G is constructed using A_b and (4.42), and G^s is constructed using A_b^s and (4.49). In both cases, the assumed time step satisfies the stability criterion of (4.46) with a margin of 10%. Fig. 4.7 locates the eigenvalues of G and G^s in the right half of the complex plane. As expected for this lossless system, the eigenvalues lie on the unit circle. However, a closer look reveals that a total of 12 eigenvalues of G occur in six complex-conjugate pairs off the unit circle. As listed in Table 4.1, three of these pairs lie inside of the unit circle, and three pairs lie outside the unit circle. Because of the existence of eigenvalues outside of the unit circle, this system ultimately goes unstable. We note that even if the time step is made progressively smaller, these eigenvalues drift closer to the unit circle but always remain outside it. Further, the addition of substantial conductive loss into the domain does not solve this problem, since the nonsymmetric case retains eigenvalues outside of the unit circle.

As indicated by the discussion of Section 4.5.1, G is ill-posed since M has complex eigenvalues. Because these eigenvalues lie very close to the unit circle, the instability occurs in the late time. For this small example, the nonsymmetric case remains stable for about 8,000 time steps. While this is well into the steady state, the late-time solution is eventually corrupted. On the other hand, G^s is stable for over 250,000 time steps and shows no signs of impending instability.

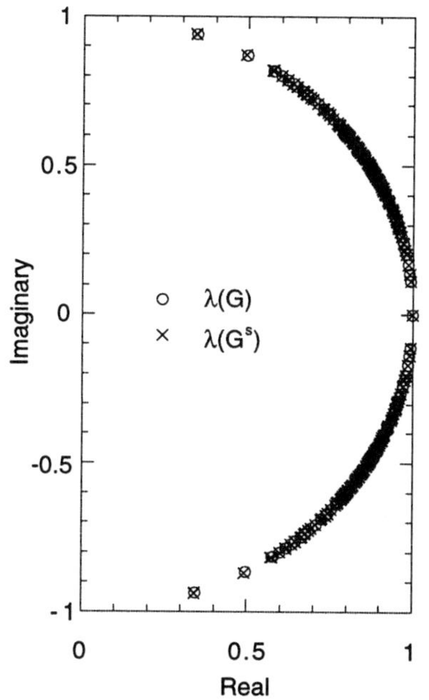

Fig. 4.7 Eigenvalues of G and G^s plotted in the complex plane for the mesh of Fig. 4.6.

TABLE 4.1

Complex-Conjugate Eigenvalue Pairs of G Off of the Unit Circle

$\lambda(G)$	$\lvert\lambda(G)\rvert$	$\angle\lambda(G)$
$0.8194 \pm j0.5736$	1.00020	$\pm 34.9952^\circ$
$0.8190 \pm j0.5734$	0.99980	$\pm 34.9952^\circ$
$0.8813 \pm j0.4735$	1.00049	$\pm 28.2490^\circ$
$0.8805 \pm j0.4731$	0.99951	$\pm 28.2490^\circ$
$0.8769 \pm j0.4821$	1.00067	$\pm 28.8021^\circ$
$0.8757 \pm j0.4815$	0.99932	$\pm 28.8021^\circ$

We note that posing A_b as symmetric is not a necessary and sufficient condition for well-posedness, as pointed out in Section 4.5.1. If A_b is poorly conditioned, then M might not have a complete set of eigenvectors, and in fact could have some complex eigenvalues. Either case leads to ill-posedness. It has been observed that if an interior angle of a grid cell is very small (near 0°) or highly oblique (near 180°), then A_b is poorly conditioned. Upon reviewing the unitary basis introduced for the N-FDTD algorithm, it is seen that the local coordinate system for this case becomes singular because the volume in (4.2) tends to zero. For the DSI algorithm, the triple scalar product in (4.34) tends to zero for this case. In practice, we find that meshes having cell interior angles smaller than about 8° or greater than about 172° can experience late-time instability. The remedy for this is to either modify the mesh or to locally alter the projection scheme.

Cavity Resonance Problem

To test the accuracy of the symmetric-projection operators, we consider the circular dielectric ring in a rectangular PEC cavity reported in [34] and shown in Fig. 4.8. Both unstructured and structured grids are used to model this problem using the DSI and N-FDTD methods, respectively. Fig. 4.9 shows a cross section of the unstructured grid.

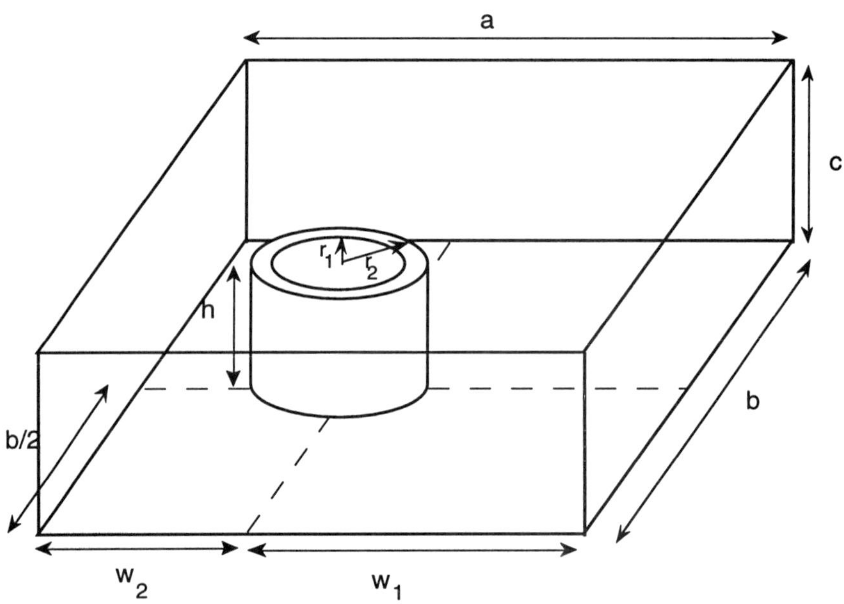

Fig. 4.8 Dielectric ring in a rectangular PEC cavity (a = 324 mm, b = 121 mm, c = 43 mm). The ring is centered at (w_1 = 207.25 mm, w_2 = 116.75 mm) in the x direction and $b/2$ in the y direction and rests on the ground plane. The ring has a height h = 39 mm, inner radius r_1 = 16.65 mm, outer radius r_2 = 26.75 mm, and ε_r = 2.06.

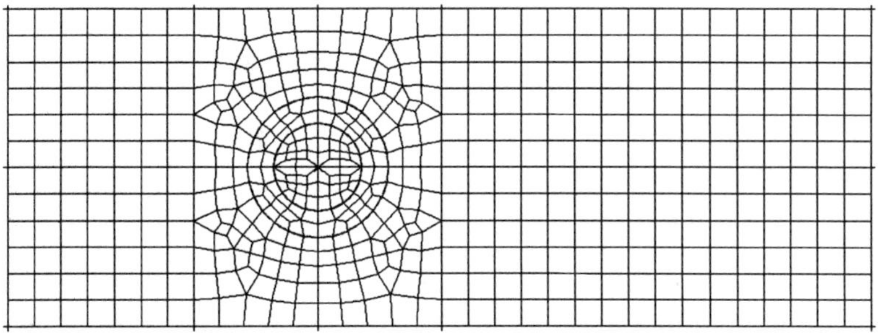

Fig. 4.9 Cross section of the hexahedral mesh (primary grid) used to model the fields within the cavity of Fig. 4.8.

The fields in the resonant cavity are simulated using: the G^{avg} and G^{sym} N-FDTD algorithms based on a structured grid; the original DSI algorithm presented in Section 4.3; and the well-posed algorithm presented in Sections 4.5.1 and 4.5.3. The cavity is excited by injecting the current density $\vec{J}(t) = -\hat{z}\,[2(t-t_0)/t_w]\,\exp[-(t-t_0)^2/t_w^2]$ at a nonsymmetric position, where $t_w = 0.2122$ ns and $t_0 = 3t_w$. The vertical field in the cavity is recorded over 25,000 time steps ($\Delta t = 4.5$ ps) and analyzed using an FFT to extract the resonant frequencies.

Table 4.2 lists the resonant frequencies for the first four cavity modes computed using the well-posed DSI and G^{sym} N-FDTD methods, and compares these results to data obtained using an implicit finite-element time-domain (FETD) method [35]. The measured resonant frequency of the dominant mode is also listed [34]. Excellent agreement within 0.8% is observed in all comparisons. Upon applying the nonsymmetric DSI and G^{avg} methods, the calculated resonant frequencies agree to within 0.1% compared with the well-posed methods.

TABLE 4.2

Resonant Frequencies of the First Four Cavity Modes of the Structure of Fig. 4.8

Mode	Measured (GHz)	Well-Posed DSI (GHz)	G^{sym} N/FDTD (GHz)	Implicit FETD (GHz)
k01	1.258	1.258	1.258	1.259
k02		1.509	1.509	1.512
k03		1.836	1.835	1.841
k04		2.158	2.161	2.175

In this example, the nonsymmetric DSI method runs for 30,000 time steps before going unstable. The symmetric DSI method runs for 250,000 time steps and shows no signs of instability.

4.6 NUMERICAL DISPERSION

The usefulness of irregular-grid techniques such as the N-FDTD and DSI/GY methods is based upon the ability of their discrete meshes to conform to nonseparable boundaries. While these methods reduce boundary-discretization error, it is understood that additional dispersive error can be introduced due to the irregularity of the lattice. One way to quantify this dispersive error is to analyze the propagation speed of a uniform plane wave within a uniform nonorthogonal mesh. This is done by deriving a dispersion relation for the associated discrete equations. Such a study provides useful insights into the convergence properties of these techniques, as well as the effects of the grid geometry. We note, however, that this type of analysis is highly controlled, and its extension to arbitrary irregular gridding is somewhat limited.

The numerical dispersion of these techniques has been studied by a number of researchers [29, 36–38]. For a uniform grid, the N-FDTD and DSI methods yield exactly the same update equations, as observed in Section 4.2. This was also deduced in [29] from the dispersion relations for these methods. In fact, the well-posed DSI algorithm and the G^{avg} and G^{sym} N-FDTD methods have the same dispersion relation for a uniform grid because their projection matrices are symmetric. Consequently, the dispersion relation need only be derived for one of the methods as it is applicable to the other methods for a uniform grid. The DSI technique is used here for this purpose. Distinctions made for irregular grids are derived based on the numerical examples in Section 4.7.

4.6.1 Dispersion Analysis Based on a Uniform Skewed Grid

The discrete dispersion relation for the N-FDTD and DSI/GY methods is based on a spectral analysis of the discrete form of Maxwell's equations. To this end, the study is based on a uniformly skewed two-dimensional lattice, as illustrated in Fig. 4.10.

We begin the analysis by discretizing Maxwell's equations using the DSI method of Section 4.3 for the uniform skewed grid of interest. A TM_Z-polarized plane wave is assumed to be propagating through the mesh with the space-time dependence

$$\exp\left(j\omega t - j\tilde{k}_x x - j\tilde{k}_y y - j\tilde{k}_z z\right) \tag{4.59}$$

Assuming lossless, homogeneous free space, (4.32) and (4.33) are used to approximate Maxwell's equations in the discrete space of Fig. 4.10. Here, the transverse magnetic fields and fluxes are located at the edge centers and the vertical electric fields are located at the grid vertices. From (4.59) and (4.33), for TM_Z-polarized fields this leads to

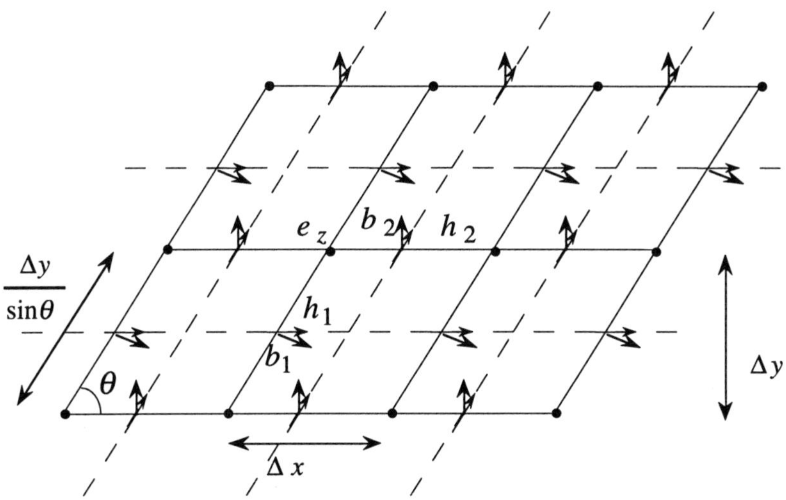

Fig. 4.10 Two-dimensional uniform skew grid with interior angle θ with vectors indicating the discrete normal flux densities and the edge-based field intensities. The dual grid is represented by the dashed lines.

$$b_1 = \frac{c\Delta t \sin\theta \Delta x}{\Delta y \Delta z} \frac{\sin\left[(\tilde{k}_x \cot\theta + \tilde{k}_y)(\Delta y/2)\right]}{\sin(\omega \Delta t/2)} e_z \tag{4.60a}$$

$$b_2 = -\frac{c\Delta t \Delta y}{\Delta x \Delta z \sin\theta} \frac{\sin\left(\tilde{k}_x \Delta x/2\right)}{\sin(\omega \Delta t/2)} e_z \tag{4.60b}$$

where b_1 and b_2 are the horizontal magnetic flux densities and the tilde denotes numerical wavenumbers. Similarly, from (4.32) we have

$$e_z = \frac{\Delta z}{\Delta y \Delta x} \frac{c\Delta t}{\sin(\omega \Delta t/2)} \left\{h_1 \sin\left[(\tilde{k}_x \cot\theta + \tilde{k}_y)(\Delta y/2)\right] - h_2 \sin\left(\tilde{k}_x \Delta x/2\right)\right\} \tag{4.61}$$

The normal-to-dual-edge projections are mapped into the spectral domain in a similar manner. From (4.34) to (4.37) it is found that

$$h_1 = b_1 \csc\theta + b_2 \cot\theta \cos\left(\tilde{k}_x \Delta x/2\right) \cos\left[(\tilde{k}_x \cot\theta + \tilde{k}_y)(\Delta y/2)\right] \tag{4.62a}$$

$$h_2 = b_2 \csc\theta + b_1 \cot\theta \cos\left(\tilde{k}_x \Delta x/2\right) \cos\left[(\tilde{k}_x \cot\theta + \tilde{k}_y)(\Delta y/2)\right] \tag{4.62b}$$

From (4.62), we observe that due to the uniformity of the grid, the projection matrix A_b is symmetric and this problem is well posed. Further, (4.62) is unchanged if $w_{i,j}$ are chosen to be unity or the triple scalar product of the normal area vectors.

The numerical dispersion relation is obtained by inserting (4.60a) and (4.60b) into (4.62a) and (4.62b), and inserting the results for h_1 and h_2 into (4.61). This leads to

$$\frac{\sin^2(\omega\Delta t/2)}{(c\Delta t)^2} = \frac{\sin^2\left(\tilde{k}_x\Delta x/2\right)}{(\Delta x\sin\theta)^2} + \frac{\sin^2\left[(\tilde{k}_x\cot\theta+\tilde{k}_y)(\Delta y/2)\right]}{(\Delta y)^2}$$
$$- \left(\frac{\Delta x\sin\theta}{\Delta y}+\frac{\Delta y}{\Delta x}\right)\left[\frac{\sin\left(\tilde{k}_x\Delta x/2\right)}{\Delta x}\right]\left\{\frac{\sin\left[(\tilde{k}_x\cot\theta+\tilde{k}_y)(\Delta y/2)\right]}{\Delta y}\right\}\times$$
$$\cos\left[\tilde{k}_x(\Delta x/2)\right]\cos\left[(\tilde{k}_x\cot\theta+\tilde{k}_y)(\Delta y/2)\right]\cos\theta \tag{4.63}$$

An identical expression results if one follows the techniques outlined in Section 4.2.

If the skew angle is 90°, (4.63) reduces to the numerical dispersion relation for an orthogonal grid [6]. Further, if we take the limit as Δx, $\Delta y/\sin\theta$, and Δt approach zero, then (4.63) reduces to the ideal dispersion relation

$$\begin{aligned}\left(\frac{\omega}{c}\right)^2 &= \frac{1}{\sin^2\theta}\left[\left(\tilde{k}_x\right)^2+\left(\tilde{k}_x\cos\theta+\tilde{k}_y\sin\theta\right)^2 - 2\tilde{k}_x\left(\tilde{k}_x\cos\theta+\tilde{k}_y\sin\theta\right)\cos\theta\right] \\ &= \frac{1}{\sin^2\theta}\left[\tilde{k}_x^2\sin^2\theta + \tilde{k}_y^2\sin^2\theta\right] = \tilde{k}_x^2+\tilde{k}_y^2\end{aligned} \tag{4.64}$$

From (4.63), we extract the numerical wavenumber from which the numerical wave velocity is computed. To this end, let

$$\tilde{k}_x = \tilde{k}\cos\phi, \qquad \tilde{k}_y = \tilde{k}\sin\phi \tag{4.65}$$

where ϕ is the angle of propagation in the grid relative to the x axis. After inserting (4.65) into (4.64), Newton's method is used to solve the nonlinear expression for $\tilde{k}$ as a function of ϕ and the mesh resolution and skew angle θ. Then, the numerical phase velocity of the discrete plane wave is computed via $\tilde{v}_p = \omega/\tilde{k}$.

Fig. 4.11 illustrates the normalized numerical phase velocity as ϕ varies from 0° to 180° for mesh skew angles $\theta = 90°$, 80°, 60°, and 45°. These results match N-FDTD dispersion data reported in [29] and DSI data reported in [38]. However, these results do not match the N-FDTD data of [38] or [36] since these formulations neglected the averaging of the contravariant terms. This was pointed out in [29]. We see from Fig. 4.11 that the numerical phase velocity depends upon the wave propagation angle, with the velocity anisotropy increasing with the skew of the mesh. For a mesh resolution $\Delta x = \Delta y/\sin\theta = \lambda/10$ (Fig. 4.11(a)), the anisotropy is on the order of 1% for 80° and 90° skew angles and over 7% for a skew angle of 45°. For a mesh resolution $\Delta x = \Delta y/\sin\theta = \lambda/20$ (Fig. 4.11(b)), the anisotropy drops by a factor of about 4:1.

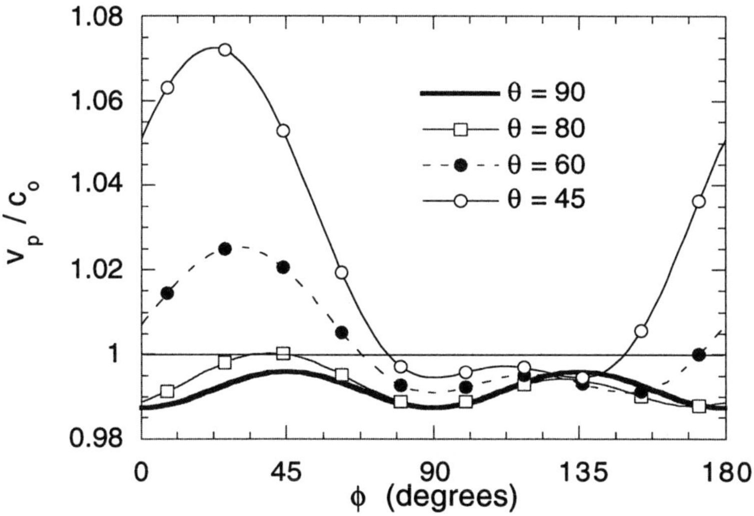

(a) $\Delta x = \Delta y / \sin\theta = \lambda/10$.

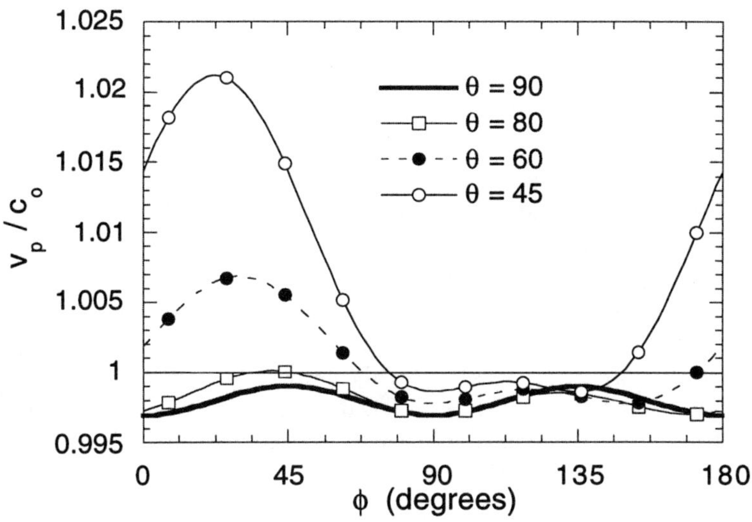

(b) $\Delta x = \Delta y / \sin\theta = \lambda/20$.

Fig. 4.11 Normalized numerical phase velocity versus propagation angle ϕ as the mesh skew angle θ in Fig. 4.10 decreases from 90° (orthogonal mesh) to 45°. For both cases, $c\Delta t = \Delta x/2$.

It can be shown for this case that the velocity anisotropy continues to drop as the mesh is refined, falling by another factor of 4:1 if the grid spacing is reduced to $\lambda/40$.

From this study, we conclude that, for the case of a uniform skewed mesh:

1. The numerical phase velocity converges in a second-order-accurate manner;
2. The mesh anisotropy increases with its skew;
3. To maintain a desired level of accuracy, a highly skewed mesh must be refined appropriately.

These results can be interpreted with respect to an irregular nonorthogonal grid in determining the quality of its mesh. It is now clear that the dispersion error is minimum if the grid is orthogonal. Thus, if one can keep the grid orthogonal throughout most of the volume and irregular only as needed near boundaries, the error can be reduced. In regions where the grid is nonorthogonal, one should ensure that the grid cells are small enough (typically $\leq \lambda/20$). Further, more local error occurs for highly oblique (long and thin) cells. This is to be expected as discussed in the previous section.

4.6.2 Example of Numerical Wave Speed

We now further investigate the analytical results derived in Section 4.6.1 by considering a canonical problem, a TEM mode propagating along a uniform microstrip transmission line. The microstrip has a width of 2.413 mm and a height of 0.795 mm above the ground plane. The substrate is assumed to be free space so that the wave speed is the free-space speed of light. Four different cases are considered to discern the effective wave speed within the numerical model, how the wave speed varies with grid skew, and how it is affected by boundary error. The four cases are summarized as follows:

Case 1. The microstrip is assumed to run parallel to the x axis so that it can be modeled using a Cartesian space lattice ($100 \times 48 \times 30$ cells; $\Delta x = \Delta y = 0.3447$ mm; $\Delta z = 0.265$ mm; and $\Delta t = 0.586$ ps). A 10-cell-thick PML ABC (see Chapter 5) is placed 10 cells from the sides and top of the microstrip and also terminates the ends of the microstrip.

Case 2. The microstrip is assumed to run at an angle of 45° relative to the x axis. It is modeled with the 45° skewed mesh shown in Fig. 4.12 that has the same dimensions as Case 1. This allows the grid to conform to the microstrip boundaries. The N-FDTD technique is used for the analysis, which for this grid is identical to the DSI method. Again, 10-cell-thick PML regions are used to terminate the exterior boundaries [39].

Case 3. The microstrip is placed in a Cartesian lattice at a 45° angle to the x axis and staircasing is used to approximate the microstrip edge. The same cell sizes and time step as Case 1 are used. However, the total grid size is much larger ($200 \times 246 \times 30$ cells) since the line is running diagonally across it. Further, the grid is extended to allow the microstrip to bend so that it enters the PML termination along the normal to the PML interface.

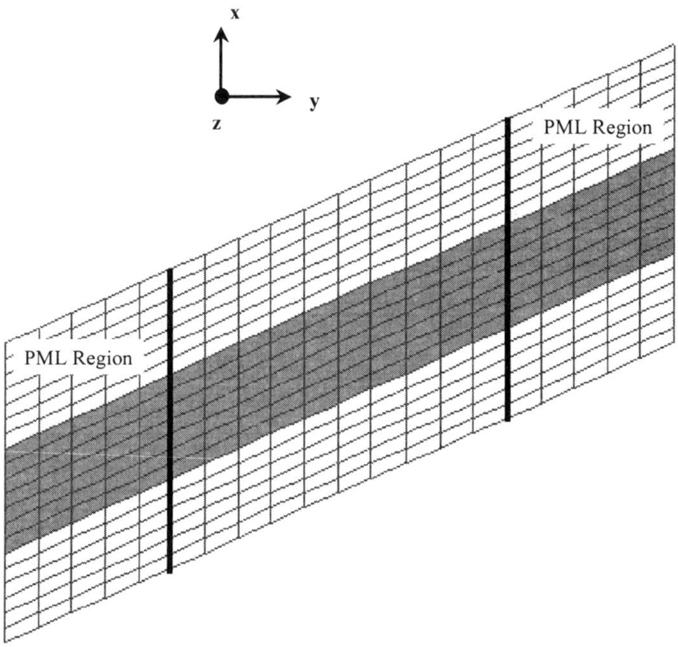

Fig. 4.12 Microstrip transmission line as gridded in a globally skewed mesh. The total mesh dimensions are not to scale.

Case 4. This is the same as Case 3, except that the cell sizes in the horizontal plane are reduced by a factor of two (e.g., $\Delta x = \Delta y = 0.17235$ mm) and Δt is reduced accordingly. The overall grid size is $400 \times 462 \times 30$ cells.

In each of these four cases, a TEM mode is excited on the microstrip. The time-dependent voltage between the upper and lower conductors is probed at two points separated axially along the strip by 9.75 mm. These data are processed using an FFT, and the numerical axial propagation constant $\tilde{\beta}$ is extracted from the phase difference of the two voltages. Finally, the numerical phase velocity is calculated as $\tilde{v} = \omega / \tilde{\beta}$.

Fig. 4.13 graphs the numerical phase velocities calculated for the four cases above compared to the exact value, $c = 2.998 \times 10^8$ m/s. Since the mode is purely TEM for the homogeneous problem, the exact phase velocity is nondispersive. For the conforming Cartesian-grid model of Case 1, the phase velocity is only about 0.67% below c. Given that the grid density for this case is about $\lambda/20$ at the highest frequency of the pulsed waveform, this agrees well with what is expected from Fig. 4.11(b). For the conforming skewed-grid model of Case 2, the wave speed exceeds c by about 2%, again agreeing well with Fig. 4.11(b).

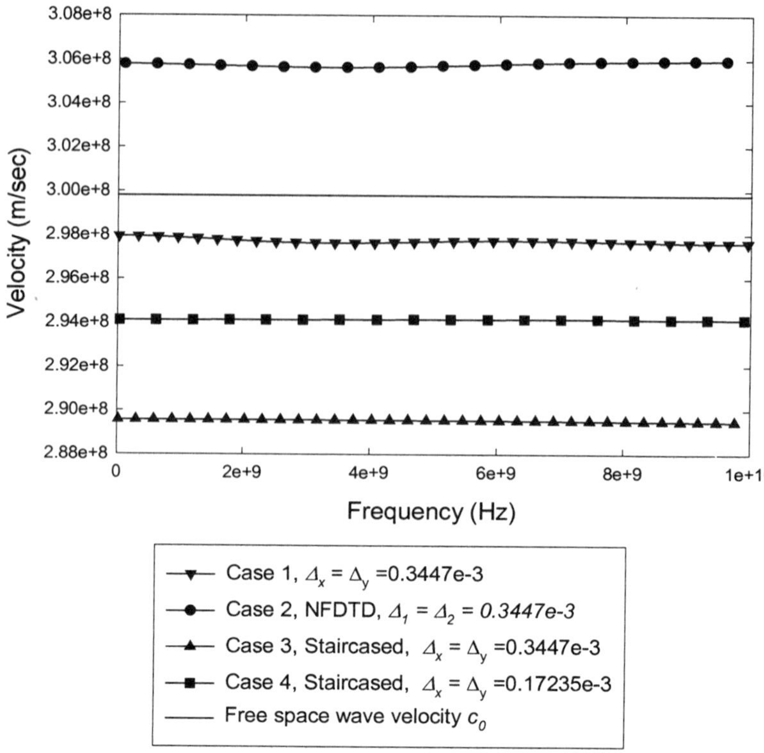

Fig. 4.13 Numerical phase velocity as calculated by the FDTD and N-FDTD techniques for a homogeneous microstrip transmission line. The microstrip width is $w = 2.413$ mm; the height is $h = 0.7995$ mm.

The staircased Cartesian-grid model of Case 3 suffers from about 75% more absolute velocity error than the conforming skewed-grid model of Case 2, yielding a wave speed about 3.5% less than c. In addition, the execution time for the enlarged grid of Case 3 is more than twice that of Case 2. We note that the discretization of Case 3 is identical to that of the conforming Cartesian-grid model of Case 1. Further, since the wave is propagating at 45° relative to the x-axis, the curves in Fig. 4.11(b) predict that the phase velocity should be improved. However, this neglects the boundary error, namely the staircasing of the edge of the microstrip in Case 3 that slows the traveling wave. We see that the boundary error due to staircasing in Case 3 well exceeds the dispersion error for the conforming Cartesian-grid model of Case 1.

The higher-resolution staircased model of Case 4 has a phase-velocity error reduced to 2% at the expense of a large increase in the grid dimensions. The execution time for Case 4 is 14 times longer than that of Case 2 for the same level of phase-velocity error.

This numerical experiment demonstrates that FDTD modeling involves both dispersive and boundary error. While the N-FDTD and DSI schemes can introduce more local dispersion error than classic Cartesian FDTD, their reduction in boundary error can well outweigh their additional dispersive error. This leads to an overall more accurate solution with a coarser grid density than using FDTD with staircased boundaries.

4.6.3 Dispersion Analysis Based on a Uniform Triangular Grid

The previous dispersion analysis was limited to quadrilateral and hexahedral cells. However, the DSI algorithm is often meshed with triangular cells in two dimensions and triangular prisms in three dimensions [14, 40]. Again, a spectral analysis is performed for such meshes, assuming uniformity. Fig. 4.14 illustrates one such two-dimensional grid composed of equilateral triangles. The secondary grid is indicated by dashed lines and is composed of hexagons. Because the triangles are equilateral, the normal and edge vectors align. Consequently, no projections are required for this grid.

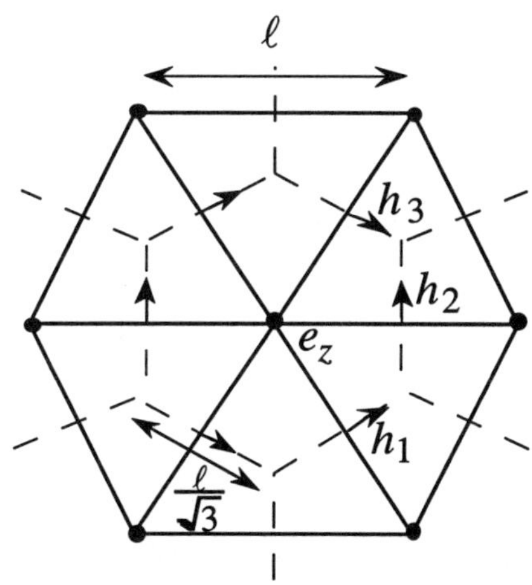

Fig. 4.14 Two-dimensional uniform unstructured mesh composed of equilateral triangular cells.

In Fig. 4.14, we assume that the transverse magnetic fields and fluxes are located at the edge centers, and the vertical electric fields are located at the vertices of the grid. Assuming that the uniform plane wave specified in (4.59) propagates in the mesh, the discrete form of Ampere's law (4.42) results in

$$e_z = \frac{2\Delta z}{\sqrt{3}\,\ell^2}\frac{c\Delta t}{\sin(\omega\Delta t/2)}\left[\begin{array}{c}-h_1\sin\left(\tilde{k}_x\dfrac{\ell}{4}-\tilde{k}_y\dfrac{\sqrt{3}\,\ell}{4}\right)-h_2\sin\left(\tilde{k}_x\dfrac{\ell}{2}\right)\\+h_3\sin\left(\tilde{k}_x\dfrac{\ell}{4}+\tilde{k}_y\dfrac{\sqrt{3}\,\ell}{4}\right)\end{array}\right] \tag{4.66}$$

Similarly, from (4.43) we have

$$h_1 = -\frac{c\,\Delta t}{\sin(\omega\Delta t/2)}\frac{1}{\sqrt{3}\,\Delta z}\sin\left(\tilde{k}_x\frac{\ell}{4} - \tilde{k}_y\frac{\sqrt{3}\,\ell}{4}\right)e_z \tag{4.67a}$$

$$h_2 = -\frac{c\,\Delta t}{\sin(\omega\Delta t/2)}\frac{1}{\sqrt{3}\,\Delta z}\sin\left(\tilde{k}_x\frac{\ell}{2}\right)e_z \tag{4.67b}$$

$$h_3 = \frac{c\,\Delta t}{\sin(\omega\Delta t/2)}\frac{1}{\sqrt{3}\,\Delta z}\sin\left(\tilde{k}_x\frac{\ell}{4} + \tilde{k}_y\frac{\sqrt{3}\,\ell}{4}\right)e_z \tag{4.67c}$$

Inserting (4.67a) to (4.67c) into (4.66) leads to the numerical dispersion relation

$$\frac{\sin^2(\omega\Delta t/2)}{(c\,\Delta t)^2} = \frac{2}{3\ell^2}\left[\sin^2\left(\tilde{k}_x\frac{\ell}{4} - \tilde{k}_y\frac{\sqrt{3}\,\ell}{4}\right) + \sin^2\left(\tilde{k}_x\frac{\ell}{2}\right) + \sin^2\left(\tilde{k}_x\frac{\ell}{4} + \tilde{k}_y\frac{\sqrt{3}\,\ell}{4}\right)\right] \tag{4.68}$$

Taking the limiting case of $\ell \to 0$, we obtain the exact dispersion relation:

$$\left(\frac{\omega}{c}\right)^2 = \frac{8}{3\ell^2}\left[\left(\tilde{k}_x\frac{\ell}{4} - \tilde{k}_y\frac{\sqrt{3}\,\ell}{4}\right)^2 + \left(\tilde{k}_x\frac{\ell}{2}\right)^2 + \left(\tilde{k}_x\frac{\ell}{4} + \tilde{k}_y\frac{\sqrt{3}\,\ell}{4}\right)^2\right] = \tilde{k}_x^2 + \tilde{k}_y^2 \tag{4.69}$$

Fig. 4.15 compares the normalized numerical phase velocity for the equilateral triangular mesh when $\ell = \lambda/10$ and $\ell = \lambda/20$ to that of a Cartesian FDTD mesh with $\Delta x = \Delta y = \ell$. We see that the velocity anisotropy of the equilateral triangular mesh is greatly reduced relative to the square mesh. Further, an analysis reveals that both the triangular and square two-dimensional meshes have the same stability limit, $c\,\Delta t \le \ell/\sqrt{2}$. Thus, both meshes are run at the same time step.

Nonuniform triangular meshes having nonequilateral-triangle cells undoubtedly have poorer dispersion properties than that given by (4.68). We further expect from the studies illustrated in Figs. 4.10 and 4.11 that triangles having very small or highly oblique interior angles suffer from relatively large dispersive error, and should therefore have edge lengths less than $\lambda/20$. Fortunately, automatic mesh generators based on a Delauney tesselation fill large regions with equilateral triangles and deform the triangles only near boundaries. For such mesh-generation schemes, excellent dispersive properties can be expected.

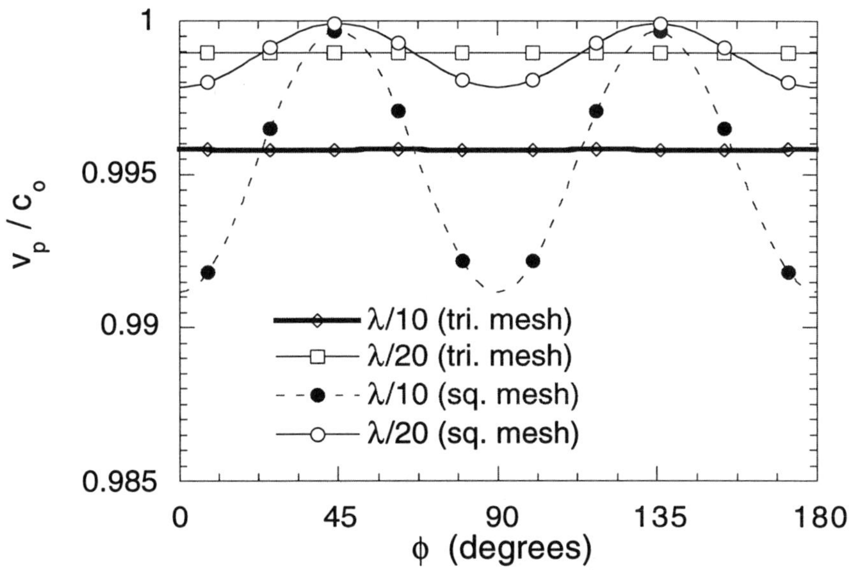

Fig. 4.15 Normalized numerical phase velocity versus propagation angle, ϕ, for the equilateral-triangle mesh and a square FDTD mesh when $\Delta x = \Delta y = \ell$ $(c\,\Delta t = \ell / \sqrt{2})$.

4.7 EXAMPLES

Sections 4.2 to 4.6 presented staggered-nonorthogonal-grid FDTD algorithms for solving Maxwell's equations. As shown in Section 4.6, these algorithms exhibit numerical dispersion that results in an anisotropic velocity for the waves propagating in the mesh. However, by utilizing grids that conform to the structural boundaries of interest, a more accurate solution of the vector fields is obtained with far fewer grid cells than are required by an orthogonal FDTD mesh. This section provides a number of modeling examples to demonstrate the effectiveness of the proposed schemes.

4.7.1 Round-Wire Transmission Line: Parallel and Twisted

The round-wire transmission line consists of wires of radius, *a,* separated by a distance, *d.* A typical application is the ribbon cable commonly found inside computers. An important figure of merit for this transmission line is its characteristic impedance, Z_0, which defines the ratio between the voltage and current on the line. Z_0 is independent of frequency for a lossless, homogeneous, ideally terminated line. However, it is dependent on near-field effects and sensitive to proper conductor discretization in numerical models. As a result, it is difficult to accurately model this line using Cartesian FDTD grids due to improper boundary representation [24]. We now analyze this geometry using the various N-FDTD techniques introduced in Section 4.2.

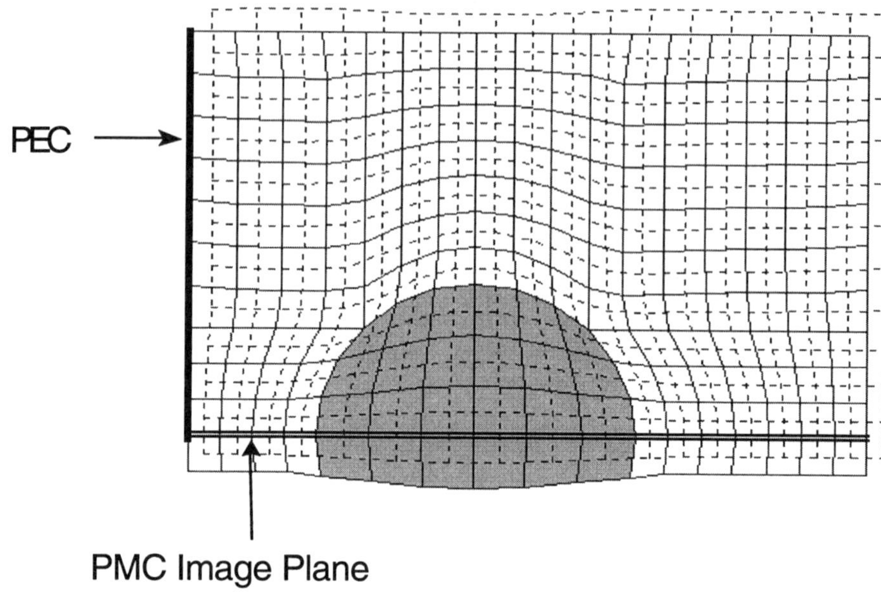

Fig. 4.16 Discretization of a two-round-wire transmission line geometry. Only one-fourth of the geometry need be discretized due to the incorporation of PEC and PMC image planes. Solid lines indicate the primary (electric field) grid; dashed lines indicate the secondary (magnetic field) grid. Wire radius $a = 0.25$ mm; wire center-to-center separation $d = 0.9$ mm.

Fig. 4.16 illustrates the transverse mesh for the round-wire geometry. An orthogonal-grid region is added to the right and top of this mesh to permit normal incorporation of a 10-cell PML ABC. In the longitudinal direction, the grid spans 100 cells with the uniform discretization $\Delta z = 0.5$ mm, and a 10-cell PML ABC is used to terminate the wire. Overall, this approach accurately approximates an infinitely long transmission line in an infinite free-space medium.

The line is excited with a Gaussian pulse having a bandwidth of about 8 GHz. To compute Z_0, the line voltage is probed at position k, and the current is probed at positions k and $k-1$ along the line, where k is chosen to be at least 30 cells from the source so that any higher-order modes excited by the source are greatly attenuated. The sampled data are processed using the FFT, and Z_0 is then computed as in [41]:

$$Z_0(\omega) = \frac{V(\omega)\, e^{-j\omega \Delta t/2}}{\sqrt{I_k(\omega)\, I_{k-1}(\omega)}} \tag{4.70}$$

Fig. 4.17 compares calculated results for Z_0 obtained using N-FDTD with the G^{avg} and G^{sym} projection techniques and also using the DN-FDTD method. The accuracy of each of these methods is measured against the exact TEM solution from [42].

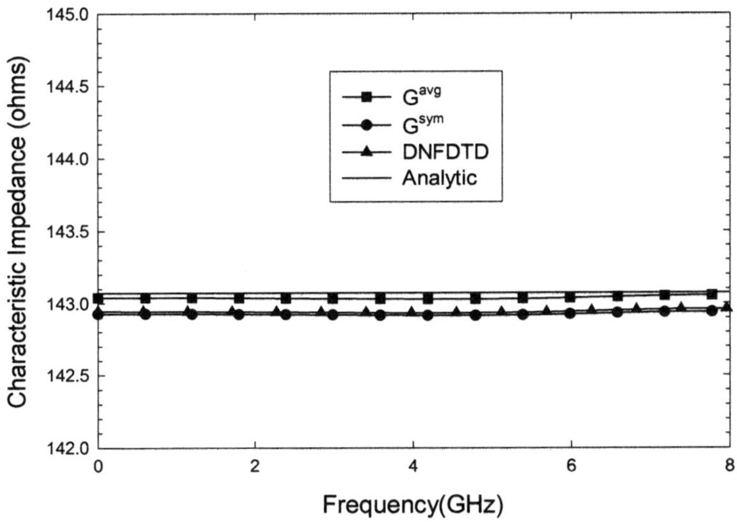

Fig. 4.17 Computed characteristic impedance for the homogeneous two-wire transmission line of Fig. 4.16.

From Fig. 4.17, we see that each of the three methods provides results in a tight 0.2Ω range just below the exact solution, with G^{avg} being the most accurate. The G^{sym} and DN-FDTD techniques provide almost identical results.

Next, the N-FDTD technique is used to analyze a fully three-dimensional nonorthogonal inhomogeneous geometry, an inhomogeneous twisted and untwisted two-wire transmission line. The geometry is constructed by applying a longitudinal twist of 1.27 turns-per-inch (315 radians/m) to the cross-section of Fig. 4.16 (corresponding to a Category-5 line). In addition, a 0.2-mm-thick dielectric coating of $\varepsilon_r = 2.58$ is added to the wire, representing insulation. While this geometry is commonly used for high-speed data communications, no analytic solution is known for its characteristic impedance.

Fig. 4.18(a) compares Z_0 for the twisted and untwisted lines. A highly expanded vertical scale is needed to observe the slight 0.075Ω reduction in Z_0 caused by the addition of twist. On the other hand, the presence of the dielectric coating causes the line to be noticeably dispersive. This justifies the common practice of using the untwisted transmission line characteristics when modeling the twisted line.

Fig. 4.18(b) compares the normalized propagation constant, β, of the twisted and untwisted lines. Here, the increased β for the twisted line is a result of the additional line length and hence the additional phase shift that results from propagation on the twisted pair. In addition, the twisting of the wires tends to increase the effective dielectric constant seen by the two wires and therefore shortens the wavelength.

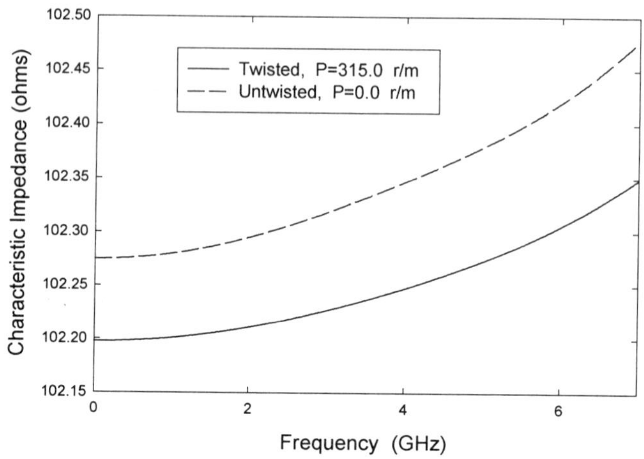

(a) Characteristic impedance.

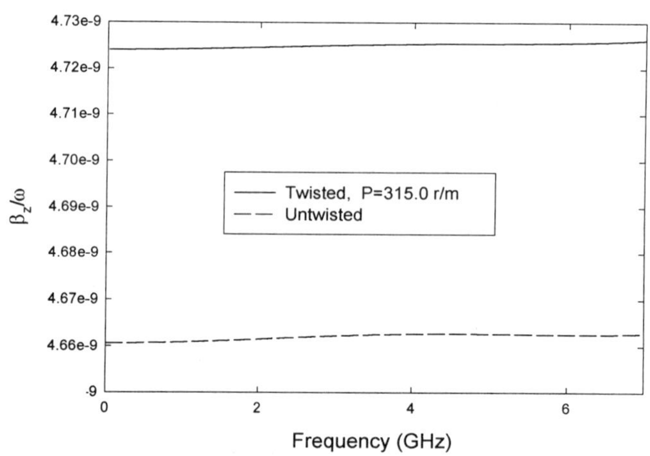

(b) Propagation constant.

Fig. 4.18 N-FDTD-calculated properties of the two-wire transmission line of Fig. 4.16. Here, each wire has a 0.2-mm-thick dielectric coating of $\varepsilon_r = 2.58$, and the wires are twisted at the rate of 1.27 turns-per-inch (315 radians/meter), which corresponds to a Category-5 line.

4.7.2 Microwave Circuit Devices

The FDTD method has been used extensively for the analysis of microwave circuit devices (see [43] for an extensive reference listing). Typically, microwave circuit devices contain intricate details that are electrically small, while the full circuit dimensions are on the order of wavelengths. While the N-FDTD and DSI algorithms are equipped to model curved boundaries, the structured grids required by N-FDTD methods can be highly restrictive when modeling circuits having complex boundary shapes. Usually, unstructured grids provides better meshing from the standpoint of accuracy and stability.

The GY code developed at the University of Kentucky, which is based on the DSI class of methods, has been applied extensively to the analysis of microwave circuit devices [14, 15]. Since a large majority of microwave circuits are printed on planar layered substrates, accurate modeling can be obtained using gridding that is unstructured in the horizontal plane and orthogonal in the vertical plane. We refer to this algorithm as the planar generalized Yee (PGY) algorithm [14]. A few examples validating these methods are presented here.

Microstrip Circular Patch Antenna

The geometry of the microstrip circular patch antenna is illustrated in Fig. 4.19.

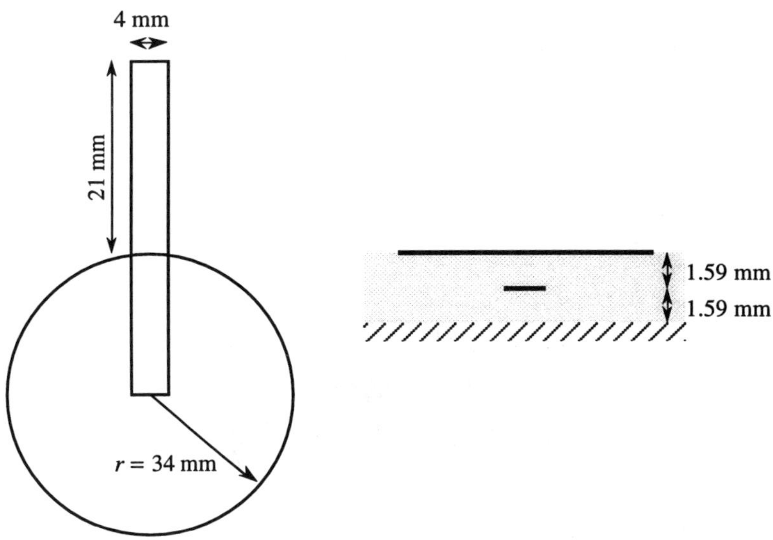

(a) (b)

Fig. 4.19 Microstrip coupled circular patch antenna. The microstrip is printed on a 1.59-mm substrate of $\varepsilon_r = 2.62$ over a ground plane, and the patch antenna is printed above a 1.59-mm superstrate of $\varepsilon_r = 2.62$. (a) top view; (b) end view.

The patch antenna of Fig. 4.19 is modeled using the PGY code as well as the frequency-domain MoM algorithm, IE3D [44]. An unstructured mesh is generated for the patch antenna that consists of 159,600 hexahdra, with the exterior mesh boundaries terminated in a 10-cell PML ABC. PML regions on the side walls and corners are composed of orthogonal cells and have field updates provided in Chapter 5. The top horizontal PML region is composed of a mesh that is orthogonal along the vertical direction, but unstructured and nonorthogonal in the transverse direction (the exception being the corner regions where the mesh is orthogonal). Since the PML interface is planar, the interface is perfectly matched. Further, the transverse field updates are simply modified as lossy field updates. The vertical field updates (or normal fields) are updated as outlined in Chapter 5.

The PGY simulation is performed using both the nonsymmetric formulation outlined in Section 4.3 and the symmetric formulation outlined in Section 4.5.3. Both simulations are run for 20,000 time steps ($\Delta t = 0.725$ ps). The nonsymmetric formulation remains stable throughout the entire computation, and no impending instability is noted. Fig. 4.20 compares the magnitude of S_{11} calculated using the symmetric and nonsymmetric PGY methods and IE3D. Both FDTD methods compare quite well with the MoM solution, exhibiting a slight downward shift in the resonant frequency. We see that enforcing symmetry in the projections introduces nearly negligible difference between the two classes of PGY results.

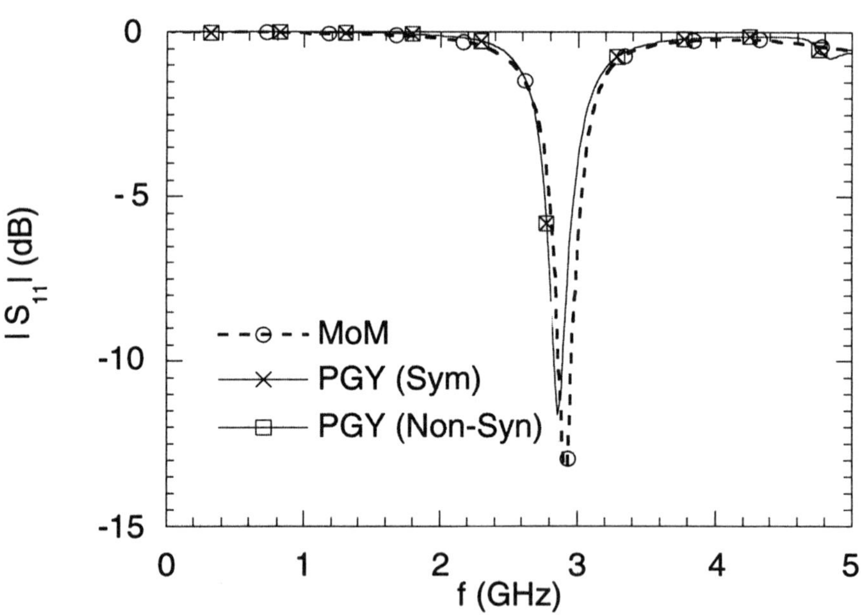

Fig. 4.20 Comparison of the reflection loss of the microstrip coupled circular patch antenna computed using the frequency-domain MoM and the symmetric and nonsymmetric PGY algorithms.

Wilkinson Power Divider

Fig. 4.21(a) shows the geometry of an eight-way Wilkinson power divider modeled using the PGY algorithm. Fig. 4.21(b) graphs the calculated S parameters, S_{11} and S_{21}.

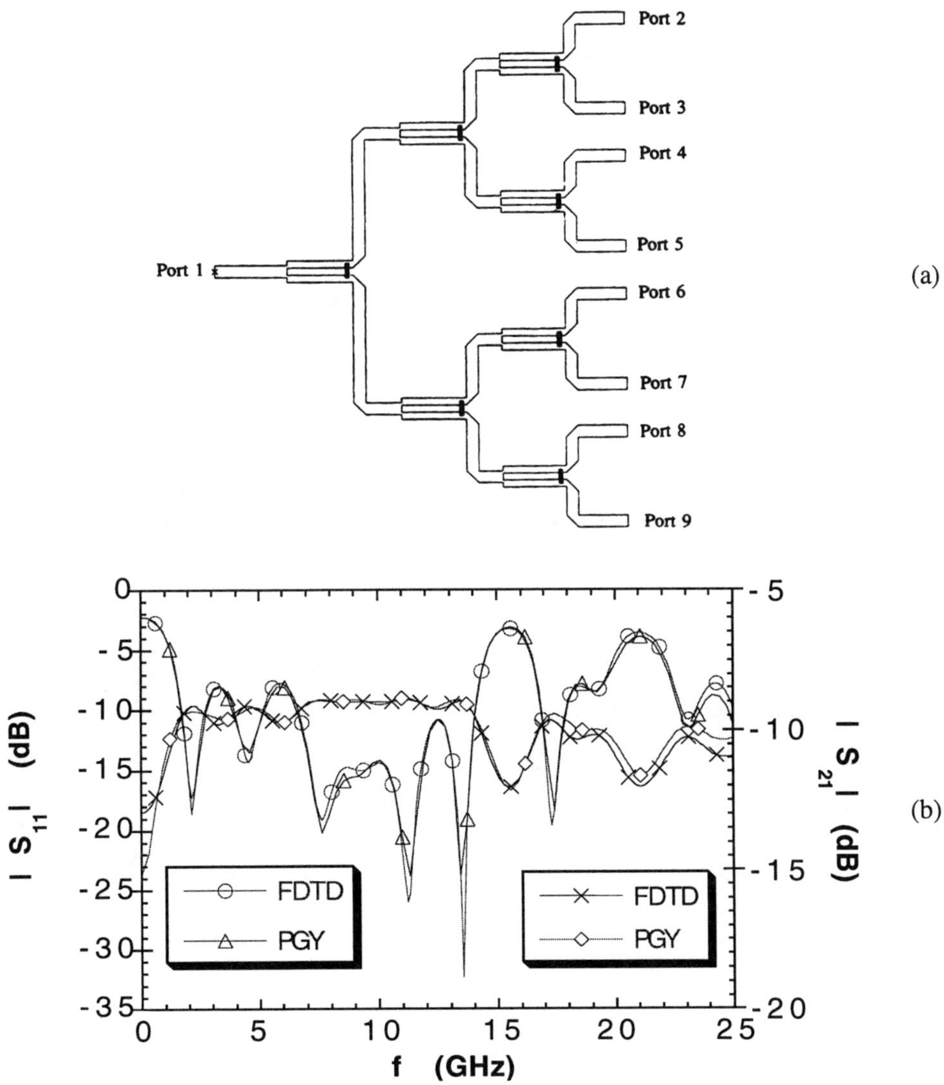

Fig. 4.21 (a) Geometry of an eight-way Wilkinson power divider; (b) S parameters calculated using the conformal PGY and staircased FDTD methods.

In Fig. 4.21(b), the PGY data are compared with results from a fine-resolution Cartesian-grid FDTD model that staircases the mitered bends in the power divider. Due to the conformal nature of its unstructured mesh, the PGY simulation is run in only one-third of the time required by FDTD to achieve the same degree of accuracy.

Transition from Coplanar Waveguide to Microstrip and Back

Fig. 4.22(a) illustrates an abrupt transition from a 50Ω coplanar waveguide (CPW) to a 50Ω microstrip and back to a 50Ω CPW. The dimensions shown are in mils.

(a)

(b)

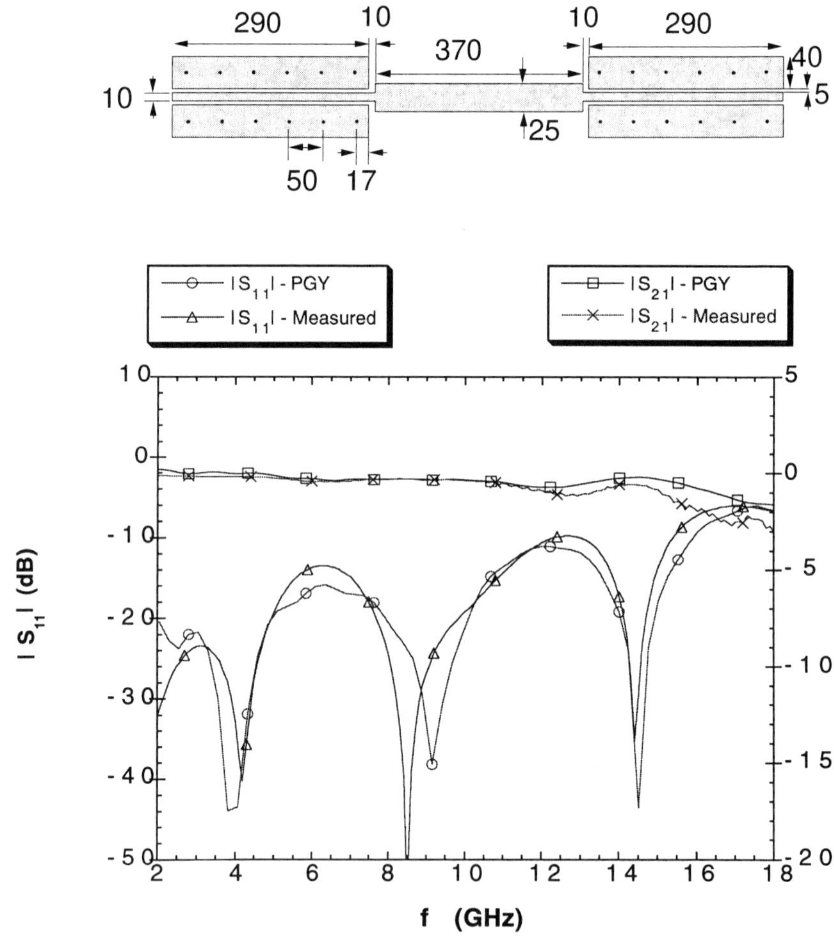

Fig. 4.22 (a) Top view of the abrupt transition from 50Ω CPW to 50Ω microstrip to 50Ω CPW printed on a 25-mil alumina substrate ($\varepsilon_r = 9.8$); (b) S parameters of the transition.

The circuit of Fig. 4.22(a) is printed on a 25-mil alumina substrate with $\varepsilon_r = 9.8$. The outer conductors of the CPW are connected to ground through 6-mil-diameter circular vias spaced 50 mils apart. Using the unstructured PGY grid, the curved surfaces of the circular vias are approximated using 6 flat facets per via. Fig. 4.22(b) compares the *S* parameters for the abrupt transition computed using the PGY code with measured data provided by Hughes. Similar results are reported in [45] using MoM.

4.7.3 Radar Cross Section of Three-Dimensional Bodies

The Electromagnetic Code Consortium (EMCC) conducted and published a series of measurements of the monostatic RCS of canonical metal targets [46]. This section reviews the results of numerical simulations of these measurements using the DSI3D code developed at Lawrence-Livermore National Laboratory. In each case, the problem mesh was generated using the TrueGrid™ software [47]. The monostatic RCS was calculated at a 10° elevation angle in the plane of the target for both horizontal and vertical incident polarizations. The second-order Higdon ABC [48] was used at the outer boundary of the mesh (the PML ABC was not available at the time). We now consider several examples.

Triangular-Shaped Flat Plates with Semicircular Back Ends

Fig. 4.23 views the DSI meshes generated for two cases of a triangular-shaped flat metal plate with a semicircular back end. In the first case, meshed as shown in Fig. 4.23(a), the vertex angle is 60°, the two straight edges are $2\lambda_0$ long, and the radius of the semicircle that defines the right boundary is $1\lambda_0$. We see that this shape is simply the union of an equilateral triangle and one-half of a circle having a diameter equal to the side length of the triangle. In the second case, meshed as shown in Fig. 4.23(b), all dimensions are the same as for the first case, but a $1\lambda_0 \times 2\lambda_0$ rectangular flat metal section is inserted between the triangle section and the half-circle section. The following differences in the meshing techniques used for these two cases are noted:

1. The lattice of Fig. 4.23(a) uses fine zoning near the vertex and edges of the target. The outer edges of the lattice join with standard rectangular Yee cells to form a computation "box" whose shape is not related to the target.
2. The lattice of Fig. 4.23(b) uses a more uniform discretization near the vertex and edges of the target. The outer edges of the lattice effectively parallel the target edges, thereby shaping the computation zone to fit the target.

Figs. 4.24 and 4.25 compare the DSI monostatic RCS calculations with measured data. Here, the units of RCS are decibels relative to one square meter (dBsm). Each figure shows results for horizontal-horizontal (H-H) polarization of the illuminating wave and the backscattered wave, and vertical-vertical (V-V) polarization of the illuminating wave and the backscattered wave. Good accuracy is seen for the H-H case; the accuracy is somewhat reduced for the V-V case.

(a)

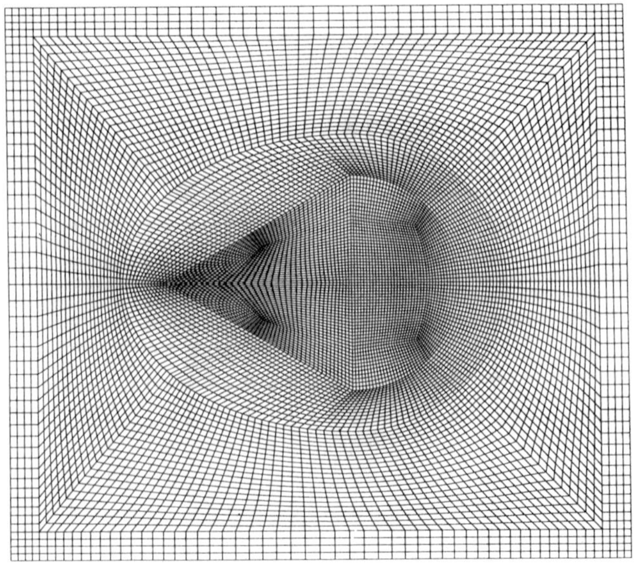

(b)

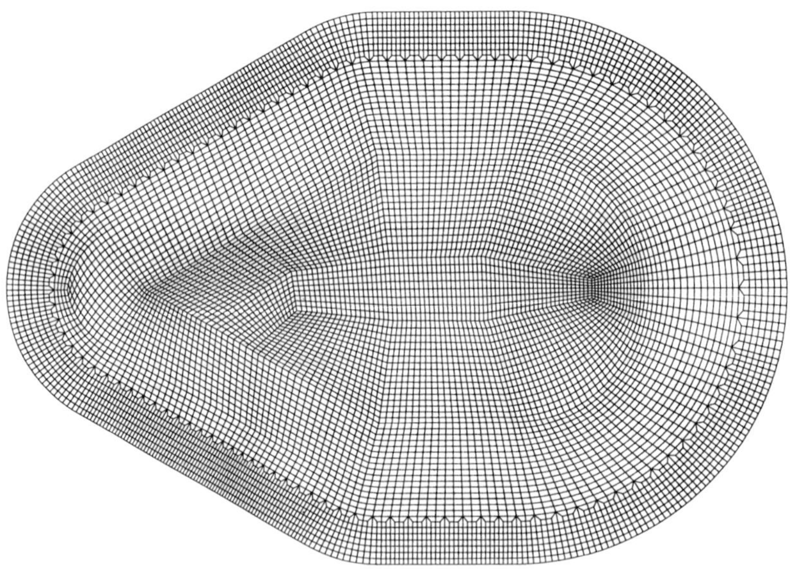

Fig. 4.23 View at each x-y plane of the DSI meshes used for the triangular-shaped flat-plate targets with semicircular back ends. The meshes do not vary in the z direction. (a) basic target; (b) basic target with a short extension between the triangle and the semicircle.

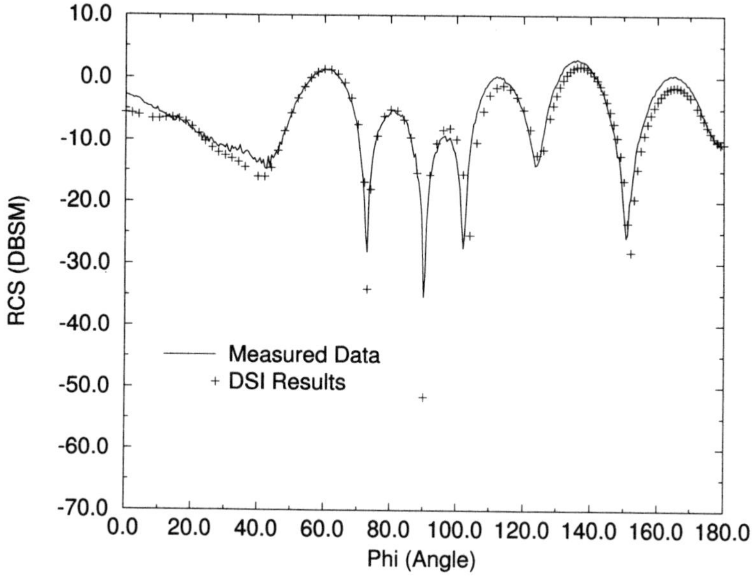

(a)

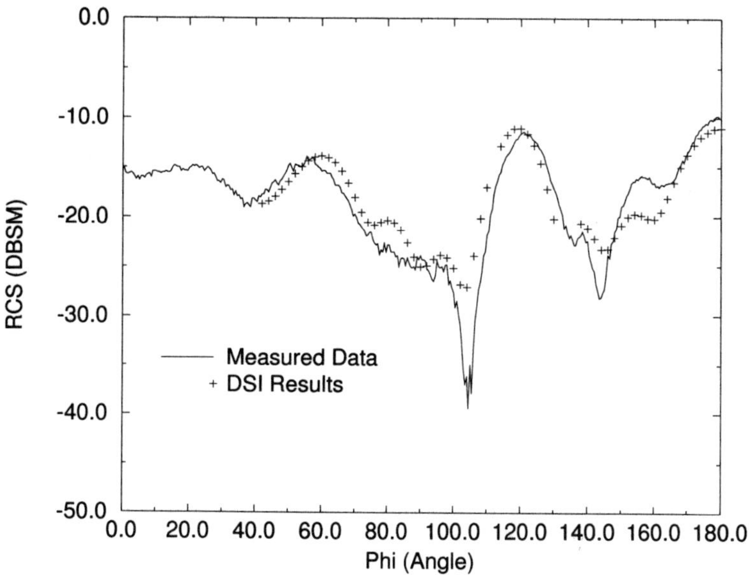

(b)

Fig. 4.24 Comparison of DSI calculations and measured data for the monostatic RCS of the basic triangular target meshed in Fig. 4.23(a): (a) H-H polarization; (b) V-V polarization.

(a)

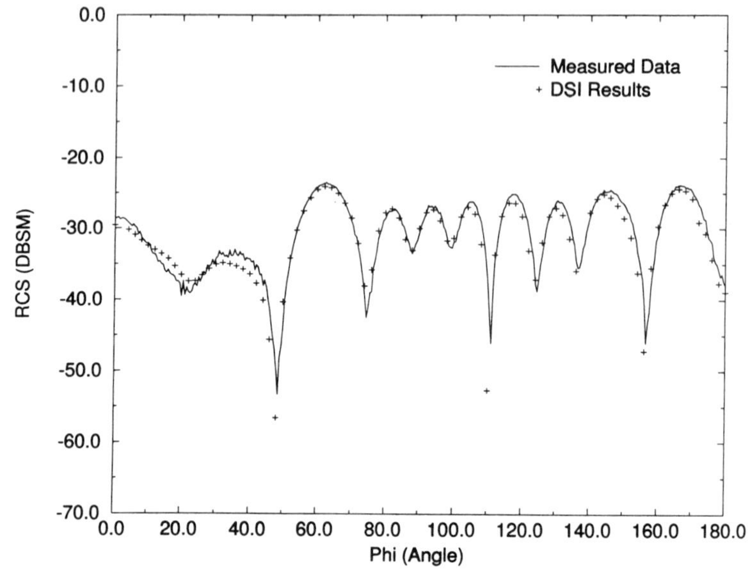

(b)

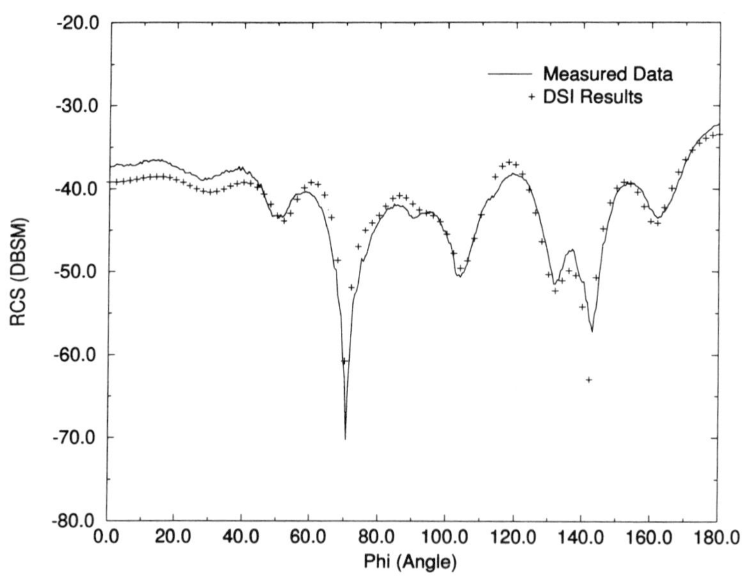

Fig. 4.25 Comparison of DSI calculations and measured data for the monostatic RCS of the extended triangular target meshed in Fig. 4.23(b): (a) H-H polarization; (b) V-V polarization.

In the cases of Figs. 4.23 to 4.25, there are a number of possible causes of the errors observed. Inadequate resolution of the object by the mesh, reflections from the ABCs used to terminate the grid, wave-propagation phase and amplitude errors within the DSI algorithm, and interpolations used in the near-to-far-field transformation are all sources of errors. Examination of each of these problems leads us to conclude that the largest source of error is due to the second-order Higdon ABCs used at the time of these modeling studies. As stated earlier, PML ABCs can be implemented in the context of the DSI algorithm so that this problem should be reduced substantially. However, the models discussed here have not yet been repeated with the PML ABC incorporated.

Almond-Shaped Target

Fig. 4.26(a) shows one-half of the DSI mesh used for computing the RCS of the 9.936-inch-long NASA almond target. Similar to the mesh used in Fig. 4.23(a), there is fine zoning near the almond, particularly near its tip, and zooming of the mesh to acquire a rectangular box outer shape. Fig. 4.26(b) views the shape of the near-to-far-field integration surface by showing a cutaway view of that surface and the meshed surface of the metallic almond residing inside.

Fig. 4.27 compares the DSI calculations and the measured data for the monostatic RCS of the almond for the H-H and V-V polarization cases. While the V-V results are in excellent agreement with the measurements, there are significant errors in the H-H results that remain unexplained. Interestingly, a very similar level of agreement with the almond's measurements for the V-V case, and disagreement for the H-H case, was reported in [49] for a conventional Cartesian FDTD code using a second-order-accurate Mur ABC and simple staircasing to approximate the surface curvature of the almond. It may be that the common factor here is the use of a second-order ABC that performs particularly poorly for the type of waves scattered outward from the almond for H polarization. Clearly, these numerical studies should be repeated using the PML ABC to reinforce or reject this possibility, and to help pin down the robustness of the DSI method for modeling arbitrary scatterers.

4.8 INTRODUCTION TO FINITE-VOLUME TIME-DOMAIN METHODS

Among the factors limiting the development of low-observables technology, radar signature measurement and prediction remain the most significant. As the RCS of large targets gets progressively smaller, the traditional methods for determining RCS become inadequate. Sources of error that once were tolerable now dominate. Neither the experimental nor computational approaches of prior years can be extended in a straightforward way to deal with these sources of error, largely because of the assumptions underlying these approaches.

An opportunity to attack these problems at a more basic level has become available through the continuing growth of computer memory and speed. The equations governing electromagnetic interactions can now be simulated without physical approximation for a

(a)

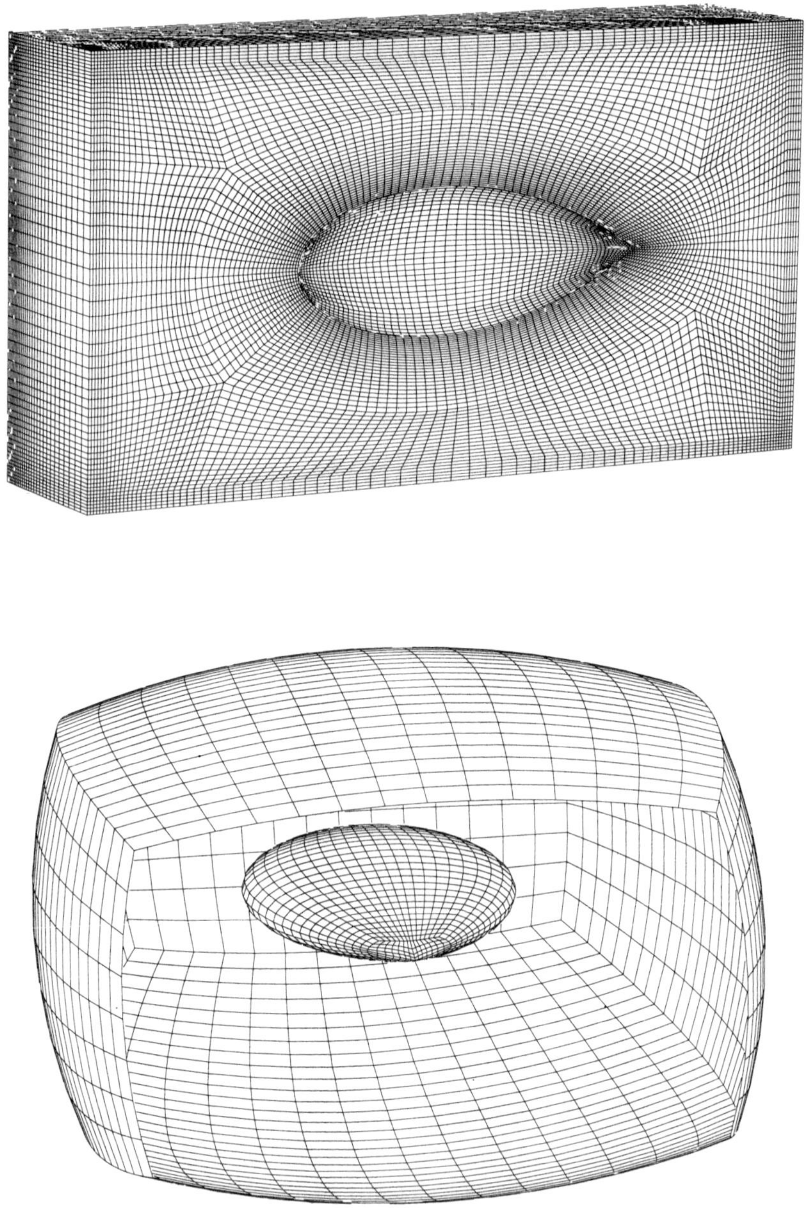

(b)

Fig. 4.26 (a) View of one-half of the mesh used for the 9.936-inch NASA almond target; (b) cutaway view of the near-to-far-field integration surface with the meshed almond surface located inside.

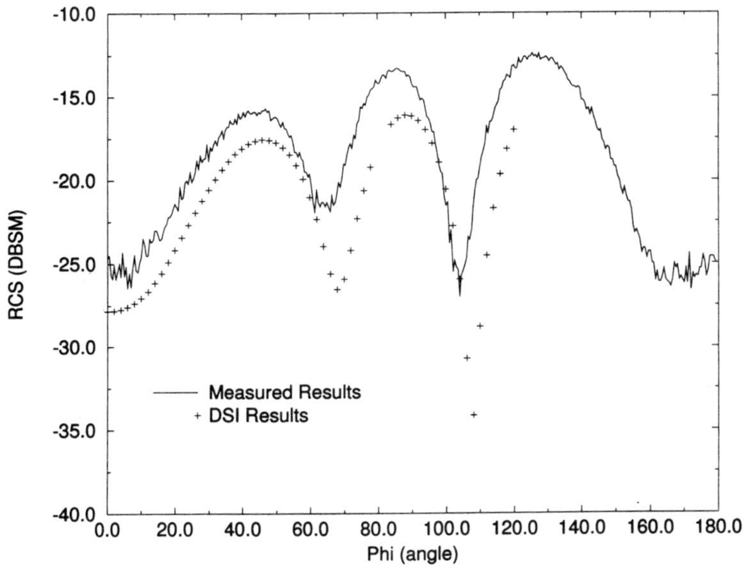

(a)

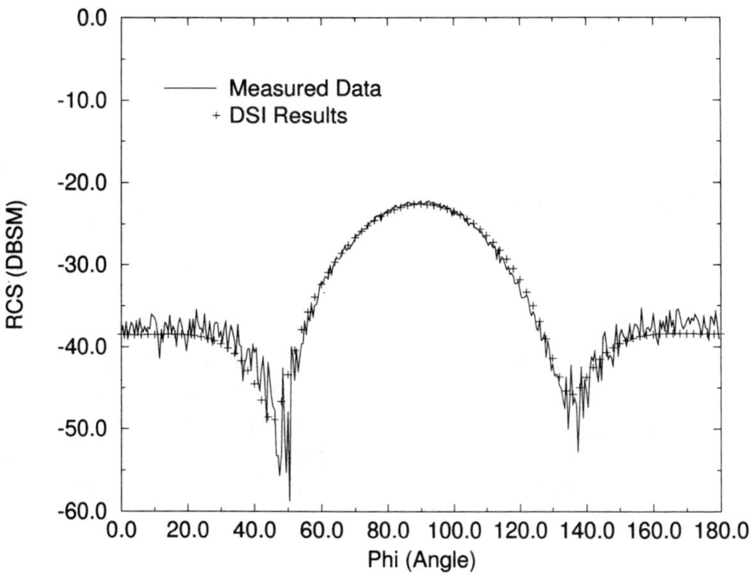

(b)

Fig. 4.27 Comparison of DSI calculations and measured data at 1.19 GHz for the monostatic RCS of the 9.936-inch NASA almond meshed in Fig. 4.26(a): (a) H-H polarization; (b) V-V polarization.

wide class of structures. This opens up possibilities for diagnosing and correcting limitations in RCS measurement techniques as well as accurately capturing subtle effects that contribute to the radar return when the ordinary scattering has been suppressed.

The large electrical size of the aerospace structures of interest places a premium on the use of efficient numerical techniques. Simply extending previous MoM approaches leads to computer memory and execution times that grow at the fourth and sixth powers, respectively, of the characteristic target dimension. (However, recently developed "fast" algorithms promise to limit this growth to the square or cube of the characteristic target dimension.) An alternative technique is the direct integration in time of a discretized form of Maxwell's equations using body-fitted grids in and around the target.

4.8.1 Background

Attempts at accurate prediction of scattering from large targets of small RCS have exposed significant difficulties for all current numerical methods, including fast integral equation solvers in the frequency domain and finite-difference, finite-volume, and finite-element schemes in the time domain. At the same time, considerable progress has been made in demonstrating the ability of these methods to solve many types of problems on massively parallel computing platforms such as the IBM SP2, Intel Paragon, and Cray T3D/E.

The most scaleable method is explicit time integration of Maxwell's equations. For example, the Rockwell code RCSMPP (developed for the EMCC under Air Force contract) has demonstrated that the time spent on communication among nodes can be made negligible. Its communications overhead is less than 3% using up to 128 nodes to generate RCS data for the General Dynamics / Lockheed VFY-218 fighter at frequencies from 100 MHz to 1 GHz. This performance is achievable because each node contains many grid cells, the vast majority of which only need to communicate among themselves (that is, within a node) to advance the solution one step in time. Good load balancing is also easy to obtain because the computations for each grid cell are nearly identical.

When the RCS prediction problem can be reduced to computing the scattering from a target surface of characteristic dimension, d, at a single frequency, frequency-domain methods have a computational advantage relative to the time-domain techniques. The frequency-domain approaches solve for a much smaller number of unknowns (the number, N, of current-vector components on the target surface) which grows as $N \sim d^2$. However, conventional MoM solutions create and attempt to invert a matrix connecting each surface unknown to every other. This leads to computer storage that grows as $N^2 \sim d^4$ and computation time that grows as $N^3 \sim d^6$. Fast methods such as the multilevel fast-multipole algorithm do not explicitly create and store this matrix, and use iteration to obtain a solution for each angle of incidence. Such techniques have achieved storage and computation times proportional to $N^\alpha \ln N$, where $1 < \alpha < 2$.

From the perspective of high-performance computing, the key problem with iterative techniques such as the fast-multipole method is that they require frequent global communications. Similar to the older MoM, each part of the surface is influenced by every other part of the surface at each iteration. As a consequence, the communications overhead grows with the number of nodes, and the overall performance becomes sensitive to the details of message passing in each type of parallel architecture (e.g., mesh, hypercube, fat tree). Low-rank approximations to blocks of the underlying matrix can reduce the computational load, but they do not affect the requirement for transfers of data from each node to every other each iteration.

Radar absorbing structures present another kind of difficulty for the fast integral-equation methods. Materials that are graded and layered in three dimensions require unknowns distributed throughout a volume with a resolution fine enough to describe the anticipated field variations. That is, $N \sim d^3$ rather than $N \sim d^2$. If a finite-element description is used for the fields inside the structure, that part of the matrix takes on a sparse banded structure, but the number of iterations required to obtain an accurate solution is generally much larger.

Time-domain methods inherently propagate electromagnetic waves through all penetrable regions of a target, and can readily deal with complex material properties involving spatial variations, time dependencies, and nonlinearities. The required computer storage is just the volume unknowns themselves, $N \sim d^3$, supplemented by the local properties of the grid. Further, results over a wide band of frequencies can be obtained from a single run with an incident pulse waveform.

The penalty associated with time-domain methods is that a substantial volume of free space around the target must be included as part of the computational domain. For the class of finite-volume techniques derived from CFD, this penalty is minimized by the use of grids that stretch rapidly away from the target. For example, the code RCSMPP rarely needs more than 15 layers of grid cells to fill the exterior volume, so the total computer storage becomes more nearly proportional to the target surface area, $N \sim d^2$, than to the volume of the domain.

Nonuniform grids do, however, affect the waves that travel through them. For waves leaving the target region, this is generally not important; but for waves bouncing between different parts of the target, the consequences can be serious. Because interactions among features separated by many wavelengths along the target can contribute substantially to the RCS, methods that explicitly propagate scattered waves over the target must satisfy stringent requirements with regard to preserving their phase and amplitude while accurately representing their interaction with each target feature.

Efforts to take advantage of proven CFD techniques have led to the formulation of a very comprehensive numerical approach for solving electromagnetic scattering and radiation problems involving complex objects [17–20, 50–65]. This technique, which we call the FVTD method, is based on the Riemann integration method [66] and recent advances in the numerical solution of partial differential equations (PDEs) in CFD [67]. It has been successfully applied to a diverse range of problems involving electrically large scatterers, inhomogeneous and layered dielectrics, cavities, frequency-dependent materials, impedance surfaces, and canonical antennas, to name just a few. The key features of the FVTD method are now summarized.

1. *Scattered-field formulation.* For RCS problems, the FVTD method casts Maxwell's equations into a scattered-field form and solves for the scattered wave. It introduces the incident wave only as a boundary condition on the conducting portions of the target, thereby avoiding propagation of the incident wave through the grid. For scattering problems involving an infinite ground plane, the problem is solved for the diffracted wave, thereby avoiding propagation of both the incident wave and the reflected wave from the ground plane. These decompositions are a direct consequence of the linearity of Maxwell's equations.
2. *Collocation of fields.* The FVTD method collocates the E and H fields in both space and time, rather than assigning them to two interleaved spatial grids and separating them one-half time step. This has the advantage of reducing the complexity in gridding, in representing material regions, and in extracting near-field data for transformation to the far field.
3. *Characteristics-based algorithms.* FVTD employs characteristics-based algorithms that contribute to its numerical stability and accuracy.
4. *Body-fitted meshes.* FVTD employs body-fitted meshes to represent the target geometry and to allow convenient implementation of the boundary conditions. Numerical gridding procedures are used to set up structured or unstructured computational cells with a prescribed variable resolution in various portions of the computational domain.

Like all explicit time-domain methods, FVTD techniques can exploit modern supercomputers having either vector or parallel architectures, thereby creating a strong capability to efficiently solve large-scale scattering and radiation problems. The use of advanced graphics workstations provides a versatile capability for performing preprocessing and postprocessing to generate informative color graphics and animations for the mesh geometries, electromagnetic fields, and power-flow visualizations.

4.8.2 Structured-Grid Solvers

Conventional time-integration schemes exhibit minimum errors in phase and amplitude on uniform grids. The classic Yee algorithm [1] and its variants [68, 69], which use two interpenetrating rectangular grids for the electric and magnetic fields and a leapfrog time-advance scheme, have been well studied in this regard. These approaches run into difficulties in accurately treating the fields at target boundaries that do not conform to the grid. Typically, they either staircase the target surface to fit the grid or truncate the grid cells through which the surface passes, resulting in a loss of accuracy in satisfying the boundary conditions. As a consequence, the surface currents they compute can introduce significant errors into the target RCS.

Another serious limitation of uniform grids for large target geometries is the storage required by the grid cells that fill the computational domain. In three dimensions, the number of cells scales as $(d/s)^3$, where d is the largest characteristic dimension of the target and s is the smallest resolvable feature, typically about $\lambda_0/20$. For the VFY-218 fighter at 1 GHz, this amounts to over a billion cells, each of which requires storage for six field components.

In CFD, time-integration schemes based on nonuniform grids were developed over 20 years ago and successfully applied to many internal and external fluid flows. Here, the basic strategy is to divide the volume around a complex geometry into zones that are independently gridded by standard techniques such as transfinite interpolation or elliptic-equation solution. The distorted Cartesian grids in neighboring zones need not match at their common interface so long as the fluxes that cross the interface are computed with sufficient accuracy. Within a zone, the grid lines can be clustered at the structure surface but diverge toward the free-stream boundary. This results in a large reduction in the total number of grid cells compared to using a uniformly spaced grid.

The Rockwell codes RCS3D and RCSMPP implement this strategy for the solution of Maxwell's equations. They demonstrate that very efficient and quite accurate RCS computations can be carried out on such grids using both vector and massively parallel supercomputing architectures.

4.8.3 Unstructured-Grid Solvers

The ease with which unstructured grids can be generated to solve problems in solid mechanics has inspired many efforts to adapt both their gridding and solution techniques to CFD and CEM. A grid consisting of hundreds of thousands of tetrahedra accurately modeling a complex shape can presently be generated on a workstation in hours. A strong theoretical basis for finite-element solution techniques has been developed over several decades [70]. However, their direct application to problems in CFD and CEM has encountered significant difficulties.

Within the CFD community, both vertex-based and cell-based schemes are implemented in a finite-volume framework [71–73]. The primary disadvantage shared by these approaches is that they require many operations on wide stencils to determine the spatial variation and time update from volume averages of the unknowns. Consequently, they are generally regarded as impractical for three-dimensional calculations.

A small-stencil finite-volume method for general grids was developed at Rockwell in the context of CEM [20]. This generalizes the upwind Lax-Wendroff algorithm in RCS3D in a natural way, reducing exactly to that form on Cartesian grids. However, it provides second-order accuracy in the solution only in that limit. To achieve acceptable accuracy in RCS predictions using this algorithm on tetrahedral grids, the number of grid nodes per wavelength must be increased substantially. Despite this limitation, the method has been successfully applied to integrating the Euler equations of fluid dynamics for flows around complex geometries, and the solutions have been validated by comparison with conventional structured-grid CFD codes [74].

The desirable features of this approach led us to investigate the sources of error in algorithms of this type [75]. Under a recently completed NASA Research Agreement, we developed a systematic means for improving the accuracy based on central, rather than upwind, estimates for solution variation inside each grid cell.

4.9 CONSERVATION FORM OF MAXWELL'S EQUATIONS AND FIELD FORMULATIONS

The major feature of the discretization approach discussed here that distinguishes it from other finite-volume and finite-difference procedures is that the E- and H-field unknowns are collocated in both space and time. These unknowns are the volume averages of the electric and magnetic fields within each cell in the space-filling lattice.

We now derive the conservation form of Maxwell's equations that leads to the collocation of the fields. The starting point is Faraday's and Ampere's laws with an impressed current source, $\vec{J}$:

$$\nabla \times \vec{E} = -\frac{\partial \vec{B}}{\partial t} \tag{4.71a}$$

$$\nabla \times \vec{H} = \frac{\partial \vec{D}}{\partial t} + \sigma \vec{E} + \vec{J} \tag{4.71b}$$

The FVTD formulation begins by integrating these equations over the volume, V, of a representative lattice cell:

$$\iiint_V \nabla \times \vec{E}\, dV = -\iiint_V \frac{\partial \vec{B}}{\partial t}\, dV \tag{4.72a}$$

$$\iiint_V \nabla \times \vec{H}\, dV = \iiint_V \frac{\partial \vec{D}}{\partial t}\, dV + \iiint_V \sigma \vec{E}\, dV + \iiint_V \vec{J}\, dV \tag{4.72b}$$

Using the Curl theorem [76] and assuming that $\vec{E}$ and $\vec{H}$ and their first derivatives satisfy certain regularity conditions, we write

$$\iiint_V \nabla \times \begin{bmatrix} \vec{E} \\ \vec{H} \end{bmatrix} dV = \oiint_S \begin{bmatrix} \hat{n} \times \vec{H} \\ \hat{n} \times \vec{E} \end{bmatrix} dS \tag{4.73}$$

where S is the surface that encloses V. We also define the volume average of the fields as

$$\vec{A}_{\text{avg}} = \frac{1}{V} \iiint_V \vec{A}\, dV \tag{4.74}$$

where $\vec{A}$ represents any of the field quantities or the current density, and $\vec{A}_{avg}$ is its volume average over the cell. For simplicity of the notation, we drop the subscript showing the average quantity. Maxwell's equations are now written as

$$\frac{\partial}{\partial t}\begin{bmatrix}\vec{D}\\ \vec{B}\end{bmatrix} = \frac{1}{V}\oiint_S \begin{bmatrix}\hat{n}\times\vec{H}\\ -\hat{n}\times\vec{E}\end{bmatrix} dS + \begin{bmatrix}-\sigma\vec{E}\\ 0\end{bmatrix} + \begin{bmatrix}-\vec{J}\\ 0\end{bmatrix} \tag{4.75}$$

4.9.1 Field Formulations

Traditionally, Maxwell's equations are arranged in terms of (and solved for) the total field in the region of interest. However, in using a differential-equation-based method to solve scattering problems, it is often advantageous to cast the problem in the scattered-field form. In this form, Maxwell's equations are solved for the scattered field (i.e., the difference between the fields in the presence and the absence of the scatterer). Moreover, when an infinite ground plane is adjacent to the scatterer, Maxwell's equations are formulated and solved for the diffracted field (i.e., the difference between the fields in the presence and the absence of both the ground plane and the scatterer). These possibilities are simply a consequence of the linearity of Maxwell's equations. If desired, the three field formulations can be cast into a single framework wherein the distinctions among these forms appear in the source terms as well as in the boundary conditions.

Consider (4.75) as the general form of Maxwell's equations. Because the source of the electromagnetic field in scattering problems is the incident wave, the impressed current source term is eliminated in our treatment. We now summarize the total-field and scattered-field formulations and their corresponding boundary conditions.

Total-Field Formulation

Governing equation:

$$\frac{\partial}{\partial t}\begin{bmatrix}\vec{D}\\ \vec{B}\end{bmatrix} = \frac{1}{V}\oiint_S \begin{bmatrix}\hat{n}\times\vec{H}\\ -\hat{n}\times\vec{E}\end{bmatrix} dS + \begin{bmatrix}-\sigma\vec{E}\\ 0\end{bmatrix} \tag{4.76}$$

Boundary conditions:

Perfect conductor: $\hat{n}\times\vec{E} = 0$ (4.77a)

Dielectric interface: $\hat{n}\times\left(\vec{E}_+ - \vec{E}_-\right) = 0, \quad \hat{n}\times\left(\vec{H}_+ - \vec{H}_-\right) = 0$ (4.77b)

Resistive sheet:

$$\hat{n}\times\left(\vec{E}_+ - \vec{E}_-\right) = 0$$
$$\hat{n}\times\left(\vec{H}_+ - \vec{H}_-\right) = -G_0\,\hat{n}\times\left(\hat{n}\times\vec{E}_\pm\right) \tag{4.77c}$$

Impedance surface: $\hat{n} \times \left(\hat{n} \times \vec{E}\right) = -\psi * \hat{n} \times \vec{H}$ (4.77d)

In (4.77), G_0 is the conductivity of the resistive sheet given by

$$G_0 = \lim_{\substack{\sigma \to \infty \\ w \to 0}} (\sigma w) \tag{4.78a}$$

ψ is the inverse Fourier transform of the surface impedance, η_s, given by

$$\psi(t) = \int_{-\infty}^{\infty} \eta_s(\omega)\, e^{j\omega t}\, d\omega \tag{4.78b}$$

and $*$ represents convolution in the time domain.

Scattered-Field Formulation

Governing equation:

$$\frac{\partial}{\partial t}\begin{bmatrix} \vec{D} \\ \vec{B} \end{bmatrix} = \frac{1}{V} \oiint_S \begin{bmatrix} \hat{n} \times \vec{H} \\ -\hat{n} \times \vec{E} \end{bmatrix} dS + \begin{bmatrix} -\sigma(\vec{E} + \vec{E}^i) \\ 0 \end{bmatrix} + \frac{\partial}{\partial t}\begin{bmatrix} (\varepsilon - \varepsilon_0)\vec{E}^i \\ (\mu - \mu_0)\vec{H}^i \end{bmatrix} \tag{4.79}$$

Boundary conditions:

Perfect conductor: $\hat{n} \times \vec{E} = -\hat{n} \times \vec{E}^i$ (4.80a)

Dielectric interface: $\hat{n} \times \left(\vec{E}_+ - \vec{E}_-\right) = 0, \quad \hat{n} \times \left(\vec{H}_+ - \vec{H}_-\right) = 0$ (4.80b)

Resistive sheet:

$$\begin{aligned} \hat{n} \times \left(\vec{E}_+ - \vec{E}_-\right) &= 0 \\ \hat{n} \times \left(\vec{H}_+ - \vec{H}_-\right) &= -G_0\, \hat{n} \times \left[\hat{n} \times (\vec{E} + \vec{E}^i)\right]_{\pm} \end{aligned} \tag{4.80c}$$

Impedance surface: $\hat{n} \times \left[\hat{n} \times (\vec{E} + \vec{E}^i)\right] = -\psi * \hat{n} \times (\vec{H} + \vec{H}^i)$ (4.80d)

In the above expressions, the *i* superscript denotes the incident field and the "+" and "-" subscripts represent the fields on the two sides of the boundary. Due to the explicit nature of the algorithm, $\vec{E}$ in the last term of the right-hand side of (4.76) and $\vec{E}$ and $\vec{H}$ in the right-hand side of (4.79) are known at any time step since their values at the previous time step are used. They can also be considered at the present time level. In that case, they must be taken to the left-hand side of (4.76) and (4.79) and the equations must be rearranged. Finally, for the diffracted-field formulation, $\vec{E}^i$ and $\vec{H}^i$ are replaced by $\vec{E}^i + \vec{E}^r$ and $\vec{H}^i + \vec{H}^r$, respectively, where the *r* superscript denotes the reflected field from the infinite ground plane.

4.9.2 Treatment of Dispersive Media

The constitutive parameters of a medium (i.e., permittivity, ε, permeability, μ, and conductivity, σ) can depend on the frequency of the applied field. Solution of Maxwell's equations in the time domain in such media is still feasible [51, 77]. However, some modification is required to cast these equations into a frame suitable for application of the FVTD method.

In the following analysis, only the permittivity is assumed to vary with frequency. Extension to the more general case where all constitutive parameters are functions of frequency is straightforward. Maxwell's equations in the frequency domain and for $\exp(-j\omega t)$ time dependence are as follows:

$$\nabla \times \vec{E}(\omega, \bar{r}) = -j\omega \vec{B}(\omega, \bar{r}) \tag{4.81a}$$

$$\nabla \times \vec{H}(\omega, \bar{r}) = \sigma \vec{E}(\omega, \bar{r}) + j\omega \vec{D}(\omega, \bar{r}) \tag{4.81b}$$

The electric displacement, $\vec{D}$, can be resolved into a fast and a slow part, $\vec{D} = \vec{D}_f + \vec{D}_s$. While the fast part is proportional to the electric field independent of the frequency of excitation, the slow part is frequency dependent and is represented in terms of a frequency-dependent susceptibility function

$$\vec{D}_s(\omega, \bar{r}) = \varepsilon_0 \chi_s(\omega, \bar{r}) \vec{E}(\omega, \bar{r}) \tag{4.82}$$

Therefore, (4.81) is written as

$$\nabla \times \vec{E}(\omega, \bar{r}) = -j\omega \vec{B}(\omega, \bar{r}) \tag{4.83a}$$

$$\nabla \times \vec{H}(\omega, \bar{r}) = \sigma \vec{E}(\omega, \bar{r}) + j\omega \varepsilon_0 \chi_s(\omega, \bar{r}) \vec{E}(\omega, \bar{r}) + j\omega \vec{D}_f(\omega, \bar{r}) \tag{4.83b}$$

Inverse Fourier transforms are now applied to (4.83) to obtain the time-dependent Maxwell's equations for the dispersive dielectric case:

$$\nabla \times \vec{E}(t, \bar{r}) = -\frac{\partial \vec{B}(t, \bar{r})}{\partial t} \tag{4.84a}$$

$$\begin{aligned} \nabla \times \vec{H}(t, \bar{r}) &= \left[\sigma + \varepsilon_0 \chi_s(0, \bar{r})\right] \vec{E}(t, \bar{r}) \\ &\quad + \varepsilon_0 \frac{\partial \chi_s(t, \bar{r})}{\partial t} * \vec{E}(t, \bar{r}) + \frac{\partial \vec{D}_f(t, \bar{r})}{\partial t} \end{aligned} \tag{4.84b}$$

where

$$\begin{bmatrix} \vec{E}(t, \bar{r}) \\ \vec{H}(t, \bar{r}) \end{bmatrix} = \int_{-\infty}^{\infty} \begin{bmatrix} \vec{E}(\omega, \bar{r}) \\ \vec{H}(\omega, \bar{r}) \end{bmatrix} e^{j\omega t}\, d\omega \tag{4.85a}$$

$$\chi_s(t, \bar{r}) = \int_{-\infty}^{\infty} \chi_s(\omega, \bar{r}) e^{j\omega t} d\omega \tag{4.85b}$$

In general, the numerical solution of (4.84) requires the complete history of $\vec{E}(t, \bar{r})$ since the second term in the right-hand side of (4.84b) is a convolution. However, for most cases of practical interest, the slow susceptibility in the frequency domain is approximated as the quotient of two rational polynomials in frequency. This susceptibility function can have simple, multiple, or complex-conjugate poles; only the case of simple poles will be dealt with here. The extension to the other two cases is straightforward. In the time domain, we have

$$\chi_s(t, \bar{r}) = \sum_{k=1}^{n} A_k(\bar{r}) e^{\omega_k(\bar{r}) t} \tag{4.86}$$

When substituted into (4.84b), this expression allows the convolution to be evaluated recursively using only two time levels of the field. Without this recursion, the memory required to store the field components at each time step would be prohibitive. After algebraic manipulation, we finally arrive at

$$\left(\frac{\partial \chi_s}{\partial t} * \vec{E}\right)^m = \sum_{k=1}^{n} e^{\omega_k \Delta t} \left[\left(\frac{\partial \chi_s}{\partial t} * \vec{E}\right)^{m-1} + \frac{\omega_k A_k \Delta t}{2} \begin{matrix} \\ + \vec{E}^{m-1} + e^{-\omega_k \Delta t} \vec{E}^m \end{matrix} \right] \tag{4.87a}$$

where the superscript indicates time $t = m\Delta t$ and

$$\left(\frac{\partial \chi_s}{\partial t} * \vec{E}\right)^0 = 0 \tag{4.87b}$$

For FVTD computation, (4.84) is written in a conservation form similar to (4.75):

$$\frac{\partial}{\partial t}\begin{bmatrix} \vec{D}_f \\ \vec{B} \end{bmatrix} = \frac{1}{V} \oiint_S \begin{bmatrix} \hat{n} \times \vec{H} \\ -\hat{n} \times \vec{E} \end{bmatrix} dS + \begin{Bmatrix} -[\sigma + \varepsilon_0 \chi_s(0, \bar{r})]\vec{E} \\ 0 \end{Bmatrix} + \begin{bmatrix} -\varepsilon_0 \dfrac{\partial \chi_s(t, \bar{r})}{\partial t} * \vec{E} \\ 0 \end{bmatrix} \tag{4.88}$$

where the convolution term is given by the recursive algorithm in (4.87).

4.10 FINITE-VOLUME DISCRETIZATION AND TIME INTEGRATION FOR STRUCTURED GRIDS

Many problems in mathematical physics are modeled by a set of linear or nonlinear partial differential equations. For example, the dynamics of a fluid are governed by the Navier-Stokes equations, and scattering and radiation phenomena of electromagnetic waves are governed by Maxwell's equations. Many of these equations lend themselves to a conservation form representation given by

$$\frac{\partial \mathbf{Q}}{\partial t} + \frac{\partial \mathbf{R}}{\partial x} + \frac{\partial \mathbf{S}}{\partial y} + \frac{\partial \mathbf{T}}{\partial z} = \mathbf{U} \tag{4.89}$$

where $\mathbf{Q}$ is the unknown vector to be computed; $\mathbf{R}$, $\mathbf{S}$, and $\mathbf{T}$ are assumed given at each instant of time; and $\mathbf{U}$ is the source of excitation or forcing function. In CFD, $\mathbf{Q}$ usually represents mass, three components of the momentum, and the energy; while $\mathbf{R}$, $\mathbf{S}$, and $\mathbf{T}$ are flux vectors in the x, y, and z coordinate directions, respectively. In electromagnetics, $\mathbf{Q}$ is identified as a vector whose components are the electric displacement and magnetic induction; and $\mathbf{R}$, $\mathbf{S}$, and $\mathbf{T}$ are magnetic and electric fields transverse to the x, y, and z coordinate directions, respectively. Application of (4.89) to many realistic problems requires a coordinate transformation to properly represent the physical domain of interest and to assist satisfying the boundary conditions.

Under the transformation of coordinates given by

$$\tau = t, \quad \xi = \xi(t, x, y, z), \quad \eta = \eta(t, x, y, z), \quad \zeta = \zeta(t, x, y, z) \tag{4.90}$$

(4.89) is recast in the conservation form given by

$$\frac{\partial Q}{\partial \tau} + \frac{\partial R}{\partial \xi} + \frac{\partial S}{\partial \eta} + \frac{\partial T}{\partial \zeta} = U \tag{4.91}$$

where

$$Q = \frac{\mathbf{Q}}{J}, \qquad U = \frac{\mathbf{U}}{J} \tag{4.92a}$$

$$R = \frac{\xi_t}{J}\mathbf{Q} + \frac{\xi_x}{J}\mathbf{R} + \frac{\xi_y}{J}\mathbf{S} + \frac{\xi_z}{J}\mathbf{T} \tag{4.92b}$$

$$S = \frac{\eta_t}{J}\mathbf{Q} + \frac{\eta_x}{J}\mathbf{R} + \frac{\eta_y}{J}\mathbf{S} + \frac{\eta_z}{J}\mathbf{T} \tag{4.92c}$$

$$T = \frac{\zeta_t}{J}\mathbf{Q} + \frac{\zeta_x}{J}\mathbf{R} + \frac{\zeta_y}{J}\mathbf{S} + \frac{\zeta_z}{J}\mathbf{T} \tag{4.92d}$$

Here, J is the Jacobian of the coordinate transformation given by

$$J = \frac{\partial(\tau, \xi, \eta, \zeta)}{\partial(x, y, z)} \tag{4.93}$$

and ξ_t, ξ_x, ξ_y, ξ_z, η_t, η_x, η_y, η_z, ζ_t, ζ_x, ζ_y, and ζ_z are the transformation metrics.

Equation (4.91) will now be discretized with respect to the spatial variables. Consider the mesh-cell geometry shown in Fig. 4.28.

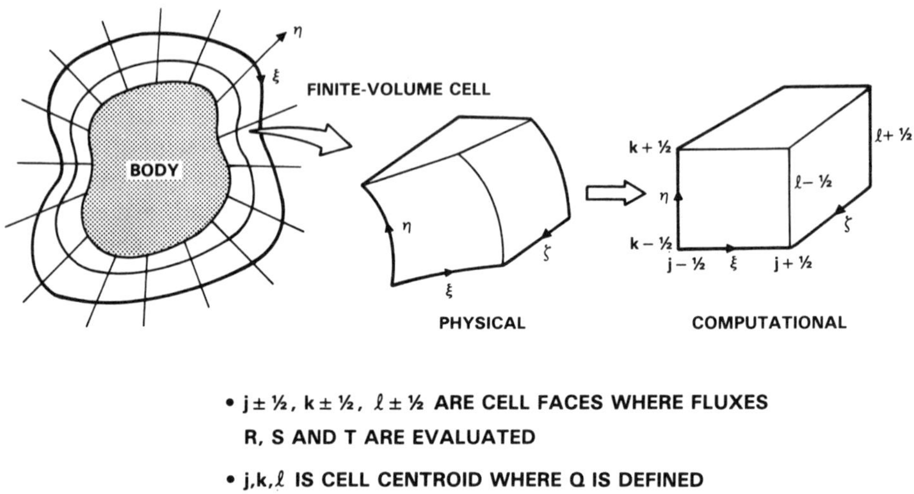

Fig. 4.28 Coordinate transformation of the mesh cell from the physical system to the computational system.

In Fig. 4.28, we associate the physical mesh cell located at (ξ, η, ζ) with the computational mesh cell located at (j, k, ℓ). The resulting discretization of (4.91) is

$$\begin{aligned} \frac{\partial Q_{j,k,\ell}}{\partial \tau} &+ \left(R_{j+\frac{1}{2},k,\ell} - R_{j-\frac{1}{2},k,\ell}\right) + \left(S_{j,k+\frac{1}{2},\ell} - S_{j,k-\frac{1}{2},\ell}\right) \\ &+ \left(T_{j,k,\ell+\frac{1}{2}} - T_{j,k,\ell-\frac{1}{2}}\right) = U_{j,k,\ell} \end{aligned} \tag{4.94}$$

where R, S, and T are the fluxes at the sides of the cell and Q is the conserved quantity inside that cell, which for convenience is assumed located at the centroid of the cell. The half-integer subscripts denote the cell sides, and the integer subscripts the cell itself or its centroid. Equation (4.94), which is the semidiscrete form of the conservation law represented by (4.91), is called a finite-volume discretization form of the conservation law. The numerical integration of this equation results in the desired solution for Q.

In the next section, an integration method for (4.94) known as the Lax-Wendroff upwind scheme is discussed. While the discussion is confined to one dimension only for the sake of clarity, extension to three dimensions is straightforward.

4.10.1 Characteristics-Based Lax-Wendroff Upwind Scheme

For one-dimensional wave motion (i.e., the telegrapher's equations), (4.91) reduces to

$$\frac{\partial Q}{\partial \tau} + \frac{\partial R}{\partial \xi} = 0 \tag{4.95}$$

In general, R could be a nonlinear function of Q. By carrying out the indicated differentiation in the second term, we obtain

$$\frac{\partial Q}{\partial \tau} + A\frac{\partial Q}{\partial \xi} = 0 \tag{4.96}$$

where $A = \partial R/\partial Q$ is the system matrix for the set of linear equations in (4.96). For nonlinear systems, (4.96) is only a quasi-linear approximation to (4.95), while for linear systems, (4.96) is an exact representation.

Across a characteristic of (4.96), the following condition known as the Rankine-Hugoniot relation (the "jump condition") is satisfied [78]:

$$-\lambda|Q| + |R| = 0 \tag{4.97}$$

where λ is an eigenvalue of the system matrix, A, and is the propagation speed of the discontinuity; and $|Q|$ and $|R|$ are jumps in the value of these quantities across a characteristic. The λ values are solutions of the eigenvalue equation

$$\det(A - \lambda I) = 0 \tag{4.98}$$

where I is an identity matrix. From (4.97), the flux, R, is expressed in terms of Q in a known neighborhood. This should become more clear when it is later applied to Maxwell's equations.

Q can now be computed from (4.97). There are many ways to solve this equation. Based on over two decades of research work in CFD, one of the best methods is the Lax-Wendroff explicit upwind scheme. This scheme is implemented in two steps, known as the predictor and corrector. At the end of the predictor step, the results are first-order accurate with respect to both time and space. Then, the corrector makes these results second-order accurate. We now review this scheme. Let

$$Q^n = Q(n\Delta\tau) \tag{4.99}$$

A Taylor series is used to expand $Q(n\Delta\tau + \Delta\tau)$ about $n\Delta\tau$:

$$Q^{n+1} = Q^n + \left.\frac{\partial Q}{\partial \tau}\right|^n \Delta\tau + \left.\frac{\partial^2 Q}{\partial \tau^2}\right|^n \frac{(\Delta\tau)^2}{2} + \cdots \tag{4.100}$$

where

$$\left.\frac{\partial^2 Q}{\partial \tau^2}\right|^n = \frac{1}{\Delta\tau}\left(\left.\frac{\partial Q}{\partial \tau}\right|^{n+1} - \left.\frac{\partial Q}{\partial \tau}\right|^n\right) + \text{higher-order terms} \tag{4.101}$$

Also, based on the characteristic theory of signal propagation, $\partial R/\partial\xi$ is resolved into the two components

$$\frac{\partial R}{\partial \xi} = \frac{\overrightarrow{\partial} \bar{R}}{\partial \xi} + \frac{\overleftarrow{\partial} \overset{+}{R}}{\partial \xi} \tag{4.102}$$

where the overbar $\rightarrow$ indicates forward differencing and the overbar $\leftarrow$ indicates backward differencing. In this way, regardless of the direction of the signal, the scheme is always an upwind one. The two derivatives are expressed as

$$\left.\frac{\overrightarrow{\partial} R^-}{\partial \xi}\right|_j = \frac{R^-_{j+1} - R^-_j}{\Delta\xi} - \frac{\Delta\xi}{2}\left.\frac{\overrightarrow{\partial}^2 R^-}{\partial \xi^2}\right|_j + \cdots \tag{4.103a}$$

$$\left.\frac{\overleftarrow{\partial} R^+}{\partial \xi}\right|_j = \frac{R^+_j - R^+_{j-1}}{\Delta\xi} + \frac{\Delta\xi}{2}\left.\frac{\overleftarrow{\partial}^2 R^+}{\partial \xi^2}\right|_j + \cdots \tag{4.103b}$$

Equations (4.103a) and (4.103b) are derived from the Taylor series expansion of R^-_{j+1} and R^+_{j-1} around j. As a result, $\partial R/\partial\xi$ at j is expressed as

$$\left.\frac{\partial R}{\partial \xi}\right|_j = \frac{R_{j+\frac{1}{2}} - R_{j-\frac{1}{2}}}{\Delta\xi} - \frac{\Delta\xi}{2}\left(\left.\frac{\overrightarrow{\partial}^2 R^-}{\partial \xi^2}\right|_j - \left.\frac{\overleftarrow{\partial}^2 R^+}{\partial \xi^2}\right|_j\right) + \cdots \tag{4.104}$$

where

$$R_{j+\frac{1}{2}} = R^-_{j+1} + R^+_j\,, \qquad R_{j-\frac{1}{2}} = R^-_j + R^+_{j-1} \tag{4.105}$$

Predictor Step

First-order accurate results with respect to the time are obtained if only two terms of the series in (4.100) are retained. It is customary to represent the current time level in the predictor as $\overline{n+1}$. The solution at the end of this step is represented by

$$Q^{\overline{n+1}} \cong Q^n - \left.\frac{\partial R}{\partial \xi}\right|^n \Delta\tau \tag{4.106}$$

and computed with first-order accuracy in space from

$$Q^{\overline{n+1}} \cong Q^n - \frac{\Delta\tau}{\Delta\xi}\left(R_{j+\frac{1}{2}} - R_{j-\frac{1}{2}}\right) \tag{4.107}$$

where only the first term of (4.104) is retained.

Corrector Step

The solution obtained in the predictor step from (4.106) is made second-order accurate by including the third term in (4.100), that is:

$$Q^{n+1} \cong Q^{\overline{n+1}} + \left.\frac{\partial^2 Q}{\partial \tau^2}\right|^n \frac{(\Delta\tau)^2}{2} \tag{4.108}$$

Substituting (4.101) in the above expression, we obtain after some manipulations

$$Q^{n+1} \cong \frac{1}{2}\left(Q^n + Q^{\overline{n+1}} - \left.\frac{\partial R}{\partial \xi}\right|^{\overline{n+1}} \Delta\tau\right) \tag{4.109}$$

Q obtained from (4.109) is second-order accurate with respect to both time and space provided that the spatial second-order truncation error is annihilated by including the second term in the right-hand side of (4.104):

$$Q_j^{n+1} = \frac{1}{2}\left(Q_j^n + Q_j^{\overline{n+1}}\right) - \frac{\Delta\tau}{2\Delta\xi}\left(R_{j+\frac{1}{2}}^{\overline{n+1}} - R_{j-\frac{1}{2}}^{\overline{n+1}}\right)$$
$$+ \frac{\Delta\tau\Delta\xi}{4}\left(\left.\frac{\overrightarrow{\partial^2 R}^{-}}{\partial \xi^2}\right|_j^n - \left.\frac{\overleftarrow{\partial^2 R}^{+}}{\partial \xi^2}\right|_j^n\right) \tag{4.110}$$

In (4.110), the second derivatives are computed from the first derivatives:

$$\left.\frac{\overset{\rightarrow}{\partial}{}^2 \overset{-}{R}}{\partial \xi^2}\right|_j = \frac{1}{\Delta\xi}\left(\left.\frac{\overset{\rightarrow}{\partial} \overset{-}{R}}{\partial \xi}\right|_{j+1} - \left.\frac{\overset{\rightarrow}{\partial} \overset{-}{R}}{\partial \xi}\right|_{j}\right) + \cdots \tag{4.111a}$$

$$\left.\frac{\overset{\leftarrow}{\partial}{}^2 \overset{+}{R}}{\partial \xi^2}\right|_j = \frac{1}{\Delta\xi}\left(\left.\frac{\overset{\leftarrow}{\partial} \overset{+}{R}}{\partial \xi}\right|_{j} - \left.\frac{\overset{\leftarrow}{\partial} \overset{+}{R}}{\partial \xi}\right|_{j-1}\right) + \cdots \tag{4.111b}$$

The first derivatives are approximated by the first terms in (4.103a) and (4.103b):

$$\left.\frac{\overset{\rightarrow}{\partial} \overset{-}{R}}{\partial \xi}\right|_j \cong \frac{R^-_{j+1} - R^-_j}{\Delta\xi} = \frac{dR^-_{j+\frac{1}{2}}}{\Delta\xi} \tag{4.112a}$$

$$\left.\frac{\overset{\leftarrow}{\partial} \overset{+}{R}}{\partial \xi}\right|_j \cong \frac{R^+_j - R^+_{j-1}}{\Delta\xi} = \frac{dR^+_{j-\frac{1}{2}}}{\Delta\xi} \tag{4.112b}$$

where the differential terms are the second-order corrections to the jump across the characteristics. For example, we have

$$dR^+_{j-\frac{1}{2}} = R^{\mathrm{R}}_j - R_{j-\frac{1}{2}} \qquad \text{or} \qquad dR^-_{j+\frac{1}{2}} = R_{j+\frac{1}{2}} - R^{\mathrm{L}}_j \tag{4.113}$$

where, as shown in Fig. 4.29, the superscripts R and L stand for the neighborhoods to the right of the positive characteristics and to the left of the negative characteristics, respectively.

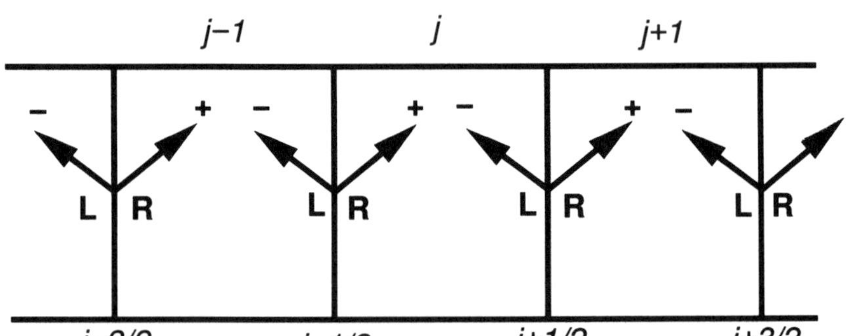

Fig. 4.29 Characteristics and neighborhoods associated with a one-dimensional wave propagation grid.

Therefore, the second derivatives in terms of the differential fluxes are

$$\left.\frac{\partial^2 \overrightarrow{R}^-}{\partial \xi^2}\right|_j \cong \frac{dR^-_{j+\frac{3}{2}} - dR^-_{j+\frac{1}{2}}}{(\Delta\xi)^2} \tag{4.114a}$$

$$\left.\frac{\partial^2 \overleftarrow{R}^+}{\partial \xi^2}\right|_j \cong \frac{dR^+_{j-\frac{1}{2}} - dR^+_{j-\frac{3}{2}}}{(\Delta\xi)^2} \tag{4.114b}$$

After substituting these expressions into (4.110), it is cast in the following form:

$$\begin{aligned} Q_j^{n+1} &= \frac{1}{2}\left(Q_j^n + Q_j^{\overline{n+1}}\right) - \frac{\Delta\tau}{2\Delta\xi}\left(R^{\overline{n+1}}_{j+\frac{1}{2}} - R^{\overline{n+1}}_{j-\frac{1}{2}}\right) \\ &\quad - \frac{\Delta\tau}{2\Delta\xi}\left(dR^+_{j-\frac{1}{2}} - dR^+_{j-\frac{3}{2}}\right)^n + \frac{\Delta\tau}{2\Delta\xi}\left(dR^-_{j+\frac{3}{2}} - dR^-_{j+\frac{1}{2}}\right)^n \end{aligned} \tag{4.115}$$

Stability Condition

Courant, Friedrichs, and Lewy demonstrated in their classical paper [79] that a necessary condition for the convergence of a finite-difference scheme is that the rate of propagation of signals in the difference scheme must be at least as large as the physical speed of propagation in the medium. Based on their work, the two-step scheme discussed here is shown [80] to be stable if and only if

$$|\lambda|_{\max} \frac{\Delta\tau}{\Delta\xi} \leq 2 \tag{4.116}$$

where $|\lambda|_{\max}$ is the largest eigenvalue of A at any point. In multidimensions, the value of $|\lambda|_{\max}$ is taken as

$$|\lambda|_{\max} = \max\left(|\lambda_\xi| + |\lambda_\eta| + |\lambda_\zeta| \right) \tag{4.117}$$

4.10.2 Application to Maxwell's Equations

Maxwell's equations (4.71) are now cast into the conservation form of (4.89):

$$\frac{\partial}{\partial t}\begin{pmatrix} \vec{D} \\ \vec{B} \end{pmatrix} + \frac{\partial}{\partial x}\begin{pmatrix} -\hat{x}\times\vec{H} \\ \hat{x}\times\vec{E} \end{pmatrix} + \frac{\partial}{\partial y}\begin{pmatrix} -\hat{y}\times\vec{H} \\ \hat{y}\times\vec{E} \end{pmatrix} + \frac{\partial}{\partial z}\begin{pmatrix} -\hat{z}\times\vec{H} \\ \hat{z}\times\vec{E} \end{pmatrix} = \begin{pmatrix} \vec{U} \\ 0 \end{pmatrix} \tag{4.118}$$

where the quantities in the second, third, and fourth terms of the left-hand side are the electric and magnetic fields tangential to, respectively, x-, y-, and z-constant faces of a cell in a Cartesian coordinate system. The body-fitted coordinate transformation previously introduced is applied to (4.118) to obtain

$$\frac{\partial}{\partial t}\begin{pmatrix}\dfrac{\vec{D}}{J}\\[2ex]\dfrac{\vec{B}}{J}\end{pmatrix}+\frac{\partial}{\partial \xi}\begin{pmatrix}\dfrac{-\vec{\xi}\times\vec{H}}{J}\\[2ex]\dfrac{\vec{\xi}\times\vec{E}}{J}\end{pmatrix}+\frac{\partial}{\partial \eta}\begin{pmatrix}\dfrac{-\vec{\eta}\times\vec{H}}{J}\\[2ex]\dfrac{\vec{\eta}\times\vec{E}}{J}\end{pmatrix}+\frac{\partial}{\partial \zeta}\begin{pmatrix}\dfrac{-\vec{\zeta}\times\vec{H}}{J}\\[2ex]\dfrac{\vec{\zeta}\times\vec{E}}{J}\end{pmatrix}=\begin{pmatrix}\dfrac{\vec{U}}{J}\\[2ex]0\end{pmatrix} \tag{4.119}$$

where J is given by (4.93) and

$$\vec{\xi} = \nabla\xi, \qquad \vec{\eta} = \nabla\eta, \qquad \vec{\zeta} = \nabla\zeta \tag{4.120}$$

As discussed earlier, (4.119) is solved numerically and $\vec{D}$ and $\vec{B}$ (or $\vec{E}$ and $\vec{H}$) are found once the tangential fields on the cell walls are obtained. This is done by applying (4.97) after the characteristics are found from (4.98). For a locally isotropic medium, in each of the three planes (t, ξ), (t, η), and (t, ζ), the matrices, A, have six eigenvalues among them, with three of them distinct. For example, in the (t, ξ) plane, we have

$$\lambda^- = -c\left|\vec{\xi}\right|, \qquad \lambda^0 = 0\,, \qquad \lambda^+ = c\left|\vec{\xi}\right| \tag{4.121}$$

where c is the speed of light in the medium. Substituting these eigenvalues in (4.97) and using the notation sketched in Fig. 4.30 results in the following expressions:

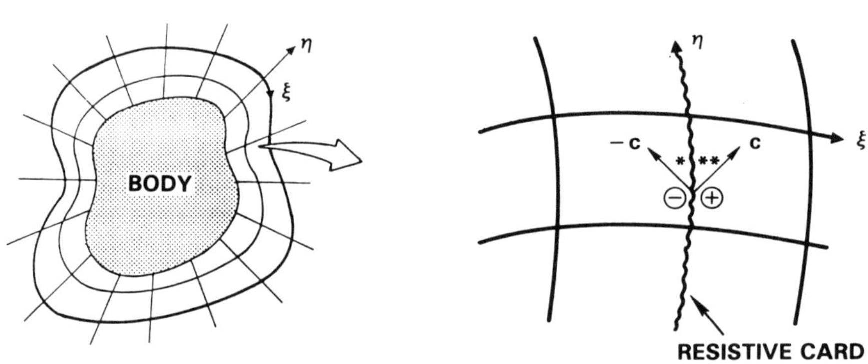

Fig. 4.30 Designation of the neighborhoods about a boundary interface between two cells.

$$Z^-\left(\vec{H}^* - \vec{H}^-\right) = -\hat{\xi} \times \left(\vec{E}^* - \vec{E}^-\right) \tag{4.122a}$$

$$Z^+\left(\vec{H}^{**} - \vec{H}^+\right) = \hat{\xi} \times \left(\vec{E}^{**} - \vec{E}^+\right) \tag{4.122b}$$

across λ^- and λ^+, respectively, and

$$\hat{\xi} \times \left(\vec{E}^{**} - \vec{E}^*\right) = 0 \tag{4.123a}$$

$$\hat{\xi} \times \left(\vec{H}^{**} - \vec{H}^*\right) = -G_0\, \hat{\xi} \times \left[\hat{\xi} \times \left(\vec{E}^* + \vec{E}^i\right)\right] \tag{4.123b}$$

across λ^0. In the above expressions, the superscript "i" represents the incident field and

$$\hat{\xi} = \frac{\nabla\xi}{|\nabla\xi|} \tag{4.124}$$

While (4.122) is valid for any of the field formulations discussed previously, (4.123) applies only to the scattered-field formulation. These equations are simply the conditions satisfied by the tangential components of the field on a constant-ξ interface between adjacent cells. Similar equations for other types of field formulations can be derived. In fact, from these expressions as well as others to be derived from them later, equivalent relations for a total-field formulation are derived by setting $\vec{E}^i = 0$. Also, upon replacing $\vec{E}^i$ by $\vec{E}^i + \vec{E}^r$, they become those of the diffracted-field formulation.

System (4.123) applies to the general case where a resistive sheet is located at the interface as depicted in Fig. 4.30. The special cases of a dielectric interface and a PEC boundary are derived in the limit as $G_0 \to 0$ and $G_0 \to \infty$, respectively. Simultaneous solution of systems (4.122) and (4.123) provides the tangential fields on the cell walls required in the numerical solution of (4.119). They are as follows:

$$\hat{\xi} \times \vec{E}^* = \hat{\xi} \times \frac{(Y\vec{E} - \hat{\xi}\times\vec{H})^- + (Y\vec{E} + \hat{\xi}\times\vec{H})^+ - G_0\vec{E}^i}{Y^- + Y^+ + G_0} \tag{4.125}$$

$$\hat{\xi} \times \vec{H}^* = \hat{\xi} \times \frac{\left[\begin{array}{c}(1 + G_0 Z^+)(Z\vec{H} + \hat{\xi}\times\vec{E})^- \\ + (Z\vec{H} - \hat{\xi}\times\vec{E})^+ + G_0 Z^+ \hat{\xi}\times\vec{E}^i\end{array}\right]}{Z^- + Z^+ + G_0 Z^- Z^+} \tag{4.126}$$

$$\hat{\xi} \times \vec{H}^{**} = \hat{\xi} \times \frac{\left[\begin{array}{c}(Z\vec{H} + \hat{\xi}\times\vec{E})^- - G_0 Z^- \hat{\xi}\times\vec{E}^i \\ + (1 + Z^- G_0)(Z\vec{H} - \hat{\xi}\times\vec{E})^+\end{array}\right]}{Z^- + Z^+ + G_0 Z^- Z^+} \tag{4.127}$$

where

$$Z^{\pm} = \sqrt{\frac{\mu^{\pm}}{\varepsilon^{\pm}}} \tag{4.128}$$

are the characteristic impedances of the media in the + and – cells, respectively, and $Y^{\pm} = 1/Z^{\pm}$ are the characteristic admittances.

For an impedance surface, it is more convenient to express the jump conditions in the frequency domain and to derive the tangential fields in that domain. Those relations are then transformed to the time domain. The end results for the scattered-field formulation are as follows:

$$\begin{aligned} \hat{\xi} \times \vec{H}^* &= \hat{\xi} \times \left\{ Z_s * \left[(Z\vec{H} + \hat{\xi} \times \vec{E})^- + (Z^- \vec{H} + \hat{\xi} \times \vec{E})^i \right] \right\} \\ &- \hat{\xi} \times \vec{H}^i \end{aligned} \tag{4.129}$$

$$\hat{\xi} \times \vec{E}^* = \hat{\xi} \times \left[(\vec{E} + Z\,\hat{\xi} \times \vec{H})^- - Z^- \hat{\xi} \times \vec{H}^* \right] \tag{4.130}$$

where

$$Z_s(\tau) = \int_{-\infty}^{\infty} \left(\frac{1}{\eta_s(\omega) + Z^-} \right) e^{j\omega\tau}\, d\omega \tag{4.131}$$

4.10.3 Outer Boundary Treatment

Equations (4.125) to (4.130) indicate that at each interface between two cells, knowledge of the field at the centroid of both cells is necessary in the evaluation of the tangential fields. At the outer boundary of the computational domain, however, depending on the grid orientation, either + or – quantities are generally unknown. Therefore, a special condition on the outer boundary needs to be established.

Usually, exact or global conditions are difficult to formulate and are demanding on computer resources but are accurate. One such method that fits the approach discussed in this chapter is to express the field everywhere outside the computational domain in terms of the tangential field on the outer boundary using either the Stratton-Chu integral relation [81] or the Franz formulation [82]. By eliminating the tangential field between (4.125) to (4.127) on one hand and the integral expression on the other hand, the unknown + quantities can be evaluated. As a result, the outer boundary is brought as close as one cell width from the target. This greatly reduces the number of unknowns in the computational domain. However, at each time step, a $2M \times 2M$ matrix must be inverted where M is the number of grid points on the outer boundary. Another exact outer boundary condition formulated for FDTD was previously proposed in [83].

Before 1994, approximate or local outer boundary conditions were less accurate than global conditions but also much less expensive in terms of computer resources. One of the best treatments of this subject is by Engquist and Majda [84]. Since 1994, Berenger's PML ABC [85] and its variants have found great utility in FDTD applications (see Chapter 5), yielding outer boundary reflection coefficients in the order of -100 dB. At present, there is no literature available on the application of the PML ABC to the characteristic-based FVTD method discussed here.

The local ABC used in the present FVTD method is equivalent to the first-order Engquist-Majda condition. The form of this boundary condition depends on the field formulation used in a particular problem. For a scattered-field formulation, the wave at the outer boundary is totally outgoing. Therefore, to derive the tangential field on the outer boundary, the incoming wave must be set equal to zero in (4.126) and (4.127) as well as $G_0 = 0$. The grid orientation determines whether the + or – term represents the incoming wave, which is set equal to zero. This leads to the following expressions for the fields tangential to the outer boundary:

$$\hat{\xi} \times \vec{E}^* = \frac{1}{2}\hat{\xi} \times (\vec{E} \pm Z\,\hat{\xi} \times \vec{H})^{\pm} \tag{4.132}$$

$$\hat{\xi} \times \vec{H}^* = \frac{1}{2}\hat{\xi} \times (\vec{H} \pm Y\,\hat{\xi} \times \vec{E})^{\mp} \tag{4.133}$$

For a total-field formulation, the incident field at the outer boundary is incoming. Thus, the appropriate + or – term must be set equal to the incoming wave. The results are as follows:

$$\hat{\xi} \times \vec{E}^* = \frac{1}{2}\hat{\xi} \times \left[(\vec{E} \pm Z\,\hat{\xi} \times \vec{H})^{\pm} + (\vec{E} - Z^{\pm}\,\hat{\xi} \times \vec{H})^{i}\right] \tag{4.134}$$

$$\hat{\xi} \times \vec{H}^* = \frac{1}{2}\hat{\xi} \times \left[(\vec{H} \pm Y\,\hat{\xi} \times \vec{E})^{\mp} + (\vec{H} + Y^{\mp}\,\hat{\xi} \times \vec{E})^{i}\right] \tag{4.135}$$

Special attention must be paid to scattering from objects over a ground plane or cavities in a ground plane, and also scattering from a half-plane or a wedge. In these problems, since only a small portion of the infinite or semi-infinite plane resides in the computational domain, it is necessary to account for the waves reflected from the part of the plane that lies outside the computational domain. This is done simply by considering those reflected waves as incoming waves and including them in the outer boundary condition. For example, if the scattered-field formulation is used, the tangential field on the outer boundary is obtained from (4.132) to (4.135), provided that the incident wave there is replaced by the reflected waves. In general, if the approximate or asymptotic behavior of the field outside the computational domain is known (such as in the half-plane problem), suitable equations similar to (4.132) to (4.135) can be formulated.

4.10.4 Multizoning and Numerical Generation of Body-Fitted Meshes

The geometry and physics of many realistic scattering and radiating structures is very complex. As a result, generating a global body-fitted mesh could be difficult, if not impossible. Dividing the computational space into less complex zones containing independent local meshes is potentially advantageous. Here, the mesh density in each zone is decided based on the local material properties. Further, the mesh lines across a zonal boundary do not have to match. Also, zones contacting the outer boundary have reduced meshing density near that boundary without significant loss of accuracy. This reduces the total number of mesh points and consequently the computer resource burden.

There are various numerical and analytical methods to construct a zonal mesh, with the optimum approach depending on the type of mesh needed. In turn, this depends on the geometry and material properties of the zone. Meshes are classified into structured (regular) and unstructured (irregular) types. Structured meshes allow convenient implementation of the FVTD algorithm and efficient processing on vector computers, but are less flexible for controlling the mesh density. Unstructured meshes are very flexible in terms of controlling the mesh density and cell shape, but involve more complicated bookkeeping to implement the numerical technique and are more suitable for parallel-processing computers. The authors believe that a hybrid approach is most useful wherein unstructured meshes are used in some zones and structured meshes in others. An example of this approach is shown in Fig. 4.31, where a hybrid grid is used to model the leading edge of an airfoil that has a complex, inhomogeneous, dielectric composition.

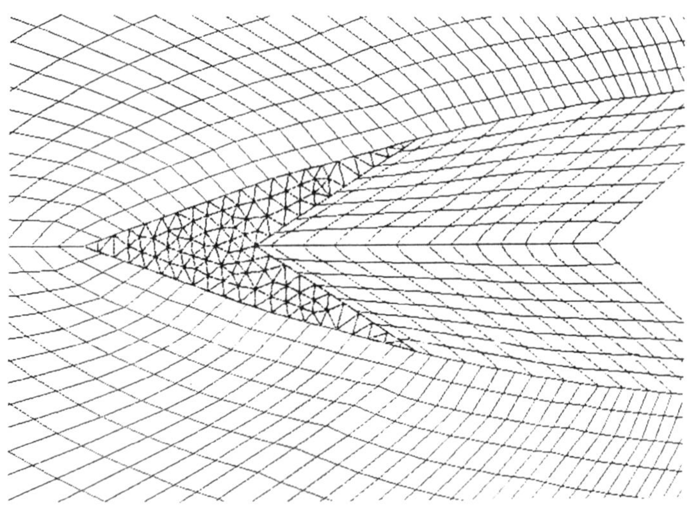

Fig.4.31 The leading edge of an airfoil where a hybrid grid is used.

The details of mesh-generation techniques are outside the scope of the present discussion. A good review of techniques for numerical generation of unstructured grids is found in [86]. Some examples of the generation of structured grids are discussed in [87–93].

4.11 FINITE-VOLUME DISCRETIZATION AND TIME INTEGRATION FOR UNSTRUCTURED MESHES

Unstructured meshes can be easily used to fill the corners, gaps, and other nearly singular regions that arise in complex aircraft geometries. The meshes in these regions do not constrain those outside in any significant way. To achieve the same freedom with structured meshes, each special region must be treated as an independent zone, and interface conditions must be enforced at every zonal boundary. Even when this is done, any substantial change in the cell size at a boundary can introduce both reflections from the boundary and changes in the character of waves that propagate across the boundary.

We note that unstructured meshing does *not* remove this source of error: rather, it *compounds* the problem because there is no relation between the dimensions of neighboring cells in the mesh. In fact, automatically generated grids typically contain a significant percentage of shards, needles, and other odd shapes that result from the final stages of volume filling. Even in regions adjoining the boundaries where resolution is specified, there is broad variation in cell aspect ratios. The burden of maintaining accuracy therefore falls on the Maxwell's equations solution algorithm.

The Lax-Wendroff upwind algorithm in Rockwell's RCSMPP code, as generalized to unstructured grids, has been found to be inadequate for modeling long-distance wave propagation through tetrahedral grids at moderate levels of resolution (10 to 20 points per wavelength). The low accuracy can be traced to how the neighbor-cell geometry enters the estimates for the spatial derivatives inside a cell. Error terms proportional to the difference in the cell dimensions appear, so that values for the fields extrapolated to the cell faces are no longer first-order accurate, and their surface integrals no longer provide a second-order-accurate update.

Including downwind as well as upwind data in estimating the spatial derivatives allows a more accurate extrapolation to be constructed. The resulting central / upwind integration algorithm does a much better job of maintaining both the phase and amplitude of waves as they propagate through distorted grids. Fig. 4.32 shows results for both the original, purely upwind algorithm and the central / upwind algorithm when used to propagate a sinusoidal wave through the stretched triangular grid shown at the top of the figure. Here, periodic boundary conditions are enforced along both the x and y directions). After the wave moves through four cycles, the purely upwind algorithm exhibits severe damping of the signal, while both the phase and amplitude are accurately propagated by the central / upwind algorithm. The extension of this algorithm to higher-order accuracy is outlined below.

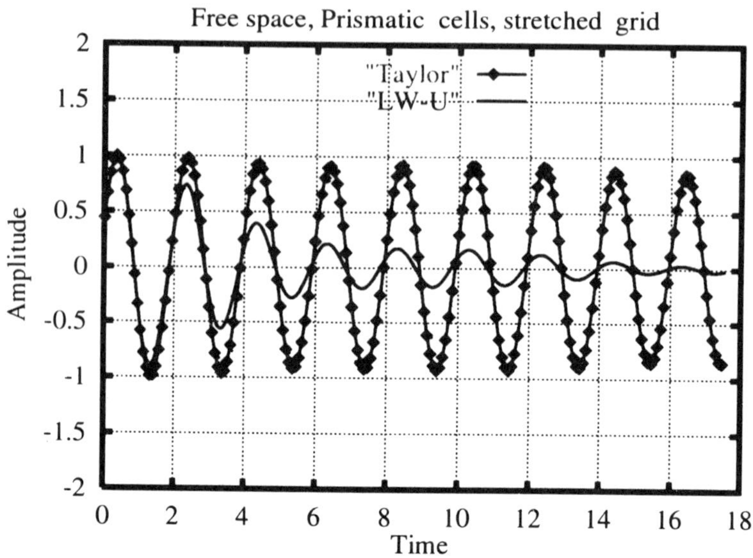

Fig.4.32 Comparison of results for the original, purely upwind FVTD algorithm and the central / upwind algorithm when used to propagate a sinusoidal wave through a stretched triangular grid.

4.11.1 An Accurate Upwind Taylor Scheme for General Grids

In developing the second-order finite-difference scheme that bears their names, Lax and Wendroff started from a Taylor series expansion of the solution $u(x, t+\Delta t)$ in the time increment, Δt. They replaced the time derivatives of u in this expansion with space derivatives of the flux, $f(u)$, where the basic partial differential equation had the form

$$\frac{\partial u}{\partial t} + \frac{\partial f(u)}{\partial x} = 0 \tag{4.136}$$

The space derivatives were then approximated by central differencing on a uniform grid. The values $u(x_j, t_n)$ could be regarded either as approximations to the point values of u or as averages over each grid cell.

Extension of this approach to unstructured grids is most naturally carried out in terms of cell averages. Integrating the multidimensional form of (4.136) given by

$$\frac{\partial \vec{u}}{\partial t} + \nabla \cdot \bar{\bar{f}} = 0 \tag{4.137}$$

over cell α in such a grid gives

$$\frac{\partial}{\partial t}\int_{\alpha} \vec{u}(\vec{r}, t)\,dV = \frac{\partial}{\partial t}\vec{U}_{\alpha}(t)V_{\alpha} = \int_{\alpha} \nabla \cdot \bar{\bar{f}}(\vec{u})\,dV = \oint_{\partial\alpha} \hat{n}\cdot\bar{\bar{f}}(\vec{u})\,dS \tag{4.138}$$

where $\vec{U}_{\alpha}$ is the cell average of the function $\vec{u}$ (which may be a vector of many components); V_{α} is the volume of the cell; $\bar{\bar{f}}(\vec{u})$ is a dyadic function; $\partial\alpha$ denotes the boundary of the cell; and $\hat{n}$ is the unit outward normal on this boundary. A Taylor series for $\vec{U}_{\alpha}(t + \Delta t)$ is written in terms of (4.138) and its higher time derivatives.

Evaluating a typical term in this series involves determining values for the function $\vec{u}$ and its spatial derivatives up to some order at the boundary of the cell. The simplest schemes of this type advance only the cell averages to the next time level and approximate or reconstruct the variation of $\vec{u}$ within a cell from these averages.

The boundary values of the reconstructed function at the interface between two cells differ in general, depending on from which side the interface is approached. Godunov suggested regarding this difference as a discontinuity in the actual solution, and proposed calculating the cell averages at the next time level by appropriately propagating this discontinuous field. This step involves solving a classical Riemann problem at each interface.

When $\bar{\bar{f}}(\vec{u})$ depends linearly on $\vec{u}$, as it does for Maxwell's equations in the limit of weak fields, the Riemann problem can be solved exactly in terms of the eigenfunctions and eigenvalues of the Jacobian matrix $\partial\bar{\bar{f}}/\partial\vec{u}$. Characteristic combinations of the elements of $\vec{u}$ can be identified that propagate as simple waves $W(\hat{n}\cdot\vec{r} \pm ct)$ across the interface. Expressing the discontinuity in $\vec{u}$ in terms of these waves allows the behavior of $\vec{u}$ at the interface for infinitesimally later times $t + \varepsilon$ to be well approximated. Because the data determining this behavior are drawn only from the waves that propagate toward the interface from either side, this process is called *upwinding*.

It is the upwind values for $\vec{u}$ and its derivatives that enter into the surface integrals of (4.138) and its derivatives in the upwind Taylor scheme. To complete the description of the scheme, we need to specify how $\vec{u}$ is to be reconstructed in each cell from its averages. The details of this procedure directly influence both the accuracy and stability of the overall algorithm, as well as its computational efficiency.

In the upwind Lax-Wendroff scheme [20], a linear variation of the solution in each cell is determined from upwind values for $\vec{u}$ at the cell faces, calculated from the average values in the two cells neighboring each interface. This procedure doubles the size of the allowable time step compared to the original Lax-Wendroff scheme, but it limits the accuracy of the reconstruction and adds considerable dissipation in cells large compared to the optimum size.

An alternative reconstruction procedure commonly used in CFD starts by assuming that the solution is smooth over a region encompassing many near neighbors of each cell, with continuous derivatives up to some order. The averages over neighboring cells are then expressed in terms of these derivatives, and a collection of these expressions is solved for the derivatives [71]. Real discontinuities in the solution such as shock waves complicate the choice of neighbors and may degrade the accuracy locally.

In electromagnetics, field discontinuities occur primarily at domain boundaries and material interfaces, the locations of which are known in advance. Appropriate boundary and interface conditions are derived from Maxwell's equations, and these conditions are then used to carry information into each cell neighboring such a boundary, provided that the boundary is smooth. Near geometrical singularities such as cracks and edges, the representation inside the cell may include singular functions. In regions away from physical boundaries, it is feasible to assume smooth behavior of the fields. This is the approach that is followed in our work.

4.11.2 Reconstruction for the Upwind Taylor Scheme

To obtain second-order accuracy in the time update for the fields on general grids, both the first and second spatial derivatives for each field component must be estimated. For this purpose, data are drawn from the averages in all cells that touch the boundary of the cell in which the variation is being reconstructed. The flux integrals over each cell face are then expressed in terms of these derivatives in combination with the first and second geometric moments of each face. The final form of the time update includes terms of up to third order in Δt, although its accuracy on nonuniform grids is only second order in the geometrical quantities defining each cell. On uniformly gridded portions of the computational domain, the accuracy of the solution increases to third order. This approach and its extension beyond second order are outlined below.

For cells that do not neighbor a physical or computational boundary, a surface that completely surrounds the cell at an infinitesimal distance from its boundary intersects every cell that shares at least one boundary point with the parent cell. It is natural to consider the data from all of these cells in any approach to reconstruct the solution variation inside the parent cell. If the solution is assumed to be analytic in this neighborhood, each cell average can be expressed in terms of the spatial derivatives of $\vec{u}$ evaluated at the centroid of the parent:

$$\begin{aligned} \vec{U}_\beta(t) &= \vec{u}(\bar{r}_\alpha, t) + (\bar{r}_\beta - \bar{r}_\alpha)\cdot\nabla u(\bar{r}_\alpha, t) \\ &+ \sum \frac{\partial^{\ell+m+n}}{\partial x^\ell\, \partial y^m\, \partial z^n}\,\vec{u}(\bar{r}_\alpha, t) \int_\beta \frac{(x'-x_\alpha)^\ell\,(y'-y_\alpha)^m\,(z'-z_\alpha)^n}{\ell!\; m!\; n!}\,\frac{dv'}{V'_\beta} \end{aligned} \tag{4.139}$$

where $\bar{r}_\alpha$ is the centroid of cell α and the integrals define the higher geometric moments $\vec{M}_\beta^{\ell,m,n}(\bar{r}_\alpha)$ of cell β about the centroid of the parent cell. The difference $\vec{U}_\beta - \vec{U}_\alpha$ is therefore expressed as a linear combination of the derivatives of $\vec{u}$ where the weights are these moments.

The number K of immediate neighbors in three dimensions is large: 26 for hexahedral meshes and about 50 for typical unstructured tetrahedral meshes. This plethora of information has led to many simplifying approaches by those pursuing unstructured-grid methods for CFD, including least-squares fitting of a small number of derivatives to the whole set of differences with various types of weighting [71]. A more consistent framework for incorporating all these data is to regard each difference as a scalar product of a derivative vector $D\vec{u}_\alpha$ and a moment vector $\vec{M}_\beta(\bar{r}_\alpha)$ in an infinite-dimensional linear vector space Ω. The K moment vectors span some subspace of Ω, and thus determine the projection of $D\vec{u}_\alpha$ on this subspace as a linear combination of the differences $\vec{U}_\beta - \vec{U}_\alpha$.

The basic question is: What does this projection tell us about the flux integrals over each face of α whose values we need in order to update $\vec{U}$ to the next time level? In fact, each face integral is another scalar product, combining the elements of $D\vec{u}_\alpha$ and $D\vec{u}_\gamma$ with geometric moments of the face, $\vec{N}_{\alpha\gamma}(\bar{r}_\alpha)$. For example, the flux integral in (4.138) is written for a linear flux function as

$$\begin{aligned}\oint_{\partial\alpha} \hat{n}\cdot\bar{\bar{f}}(\vec{u}^*)\,ds &= \oint_{\partial\alpha} \hat{n}\cdot\left[\bar{\bar{f}}(0) + \sum_i \vec{u}_i \left.\frac{\partial\bar{\bar{f}}(\vec{u}_i)}{\partial\vec{u}_i}\right|_{\vec{u}=0}\right] ds \\ &= \sum_i \frac{\partial\bar{\bar{f}}(\vec{u}_i)}{\partial\vec{u}_i}\cdot\sum_\gamma \hat{n}_{\alpha\gamma}\int_{\alpha\gamma} \vec{u}^*\,ds\end{aligned} \tag{4.140}$$

where $\vec{u}^*$ is the linear combination of $\vec{u}_\alpha(\bar{r},t)$ and $\vec{u}_\gamma(\bar{r},t)$ determined by solving the Riemann problem at $\bar{r}$, and $\alpha\gamma$ is the face boundary between cells α and γ. The function $\vec{u}_\alpha$ is the Taylor series determined by $D\vec{u}_\alpha$, and similarly for $\vec{u}_\gamma$. The integral of $\vec{u}_\alpha$ over the face $\alpha\gamma$ is thus written as a sum of integrals of powers of $(\bar{r} - \bar{r}_\alpha)$.

In general, the vector $\vec{N}_{\alpha\gamma}$ does not lie completely in the subspace spanned by the vectors $\vec{M}_\beta$. Consequently, its scalar product with $D\vec{u}_\alpha$ is only partly determined because combinations of the components of $D\vec{u}_\alpha$ that are orthogonal to this subspace are unknown. Setting these combinations to zero results in a definite prescription for estimating the flux integrals. This prescription gives a weighted sum of the differences $\vec{U}_\beta - \vec{U}_\alpha$ for each integral, with weights $w_{\alpha\gamma}^\beta$ determined from $\vec{M}_\beta$ and $\vec{N}_{\alpha\gamma}$. Constraints such as the divergence conditions on $\vec{D}$ and $\vec{B}$ are also applied to better estimate the orthogonal components of $D\vec{u}_\alpha$.

An important feature of this approach is that only the immediate neighbors of cells α and γ are involved in the time update. Every face integral has a stencil that extends only as far as the immediate neighbors of γ. This contrasts with the standard Runge-Kutta schemes popular in CFD that spread the stencil to as many extra layers of neighbors as there are stages in the time update.

The storage required at each face to implement this scheme can be rather large: K coefficients $w_{\alpha\gamma}^{\beta}$ are needed to incorporate all of the geometrical data describing the surrounding grid. In practice, only a few of these are required to provide second-order accuracy in the grid variables.

4.11.3 Singularities and High-Order Boundary Conditions

Cracks, edges, and thin wires are important features of radar targets that pose special problems for any exact solution method. Highly curved boundaries and interfaces share many of the same difficulties. Brute-force high-resolution gridding can be used to obtain accuracy, but poses generally prohibitive computer resource burdens.

A better approach is to incorporate the known analytic behavior of each field component into the description of its variation within any cell neighboring geometrically singular boundaries. Rather than assuming that a Taylor series can describe the fields, we can derive from Maxwell's equations an expansion in singular functions that provides any desired degree of accuracy inside the cell. The coefficients in this expansion are determined in a manner analogous to the reconstruction procedure described earlier. This involves integrals of the singular functions over each neighbor cell and each cell face that are evaluated before the simulation is begun.

At a highly curved boundary, any expressions derived by treating the boundary as locally flat can be seriously in error. Maxwell's equations provide relations not only among the tangential fields at a boundary, but also among the tangential and normal derivatives of the fields. These derivatives and their higher-order counterparts explicitly depend on the boundary shape, and terms such as the local curvature mix tangential and normal derivatives in the boundary conditions. For this reason, the reconstruction procedure for the new algorithm must be extended beyond the use of "ghost" cells in the conductor to incorporate the actual shape-dependent conditions on the higher derivatives inside the cell. Results on this currently active research item will be reported later.

4.11.4 Runge-Kutta Integration Schemes for Maxwell's Equations

The dissipation that is an inescapable part of upwinding can effectively be avoided by using central estimates for spatial derivatives. While Yee's scheme is the method of this type most well-known to the electromagnetics community, Runge-Kutta central schemes have been widely exploited in CFD [94].

The essential difference between Lax-Wendroff and Runge-Kutta schemes can be understood with reference to (4.138), which expresses the time derivative of the volume average, $\vec{U}_\alpha$, in terms of a surface integral of the flux. Rather than expanding $\vec{U}_\alpha$ in a Taylor series in time, we regard (4.138) as a first-order system of ordinary differential equations for the unknown $\vec{U}$s. The estimate for $\vec{u}(\vec{r}, t)$ at the cell interface is expressed as a linear combination of the surrounding $\vec{U}$s, closing the system of equations, which is now written as

$$\frac{d\vec{U}}{dt} = \vec{H}\left(\vec{U}\right) \tag{4.141}$$

where $\vec{H}$ is a sum over the flux from each face of the cell. The lowest-order central estimate for $\vec{u}$ at a cell face $\alpha\beta$ is just $(\vec{U}_\alpha + \vec{U}_\beta)/2$. In contrast to the upwind Lax-Wendroff scheme, this uses no information about the characteristics of the flux.

The Runge-Kutta schemes for integrating a first-order system of the type of (4.141) are explicit single-step methods that involve constructing a number of intermediate estimates $\vec{U}^{(\nu)}$ at time levels between n and $n+1$. Each intermediate stage has the form

$$\vec{U}^{(\nu)} = \vec{U}^n + \mu_\nu \Delta t \vec{H}\left[\lambda_{\nu 1}\vec{U}^{(\nu-1)} + \lambda_{\nu 2}\vec{U}^{(\nu-2)} + \cdots\right] \tag{4.142}$$

with $\vec{U}^{(0)} = \vec{U}^n$. The final update is written as

$$\vec{U}^{n+1} = \vec{U}^n + \Delta t\left(\alpha_0 \vec{H}_0 + \alpha_1 \vec{H}_1 + \cdots + \alpha_{\kappa-1}\vec{H}_{\kappa-1}\right) \tag{4.143a}$$

where

$$\vec{H}_\nu = \vec{H}\left[\lambda_{\nu 1}\vec{U}^{(\nu-1)} + \lambda_{\nu 2}\vec{U}^{(\nu-2)} + \cdots\right] \tag{4.143b}$$

and κ is the number of stages in the scheme. Time accuracy up to the order of the number of stages is obtained with appropriate conditions on the coefficients μ, λ, and α.

The advantage of Runge-Kutta schemes of third order and higher for wave propagation is that they need not damp oscillatory solutions: their region of stability includes purely real frequencies. When $\vec{H}$ is a linear function of $\vec{U}$, as it is for Maxwell's equations, each eigenvector $\vec{E}_\lambda$ of $\vec{H}$ satisfies (4.141) with a time dependence $\exp(j\omega t)$:

$$j\omega \vec{E}_\lambda = \vec{H}(\vec{E}_\lambda) = \lambda \vec{E}_\lambda \tag{4.144}$$

where the eigenvalues, λ, are determined by the spatial discretization and the boundary conditions. For a central discretization on a uniform grid with periodic boundary conditions, all of the eigenvalues are imaginary and the eigenfunctions are in fact simple sinusoidal waves. Also, the set of eigenvectors of $\vec{H}$ spans the complete space of solutions for $\vec{U}$, so that any initial condition on $\vec{U}$ can be represented by them.

For a general unstructured grid, the eigenvectors of $\vec{H}$ are not sinusoids, but a central discretization can still produce an imaginary eigenspectrum. In such a case, the Runge-Kutta methods preserve the amplitude of any initial sinusoidal wave indefinitely. However, the phase does not track that of the exact solution except at long wavelengths compared to the size of the grid cells. Further, the wave does not remain sinusoidal as is propagates. This is because the various eigenvectors composing it have different characteristic frequencies. We note that these negative effects are mitigated by using high-order central discretizations.

When high spatial frequencies are present in the initial conditions (for example, in the scattered-field formulation that has zero initial fields except at the target boundary), the lack of dissipation for high-frequency eigensolutions poses a problem. Although the behavior of these solutions bears very little resemblance to exact solutions of Maxwell's equations at these frequencies, they continue to add unwanted structure to the solution indefinitely. To remove this structure, filtering procedures of various forms have been applied, including the addition of damping terms in (4.141) proportional to some high spatial derivative of the solution. Filters with very little damping at long wavelengths can be implemented in this way.

4.11.5 Mixed and Merged Cells

A very useful feature of the finite-volume algorithms discussed here is that they apply to general collections of grid elements: a tetrahedron may neighbor a prism, a hexahedron, and several other tetrahedra; or a region of tetrahedra may interface with a region of uniform cubes. The reconstruction procedure involves only weighted averages of the nearest-neighbor data, with fixed weights determined by the local mesh geometry. The flux integrals are evaluated separately for each face, so the total number of faces bounding a cell does not affect the form of the algorithm.

We can take advantage of this generality in two different ways. First, we can maximize the region in which higher-order accuracy is maintained. Second, we can eliminate cells having high aspect ratios (e.g., flat shards or needles) by merging them with their neighbors.

A complex target typically needs tetrahedral and prismatic cells covering its surfaces. These cells are extensions of triangular surface grids constructed to capture all important details of the geometry. Just beyond these surface layers is the region through which long-range wave interactions pass from one feature of the target to another. A nearly uniform hexahedral grid filling most of this region needs many fewer cells to achieve the desired accuracy compared to prisms or tetrahedra. Outside of this interaction region, the mesh can stretch rapidly to the boundary of the computational domain, and the most efficient cell geometry can be chosen to fill this remaining space. Depending on the target, this could be sections of polar gridding, blocks of cubes, or sparsely packed tetrahedra. In all cases, the objective is to minimize the number of operations and the storage required to obtain a solution of the desired accuracy.

To avoid having to repeatedly regenerate portions of volume meshes in order to get rid of cells that reduce the accuracy and stability of the algorithm, each isolated "bad" cell is merged with a selected neighbor to create a larger cell having more faces but less variation among its dimensions. A criterion such as the ratio of circumscribed sphere size to inscribed sphere size is used to judge the quality of the merged cell.

4.12 NUMERICAL RESULTS AND VALIDATION

In computing electromagnetic fields using the FVTD methods discussed here, two issues need to be addressed. The first issue is the required resolution in meshing the object and its surrounding. Our experience has been that, for a second-order-accurate method, an average resolution of 10 to 15 grid lines per wavelength for the object itself results in a level of accuracy sufficient for most problems. The computational space around the object, however, does not require this resolution. As previously discussed, the density of the FVTD mesh drops (i.e., the grid declusters) as one moves away from the surface of the object toward the outer boundary of the computational domain. There, on the average, only 4 to 5 grid lines per wavelength are sufficient. (Note that these criteria are not applicable for electrically small objects where far more than 10 grid lines per wavelength may be needed to properly define the object shape.)

The other issue is the location of the outer boundary. How far the outer boundary should be located from the object, among other things, depends on the order of the ABC used. The findings discussed here are for a first-order ABC used in the context of FVTD. For electrically small bodies, the distance between the object and the outer boundary should be much larger than the characteristic dimension of the object. For electrically large objects, this distance should be much larger than a wavelength. In the intermediate range, the two criteria tend to coincide. As a rule of thumb, a span of 10 to 15 cells from the body surface to the outer mesh boundary is adequate, assuming a suitable choice for the height of the first grid cell and a proper declustering rate. The application of higher-order ABCs is expected to relax these requirements while preserving the accuracy of the computation.

Based on the algorithms discussed in the previous sections, computer codes have been developed for both vector and parallel machines that implement structured and unstructured grids in two- and three-dimensional spaces. All codes have a modular construction comprised of a geometry and material package, the main solver, and a preprocessor and postprocessor. The geometry and material package contains meshing and material data for each zone, zonal boundary specifications, incident-wave parameters, and grid diagnostics. The main solver numerically implement the time-stepping algorithms and the boundary conditions. Finally, the preprocessors and postprocessors implement FFTs; provide graphic tools for plotting the grid, field components, power flow, and scattering or radiation patterns; and provide means to generate animations.

Sample three-dimensional results obtained using the FVTD technique are presented next. Most of these results are achieved using a mesh resolution of 10 grid lines per wavelength for the object.

4.12.1 Structured-Grid Examples

The results presented here are generated by Rockwell's RCS3D code that uses the Lax-Wendroff algorithm and structured grids. The examples chosen have analytical solutions so that the issue of uncertainty resulting from measurement errors is avoided.

The first case, shown in Fig. 4.33, is the bistatic RCS of a dielectric sphere of radius a, electrical size $k_0 a = 1$ (k_0 is the wavenumber in free space), and relative permittivity $\varepsilon_r = 2.56$. The FVTD-computed results for both the E and H planes show excellent agreement with the exact Mie series.

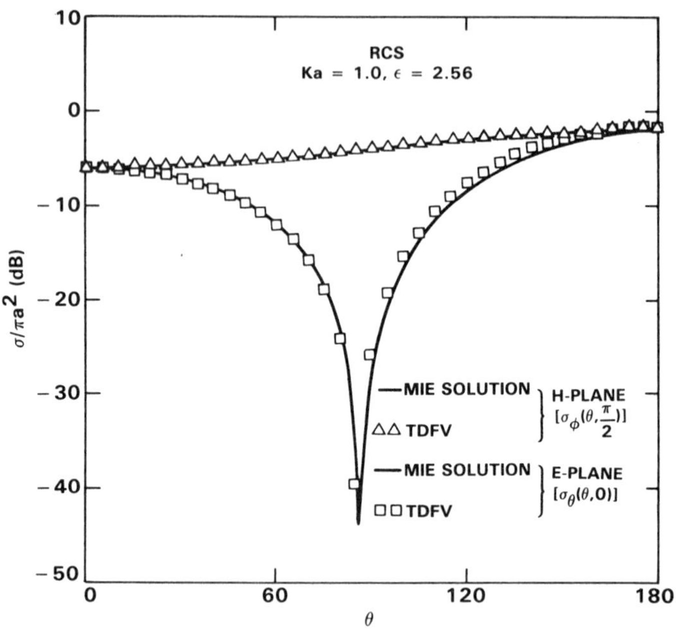

Fig. 4.33 Comparison of FVTD and Mie solutions for the bistatic radar cross section of a dielectric sphere of size $k_0 a = 1$ and $\varepsilon_r = 2.56$ in both the E and H planes.

The second case, shown in Fig. 4.34, is the bistatic RCS of a circular PEC disk of size $k_0 a = 10$ illuminated by a plane wave traveling along its axis. Again, the agreement between the FVTD and analytic solutions is very good.

4.12.2 Unstructured-Grid Examples

The results presented here are obtained using Rockwell's unstructured parallel code, UPRCS, which is based on the new fourth-order Runge-Kutta algorithm discussed in the previous sections. The examples chosen here do not have analytical solutions, but have been the subject of detailed measurements commissioned by the EMCC.

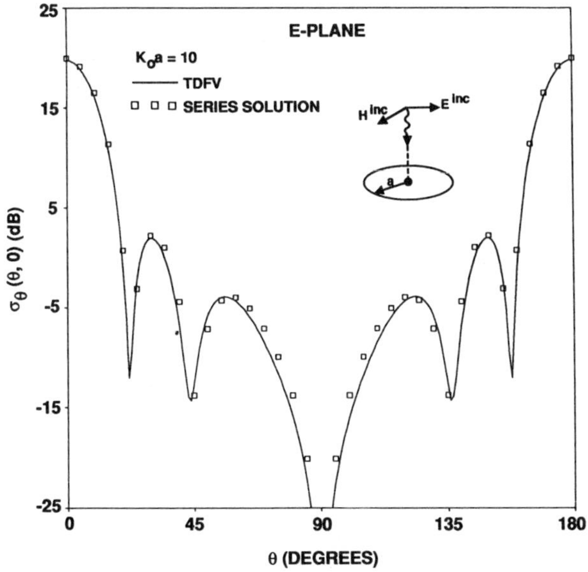

(a)

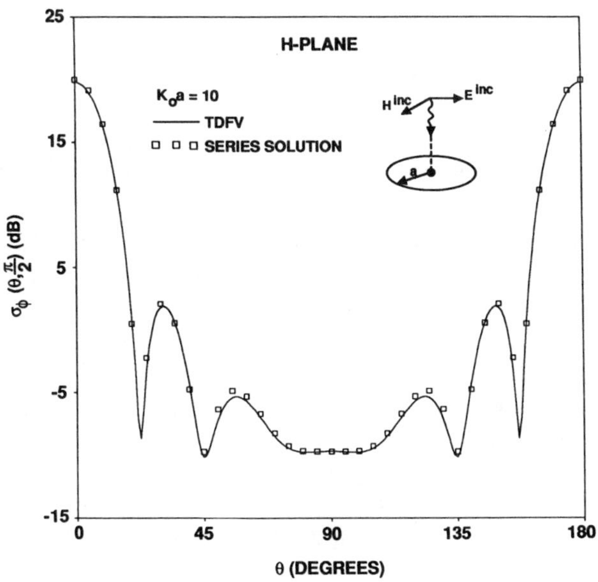

(b)

Fig. 4.34 Comparison of FVTD and analytic solutions for the bistatic RCS of a circular PEC disk of size $k_0 a = 10$ due to an axially incident plane wave: (a) E plane; (b) H plane.

The first case, shown in Fig. 4.35, is the monostatic RCS of the EMCC's "business card," a thin, flat, rectangular PEC plate of electrical size $3.5\lambda_0 \times 2.0\lambda_0$. This seemingly simple geometry is deceptively complex and challenging due to the presence of traveling waves along its surface.

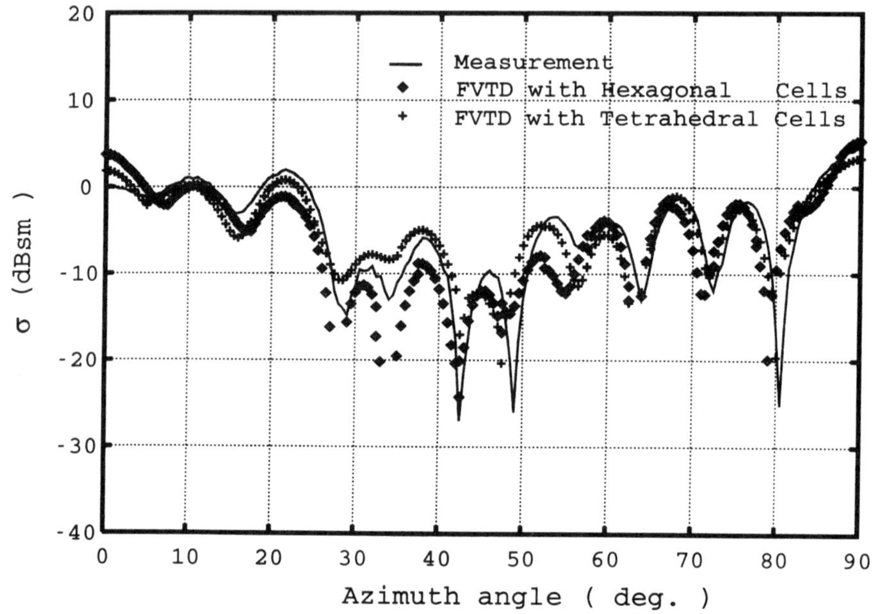

Fig. 4.35 Comparison of FVTD and measured data for the monostatic RCS of the EMCC "business card" scatterer, a thin, flat, rectangular PEC plate of electrical size $3.5\lambda_0 \times 2.0\lambda_0$.

Fig. 4.35 compares FVTD results with measurements for two different grids, one using hexagonal cells and the other using tetrahedral cells. While the results from the two grids are not identical, they are in reasonable agreement with the measurements. We note that, to conserve computer time, only the results for 0°, 10°, and so forth, are computed from the FVTD code. The RCS for angles between these values is obtained from an approximate technique known as the bistatic-to-monostatic interpolation [95]. We further note that RCS results for canonical geometries such as a sphere or cylinder are not sensitive to the cell shape and they usually match extremely well with benchmark data.

The second unstructured-grid example involves the monostatic RCS of the General Dynamics / Lockheed VFY-218 prototype fighter plane. Measurements of a scale-model of this plane are available again due to the work of the EMCC. Fig. 4.36 illustrates how the unstructured mesh realizes the airplane's geometrical features. Fig. 4.37 compares the FVTD-computed monostatic RCS for vertical and horizontal polarizations of the incident wave at 100 MHz with experimental data. The agreement is very good for the

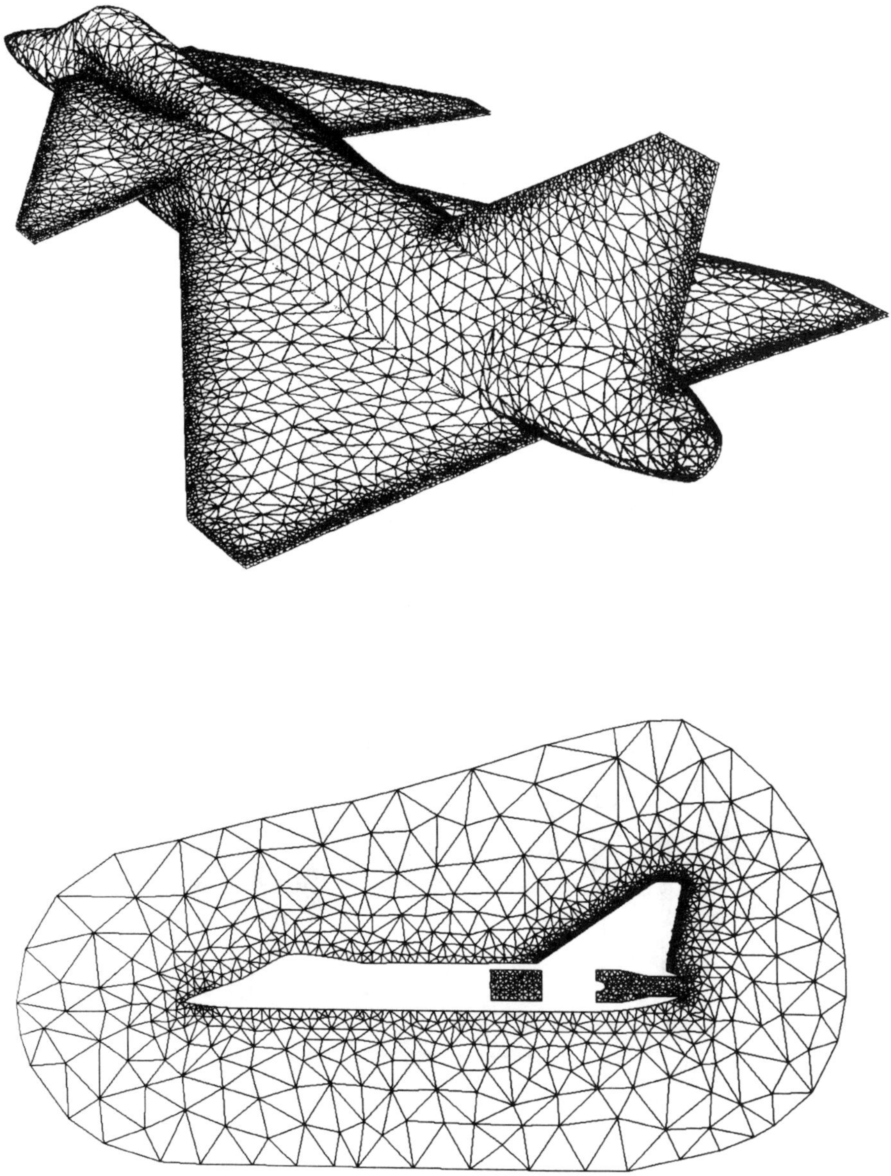

(a)

(b)

Fig. 4.36 The unstructured FVTD mesh for the VFY-218 prototype fighter airplane: (a) on the surface of the plane; (b) on a cross section through the computational domain.

(a)

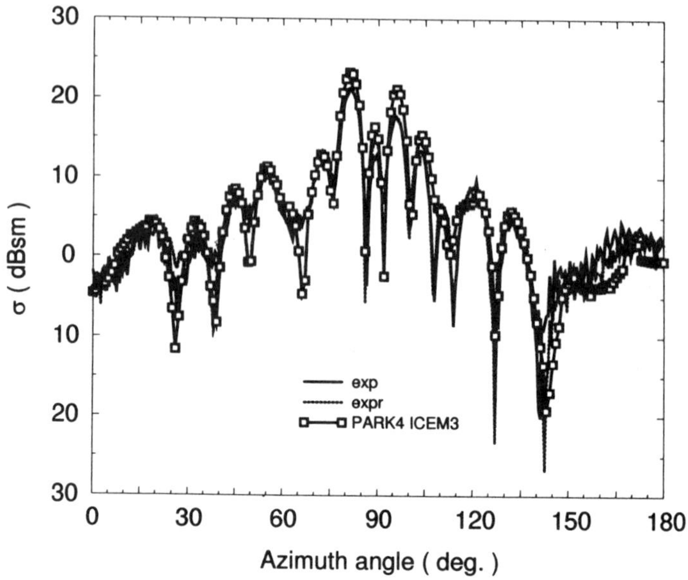

(b)

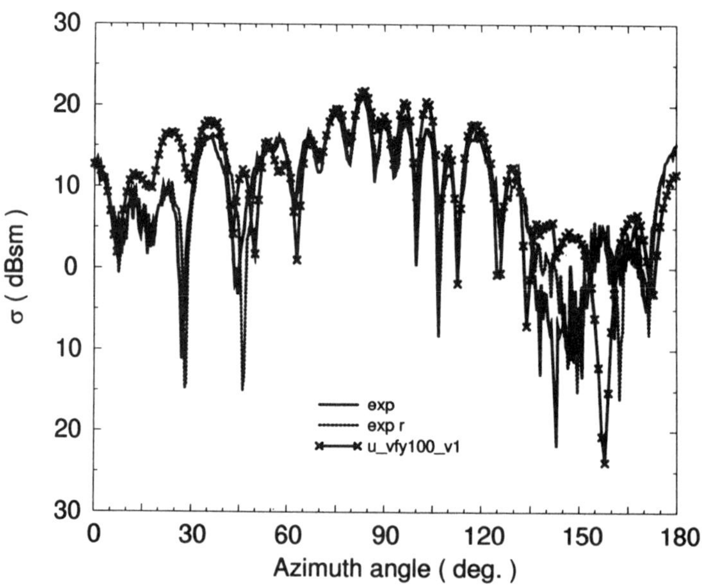

Fig. 4.37 Comparison of FVTD and measured data for the monostatic RCS of the VFY-218 prototype fighter airplane of Fig. 4.36 at 100 MHz: (a) V-V polarization; (b) H-H polarization.

vertical-polarization (V-V) case, and reasonably good for the horizontal-polarization (H-H) case. A color graphic of the FVTD-calculated near field is shown in the set of color plates elsewhere in this book.

4.13 PERFORMANCE OF THE UNSTRUCTURED PARALLEL FVTD CODE UPRCS

For computer software and architectures to be called "scaleable," they must efficiently make use of parallelization in two measurable ways. In the first, called *scale-up* (or sometimes scaled speed-up), a constant amount of computational work is given to each processor while the number of processors being utilized is increased. While the total size of the problem being solved scales upwards linearly with the number of processors, in the ideal case the computation time should remain constant. The second measure of scaleability is called *speed-up*. In this case, the total size of the problem is kept constant while it is spread over more and more processors. In this case, the overall computation time should ideally scale down linearly with the number of processors.

In this section, we report measurements of the scaleability of the unstructured, parallel FVTD code UPRCS implemented on an IBM SP system consisting of 256 processors each having 512 MB of random access memory (RAM). Scaleability experiments have been performed for a simple spherical scatterer and for the General Dynamics / Lockheed VFY-218 prototype fighter.

4.13.1 Testing Scaleability Using Sphere Meshes

Construction of four meshes of different sizes for a sphere allows direct testing of both measures of scaleability for a simple target. The meshes incorporate hexahedral cells and have the following total cell count: small, 76,800; medium, 307,200; large, 1,228,800; and extra large, 4,915,200 cells. (The extra-large mesh needs a total of 1.7 GB of memory to run the problem.) All tests involve running the code for 500 time steps.

Scale-Up Performance

The first group of tests involves loading variable numbers of processors with a fixed number of cells per processor: 19,200, 38,400, or 76,800. Fig. 4.38(a) shows the results of this study. We see that the overall increase in running time in going from the small grid to the large grid is least when each processor is most heavily loaded. This is due to the communications overhead between processors, which is proportional to the surface-to-volume ratio of the mesh zones allocated to the processors. Clearly, this ratio is minimized when the volume of each mesh zone loading each processor is maximized. Fig. 4.38(a) also plots the ideal situation where no additional time is needed as the problem scales up. The case of 76,800 cells per processors is seen to approach the ideal, yielding less than a 10% running-time increase when expanding the problem size by 64:1.

(a)

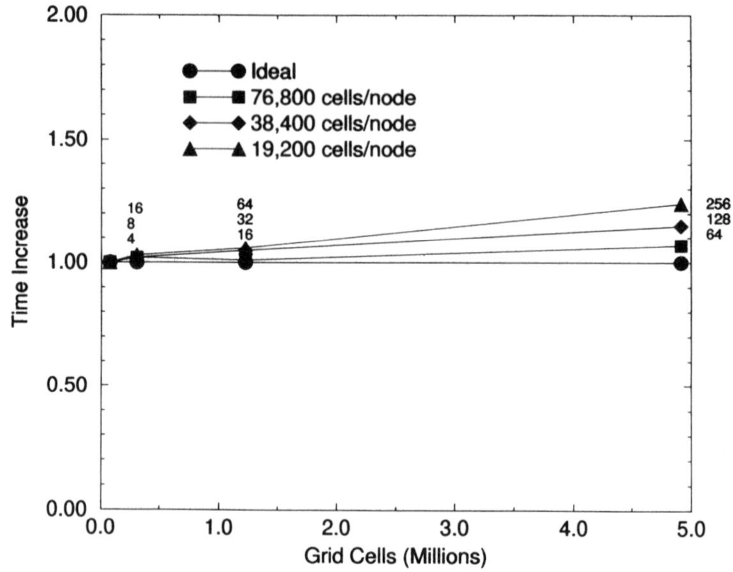

(b)

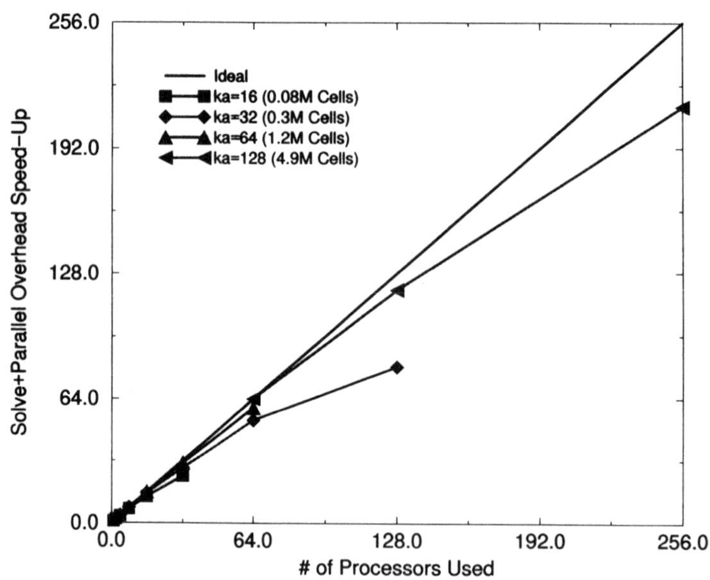

Fig. 4.38 Measured scaleability of the unstructured parallel FVTD code UPRCS for sphere meshes: (a) scale-up performance; (b) speed-up performance.

Speed-Up Performance

Fig. 4.38(b) shows the speed-up results for the four different sphere meshes when spread over up to 256 processors. The largest mesh exhibits the best speed-up, about 210:1, when spread over 256 processors. However, the speed-up departs markedly from the ideal even for this case when more than 128 processors are used. There is evidently a jump in the parallel overhead in this regime.

We note that the parallel overhead is due primarily to load imbalances in the number of boundary condition faces on each processor. The use of faster processors would probably reduce the amount of these load imbalances; this is evidenced by preliminary studies using an IBM SP system having 160-MHz versus 120-MHz processors. Here, the extra large grid shows improved parallel performance on the faster processors, with the parallel overhead decreasing from 22% to 13%. Similar results are seen in the next section for the large fighter airplane grid.

4.13.2 Scaleable Performance for Fighter Aircraft Models

We now consider the results of running two differently gridded models of the General Dynamics / Lockheed VFY-218 prototype fighter of Fig. 4.36 to assess the speed-up of the UPRCS code for more realistic problems. Unlike the sphere test cases, the fighter meshes are more likely to consist of tetrahedral cells and have load imbalances due to uneven boundary-condition loading across processors.

Small Fighter Grid

This mesh consists of 206,436 tetrahedral cells and is suitable for calculating the RCS of the VFY-218 in the frequency range 100 to 300 MHz. Fig. 4.39(a) shows that the speed-up for this case is nearly ideal up to 32 processors. This occurs even though the parallel overhead increases from 1.5% to 12% of the total time. The increase in parallel overhead is apparently being matched by an increase in computational efficiency.

Large Fighter Grid

This mesh consists of 1,352,740 tetrahedral cells and is suitable for calculating the RCS of the VFY-218 in the frequency range 300 MHz to 1 GHz. Using 120-MHz IBM SP processors, an average solution for this grid (marched 5,000 time steps) can be obtained in 108 min on 8 processors, and only 5.3 min on 256 processors. Using 160-MHz processors, these times are reduced to 81 min and 3.4 min, respectively.

Fig. 4.39(b) shows the speed-up for this case. We see that the departure from the ideal become quite evident when using more than 64 processors. As a result, running this model on 256 120-MHz processors is probably not worthwhile in production runs. Due to the light computational loading at this scale (only 5,300 cells/processor), parallel overhead is up at 33.5% and the speed-up relative to the use of 128 processors is only

(a)

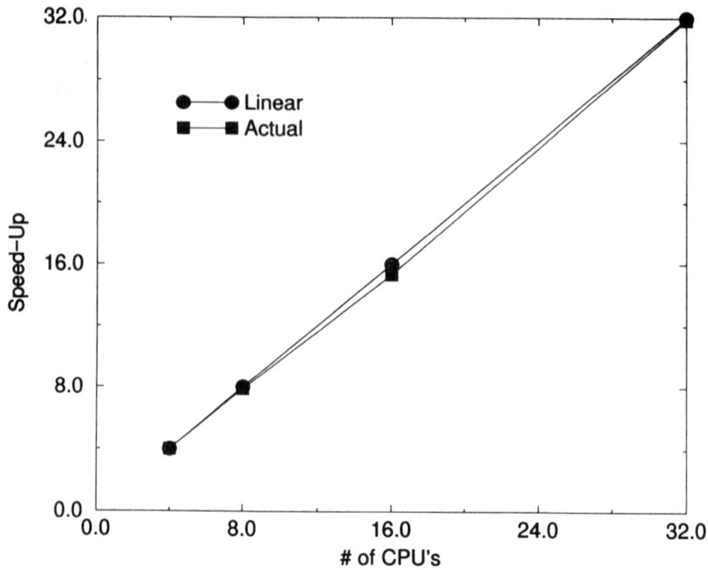

(b)

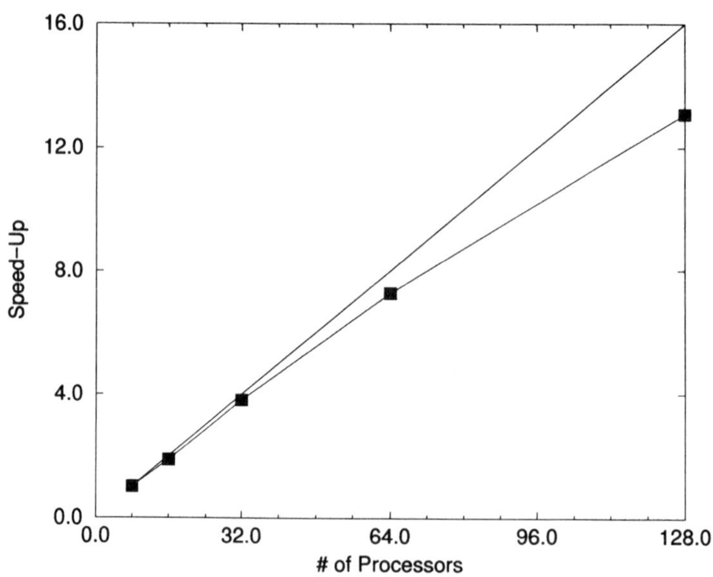

Fig. 4.39 Measured speed-up of the unstructured parallel FVTD code UPRCS for the VFY-218 fighter meshes: (a) small mesh, RCS models to 300 MHz; (b) large mesh, RCS models to 1 GHz.

1.57. However, the parallel speed-up improves when using 256 160-MHz processors, which exhibit a respectable 1.82 speed-up over the 128-processor run.

For the 160-MHz-processor runs, parallel overhead might be expected to increase as a percentage of total solve time due to the solve times being reduced (faster clock). However, the 160-MHz-processor runs almost all have *smaller* parallel overheads. This is probably due to a reduced contribution to the parallel overhead from load imbalances, the major source of parallel overhead for the UPRCS code. Most of the time attributed to actual communications, that is, sending data between processors, is hidden behind the solve time in the UPRCS code. In general, the UPRCS solve time for the 160-MHz processors is 25% faster than with the 120-MHz processors, as expected.

4.14 SUMMARY AND CONCLUSIONS

The FDTD method has been very successful in solving a wide variety of electromagnetic wave interaction problems. However, a principal limitation of FDTD in its original form is its restriction to orthogonal grids. This chapter focused on two powerful means to address this limitation: nonorthogonal staggered-grid generalizations of FDTD, including the N-FDTD and DSI methods (Sections 6.2 – 6.7); and FVTD techniques (Sections 6.8 – 6.13). Key aspects of these approaches were discussed, including numerical stability, well-posedness, sources of error, accuracy, and computational efficiency. The intent of the chapter has been to provide both a broad and comprehensive review of these emerging computational electromagnetics techniques.

REFERENCES

[1] Yee, K. S., "Numerical solution of initial boundary value problems in isotropic media," *IEEE Trans. Antennas and Propagation*, Vol. 14, 1966, pp. 302–307.

[2] Umashankar, K. R., A. Taflove, and B. Beker, "Calculation and experimental validation of induced currents on coupled wires in an arbitrary shaped cavity," *IEEE Trans. Antennas and Propagation*, Vol. 35, 1987, pp. 1248–1257.

[3] Taflove, A., K. Umashankar, B. Beker, F. Harfoush, and K. Yee, "Detailed FDTD analysis of electromagnetic fields penetrating narrow slots and lapped joints in thick conducting screens," *IEEE Trans. Antennas and Propagation,* Vol. 36, 1988, pp. 247–257.

[4] Jurgens, T. G., A. Taflove, K. R. Umashankar, and T. G. Moore, "Finite-difference time-domain modeling of curved surfaces," *IEEE Trans. Antennas and Propagation*, Vol. 40, 1992, pp. 357–366.

[5] Craddock, I. J., C. J. Railton, and J. P. McGeehan, "Derivation and application of a passive equivalent circuit for the finite-difference time-domain algorithm," *IEEE Microwave and Guided Wave Lett.*, Vol. 6, 1996, pp. 40–42.

[6] Taflove, A., *Computational Electrodynamics: The Finite-Difference Time-Domain Method*, Norwood, MA: Artech House, 1995.

[7] Holland, R., "Finite-difference solution of Maxwell's equations in generalized nonorthogonal coordinates," *IEEE Trans. Nuclear Science*, Vol. NS-30, 1983, pp. 4589–4591.

[8] Lee, J.-F., "Finite difference time domain algorithm for non-orthogonal grids and its application to the solution of electromagnetic scattering problems," *Archiv fur Elektronik und Uebertragungstechnik,* Vol. 46, 1992, pp. 328–335.

[9] Lee, J.-F., "Numerical solutions of TM scattering using an obliquely Cartesian finite difference time domain algorithm," *IEE Proc. H (Microwaves, Antennas and Propagation)*, Vol. 140, 1992, pp. 23–28.

[10] Lee, J.-F., R. Palendech, and R. Mittra, "Modeling three-dimensional discontinuities in waveguides using nonorthogonal FDTD algorithm," *IEEE Trans. Microwave Theory and Techniques,* Vol. 40, 1992, pp. 346–352.

[11] Madsen, N., "Divergence preserving discrete surface integral methods for Maxwell's equations using nonorthogonal unstructured grids," Technical Report UCRL-JC-109787, Lawrence-Livermore National Laboratory, Livermore, CA, 1992.

[12] Madsen, N., "Divergence preserving discrete surface integral methods for Maxwell's equations using nonorthogonal unstructured grids," *J. Computational Physics*, Vol. 119, 1995, pp. 34–45.

[13] Gedney, S., and F. Lansing, "Full wave analysis of printed microstrip devices using a generalized Yee algorithm," *Proc. IEEE Antennas and Propagation Soc. Int. Symp.*, Ann Arbor, MI, 1993, pp. 1179–1182.

[14] Gedney, S., and F. Lansing, "A parallel planar generalized Yee algorithm for the analysis of microwave circuit devices," *Int. J. Numerical Modeling (Electronic Networks, Devices and Fields)*, Vol. 8, 1995, pp. 265–276.

[15] Gedney, S., F. Lansing and D. Rascoe, "A full-wave analysis of passive monolithic integrated circuit devices using a generalized Yee algorithm," *IEEE Trans. Microwave Theory and Techniques*, Vol. 44, 1996, pp. 1393–1400.

[16] Madsen, N. K., and R. W. Ziolkowski, "A three-dimensional modified finite volume technique for Maxwell's equations," *Electromagnetics,* Vol. 10, 1990, pp. 147–161.

[17] Mohammadian, A. H., V. Shankar, and W. F. Hall, "Computation of electromagnetic scattering and radiation using a time-domain finite-volume discretization procedure," *Computer Physics Communication,* Vol. 68, 1991, pp. 175–185.

[18] Shankar, V., A. Mohammadian, and W. Hall, "A time-domain finite-volume treatment for the Maxwell equations," *Electromagnetics,* Vol. 10, 1990, pp. 147–162.

[19] Shang, J. S., "Characteristic-based algorithms for solving the Maxwell's equations in the time domain," *IEEE Antennas and Propagation Magazine,* Vol. 37, 1995, pp. 15–23.

[20] Rowell, C., V. Shankar, W. F. Hall and A. H. Mohammadian, "Algorithmic aspects and computing trends in computational electromagnetics using massively parallel architectures," *Proc. IEEE Conf. on Algorithms and Architectures for Parallel Processing*, Brisbane, Australia, 1995.

[21] Riley, D. J., and C. D. Turner, "VOLMAX: a solid-model-based, transient volumetric Maxwell solver using hybrid grids," *IEEE Antennas and Propagation Magazine,* Vol. 39, Feb. 1997, pp. 20–33.

[22] Yee, K. S., and J. S. Chen, "The finite-difference time-domain (FDTD) and finite-volume time-domain (FVTD) methods in solving Maxwell's equations," *IEEE Trans. Antennas and Propagation,* Vol. 45, 1997, pp. 354–363.

[23] Stratton, J. A., *Electromagnetic Theory*, New York: McGraw-Hill, 1941.

[24] Roden, J. A., *Broadband Electromagnetic Analysis of Complex Structures with the Finite-Difference Time-Domain Technique in General Curvilinear Coordinates*, Ph.D. dissertation, University of Kentucky, 1997.

[25] Igarashi, H., and T. Honma, "BE and FE solutions to open boundary potential problems with helical symmetry," *IEEE Trans. Magnetics*, Vol. 28, 1992.

[26] Becker, W. D., P. Harms, and R. Mittra, "Time domain electromagnetic analysis of interconnects in a computer chip package," *IEEE Trans. Microwave Theory and Techniques*, Vol. 40, 1992, pp. 2155–2163.

[27] Gedney, S., and F. Lansing, "Time domain analysis of planar microstrip devices using a generalized Yee algorithm based on unstructured grids," *Proc. 1993 Progress In Electromagnetics Research Symp.*, Pasadena, CA, 1993, p. 561.

[28] Gedney, S. D., and J. A. Roden, *Well Posed Nonorthogonal FDTD Methods*, Tech. Report EMG-98-01, University of Kentucky, 1998.

[29] Xiao, F., and H. Yabe, "Numerical dispersion relation for FDTD method in general curvilinear coordinates," *IEEE Microwave and Guided Wave Lett.*, Vol. 7, 1997, pp. 48-50.

[30] Lee, J. F., R. Lee and A. Cangellaris, "Time-domain finite-element methods," *IEEE Trans. Antennas and Propagation*, Vol. 45, 1997, pp. 430–442.

[31] Gustafsson, B., H.-O. Kreiss, and J. Oliger, *Time-Dependent Problems and Difference Methods*, New York: Wiley, 1995.

[32] Trefethen, L. N., "Lax-stability vs. eigenvalue stability of spectral methods," *Numerical Methods for Fluid Dynamics III* , M. J. Baines, ed., Oxford: Clarendon Press, 1988, pp. 237-253.

[33] Trefethen, L. N., "Psuedospectra of matrices," *Numerical Analysis 1991*, G. A. Watson, ed., Essex, UK: Longman Scientific, 1992, pp. 234–266.

[34] Bardi, I., O. Biro, K. Preis, and B. R. Vrisk, "Nodal and edge element analysis of inhomogeneous loaded 3D cavities," *IEEE Trans. Magnetics*, Vol. 28, 1992, pp. 1142–1145.

[35] Gedney, S., and U. Navsariwala, "An unconditionally stable implicit finite-element time-domain solution of the vector wave equation," *IEEE Microwave and Guided Wave Lett.*, Vol. 5, 1995, pp. 332–334.

[36] Ray, S. L., "Numerical dispersion and stability characteristics of time-domain methods on nonorthogonal meshes," *IEEE Trans. Antennas and Propagation*, Vol. 41, 1993, pp. 233–235.

[37] Navarro, E. A., C. Wu, P. Y. Chung, and J. Litva, "Some considerations about the finite difference time domain method in general curvilinear coordinates," *IEEE Microwave and Guided Wave Lett.*, Vol. 4, 1994, pp. 396–398.

[38] Shi, H., and J. L. Drewniak, "Dispersion comparison for DSI- and tensor-based nonorthogonal FDTD," *IEEE Microwave and Guided Wave Lett.*, Vol. 6, 1996, pp. 193–195.

[39] Roden, J. A., and S. D. Gedney, "Efficient implementation of the uniaxial based PML media in three-dimensional nonorthogonal coordinates using the FDTD technique," *Microwave and Optical Technology Lett.*, Vol. 10, 1997, pp. 71–75.

[40] Gedney, S., and F. Lansing, "Explicit time-domain solution of Maxwell's equations using nonorthogonal and unstructured grids," Chap. 11 in *Computational Electrodynamics: The Finite-Difference Time-Domain Method*, A. Taflove, ed., Norwood, MA: Artech House, 1995.

[41] Fang, J., and D. Xeu, "Numerical errors in the computation of impedances by FDTD method and ways to eliminate them," *IEEE Microwave and Guided Wave Lett.*, Vol. 5, 1995, pp. 1–6.

[42] Ramo, S., J. Whinnery, and T. Van Duzer, *Fields and Waves in Communication Electronics*, 3rd ed., New York: Wiley, 1994.

[43] Itoh, T., and B. Houshmand, eds., *Time-Domain Methods for Microwave Structures*, New York: IEEE Press, 1998.

[44] Zeland Software, 39120 Argonaut Way, Suite 499, Freemont, CA 94538.

[45] Tsai, M. J., C. Chen, N. G. Alexopoulos and T.-S. Horng, "Multiple arbitrary shape via-hole and air-bridge transitions in multilayered structures," *IEEE Trans. Microwave Theory and Techniques*, Vol. 44, 1996, pp. 2504–2511.

[46] Woo, A. C., H. T. G. Wang, M. J. Schuh, and M. L. Sanders, *Benchmark Radar Targets for the Validation of Computational Electromagnetics Programs*, NASA Tech. Memorandum 108829, Ames Research Center, Moffett Field, CA, 1994.

[47] Technical Report True Grid, XYZ Scientific Applications, Inc., 1324 Concannon Blvd, Livermore, CA 94550,

[48] Higdon, R. L., "Numerical absorbing boundary conditions for difference approximations to the multi-dimensional wave equation," *Mathematics of Computation,* Vol. 47, 1987, pp. 437–459.

[49] Taflove, A., and S. Barnard, *Application of Massively Parallel Supercomputing to 3-D RCS Methods and Modeling of Complex Materials*, Final Report, NASA Contract NAS2-13890, Ames Research Center, Moffett Field, CA, 1994.

[50] Shankar, V., and W. F. Hall, "A time-domain differential solver for electromagnetic scattering," *Proc. National Radio Science Meeting*, Boulder, CO, 1988.

[51] Mohammadian, A. H., V. Shankar, and W. F. Hall, "Time-domain solution of Maxwell's equations in frequency-dependent media," *Proc. National Radio Science Meeting*, Boulder, CO, 1989.

[52] Mohammadian, A. H., V. Shankar, and W. F. Hall, "Numerical solution to scattering from 2-D infinite bodies in finite computational domain," *Proc. 5th Annual Review of Progress in Applied Computational Electromagnetics*, Monterey, CA, 1989.

[53] Shankar, V., W. F. Hall, and A. H. Mohammadian, "A time domain differential solver for electromagnetic scattering problems," *Proc. IEEE* , Vol. 77 1989, pp. 709–721.

[54] Mohammadian, A. H., V. Shankar, and W. F. Hall, "Toward comprehensive validation of the Rockwell differential solver for Maxwell's equations," *Proc. IEEE Antennas and Propagation Society Int. Symp. and URSI Radio Science Meeting*, San Jose, CA, 1989.

[55] Mohammadian, A. H., V. Shankar, and W. F. Hall, " Validation of a CFD-based Maxwell solver using canonical scattering problems," in *Proc. Workshop on Software Validation*, San Jose, CA, 1989.

[56] Shankar, V., W. F. Hall, and A. H. Mohammadian, "A CFD-based finite-volume procedure for computational electromagnetics – interdisciplinary applications of CFD methods," in *Proc. AIAA 9th Computational Fluid Conference*, Buffalo, NY, 1989.

[57] Mohammadian, A. H., V. Shankar, and W. F. Hall, "Application of the time-domain differential solver to 2-D electromagnetic penetration problems," *IEEE Trans. Magnetics*, Vol. 25, 1989, pp. 2875–2877.

[58] Shankar, V., W. F. Hall, and A. H. Mohammadian, "A three-dimensional Maxwell's equation solver for computation of scattering from layered media," *IEEE Trans. Magnetics,* Vol. 25, 1989, pp. 3098–3103.

[59] Shankar, V., W. F. Hall, and A. H. Mohammadian, "A multizone time-domain finite-volume procedure for electromagnetics computation," *Proc. 1989 Int. Symp. on Antennas and Propagation (ISAP)*, Tokyo, Japan, 1989.

[60] Mohammadian, A. H., V. Shankar, and W. F. Hall, "Simulation of EM radiation from antenna elements and arrays using CFD-based FVTD method," *Proc. National Radio Science Meeting*, Boulder, CO, 1990.

[61] Mohammadian, A. H., V. Shankar, and W. F. Hall, "Some preliminary results for scattering and radiation using time-domain finite-volume method," *Proc. 6th Annual Review of Progress in Computational Electromagnetics*, Monterey, CA, 1990.

[62] Mohammadian, A. H., V. Shankar,, and W. F. Hall, "Application of the time-domain finite-volume method to antenna and array problems," *Proc. IEEE Antennas and Propagation Soc. Int. Symp.*, Dallas, Texas, 1990.

[63] Shankar, V., W. F. Hall, and A. H. Mohammadian, "Development of computational fluid dynamics (CFD) based time-domain algorithm for Maxwell's equations," *Proc. IEEE Antennas and Propagation Soc. Int. Symp.*, Dallas, Texas, 1990.

[64] Shankar, V., W. F. Hall, A. H. Mohammadian, and S. Chakravarthy, "Application of CFD based methods to problems in computational science," *Comp. Syst. Eng.* Vol. 1, 1990, pp. 7–22.

[65] Shankar, V., W. F. Hall, A. H. Mohammadian, and C. Rowell, "Development of a CFD based algorithm for computational electromagnetics (CEM)," *Computational Fluid Dynamics Review 1995*, M. Hafez and K. Oshima, eds., New York: Wiley 1995.

[66] Sommerfeld, A., *Partial Differential Equations in Physics*, New York: Academic Press, 1949.

[67] Shankar, V., S. Chakravarthy, and K.Y. Szema, "Development and application of CFD methods to problems in computational science, *Proc. Workshop on CFD for Aerospace Problems*, Univ. Tennessee Space Inst., UTSI Pub. no. E02-4005-013-88, 1988.

[68] Taflove, A., and K. R. Umashankar, "Review of FDTD numerical modeling of electromagnetic wave scattering and radar cross section," *Proc. IEEE*, Vol. 77, 1989, pp. 682–699.

[69] Petropoulous, P. G., "Phase error control for FDTD methods of second and fourth order accuracy," *IEEE Trans. Antennas and Propagation*, Vol. 42, 1994, pp. 859–862.

[70] Hughes, T. J. R., *The Finite Element Method,* Englewood Cliffs: Prentice-Hall, 1987.

[71] Barth, T. J., "Recent developments in high order k-exact reconstruction on unstructured meshes," AIAA paper 93–0668, Jan. 1993.

[72] Frink, N. T., "Recent progress toward a three-dimensional unstructured Navier-Stokes flow solver," AIAA paper 94–0061, Jan. 1994.

[73] Mavriplis, D. J., "Three-dimensional multigrid Reynolds-averaged Navier-Stokes solver for unstructured grids," *AIAA Journal*, Vol. 33, 1995, pp. 445–453.

[74] Szema, K. Y., S.V. Ramakrishnan, and V. V. Shankar, "Application of a generalized Lax-Wendroff scheme for unstructured Euler computations," *Proc. 14th AIAA Applied Aerodynamics Conf.*, New Orleans, 1996.

[75] Palaniswamy, S., W. F. Hall, and V. Shankar, "Numerical solution to Maxwell's equations in the time domain on nonuniform grids," *Radio Science*, Vol. 31, 1996, pp. 905–912.

[76] Tai, C.-T., *Generalized Vector and Dyadic Analysis*, New York: IEEE Press, 1992.

[77] Holland, R., "Time-domain treatment of Maxwell's equations in frequency dependent media," *Proc. 2nd Annual Review of Progress in Applied Computational Electromagnetics*, Monterey, CA, 1986.

[78] Lax, R., and B. Wendroff, "Systems of conservation laws," *Commun. Pure Appl. Math.*, Vol. 13, 1960, pp. 217–237.

[79] Courant, R., K. Friedrichs, and H. Lewy, "On the partial difference equations of mathematical physics, *IBM Journal*, Vol. 11, 1967, pp. 215–237.

[80] Warming, R. F., and R. M. Beam, "Upwind second-order difference schemes and applications in aerodynamics flows, *AIAA Journal*, Vol. 14, 1976, pp. 1241–1249.

[81] Stratton, J. A., *Electromagnetic Theory*: New York, McGraw-Hill, 1941.

[82] Von Franz, W., "Zur formulierung des Huygensschen prinzips," *Z. Naturforsch. A*, Vol. 3a, 1948, pp. 500–506.

[83] Ziolkowski, R. W., N. K. Madsen, and R. C. Carpenter, "Three-dimensional computer modeling of electromagnetic fields: a global lookback lattice truncation scheme," *J. Comput. Phys.*, Vol. 50, 1983, pp. 360–408.

[84] Engquist, B., and A. Majda, "Absorbing boundary conditions for the numerical simulation of waves," *Math Comput.,* Vol. 31, 1977, pp. 629–651.

[85] Berenger, J.-P., "A perfectly matched layer for the absorption of electromagnetic waves," *J. Comput. Phys.*, Vol. 114, 1994, pp. 185–200.

[86] Thacker, W. C., "A brief review of techniques for generating irregular computational grids," *Int. J. Num. Meth. Eng.*, Vol. 15, 1980, pp. 1335–1341.

[87] Thompson, J. F., F. C. Thames, and C. W. Mastin, "Automatic numerical generation of body-fitted curvilinear coordinate system for field containing any number of arbitrary two-dimensional bodies," *J. Comput. Phys.*, Vol. 15, 1974, pp. 299–319.

[88] Steger, J. L., and R.L. Sorenson, "Automatic mesh-point clustering near a boundary in grid generation with elliptic partial differential equations," *J. Comput. Phys.*, Vol. 33, 1979, pp. 405–410.

[89] Thomas, P. D., and J. F. Middlecoff, "Direct control of the grid point distribution in meshes generated by elliptic equations," *AIAA Journal*, Vol. 18, 1980, pp. 652–656.

[90] Kreis, R. I., F. C. Thames, and H. A. Hassan, "Application of a variation method for generating adaptive grids," *AIAA Journal*, Vol. 24, 1986, pp. 408–410.

[91] Nakahashi, K., and G. S. Deiwert, "Three-dimensional adaptive grid method," *AIAA Journal*, Vol. 24, 1986, pp. 948–954.

[92] Kennon, S. R., and G .S. Dulikravich, "Generation of computational grids using optimization," *AIAA Journal*, Vol. 24, 1986, pp. 1069–1073.

[93] Shankar, V., S. Rudy, and K. Y. Szema, "Application of a two-dimensional grid solver for three-dimensional problems," *AIAA Journal*, Vol. 23, 1985, pp. 326–327.

[94] Jameson, A., "Successes and challenges in computational aerodynamics," AIAA paper 87–1184, *Proc. 8th AIAA Computational Fluid Dynamics Conference*, 1987.

[95] Schuh, M. J., A. C. Woo, and M. P. Simon, "The monostatic/bistatic approximation," *IEEE Antennas & Propagation Magazine*, Vol. 36, no. 4, 1994, pp. 76–78.

Chapter 5

The Perfectly Matched Layer Absorbing Medium

Stephen D. Gedney

5.1 INTRODUCTION

One of the greatest challenges of the FDTD method has been the efficient and accurate solution of electromagnetic wave interaction problems in unbounded regions. For such problems, an absorbing boundary condition (ABC) must be introduced at the outer lattice boundary to simulate the extension of the lattice to infinity. Over the years, a number of techniques have been introduced to achieve this. An excellent compendium of the development of ABCs is provided in [1].

As shown in Fig. 5.1, one approach to realize an ABC is to terminate the outer boundary of the space lattice in an absorbing material medium. This is analogous to the physical treatment of the walls of an anechoic chamber. Ideally, the absorbing medium is only a few lattice cells thick, reflectionless to all impinging waves over their full frequency spectrum, highly absorbing, and effective in the near field of a source or a scatterer. An early attempt at implementing such an absorbing material boundary condition was reported by Holland [2], who utilized a conventional lossy dispersionless absorbing medium. The difficulty with this tactic is that such an absorbing layer is matched only to normally incident plane waves. As a result, the entire category of lossy-material ABCs had had only limited application in electromagnetics.

Since 1994, a new fervor in this area has been created by J.-P. Berenger's introduction of a highly effective absorbing-material ABC designated the perfectly matched layer, or PML [3]. The innovation of Berenger's PML is that plane waves of arbitrary incidence, polarization, and frequency are matched at the boundary.

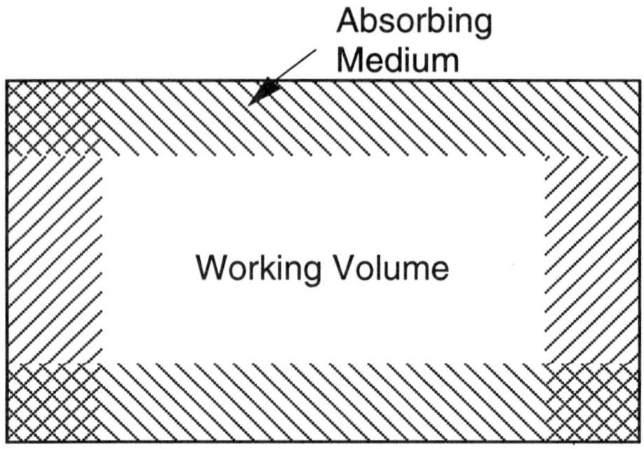

Fig. 5.1 Anechoic chamber analogy of terminating an FDTD lattice using an absorbing material medium.

To this end, Berenger derived a *split-field* formulation of Maxwell's equations. Namely, each vector field component is split into two orthogonal components. Each of the 12 orthogonal components is then expressed as satisfying a coupled set of hyperbolic equations. By choosing loss parameters consistent with a dispersionless medium, a perfectly matched planar interface is derived.

In the continuous space, the PML absorber and the host medium are perfectly matched. However, in the discrete FDTD space, large reflections can result. To keep the PML absorber thin, the conductivity of the medium must be large to be an effective absorber. However, in the discrete space, the electric and magnetic material parameters are represented in a piecewise constant manner. As well, the electric and magnetic material parameters are spatially staggered. Consequently, large spurious reflections can result from discretization errors. Berenger proposed spatial scaling of the PML material parameters to reduce the discretization error due to material discontinuities [3]. He also reported means to reduce the discretization errors for the second-order FDTD scheme, and to reduce the PML thickness for efficient implementation [4–8].

Following Berenger's seminal work, many papers appeared validating his technique [9–18] as well as applying FDTD with the PML medium [19–29]. A number of modifications of the PML were also proposed to enhance its performance [4, 7, 15, 30–32]. The original split-field PML concept has also been restated in a stretched-coordinate form [10, 33]. This form has extended the use of the PML to cylindrical and spherical coordinate systems [34] through a coordinate mapping of the metrics.

The split-field PML has also been posed as a uniaxial anisotropic medium [35–37]. Specifically, a uniaxial anisotropic medium with both magnetic and electric permittivity tensors was proposed that is perfectly matched to an arbitrary half space.

Introducing loss into the tensors results in a perfectly matched absorbing medium. The uniaxial PML (UPML) was first applied within the FDTD method by Gedney [36, 37]. The UPML is intriguing because it is based on a Maxwellian formulation rather than a mathematical model. In fact, an attempt to realize a physical UPML medium has been reported [38].

This chapter reviews the theoretical foundations of Berenger's PML, its stretched-coordinate representation, and the uniaxial PML. The relationships between the three methods are highlighted. The efficient implementation of the uniaxial PML within an FDTD lattice is presented. Further, the numerical discretization error is discussed, and various PML parameters and scaling functions are evaluated. It is also demonstrated that PML is applicable to inhomogeneous media and can also be applied in non-Cartesian coordinate systems.

5.2 BERENGER'S SPLIT-FIELD FORMULATION OF PML

5.2.1 Dispersionless Medium

In the exploration of a reflectionless material interface, consider a plane wave incident from a free space ($x < 0$) onto a material half-space ($x > 0$) (Fig. 5.2). The planar interface separating the two media is assumed to be the $x = 0$ plane. The material half-space is assumed to have both a magnetic conductivity, σ^*, and an electric conductivity, σ. Consider a TE$_z$-polarized uniform plane wave incident upon this interface, $\vec{H}^{\text{inc}} = \hat{z} H_0 \exp(-j\beta_x^i x - j\beta_y^i y)$, propagating with angle θ^i relative to the normal x-axis. The fields in the two regions are then posed as:

$$\left.\begin{aligned} \vec{H}_1 &= \hat{z}\, H_0\left(1 + \Gamma e^{2j\beta_x^i x}\right) e^{-j\beta_x^i x - j\beta_y^i y} \\ \vec{E}_1 &= \left[-\hat{x}\frac{\beta_y^i}{\omega\varepsilon_1}\left(1 + \Gamma e^{2j\beta_x^i x}\right) + \hat{y}\frac{\beta_x^i}{\omega\varepsilon_1}\left(1 - \Gamma e^{2j\beta_x^i x}\right)\right] H_0\, e^{-j\beta_x^i x - j\beta_y^i y} \end{aligned}\right\} \quad (x<0) \qquad (5.1a)$$

$$\left.\begin{aligned} \vec{H}_2 &= \hat{z}\, H_0\, \tau\, e^{-j\beta_x^t x - j\beta_y^t y} \\ \vec{E}_2 &= \left[-\hat{x}\frac{\beta_y^t}{\omega\varepsilon_2\left(1 + \frac{\sigma}{j\omega\varepsilon_2}\right)} + \hat{y}\frac{\beta_x^t}{\omega\varepsilon_2\left(1 + \frac{\sigma}{j\omega\varepsilon_2}\right)}\right] H_0\, \tau\, e^{-j\beta_x^t x - j\beta_y^t y} \end{aligned}\right\} \quad (x>0) \qquad (5.1b)$$

where Γ and τ are the reflection and transmission coefficients, respectively, and from the dispersion relationships:

$$\beta_x^i = k_1 \cos\theta^i\,, \qquad \beta_y^i = k_1 \sin\theta^i \quad \Big\}\ (x<0) \qquad (5.2a)$$

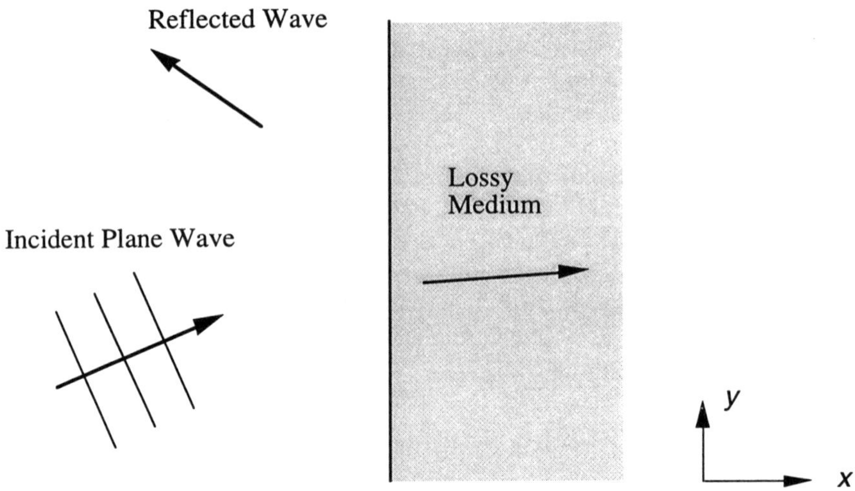

Fig. 5.2 Plane wave impinging on a lossy half-space.

$$\beta_x^t = \sqrt{(k_2)^2\left(1+\frac{\sigma}{j\omega\varepsilon_2}\right)\left(1+\frac{\sigma^*}{j\omega\mu_2}\right) - \left(\beta_y^t\right)^2} \quad \Bigg\} \; (x>0) \tag{5.2b}$$

where $k_i = \omega\sqrt{\varepsilon_i \mu_i}$ $(i = 1,2)$. Enforcing the continuity of the tangential fields across the boundary interface $(x = 0)$ leads to

$$\Gamma = \frac{\dfrac{\beta_x^i}{\omega\varepsilon_1} - \dfrac{\beta_x^t}{\omega\varepsilon_2(1+\sigma/j\omega\varepsilon_2)}}{\dfrac{\beta_x^i}{\omega\varepsilon_1} + \dfrac{\beta_x^t}{\omega\varepsilon_2(1+\sigma/j\omega\varepsilon_2)}} \; ; \qquad \tau = 1 + \Gamma \tag{5.3}$$

and $\beta_y^t = \beta_y^i = k_1 \sin\theta^i$. If the wave is normally incident ($\theta^i = 0$), we find that

$$\Gamma = \frac{\eta_1 - \eta_2}{\eta_1 + \eta_2} \tag{5.4}$$

where

$$\eta_1 = \sqrt{\frac{\mu_1}{\varepsilon_1}}, \qquad \eta_2 = \sqrt{\frac{\mu_2(1+\sigma^*/j\omega\mu_2)}{\varepsilon_2(1+\sigma/j\omega\varepsilon_2)}} \tag{5.5}$$

Subsequently, if $\mu_2 = \mu_1$, $\varepsilon_2 = \varepsilon_1$, and

$$\frac{\sigma^*}{\mu_1} = \frac{\sigma}{\varepsilon_1} \tag{5.6}$$

then $\Gamma = 0$. Second, from (5.2b), we find that

$$\beta_x^t = \left(1 + \frac{\sigma}{j\omega\varepsilon_1}\right)k_1 = k_1 - j\sigma\eta_1 \tag{5.7}$$

The real part of (5.7) is the propagating component of the wave, and the imaginary part is the attenuating component. From the real component, we observe that the wave speed in this medium is identical to that in medium 1. Further, despite being a lossy medium, the wave is dispersionless — that is, the wave speed is frequency-independent for an axially propagating wave. Hence, this type of medium has been quantified as a dispersionless medium.

From (5.1b), the resultant fields in the dispersionless medium excited by a normally incident wave are expressed as

$$\vec{E}_2 = \hat{y}\,\eta_1 H_0 \exp(-jk_1 x)\exp(-\sigma\eta_1 x) \tag{5.8a}$$

$$\vec{H}_2 = \hat{z}\,H_0 \exp(-jk_1 x)\exp(-\sigma\eta_1 x) \tag{5.8b}$$

In summary, we see that given a medium with magnetic and electric conductivity defined by (5.6), the medium is matched across a planar boundary for all normally incident waves. Further, the wave propagating in the dispersionless medium has the same propagation characteristics as the incident wave, but attenuates along the normal direction.

5.2.2 Berenger's PML Medium

The dispersionless medium discussed in the previous section has been used with limited success to terminate FDTD lattices [2, 39]. The obvious limitation of this approach is that the dispersionless medium is matched to the interior region only for normally incident waves. Therefore, oblique incident waves partially reflect back into the computation region and corrupt the solution. As a result, the absorbing medium must be placed sufficiently far away from any source or discontinuity such that the impinging waves are planar and near-normally incident. This is impractical for large or elongated problem domains.

Berenger recognized that by adding an additional degree of freedom through splitting the transverse fields into two orthogonal components, the medium can be matched even at oblique incidence. Consider again the geometry introduced in Fig. 5.2

wherein a TE_z-polarized plane wave is assumed to impinge on the planar interface of the lossy medium. Within the lossy medium, Maxwell's curl equations are expressed in their time-harmonic form as

$$j\omega\varepsilon_2\left(1 + \frac{\sigma}{j\omega\varepsilon_0}\right)E_x = \frac{\partial H_z}{\partial y}, \qquad j\omega\varepsilon_2\left(1 + \frac{\sigma}{j\omega\varepsilon_0}\right)E_y = -\frac{\partial H_z}{\partial x} \tag{5.9a}$$

$$j\omega\mu_2\left(1 + \frac{\sigma^*}{j\omega\mu_0}\right)H_z = \frac{\partial E_x}{\partial y} - \frac{\partial E_y}{\partial x} \tag{5.9b}$$

Note that the electric and magnetic conductivities in (5.9a) and (5.9b) are intentionally normalized by the relative permittivity and permeability, respectively. This is in anticipation of matching to inhomogeneous media as discussed later in Section 5.4.4. These equations are then modified by splitting the transverse magnetic field into two orthogonal terms:

$$H_y = H_{yx} + H_{yz} \tag{5.10}$$

such that (5.9b) is rewritten as

$$j\omega\mu_2\left(1 + \frac{\sigma_x^*}{j\omega\mu_0}\right)H_{zx} = -\frac{\partial E_y}{\partial x} \tag{5.11a}$$

$$j\omega\mu_2\left(1 + \frac{\sigma_y^*}{j\omega\mu_0}\right)H_{zy} = \frac{\partial E_x}{\partial y} \tag{5.11b}$$

The notation is further reduced by introducing the variables:

$$s_k = \left(1 + \frac{\sigma_k}{j\omega\varepsilon_0}\right), \qquad s_k^* = \left(1 + \frac{\sigma_k^*}{j\omega\mu_0}\right); \qquad k = x,\, y \tag{5.12}$$

Then, (5.11a) and (5.11b) are rewritten as

$$j\omega\varepsilon_2\, s_y\, E_x = \frac{\partial\left(H_{zx} + H_{zy}\right)}{\partial y} \tag{5.13a}$$

$$j\omega\varepsilon_2\, s_x\, E_y = -\frac{\partial\left(H_{zx} + H_{zy}\right)}{\partial x} \tag{5.13b}$$

The next step is to derive plane-wave solutions within the lossy material region. To this end, (5.13a) is differentiated with respect to y and (5.13b) with respect to x. Then, substituting the expressions for $\partial E_x/\partial y$ and $\partial E_y/\partial x$ into (5.11a) and (5.11b) leads to

$$-\omega^2 \mu_2 \varepsilon_2 H_{zx} = -\frac{1}{s_x^*}\frac{\partial}{\partial x}\frac{1}{s_x}\frac{\partial}{\partial x}\left(H_{zx} + H_{zy}\right) \tag{5.14a}$$

$$-\omega^2 \mu_2 \varepsilon_2 H_{zy} = -\frac{1}{s_y^*}\frac{\partial}{\partial y}\frac{1}{s_y}\frac{\partial}{\partial y}\left(H_{zx} + H_{zy}\right) \tag{5.14b}$$

Adding these together and using (5.10), leads the representative wave equation:

$$\frac{1}{s_x^*}\frac{\partial}{\partial x}\frac{1}{s_x}\frac{\partial}{\partial x}H_z + \frac{1}{s_y^*}\frac{\partial}{\partial y}\frac{1}{s_y}\frac{\partial}{\partial y}H_z + \omega^2 \mu_2 \varepsilon_2 H_z = 0 \tag{5.15}$$

This wave equation supports the solutions

$$H_z = H_0 \tau e^{-j\sqrt{s_x s_x^*}\,\beta_x x \,-\, j\sqrt{s_y s_y^*}\,\beta_y y} \tag{5.16}$$

with the dispersion relationship

$$\left(\beta_x\right)^2 + \left(\beta_y\right)^2 = \left(k_2\right)^2 \tag{5.17}$$

Then, from (5.13a), (5.13b), and (5.10), we have

$$E_x = -H_0 \tau \frac{\beta_y}{\omega \varepsilon_2}\sqrt{\frac{s_y^*}{s_y}}\, e^{-j\sqrt{s_x s_x^*}\,\beta_x x \,-\, j\sqrt{s_y s_y^*}\,\beta_y y} \tag{5.18a}$$

$$E_y = H_0 \tau \frac{\beta_x}{\omega \varepsilon_2}\sqrt{\frac{s_x^*}{s_x}}\, e^{-j\sqrt{s_x s_x^*}\,\beta_x x \,-\, j\sqrt{s_y s_y^*}\,\beta_y y} \tag{5.18b}$$

The object of this exercise is to determine the characteristics of s_x, s_x^*, s_y, and s_y^* such that the boundary interface is reflectionless. Despite the field splitting, the tangential fields must still be continuous across the $x = 0$ interface. A plane wave is assumed to be incident on this hypothetical material medium (Fig. 5.1) rendering total fields (5.1a) in region 1 and transmitted fields (5.16) and (5.18) in region 2. The reflection and transmission coefficients are derived by enforcing the continuity of the tangential fields at $x = 0$, leading to

$$\Gamma = \frac{\dfrac{\beta_x^i}{\omega \varepsilon_1} - \dfrac{\beta_x}{\omega \varepsilon_2}\sqrt{\dfrac{s_x^*}{s_x}}}{\dfrac{\beta_x^i}{\omega \varepsilon_1} + \dfrac{\beta_x}{\omega \varepsilon_2}\sqrt{\dfrac{s_x^*}{s_x}}}; \qquad \tau = 1 + \Gamma \tag{5.19}$$

and $\beta_y = \beta_y^i = k_1 \sin\theta^i$. From the latter relation, we see that $s_y = s_y^* = 1$. We observe that if $\varepsilon_2 = \varepsilon_1$ and $s_x = s_x^*$ (or $\sigma_x / \varepsilon_0 = \sigma_x^* / \mu_0$), then $\beta_x = \beta_x^i$ and $\Gamma = 0$. This is true for all incident angles θ^i. The resultant TE_z field transmitted into the absorbing medium is then

$$H_z = H_0 \, e^{-js_x\beta_x^i x - j\beta_y^i y} = H_0 \, e^{-j\beta_x^i x - j\beta_y^i y} \, e^{-\sigma \eta_1 \varepsilon_{r1} \cos\theta^i x} \tag{5.20a}$$

$$E_x = -H_0 \, \eta_1 \sin\theta^i \, e^{-j\beta_x^i x - j\beta_y^i y} \, e^{-\sigma \eta_1 \varepsilon_{r1} \cos\theta^i x} \tag{5.20b}$$

$$E_y = H_0 \, \eta_1 \cos\theta^i \, e^{-j\beta_x^i x - j\beta_y^i y} \, e^{-\sigma \eta_1 \varepsilon_{r1} \cos\theta^i x} \tag{5.20c}$$

Thus, we observe that the purely transmitted field propagates with the same wave speed as the impinging field while attenuating along the normal direction with an attenuation constant of $\sigma \eta_1 \varepsilon_{r1} \cos\theta^i$. Unlike the dispersionless material in the previous section, this property applies to all wave angles of incidence. Note that this exercise can be repeated for the dual-polarization splitting, $E_z = E_{zx} + E_{zy}$. If $\mu_2 = \mu_1$ and $s_x = s_x^*$, then $\beta_x = \beta_x^i$ and $\Gamma = 0$ for all θ^i.

In summary, if a wave impinges on a half-space of this perfectly matched absorbing material, it is perfectly transmitted for *all* polarizations, *all* frequencies, and *all* angles of incidence. Further, the transmitted wave propagates with the same wave speed as the incident wave, has the same characteristic impedance as the incident wave, but attenuates along the normal direction. Finally, this hypothetical absorbing material can be used within the framework of a discrete numerical method such as FDTD to terminate all outward traveling waves.

5.2.3 Corner Region — Matching a PML to a PML

The PML has compelling advantages to serve as an absorbing medium to terminate FDTD lattices. Referring to Fig. 5.1, exterior boundaries of the rectangular lattice can be terminated with a PML absorbing layer. However, for FDTD lattice computations, there is not a single planar boundary, but rather multiple planar boundaries. These overlap in the corner regions.

Originally, there was some question as to whether the PML is actually matched in the corner region [11]. In fact it is, and through a systematic analysis this can be proven. To demonstrate this, we need to derive a PML that is matched along an orthogonal axis to another PML media.

Consider the PML medium constructed in Section 5.2.3 for an x-normal boundary. This medium is to be matched to a planar half-space across a y-normal interface. The incident field impinging on the boundary is provided in (5.20). Assuming a reflected field with amplitude Γ, the total field in the region ($y < 0$) is:

$$\vec{H}_2 = \hat{z} H_0 \left(1 + \Gamma e^{j2\beta_y^i y}\right) e^{-js_x \beta_x^i x - j\beta_y^i y} \tag{5.21a}$$

$$\vec{E}_2 = \left[-\hat{x} \frac{\beta_y^i}{\omega \varepsilon_2}\left(1 - \Gamma e^{2j\beta_y^i y}\right) + \hat{y} \frac{s_x \beta_x^i}{\omega s_x \varepsilon_2}\left(1 + \Gamma e^{2j\beta_y^i y}\right)\right] H_0\, e^{-js_x\beta_x^i x - j\beta_y^i y} \tag{5.21b}$$

The magnetic field in the region $y > 0$ is again split as in (5.10). Then Ampere's and Faraday's laws are expressed as in (5.11a) to (5.13b). Again the wave equation in (5.15) is derived; the exception is that $s_x^* = s_x$. This leads to posed solutions in the region ($y > 0$):

$$H_{z_3} = H_0 \tau e^{-js_x \beta_x x - j\sqrt{s_y s_y^*}\, \beta_y y} \tag{5.22a}$$

$$E_{x_3} = -H_0 \tau \frac{\beta_y}{\omega \varepsilon_3} \sqrt{\frac{s_y^*}{s_y}}\, e^{-js_x \beta_x x - j\sqrt{s_y s_y^*}\, \beta_y y} \tag{5.22b}$$

$$E_{y_3} = H_0 \tau \frac{\beta_x}{\omega \varepsilon_3} e^{-js_x \beta_x x - j\sqrt{s_y s_y^*}\, \beta_y y} \tag{5.22c}$$

with the dispersion relationship $(\beta_x)^2 + (\beta_y)^2 = (k_3)^2$. Again, enforcing the continuity of the tangential fields across the interface leads to

$$\Gamma = \frac{\dfrac{\beta_y^i}{\omega \varepsilon_2} - \dfrac{\beta_y}{\omega \varepsilon_3}\sqrt{\dfrac{s_y^*}{s_y}}}{\dfrac{\beta_y^i}{\omega \varepsilon_2} + \dfrac{\beta_y}{\omega \varepsilon_3}\sqrt{\dfrac{s_y^*}{s_y}}}, \qquad \tau = 1 + \Gamma \tag{5.23}$$

and $s_x\beta_x = s_x\beta_x^i$. If $\varepsilon_3 = \varepsilon_2$, s_x is continuous across the interface; and if $s_y = s_y^*$, then $\beta_y = \beta_y^i$ and $\Gamma = 0$. This is true for all incident angles $\phi^i = 90^\circ - \theta^i$

(off the y-normal axis). Finally, the resultant TE$_Z$ field transmitted into the absorbing medium in the overlapping region is

$$H_z = H_0 e^{-js_x\beta_x^i x - js_y\beta_y^i y} = H_0 e^{-j\beta_x^i x - j\beta_y^i y} e^{-\sigma\eta_1 \varepsilon_{r1}\cos\theta^i x} e^{-\sigma\eta_1 \varepsilon_{r1}\sin\theta^i y} \quad (5.24a)$$

$$E_x = -H_0 \eta_1 \sin\theta^i e^{-j\beta_x^i x - j\beta_y^i y} e^{-\sigma\eta_1 \varepsilon_{r1}\cos\theta^i x} e^{-\sigma\eta_1 \varepsilon_{r1}\sin\theta^i y} \quad (5.24b)$$

$$E_y = H_0 \eta_1 \cos\theta^i e^{-j\beta_x^i x - j\beta_y^i y} e^{-\sigma\eta_1 \varepsilon_{r1}\cos\theta^i x} e^{-\sigma\eta_1 \varepsilon_{r1}\sin\theta^i y} \quad (5.24c)$$

(Note that θ^i is referenced to the $x = 0$ plane.) Thus, the impinging wave is purely transmitted into the overlapping medium and propagates with its same wave speed while attenuating along *both* normal directions with the factors $\alpha_x = \sigma\eta_1\varepsilon_{r1}\cos\theta^i$ and $\alpha_y = \sigma\eta_1\varepsilon_{r1}\sin\theta^i$. Again, if this were repeated for the TM$_Z$-polarized case, requiring a splitting of $E_z = E_{zx} + E_{zy}$, we would determine that if $\mu_3 = \mu_2$ and $s_y = s_y^*$, then $\beta_y = \beta_y^i$ and $\Gamma = 0$. Thus the corner regions are perfectly matched as well.

5.2.4 Berenger's PML in Three Dimensions

The analogies used in the previous section can be extended to a third overlapping region with a z-normal boundary. In this case, a general three-dimensional expression for the split-field PML is used [6, 9, 13]. The resulting modified curl equations in time-dependent form are given by

$$\begin{aligned}
\left(\varepsilon_0\varepsilon_r\frac{\partial}{\partial t} + \sigma_y\varepsilon_r\right)E_{xy} &= \frac{\partial}{\partial y}(H_{zx} + H_{zy}) \\
\left(\varepsilon_0\varepsilon_r\frac{\partial}{\partial t} + \sigma_z\varepsilon_r\right)E_{xz} &= -\frac{\partial}{\partial z}(H_{yx} + H_{yz}) \\
\left(\varepsilon_0\varepsilon_r\frac{\partial}{\partial t} + \sigma_z\varepsilon_r\right)E_{yz} &= \frac{\partial}{\partial z}(H_{xy} + H_{xz}) \\
\left(\varepsilon_0\varepsilon_r\frac{\partial}{\partial t} + \sigma_x\varepsilon_r\right)E_{yx} &= -\frac{\partial}{\partial x}(H_{zx} + H_{zy}) \\
\left(\varepsilon_0\varepsilon_r\frac{\partial}{\partial t} + \sigma_x\varepsilon_r\right)E_{zx} &= \frac{\partial}{\partial x}(H_{yx} + H_{yz}) \\
\left(\varepsilon_0\varepsilon_r\frac{\partial}{\partial t} + \sigma_y\varepsilon_r\right)E_{zy} &= -\frac{\partial}{\partial y}(H_{xy} + H_{xz})
\end{aligned} \quad (5.25a)$$

for the modified Ampere's law, and

$$
\begin{aligned}
\left(\mu_0\,\mu_r\frac{\partial}{\partial t} + \sigma_y\,\eta_0^2\,\mu_r\right)H_{xy} &= -\frac{\partial}{\partial y}(E_{zx} + E_{zy}) \\
\left(\mu_0\,\mu_r\frac{\partial}{\partial t} + \sigma_z\,\eta_0^2\,\mu_r\right)H_{xz} &= \frac{\partial}{\partial z}(E_{yx} + E_{yz}) \\
\left(\mu_0\,\mu_r\frac{\partial}{\partial t} + \sigma_z\,\eta_0^2\,\mu_r\right)H_{yz} &= -\frac{\partial}{\partial z}(E_{xy} + E_{xz}) \\
\left(\mu_0\,\mu_r\frac{\partial}{\partial t} + \sigma_x\,\eta_0^2\,\mu_r\right)H_{yx} &= \frac{\partial}{\partial x}(E_{zx} + E_{zy}) \qquad (5.25b) \\
\left(\mu_0\,\mu_r\frac{\partial}{\partial t} + \sigma_x\,\eta_0^2\,\mu_r\right)H_{zx} &= -\frac{\partial}{\partial x}(E_{yx} + E_{yz}) \\
\left(\mu_0\,\mu_r\frac{\partial}{\partial t} + \sigma_y\,\eta_0^2\,\mu_r\right)H_{zy} &= \frac{\partial}{\partial y}(E_{xy} + E_{xz})
\end{aligned}
$$

for the modified Faraday's law. Note that the relative permittivities and permeabilities are included in these equations, which are posed in a general form so that they are valid for inhomogeneous as well as homogeneous media [36, 37].

5.3 A STRETCHED-COORDINATE FORMULATION

A more compact form of the split-field equations of (5.25a) and (5.25b) was introduced by Chew and Weedon [10] and independently by Rappaport [33]. Here, the split-field equations are re-posed in a nonsplit form that maps Maxwell's equations into a complex coordinate space. To this end, a coordinate mapping into a complex space is introduced [10]:

$$
\tilde{x} \to \int_0^x s_x(x')\,dx', \qquad \tilde{y} \to \int_0^y s_y(y')\,dy', \qquad \tilde{z} \to \int_0^z s_z(z')\,dz' \qquad (5.26a)
$$

where it is assumed that the PML parameters, s_i, are continuous functions along the axial directions. The partial derivatives in the stretched coordinate space are then

$$
\frac{\partial}{\partial \tilde{x}} = \frac{1}{s_x}\frac{\partial}{\partial x}, \qquad \frac{\partial}{\partial \tilde{y}} = \frac{1}{s_y}\frac{\partial}{\partial y}, \qquad \frac{\partial}{\partial \tilde{z}} = \frac{1}{s_z}\frac{\partial}{\partial z} \qquad (5.26b)
$$

Thus, the ∇ operator in the mapped space is defined as

$$\tilde{\nabla} = \hat{x}\frac{\partial}{\partial \tilde{x}} + \hat{y}\frac{\partial}{\partial \tilde{y}} + \hat{z}\frac{\partial}{\partial \tilde{z}} = \hat{x}\frac{1}{s_x}\frac{\partial}{\partial x} + \hat{y}\frac{1}{s_y}\frac{\partial}{\partial y} + \hat{z}\frac{1}{s_z}\frac{\partial}{\partial z} \tag{5.26c}$$

Maxwell's equations in the complex-coordinate stretched space are then expressed as

$$\begin{aligned} j\omega\varepsilon\vec{E} &= \tilde{\nabla}\times\vec{H} \\ &= \hat{x}\left(\frac{1}{s_y}\frac{\partial}{\partial y}H_z - \frac{1}{s_z}\frac{\partial}{\partial z}H_y\right) + \hat{y}\left(\frac{1}{s_z}\frac{\partial}{\partial z}H_x - \frac{1}{s_x}\frac{\partial}{\partial x}H_z\right) \\ &+ \hat{z}\left(\frac{1}{s_x}\frac{\partial}{\partial x}H_y - \frac{1}{s_y}\frac{\partial}{\partial y}H_x\right) \end{aligned} \tag{5.27a}$$

$$\begin{aligned} -j\omega\mu\vec{H} &= \tilde{\nabla}\times\vec{E} \\ &= \hat{x}\left(\frac{1}{s_y}\frac{\partial}{\partial y}E_z - \frac{1}{s_z}\frac{\partial}{\partial z}E_y\right) + \hat{y}\left(\frac{1}{s_z}\frac{\partial}{\partial z}E_x - \frac{1}{s_x}\frac{\partial}{\partial x}E_z\right) \\ &+ \hat{z}\left(\frac{1}{s_x}\frac{\partial}{\partial x}E_y - \frac{1}{s_y}\frac{\partial}{\partial y}E_x\right) \end{aligned} \tag{5.27b}$$

A direct relationship can be drawn between the stretched-coordinate form of Maxwell's equations and Berenger's split-field PML. To demonstrate this, first rewrite the split-field equations of (5.25a) for the time-harmonic case:

$$j\omega\varepsilon s_y E_{xy} = \frac{\partial}{\partial y}(H_{zx} + H_{zy}) \tag{5.28a}$$

$$j\omega\varepsilon s_z E_{xz} = -\frac{\partial}{\partial z}(H_{yx} + H_{yz}) \tag{5.28b}$$

$$j\omega\varepsilon s_z E_{yz} = \frac{\partial}{\partial z}(H_{xy} + H_{xz}) \tag{5.28c}$$

$$j\omega\varepsilon s_x E_{yx} = -\frac{\partial}{\partial x}(H_{zx} + H_{zy}) \tag{5.28d}$$

$$j\omega\varepsilon s_x E_{zx} = \frac{\partial}{\partial x}(H_{yx} + H_{yz}) \tag{5.28e}$$

$$j\omega\varepsilon s_y E_{zy} = -\frac{\partial}{\partial y}(H_{xy} + H_{xz}) \tag{5.28f}$$

Then, add (5.28a) + (5.28b); (5.28c) + (5.28d); and (5.28e) + (5.28f); and use the relationships $E_x = E_{xy} + E_{xz}$, $E_y = E_{yx} + E_{yz}$, and $E_z = E_{zx} + E_{zy}$. This yields

$$j\omega\varepsilon E_x = \frac{\partial}{s_y \partial y} H_z - \frac{\partial}{s_z \partial z} H_y \tag{5.29a}$$

$$j\omega\varepsilon E_y = \frac{\partial}{s_z \partial z} H_x - \frac{\partial}{s_x \partial x} H_z \tag{5.29b}$$

$$j\omega\varepsilon E_z = \frac{\partial}{s_x \partial x} H_y - \frac{\partial}{s_y \partial y} H_x \tag{5.29c}$$

which is identical to (5.27a). This procedure is repeated for the modified Faraday's law in (5.25b), leading exactly to (5.27b). Specifically, we see that the stretched-coordinate form of the PML is equivalent to the split-field PML; however, it re-poses it in a nonsplit form.

The principal advantage the complex stretched-coordinate formulation offers is the ease of mathematically manipulating the PML equations, thereby simplifying the understanding of the behavior of the PML. It also provides a pathway to mapping the PML into other coordinate systems such as cylindrical and spherical coordinates [34, 40], as well as utilizing the split-field PML in frequency-domain finite-element methods based on unstructured discretizations [33, 41].

5.4 AN ANISOTROPIC PML ABSORBING MEDIUM

The split-field PML introduced by Berenger is a hypothetical medium based on a mathematical model. Neither it nor its coordinate-stretched analog is a physical medium. Due to the coordinate dependence of the loss terms, if such a physical medium were to exist, it must be anisotropic.

Indeed, an anisotropic perfectly matched medium does exist, and was first discussed by Sacks et al. [35]. For a single interface, the anisotropic medium is uniaxial and is composed of both electric and magnetic permittivity tensors. This formulation performs as well as Berenger's PML and avoids the field splitting. In this section, the theoretical development of the uniaxial PML is introduced. Its comparison with Berenger's PML and the coordinate-stretched PML formulation is discussed. Later, in Section 5.6, its effective implementation within the FDTD method is presented.

5.4.1 Perfectly Matched Uniaxial Medium

Consider an arbitrarily polarized time-harmonic plane wave, $\vec{H}^{inc} = \vec{H}_0 e^{-j\beta_x^i x - j\beta_y^i y}$, propagating in an isotropic medium that is incident on a material half-space described as a uniaxial anisotropic medium. The interface between the two media is the $x = 0$ plane. The fields excited within the uniaxial medium are also plane-wave in nature and satisfy Maxwell's equations. For the plane wave in the material half-space, the curl equations yield

$$\vec{\beta}^a \times \vec{E} = \omega \overline{\overline{\mu}} \vec{H}, \qquad \vec{\beta}^a \times \vec{H} = -\omega \overline{\overline{\varepsilon}} \vec{E} \tag{5.30}$$

where $\beta^a = \hat{x}\beta_x^a + \hat{y}\beta_y^a$ and the permittivity and permeability are the uniaxial tensors

$$\overline{\overline{\varepsilon}} = \varepsilon_1 \begin{bmatrix} a & 0 & 0 \\ 0 & b & 0 \\ 0 & 0 & b \end{bmatrix}, \qquad \overline{\overline{\mu}} = \mu_1 \begin{bmatrix} c & 0 & 0 \\ 0 & d & 0 \\ 0 & 0 & d \end{bmatrix} \tag{5.31}$$

having $\varepsilon_{yy} = \varepsilon_{zz}$ and $\mu_{yy} = \mu_{zz}$ (since the medium is expected to be rotationally symmetric about the x axis). The wave equation is then derived from the coupled curl equations:

$$\vec{\beta}^a \times \left(\overline{\overline{\varepsilon}}^{-1} \vec{\beta}^a\right) \times \vec{H} + \omega^2 \overline{\overline{\mu}} \vec{H} = 0 \tag{5.32}$$

Expressing the cross products as matrix operators, the wave equation is expressed more suitably in matrix form as

$$\begin{bmatrix} k^2 c - \left(\beta_y^a\right)^2 b^{-1} & \beta_x^a \beta_y^a b^{-1} & 0 \\ \beta_x^a \beta_y^a b^{-1} & k^2 d - \left(\beta_x^a\right)^2 b^{-1} & 0 \\ 0 & 0 & k^2 d - \left(\beta_x^a\right)^2 b^{-1} - \left(\beta_y^a\right)^2 a^{-1} \end{bmatrix} \begin{bmatrix} H_x \\ H_y \\ H_z \end{bmatrix} = 0 \tag{5.33}$$

where $k^2 = \omega^2 \mu_1 \varepsilon_1$. The dispersion relationship for the uniaxial medium is derived from the determinant of the matrix operator. Solving for β_x^a, we find that there are four eigenmode solutions. Conveniently, these solutions can be decoupled into forward and backward $\mathrm{TE_z}$ and $\mathrm{TM_z}$ modes, which satisfy the dispersion relationships

$$\begin{aligned} k^2 - \left(\beta_x^a\right)^2 b^{-1} d^{-1} - \left(\beta_y^a\right)^2 a^{-1} d^{-1} &= 0 \qquad \mathrm{TE_z} \\ k^2 - \left(\beta_x^a\right)^2 b^{-1} d^{-1} - \left(\beta_y^a\right)^2 b^{-1} c^{-1} &= 0 \qquad \mathrm{TM_z} \end{aligned} \tag{5.34}$$

The reflection coefficient at the interface of the two half-spaces can now be computed. Let us assume a TE_z incident wave. Then, in the isotropic region $x < 0$, the fields are expressed as a superposition of the incident and reflected fields as

$$\vec{H}_1 = \hat{z} H_0 \left(1 + \Gamma e^{2j\beta_x^i x}\right) e^{-j\beta_x^i x - j\beta_y^i y}$$
$$\vec{E}_1 = \left[-\hat{x} \frac{\beta_y^i}{\omega \varepsilon_1}\left(1 + \Gamma e^{2j\beta_x^i x}\right) + \hat{y} \frac{\beta_x^i}{\omega \varepsilon_1}\left(1 - \Gamma e^{2j\beta_x^i x}\right)\right] H_0 e^{-j\beta_x^i x - j\beta_y^i y} \tag{5.35}$$

The wave transmitted into the anisotropic half-space $x > 0$ is also TE_z with propagation characteristics governed by dispersion relationship (5.34), and is expressed as

$$\vec{H}_2 = \hat{z} H_0 \tau e^{-j\beta_x^a x - j\beta_y^a y}$$
$$\vec{E}_2 = \left(-\hat{x} \frac{\beta_y^a}{\omega \varepsilon_1 a} + \hat{y} \frac{\beta_x^a}{\omega \varepsilon_1 b}\right) H_0 \tau e^{-j\beta_x^a x - j\beta_y^a y} \tag{5.36}$$

Γ and τ are the reflection and transmission coefficients, respectively, and are computed by enforcing the continuity of the tangential fields across the boundary interface:

$$\Gamma = \frac{\beta_x^i - \beta_x^a b^{-1}}{\beta_x^i + \beta_x^a b^{-1}}, \qquad \tau = 1 + \Gamma = \frac{2\beta_x^i}{\beta_x^i + \beta_x^a b^{-1}} \tag{5.37}$$

and $\beta_y^a = \beta_y^i$ due to phase matching. Observing (5.37), we see that if

$$\beta_x^a = \beta_x^i b \tag{5.38}$$

then $\Gamma = 0$ for all angles of incidence. From (5.34), for a TE_z-polarized wave:

$$\beta_x^a = \sqrt{k^2 b d - \left(\beta_y^i\right)^2 a^{-1} b} \tag{5.39}$$

Then, if $d = b$ and $a^{-1} = b$, we have

$$\beta_x^a = \sqrt{k^2 b^2 - \left(\beta_y^i\right)^2 b^2} = b\sqrt{k^2 - \left(\beta_y^i\right)^2} \equiv b\beta_x^i \tag{5.40}$$

Consequently, $\Gamma = 0$ for all β_x^i (reflectionless).

The above exercise can be repeated for TM_z polarization. The reflection coefficient for the TM_z-polarized wave is the dual of (5.37) and is found by replacing b with d (and vice versa), and a with c. For this case, the reflectionless condition holds if $b = d$ and $c^{-1} = d$.

In summary, given a plane wave incident on a half-space with an interface in an x = constant plane and composed of a uniaxial medium with the permittivity and permeability tensors:

$$\bar{\bar{\varepsilon}} = \varepsilon_1 \bar{\bar{s}}, \qquad \bar{\bar{\mu}} = \mu_1 \bar{\bar{s}}, \qquad \bar{\bar{s}} = \begin{bmatrix} s_x^{-1} & 0 & 0 \\ 0 & s_x & 0 \\ 0 & 0 & s_x \end{bmatrix} \tag{5.41}$$

the plane wave is purely transmitted into the uniaxial medium. This reflectionless property is completely independent of the angle of incidence, polarization, and frequency of the incident wave. Further, from (5.34), the propagation characteristics of the TE- and TM-polarized waves are identical. We call this medium a UPML in recognition of its uniaxial anisotropy.

Similarly to Berenger's PML, the reflectionless property of the UPML is valid for any s_x. For example, choose $s_x = 1 + \sigma_x / j\omega\varepsilon_0$. Then, from (5.38) we have

$$\beta_x^a = \left(1 - j\sigma/\omega\varepsilon_0\right)\beta_x^i \tag{5.42}$$

It is interesting to note that the real part of β_x^a is identical to that of the incident wave, implying that the phase velocities are identical. The axial characteristic wave impedance is also identical to the incident wave, which is a consequence of the fact that the media are perfectly matched. We note further that the transmitted wave attenuates along the x direction.

Finally, given a TE_z incident wave, the field intensities in the UPML are given by

$$\begin{aligned} \vec{H}_2 &= \hat{z} H_0 e^{-js_x\beta_x^i x - j\beta_y^a y} = \hat{z} H_0 e^{-j\beta_x^i x - j\beta_y^a y} e^{-\sigma_x \eta_1 \varepsilon_{r1} \cos\theta^i x} \\ \vec{E}_2 &= \left(-\hat{x} s_x \eta_1 \sin\theta^i + \hat{y}\eta_1 \cos\theta^i\right) H_0 e^{-j\beta_x^i x - j\beta_y^i y} e^{-\sigma_x \eta_1 \varepsilon_{r1} \cos\theta^i x} \end{aligned} \tag{5.43}$$

where θ^i is the angle of incidence relative to the x-axis. Thus, along the normal axis, the wave propagates with the same phase velocity as the incident while simultaneously being attenuated. The attenuation factor is independent of frequency, although it is dependent on the angle of incidence and the conductivity of the medium.

5.4.2 Relationship to Berenger's Split-Field PML

Comparing the TE_z fields transmitted into the UPML in (5.43) with the corresponding fields for Berenger's split-field PML in (5.20), we observe identical propagation characteristics. Further examination of (5.15) and (5.33) reveals that the two methods result in the same wave equation. Consequently, the plane waves satisfy the same dispersion relation. Comparing the tangential electric and magnetic field components in (5.43) and (5.20), we see that they, too, are identical.

However, upon comparing the normal electric fields, E_x, we note that they differ by a factor of s_x. In fact, by comparing with the incident fields, it is clear that in the split-field formulation, E_x is continuous across the $x = 0$ boundary. However, for the UPML, E_x is discontinuous while $D_x = s_x^{-1} E_x$ is continuous. This implies that the two methods host different divergence theorems.

Within the UPML, Gauss' law for the electric field is explicitly written as

$$\nabla \cdot \vec{D} \; = \; \nabla \cdot \left(\varepsilon \bar{\bar{s}} \, \vec{E} \right) \; = \; \frac{\partial}{\partial x}\left(\varepsilon \, s_x^{-1} E_x \right) + \frac{\partial}{\partial y}\left(\varepsilon \, s_x E_y \right) + \frac{\partial}{\partial z}\left(\varepsilon \, s_x E_z \right) \; = \; 0 \qquad (5.44)$$

This implies that $D_x = \varepsilon \, s_x^{-1} E_x$ must be continuous across the $x = 0$ boundary interface since no sources are assumed on this boundary. It was shown that ε must be continuous across the boundary for a perfectly matched condition. Thus, D_x and $s_x^{-1} E_x$ must be continuous across $x = 0$. Comparing (5.43) with (5.35), this is indeed true for the TE_z-polarized wave.

Next, consider Gauss' law for Berenger's split-field PML formulation. The del operator in the stretched-coordinate space is defined from (5.29c) as

$$\nabla \; = \; \hat{x} \, \frac{\partial}{s_x^{-1} \, \partial x} + \hat{y} \, \frac{\partial}{\partial y} + \hat{z} \, \frac{\partial}{\partial z} \qquad (5.45)$$

Therefore, we can express the divergence of the electric flux density as

$$\frac{\partial}{s_x \, \partial x}\left(\varepsilon E_x \right) + \frac{\partial}{\partial y}\left(\varepsilon E_y \right) + \frac{\partial}{\partial z}\left(\varepsilon E_z \right) \; = \; 0 \qquad (5.46)$$

Since ε is continuous across the boundary and s_x^{-1} occurs outside the derivative, both E_x and D_x are continuous.

In summary, we demonstrated that Berenger's split-field PML and the UPML have the same propagation characteristics since they both result in the same wave equation. The two formulations do have different Gauss' laws. Hence, the electric and magnetic vector field intensity components that are normal to the PML interface are different. This was identified in [37, 42] where it was shown that Berenger's PML could thus be derived from the UPML through a scaling of the normal field intensities.

5.4.3 Corner Regions

Each planar boundary of the FDTD lattice can be terminated in a UPML slab. However, as shown in Fig. 5.3, there are corner regions where the UPML media overlap. As found with Berenger's PML, the constitutive parameters of the corner regions are determined by matching a plane wave propagating in a UPML medium characterized by (5.41) to a UPML boundary having its normal along the x or y directions.

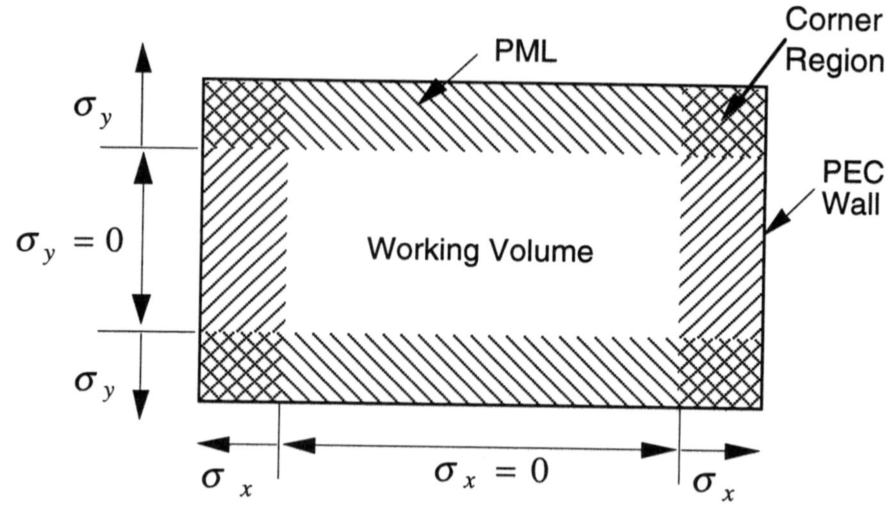

Fig. 5.3 FDTD lattice terminated by PML slabs. The interior or working volume is the region where the physical fields are to be modeled. The PML regions absorb outgoing scattered or radiated numerical electromagnetic wave analogs.

The analysis is quite similar to that presented in Section 5.2.3 and follows (5.32) to (5.43). For the sake of illustration, assume that the new UPML medium has a normal aligned along the y axis. For a matched condition, ε and μ must again be continuous across the interface, and the electric and magnetic tensors must be

$$\bar{\bar{\varepsilon}} = \bar{\bar{\mu}} = \begin{bmatrix} s_x^{-1} & 0 & 0 \\ 0 & s_x & 0 \\ 0 & 0 & s_x \end{bmatrix} \begin{bmatrix} s_y & 0 & 0 \\ 0 & s_y^{-1} & 0 \\ 0 & 0 & s_y \end{bmatrix} = \begin{bmatrix} s_x^{-1} s_y & 0 & 0 \\ 0 & s_x s_y^{-1} & 0 \\ 0 & 0 & s_x s_y \end{bmatrix} \tag{5.47}$$

Repeating this for a z-normal boundary, the general tensor is determined to be

$$\bar{\bar{\varepsilon}} = \bar{\bar{\mu}} = \begin{bmatrix} \dfrac{s_y s_z}{s_x} & 0 & 0 \\ 0 & \dfrac{s_x s_z}{s_y} & 0 \\ 0 & 0 & \dfrac{s_x s_y}{s_z} \end{bmatrix} = \bar{\bar{s}} \tag{5.48}$$

In general, Maxwell's curl equations in the PML regions are written as

$$\nabla \times \vec{E} = -j\omega\mu\bar{\bar{s}}\vec{H}, \qquad \nabla \times \vec{H} = j\omega\varepsilon\bar{\bar{s}}\vec{E}. \tag{5.49}$$

We could then define the FDTD space as being a PML medium in all regions. Using the definitions for s_x, s_y, and s_z given in (5.12) in the interior FDTD space shown in Fig. 5.3, σ_x, σ_y, and σ_z are then simply zero and (5.48) reduces to Maxwell's equations in the original isotropic space. Consequently, $\bar{\bar{s}}$ is the identity tensor in this region. Similarly, in regions where σ_y and σ_z both equal zero and σ_x is nonzero, $\bar{\bar{s}}$ reduces to (5.41). Thus, (5.48) provides a general tensor that is defined throughout the FDTD lattice and explicitly treats the interior and corner regions simultaneously.

Defining the electric and magnetic tensors as in (5.48), we observe that the generalized tensors are no longer uniaxial by strict definition. However, the anisotropic PML is still referenced as being uniaxial, since it is uniaxial in the nonoverlapping regions.

5.4.4 Matching the UPML to Generalized Media

In the previous sections, the PML was matched to a homogeneous medium supporting plane-wave solutions impinging on the PML boundary. Because the PML is matched for all angles of incidence and polarizations, we could generalize this to arbitrary finite sources or to scattered fields emanating from complex geometries by using plane-wave expansions of the fields. However, a more general proof is needed to demonstrate its effectiveness for matching to waves propagating in inhomogeneous media. This is the focus of this section.

Consider a wave propagating in an inhomogeneous medium along an axial direction chosen arbitrarily to be the z axis. We assume that the material medium is inhomogeneous along the transverse axes (x, y) and invariant along the axial direction. The electromagnetic waves supported by such a medium can be decomposed into TE_z- and TM_z-polarized waves. In this medium, we pose time-harmonic solutions to the wave equation of the form

$$\text{TE}_z: \quad H_z = h_z(x, y)\, e^{-\gamma_z z} \tag{5.50a}$$

$$\text{TM}_z: \quad E_z = e_z(x, y) e^{-\gamma_z z} \tag{5.50b}$$

where $E_z = 0$ for the TE_z-polarized waves and $H_z = 0$ for the TM_z-polarized waves. From Maxwell's equations for the TE_z-polarization case, the remaining fields are derived from the axial fields as

$$E_x = \frac{-j\omega\mu}{k^2+\gamma_z^2}\frac{\partial H_z}{\partial y}, \qquad H_x = \frac{-\gamma_z}{k^2+\gamma_z^2}\frac{\partial H_z}{\partial x}$$

$$E_y = \frac{j\omega\mu}{k^2+\gamma_z^2}\frac{\partial H_z}{\partial x}, \qquad H_y = \frac{-\gamma_z}{k^2+\gamma_z^2}\frac{\partial H_z}{\partial y} \tag{5.51}$$

$$E_z = 0$$

The TM_z fields are derived from (5.50b), leading to a dual expression of (5.51).

A PML half-space is interfaced with the inhomogeneous medium at the $z = 0$ plane. We assume that the transversely inhomogeneous material profile extends through the PML. From (5.48), the PML has the following properties:

$$\bar{\bar{\varepsilon}} = \varepsilon(x,y)\begin{bmatrix} s_z & 0 & 0 \\ 0 & s_z & 0 \\ 0 & 0 & s_z^{-1} \end{bmatrix}, \qquad \bar{\bar{\mu}} = \mu(x,y)\begin{bmatrix} s_z & 0 & 0 \\ 0 & s_z & 0 \\ 0 & 0 & s_z^{-1} \end{bmatrix} \tag{5.52}$$

where ε and μ can be complex. The generalized wave equation is then derived, and the axial field solutions are obtained for the dual polarizations:

$$\text{TE}_z\text{:} \qquad H_z = s_z h_z(x, y)e^{-s_z \gamma_z z} \tag{5.53a}$$

$$\text{TM}_z\text{:} \qquad E_z = s_z e_z(x, y)e^{-s_z \gamma_z z} \tag{5.53b}$$

where $h_z(x,y)$ and $e_z(x,y)$ are identical to those in (5.50a) and (5.50b), respectively. Note that the s_z weighting the normal fields is consistent with Gauss' laws defined in (5.44). For the TE_z-polarized case, the remaining fields are again derived from Maxwell's equations in the uniaxial medium, leading to

$$E_x = \frac{-j\omega\mu}{k^2+\gamma_z^2}\frac{1}{s_z}\frac{\partial H_z}{\partial y}, \qquad H_x = \frac{-\gamma_z}{k^2+\gamma_z^2}\frac{1}{s_z}\frac{\partial H_z}{\partial x}$$

$$E_y = \frac{j\omega\mu}{k^2+\gamma_z^2}\frac{1}{s_z}\frac{\partial H_z}{\partial x}, \qquad H_y = \frac{-\gamma_z}{k^2+\gamma_z^2}\frac{1}{s_z}\frac{\partial H_z}{\partial y} \tag{5.54}$$

$$E_z = 0$$

The TM_z fields are simply the dual of (5.54).

Upon inserting the axial fields of (5.53a) into (5.54), we see that the transverse fields are continuous across the $z = 0$ plane. In addition, the wave transmitted into the PML medium has the same characteristic impedance relative to the axial direction as that of the impinging wave. Consequently, the uniaxial medium is perfectly matched to the inhomogeneous space. The same can be shown to be true for the dual TM_z case.

5.5 THEORETICAL PERFORMANCE OF THE PML

5.5.1 The Continuous Space

The preceding sections developed the theoretical foundation for perfectly matched absorbing media suitable for terminating FDTD space lattices. However, in all of the analyses, the PML media were assumed to be half-spaces. When used to truncate the FDTD lattice, the PML must have finite thickness and be terminated by a boundary. If the boundary is assumed to be a PEC wall, as shown in Fig. 5.3, power reflects back into the interior FDTD region. The reflection is computed using a simple transmission line analysis and has an amplitude of

$$R(\theta) = \exp\left(-2\sigma\eta\varepsilon_r d\cos\theta\right) \tag{5.55}$$

where θ is the angle of incidence, d is the thickness of the PML slab, η is the characteristic impedance of the reference material, and σ is the conductivity of the PML. Within the context of an FDTD simulation, $R(\theta)$ is referred to as the "reflection error" since it is a nonphysical reflection due to the PEC-backed PML slab. This reflection error decays exponentially with the thickness of the PML as well as with the conductivity of the PML. However, the cosine dependence of θ dictates that at larger angles of incidence, the reflection error increases for a fixed σ and d. To be effective within an FDTD simulation, it is desirable for the PML medium to be as thin as possible. Thus, for a small d, we must have a large conductivity to reduce $R(\theta)$ to an acceptably small level.

In this discussion, we have assumed that the wave impinging on the PML medium is purely propagating. However, there are important situations where the PML could be used to terminate evanescent waves. One such case would be the termination of a hollow rectangular waveguide that supports both evanescent and propagating modes. From (5.38), we derive [25]

$$\gamma_x^a = s_x\gamma_x^i = s_x\left(\alpha_x^i + j\beta_x^i\right) \tag{5.56a}$$

If $s_x = 1 + \sigma_x / j\omega\varepsilon_0$, then

$$\gamma_x^a = \left(\alpha_x^i + \beta_x^i\frac{\sigma_x}{\omega\varepsilon_0}\right) + j\left(\beta_x^i - \alpha_x^i\frac{\sigma_x}{\omega\varepsilon_0}\right) \tag{5.56b}$$

If $\beta_x^i = 0$, we see that the wave attenuates at the same rate as the evanescent wave. Here, the PML is useless unless it is placed sufficiently far from the source of the evanescent wave such that it decays on its own. One simple solution to this problem is to modify the real part of s_x. Specifically, let

$$s_x = \kappa_x + \sigma_x / j\omega\varepsilon_0 \tag{5.57a}$$

Then, we have

$$\gamma_x^a = s_x \gamma_x^i = \left(\kappa_x \alpha_x^i + \beta_x^i \frac{\sigma_x}{\omega\varepsilon_0}\right) + j\left(\kappa_x \beta_x^i - \alpha_x^i \frac{\sigma_x}{\omega\varepsilon_0}\right) \tag{5.57b}$$

Choosing $\kappa_x > 0$ increases the decay rate of the evanescent waves.

5.5.2 The Discrete Space

Spatial Scaling of the PML Material Parameters

The PML interface presents a discontinuity in both the electric and magnetic conductivity. In the discrete-space representation of Maxwell's curl equations via the FDTD method, these discontinuities are modeled using a linear approximation [1]. Further, σ and σ^* are staggered by one-half lattice cell due to Yee's interleaving of the electric and magnetic fields. Consequently, discontinuities in σ and σ^* can lead to significant discretization error that is manifested as a spurious reflection. To reduce this reflection error, Berenger proposed to spatially scale the conductivity profile along the normal axis [3]. It can be shown that as long as the conductivity is scaled along normal axes, the material remains perfectly matched. This is easily seen from the stretched-coordinate analog presented in Section 5.3.

In the continuous space, assume that σ is a function of x. Assuming a PEC-backed PML slab of thickness d, we see from (5.26a) and (5.24) or (5.43) that the reflection due to a purely propagating wave impinging at angle θ is

$$R(\theta) = \exp\left[-2\eta\varepsilon_r \cos\theta \int_0^d \sigma(x)\,dx\right] \tag{5.58}$$

Several profiles have been suggested for scaling σ. The most successful use polynomial scaling or geometric scaling [4]. Polynomial scaling is simply

$$\sigma(x) = \left(\frac{x}{d}\right)^m \sigma_{\max} \tag{5.59}$$

This scales the conductivity from zero at the interface of the PML and the working volume to $\sigma_{\max}$ at the PEC boundary. The resultant reflection coefficient is

$$R(\theta) = \exp\left[-2\eta\,\varepsilon_r \sigma_{\max} d \cos\theta / (m+1)\right] \tag{5.60}$$

Polynomial scaling provides two parameters for a fixed PML thickness: $\sigma_{\max}$ and m. For large m, the conductivity distribution is relatively flat near the interface of the PML and the working volume. However, deeper into the PML, the conductivity increases more rapidly than for small m. In this region, the field amplitudes are substantially decayed and reflections due to the discretization error contribute less. Typically, m in the range between 3 and 4 has been found to be suitable [4, 7, 15, 37].

If the real part of s is greater than one as in (5.57a), then it can also be spatially scaled using polynomial scaling. Here, κ ranges from one at the PML surface to a maximum value $\kappa_{max} \geq 1$ at the PML/PEC boundary. To this end, we let

$$\kappa(x) = 1 + (\kappa_{max} - 1)\left(\frac{x}{d}\right)^m \tag{5.61}$$

For polynomial scaling, the PML parameters can be readily determined for a given error estimate. For example, let m, d, and the desired reflection error $R(0)$ be known. Then, from (5.60), σ_{max} is computed as

$$\sigma_{max} = -\frac{(m+1)\ln[R(0)]}{2\eta\varepsilon_r d} \tag{5.62}$$

For geometric scaling, the PML conductivity profile is defined as [7]

$$\sigma(x) = \sigma_0\left(g^{1/\Delta x}\right)^x \tag{5.63}$$

where σ_0 is the PML conductivity in the $x = 0$ plane (the PML/working-volume interface); g is the scaling factor; Δx is the FDTD space increment; and $d = N\Delta x$ is the PML thickness. Here, the PML conductivity scales from σ_0 at $x = 0$ to a maximum value of $g^N \sigma_0$ at the PML/PEC boundary. From (5.58), this results in

$$R(\theta) = \exp\left[-2\eta\sigma_0\Delta x(g^N - 1)\cos\theta / \ln g\right] \tag{5.64}$$

For geometric scaling, there are two parameters to choose for a fixed d: g and σ_0. σ_0 must be small to minimize the initial discretization error. The metric $g > 1$ governs the rate of increase of the conductivity within the PML. Large values of g flatten the conductivity profile near $x = 0$ and steepen it deeper into the PML. Similarly, if $\kappa > 1$, it can be geometrically scaled as

$$\kappa(x) = \left(q^{1/\Delta x}\right)^x \tag{5.65}$$

Then, κ ranges from 1 at $x = 0$ to a maximum value of q^N at the PML/PEC boundary. For geometric scaling, we would normally predetermine g, d, and $R(0)$. Then, we obtain

$$\sigma_0 = -\frac{\ln[R(0)]\ln(g)}{2\eta\varepsilon_r\Delta x\left(g^N - 1\right)} \tag{5.66}$$

Discretization Error

The design of an effective PML requires balancing the theoretical reflection error, $R(\theta)$, and the numerical discretization error. For example, (5.62) provides $\sigma_{\max}$ for a spatially scaled conductivity given a predetermined $R(\theta)$ and m. If $\sigma_{\max}$ is small, the primary reflection from the PML is due to its PEC backing. Equation (5.58) provides a fairly accurate approximation of the reflection error for this case. However, we normally choose $\sigma_{\max}$ to be as large as possible to minimize $R(\theta)$. Unfortunately, if $\sigma_{\max}$ is too large, the discretization error due to the FDTD approximation dominates, and the actual reflection error is potentially orders of magnitude higher than what (5.58) predicts. Consequently, there is an optimal choice for $R(\theta)$ that balances reflection from the PEC wall and discretization error.

It was postulated by Berenger [4, 7] that the largest discretization error manifested as reflection error occurs at $x=0$, the PML / working-volume interface. Energy penetrating further into the PML that is subsequently reflected is attenuated before leaving the PML, and typically is not as large a contribution. Thus, it is desirable to minimize the step discontinuity at $x=0$. As discussed earlier in this section, one way to achieve this is by spatially scaling the PML conductivity to flatten its profile near $x=0$. However, if the subsequent rise of conductivity with depth is too rapid, reflections from deeper within the PML can dominate.

For geometric scaling, the optimal g is typically between 2 and 3. For polynomial scaling, the optimal m is typically between 3 and 4. Through extensive experimental study, it was demonstrated in [37, 43] that, for a broad range of applications, an optimal choice for a 10-cell-thick PML is $R(0) \approx e^{-16}$. For a 5-cell thick PML, $R(0) \approx e^{-8}$ is optimal. From (5.62), this leads to an optimal choice for $\sigma_{\max}$:

$$\sigma_{\text{opt}} = \frac{m+1}{150\pi\sqrt{\varepsilon_r}\,\Delta x} \tag{5.67}$$

This expression has proven to be quite robust for many applications, as will be illustrated in Section 5.7. However, its value may be too large when the PML terminates highly elongated resonant structures or sources with a very long time duration (such as a unit step). For the latter case, large reflection errors can occur in the late time. This can be seen in the following discussion.

Consider the reflection at the air / PML interface of a normally incident TE_z-polarized wave. From (5.37), this is

$$\Gamma_0 = \left(1 - \sqrt{\frac{s_x^*(0)}{s_x(0)}}\right) \Bigg/ \left(1 + \sqrt{\frac{s_x^*(0)}{s_x(0)}}\right) \tag{5.68}$$

where s_x^* is from the magnetic tensor. In the discrete space, the magnetic tensor and the electric tensor parameters are staggered by one half of a cell. Therefore, at the air / PML interface, $s_x^* = 1$. Thus, we have

$$\Gamma = \frac{\sqrt{s_x(0)} - 1}{\sqrt{s_x(0)} + 1} \tag{5.69}$$

Given s_x from (5.57a), this further reduces to

$$\Gamma = \frac{\sqrt{1 + \left[\sigma_x(0)/j\omega\varepsilon_0\right]} - 1}{\sqrt{1 + \left[\sigma_x(0)/j\omega\varepsilon_0\right]} + 1} = \frac{\sqrt{\omega - \left[j\sigma_x(0)/\varepsilon_0\right]} - \sqrt{\omega}}{\sqrt{\omega - \left[j\sigma_x(0)/\varepsilon_0\right]} + \sqrt{\omega}} \tag{5.70}$$

where $\kappa_x(0) = 1$. In the region that $\omega >> \sigma_x(0)/\varepsilon_0$, then $\Gamma \approx 0$. However, at low frequencies where $\omega << \sigma_x(0)/\varepsilon_0$, then $\Gamma \approx 1$. Therefore, there is a low-pass filtering of the reflected wave. In the time domain, this is convolutional so that narrow pulses impinging on the interface have small reflections while broad pulses encounter relatively large reflections [4, 7]. A cutoff frequency is defined for this low-pass filter as [4]

$$f_c = \frac{\sigma_x(0)}{2\pi\varepsilon_0} = 1.8 \times 10^{10}\,\sigma_x(0) \tag{5.71}$$

Thus, when $f > f_c$ most of the power is transmitted into the PML, and when $f < f_c$ most of the power is reflected. The time-dependent expression for the reflected field is the convolution of Γ with the transverse electric field. In the time domain, this allows for small reflections from the interface up to a time of

$$t_c = \frac{1}{f_c} = \frac{1}{1.8 \times 10^{10}\,\sigma_x(0)} \tag{5.72}$$

Beyond this time, larger reflections accumulate. This was verified experimentally by Berenger in [4].

Observing (5.72), we see that the "time exposure" of the PML interface is increased by decreasing $\sigma_x(0)$. However, this is at the expense of reducing $R(\theta)$. Thus, again balancing the discretization error and the theoretical reflection error, we expect that in such a case, the optimal value for $\sigma_{\max}$ is smaller than that predicted by (5.67).

This filtering effect of the discretization error is due to the choice of s_x in (5.57a). It is understood that the PML is matched for any finite value of s_x. Of course, stability and causality limit the location of s_x in the complex plane to the lower right quadrant. We note that the choice $s_x = \kappa_x + \sigma_x / j\omega\varepsilon_0$ leads to a causal and weakly stable formulation. It was originally claimed that this choice is not causal because of the pole at the origin of the complex-frequency plane. However, causality can be shown with the Kramers-Kronig relations by employing a Cauchy principal-value integral around a small circle about the origin. In fact, causality is expected since a conducting medium is a causal medium. The reader is referred to [44] for more details. An alternate choice for s_x that may have some benefit is

$$s_x = \kappa_x + \frac{\sigma_x}{\alpha_x + j\omega\varepsilon_0} \tag{5.73}$$

which was originally proposed in [45]. This form of s_x is referred to here as the complex-frequency-shifted (CFS) s_x. Assuming $\kappa_x > 1$, $\sigma_x \geq 0$, and $\alpha_x > 0$, this choice for s_x is causal and stable [44,45]. It also offers some unique advantages over the previous choice for s_x in that the discrete boundary does not have a low-pass filtering effect.

Returning to (5.69), using s_x defined in (5.73) we have

$$\Gamma = \frac{\sqrt{1 + \left[\sigma_x(0)/\left(a_x + j\omega\varepsilon_0\right)\right]} - 1}{\sqrt{1 + \left[\sigma_x(0)/\left(a_x + j\omega\varepsilon_0\right)\right]} + 1} \tag{5.74a}$$

In the limit that $\omega >> \alpha_x$ and $\omega >> \sigma_x(0)$, $\Gamma \to 0$. However, for $\omega << \alpha_x$, we have

$$\Gamma \approx \frac{\sqrt{1 + \left[\sigma_x(0)/a_x\right]} - 1}{\sqrt{1 + \left[\sigma_x(0)/a_x\right]} + 1} \tag{5.74b}$$

Assuming that $\sigma_x(0)/\alpha_x << 1$, then Γ is very small. Thus, we see that by pushing the pole off the real axis of the complex-frequency plane, the low-pass filtering effect of the discretization error is alleviated. In fact, it was shown in [46] that this formulation can be extended to the static analysis ($\omega = 0$) of electric or magnetic fields without special treatment of the pole at the origin. Specifically, in the static case, s_x is purely real. Following the analysis of Section 5.3, this is analogous to stretching the mesh along the normal coordinate axis.

Employing (5.73) has an advantage for electromagnetic wave interaction problems having long simulation times and dominant low-frequency excitations. However, it is shown in the next section that implementing (5.73) in an FDTD code can result in a large increase in memory requirements, and should only be used as needed.

5.6 EFFICIENT IMPLEMENTATION OF PML IN FDTD

This section discusses mapping the UPML presented in Section 5.4 into the discrete FDTD space. The FDTD approximation is derived from Maxwell's curl equations within the uniaxial medium as defined in (5.48) and (5.49).

5.6.1 Derivation of the Finite-Difference Expressions

Starting with the UPML equations (5.48) and (5.49), Ampere's law is expressed as

$$\begin{bmatrix} \frac{\partial}{\partial y}H_z - \frac{\partial}{\partial z}H_y \\ \frac{\partial}{\partial z}H_x - \frac{\partial}{\partial x}H_z \\ \frac{\partial}{\partial x}H_y - \frac{\partial}{\partial y}H_x \end{bmatrix} = j\omega\varepsilon_0\varepsilon_r \begin{bmatrix} \frac{s_y s_z}{s_x} & 0 & 0 \\ 0 & \frac{s_x s_z}{s_y} & 0 \\ 0 & 0 & \frac{s_x s_y}{s_z} \end{bmatrix} \begin{bmatrix} E_x \\ E_y \\ E_z \end{bmatrix} \tag{5.75a}$$

where ε_r is the relative permittivity of the medium, and it is assumed that

$$s_x = \kappa_x + \frac{\sigma_x}{j\omega\varepsilon_0}, \qquad s_y = \kappa_y + \frac{\sigma_y}{j\omega\varepsilon_0}, \qquad s_z = \kappa_z + \frac{\sigma_z}{j\omega\varepsilon_0} \tag{5.75b}$$

Inserting (5.75b) into (5.75a) and then transforming into the time domain leads to a convolution between the tensor coefficients and the electric field intensity. We note that implementing this convolution directly within FDTD would be computationally intensive. A much more efficient approach is to define the proper constitutive relationship to decouple the frequency-dependent terms [36, 37]. Specifically, let

$$D_x = \varepsilon_0\varepsilon_r\frac{s_z}{s_x}E_x \tag{5.76a}$$

$$D_y = \varepsilon_0\varepsilon_r\frac{s_x}{s_y}E_y \tag{5.76b}$$

$$D_z = \varepsilon_0\varepsilon_r\frac{s_y}{s_z}E_z \tag{5.76c}$$

Then, (5.75a) is rewritten as

$$\begin{bmatrix} \frac{\partial}{\partial y}H_z - \frac{\partial}{\partial z}H_y \\ \frac{\partial}{\partial z}H_x - \frac{\partial}{\partial x}H_z \\ \frac{\partial}{\partial x}H_y - \frac{\partial}{\partial y}H_x \end{bmatrix} = j\omega \begin{bmatrix} s_y & 0 & 0 \\ 0 & s_z & 0 \\ 0 & 0 & s_x \end{bmatrix} \begin{bmatrix} D_x \\ D_y \\ D_z \end{bmatrix} \tag{5.77}$$

From (5.75a), this is transformed into the time domain using the identity $j\omega f(\bar{r}, \omega) \rightarrow (\partial/\partial t) f(\bar{r}, t)$

$$\begin{bmatrix} \frac{\partial}{\partial y} H_z - \frac{\partial}{\partial z} H_y \\ \frac{\partial}{\partial z} H_x - \frac{\partial}{\partial x} H_z \\ \frac{\partial}{\partial x} H_y - \frac{\partial}{\partial y} H_x \end{bmatrix} = \frac{\partial}{\partial t} \begin{bmatrix} \kappa_y & 0 & 0 \\ 0 & \kappa_z & 0 \\ 0 & 0 & \kappa_x \end{bmatrix} \begin{bmatrix} D_x \\ D_y \\ D_z \end{bmatrix} + \frac{1}{\varepsilon_0} \begin{bmatrix} \sigma_y & 0 & 0 \\ 0 & \sigma_z & 0 \\ 0 & 0 & \sigma_x \end{bmatrix} \begin{bmatrix} D_x \\ D_y \\ D_z \end{bmatrix} \tag{5.78}$$

Using central-difference approximations for the space and time derivatives and time-averaging the lossy terms, standard FDTD expressions are derived for (5.78) [1].

Next, we focus on (5.76a) to (5.76c). Starting with (5.76a), multiply both sides by s_x. Then, from (5.75b):

$$\left(\kappa_x + \frac{\sigma_x}{j\omega\varepsilon_0} \right) D_x = \varepsilon_0 \varepsilon_r \left(\kappa_z + \frac{\sigma_z}{j\omega\varepsilon_0} \right) E_x \tag{5.79}$$

Multiplying both sides by $j\omega$ and transforming into the time domain leads to

$$\frac{\partial}{\partial t}\left(\kappa_x D_x\right) + \frac{\sigma_x}{\varepsilon_0} D_x = \varepsilon_0 \varepsilon_r \left[\frac{\partial}{\partial t}\left(\kappa_z E_x\right) + \frac{\sigma_z}{\varepsilon_0} E_x \right] \tag{5.80a}$$

Similarly, from (5.76b) and (5.76c), we obtain

$$\frac{\partial}{\partial t}\left(\kappa_y D_y\right) + \frac{\sigma_y}{\varepsilon_0} D_y = \varepsilon_0 \varepsilon_r \left[\frac{\partial}{\partial t}\left(\kappa_x E_y\right) + \frac{\sigma_x}{\varepsilon_0} E_y \right] \tag{5.80b}$$

$$\frac{\partial}{\partial t}\left(\kappa_z D_z\right) + \frac{\sigma_z}{\varepsilon_0} D_z = \varepsilon_0 \varepsilon_r \left[\frac{\partial}{\partial t}\left(\kappa_y E_z\right) + \frac{\sigma_y}{\varepsilon_0} E_z \right] \tag{5.80c}$$

The FDTD solution for the fields in the PML is a two-step update based on (5.78) and (5.80a) to (5.80c). Consistent with the usual FDTD format, the fields are discretized on a dual-staggered grid where the spatial derivatives in (5.78) are realized using second-order-accurate central differences. Similarly, the time derivatives are implemented using second-order-accurate central-difference (leapfrog) staggering of the E and H fields in time. Here, the loss terms are averaged in time. This leads to explicit field-update expressions, illustrated here for the first row of (5.78):

$$D_x\Big|_{i+\frac{1}{2},j,k}^{n+\frac{1}{2}} = \left(\frac{\dfrac{\kappa_y}{\Delta t} - \dfrac{\sigma_y}{2\varepsilon_0}}{\dfrac{\kappa_y}{\Delta t} + \dfrac{\sigma_y}{2\varepsilon_0}}\right) D_x\Big|_{i+\frac{1}{2},j,k}^{n-\frac{1}{2}} + \left(\frac{1}{\dfrac{\kappa_y}{\Delta t} + \dfrac{\sigma_y}{2\varepsilon_0}}\right) \times$$

$$\left(\frac{H_z\Big|_{i+\frac{1}{2},j+\frac{1}{2},k}^{n} - H_z\Big|_{i+\frac{1}{2},j-\frac{1}{2},k}^{n}}{\Delta y} - \frac{H_y\Big|_{i+\frac{1}{2},j,k+\frac{1}{2}}^{n} - H_y\Big|_{i+\frac{1}{2},j,k-\frac{1}{2}}^{n}}{\Delta z}\right) \tag{5.81}$$

Similar expressions can be derived for D_y and D_z.

Given knowledge of $D_x^{n+1/2}$, then $E_x^{n+1/2}$ can be derived from (5.80a). Again using leapfrog central differences for the time derivatives and time-averaging the loss terms, we obtain the second-order explicit update expression:

$$E_x\Big|_{i+\frac{1}{2},j,k}^{n+\frac{1}{2}} = \left(\frac{\kappa_z - \dfrac{\sigma_z \Delta t}{2\varepsilon_0}}{\kappa_z + \dfrac{\sigma_z \Delta t}{2\varepsilon_0}}\right) E_x\Big|_{i+\frac{1}{2},j,k}^{n-\frac{1}{2}} + \left[\frac{1}{\left(\kappa_z + \dfrac{\sigma_z \Delta t}{2\varepsilon_0}\right)\varepsilon_0\,\varepsilon_r}\right] \times$$

$$\left[\left(\kappa_x + \frac{\sigma_x \Delta t}{2\varepsilon_0}\right) D_x\Big|_{i+\frac{1}{2},j,k}^{n+\frac{1}{2}} - \left(\kappa_x - \frac{\sigma_x \Delta t}{2\varepsilon_0}\right) D_x\Big|_{i+\frac{1}{2},j,k}^{n-\frac{1}{2}}\right] \tag{5.82}$$

Similar update equations are derived for the remaining electric field components.

Starting with Faraday's law in (5.49) and (5.48), similar expressions are derived for the magnetic field updates. For example, the update for H_x is given by the two-step procedure:

$$B_x\Big|_{i,j+\frac{1}{2},k+\frac{1}{2}}^{n+1} = \left(\frac{\dfrac{\kappa_y}{\Delta t} - \dfrac{\sigma_y}{2\varepsilon_0}}{\dfrac{\kappa_y}{\Delta t} + \dfrac{\sigma_y}{2\varepsilon_0}}\right) B_x\Big|_{i,j+\frac{1}{2},k+\frac{1}{2}}^{n} - \left(\frac{1}{\dfrac{\kappa_y}{\Delta t} + \dfrac{\sigma_y}{2\varepsilon_0}}\right) \times$$

$$\left(\frac{E_z\Big|_{i,j+1,k+\frac{1}{2}}^{n+\frac{1}{2}} - E_z\Big|_{i,j,k+\frac{1}{2}}^{n+\frac{1}{2}}}{\Delta y} - \frac{E_y\Big|_{i,j+\frac{1}{2},k+1}^{n+\frac{1}{2}} - E_y\Big|_{i,j+\frac{1}{2},k}^{n+\frac{1}{2}}}{\Delta z}\right) \tag{5.83a}$$

$$H_x\Big|_{i,j+\frac{1}{2},k+\frac{1}{2}}^{n+1} = \left(\frac{\kappa_z - \frac{\sigma_z \Delta t}{2\varepsilon_0}}{\kappa_z + \frac{\sigma_z \Delta t}{2\varepsilon_0}}\right) H_x\Big|_{i,j+\frac{1}{2},k+\frac{1}{2}}^{n} + \left[\frac{1}{\left(\kappa_z + \frac{\sigma_z \Delta t}{2\varepsilon_0}\right)\mu_0 \mu_r}\right] \times$$

$$\left[\left(\kappa_x + \frac{\sigma_x \Delta t}{2\varepsilon_0}\right) B_x\Big|_{i,j+\frac{1}{2},k+\frac{1}{2}}^{n+1} - \left(\kappa_x - \frac{\sigma_x \Delta t}{2\varepsilon_o}\right) B_x\Big|_{i,j+\frac{1}{2},k+\frac{1}{2}}^{n}\right] \tag{5.83b}$$

where the dual constitutive relation $B_x = \mu_0 \mu_r (s_z/s_x) H_x$ is chosen. Similar expressions can be derived for the remaining two magnetic field components.

It can be shown that the explicit updates defined for all six field components based on (5.81) to (5.83) are stable within the Courant limit [14]. Further, it can be shown that the discrete fields satisfy Gauss' laws and that the UPML is well posed [47].

5.6.2 Computer Implementation of the UPML

The time-dependent electric and magnetic fields within the UPML are computed using an explicit time-marching scheme as derived in (5.81) and (5.82) for E_x and in (5.83a) and (5.83b) for H_x. Based on these updates, we see that the UPML is easily and efficiently implemented within the framework of an existing FDTD code. For example, from (5.81) and (5.82) , the discrete field updates are expressed as a triply nested loop (illustrated here in FORTRAN):

```
do 10 k=2,nz-1
   do 10 j=2,ny-1
      do 10 i=1,nx-1
         ds = dx(i,j,k)
         dx(i,j,k) = ay(j)*dx(i,j,k)
                   + by(j)*(hz(i,j,k) - hz(i,j-1,k) -
                            hy(i,j,k) + hy(i,j,k-1))
         ex(i,j,k) = az(k)*ex(i,j,k)
                   + bz(k)*(ax(i)*dx(i,j,k) - bx(i)*ds)*erx(i,j,k)
10 continue
```

(5.84)

where

$$\text{ax(i)} = \kappa_{x_i} + \frac{\sigma_{x_i} \Delta t}{2\varepsilon_0} \tag{5.85a}$$

$$\texttt{ay(j)} = \frac{2\varepsilon_0 \kappa_{y_j} - \sigma_{y_j} \Delta t}{2\varepsilon_0 \kappa_{y_j} + \sigma_{y_j} \Delta t} \tag{5.85b}$$

$$\texttt{az(k)} = \frac{2\varepsilon_0 \kappa_{z_i} - \sigma_{z_i} \Delta t}{2\varepsilon_0 \kappa_{z_i} + \sigma_{z_i} \Delta t} \tag{5.85c}$$

$$\texttt{bx(i)} = \kappa_{x_i} - \frac{\sigma_{x_i} \Delta t}{2\varepsilon_0} \tag{5.85d}$$

$$\texttt{by(j)} = \frac{2\varepsilon_0 \Delta t}{2\varepsilon_0 \kappa_{y_j} + \sigma_{y_j} \Delta t} \cdot \frac{\Delta x}{\Delta y \Delta z} \tag{5.85e}$$

$$\texttt{bz(k)} = \frac{2\varepsilon_0}{2\varepsilon_0 \kappa_{z_i} + \sigma_{z_i} \Delta t} \tag{5.85f}$$

$$\texttt{erx(i,j,k)} = \frac{1}{\varepsilon_0 \varepsilon_r(i, j, k)} \tag{5.85g}$$

and the fields are scaled by their edge length (e.g., $E_x = E_x \Delta x$). The array `erx(i,j,k)` permits modeling of three-dimensional inhomogeneities in the permittivity.

In the interior working volume, we assume that $\sigma_i = 0$ and $\kappa_i = 1$. In the PML regions, these parameters are assumed to have a polynomial- or geometric-scaled spatial variation along the respective axes of the PML slabs such as that given in (5.59) and (5.61) or in (5.63) and (5.65). As a result, the update coefficients in (5.85) vary simply in one dimension.

Assuming PML throughout the entire FDTD lattice in this manner has the limitation that the flux densities, D_x and B_x, must be stored everywhere in the grid. However, this approach offers the significant advantage of simplifying the modification of existing FDTD codes. An alternative is to write a triply nested loop for the interior fields and separate loops for the various PML slabs (segregating the corner regions). In this case, the auxiliary variables need to be stored only in the PML region, leading to memory savings. Further, in this circumstance, the UPML requires considerably less storage than Berenger's PML since only the normal fields require dual storage as opposed to the two tangential fields. The memory requirement for the UPML truncation on all six outer lattice boundaries is

$$\begin{gathered} 6N_xN_yN_z + 8N_{\text{pml}}(N_xN_y + N_yN_z + N_zN_x) - \\ 16N_{\text{pml}}(N_x + N_y + N_z) + \left(24N_{\text{pml}}\right)^2 \end{gathered} \tag{5.86}$$

real numbers as compared to $6N_xN_yN_z$ real numbers required by a FDTD method with a local ABC. A comparison of the storage requirements is presented in Fig. 5.4. This illustrates the percentage of additional memory required to implement a 4-cell-thick

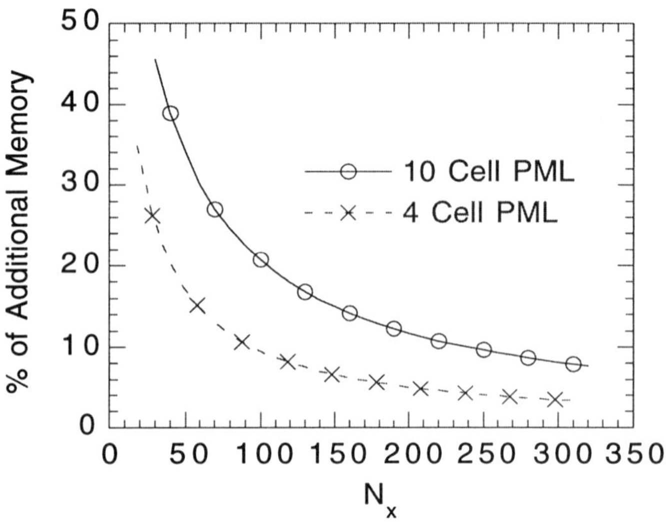

Fig. 5.4 Percentage of additional memory required by the UPML algorithm for a 4-cell-thick and a 10-cell-thick ABC given $N_x = N_y = N_z$.

and a 10-cell-thick UPML ABC in a cubic lattice ($N_x = N_y = N_z$) compared to that required by an FDTD lattice of the same dimensions using a local ABC . We see that the storage penalty for the 10-cell ABC drops below 10% for $N_x > 240$. Note also that local ABCs must be placed *much* further out than the PML ABC, requiring even larger lattices. Consequently, the UPML ABC can lead to an overall decrease in the required memory to achieve a given desired outer-boundary reflectivity.

If we assume PML throughout the entire FDTD lattice as in (5.84), the memory requirement is $12N_xN_yN_z$ real numbers. This option is suitable when coding simplicity is desired and memory is not a constraint.

5.6.3 The Complex Frequency-Shifted Tensor

In Section 5.5.2, we demonstrated that one limitation of using (5.75b) is that signals of long duration can suffer from large reflections off the PML interface. This effect can be circumvented by using the CFS tensor. We let

$$s_x = \kappa_x + \frac{\sigma_x}{\alpha_x + j\omega\varepsilon_0}, \quad s_y = \kappa_y + \frac{\sigma_y}{\alpha_y + j\omega\varepsilon_0}, \quad s_z = \kappa_z + \frac{\sigma_z}{\alpha_z + j\omega\varepsilon_0} \tag{5.87}$$

To derive the time-dependent formulation for Ampere's law from (5.75a) and (5.87), consider for example an x-normal boundary ($s_y = s_z = 1$). In the previous case where $\alpha_x = 0$, we introduced a constitutive relation that results in an auxiliary variable for D_x, as in (5.76a). The transverse fields did not need auxiliary variables when $s_y = s_x = 1$.

This is not the case for a CFS tensor due to α_x, which requires the following auxiliary variables:

$$D_x = \varepsilon_0 \varepsilon_r \frac{1}{s_x} E_x, \qquad D_y = \varepsilon_0 \varepsilon_r s_x E_y, \qquad D_z = \varepsilon_0 \varepsilon_r s_x E_z \tag{5.88}$$

Then, (5.75a) is rewritten in its time-dependent form as

$$\frac{\partial}{\partial t} \vec{D} = \nabla \times \vec{H} \tag{5.89}$$

The time-dependent electric field intensities are then derived from the constitutive relationships in (5.87) and (5.88) as

$$\varepsilon_r \varepsilon_0 \left(\alpha_x E_x + \varepsilon_0 \frac{\partial}{\partial t} E_x \right) = (\kappa_x \alpha_x + \sigma_x) D_x + \kappa_x \varepsilon_0 \frac{\partial}{\partial t} D_x \tag{5.90}$$

for E_x and

$$\varepsilon_r \varepsilon_0 \left[(\kappa_x \alpha_x + \sigma_x) E_y + \kappa_x \varepsilon_0 \frac{\partial}{\partial t} E_y \right] = \alpha_x D_y + \varepsilon_0 \frac{\partial}{\partial t} D_y \tag{5.91}$$

for E_y. A similar expression is derived for E_z.

Thus, within the PML, field updates for the flux density $\vec{D}$ are derived from (5.89). Then, the field updates for the electric field vector are derived from (5.90) for the normal field component and from (5.91) for the transverse fields. This leads to a second-order-accurate field updating. The additional cost over the previous method is having to store auxiliary variables for the normal and the transverse fields.

In the corner regions, the PML update expressions are a bit more complex. However, a stable formulation can nevertheless be derived. To illustrate this process, only the field update for E_x will be derived. (The remaining corner fields can be updated in an analogous manner.) In the corner region, the constitutive relations are

$$D_x = \varepsilon_0 \varepsilon_r \frac{1}{s_x} E_x \tag{5.92a}$$

$$P_{x_1} = s_y D_x \tag{5.92b}$$

$$P_{x_2} = s_z P_{x_1} \tag{5.92c}$$

and similarly for the remaining field components. Thus, from Ampere's law,

$$\frac{\partial}{\partial t} P_{x_2} = \frac{\partial}{\partial y} H_z - \frac{\partial}{\partial z} H_y \tag{5.93}$$

Then, from (5.92c), (5.92b), and (5.92a), respectively, we derive the equations:

$$(\kappa_z \alpha_z + \sigma_z) P_{x_1} + \kappa_z \varepsilon_0 \frac{\partial}{\partial t} P_{x_1} = \alpha_z P_{x_2} + \varepsilon_0 \frac{\partial}{\partial t} P_{x_2} \tag{5.94a}$$

$$(\kappa_y \alpha_y + \sigma_y) D_x + \kappa_y \varepsilon_0 \frac{\partial}{\partial t} D_x = \alpha_y P_{x_1} + \varepsilon_0 \frac{\partial}{\partial t} P_{x_1} \tag{5.94b}$$

$$\varepsilon_r \varepsilon_0 \left(\alpha_x E_x + \varepsilon_0 \frac{\partial}{\partial t} E_x \right) = (\kappa_x \alpha_x + \sigma_x) D_x + \kappa_x \varepsilon_0 \frac{\partial}{\partial t} D_x \tag{5.94c}$$

Subsequently, (5.93) is used to predict P_{x_2} given H and P_{x_2} from the previous time step. Then, (5.94a) is used to predict P_{x_1} given the current and past value of P_{x_2}. Equation (5.94b) is used to predict D_x given the current and past values of P_{x_1}. Finally, (5.94c) is used to predict E_x given the current and past values of D_x.

Comparing this approach with (5.75) to (5.83), we see that this method requires the introduction of three auxiliary expressions and variables compared to one required by the previous method. This results in an increase in computer time and memory. Thus, while this formulation offers some merit, its use is reserved for special cases.

5.7 NUMERICAL EXPERIMENTS WITH PML

In this section, the PML termination of FDTD lattices is presented for a number of sample applications. The goal is to provide an understanding of the effect of the PML parameters and scaling functions on PML effectiveness. In this manner, the reader can better understand how to properly choose these parameters.

5.7.1 Parallel-Plate Waveguide, Pulsed TEM Mode

Consider a two-dimensional parallel-plate waveguide propagating a pulsed TEM mode in the z direction. The top and bottom plates of the waveguide are assumed to be PEC boundaries separated by a y distance of 15 mm. A Yee FDTD grid with 1-mm square cells and a time step of 0.98 of the Courant limit is used. The pulsed source has a Gaussian time waveform with a 12-GHz bandwidth, and is located only 10 cells from the grid truncation formed by a UPML absorber backed by a PEC wall. Both 5-cell and 10-cell PMLs are tested in comparison with a reference computation where the terminating grid boundary is 1000 cells from the source. The maximum relative error in the electric field computed throughout the mesh is given by

$$\text{Maximum Relative Error} = \sup\left(\left| E_y\big|_{i,j}^{n} - E_{y,\text{ref}}\big|_{i,j}^{n} \right| \Big/ \left| E_{y,\max}\big|_{i,j} \right| \right) \tag{5.95}$$

where $|E_{y,\max}|_{i,j}|$ is the maximum amplitude of the reference electric field at grid coordinate location (i, j) over the 1000 time steps of the simulation.

Fig. 5.5 illustrates the maximum relative error for this case. The results are also compared with the theoretical reflection error for a normally incident TEM wave, or $R(0)$ from (5.60). For small values of $\sigma_{\max}$, the actual error and the predicted error follow quite closely. However, as $\sigma_{\max}$ increases, the reflection error flattens out and then ultimately increases. As discussed earlier, this is due to the discretization error of the grid. On the average, when $\sigma_{\max} \approx \sigma_{opt}$ (as defined by (5.67)), the 10-cell-thick PML results in approximately 1/100 of the reflection error of the 5-cell-thick PML.

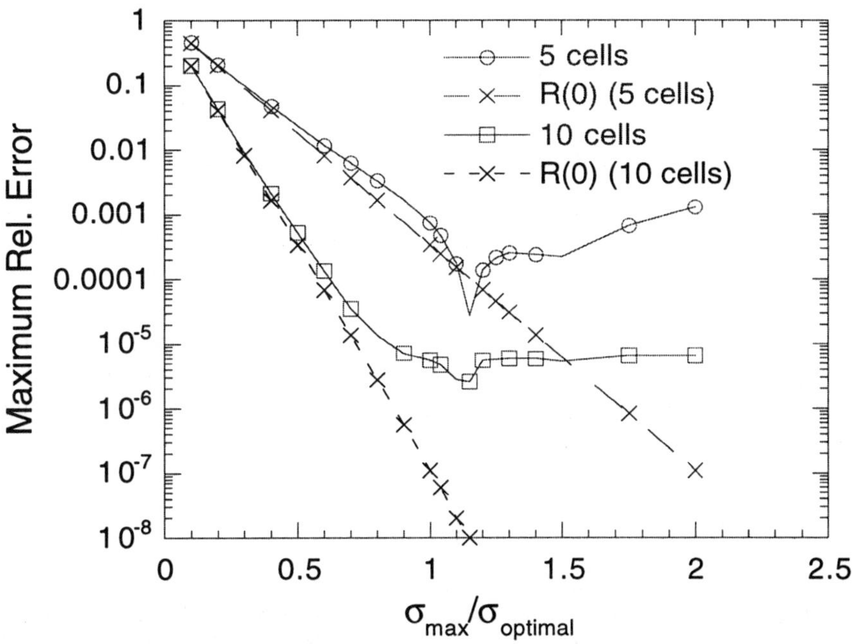

Fig. 5.5 Maximum relative error due to a 5-cell and a 10-cell UPML termination of a parallel-plate waveguide propagating a pulsed TEM mode over 1000 time steps. The results are a function of $\sigma_{\max}/\sigma_{opt}$ and are also compared with the theoretical reflection, $R(0)$. Polynomial spatial scaling is used with $m = 4$.

5.7.2 Current Source Radiating in an Unbounded Region

The TEM excitation of the parallel-plate waveguide is a fairly simple problem since the wave is purely propagating and is normally incident on the PML. Hence, optimal performance is expected. The next example is that of a current source, $\vec{J}$, excited in a square FDTD grid, as illustrated in Fig. 5.6. The source is vertically directed and invariant along the axial direction. Hence, it radiates two-dimensional TE waves.

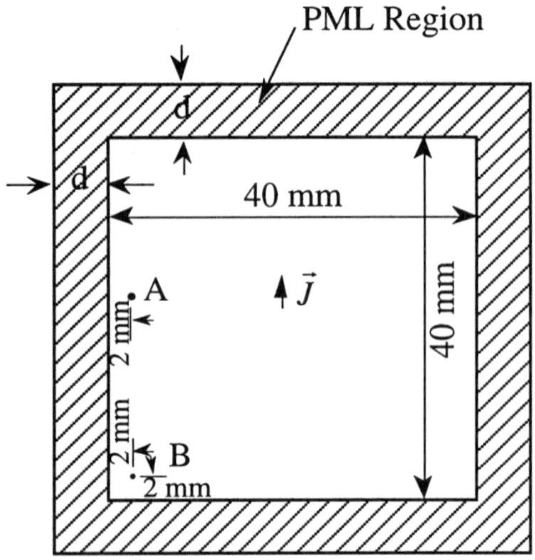

Fig. 5.6 TE-polarized wave excited by a vertically directed electric current source in a 2-D region. The working volume is 40 x 40 mm and is surrounded by UPML of thickness d. The source is located at the center of the region. The fields are probed at two observation points A and B.

A two-dimensional adaptation of the algorithm of (5.81) to (5.83) is implemented in a 40 x 40-cell Yee grid with 1-mm square cells and a time step of 0.98 of the Courant limit. As shown in Fig. 5.6, the current source is centered in the computational domain. It has the time signature of a differentiated Gaussian pulse:

$$J_y(x_0, y_0, t) = -2\left[(t-t_0)/t_w\right]\exp\left\{-\left[(t-t_0)/t_w\right]^2\right\} \tag{5.96}$$

where $t_w = 26.53$ ps and $t_0 = 4t_w$. (Note that the amplitude is scaled by t_w.) The electric field is probed at two points, A and B, as shown in the figure. Point A is in the same plane as the source and two cells from the PML, and point B is two cells from the bottom and side PMLs. Time-stepping runs over 1000 iterations, well past the steady-state response. Both 5-cell and 10-cell UPML ABCs are used with polynomial scaling ($m = 4$, $\sigma_{max} = \sigma_{opt}$, and $\kappa_{max} = 1$).

For this case, the reference solution is obtained using a very large 1240 x 1240-cell grid. The current source is centered within this grid, and the field-observation points, A and B, are at the same positions relative to the source as in the test grid. The reference grid is sufficiently large such that there are no reflections from its outer boundaries during the 1000 time steps.

Fig. 5.7 illustrates the relative error computed at points A and B as a function of time. This error is computed in a manner similar to (5.95). We see that the maximum error occurs in the late time, and is predominately a low-frequency error as anticipated by (5.72). As time advances further, this error gradually decreases. We also note that

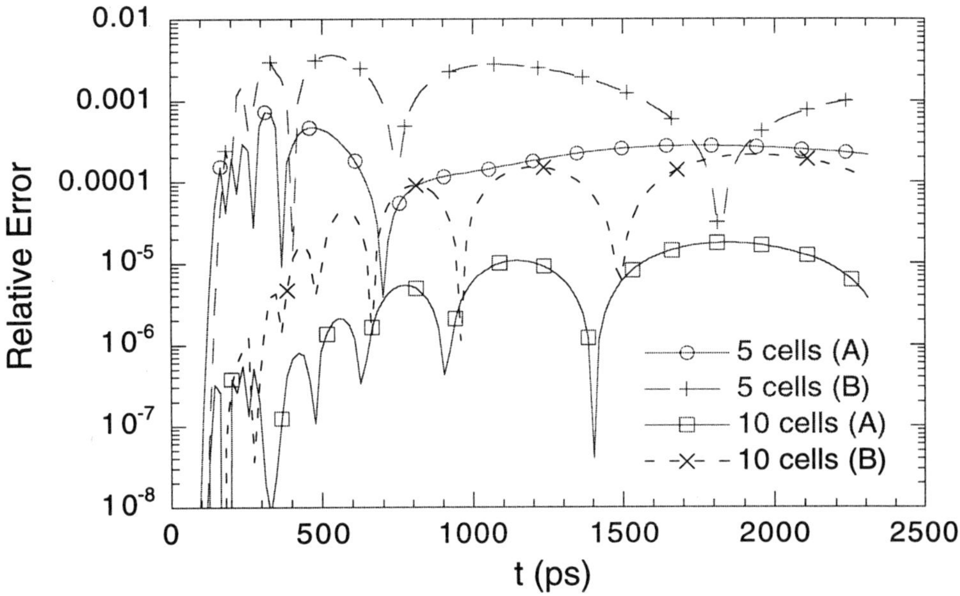

Fig. 5.7 Relative error at points *A* and *B* of Fig. 5.6 over 1000 time steps for 5-cell and 10-cell UPMLs with $\sigma_{max} = \sigma_{opt}$ and $\kappa = 1$.

the error at point *A* is always less than that at point *B*. This is because the wave impinging on the PML near *A* is nearly normally incident and undergoes maximum absorption. At *B*, while the amplitude of the outgoing wave is smaller due to the radiation pattern of the source, the wave impinges on the PML more obliquely at 45°.

The relative error in Fig. 5.7 is calculated assuming a fixed σ_{max} equal to σ_{opt}. Additional information is gained by observing the maximum error over 1000 time steps recorded at points *A* and *B* as a function of σ_{max}. This is illustrated in Fig. 5.8 where σ_{max} is varied from zero to $3\sigma_{opt}$ for the 5-cell and 10-cell PMLs. As before, the same σ_{max} is used for each of the four PML slabs and $m = 4$. We see that the optimal choice for σ_{max} is quite close to σ_{opt}. Again, the maximum error at point *B* is about an order of magnitude larger than that at point *A*.

Fig. 5.9 illustrates the corresponding error for a 15-cell PML. Here, we see that for $\sigma_{max} = \sigma_{opt}$, the errors at both points *A* and *B* are the same and in the order of 10^{-5}.

Figs. 5.10(a) and 5.10(b) are contour plots resulting from a comprehensive parametric study of this case. The maximum relative error at points *A* and *B*, respectively, recorded over 1000 time steps is plotted versus both $\sigma_{max}/\sigma_{opt}$ and polynomial order, *m* (not necessarily an integer), in the range from 1 to 5 for a 10-cell UPML. The minimum error is found for $3 < m < 4$ and $\sigma_{max} \cong 0.75\sigma_{opt}$. For $\sigma_{max} = \sigma_{opt}$, the error is on the order of -90 dB at point *A* and -75 dB at point *B*.

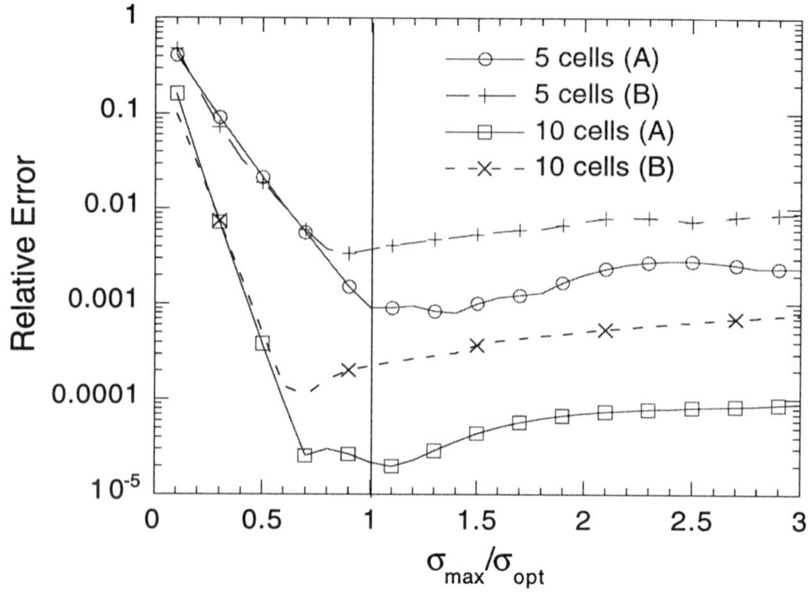

Fig. 5.8 Maximum relative error due to 5-cell and 10-cell UPML ABCs in the grid of Fig. 5.6 versus $\sigma_{max}/\sigma_{opt}$ over a 1000 time-step observation period.

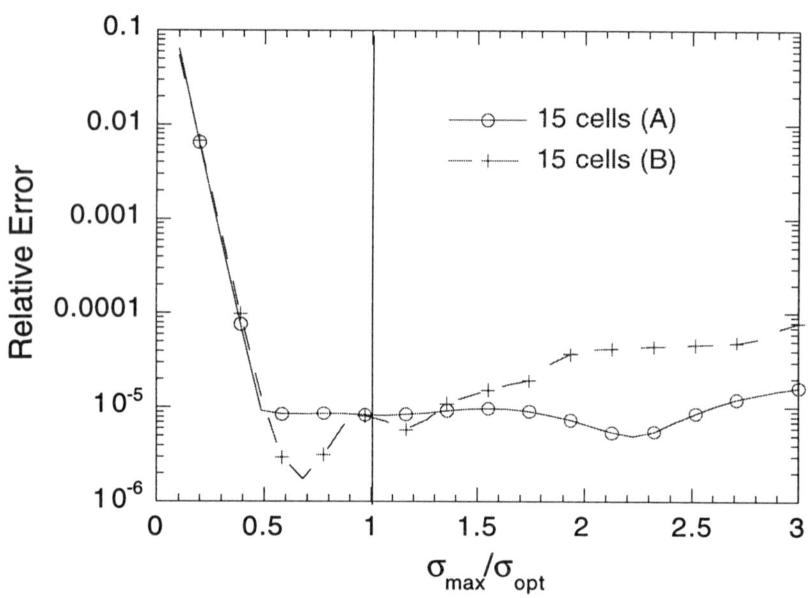

Fig. 5.9 Maximum relative error due to a 15-cell UPML ABC in the grid of Fig. 5.6 versus $\sigma_{max}/\sigma_{opt}$ over a 1000 time-step observation period.

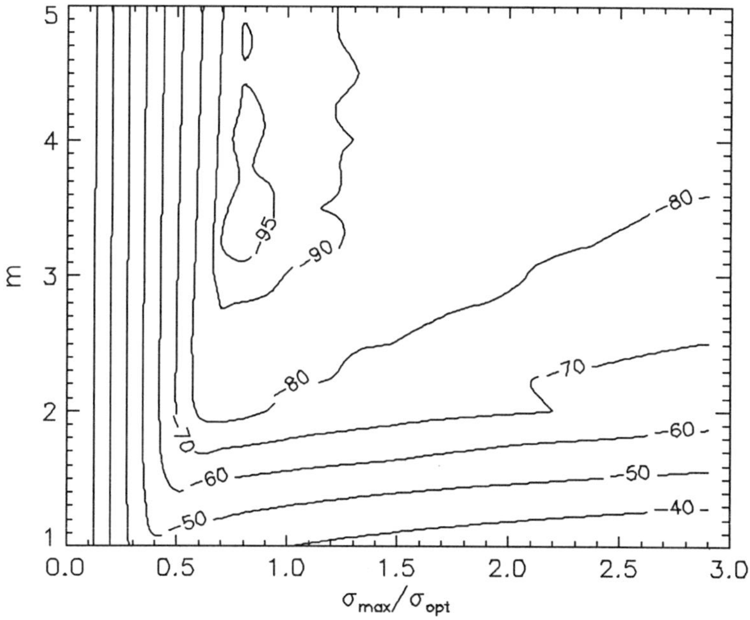

(a)

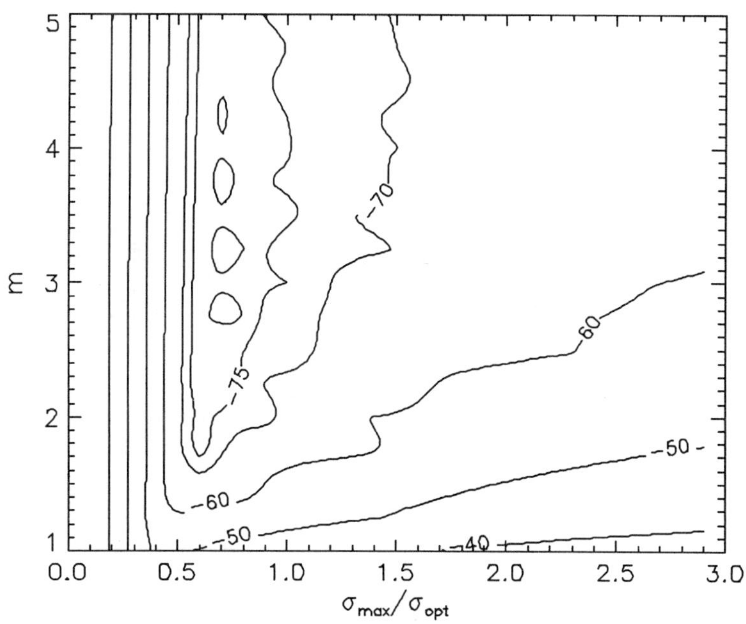

(b)

Fig. 5.10 Contour plots of the maximum relative error in dB in the grid of Fig. 5.6 versus $\sigma_{max}/\sigma_{opt}$ and polynomial order, m, for a 10-cell UPML: (a) at point A; (b) at point B.

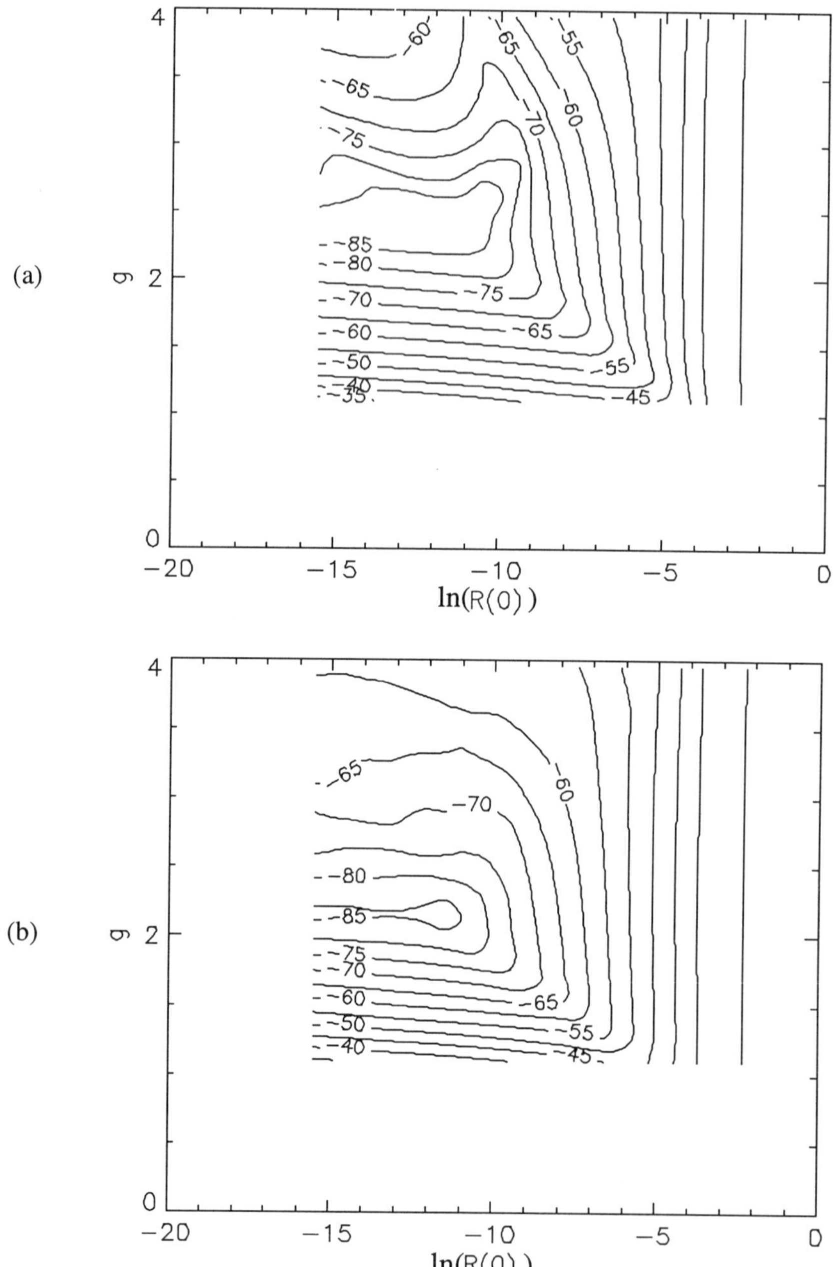

Fig. 5.11 Contour plots of the maximum relative error in dB in the grid of Fig. 5.6 versus g and $\ln[R(0)]$ for a geometrically scaled 10-cell UPML: (a) at point A; (b) at point B.

A similar study is performed with geometric scaling applied to the UPML parameters. Fig. 5.11 is a contour plot of the maximum relative error over 1000 time steps at points A and B versus $R(0)$ and g (see (5.63) and (5.64)) for a 10-cell-thick UPML. Interestingly, for $g = 2.2$, about -85 dB of reflection error is realized at both points A and B when $\ln[R(0)]$ is between -12 and -16. We also observe that the effectiveness of the UPML is quite sensitive to the choice of g.

5.7.3 Highly Elongated Domains

The previous example involves a square grid with a source placed at its center. For such a case, the incident angle of the wave impinging on the PML boundaries never exceeds 45° and the PML performs extremely well. However, if electromagnetic waves impinge on the PML at highly oblique angles, the performance of the PML is expected to degrade since $\ln[R(0)]$ is proportional to the cosine of the incident angle.

Fig. 5.12 illustrates a sample problem involving a current source, $\vec{J}$, radiating TE-polarized waves in a highly elongated FDTD grid. The relative error in the electric field is computed at point A (10 mm from the source), point B (50 mm from the source), point C (100 mm from the source), point D (200 mm from the source), and point E (400 mm from the source). We note that the source is polarized such that its radiated signal impinging on the long grid boundary is maximum in amplitude. From the problem geometry, it is apparent that the specularly reflected wave from the long-grid-boundary PML arriving at point E is incident on the PML at 89°. Thus, we expect that the reflection error at point E is greatly degraded from that observed at point A.

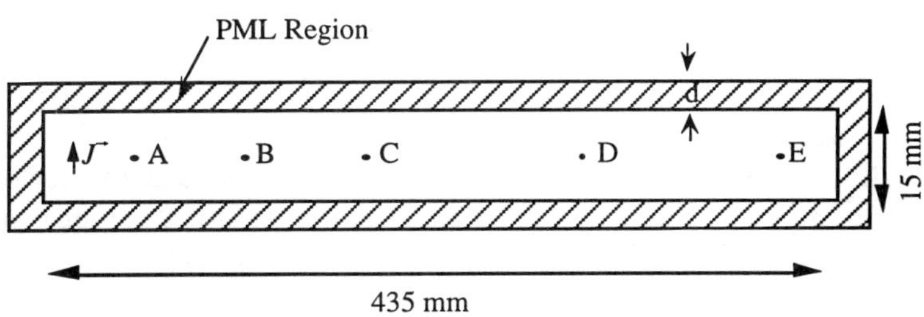

Fig. 5.12 Current element $\vec{J}$ radiating in an elongated FDTD grid (not to scale) terminated by PML. Distance of each observation point from the source: A, 10 mm; B, 50 mm; C, 100 mm; D, 200 mm; and E, 400 mm.

The model of Fig. 5.12 uses a 435 x 15-cell (working volume) grid having 1-mm square cells and Δt at 0.98 times the Courant limit. The source has the same differentiated Gaussian-pulse waveform used previously (see (5.96)). A 10-cell UPML absorber with polynomial spatial scaling ($m = 4$ and $\kappa_{\max} = 1$) is used on all sides.

Fig. 5.13 illustrates the maximum relative error recorded at each of the observation points over 1000 time steps as a function of $\sigma_{\max}$. Again, the error is referenced to a

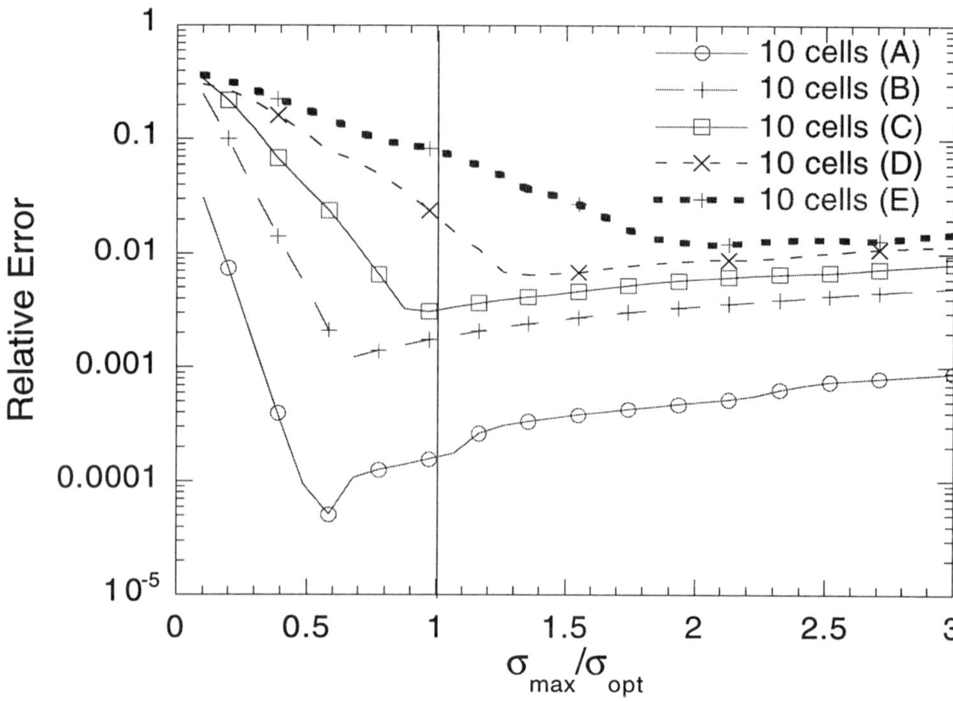

Fig. 5.13 Maximum relative error at points A, B, C, D, and E of the grid of Fig. 5.12 due to a polynomial-scaled 10-cell PML vs. $\sigma_{max}/\sigma_{opt}$. Parameters $m = 4$ and $\kappa = 1$ are fixed.

large-domain simulation. The error at point A is less than 0.0001, or -80 dB. However, at points B through E, the error gets progressively worse. From (5.60), we expect that larger values of σ_{max} are required to reduce the reflection error since σ_{opt} is referenced to normal incidence. The caveat is that larger values of σ_{max} lead to larger step discontinuities in the material profile and hence, a larger discretization error.

The radiation due to the current source is characterized by a wide spectrum of waves containing evanescent as well as propagating modes. However, the evanescent modes are not absorbed by the PML when $\kappa = 1$. Increasing κ should help this situation. Fig. 5.14 plots contours of the maximum relative error as a function of $\sigma_{max}/\sigma_{opt}$ and κ_{max} at points A and E. Indeed, we see that increasing κ_{max} causes the reflection error at point E to decrease. In fact, in the range of $\sigma_{max} \approx 1.6\sigma_{opt}$ and $\kappa_{max} \approx 25$, the reflection error drops two orders of magnitude to less than -65 dB. Unfortunately, increasing σ_{max} and κ_{max} degrades the reflection error at point A. However, from Fig. 5.14, the reflection error at point A is still less than -65 dB.

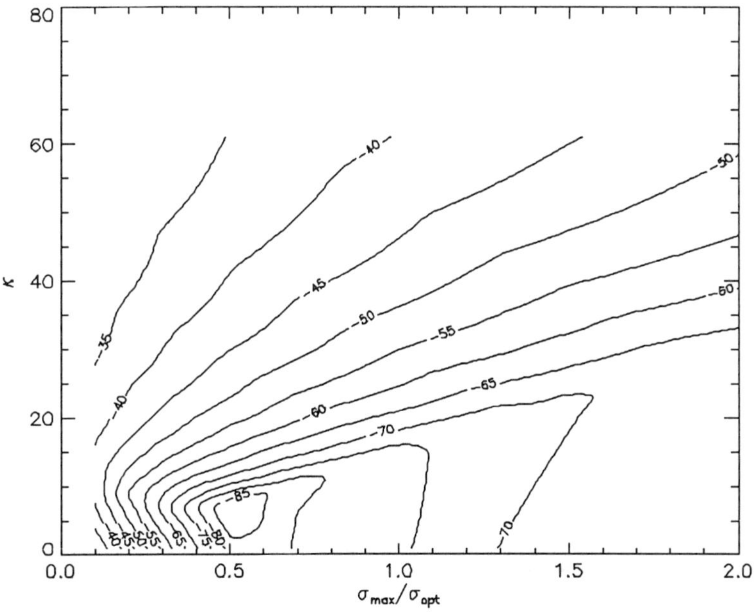

(a)

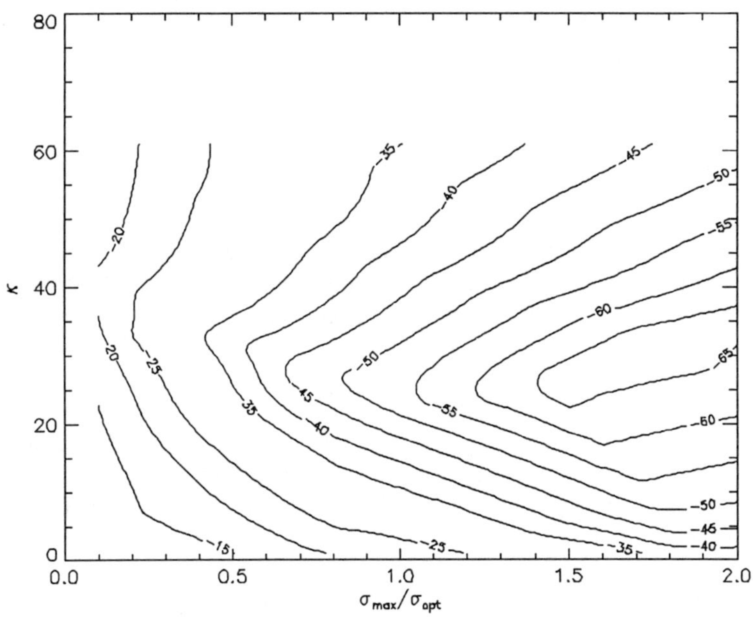

(b)

Fig. 5.14 Contour plots of the maximum relative error in dB in the grid of Fig. 5.12 vs. $\kappa_{y-\max}$ and $\sigma_{y-\max} / \sigma_{\text{opt}}$ for a polynomial-scaled 10-cell UPML: (a) at point A; (b) at point E.

5.7.4 TE Excitation of a Long Thin PEC Strip

This example considers the performance of PML in FDTD modeling of TE-polarized electromagnetic wave interactions with a long, thin PEC strip. Here, a perpendicular current source, $\vec{J}$, excites creeping waves that propagate around the strip. This potentially requires a longer simulation time to reach a steady state.

Fig. 5.15 shows the FDTD grid geometry for the strip problem. The space and time increments are unchanged from the previous examples, and the grid termination is realized by a 10-cell UPML having $m = 4$ polynomial profile scaling. Initially, the PML interface is placed only 3 cells from each side of the strip, leaving a working volume of 106 x 6 cells. The vertical (E_y) and horizontal (E_x) electric fields are observed at the very edge of the strip where the diffracted field is singular, and also near the source as a reference point. The source has the same differentiated-Gaussian time signature given by (5.96).

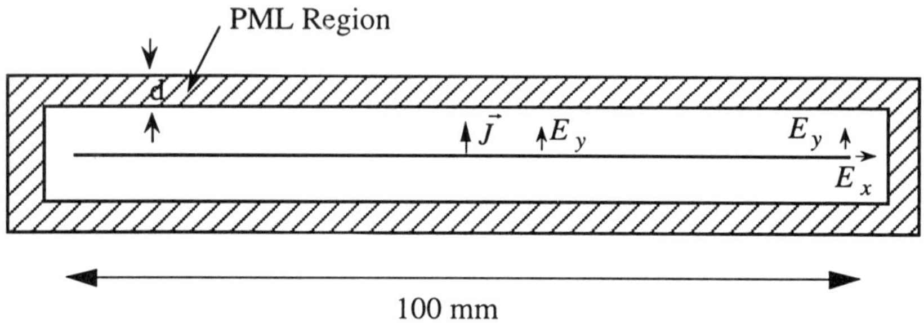

Fig. 5.15 Perpendicular current source $\vec{J}$ radiating in the presence of a long PEC strip. E_y is observed 10 mm from the current source and at the edge of the strip, and E_x is also probed at the edge.

Fig. 5.16 plots contours of the maximum relative error in E_y at an observation point 10 mm from the source, and the maximum relative error in E_x at the strip edge. The contours are plotted as a function of $\sigma_{\max}/\sigma_{\text{opt}}$ and $\kappa_{\max}$ for an observation period of 1000 time steps. For this case, the reference fields are computed on a large grid having PML outer boundaries 600 cells away from the strip on all sides.

From Fig. 5.16, the minimum reflection error is on the order of -60 dB near the source and -30 dB near the strip edge, with the error in E_y at the edge nearly the same as that for E_x. This error is unacceptably high. It is due to the elongated domain and the propagation of creeping waves around the strip for an extended period of time. Observations indicate that the error is largest after the initial pulse passes by. This is due to the low-pass filtering effect of the PML interface. From Section 5.5.2, the error can be reduced by: (1) reducing $\sigma_n(0)$, where n is the normal axis; (2) placing the PML interface further away from the strip; or (3) using the CFS tensor derived from (5.87).

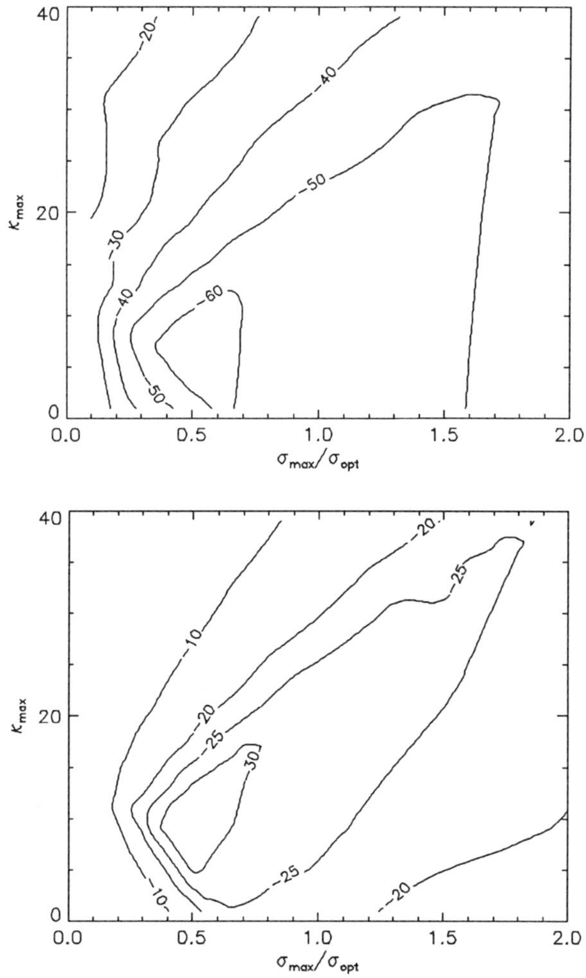

(a)

(b)

Fig. 5.16 Contour plots of the maximum relative error in dB vs. κ_{max} and $\sigma_{max} / \sigma_{opt}$ for the thin-strip model of Fig. 5.15. Polynomial-scaled ($m = 4$) 10-cell UPML is placed only 3 cells from each strip edge. (a) E_y field 10 mm from the source; (b) E_x field at the strip edge.

Moving the PML interface to 12 cells from each side of the strip results in the error contour plots of Fig. 5.17. The error near the source and the strip edge is diminished to -85 dB and -50 dB, respectively. However, to reduce the error near the strip edge to an acceptable level, the PML interface still needs to be pushed another 5 to 10 cells away from the strip. We also see that the minimum error near the edge is realized when $\sigma_{max} = 0.6\sigma_{opt}$ and $\kappa_{max} = 7$. The smaller σ_{max} is contradictory to that seen in the previous example of a homogeneous space. However, reducing σ_{max} reduces $\sigma(0)$, which delays the late-time reflection error (see (5.68) to (5.72)).

(a)

(b)

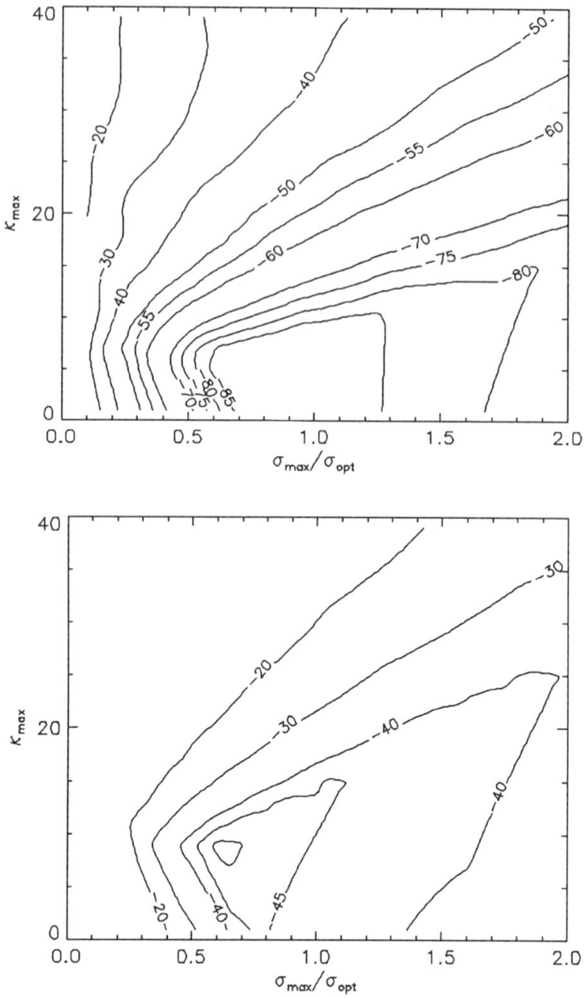

Fig. 5.17 Contour plots of the maximum relative error in dB vs. κ_{max} and $\sigma_{max} / \sigma_{opt}$ for the thin-strip model of Fig. 5.15. Polynomial-scaled ($m = 4$) 10-cell UPML is placed 12 cells from each strip edge. (a) E_y field 10 mm from the source; (b) E_x field at the strip edge.

An alternative to increasing the distance from the scatterer to the PML is to increase the PML thickness. This further decreases $\sigma(0)$ since it can be scaled over more cells. Fig. 5.18 provides contour plots of the reflection error for a 15-cell-thick UPML placed only 3 cells away from the strip in all directions. Relative to the results of Fig. 5.16, the error is decreased by over 25 dB near the source and near the edge.

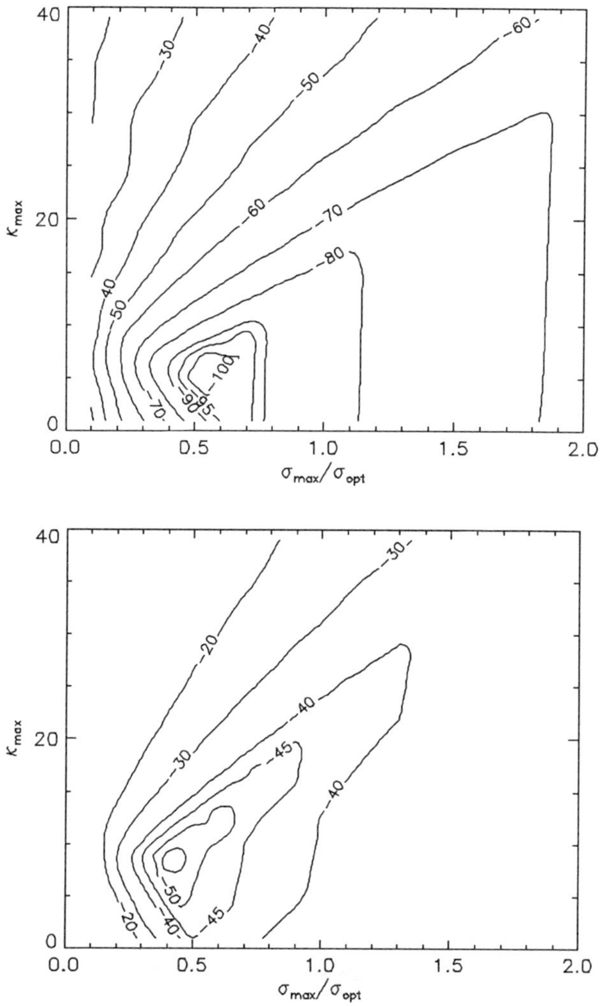

(a)

(b)

Fig. 5.18 Contour plots of the maximum relative error in dB vs. κ_{max} and $\sigma_{max} / \sigma_{opt}$ for the thin-strip model of Fig. 5.15. Polynomial-scaled (m = 4) 15-cell UPML is placed only 3 cells from each strip edge. (a) E_y field 10 mm from the source; (b) E_x field at the strip edge.

Finally, the CFS tensor is employed. A 10-cell UPML designed using (5.87) to (5.94) is placed only 3 cells away from the strip edge on all sides. Fig. 5.19 illustrates the resulting error as a function of κ_{max} and $\sigma_{max}/\sigma_{opt}$. Here, the error drops below -65 dB near the strip edge and -90 dB in the center, and exhibits no slow oscillations in the late time. The optimal σ_{max} is quite close to σ_{opt} as in the free-space case. However, the cost of implementing this PML is much higher than the previous cases.

(a)

(b)

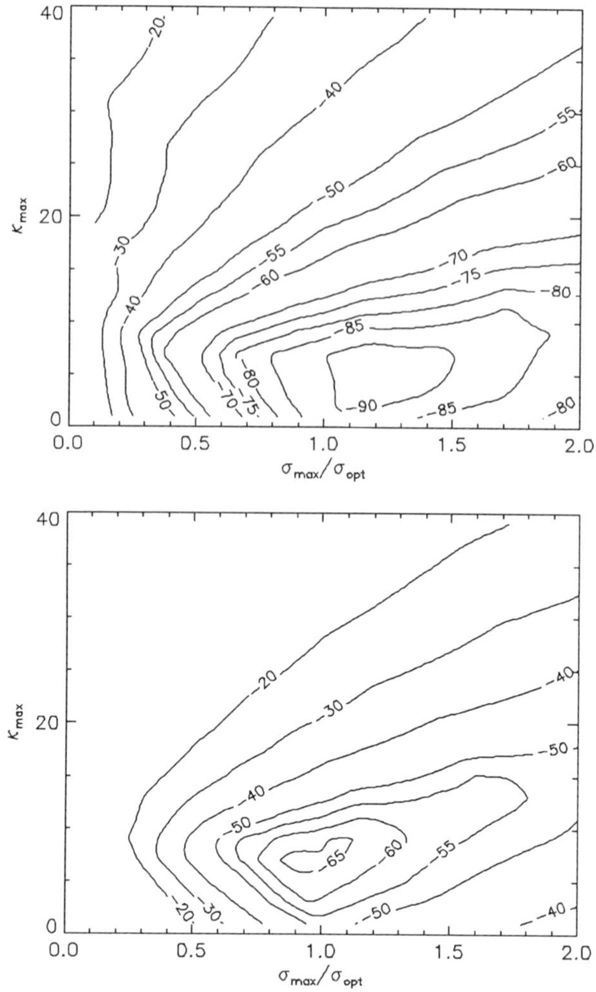

Fig. 5.19 Contour plots of the maximum relative error in dB vs. $\kappa_{\max}$ and $\sigma_{\max} / \sigma_{\text{opt}}$ for the thin-strip model of Fig. 5.15. Polynomial-scaled ($m = 4$) 10-cell UPML is placed only 3 cells from each strip edge. CFS tensor is used with $\alpha_x = \alpha_y = 0.069$. (a) E_y field 10 mm from the source; (b) E_x field at the strip edge.

5.7.5 Printed Microwave Circuits and Antennas

In Section 5.4.4, we showed theoretically that the PML ABC could be used to terminate FDTD lattices modeling inhomogeneous materials of infinite extent. This is validated here through some simple modeling examples involving printed microwave circuits and antennas. Often, when modeling printed microwave circuits, we simply assume that the dielectric substrate and superstrate are infinite in extent. Typically, this is a valid approximation if the dielectric substrate edge is sufficiently far away from the metal signal trace. Rather than extending the FDTD lattice out beyond the dielectric substrate edge, the lattice is terminated as close as possible to the signal trace using an absorbing medium such as the PML.

Fig. 5.20 depicts the cross-section geometry of the first example, a uniform 50-Ω microstrip line (10-mil width, zero thickness) printed on a 10-mil-thick alumina substrate ($\varepsilon_r = 9.8$). The region above the substrate is assumed to be air. Spatially scaled UPML is used to terminate the outer grid boundaries as shown in the figure. The microstrip line is excited at one end by a voltage source with a Gaussian profile and a 40-GHz bandwidth. At the other end, the metal trace and the dielectric substrate extend through the PML and terminate in the PEC wall backing the PML.

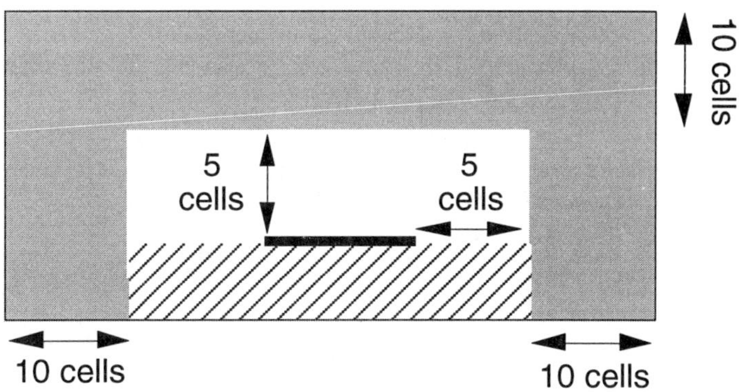

Fig. 5.20 Cross section of a 50-Ω microstrip line printed on a 10-mil alumina substrate terminating in PML. FDTD lattice sidewalls are also terminated by PML.

A number of studies with this model are now discussed to characterize the effects of the PML thickness, conductivity, and order of spatial variation, m, on its absorbing properties. We first recall from (5.67) that σ_{opt} varies as $(\varepsilon_r)^{-1/2}$. However, for the microstrip geometry, ε_r is inhomogeneous, causing an ambiguity. We resolve this by applying quasistatic theory to determine the effective dielectric constant of the microstrip line, which is $\varepsilon_{eff} = 6.62$ for this example. Then, we use this value to estimate σ_{opt} for several values of m, obtaining $\sigma_{\text{opt}} = 21$ S/m ($m = 2$); 28 S/m ($m = 3$); 35 S/m ($m = 4$); and 42 S/m ($m = 5$). This theory is supported by the results of numerical experiments shown in Fig. 5.21, where the reflection error due to a 10-cell, $m = 4$ UPML termination of the microstrip line is minimum when $\sigma_{\max} = 40$. Fig. 5.22 illustrates the reflection error due to the 10-cell UPML termination for various values of m, where for each case, $\sigma_{\max}$ is chosen close to σ_{opt}. This figure shows that $m = 4$ provides the smallest reflection error over the 0 to 50 GHz range considered.

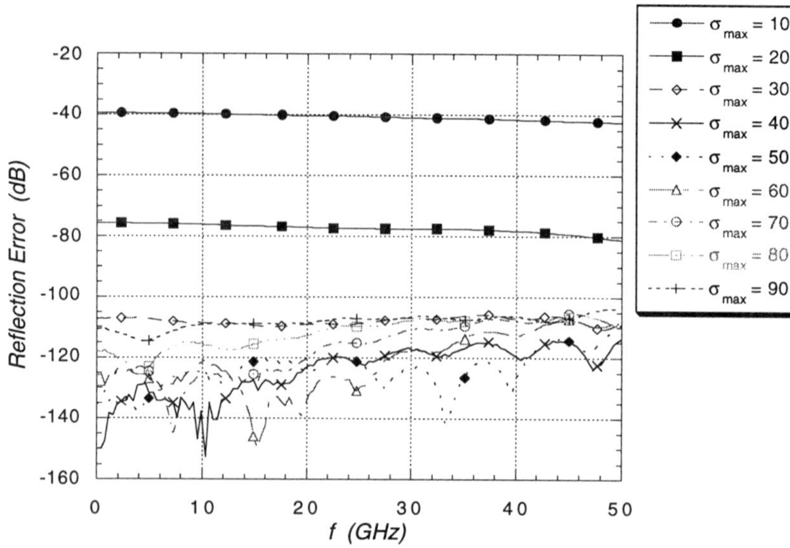

Fig. 5.21 Reflection error due to a 10-cell UPML termination of the 50-Ω microstrip line of Fig. 5.20 for various values of $\sigma_{\max}$. For this case ($m = 4$), $\sigma_{\text{opt}} = 35$. *Source*: Gedney, *IEEE Trans. Antennas and Propagation*, 1996, pp. 1630–1639, © 1996 IEEE.

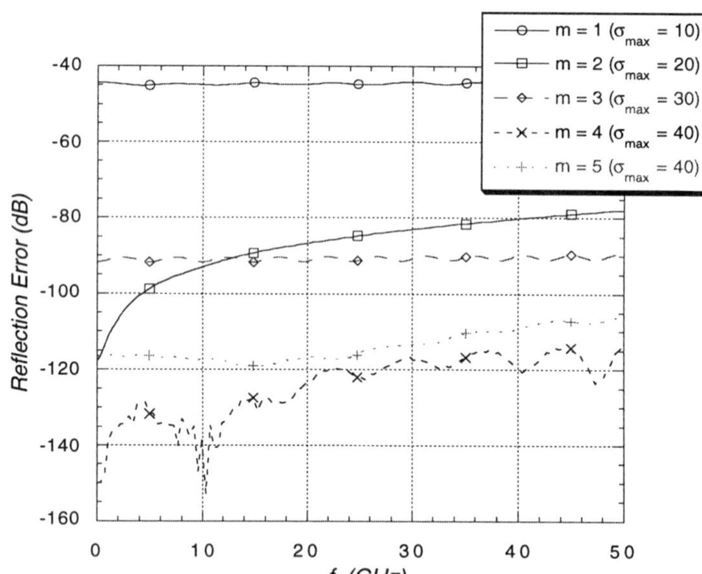

Fig. 5.22 Reflection error corresponding to Fig. 5.21, but for various values of m. $\sigma_{\max}$ is approximately equal to σ_{opt} in each case. *Source*: Gedney, *IEEE Trans. Antennas and Propagation*, 1996, pp. 1630–1639, © 1996 IEEE.

Fig. 5.23 provides the results of numerical experiments for the reflection error versus σ_{max} for a 10-cell PML with $m = 2$ and $m = 4$, and a 5-cell PML with $m = 4$. These results are also compared with the theory of (5.60), assuming $\varepsilon_r = \varepsilon_{eff} = 6.62$ and $\theta = 0$. The numerical reflection coefficient is closely approximated by (5.60) for small values of σ_{max}, since the reflection is predominately due to the terminating PEC wall. However, the reflection error deviates from the predicted value as the discretization error increases. This error is minimum in the near vicinity of σ_{opt}, and then increases as σ_{max} increases further.

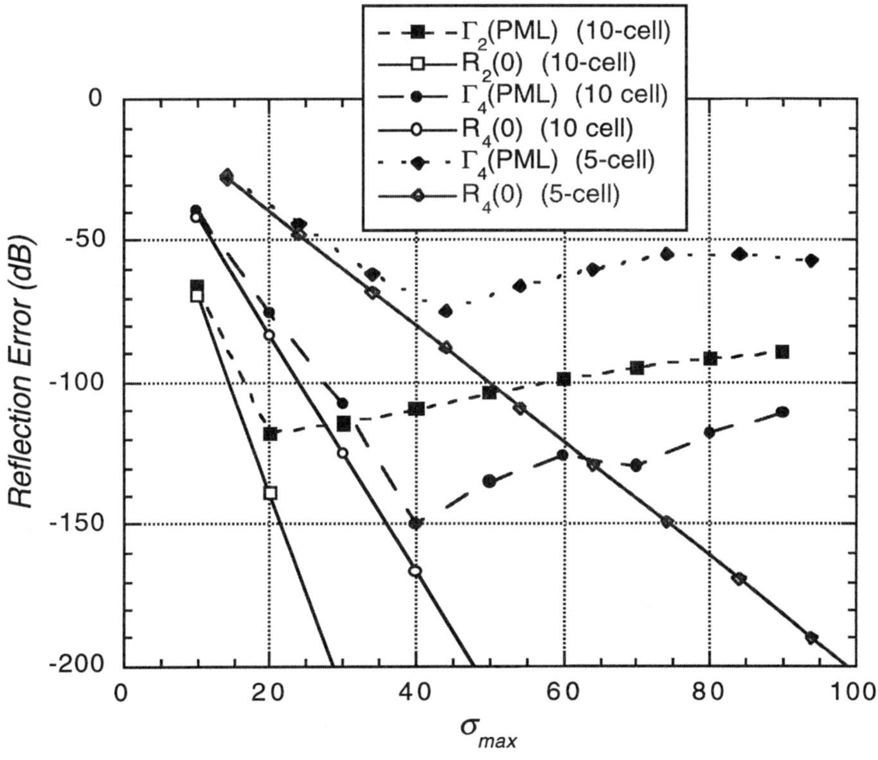

Fig. 5.23 Comparison of the reflection error computed using the split-field PML with the theoretical values. Cases considered: 10-cell PML termination of the microstrip line of Fig. 5.20 for $m = 2$ and $m = 4$; 5-cell PML termination for $m = 4$. *Source*: Gedney, *IEEE Trans. Antennas and Propagation*, 1996, pp. 1630–1639, © 1996 IEEE.

It is also important to study the FDTD lattice side terminations, upon which are incident highly oblique waves and surface-wave modes. Fig. 5.24 illustrates the reflection error due to the side terminations for a 10-cell PML ($m = 4$) for various

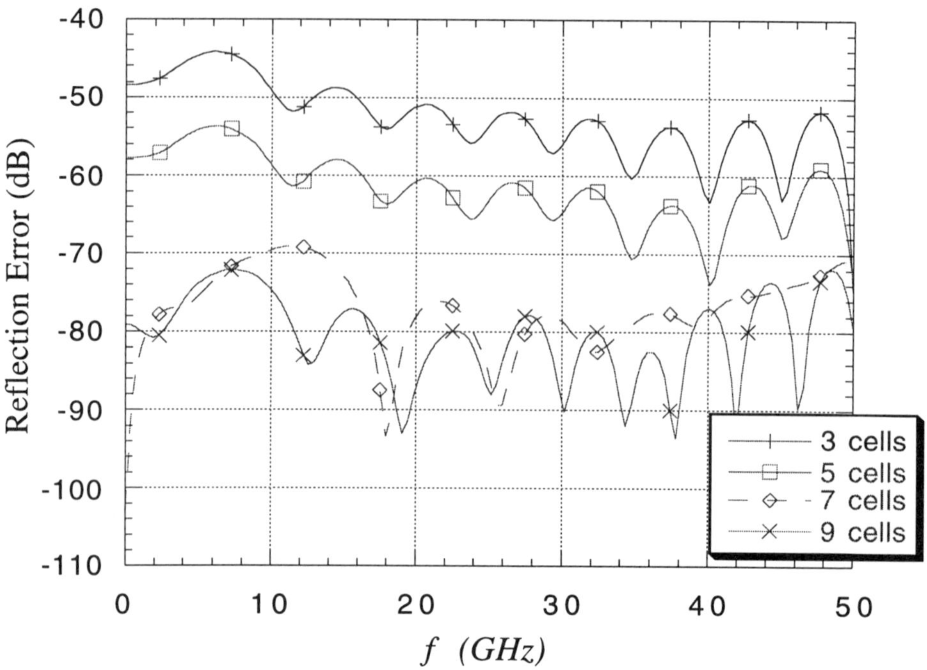

Fig. 5.24 Reflection error due to 10-cell UPML used for side terminations in Fig. 5.20 as a function of the distance from the edge of the microstrip to the PML interface. *Source*: Gedney, *IEEE Trans. Antennas and Propagation*, 1996, pp. 1630–1639, © 1996 IEEE.

distances between the PML interface and the edge of the microstrip. As expected, the reflection error here is larger than for the end-wall PML terminations. If better than -70 dB reflection is required across the band, the interface must be at least 7 cells from the edge of the microstrip.

Numerical experiments show that as ε_r decreases, the PML interface can be placed closer to the edge of the microstrip. For example, if $\varepsilon_r = 2.2$, only 3 cells from the strip edge is sufficient. We suspect that this is due to weaker coupling into surface-wave modes as ε_r decreases. Increasing $\kappa_{\max}$ should reduce the reflections from the side PMLs.

Finally, Fig. 5.25 provides FDTD-computed results for $|S_{11}|$ of a microstrip-fed patch antenna printed on a 31.25-mil Duroid substrate of $\varepsilon_r = 2.2$ [48,49]. The discretization used for this problem is $\Delta x = 0.389$ mm, $\Delta y = 0.4$ mm, $\Delta z = 0.1588$ mm, and $\Delta t = 0.441$ ps. The source microstrip line is 50 cells long and is excited by a Gaussian-pulse voltage source having a 30-GHz bandwidth. The FDTD lattice is terminated by a 10-cell UPML ABC placed either 3 cells from the edges of the patch antenna and 5 cells above the antenna surface, or 10 cells from the edges of the patch.

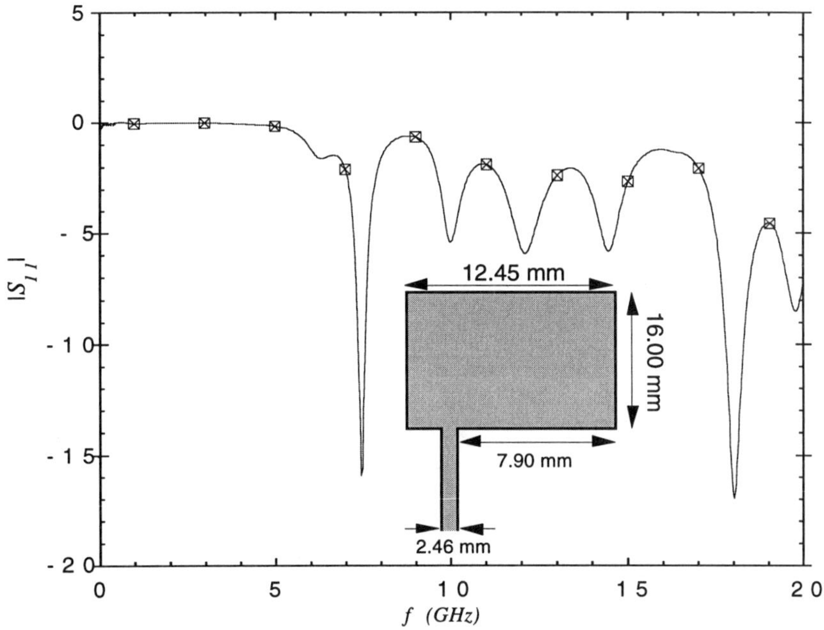

Fig. 5.25 $|S_{11}|$ of a microstrip-fed patch antenna computed using FDTD. The lattice is terminated by a 10-cell UPML placed either 3 cells from the edge of the patch and 5 cells above the patch (—✱—) or 10 cells from the edges of the patch (—⊠—). *Source*: Gedney, *IEEE Trans. Antennas and Propagation*, 1996, pp. 1630–1639, © 1996 IEEE.

Spatial scaling of the PML is used with $m = 4$ and σ_{opt} from (5.67). Fig. 5.25 shows negligible difference in $|S_{11}|$ over most of the frequency band for the two locations of the PML lattice termination. However, as expected from the discussion in 5.5.2, there is a small difference at very low frequencies.

5.8 PML TERMINATION FOR CONDUCTIVE MEDIA

For many applications, it is necessary to simulate electromagnetic wave interactions within lossy media of significant spatial extent. Examples include wave propagation within microwave circuits printed on lossy dielectric substrates, and impulsive electromagnetic scattering by objects buried in a lossy half-space such as the ground. In such cases, it is desirable to simulate the lossy material extending to infinity through the use of a PML absorbing boundary. Following the theory in Section 5.4.4, it can be easily shown that PML is inherently matched to a lossy medium [25–28]. In this section, a UPML for terminating inhomogeneous lossy media and its time-dependent representation is presented along with numerical verification.

5.8.1 Theory

It is desired to match a UPML along a planar boundary to a lossy half-space of parameters ε_r and σ. To match the lossy medium, it must extend through the PML with the same ε_r and σ. Hence, Ampere's law in the PML can be expressed as

$$\begin{bmatrix} \frac{\partial}{\partial y}H_z - \frac{\partial}{\partial z}H_y \\ \frac{\partial}{\partial z}H_x - \frac{\partial}{\partial x}H_z \\ \frac{\partial}{\partial x}H_y - \frac{\partial}{\partial y}H_x \end{bmatrix} = j\omega\varepsilon_0\left(\varepsilon_r + \frac{\sigma}{j\omega\varepsilon_r}\right)\begin{bmatrix} \frac{s_y s_z}{s_x} & 0 & 0 \\ 0 & \frac{s_x s_z}{s_y} & 0 \\ 0 & 0 & \frac{s_x s_y}{s_z} \end{bmatrix}\begin{bmatrix} E_x \\ E_y \\ E_z \end{bmatrix} \tag{5.97}$$

where s_x, s_y, and s_z are given in (5.75b). The next step is to derive a time-dependent representation. To this end, a mapping is performed to pose (5.97) as a first-order partial differential equation in time. Auxiliary variables are introduced in a manner similar to that of Section 5.6. However, due to the frequency dependence of the loss term, an additional auxiliary variable must be introduced. Thus, we define

$$P_x = \frac{s_z}{s_x}E_x, \quad P_y = \frac{s_x}{s_y}E_y, \quad P_z = \frac{s_y}{s_z}E_z \tag{5.98a}$$

$$P_x' = s_y P_x, \quad P_y' = s_z P_y, \quad P_z' = s_x P_z \tag{5.98b}$$

Using these expressions, (5.97) is decoupled into the set of differential equations:

$$\begin{bmatrix} \frac{\partial}{\partial y}H_z - \frac{\partial}{\partial z}H_y \\ \frac{\partial}{\partial z}H_x - \frac{\partial}{\partial x}H_z \\ \frac{\partial}{\partial x}H_y - \frac{\partial}{\partial y}H_x \end{bmatrix} = \frac{\partial}{\partial t}\varepsilon_0\varepsilon_r\begin{bmatrix} P_x' \\ P_y' \\ P_z' \end{bmatrix} + \sigma\begin{bmatrix} P_x' \\ P_y' \\ P_z' \end{bmatrix} \tag{5.99a}$$

$$\frac{\partial}{\partial t}\begin{bmatrix} P_x' \\ P_y' \\ P_z' \end{bmatrix} = \frac{\partial}{\partial t}\begin{bmatrix} \kappa_y & 0 & 0 \\ 0 & \kappa_z & 0 \\ 0 & 0 & \kappa_x \end{bmatrix}\begin{bmatrix} P_x \\ P_y \\ P_z \end{bmatrix} + \frac{1}{\varepsilon_0}\begin{bmatrix} \sigma_y & 0 & 0 \\ 0 & \sigma_z & 0 \\ 0 & 0 & \sigma_x \end{bmatrix}\begin{bmatrix} P_x \\ P_y \\ P_z \end{bmatrix} \tag{5.99b}$$

$$\frac{d}{dt}(\kappa_x P_x) + \frac{\sigma_x}{\varepsilon_0}P_x = \frac{d}{dt}(\kappa_z E_x) + \frac{\sigma_z}{\varepsilon_0}E_x \tag{5.99c}$$

$$\frac{d}{dt}\left(\kappa_y P_y\right) + \frac{\sigma_y}{\varepsilon_0} P_y = \frac{d}{dt}\left(\kappa_x E_y\right) + \frac{\sigma_x}{\varepsilon_0} E_y \tag{5.99d}$$

$$\frac{d}{dt}\left(\kappa_z P_z\right) + \frac{\sigma_z}{\varepsilon_0} P_z = \frac{d}{dt}\left(\kappa_y E_z\right) + \frac{\sigma_y}{\varepsilon_0} E_z \tag{5.99e}$$

Equations (5.99b) to (5.99e) are discretized in the same fashion as presented in (5.82). Equation (5.99a) is discretized as a standard lossy update similar to (5.81), applying central-difference approximations to space and time derivatives and averaging the loss term in time. This yields a second-order update for the electric field. The magnetic field update within the PML is identical to that presented in (5.83a) and (5.83b).

The advantage of the formulation presented in (5.99) is that within the working volume where $\sigma_x = \sigma_y = \sigma_z = 0$ and $\kappa_x = \kappa_y = \kappa_z = 1$, we have $P_x' = P_x = E_x$, and so forth, and (5.99a) reduces to Ampere's law in a lossy isotropic medium. Further, in this region, the updates (5.99b) to (5.99e) are not needed. Also, in nonoverlapping PML regions, only a single auxiliary variable need be introduced for the electric field.

5.8.2 Numerical Example

Consider the simulated lossy half-space geometry of Fig. 5.26 shown in a two-dimensional TE-polarized FDTD grid with field components E_x, E_y, and H_z:

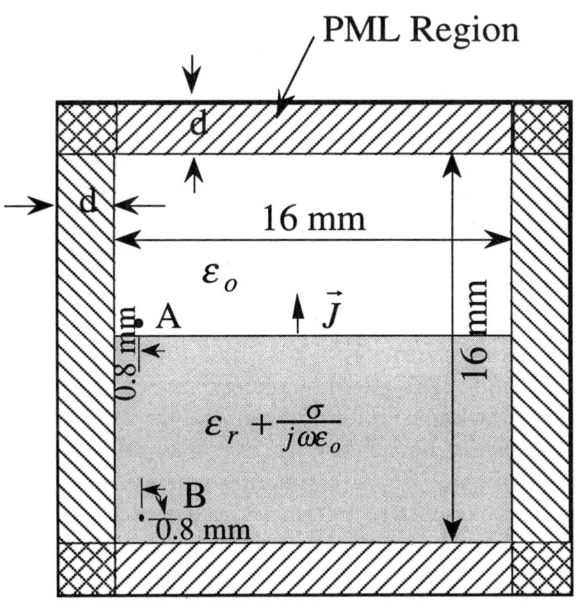

Fig. 5.26 Vertical electric current source above a lossy half-space. The working area is 40 × 40 cells ($\Delta x = \Delta y = 0.4$ mm), and the PML is 10 cells thick.

In the example shown in Fig. 5.26, the working area of the FDTD grid has 40×40 cells with $\Delta x = \Delta y = 0.4$ mm. The time step is set at 0.98 of the Courant limit ($\Delta t = 0.9246$ ps). A 10-cell polynomial-scaled UPML is used to terminate each exterior boundary of the grid. The lossy half-space of parameters $\varepsilon_r = 10$ and $\sigma = 0.3$ S/m extends completely through the PML to its PEC outer boundary. Here, the two-dimensional FDTD implementation based on (5.99a) to (5.99e) is used. On the left-side and right-side PMLs where ε_r is inhomogeneous, an average value of $\varepsilon_r = 5.5$ is assumed to estimate σ_{opt}. Despite σ_{opt} being different for each PML, the ratio of $\sigma_{max} / \sigma_{opt}$ is kept the same on all PML absorbers for each simulation.

We excite the grid of Fig. 5.26 by applying a vertical electric current source having the differentiated-Gaussian-pulse time signature

$$J_y(x_0, y_0, t) = -2\left(\frac{t - t_0}{t_w}\right) e^{-[(t - t_0)/t_w]^2} \Bigg\} \; t_w = 26.53 \text{ ps}, \;\; t_0 = 4t_w \tag{5.100}$$

two cells above the lossy half-space. Field data are recorded at point A, two cells above the lossy half-space and two cells from the PML; and at point B, two cells from the bottom and side PMLs in the lossy medium.

Fig. 5.27 plots contours of the maximum relative error due to outer boundary reflections in Fig. 5.26 over 1000 time steps at observation points A and B. The contours are plotted versus the PML polynomial order, m, and the parameter $\sigma_{max} / \sigma_{opt}$. In all of these results, $\kappa = 1$. We note that the simulation time goes well past the steady-state response. From this figure, the minimum reflection error occurs for $3 < m < 4$ and $\sigma_{max} / \sigma_{opt} \approx 1.1$. The error is minimum for a slightly larger value of σ_{max} at point B than that at point A. This is due to the fact that the wave impinges on the PML at a steeper angle at point B.

We expect that increasing κ should improve the performance of the PML by attenuating evanescent waves. Fig. 5.28 plots contours of the reflection errors at points A and B as a function of κ_{max} and $\sigma_{max} / \sigma_{opt}$ for $m = 3.5$. Within the PML, κ is scaled from 1 to κ_{max} using (5.61). We observe that increasing κ_{max} to approximately 7 indeed improves the reflection error at point B. Again, minimum reflection error is realized when $\sigma_{max} / \sigma_{opt} \approx 1.1$ for this example.

We noted earlier that the material profile is inhomogeneous along the vertical direction on the left and right grid boundaries in Fig. 5.26. However, the PML parameters, σ_x, are constant along the vertical direction but scaled along the x direction. Thus, σ_x must be tuned to an averaged rather than a specific medium. We may be tempted to vary σ_x as a function of y to match the free-space region separately from the lossy-dielectric region. However, if σ_x is a function of y, Gauss' law presented in Section 5.4.2 is violated and spurious charge is introduced at the discontinuity. This leads to algorithmic instability. The mistake in this line of thought is that the PML boundary is not matched to the free-space region and the lossy medium separately, but rather to the entire planar boundary of the inhomogeneous medium.

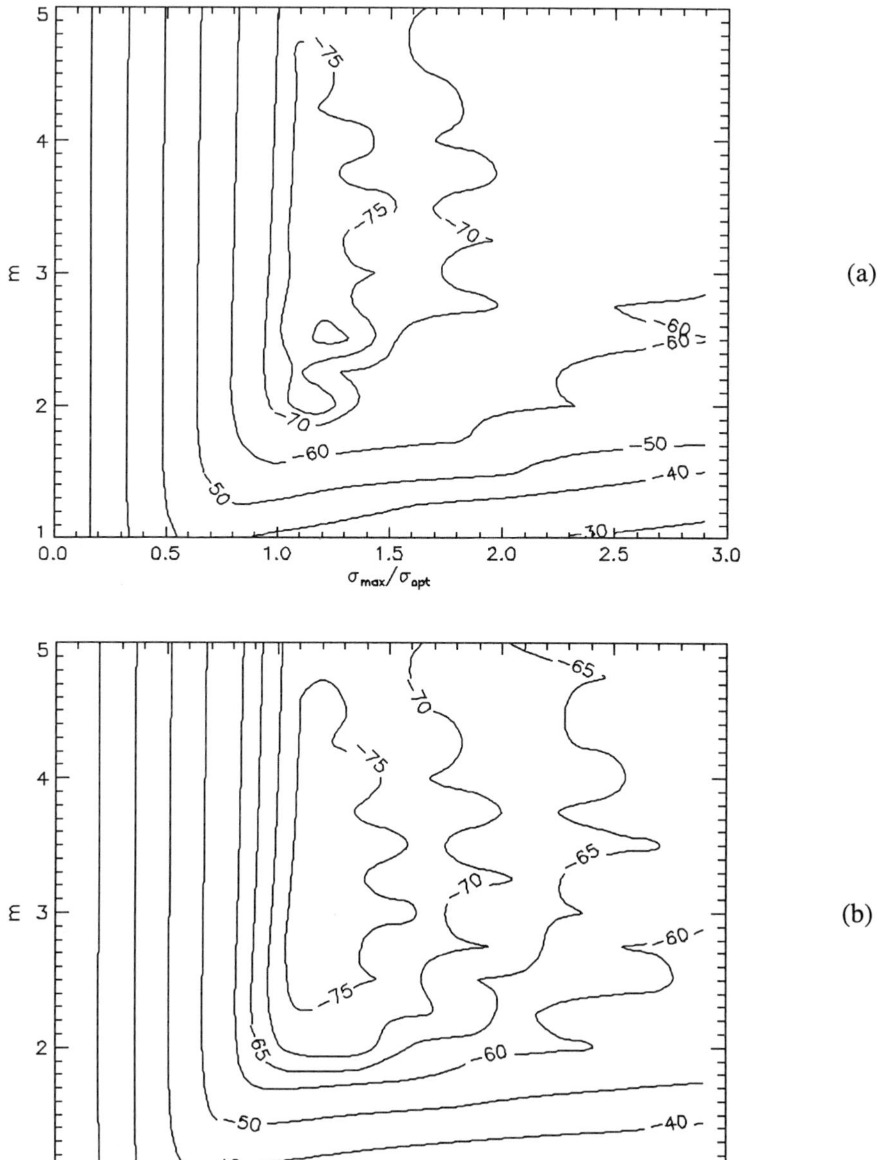

Fig. 5.27 Contour plots of the maximum relative error in dB due to outer boundary reflections in Fig. 5.26 over 1000 time steps. The performance of the 10-cell-thick UPML is given as a function of m and $\sigma_{max} / \sigma_{opt}$, with $\kappa = 1$: (a) at point A; (b) at point B.

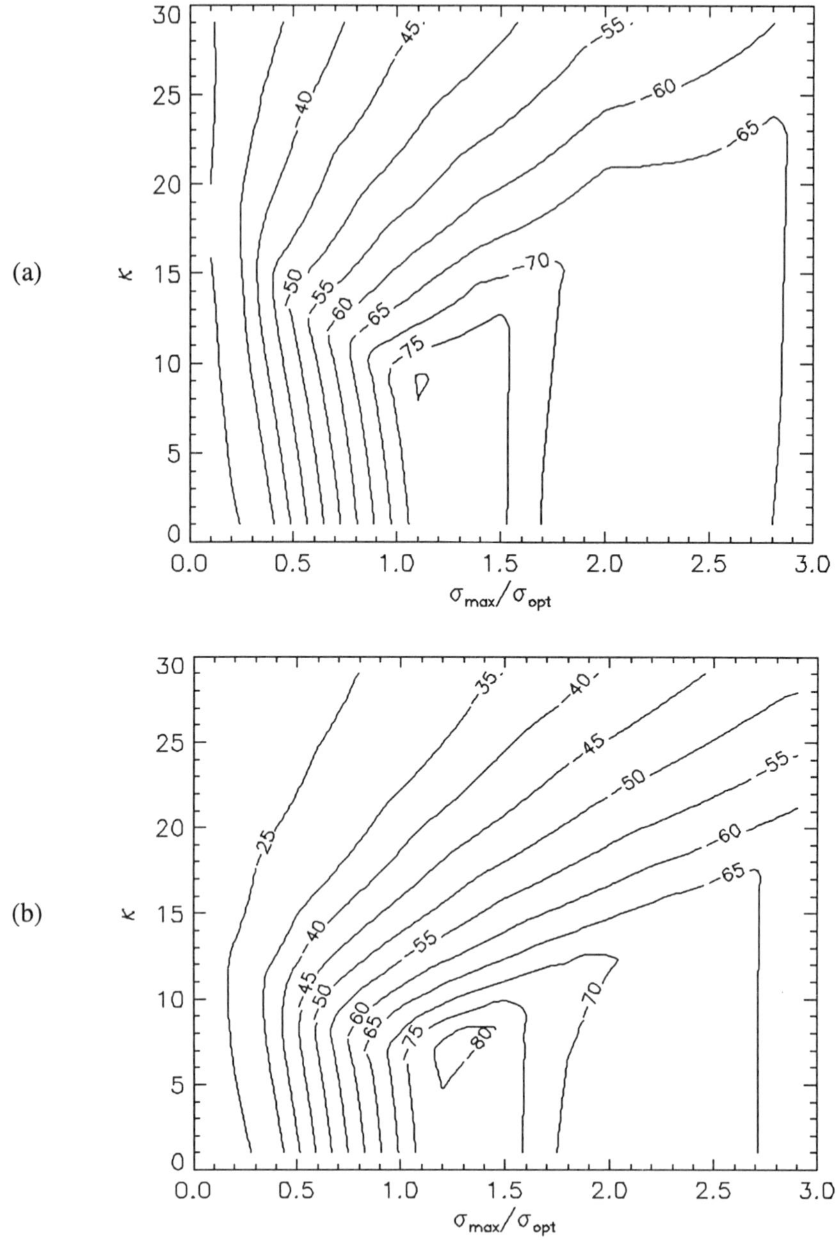

Fig. 5.28 Contour plots of the maximum relative error in dB due to outer boundary reflections in Fig. 5.26 over 1000 time steps. The performance of the 10-cell-thick UPML is given as a function of κ and $\sigma_{max} / \sigma_{opt}$, with $m = 3.5$: (a) at point A; (b) at point B.

Finally, Fig. 5.29 graphs the time waveforms at points *A* and *B* when the FDTD grid is terminated on all sides by an optimal 10-cell UPML ($\sigma_{max} = 1.1\,\sigma_{opt}$, $m = 3.5$, $\kappa_{max} = 7$). The results are compared with the waveforms calculated on an extremely large grid. Since the reflection error is about -80 dB, the results are indistinguishable.

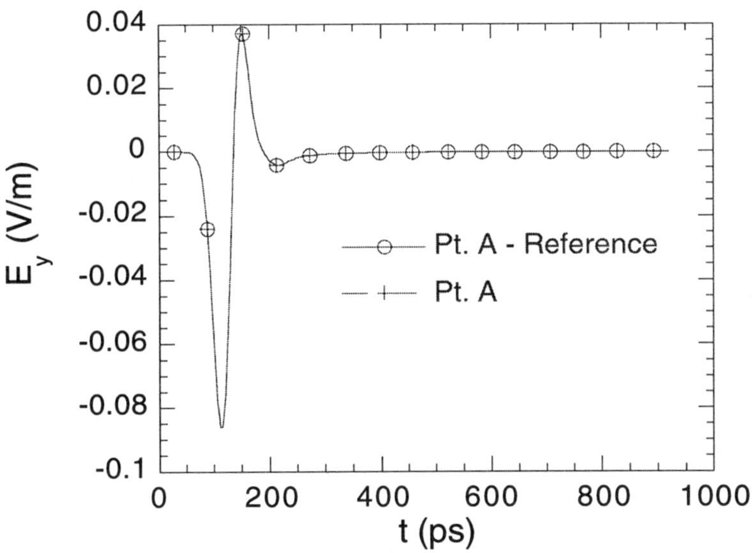

(a)

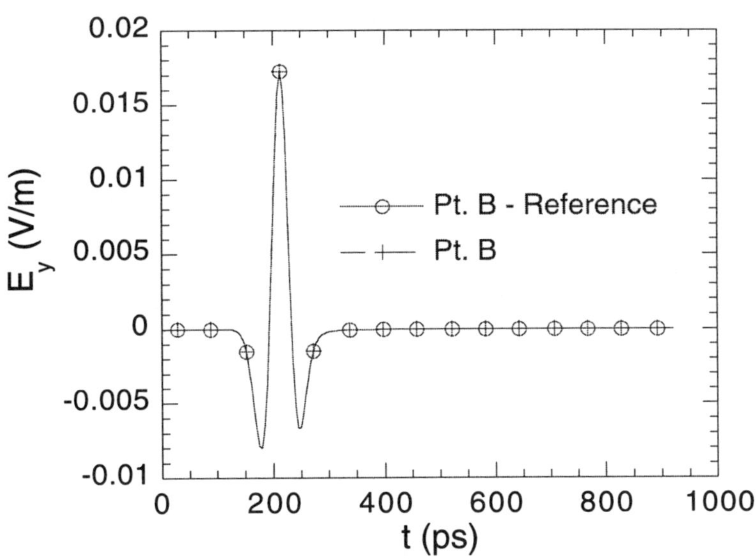

(b)

Fig. 5.29 Time waveforms of E_y in Fig. 5.26 with optimized 10-cell UPML compared with reference solution computed on an extremely large grid: (a) at point *A*; (b) at point *B*.

5.9 PML TERMINATION FOR DISPERSIVE MEDIA

The FDTD method has provided accurate simulations of electromagnetic phenomena in linear and nonlinear dispersive material media for a number of applications [1,50,51]. However, there are no pseudo-differential ABCs that provide broadband absorption for FDTD simulations of highly dispersive materials. Classical outer-boundary operators such as those of Mur and Higdon must assume an effective phase velocity of the incident wave. If a medium is highly dispersive and lossy, these boundary operators perform very poorly when terminating a broadband pulse. In this section, a PML is derived for a second-order (Lorentz) dispersive medium [25]. Based on this approach, PML absorbers can be developed for other media having Debye and multiple Lorentz resonances, or even more general dispersive behavior [1].

5.9.1 Theory

Consider a medium having a complex relative dielectric permittivity with a second-order (Lorentz) dispersion:

$$\varepsilon_r(\omega) = \varepsilon_\infty + \frac{(\varepsilon_s - \varepsilon_\infty)\,\omega_0^2}{\omega_0^2 + 2j\omega\delta - \omega^2} \tag{5.101}$$

where ω_0 is the resonant radial frequency and δ is a damping constant. The FDTD algorithm used here is that originally proposed by Joseph et al. [50], which introduces a second-order difference equation to update the electric field intensity given the flux density. The relative permittivity defines the constitutive relationship between the electric flux density and the electric field:

$$\vec{D} = \varepsilon_r(\omega)\vec{E} \tag{5.102}$$

Assume the second-order Lorentz medium of (5.101) for (5.102). Then, multiplying both sides by the denominator leads to

$$(\omega_0^2 + 2j\omega\delta - \omega^2)D_i = (\omega_0^2\,\varepsilon_s + 2j\omega\delta\,\varepsilon_\infty - \omega^2\,\varepsilon_\infty)E_i \tag{5.103}$$

where $i = x, y,$ or z. This is subsequently transformed into the time domain, yielding the differential equation

$$\left(\omega_0^2 + 2\delta\frac{\partial}{\partial t} - \frac{\partial^2}{\partial t^2}\right)D_i = \left(\omega_0^2\,\varepsilon_s + 2\delta\,\varepsilon_\infty\frac{\partial}{\partial t} - \varepsilon_\infty\frac{\partial^2}{\partial t^2}\right)E_i \tag{5.104}$$

This equation is numerically implemented with second-order accuracy by using central-difference approximations of the first-order and second-order time derivatives, and a trapezoidal time-integration of the constant terms. This results in the update relation:

$$E_i^{n+\frac{1}{2}} = \frac{c_1 E_i^{n-\frac{1}{2}} + c_2 E_i^{n-\frac{3}{2}} + c_3 D_i^{n+\frac{1}{2}} + c_4 D_i^{n-\frac{1}{2}} + c_5 D_i^{n-\frac{3}{2}}}{c_6} \tag{5.105}$$

where

$$c_1 = -\left[\frac{\omega_0^2\,\varepsilon_s}{2} - \frac{2\varepsilon_\infty}{(\Delta t)^2}\right], \qquad c_2 = -\left[\frac{\omega_0^2\,\varepsilon_s}{4} - \frac{\delta\,\varepsilon_\infty}{\Delta t} + \frac{\varepsilon_\infty}{(\Delta t)^2}\right] \tag{5.106a}$$

$$c_3 = \left[\frac{\omega_0^2}{4} + \frac{\delta}{\Delta t} + \frac{1}{(\Delta t)^2}\right], \qquad c_4 = \left[\frac{\omega_0^2}{2} - \frac{2}{(\Delta t)^2}\right] \tag{5.106b}$$

$$c_5 = \left[\frac{\omega_0^2}{4} - \frac{\delta}{\Delta t} + \frac{1}{(\Delta t)^2}\right], \qquad c_6 = \left[\frac{\omega_0^2\,\varepsilon_s}{4} + \frac{\delta\,\varepsilon_\infty}{\Delta t} + \frac{\varepsilon_\infty}{(\Delta t)^2}\right] \tag{5.106c}$$

Within the dispersive region, the displacement flux densities are updated using the traditional FDTD update expressions as proposed in [50] .

The intent of this study is to terminate the FDTD lattice loaded with a dispersive material such that the simulation emulates the dispersive medium extending to infinity. To do this, the dispersive medium must extend through the PML. In previous sections, it was demonstrated that PML is perfectly matched to a medium provided that the permittivity of the medium extends normally through the PML. The formulation makes no distinction regarding the permittivity being frequency dependent. Therefore, the PML can be matched to a dispersive medium.

In this section, the PML is represented as a uniaxial medium. The implementation follows closely that presented in the previous section. Within the PML, Ampere's law is expressed as

$$\begin{bmatrix} \frac{\partial}{\partial y}H_z - \frac{\partial}{\partial z}H_y \\ \frac{\partial}{\partial z}H_x - \frac{\partial}{\partial x}H_z \\ \frac{\partial}{\partial x}H_y - \frac{\partial}{\partial y}H_x \end{bmatrix} = j\omega\varepsilon_0\,\hat{\varepsilon}_r(\omega) \begin{bmatrix} \frac{s_y s_z}{s_x} & 0 & 0 \\ 0 & \frac{s_x s_z}{s_y} & 0 \\ 0 & 0 & \frac{s_x s_y}{s_z} \end{bmatrix} \begin{bmatrix} E_x \\ E_y \\ E_z \end{bmatrix} \tag{5.107}$$

where s_x, s_y, and s_z are expressed in (5.75b). Again, an auxiliary variable is introduced to pose the curl equation as first-order in time. However, due to the frequency dependence of ε_r, an additional auxiliary variable must be introduced. Defining

$$D_x = \varepsilon_0 \frac{s_z}{s_x} E_x\,, \qquad D_y = \varepsilon_0 \frac{s_x}{s_y} E_y\,, \qquad D_z = \varepsilon_0 \frac{s_y}{s_z} E_z \tag{5.108a}$$

$$P_x = \hat{\varepsilon}_r(\omega) D_x\,, \qquad P_y = \hat{\varepsilon}_r(\omega) D_y\,, \qquad P_z = \hat{\varepsilon}_r(\omega) D_z \tag{5.108b}$$

we obtain the following set of differential equations:

$$\begin{bmatrix} \frac{\partial}{\partial y} H_z - \frac{\partial}{\partial z} H_y \\ \frac{\partial}{\partial z} H_x - \frac{\partial}{\partial x} H_z \\ \frac{\partial}{\partial x} H_y - \frac{\partial}{\partial y} H_x \end{bmatrix} = \frac{\partial}{\partial t} \begin{bmatrix} \kappa_y & 0 & 0 \\ 0 & \kappa_z & 0 \\ 0 & 0 & \kappa_x \end{bmatrix} \begin{bmatrix} P_x \\ P_y \\ P_z \end{bmatrix} + \frac{1}{\varepsilon_0} \begin{bmatrix} \sigma_y & 0 & 0 \\ 0 & \sigma_z & 0 \\ 0 & 0 & \sigma_x \end{bmatrix} \begin{bmatrix} P_x \\ P_y \\ P_z \end{bmatrix} \tag{5.109a}$$

$$\left(\omega_0^2 + 2\delta \frac{\partial}{\partial t} - \frac{\partial^2}{\partial t^2} \right) \begin{bmatrix} P_x \\ P_y \\ P_z \end{bmatrix} = \left(\omega_0^2 \varepsilon_s + 2\delta \varepsilon_\infty \frac{\partial}{\partial t} - \varepsilon_\infty \frac{\partial^2}{\partial t^2} \right) \begin{bmatrix} D_x \\ D_y \\ D_z \end{bmatrix} \tag{5.109b}$$

$$\frac{d}{dt}\left(\kappa_x D_x\right) + \frac{\sigma_x}{\varepsilon_0} D_x = \varepsilon_0 \left[\frac{d}{dt}\left(\kappa_z E_x\right) + \frac{\sigma_z}{\varepsilon_0} E_x \right] \tag{5.109c}$$

$$\frac{d}{dt}\left(\kappa_y D_y\right) + \frac{\sigma_y}{\varepsilon_0} D_y = \varepsilon_0 \left[\frac{d}{dt}\left(\kappa_x E_y\right) + \frac{\sigma_x}{\varepsilon_0} E_y \right] \tag{5.109d}$$

$$\frac{d}{dt}\left(\kappa_z D_z\right) + \frac{\sigma_z}{\varepsilon_0} D_z = \varepsilon_0 \left[\frac{d}{dt}\left(\kappa_y E_z\right) + \frac{\sigma_y}{\varepsilon_0} E_z \right] \tag{5.109e}$$

Equations (5.109a) and (5.109c) to (5.109e) are discretized in the same way as in (5.81) and (5.82). Equation (5.109b) is discretized according to (5.105). This combines to provide a second-order update expression for the electric field. The magnetic field update within the PML is identical to that presented in (5.83a) and (5.83b).

5.9.2 Numerical Example: Scattering by a Lorentz Medium

To demonstrate the effectiveness of this technique, we consider the case of a TE-polarized plane-wave pulse normally incident from free space onto a dispersive dielectric half-space. The dielectric is assumed to have the second-order Lorentz dispersion of (5.101) with the parameters reported in [50]: $\varepsilon_s = 2.25$, $\varepsilon_\infty = 1$, $\omega_0 = 4\times10^{16}$ rad/sec, and $\delta = 0.28\times10^{16}$ sec^{-1}. Note that the dielectric is very lossy at the resonant frequency, ω_0.

For this study, $\Delta x = 1\times10^{-10}$ m and $\Delta t = 2\times10^{-19}$ sec. The plane wave has a Gaussian-pulse time dependence with a bandwidth of 3×10^{16} Hz. Both the free-space region and the dielectric half-space are terminated in a 10-cell UPML.

Fig. 5.30(a) illustrates the reflection error due to the PML observed just in front of the air / dispersive-dielectric interface for several values of σ_{max} and for $\kappa_{max} = 1$. (Note from (5.67) that $\sigma_{opt} = 1.1\times10^8$ S/m.) We observe that there is a sharp rise in the error near 0.8×10^{16} Hz. This is in the frequency range where the wave in the dispersive medium is nearly purely attenuating. Away from this region, reflection errors drop to the order of -90 dB. At dc, a reflection error of -70 dB is realized.

Fig. 5.30(b) illustrates the reflection error due to the PML termination as κ_{max} is varied and $\sigma_{max} = 2\times10^8$ S/m. Increasing κ_{max} from 1 to 60 reduces the peak reflection error from -53 dB to -68 dB. However, further increasing κ_{max} causes the error in the propagating region to increase, as well.

It is clear that the PML is highly effective in terminating a lossy dispersive medium across the entire spectrum. This is significant since prior to the PML approach, there was no absorbing boundary available that could provide such a broadband termination. The only alternative was to greatly extend the FDTD lattice.

5.10 PML IN CYLINDRICAL AND SPHERICAL COORDINATES

For specialized problems, the FDTD method can be posed in cylindrical or spherical coordinates. Specifically, if the boundaries of the problem geometry are separable within one of these coordinate systems, the boundary discretization error can be greatly reduced. It is thus desirable to extend the PML to a these spaces. The previous sections of this chapter have assumed PML layers with planar boundaries. Unfortunately, neither the UPML nor Berenger's PML can be directly implemented in cylindrical or spherical coordinates as posed [34, 52], except in the limit as the radius of the boundary interface tends to infinity.

5.10.1 Stretched-Coordinate Formulation

A specialized PML can be designed for the cylindrical-coordinate space. This was demonstrated by Maloney et al. [53] using a graphical solution based on the UPML, and by Chew et al. [34, 52] using a stretched-coordinate approach. In this section, the stretched-coordinate form of the PML is derived, and then a one-to-one relationship with the UPML is presented.

(a)

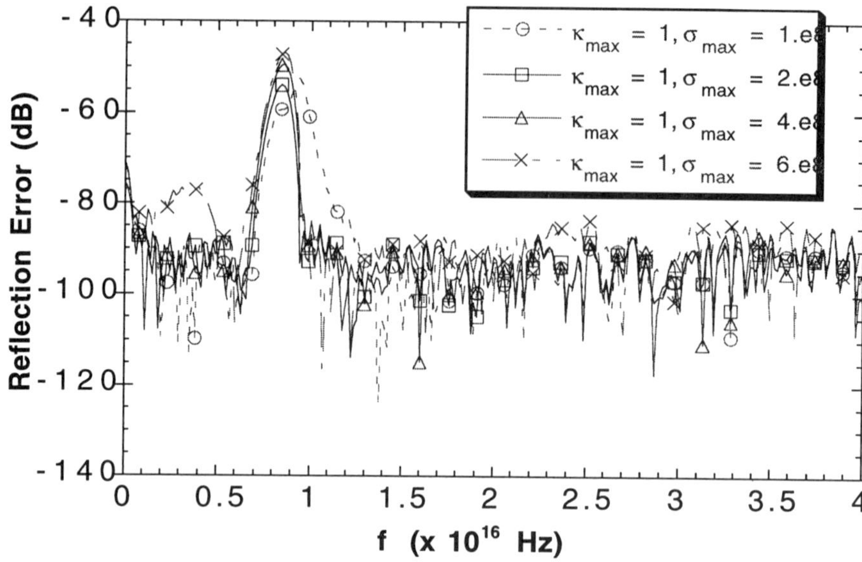

(b)

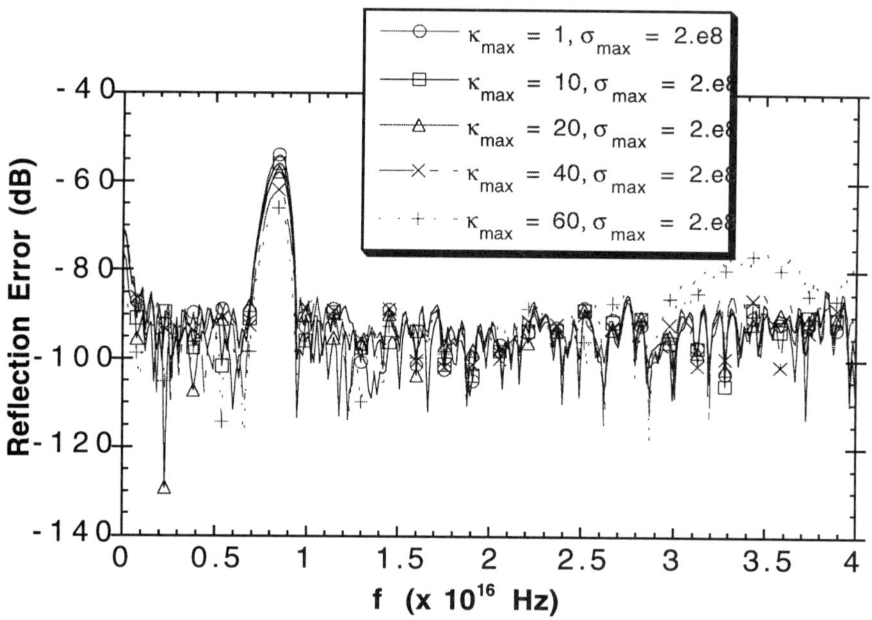

Fig. 5.30 Reflection error due to a 10-cell PML terminating an FDTD grid containing a dielectric half-space characterized by a second-order (Lorentz) dispersion for several values of: (a) σ_{max}; (b) κ_{max}. *Source:* Gedney, *Electromagnetics*, 1996, pp. 399–415.

In the cylindrical-coordinate space, Ampere's law is expressed as

$$j\omega\varepsilon_0 E_\rho = \frac{1}{\rho}\frac{\partial H_z}{\partial \phi} - \frac{\partial H_\phi}{\partial z} \tag{5.110a}$$

$$j\omega\varepsilon_0 E_\phi = \frac{\partial H_\rho}{\partial z} - \frac{\partial H_z}{\partial \rho} \tag{5.110b}$$

$$j\omega\varepsilon_0 E_z = \frac{1}{\rho}\frac{\partial(\rho H_\phi)}{\partial \rho} - \frac{1}{\rho}\frac{\partial H_z}{\partial \phi} \tag{5.110c}$$

Faraday's law is derived by duality. Define the region $\rho < \rho_0$ to be free space and the region $\rho > \rho_0$ to be a PML. The region is also terminated in the z-direction, defining $z < z_0$ to be free space and $z > z_0$ to be PML. It is desired to map Maxwell's equations in the cylindrical-coordinate space into a complex stretched-coordinate space. Following (5.26a) to (5.26b), we define the stretched-coordinate variables for the cylindrical-coordinate space:

$$\widehat{\rho} \rightarrow \rho_0 + \int_{\rho_0}^{\rho} s_\rho(\rho')d\rho', \qquad \widehat{z} \rightarrow z_0 + \int_{z_0}^{z} s_z(z')dz' \tag{5.111}$$

Note that the leading ρ_0 and z_0 terms arise from the fact that over the ranges $\{0 \le \rho \le \rho_0\}$ and $\{0 \le z \le z_0\}$, $s_\rho = 1$ and $s_z = 1$, respectively. Further,

$$\frac{\partial}{\partial\widehat{\rho}} = \frac{1}{s_\rho}\frac{\partial}{\partial\rho}, \qquad \frac{\partial}{\partial\widehat{z}} = \frac{1}{s_z}\frac{\partial}{\partial z} \tag{5.112}$$

Subsequently, (5.110) is mapped into the cylindrical, complex stretched-coordinate space:

$$j\omega\varepsilon_0 E_\rho = \frac{1}{\widehat{\rho}}\frac{\partial H_z}{\partial \phi} - \frac{\partial H_\phi}{\partial \widehat{z}} \tag{5.113a}$$

$$j\omega\varepsilon_0 E_\phi = \frac{\partial H_\rho}{\partial \widehat{z}} - \frac{\partial H_z}{\partial \widehat{\rho}} \tag{5.113b}$$

$$j\omega\varepsilon_0 E_z = \frac{1}{\widehat{\rho}}\frac{\partial(\widehat{\rho} H_\phi)}{\partial \widehat{\rho}} - \frac{1}{\widehat{\rho}}\frac{\partial H_\rho}{\partial \phi} \tag{5.113c}$$

This is repeated for Faraday's law.

To demonstrate the perfectly matched condition of the wave in the cylindrical-coordinate space, we assume a TM-polarized cylindrical wave incident on the PML. For this polarization, Maxwell's equations in the stretched-coordinate space reduce to

$$j\omega\mu_0 H_\rho = -\frac{1}{\hat{\rho}}\frac{\partial E_z}{\partial \phi} \tag{5.114a}$$

$$j\omega\mu_0 H_\phi = \frac{\partial E_z}{\partial \hat{\rho}} \tag{5.114b}$$

$$j\omega\varepsilon_0 E_z = \frac{1}{\hat{\rho}}\frac{\partial(\hat{\rho} H_\phi)}{\partial \hat{\rho}} - \frac{1}{\hat{\rho}}\frac{\partial H_\rho}{\partial \phi} \tag{5.114c}$$

Combining these coupled first-order equations into a second-order partial differential equation with respect to E_z leads to the well-known cylindrical wave equation

$$\left[\frac{1}{\hat{\rho}}\frac{\partial}{\partial \hat{\rho}}\left(\hat{\rho}\frac{\partial}{\partial \hat{\rho}}\right) + \frac{1}{\hat{\rho}^2}\frac{\partial^2}{\partial \phi^2} + k^2\right] E_z(\hat{\rho}, \phi) = 0 \tag{5.115}$$

which has Bessel function solutions of the form

$$E_{z_n}(\hat{\rho}, \phi) = A_n B_n(k\hat{\rho}) \exp(jn\phi) \tag{5.116}$$

where B_n is a general nth-order Bessel function. In the PML ($\rho > \rho_0$), the wave is assumed to be outward traveling. Hence, the fields have the solutions

$$E_{z_n}(\hat{\rho}, \phi) = A_n H_n^{(2)}(k\hat{\rho}) \exp(jn\phi) \tag{5.117a}$$

$$H_{\rho_n}(\hat{\rho}, \phi) = -\frac{A_n}{\omega\mu_0}\frac{n}{\hat{\rho}} H_n^{(2)}(k\hat{\rho}) \exp(jn\phi) \tag{5.117b}$$

$$H_{\phi_n}(\hat{\rho}, \phi) = \frac{kA_n}{j\omega\mu_0} H_n^{(2)\prime}(k\hat{\rho}) \exp(jn\phi) \tag{5.117c}$$

where $H_n^{(2)}$ is the nth-order Hankel function of the second kind and $H_n^{(2)\prime}$ is the derivative of $H_n^{(2)}$ with respect to the argument. Thus, the transverse wave impedance relative to the radial direction is

$$Z_{\hat{\rho}_n}(\rho) = \frac{E_{z_n}}{H_{\phi_n}} = \frac{j\omega\mu_0}{k}\frac{H_n^{(2)}(k\hat{\rho})}{H_n^{(2)\prime}(k\hat{\rho})} \tag{5.118}$$

In the region $\rho < \rho_0$, the impedance is similarly derived as

$$Z_{\rho_n}(\rho) = \frac{E_{z_n}}{H_{\phi_n}} = \frac{j\omega\mu_0}{k} \frac{H_n^{(2)}(k\rho)}{H_n^{(2)\prime}(k\rho)} \tag{5.119}$$

At the radial interface $\rho = \rho_0$, we have $\hat{\rho} = \rho_0$ as derived from (5.111). As a result, $Z_{\rho_n}(\rho_0) = Z_{\hat{\rho}_n}(\rho_0)$. Due to this impedance match, the wave is purely transmitted. We note that the impedance match is independent of frequency and azimuthal order, n. This can also be demonstrated via duality for the TE-polarization case. Thus, we conclude that the two media are perfectly matched. Further, by choosing s_ρ to be complex, the wave attenuates along the radial direction.

Finally, from (5.112) and (5.113), Faraday's law is expressed in the stretched-coordinate space as:

$$j\omega\varepsilon_0 E_\rho = \frac{1}{\hat{\rho}}\frac{\partial H_z}{\partial \phi} - \frac{\partial H_\phi}{s_z \partial z} \tag{5.120a}$$

$$j\omega\varepsilon_0 E_\phi = \frac{\partial H_\rho}{s_z \partial z} - \frac{\partial H_z}{s_\rho \partial \rho} \tag{5.120b}$$

$$j\omega\varepsilon_0 E_z = \frac{1}{\hat{\rho}}\frac{\partial(\hat{\rho} H_\phi)}{s_\rho \partial \rho} - \frac{1}{\hat{\rho}}\frac{\partial H_\rho}{\partial \phi} \tag{5.120c}$$

where $\hat{\rho}$ is given in (5.111). Ampere's law is the dual of these equations.

5.10.2 UPML in Cylindrical Coordinates

Maloney et al. also derived a UPML for the cylindrical-coordinate space [53]. Their development involved using a graphical approach. From the discussion in 5.4.2, it is expected that there is a direct relationship between the stretched-coordinate formulation and that derived directly for a uniaxial medium. This was pointed out by Teixeira and Chew in [54]. To this end, multiply (5.120a) by $s_z \rho / \hat{\rho}$; (5.120b) by $s_z s_\rho$; and (5.120c) by $s_\rho \hat{\rho} / \rho$. This leads to

$$j\omega\varepsilon_0 s_z \frac{\hat{\rho}}{\rho} E_\rho = \frac{1}{\rho}\frac{\partial (s_z H_z)}{\partial \phi} - \frac{\partial}{\partial z}\left(\frac{\hat{\rho}}{\rho} H_\phi\right) \tag{5.121a}$$

$$j\omega\varepsilon_0 s_z s_\rho E_\phi = \frac{\partial\left(s_\rho H_\rho\right)}{\partial z} - \frac{\partial (s_z H_z)}{\partial \rho} \tag{5.121b}$$

$$j\omega\varepsilon_0 s_\rho \frac{\hat{\rho}}{\rho} E_z = \frac{1}{\rho}\frac{\partial}{\partial\rho}\left(\hat{\rho} H_\phi\right) - \frac{1}{\rho}\frac{\partial\left(s_\rho H_\rho\right)}{\partial\phi} \tag{5.121c}$$

Finally, the fields are normalized as

$$E'_\rho = s_\rho E_\rho, \qquad E'_\phi = \frac{\hat{\rho}}{\rho} E_\phi, \qquad E'_z = s_z E_z \tag{5.122a}$$

$$H'_\rho = s_\rho H_\rho, \qquad H'_\phi = \frac{\hat{\rho}}{\rho} H_\phi, \qquad H'_z = s_z H_z \tag{5.122b}$$

which leads to

$$j\omega\varepsilon_0 \frac{s_z \hat{\rho}}{s_\rho \rho} E'_\rho = \frac{1}{\rho}\frac{\partial H'_z}{\partial\phi} - \frac{\partial H'_\phi}{\partial z} \tag{5.123a}$$

$$j\omega\varepsilon_0 \frac{s_z s_\rho \rho}{\hat{\rho}} E'_\phi = \frac{\partial H'_\rho}{\partial z} - \frac{\partial H'_z}{\partial\rho} \tag{5.123b}$$

$$j\omega\varepsilon_0 \frac{s_\rho \hat{\rho}}{s_z \rho} E'_z = \frac{1}{\rho}\frac{\partial}{\partial\rho}\left(\rho H'_\phi\right) - \frac{1}{\rho}\frac{\partial H'_\rho}{\partial\phi} \tag{5.123c}$$

Repeating this procedure for Faraday's law, Maxwell's curl equations in the uniaxial medium are expressed as

$$j\omega\varepsilon_0 \bar{\bar{s}}_c \vec{E} = \nabla\times\vec{H} \tag{5.124a}$$

$$j\omega\mu_0 \bar{\bar{s}}_c \vec{H} = \nabla\times\vec{E} \tag{5.124b}$$

where

$$\bar{\bar{s}}_c = \begin{bmatrix} \dfrac{s_z\hat{\rho}}{s_\rho\rho} & 0 & 0 \\ 0 & \dfrac{s_z s_\rho \rho}{\hat{\rho}} & 0 \\ 0 & 0 & \dfrac{s_\rho\hat{\rho}}{s_z\rho} \end{bmatrix} \tag{5.125}$$

and $\hat{\rho}$ is defined in (5.111). This is identical to that derived by Maloney et al. using a graphical solution [53] with the added generality of s_z.

Consider the TM-polarized case. From (5.124) and (5.125) we have

$$j\omega\mu_0 \frac{\widehat{\rho}}{s_\rho \rho} H_\rho = -\frac{1}{\rho}\frac{\partial E_z}{\partial \phi} \tag{5.126a}$$

$$j\omega\mu_0 \frac{s_\rho \rho}{\widehat{\rho}} H_\phi = \frac{\partial E_z}{\partial \rho} \tag{5.126b}$$

$$j\omega\varepsilon_0 \frac{s_\rho \widehat{\rho}}{\rho} E_z = \frac{1}{\rho}\frac{\partial(\rho H_\phi)}{\partial \rho} - \frac{1}{\rho}\frac{\partial H_\rho}{\partial \phi} \tag{5.126c}$$

Again, combining terms by substituting H_ρ and H_ϕ from (5.126a) and (5.126b) into (5.126c), the scalar wave equation is derived:

$$\left[\frac{1}{\widehat{\rho}}\frac{\partial}{s_\rho \partial \rho}\left(\widehat{\rho}\frac{\partial}{s_\rho \partial \rho}\right) + \frac{1}{\widehat{\rho}^2}\frac{\partial^2}{\partial \phi^2} + k^2\right] E_z(\rho, \phi) = 0 \tag{5.127}$$

which is identical to that derived using the stretched-coordinate formulation of (5.115). Hence, E_z has the same set of solutions as (5.116). From (5.126a) and (5.126b), we have in the uniaxial region

$$E_{z_n}(\widehat{\rho}, \phi) = A_n H_n^{(2)}(k\widehat{\rho}) \exp(jn\phi) \tag{5.128a}$$

$$H_{\rho_n}(\widehat{\rho}, \phi) = -\frac{A_n}{\omega\mu_0}\frac{n}{\widehat{\rho}} s_\rho H_n^{(2)}(k\widehat{\rho}) \exp(jn\phi) \tag{5.128b}$$

$$H_{\phi_n}(\widehat{\rho}, \phi) = \frac{k A_n}{j\omega\mu_0}\frac{\widehat{\rho}}{\rho} H_n^{(2)\prime}(k\widehat{\rho}) \exp(jn\phi) \tag{5.128c}$$

Comparing this with the fields in (5.117) derived in the stretched-coordinate medium, we see that H_ρ and H_ϕ are scaled according to (5.122b), as expected. Further, the fields are still divergence free, even at the interface. This is demonstrated following the argument in Section 5.4.2.

At the $\rho = \rho_0$ boundary, E_{z_n} and H_{ϕ_n} are both continuous since $\widehat{\rho} = \rho_0$ there. This is further seen by the characteristic wave impedance

$$Z_{\rho_n\text{-UPML}}(\rho) = \frac{E_{z_n}}{H_{\phi_n}} = \frac{j\omega\mu_0}{k}\frac{\rho}{\widehat{\rho}}\frac{H_n^{(2)}(k\widehat{\rho})}{H_n^{(2)\prime}(k\widehat{\rho})} \tag{5.129}$$

Comparing (5.129) with (5.119), it is observed that $Z_{\rho_n\text{-UPML}}(\rho_0) = Z_{\rho_n}(\rho_0)$. It is interesting to note that, due to the scaling of H_ϕ introduced in (5.122), the characteristic impedance of the transmitted wave is different from that of the stretched-coordinate counterpart when $\rho > \rho_0$.

Again, this can be repeated for the case of TE polarization. Thus, a matched condition is achieved for both polarizations for all orders, n, and for all frequencies.

Based on this analysis, an additional conclusion can now be drawn. In the early developments of PML theory, it was suggested to terminate the radial boundary with a tensor analogous to the Cartesian coordinate system, but replacing s_x with s_ρ, and so forth. However, it should now be clear that this does not work. Specifically, in Cartesian coordinates, the characteristic impedance of the propagating wave is position-independent. However, in cylindrical coordinates, the characteristic impedance of the wave is radially dependent. Consequently, the tensor must also be radially dependent to be matched, as we have derived from the stretched-coordinate representation.

5.10.3 Time-Dependent UPML in Cylindrical Coordinates

The intent of this study is to develop a PML for the truncation of FDTD lattices in the cylindrical-coordinate space. Therefore, time-dependent equations for the explicit field updates need to be derived. In this section, a derivation based on the UPML formulation of Section 5.5.2 is presented. A similar set of equations can be derived based on field-splitting using the stretched-coordinate formulation of Section 5.10.1. This is left to the reader.

Initially, we assume that s_z and s_ρ are complex and spatially dependent:

$$s_z = 1 - j\frac{\sigma_z(z)}{\omega\varepsilon_0}, \qquad s_\rho = 1 - j\frac{\sigma_\rho(\rho)}{\omega\varepsilon_0} \tag{5.130}$$

where $\sigma_z(z) = 0$ for $z < z_0$ and $\sigma_\rho(\rho) = 0$ for $\rho < \rho_0$. From (5.111), we have

$$\hat{\rho} = \begin{cases} \rho & \rho < \rho_0 \\ \rho - j\dfrac{1}{\omega\varepsilon_0}\displaystyle\int_{\rho_0}^{\rho}\sigma_\rho(\rho')d\rho' & \rho > \rho_0 \end{cases} \tag{5.131}$$

Due to the frequency dependence of s_z and s_ρ, the curl equations in the uniaxial medium as defined by (5.124) and (5.125) have higher-order time derivatives. This is undesirable for an FDTD representation since it can lead to instability. Consequently, auxiliary variables are introduced to formulate a coupled set of first-order equations:

$$P_\rho = \frac{\hat{\rho}}{s_\rho \rho}E_\rho, \qquad P_\phi = \frac{s_z \rho}{\hat{\rho}}E_\phi, \qquad P_z = \frac{\hat{\rho}}{s_z \rho}E_z \tag{5.132a}$$

$$Q_\rho = \frac{\hat{\rho}}{s_\rho \rho} H_\rho, \qquad Q_\phi = \frac{s_z \rho}{\hat{\rho}} H_\phi, \qquad Q_z = \frac{\hat{\rho}}{s_z \rho} H_z \tag{5.132b}$$

Following a procedure similar to that presented in Section 5.6, we obtain the coupled set of time-dependent equations:

$$\frac{\partial}{\partial t}\left(\varepsilon_0 P_\rho\right) + \sigma_z(z) P_\rho = \frac{1}{\rho}\frac{\partial}{\partial \phi} H_z - \frac{\partial}{\partial z} H_\phi \tag{5.133a}$$

$$\frac{\partial}{\partial t}\left(\varepsilon_0 P_\phi\right) + \sigma_\rho(\rho) P_\phi = \frac{\partial}{\partial z} H_\rho - \frac{\partial}{\partial \rho} H_z \tag{5.133b}$$

$$\frac{\partial}{\partial t}\left(\varepsilon_0 P_z\right) + \sigma_\rho(\rho) P_z = \frac{1}{\rho}\frac{\partial}{\partial \rho}\left(\rho H_\phi\right) - \frac{1}{\rho}\frac{\partial}{\partial \phi} H_\rho \tag{5.133c}$$

$$\frac{\partial}{\partial t} P_\rho + \frac{\sigma_\rho(\rho)}{\varepsilon_0} P_\rho = \frac{\partial}{\partial t} E_\rho + \frac{1}{\rho \varepsilon_0}\left[\int_{\rho_0}^{\rho} \sigma_\rho(\rho') d\rho'\right] E_\rho \tag{5.133d}$$

$$\frac{\partial}{\partial t} P_\phi + \frac{1}{\rho \varepsilon_0}\left[\int_{\rho_0}^{\rho} \sigma_\rho(\rho') d\rho'\right] P_\phi = \frac{\partial}{\partial t} E_\phi + \frac{\sigma_z(z)}{\varepsilon_0} E_\phi \tag{5.133e}$$

$$\frac{\partial}{\partial t} P_z + \frac{\sigma_z(z)}{\varepsilon_0} P_z = \frac{\partial}{\partial t} E_z + \frac{1}{\rho \varepsilon_0}\left[\int_{\rho_0}^{\rho} \sigma_\rho(\rho') d\rho'\right] E_z \tag{5.133f}$$

A dual expression can be derived for Faraday's law.

Explicit FDTD update equations can be derived from (5.133a) to (5.133f). To this end, (5.133a) to (5.133c) are discretized using the usual staggered-grid scheme in the cylindrical-coordinate space [55]. The time derivatives in (5.133d) to (5.133f) are approximated using central differences, and the loss terms are averaged in time (trapezoidal time integration). This leads to difference equations similar to (5.82).

The conductivity in the PML region must be scaled to reduce discretization error. Spatial or geometric scaling as defined in Section 5.5.2 can be used. For example, assuming spatial scaling, let

$$s_\rho = \begin{cases} 1 & \rho < \rho_0 \\ 1 - j\dfrac{\sigma_{\max}}{\omega \varepsilon_0}\left(\dfrac{\rho - \rho_0}{\rho_1 - \rho_0}\right)^m & \rho_0 < \rho < \rho_1 \end{cases} \tag{5.134}$$

Then, from (5.95) we have

$$\widehat{\rho} = \begin{cases} \rho & \rho < \rho_0 \\ \rho - j\dfrac{\sigma_{\max}}{(m+1)\,\omega\varepsilon_0}\dfrac{(\rho - \rho_0)^{m+1}}{(\rho_1 - \rho_0)^m} & \rho_1 < \rho < \rho_0 \end{cases} \tag{5.135}$$

The attenuation of the outward traveling wave for the nth mode is then realized by the imaginary argument of the second-order Hankel function $H_0^{(2)}(k\widehat{\rho})$. Due to the radial dependence of the attenuation, the optimal choices for m and $\sigma_{\max}$ are a function of ρ_0.

5.10.4 PML for a Cylindrical Shell

For some geometries that are separable in a cylindrical-coordinate space, the boundaries are contained within a thin shell, as illustrated in Fig. 5.31. One example is a patch antenna on a curved surface. Here, it would be useful to implement a conformal FDTD grid boundary. This problem poses two additional challenges: terminating the ϕ-normal boundaries, and terminating the interior ρ boundary.

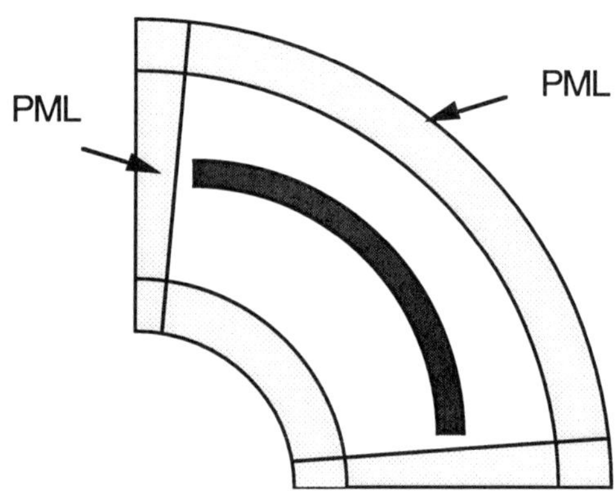

Fig. 5.31 Cylindrical-shell domain bounded with upper and lower ρ and ϕ limits terminated into finite-thickness PML.

Here, from the stretched-coordinate formulation, define the stretched ϕ coordinate

$$\widehat{\phi} \rightarrow \phi_0 + \int_{\phi_0}^{\phi} s_\phi(\phi')\,d\phi' \tag{5.136}$$

Then, it follows that

$$\frac{\partial}{\partial \widehat{\phi}} = \frac{1}{s_\phi}\frac{\partial}{\partial \phi} \tag{5.137}$$

This leads to a simple modification of (5.120).

This can be further carried over to the uniaxial formulation. To this end, we follow the same procedure outlined in (5.121) to (5.125), with the exception that we normalize the azimuthal terms as

$$E'_\phi = \frac{\widehat{\rho}}{\rho} s_\phi E_\phi , \qquad H'_\phi = \frac{\widehat{\rho}}{\rho} s_\phi H_\phi \tag{5.138}$$

This leads to the set of curl equations in (5.125) with the tensor

$$\overline{\overline{s}}_c = \begin{bmatrix} \dfrac{s_z s_\phi \widehat{\rho}}{s_\rho \rho} & 0 & 0 \\ 0 & \dfrac{s_z s_\rho \rho}{s_\phi \widehat{\rho}} & 0 \\ 0 & 0 & \dfrac{s_\rho s_\phi \widehat{\rho}}{s_z \rho} \end{bmatrix} \tag{5.139}$$

Again, a set of first-order equations can be derived similar to those proposed in the previous section. However, we observe that an additional storage variable must be introduced because of s_ϕ.

The second challenge posed by the geometry of Fig. 5.31 is the interior radial boundary. Specifically, the waves incident on this boundary are inward traveling. One must choose s_ρ such that the wave in the PML attenuates and the formulation is stable. For example, for the TM-polarized solutions (5.116), the waves are of the form

$$E_{z_n}(\widehat{\rho}, \phi) = A_n H_n^{(1)}(k\widehat{\rho}) \exp(jn\phi) \tag{5.140}$$

where $H_n^{(1)}$ is a Hankel function of the first kind. From the large-argument theory of Bessel functions, we have the relationship that

$$\lim_{x \to +\infty} H_n^{(1)}(jx) \approx \sqrt{\frac{2}{\pi}}\, j^{-(n+1)} \exp(-x) \tag{5.141}$$

Therefore, on the convex interior boundary, we need to have the imaginary part of $\widehat{\rho}$ positive and increasing as ρ decreases. At the same time, $s_\rho(\rho)$ must be negative-imaginary to obey causality [44]. Thus, choosing

$$s_\rho = 1 - j\,\frac{\sigma_\rho(\rho)}{\omega\varepsilon_0} \tag{5.142}$$

where s_ρ is increasing for $\rho < \rho_1$, and then defining

$$\hat{\rho} = \begin{cases} \rho & \rho > \rho_1 \\ \rho_1 - \int\limits_{\rho}^{\rho_1} s_\rho(\rho')d\rho' & \rho < \rho_1 \end{cases} \tag{5.143}$$

we see that the imaginary part of $\hat{\rho}$ is positive. For example, assuming a spatial scaling for s_ρ of the form

$$\sigma_\rho(\rho) = \begin{cases} \rho & \rho > \rho_1 \\ \dfrac{(\rho_1 - \rho)^m}{(\rho_1 - \rho_2)^m} & \rho_2 < \rho < \rho_1 \end{cases} \tag{5.144}$$

where ρ_2 is the inner PEC boundary backing the PML and ρ_1 is the boundary of the air / PML interface. Then, from (5.143), we have

$$\hat{\rho} = \begin{cases} \rho & \rho > \rho_1 \\ \rho + j\,\dfrac{\sigma_{\max}}{\omega\varepsilon_0}\,\dfrac{1}{(m+1)}\,\dfrac{(\rho_1 - \rho)^{m+1}}{(\rho_1 - \rho_2)^m} & \rho_2 < \rho < \rho_1 \end{cases} \tag{5.145}$$

Therefore, in the region $\rho_2 < \rho < \rho_1$, we have $\mathrm{Im}(s_\rho) < 0$ and $\mathrm{Im}(\tilde{\rho}) > 0$, and the formulation is stable. The stability and effectiveness of the PML for the radial convex boundary was demonstrated for the $m = 0$ case by Maloney and Kessler [56], who used a cylindrical FDTD code to simulate a radial waveguide excited with high-order modes. Note that choosing optimal values for $\sigma_{\max}$ and m is quite dependent on ρ_1, and numerical experiments are needed.

5.10.5 PML in Spherical Coordinates

Using a stretched-coordinate mapping, a PML can also be posed for a spherical-coordinate space [34, 52] following arguments similar to those presented in the previous section. In the spherical-coordinate space, Ampere's law is expressed as

$$j\omega\varepsilon_0 E_r = \frac{1}{r\sin\theta}\left[\frac{\partial}{\partial\theta}\left(H_\phi \sin\theta\right) - \frac{\partial}{\partial\phi}H_\theta\right] \tag{5.146a}$$

$$j\omega\varepsilon_0 E_\theta = \frac{1}{r}\left[\frac{1}{\sin\theta}\frac{\partial}{\partial\phi}H_r - \frac{\partial}{\partial r}\left(r H_\phi\right)\right] \tag{5.146b}$$

$$j\omega\varepsilon_0 E_\phi = \frac{1}{r}\left[\frac{\partial}{\partial r}(r H_\theta) - \frac{\partial}{\partial\theta}H_r\right] \tag{5.146c}$$

We assume that the PML is only placed exterior to the radial boundary $r = r_0$. In this region, the radial stretched coordinate is defined as

$$\hat{r} = r_0 + \int_{r_0}^{r} s_r(r')\,dr' \tag{5.147a}$$

which leads to the relation

$$\frac{\partial}{\partial\hat{r}} = \frac{1}{s_r}\frac{\partial}{\partial r} \tag{5.147b}$$

Subsequently, in the stretched-coordinate system, (5.146) becomes

$$j\omega\varepsilon_0 E_r = \frac{1}{\hat{r}\sin\theta}\left[\frac{\partial}{\partial\theta}\left(H_\phi\sin\theta\right) - \frac{\partial}{\partial\phi}H_\theta\right] \tag{5.148a}$$

$$j\omega\varepsilon_0 E_\theta = \frac{1}{\hat{r}}\left[\frac{1}{\sin\theta}\frac{\partial}{\partial\phi}H_r - \frac{\partial}{s_r\,\partial r}\left(\hat{r} H_\phi\right)\right] \tag{5.148b}$$

$$j\omega\varepsilon_0 E_\phi = \frac{1}{\hat{r}}\left[\frac{\partial}{s_r\,\partial r}(\hat{r} H_\theta) - \frac{\partial}{\partial\theta}H_r\right] \tag{5.148c}$$

From the stretched-coordinate form, a uniaxial medium is derived [54]. To this end, multiply (5.148a) by $(\hat{r}/r)^2$; (5.148b) by $s_r(\hat{r}/r)$; and (5.148c) by $s_r(\hat{r}/r)$, leading to

$$j\omega\varepsilon_0\left(\frac{\hat{r}}{r}\right)^2 E_r = \frac{1}{r\sin\theta}\left[\frac{\partial}{\partial\theta}\left(\frac{\hat{r}}{r}H_\phi\sin\theta\right) - \frac{\partial}{\partial\phi}\left(\frac{\hat{r}}{r}H_\theta\right)\right] \tag{5.149a}$$

$$j\omega\varepsilon_0\, s_r\frac{\hat{r}}{r}E_\theta = \frac{1}{r}\left[\frac{1}{\sin\theta}\frac{\partial}{\partial\phi}\left(s_r H_r\right) - \frac{\partial}{\partial r}\left(\hat{r} H_\phi\right)\right] \tag{5.149b}$$

$$j\omega\varepsilon_0\, s_r\frac{\hat{r}}{r}E_\phi = \frac{1}{r}\left[\frac{\partial}{\partial r}(\hat{r} H_\theta) - \frac{\partial}{\partial\theta}\left(s_r H_r\right)\right] \tag{5.149c}$$

The fields are then normalized as [54]

$$E'_r = s_r E_r, \qquad E'_\theta = \frac{\hat{r}}{r} E_\theta, \qquad E'_\phi = \frac{\hat{r}}{r} E_\phi \tag{5.150a}$$

$$H'_r = s_r H_r, \qquad H'_\theta = \frac{\hat{r}}{r} H_\theta, \qquad H'_\phi = \frac{\hat{r}}{r} H_\phi \tag{5.150b}$$

This procedure is repeated for Faraday's law. Finally, Maxwell's curl equations in the uniaxial medium defined for the spherical-coordinate space are expressed as

$$j\omega\varepsilon_0 \bar{\bar{s}}_s \vec{E} = \nabla \times \vec{H} \tag{5.151a}$$

$$j\omega\mu_0 \bar{\bar{s}}_s \vec{H} = \nabla \times \vec{E} \tag{5.151b}$$

where

$$\bar{\bar{s}}_s = \begin{bmatrix} \frac{1}{s_r}\left(\frac{\hat{r}}{r}\right)^2 & 0 & 0 \\ 0 & s_r & 0 \\ 0 & 0 & s_r \end{bmatrix} \tag{5.152}$$

The wave equation derived for a uniaxial medium with this tensor is identical to that derived via the stretched-coordinate space. Thus, the waves in the stretched-coordinate space and the uniaxial medium satisfy the same dispersion relationship. Further, the waves also have the same characteristic impedance.

It can be shown that the characteristic wave impedance is perfectly matched at the $r = r_0$ boundary for all spherical modes and for all frequencies, given $\hat{r}$ as defined in (5.147a). Thus, the tensor in (5.152) combined with (5.147) constitutes a PML medium for the $r = r_0$ boundary.

The time-dependent form of Maxwell's curl equations is derived directly from (5.151) and (5.152). To this end, first define

$$s_r = 1 - j\frac{\sigma_r(r)}{\omega\varepsilon_0} \tag{5.153a}$$

$$\hat{r} = \begin{cases} r & r < r_0 \\ r - j\dfrac{1}{\omega\varepsilon_0}\displaystyle\int_{r_0}^{r} \sigma_r(r')\,dr' & r > r_0 \end{cases} \tag{5.153b}$$

Then, we introduce the auxiliary variables

$$P_r = \frac{1}{s_r}\left(\frac{\hat{r}}{r}\right)E_r, \qquad Q_r = \frac{1}{s_r}\left(\frac{\hat{r}}{r}\right)H_r \tag{5.154}$$

Again, following the procedure outlined in Section 5.10.3, the time-dependent Maxwell's curl equations lead to the following coupled set of equations:

$$\frac{\partial}{\partial t}\left(\varepsilon_0 P_r\right) + \frac{1}{r}\int_{r_0}^{r} \sigma_r(r')\,dr' = \frac{1}{r\sin\theta}\left[\frac{\partial}{\partial\theta}\left(H_\phi \sin\theta\right) - \frac{\partial}{\partial\phi}H_\theta\right] \tag{5.155a}$$

$$\frac{\partial}{\partial t}\left(\varepsilon_0 E_\theta\right) + \sigma_r(r)E_\theta = \frac{1}{r}\left[\frac{1}{\sin\theta}\frac{\partial}{\partial\phi}H_r - \frac{\partial}{\partial r}\left(rH_\phi\right)\right] \tag{5.155b}$$

$$\frac{\partial}{\partial t}\left(\varepsilon_0 E_\phi\right) + \sigma_r(r)E_\phi = \frac{1}{r}\left[\frac{\partial}{\partial r}(rH_\theta) - \frac{\partial}{\partial\theta}H_r\right] \tag{5.155c}$$

$$\frac{\partial}{\partial t}P_r + \frac{\sigma_r(r)}{\varepsilon_0}P_r = \frac{\partial}{\partial t}E_r + \left[\frac{1}{\varepsilon_0 r}\int_{r_0}^{r}\sigma_r(r')\,dr'\right]E_r \tag{5.155d}$$

A dual expression can be derived for Faraday's law.

Explicit FDTD update equations are derived from (5.155a) to (5.155c) and the dual field equations. To this end, the fields are discretized using the usual staggered-grid scheme in the spherical-coordinate space [57]. The time derivatives in (5.155d) are also approximated using central differences, and the loss terms are averaged in time (trapezoidal time-integration). This leads to second-order expressions similar to (5.82).

5.11 SUMMARY AND CONCLUSIONS

This chapter provided a comprehensive review of the theoretical foundation and numerical implementation of PML ABCs in the FDTD method. Detailed parametric studies of PML performance under a variety of conditions were reported. The review included:

- Berenger's original split-field PML concept;
- The stretched-coordinate PML formulation;
- Theoretical development of the uniaxial anisotropic PML;
- Theoretical performance of the PML;
- Efficient implementation of PML in FDTD;

- Numerical experiments to provide an understanding of the effect of the PML parameters and scaling functions on PML effectiveness;
- PML termination for conductive media;
- PML termination for dispersive media;
- PML in cylindrical and spherical coordinates.

We conclude that the PML concept represents one of the most significant advances in the historical development of the FDTD method. PML technology has expanded the potential computational dynamic range of FDTD models by orders of magnitude relative to that provided by previous ABCs. Currently, there exists sufficient understanding of the theoretical basis, numerical implementation, and parameter optimization of PML absorbers to permit their reliable and efficient usage in the context of most (if not all) linear FDTD Maxwell's equations models. An interesting future challenge is the development of PML for terminating nonlinear media, which may be of increased interest in the FDTD community as researchers apply full-wave Maxwell's equations models to study the dynamics of passive and active micron-scale optical structures.

REFERENCES

[1] Taflove, A., *Computational Electrodynamics: The Finite-Difference Time-Domain Method*, Norwood, MA: Artech House, 1995.

[2] Holland, R. and J. Williams, "Total-field versus scattered-field finite-difference," *IEEE Trans. Nuclear Science,* Vol. 30, 1983, pp. 4583–4587.

[3] Berenger, J. P., "A perfectly matched layer for the absorption of electromagnetic waves," *J. Computational Physics,* Vol. 114, 1994, pp. 185–200.

[4] Berenger, J. P., "Perfectly matched layer for the FDTD solution of wave-structure interaction problems," *IEEE Trans. Antennas and Propagation*, Vol. 51, 1996, pp. 110–117.

[5] Berenger, J. P., "A perfectly matched layer for free-space simulations in finite-difference computer codes," *Annales des Telecommunications,* Vol. 51, 1996, pp. 39–46.

[6] Berenger, J. P., "Three-dimensional perfectly matched layer for the absorption of electromagnetic waves," *J. Computational Physics*, Vol. 127, 1996, pp. 363–379.

[7] Berenger, J. P., "Improved PML for the FDTD solution of wave-structure interaction problems," *IEEE Trans. Antennas and Propagation*, Vol. 45, 1997, pp. 466–473.

[8] Berenger, J. P., "Numerical reflection of evanescent waves from perfectly matched layers," *Proc. IEEE Antennas and Propagat. Soc. Int. Symposium*, Montreal, Canada, Vol. 3, pp. 1888–1891, 1997.

[9] Katz, D. S., E. T. Thiele, and A. Taflove, "Validation and extension to three-dimensions of the Berenger PML absorbing boundary condition for FDTD meshes," *IEEE Microwave and Guided Wave Letters,* Vol. 4, 1994, pp. 268–270.

[10] Chew, W. C., and W. H. Weedon, "A 3D perfectly matched medium from modified Maxwell's equations with stretched coordinates," *IEEE Microwave and Guided Wave Letters,* Vol. 7, 1994, pp. 599–604.

[11] Mittra, R., and U. Pekel, "A new look at the perfectly matched layer (PML) concept for the reflectionless absorption of electromagnetic waves," *IEEE Microwave and Guided Wave Letters,* Vol. 5, 1995, pp. 84–87.

[12] Andrew, W. V., C. A. Balanis, and P. A. Tirkas, "A comparison of the Berenger perfectly matched layer and the Lindman higher-order ABC's for the FDTD method," *IEEE Microwave and Guided Wave Letters,* Vol. 5, 1995, pp. 192–194.

[13] Cai, M., and H. Zhou, "Proof of perfectly matched layer conditions in three dimensions," *Electronics Letters,* Vol. 31, 1995, pp. 1675–1676.

[14] Nehrbass, J. W., J.-F. Lee, and R. Lee, "Stability analysis for perfectly matched layered absorbers," *Electromagnetics,* Vol. 16, 1996, pp. 385–389.

[15] Wu, Z., and J. Fang, "High-performance PML algorithms," *IEEE Microwave and Guided Wave Letters,* Vol. 6, 1996, pp. 335–337.

[16] Wittwer, D. C., and R. W. Ziolkowski, "How to design the imperfect Berenger PML," *Electromagnetics,* Vol. 16, 1996, pp. 465–468.

[17] Veihl, J. C., and R. Mittra, "An efficient implementation of Berenger's perfectly matched layer (PML) for finite-difference time-domain mesh truncation," *IEEE Microwave and Guided Wave Letters,* Vol. 6, 1996, pp. 94–96.

[18] De Moerloose, J., and M. A. Stuchley, "Reflection analysis of PML ABC's for low-frequency applications," *IEEE Microwave and Guided Wave Letters,* Vol. 6, 1996, pp. 177–179.

[19] Reuter, C. E., R. M. Joseph, E. T. Thiele, D. S. Katz, and A. Taflove, "Ultrawideband absorbing boundary condition for termination of waveguiding structures in FDTD simulations," *IEEE Microwave and Guided Wave Letters,* Vol. 4, 1994, pp. 344–345.

[20] Chen, J. C., and K. Li, "Quartic perfectly matched layers for dielectric waveguides and gratings," *Microwave and Optical Technology Letters,* Vol. 10, 1995, pp. 319–323.

[21] Wu, Z., and J. Fang, "Numerical implementation and performance of perfectly matched layer boundary condition for waveguide structures," *IEEE Trans. Microwave Theory and Techniques,* Vol. 43, 1995, pp. 2676–2683.

[22] Fang, J., and Z. Wu, "Generalized perfectly matched layer—an extension of Berenger's perfectly matched layer boundary condition," *IEEE Microwave and Guided Wave Letters,* Vol. 5, 1995, pp. 451–453.

[23] Bernabeu Verdu, J., R. Gillard, K. Moustadir, and J. Citerne, "An extension of the PML technique to the FDTD analysis of multilayer planar circuits and antennas," *Microwave and Optical Technology Letters,* Vol. 10, 1995, pp. 323–327.

[24] Pasik, M. F., G. Aguirre, and A. C. Cangellaris, "Application of the PML absorbing boundary condition to the analysis of patch antennas," *Electromagnetics,* Vol. 16, 1996, pp. 435–449.

[25] Gedney, S. D., "An anisotropic PML absorbing media for FDTD simulation of fields in lossy dispersive media," *Electromagnetics,* Vol. 16, 1996, pp. 399–415.

[26] Lau, Y. C., M. S. Leong, and P. S. Kooi, "Extension of Berenger's PML boundary condition in matching lossy medium and evanescent waves," *Electronics Letters,* Vol. 32, 1996, pp. 974–976.

[27] Fang, J., and Z. Wu, "Generalized perfectly matched layer for the absorption of propagating and evanescent waves in lossless and lossy media," *IEEE Trans. Microwave Theory and Techniques,* Vol. 14, 1996, pp. 2216–2222.

[28] Liu, Q. H., "An FDTD algorithm with perfectly matched layers for conductive media," *Microwave and Optical Technology Letters,* Vol. 14, 1997, pp. 134–137.

[29] Roden, J. A., S. D. Gedney, M. P. Kesler, J. G. Maloney, and P. H. Harms, "Time-domain analysis of periodic structures at oblique incidence: Orthogonal and nonorthogonal FDTD implementations," *IEEE Trans. Microwave Theory and Techniques*, vol. 46, 1998, pp. 420-427.

[30] Fang, J., and Z. Wu, "Closed-form expression of numerical reflection coefficient at PML interfaces and optimization of PML performance," *IEEE Microwave and Guided Wave Letters,* Vol. 6, 1996, pp. 332–334.

[31] Sullivan, D. M., "A simplified PML for use with the FDTD method," *IEEE Microwave and Guided Wave Letters,* Vol. 6, 1996, pp. 97–99.

[32] Sullivan, D. M., "An unsplit step three-dimensional PML for use with the FDTD," *IEEE Microwave and Guided Wave Letters,* Vol. 7, 1997, pp. 184–186.

[33] Rappaport, C. M., "Perfectly matched absorbing boundary conditions based on anisotropic lossy mapping of space," *IEEE Microwave and Guided Wave Letters,* Vol. 5, 1995, pp. 90–92.

[34] Teixeira, F. L., and W. C. Chew, "PML-FDTD in cylindrical and spherical coordinates," *IEEE Microwave and Guided Wave Letters,* Vol. 7, 1997, pp. 285–287.

[35] Sacks, Z. S., D. M. Kingsland, R. Lee, and J. F. Lee, "A perfectly matched anisotropic absorber for use as an absorbing boundary condition," *IEEE Trans. Antennas and Propagation*, Vol. 43, 1995, pp. 1460–1463.

[36] Gedney, S. D., "An anisotropic perfectly matched layer absorbing media for the truncation of FDTD lattices," Technical Report EMG-95-006, Department of Electrical Engineering, University of Kentucky, March 1995.

[37] Gedney, S. D., "An anisotropic perfectly matched layer absorbing media for the truncation of FDTD lattices," *IEEE Trans. Antennas and Propagation*, Vol. 44, 1996, pp. 1630–1639.

[38] Ziolkowski, R. W., "Time-derivative Lorentz materials and their utilization as electromagnetic absorbers," *Physical Review E,* Vol. 55, 1997, pp. 7696–7703, pt. B.

[39] Berenger, J. P., "Calcul de la diffraction a l'aide d'une méthode aux différence finies," Actes du colloque CEM, CNFRS-URSI, Trégastel France, June 1983.

[40] Teixeira, F. L., and W. C. Chew, "Perfectly matched layer in cylindrical coordinates," *Proc. IEEE Antennas and Propagation Society International Symposium*, Montreal, Canada, Vol. 3, pp. 1908–1911, 1997.

[41] Chew, W. C., and J. M. Jin, "Perfectly matched layers in the discretized space: An analysis and optimization," *Electromagnetics,* Vol. 16, 1996, pp. 325–340.

[42] Gedney, S., "An anisotropic perfectly matched layer absorbing media for the truncation of FDTD lattices," Tech. Report EMG-95-006, University of Kentucky, March, 1995.

[43] He, L., "FDTD—Advances in sub-sampling methods, UPML, and higher-order boundary conditions," M. S. thesis, University of Kentucky, Lexington, KY, May 1997.

[44] Teixer, F. L., and W. C. Chew, "Causality and dynamical stability of perfectly matched layers for FDTD simulation," Technical Report CCEM-27-97, Center for Computational Electromagnetics, University of Illinois, Urbana, IL, Sept., 1997.

[45] Kuzuoglu, M., and R. Mittra, "Frequency dependence of the constitutive parameters of causal perfectly matched anisotropic absorbers," *IEEE Microwave and Guided Wave Letters,* Vol. 6, 1996, pp. 447–449.

[46] Veremey, V., and R. Mittra, "A technique for fast calculations of capacitance matrices of interconnect structures," IEEE 6th Topical Meeting on Electronic Performance of Electronic Packaging, San Jose, CA, Vol. 1, pp. 244–247, 1997.

[47] Abarbanel, S., and D. Gottlieb, "A mathematical analysis of the PML method," *J. Computational Physics,* Vol. 134, 1997, pp. 357–363.

[48] Sheen, D., S. Ali, M. Abouzahra, and J. A. Kong, "Application of the three-dimensional finite-difference time-domain method to the analysis of planar microstrip circuits," *IEEE Trans. Microwave Theory and Techniques,* Vol. 38, 1990, pp. 849–856.

[49] Tulintseff, A., "The finite-difference time-domain method and computer program description applied to multi-layered microstrip antenna and circuit configurations," Technical Report AMT: 336.5-92-041, Jet Propulsion Laboratory, Pasadena, CA, May 1992.

[50] Joseph, R. M., S. C. Hagness, and A. Taflove, "Direct time integration of Maxwell's equations in linear dispersive media with absorption for scattering and propagation of femtosecond electromagnetic pulses," *Optics Letters,* Vol. 16, 1991, pp. 1412–1414.

[51] Luebbers, R. J., and F. Hunsberger, "FDTD for Nth-order dispersive media," *IEEE Trans. Antennas and Propagation,* Vol. 40, 1992, pp. 1297–1301.

[52] Chew, W. C., J. M. Jin, and E. Michielssen, "Complex coordinate stretching as a generalized absorbing boundary condition," *Microwave and Optical Technology Letters,* Vol. 15, 1997, pp. 363–369.

[53] Maloney, J., M. Kesler, and G. Smith, "Generalization of PML to cylindrical geometries," *Proc. 13th Annual Review of Progress in Applied Computational Electromagnetics*, Monterey, CA, March 17–21, 1997, Vol. 2, pp. 900–908.

[54] Teixeira, F. L., and W. C. Chew, "Conformal perfectly matched absorber for electromagnetic media," Technical Report CCEM-29-97, Center for Computational Electromagnetics, University of Illinois, Urbana, IL, Sept. 1997.

[55] Fusco, M., "FDTD algorithm in curvilinear coordinates," *IEEE Trans. Antennas and Propagation,* Vol. 38, 1990, pp. 76–89.

[56] Maloney, J., and M. Kesler, personal communication, 1997.

[57] Holland, R., "THREDS: A finite-difference time-domain EMP code in 3D spherical coordinates," *IEEE Trans. Nuclear Science,* Vol. 30, 1983, pp. 4592-4595.

Chapter 6

Analysis of Periodic Structures

James G. Maloney and Morris P. Kesler

6.1 INTRODUCTION

Many structures of electromagnetic interest possess a periodicity in one or more dimensions. For example, a frequency selective surface (FSS) is commonly used in a radome to control the energy that reaches an antenna. A typical FSS consists of one or more layers of material, each of which is formed from an element periodically replicated in two dimensions. Another periodic structure that has received considerable attention in recent years is the photonic bandgap (PBG) structure. A PBG is a type of periodic dielectric structure that has frequency regions in which electromagnetic propagation is forbidden. A third type of structure that can be considered periodic is an antenna array. If large enough, many of the important parameters of an array can be analyzed by assuming that the structure is periodic.

The FDTD technique can be applied to the analysis of periodic structures such as the ones mentioned above. Such structures often have fine details at the element level that must be accurately modeled in order to predict the correct electromagnetic behavior. Coupled with the fact that the overall structure may consist of many replicas of the basic element, this level of detail leads to computational problems that are unmanageable. One way to alleviate the computational burden is to model only the individual element and use boundary conditions to simulate the effect of periodic replication.

As an example of a periodic structure, consider the two-dimensional, electromagnetic screen shown in Fig. 6.1. The screen consists of infinitely long, parallel,

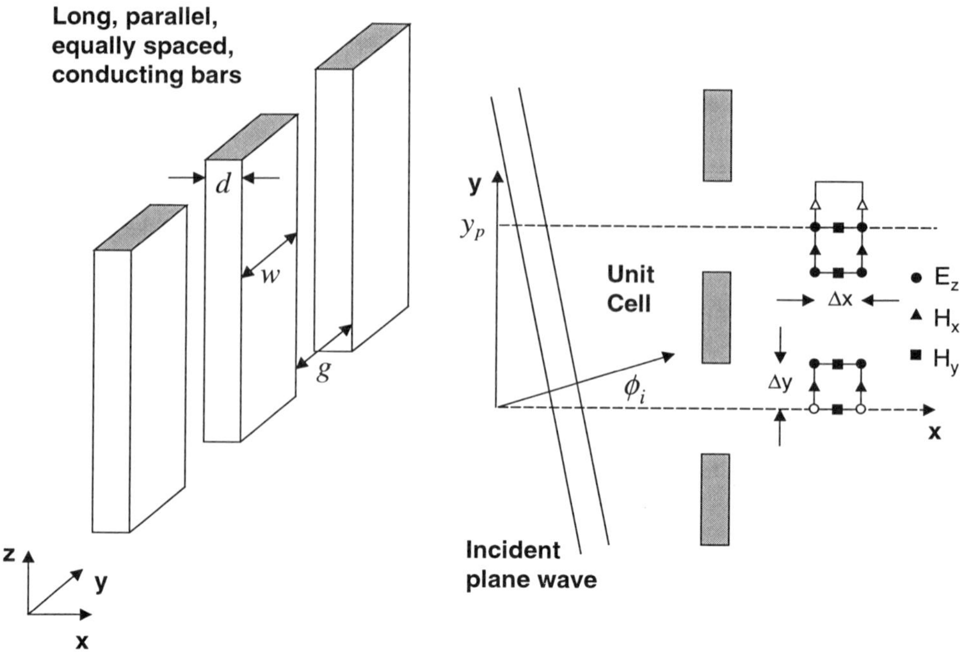

Fig. 6.1 Two-dimensional electromagnetic screen (left) and a top view of the structure showing the field components at the edges of the unit cell.

conducting bars of width w; thickness d; and separated by a gap g. The illumination is an electromagnetic plane wave with the electric field parallel to the bars (z polarized), and the direction of propagation is at angle ϕ_i with respect to the x axis as shown in the figure. This is a two-dimensional electromagnetic problem; only the TE fields (E_z, H_x, and H_y) are nonzero. Because of the periodicity inherent in the geometry, only the fields in the "unit cell" are unique and need to be determined. Dashed lines in the figure denote the edges of the unit cell.

Now, consider discretizing the solution space within the unit cell using a traditional Yee lattice. Field components that can be updated using the fields in the unit cell are denoted by solid black symbols, while those that cannot be updated are denoted by white symbols. The field components that cannot be updated can instead be determined using periodic boundary conditions that relate the fields on one side of the unit cell to those on the other side. Of course, this relationship depends on the angle at which the incident field impinges on the structure. For the geometry shown, the relationship is expressed in the frequency domain by

$$H_x\left(x, y = y_p + \Delta y/2\right) = H_x\left(x, y = \Delta y/2\right)\exp\left(-jk_y y_p\right) \tag{6.1a}$$

$$E_z(x, y = 0) = E_z\left(x, y = y_p\right)\exp\left(+jk_y y_p\right) \tag{6.1b}$$

where $k_y = k_0 \sin\phi_i$ and k_0 is the free-space wavenumber. FDTD requires the boundary conditions to be expressed in the time domain. In the time domain, (6.1) becomes

$$H_x\left(x, y = y_p + \Delta y/2, t\right) = H_x\left(x, y = \Delta y/2, t - y_p \sin\phi_i\right) \tag{6.2a}$$

$$E_z(x, y = 0, t) = E_z\left(x, y = y_p, t + y_p \sin\phi_i\right) \tag{6.2b}$$

Equation (6.2) shows that the periodic boundary condition involves using both previous as well as future field components. Specifically, the H_x values at $y = y_p + \Delta y/2$ are equal to previous values of H_x at $y = \Delta y/2$, while the E_z values at $y = 0$ are equal to future values of E_z at $y = y_p$. Previous values can simply be saved using additional computational storage. However, there is no simple solution for this need for future values of E_z.

This chapter discusses several of the techniques that have been proposed for dealing with this problem. We have divided the techniques into two classes, as indicated in Fig. 6.2. The techniques listed on the right side of this figure all use a field transformation to eliminate the need for time-advanced data. The transformed field equations are then discretized and solved using FDTD techniques. Each technique uses a different approach in handling the extra terms introduced by the field transformation. Numerical stability is an important consideration in each of these approaches. All of the other techniques are listed on the left under "Direct Field Methods." These methods work directly with Maxwell's equations with periodic boundary conditions.

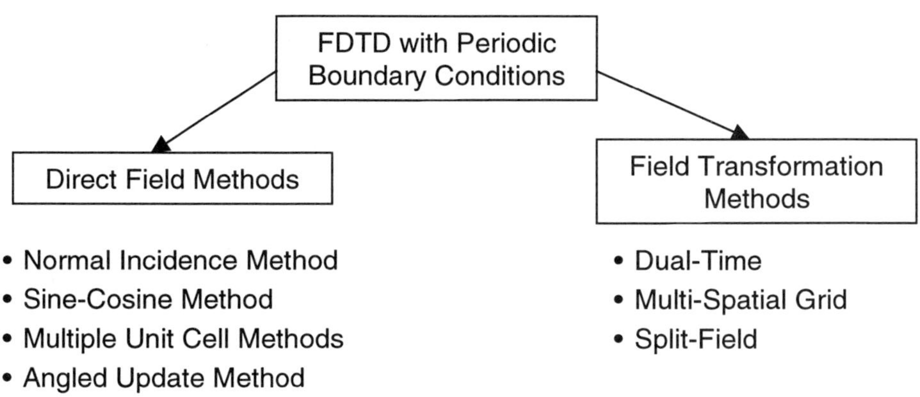

Fig. 6.2 The techniques used to perform FDTD with periodic boundary conditions are divided into two classes. Each technique is discussed in the chapter.

Another type of periodic problem is one in which the periodicity is in the direction of propagation. One example of this type is a Bragg filter. Several authors have investigated this class of problem using an FDTD approach [1–3]. In this chapter, we will focus on geometries that are periodic in the transverse directions and will not consider longitudinal periodicity.

The chapter is organized as follows. Section 6.2 contains a brief overview of scattering from periodic geometries. Section 6.3 contains a summary and brief description of the direct-field methods listed on the left side of Fig. 6.2, while Section 6.4 focuses on the field-transformation approaches. Section 6.5 looks at applications of the split-field technique (one of the field-transformation approaches) on three types of periodic structures. A summary is contained in Section 6.6.

6.2 REVIEW OF SCATTERING FROM PERIODIC STRUCTURES

Before we delve into the intricacies of applying the FDTD method to the analysis of scattering from periodic structures, a brief discussion of this topic in general is appropriate. For simplicity, consider the case of a planar scatterer having one dimension of periodicity, as shown in Fig. 6.3. The geometry is periodic in the y direction with period y_p. The illumination (shown bold and denoted I in the figure) is an electromagnetic plane wave at angle ϕ_i. The traditional far-field transmitted and reflected waves are at angles $\phi_t = \phi_i$ and $\phi_r = 180° - \phi_i$ (shown bold and denoted T_0 and R_0 in the figure).

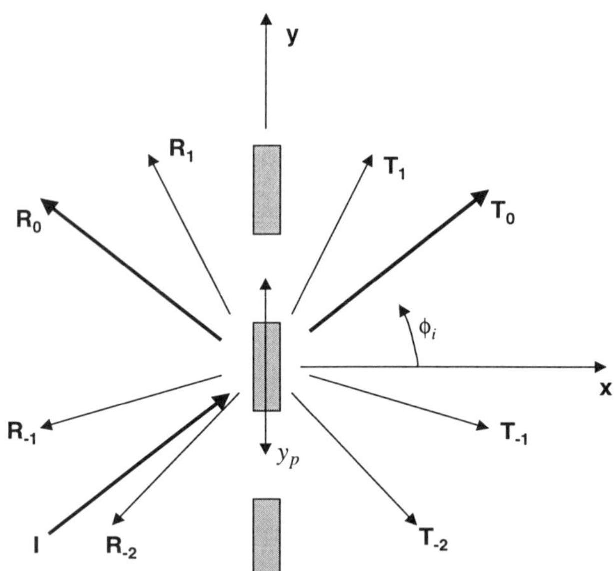

Fig. 6.3 Schematic representation of scattering from periodic media.

The presence of periodicity in the scatterer can lead to the appearance of additional angles of far-field transmission and reflection. These are often referred to as Floquet modes or grating lobes. The set of angles, ϕ_t^n, are the roots of the following equation:

$$\sin\phi_t^n = \left(\frac{c}{y_p f}\right) n + \sin\phi_i \tag{6.3}$$

The angles of reflection, ϕ_r^n, are simply related to the angles of transmission by

$$\phi_r^n = 180^\circ - \phi_t^n \tag{6.4}$$

To illustrate the appearance and behavior of these additional angles of transmission and reflection, consider the case of $\phi_i = 30^\circ$. The angles of transmission, ϕ_t^n, are shown as a function of the normalized frequency, $y_p f/c$, in Fig. 6.4. As the frequency is increased, new solutions appear. The angles of these new solutions begin at ±90°.

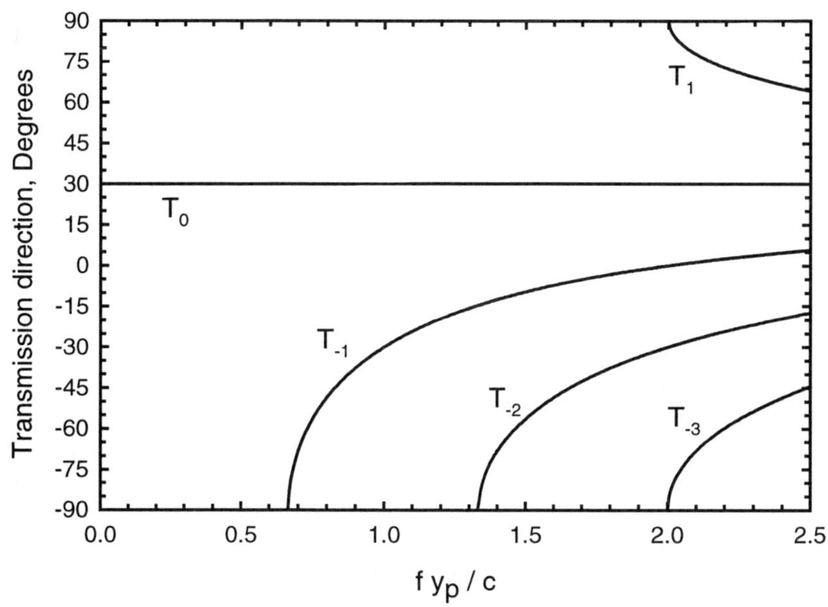

Fig. 6.4 Transmission directions, ϕ_t^n, of the periodic structure as a function of frequency for the case of $\phi_i = 30^\circ$.

Note that the number of additional angles is dependent on the frequency and the angle of incidence. For positive ϕ_i, the following equations give the turn-on frequency for the first few orders:

$$f_{\text{turn-on}} = 0 \qquad n = 0 \tag{6.5a}$$

$$f_{\text{turn-on}} = \frac{c}{y_p(1 \mp \sin\phi_i)} \qquad n = \pm 1 \tag{6.5b}$$

$$f_{\text{turn-on}} = \frac{2c}{y_p(1 \mp \sin\phi_i)} \qquad n = \pm 2 \tag{6.5c}$$

We see that for low frequencies $f < c/[y_p(1+\sin\phi_i)]$, only one root exists, namely $\phi_t^0 = \phi_i$. This is in the traditional transmitted direction.

The fields on a planar surface in front of and behind the scatterer can be used to determine the amplitude and phase of the signals reflected and transmitted in each order, n. Because of the periodic nature of the geometry, only the fields in the unit cell need be considered in performing this calculation. In the frequency domain, the far field for order n is given by

$$E_{\text{ff}}^n(\omega) = \frac{1}{y_p}\int_0^{y_p} E(\omega, y)\exp\left[j\left(2\pi n y/y_p + k_0 y\sin\phi_i\right)\right]dy \tag{6.6}$$

Thus, the far field for a given ray is found by summing up the fields across the unit cell with the appropriate phase taper. Although (6.6) is written in terms of the electric field in the frequency domain, this sum can also be performed using the time-domain fields directly. Transforming (6.6) to the time-domain and discretizing yields

$$E_{\text{ff}}^n(t) = \frac{1}{N_y}\sum_{i=1}^{N_y} E\left(t + \frac{i\Delta y\sin\phi_i}{c},\ i\Delta y\right)\exp\left(j2\pi n i/N_y\right) \tag{6.7}$$

where N_y is the number of cells across the unit cell.

Note that in general this sum produces a complex result. This occurs because the time-domain result is not a physically measurable quantity, since the angle at which ray n propagates is frequency dependent. When the sum is transformed into the frequency domain, physical significance can be assigned to the individual frequency components.

We also note that in (6.7), a time gradient $i\Delta y\sin\phi_i/c$ is applied to the field across the unit cell before the sum is performed. In other words, the sum is performed using the field at different times across the grid. This time-shifted field is exactly equal to a variable, P, that is introduced in Section 6.4 in the discussion of the field transformation methods. In terms of P, the sum is simply given by

$$E_{\text{ff}}^n(t) = \frac{1}{N_y}\sum_{i=1}^{N_y} P(t,\ i\Delta y)\exp\left(j2\pi n i/N_y\right) \tag{6.8}$$

For illustration, consider the two-dimensional lossy electromagnetic screen shown in Fig. 6.1 where $w/g = 1$, $w/d = 4$, $1/\sigma d = 63\ \Omega$, and $\phi_i = 30°$. Fig. 6.5 shows the relative amplitudes of the various transmitted waves calculated using a frequency-domain, mode-matching technique [4]. Comparisons to these data are made throughout the chapter to illustrate the capabilities of the methods to be discussed.

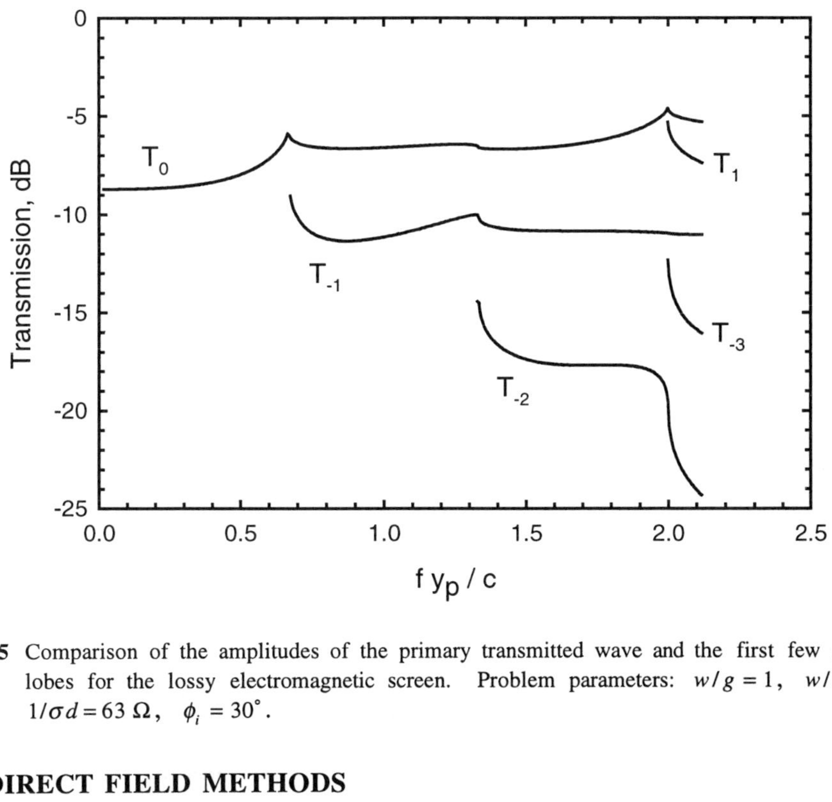

Fig. 6.5 Comparison of the amplitudes of the primary transmitted wave and the first few grating lobes for the lossy electromagnetic screen. Problem parameters: $w/g = 1$, $w/d = 4$, $1/\sigma d = 63\ \Omega$, $\phi_i = 30°$.

6.3 DIRECT FIELD METHODS

There have been several direct field methods proposed to implement periodic boundary conditions in FDTD. This section contains a description of these techniques.

6.3.1 Normal Incidence Case

The difficulty associated with the periodic boundary condition in the time domain does not exist when the excitation is normally incident. Consider the two-dimensional configuration shown in Fig. 6.6. The wavefront has no spatial tilt, and therefore, no time delay or advance is required in the periodic boundary condition. Equation (6.2) for $\phi_i = 0$ becomes

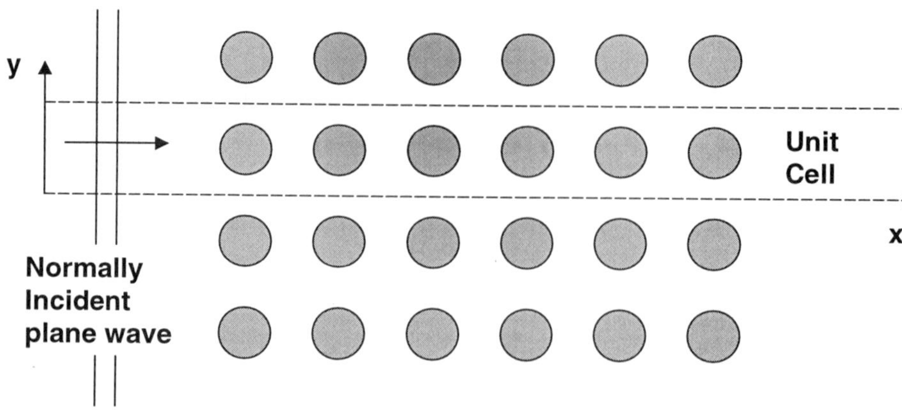

Fig. 6.6 Top view of a plane wave incident on a periodic structure.

$$H_x\left(x, y = y_p + \Delta y/2, t\right) = H_x\left(x, y = \Delta y/2, t\right) \tag{6.9a}$$

$$E_z(x, y = 0, t) = E_z\left(x, y = y_p, t\right) \tag{6.9b}$$

Because this method is just the usual FDTD technique and the boundary condition (6.9) relies only on values at the current time level, the stability of the method is unchanged; that is

$$\frac{c\Delta t}{\Delta s} \leq \frac{1}{\sqrt{N}} \tag{6.10}$$

for N-dimensional problems. Also, the grid dispersion remains unchanged and the usual rule of thumb of 20 cells per wavelength should remain as valid as for the traditional FDTD method.

The normal-incidence periodic boundary condition for FDTD has been successfully applied to a number of scattering and radiation problems. In [5], the scattering from parallel strips and the radiation from parallel slots compared favorably with published theoretical results. FDTD computations for scattering from parallel strips compared well with MoM results in [6]. Wideband results for a multilayer, quasi-optical rotator array compared well to experimental measurements in [7]. Rough-surface scattering was studied in [8].

We have successfully used the normal-incidence periodic boundary condition in FDTD modeling of PBG materials [9, 10]. For example, consider a PBG material composed of six rows of 4-mm-diameter Pyrex rods ($\varepsilon_r = 4.2$) on a 9-mm square lattice, as shown in Fig. 6.6. For the FDTD grid, 0.25-mm spatial cells are used. Thus, the rods are 16 cells in diameter, and the unit cell is 36 cells wide.

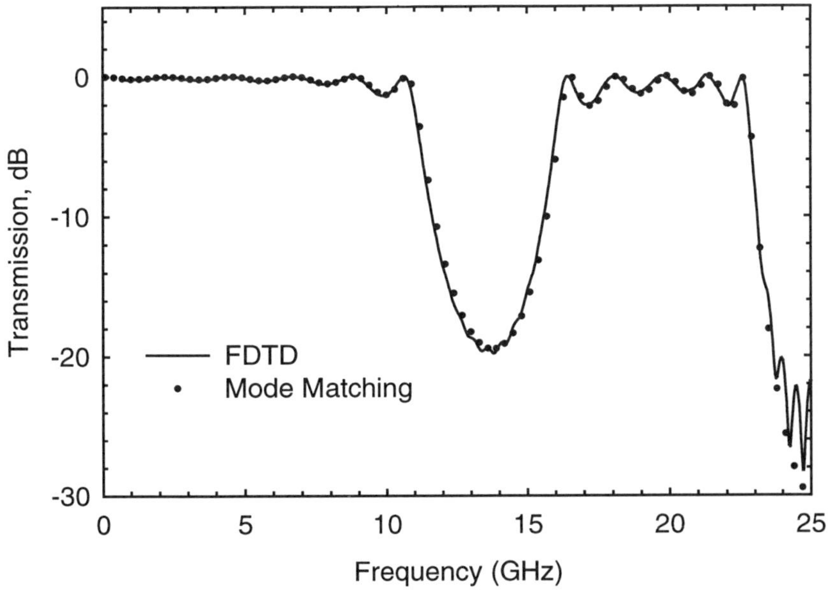

Fig. 6.7 Magnitude of transmission through a 6-row PBG crystal composed of 4-mm-diameter Pyrex rods ($\varepsilon_r = 4.2$) on a 9-mm square lattice.

Fig. 6.7 shows the magnitude of the transmission coefficient for the periodic array of Pyrex rods. The FDTD results for normal incidence (solid line) and the mode-matching result (dots) are seen to be in excellent agreement. Comparisons of FDTD modeling and experimental data for PBG materials have also indicated good agreement [9, 10].

6.3.2 Multiple Unit Cells

Several research groups have tried to solve the conundrum of needing future fields through the use of multiple unit cells. Fig. 6.8 illustrates the basic strategy for a two-dimensional problem; the extension to three dimensions is straightforward. The portion of the periodic boundary condition involving time delays is used to truncate one edge of the grid; that is,

$$H_x(x,\ y=B,\ t) \ = \ H_x\left(x,\ y=A,\ t-y_p\sin\phi_i\right) \tag{6.11}$$

where $\phi_i \geq 0$. The other three edges of the solution space are truncated with absorbing boundary conditions (ABCs). The plane wave is injected at a surface on the left.

One problem associated with this method is immediately apparent. The ABC along the bottom of the grid interacts strongly with the incident wave that is propagating along

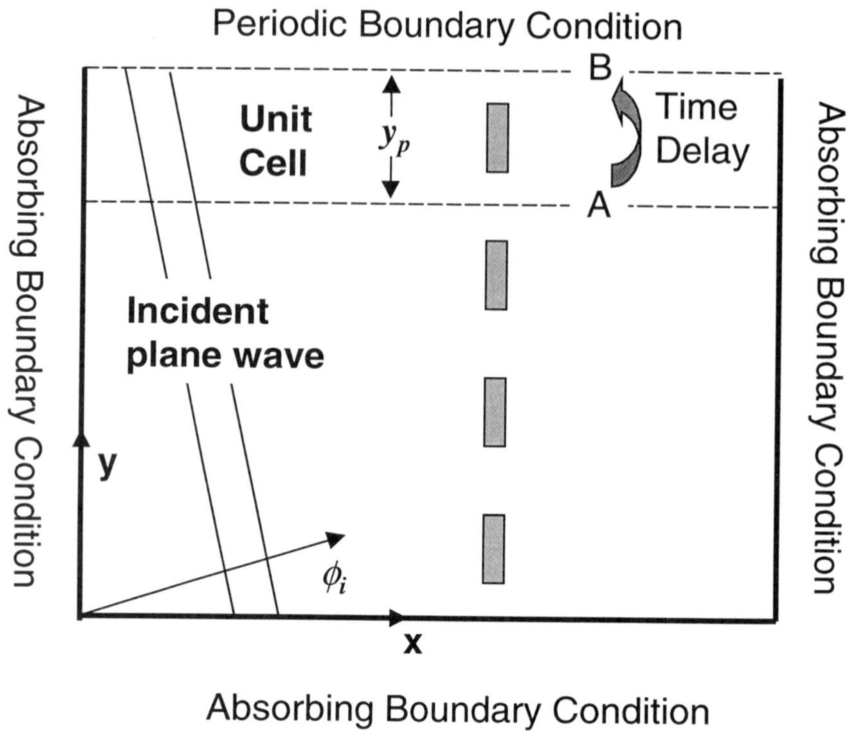

Fig. 6.8 Cross section of the grid for the multiple unit cell FDTD algorithm.

and away from the bottom boundary. This source of error is apparently mitigated through the use of additional unit cells. At the completion of the simulation, only the fields in the unit cell at the top of the figure are considered the solution.

Numerical experiments using this approach indicate that scattering from a singly periodic array of strips using 3 unit cells compares well with a large array simulation of 11 unit cells. Radiation from a doubly periodic array of rectangular apertures using 3×3 unit cells also compares well with a modal analysis for angles out to 80° from normal [11]. Input impedance calculations of a doubly periodic array of patch antennas using 5×5 unit cells compare reasonably with spectral-domain MoM results [12].

This method appears to retain the usual FDTD stability limit (6.10). Also, its level of grid dispersion should remain unchanged from the usual. The drawbacks to this method are the increased computational storage requirements of multiple unit cells, and the inherent finite-array approximation that is utilized to avoid the time-advanced field conundrum. We expect the level of accuracy to be comparable to assuming that the result from a unit cell in the center of a small finite array is equivalent to the infinite-array result.

6.3.3 Sine-Cosine Method

The sine-cosine method approaches the problem of modeling periodic boundaries from a single-frequency perspective. Thus, it loses the wideband capability of FDTD modeling but retains all of its other characteristics. The technique uses two separate grids computed simultaneously, one excited with a $\cos \omega t$ time dependence and the other with $\sin \omega t$. Fig. 6.9 shows a schematic representation of a two-dimensional configuration.

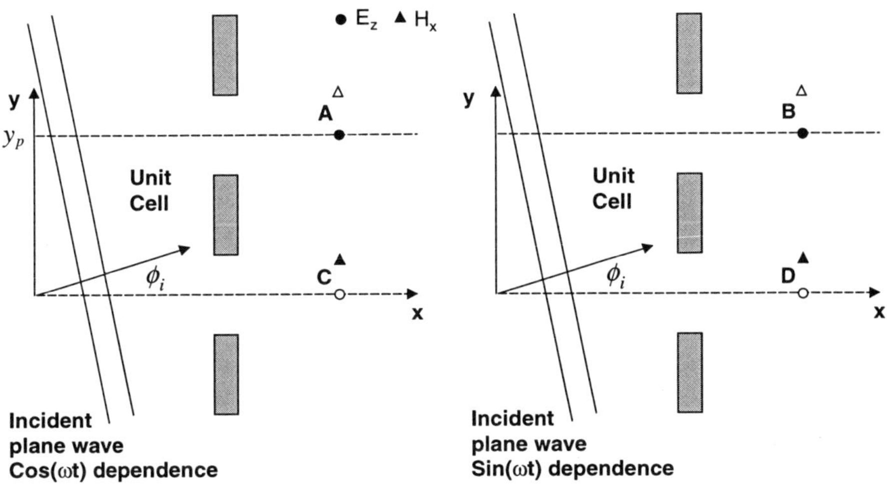

Fig. 6.9 Periodic unit cell for the sine-cosine single-frequency technique in two dimensions.

The sine-cosine method uses the following procedure to implement the periodic boundary condition. The instantaneous fields at the known boundary elements are combined to form a complex phasor; for example, $E_z(A) + jE_z(B)$. Next, the phase shifts in the frequency-domain periodic boundary condition (6.1) are used to compute the complex field at the unknown locations. Finally, the real part is used to truncate the $\cos \omega t$ grid, and the imaginary part is used to truncate the $\sin \omega t$ grid. For this example,

$$E_z(C) = \mathrm{Re}\left\{ \left[E_z(A) + jE_z(B) \right] \exp\left(jk_y y_p \right) \right\} \tag{6.12a}$$

$$E_z(D) = \mathrm{Im}\left\{ \left[E_z(A) + jE_z(B) \right] \exp\left(jk_y y_p \right) \right\} \tag{6.12b}$$

and

$$H_x(A) = \mathrm{Re}\left\{ \left[H_x(C) + jH_x(D) \right] \exp\left(-jk_y y_p \right) \right\} \tag{6.13a}$$

$$H_x(B) = \operatorname{Im}\left\{\left[H_x(C) + jH_x(D)\right] \exp\left(-jk_y y_p\right)\right\} \tag{6.13b}$$

This approach has been successfully employed to model various periodic structures [13–16]. A finite-volume time-domain technique incorporated an approach similar to this for modeling periodic boundaries in [17]. The stability of this approach is the same as for the usual FDTD algorithm. The major drawback is that only a single frequency is calculated per simulation.

To illustrate the sine-cosine technique, the transmission through the lossy electromagnetic screen discussed earlier is calculated. Fig. 6.10 shows the results for an angle of incidence of 30°, along with the results from the frequency-domain mode-matching technique. There is excellent agreement between the two techniques.

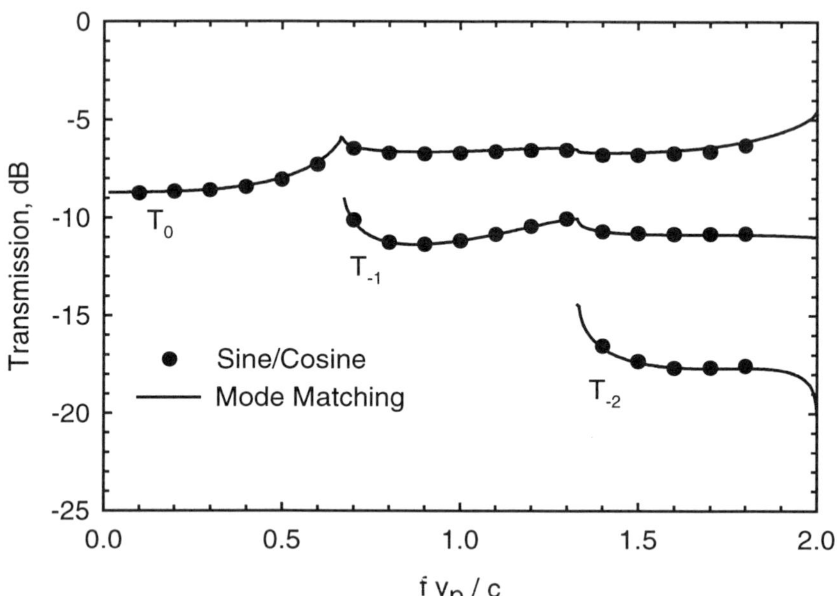

Fig. 6.10 Transmission through the lossy screen computed using the sine-cosine technique. The angle of incidence is 30°.

6.3.4 Angled-Update Method

The last direct field method to be discussed is the angled-update method. The formulation of this method is based on the observation that field components in the solution space do not all have to be at the same time level. In fact, this same observation is the basis for the efficient out-of-core FDTD solvers [18].

Before the mechanics of the angled-update method are presented, the development of a time gradient across a numerical grid is described. A two-dimensional grid is utilized to illustrate the method. Consider the field components in a slice of constant x across the grid, as shown in Fig. 6.11, top.

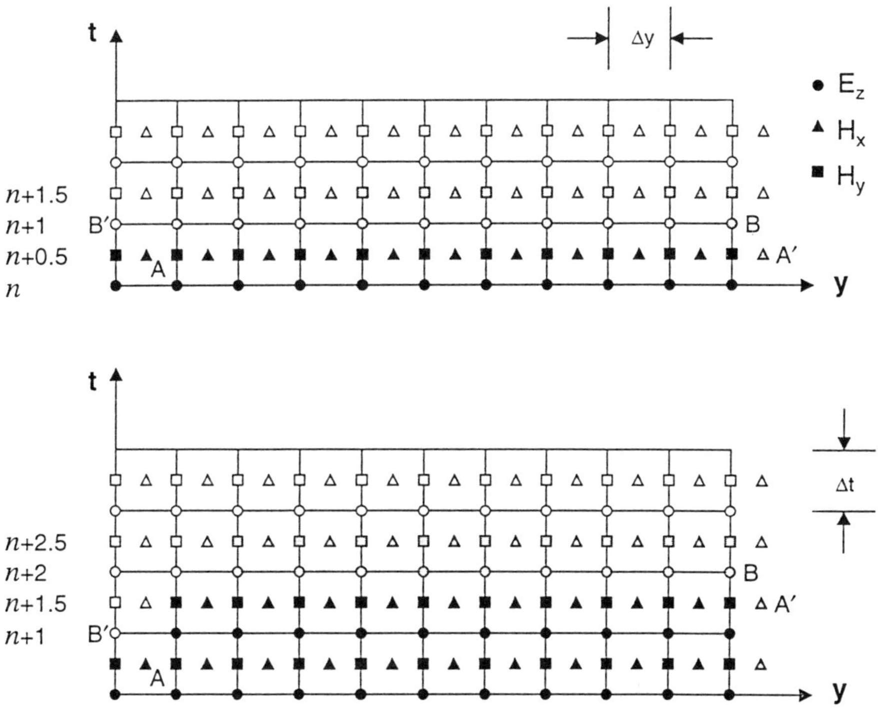

Fig. 6.11 Schematic representation of the field components in a slice of constant x of the two-dimensional FDTD grid. Top: initial stage of building the time gradient across the grid; bottom: the second stage.

Initially, the electric field components at time, n, and the magnetic field components at $n + 0.5$ are known. These are shown as black symbols. The time-domain periodic boundary condition (6.2) is used to relate the fields at A′ and B′ (that cannot be updated using the Yee algorithm) to the fields at A and B, respectively. The H_x component at A′ is determined from the saved, previous values of H_x at A. Once H_x at A′ is determined from the periodic boundary condition, the E_z components at $n+1$ can be updated. We note that the E_z field at B′ cannot be updated directly nor computed yet using the periodic boundary condition. Continuing, most of the H_x and H_y fields at $n+1.5$ can be updated using the recently updated E_z fields at $n+1$. The components computed in this sequence are denoted by gray symbols in Fig. 6.11, top.

This sequence can be repeated as shown in Fig. 6.11, bottom. Again, the H_x field at A′ is determined from the saved previous values of H_x at A. This allows most of the E_z fields at time $n+2$ to be updated. Now, most of the H_x and H_y components at $n+2.5$ can be updated using the recently updated E_z fields at $n+2$. As before, the components computed in this sequence are denoted by gray symbols.

This process can be repeated up to $N_y - 1$ times, where N_y is the number of cells across the grid in the y direction. Eventually, the E_z field at B is known $N_y - 1$ time steps in the future relative to the time level of the unknown E_z field at B′. For moderate plane-wave steering angles, this sequence is stopped after

$$N_{\text{slant}} = \text{ceil}\left[(N_y - 1)\sqrt{2}\sin\phi_i\right] \le N_y - 1 \tag{6.14}$$

stages where "ceil" is a function that rounds its argument to the next higher integer. The sequence is stopped at this point so that the largest possible Δt can be used. This general case is shown in Fig. 6.12.

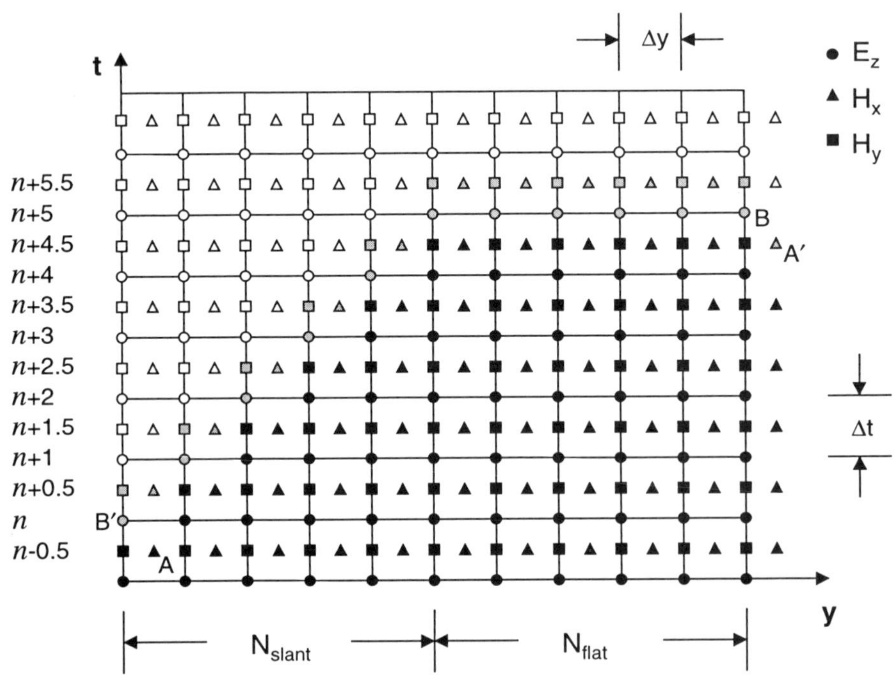

Fig. 6.12 Schematic representation of the field components on a slice of constant x of the two-dimensional FDTD grid. The time gradient across the numerical grid utilized in the angled update method is evident. Black symbols denote field components that are known; gray indicate those that will be updated in the next cycle; and white are unknown.

Now, the time step is chosen as

$$\frac{c\Delta t}{\Delta y} = \frac{N_y \sin \phi_i}{N_{\text{slant}}} \leq \frac{1}{\sqrt{2}} \tag{6.15}$$

so that the time differential between points A–A′ and between B–B′ is identical to the time delay and time advance required in the periodic boundary condition (6.2). We note that continuing the above sequence beyond the point indicated in (6.14) would require a smaller time step according to (6.15).

Because all fields in the FDTD grid are zero at the start of the simulation, the angled-update method can begin the simulation with the time gradient illustrated in Fig. 6.12 already established. Only computer storage for the most recent magnetic and electric field component is required. A single cycle of the update procedure is as follows:

1. Assuming N_{slant} is set by (6.14) and Δt is set by (6.15), then the time gradient established across the grid allows $H_y^{n+4.5}(\text{A}')$ to be determined from the simplified periodic boundary condition, $H_y^{n+4.5}(\text{A}') = H_y^{n-0.5}(\text{A})$, for all values of x.
2. Update $E_z^{n+5}(\text{B})$ for all values of x using the Yee algorithm.
3. Determine $E_z^{n}(\text{B}')$ from the simplified periodic boundary condition, $E_z^{n}(\text{B}') = E_z^{n+5}(\text{B})$, for all values of x.
4. Update fields along the slant portion of the grid:
   a. Update $H_x^{n+0.5}$ and $H_y^{n+0.5}$ for all values of x using the Yee algorithm.
   b. Update E_z^{n+1} for all values of x using the Yee algorithm.
   c. Repeat (a) and (b) sequentially.
5. Update E_z^{n+5} for all cells along the flat portion and for all values of x using the Yee algorithm.
6. Finally, update $H_x^{n+5.5}$ and $H_y^{n+5.5}$ for all cells along the flat portion and for all values of x using the Yee algorithm.

The field components updated in one iteration are shown as gray symbols in Fig. 6.12. We note that the above procedure is still a leapfrog method.

Because the angled-update method only rearranges the order in which field components are updated and does not change the regular FDTD update equations, the stability requirement is unchanged from (6.10). Further, the grid dispersion is unchanged, and thus the usual rule of thumb of at least 20 cells per wavelength applies.

The major limitation of this method is that the maximum steering angle is limited to

$$\phi_{\max} = \sin^{-1}\left(\frac{N_y - 1}{\sqrt{2}\, N_y}\right) < 45^\circ \tag{6.16}$$

because the maximum time gradient across the grid is $(N_y - 1)\Delta t$, whereas the required time gradient is $N_y \Delta y \sin\phi / c$. Since in the two-dimensional case, stability limits the maximum time-step to $\Delta t \le \Delta y / c\sqrt{2}$, then (6.16) follows. Note that for the three-dimensional case, the maximum steering angle is approximately 35°.

As an illustration of the angled-update technique, the transmission through the lossy screen example discussed earlier is computed for an angle of incidence of 30° and compared to mode-matching results. As seen in Fig. 6.13, there is very good agreement.

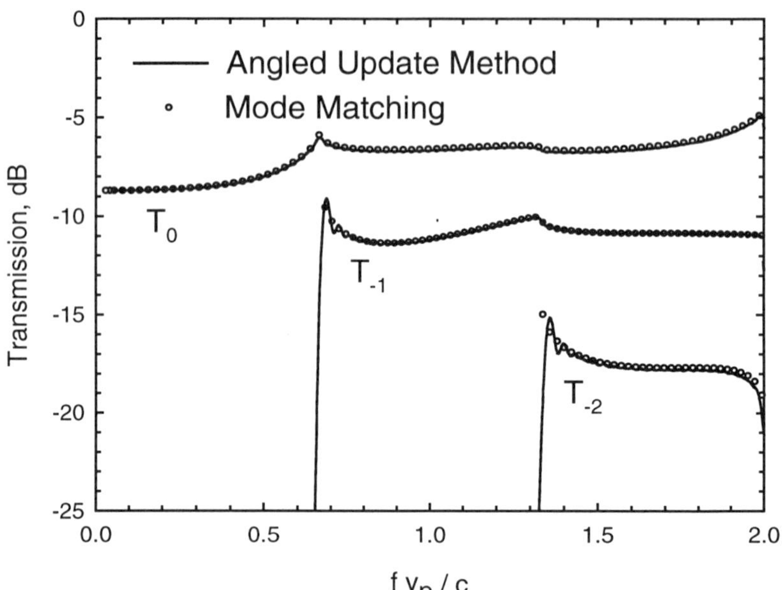

Fig. 6.13 Comparison of transmission through the lossy screen for an angle of incidence of 30° calculated using the angled-update technique and mode matching.

6.4 FIELD-TRANSFORMATION APPROACH

The second class of techniques for analyzing periodic structures uses a field transformation to remove the time gradient across the grid [19], in contrast to the numerical time gradient introduced in the angled-update method. The resulting equations are then discretized and solved using an FDTD approach. However, the transformed equations have additional terms that require special treatment. In this section, a detailed derivation of the transformed field approach is given. We start with Maxwell's equations in two dimensions to keep the analysis simple. The full three-dimensional case is discussed in a later section.

In the frequency domain, Maxwell's equations for the two-dimensional case of interest are given by

$$\frac{\partial E_z}{\partial y} = -j\omega\mu H_x \tag{6.17a}$$

$$\frac{\partial E_z}{\partial x} = j\omega\mu H_y \tag{6.17b}$$

$$\frac{\partial H_y}{\partial x} - \frac{\partial H_x}{\partial y} = j\omega\varepsilon E_z \tag{6.17c}$$

To remove the phase that is accumulated across the grid, a new set of field variables is introduced:

$$P_z = E_z \exp\left(jk_y y\right) \tag{6.18a}$$

$$Q_x = \eta_0 H_x \exp\left(jk_y y\right) \tag{6.18b}$$

$$Q_y = \eta_0 H_y \exp\left(jk_y y\right) \tag{6.18c}$$

The periodic boundary conditions (6.1) for the transformed field variables are now simply

$$P_z(x,\ y=0) = P_z(x,\ y=y_p) \tag{6.19a}$$

$$Q_x(x,\ y=0) = Q_x(x,\ y=y_p) \tag{6.19b}$$

Substituting the relations of (6.18) into Maxwell's equations (6.17) yields

$$\frac{j\omega\mu_r}{c} Q_x = -\frac{\partial P_z}{\partial y} + \left\{jk_y P_z\right\} \tag{6.20a}$$

$$\frac{j\omega\mu_r}{c} Q_y = \frac{\partial P_z}{\partial x} \tag{6.20b}$$

$$\frac{j\omega\varepsilon_r}{c} P_z = \frac{\partial Q_y}{\partial x} - \frac{\partial Q_x}{\partial y} + \left\{jk_y Q_x\right\} \tag{6.20c}$$

The terms in brackets are extra terms resulting from the transformation (6.18). Expressing (6.20) in the time domain, we have

$$\frac{\mu_r}{c}\frac{\partial Q_x}{\partial t} = -\frac{\partial P_z}{\partial y} + \left\{\frac{\sin\phi}{c}\frac{\partial P_z}{\partial t}\right\} \tag{6.21a}$$

$$\frac{\mu_r}{c}\frac{\partial Q_y}{\partial t} = \frac{\partial P_z}{\partial x} \tag{6.21b}$$

$$\frac{\varepsilon_r}{c}\frac{\partial P_z}{\partial t} = \frac{\partial Q_y}{\partial x} - \frac{\partial Q_x}{\partial y} + \left\{\frac{\sin\phi}{c}\frac{\partial Q_x}{\partial t}\right\} \tag{6.21c}$$

where c is the speed of light in free space, and the wave is assumed to be incident on the periodic structure from free space. Combining (6.21a) and (6.21c) to eliminate the time derivatives on the right-hand side results in the basic equations to be discretized:

$$\left(\frac{\varepsilon_r\mu_r - \sin^2\phi}{c}\right)\frac{\partial Q_x}{\partial t} = -\varepsilon_r\frac{\partial P_z}{\partial y} + \left(\frac{\partial Q_y}{\partial x} - \frac{\partial Q_x}{\partial y}\right)\sin\phi \tag{6.22a}$$

$$\frac{\mu_r}{c}\frac{\partial Q_y}{\partial t} = \frac{\partial P_z}{\partial x} \tag{6.22b}$$

$$\left(\frac{\varepsilon_r\mu_r - \sin^2\phi}{c}\right)\frac{\partial P_z}{\partial t} = -\frac{\partial P_z}{\partial y}\sin\phi + \mu_r\left(\frac{\partial Q_y}{\partial x} - \frac{\partial Q_x}{\partial y}\right) \tag{6.22c}$$

It is important to note the differences between (6.22) and the normal FDTD update equations, which is found by letting $\phi = 0$.

1. The presence of a space and time derivative of the same variable in an equation means that the usual Yee leapfrog updating must be modified. For example, in (6.22a) with $\phi = 0$, Q_x at time step $n+1/2$ is determined by the derivative of P_z at n and the previous value of Q_x at $n-1/2$. However, with $\phi \neq 0$, there are spatial derivatives of Q_x and Q_y present on the right-hand side as well. These derivatives must be handled properly before this technique can be made stable.
2. The usual Courant stability criterion must be modified because of the extra terms and the factor multiplying the time derivatives in (6.22a) and (6.22c).

Before discussing techniques for discretizing (6.22), it is instructive to examine the properties of the continuous equations. In particular, the dispersion relation is of interest since we will be looking at the numerical dispersion of the different techniques used to implement (6.22). To find the dispersion relation, assume the fields are of the form

$$Q_x,\ Q_y,\ P_z \propto \exp\left(j\omega t - jk_x^s x - jk_y^s y\right) \tag{6.23}$$

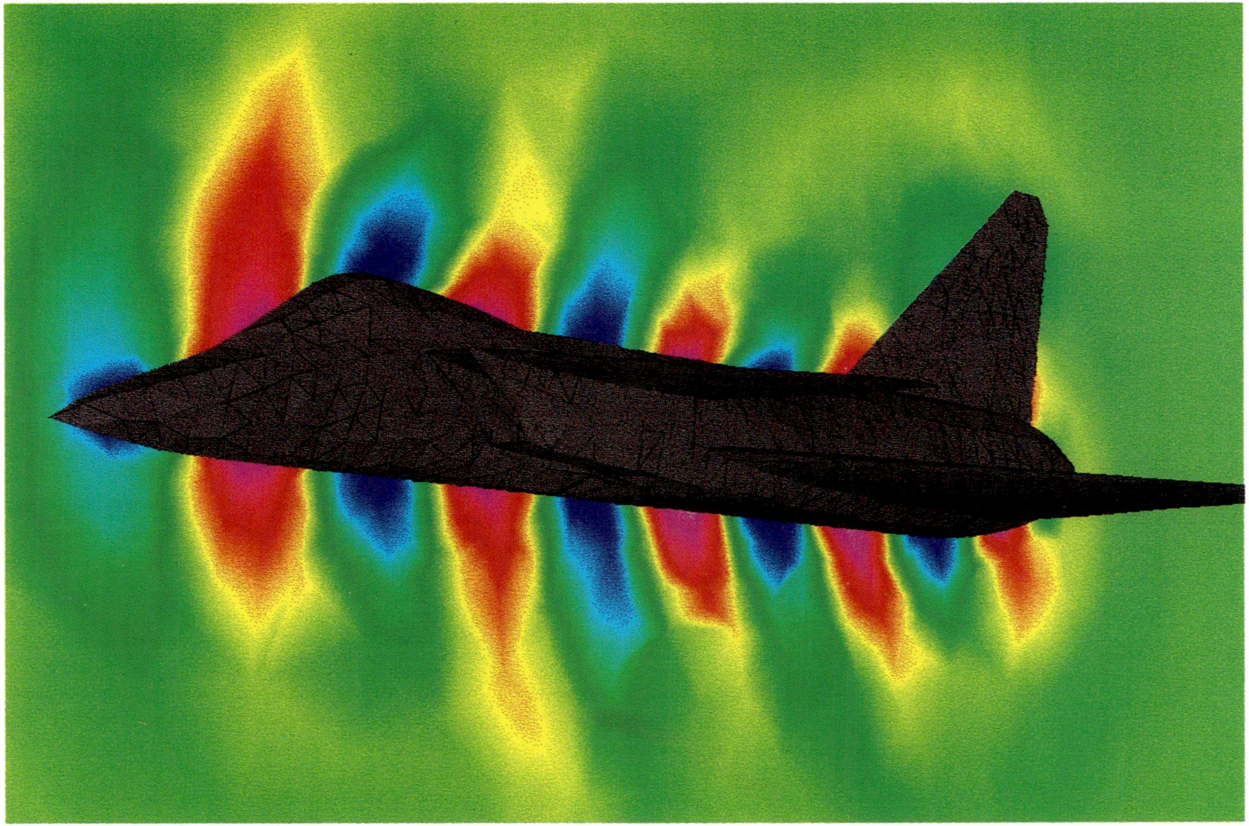

Plate 1. Visualization of the faceted surface and the surrounding low-frequency electromagnetic field for the FVTD model of the full-scale VFY-218 prototype fighter airplane. (See also Fig. 4.36.)

Plate 2. Detailed view of the engine inlet of the fighter airplane model of Plate 1, showing the internal geometry and the field enhancement at the edges of the inlet.

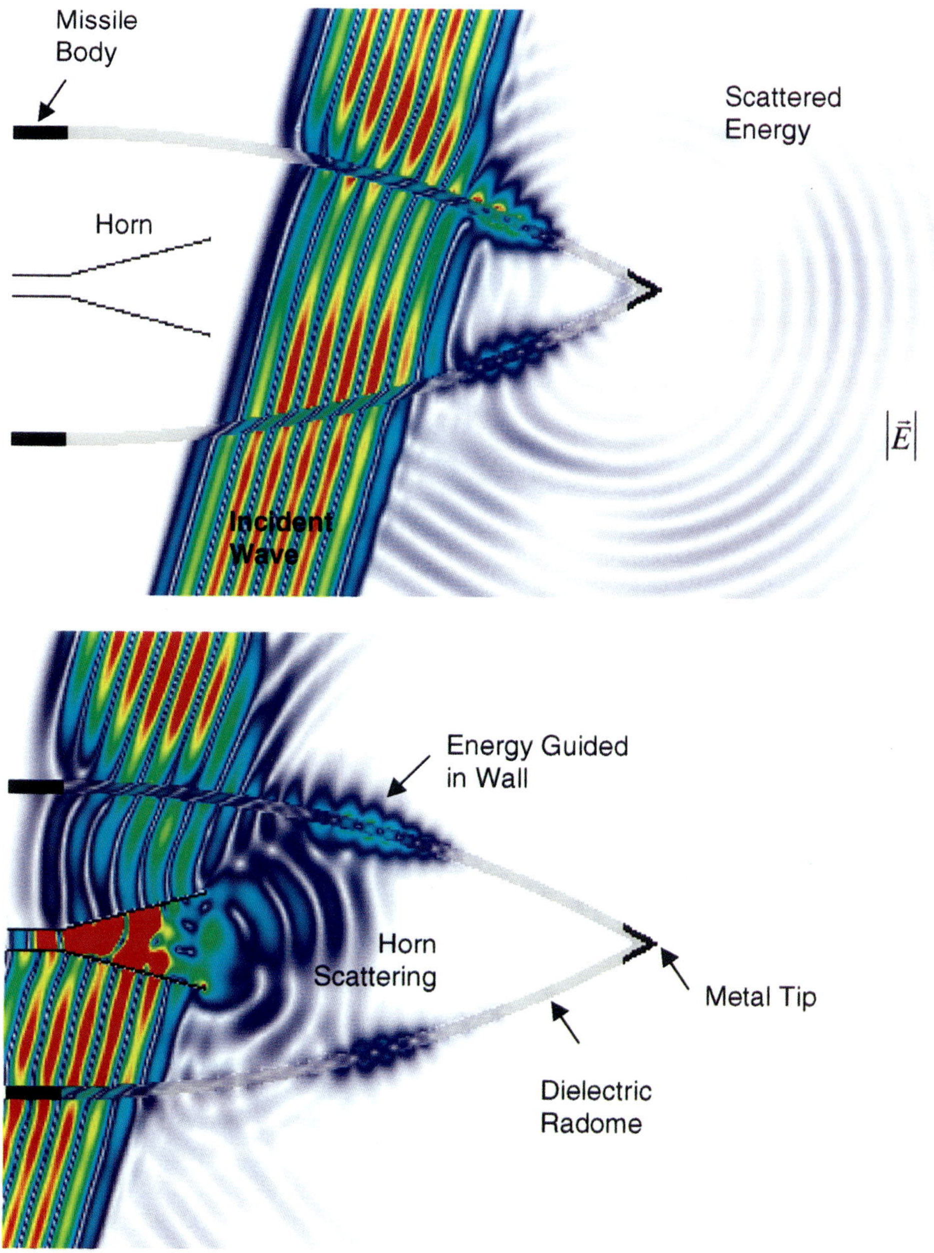

Plate 3. Visualization of plane-wave penetration and scattering for the FDTD model of the radome of a missile containing a horn antenna. The plane wave propagates from right to left and is obliquely incident at 15° from boresight. (See also Fig. 7.25.)

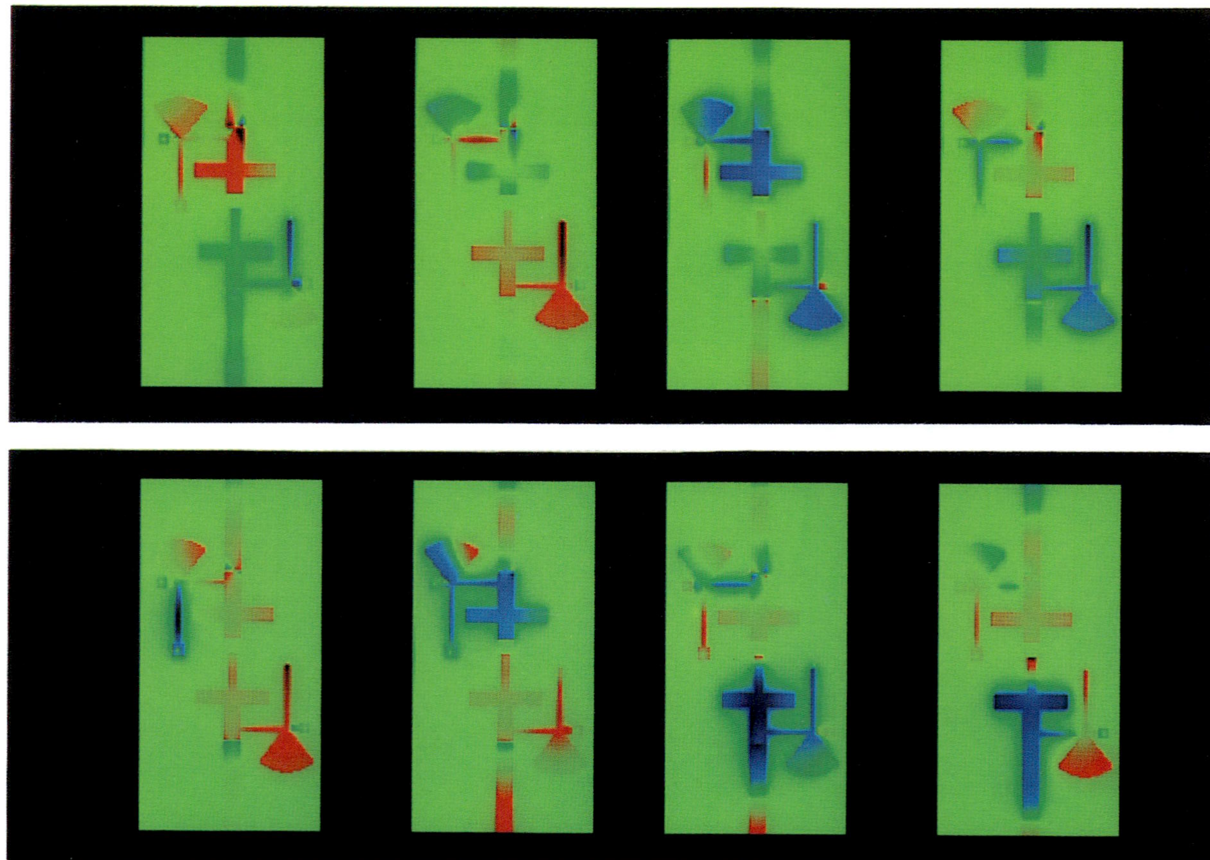

Plate 4. Visualization of the FDTD-computed electromagnetic field within a microwave amplifier. The sequence of snapshots in time goes from top left to bottom right. (See also Fig. 8.15.)

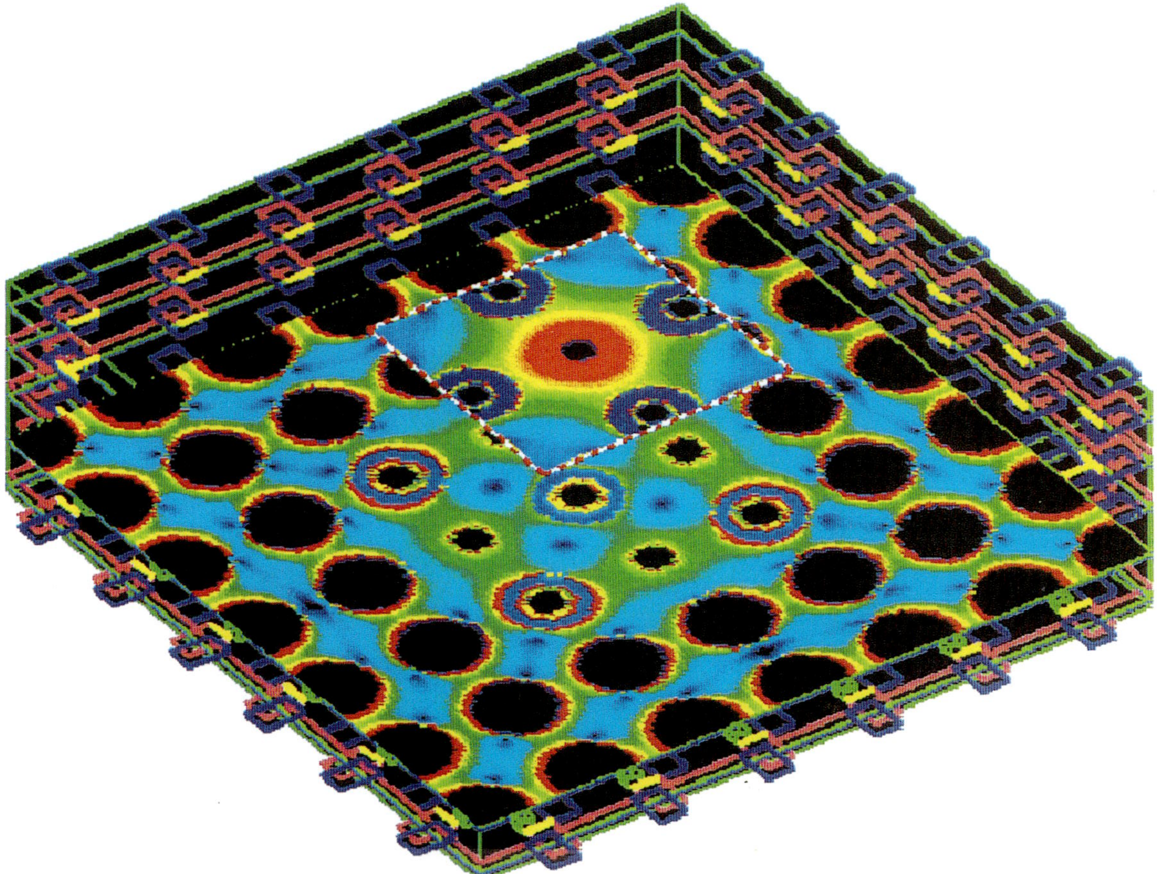

Plate 5. Visualization of the FDTD-computed current density for an eight-layer power distribution structure, showing the meshed printed-circuit-board ground plane (source). The multichip module (load) ground layer is located within the dotted rectangle. (See also Fig. 8.22.)

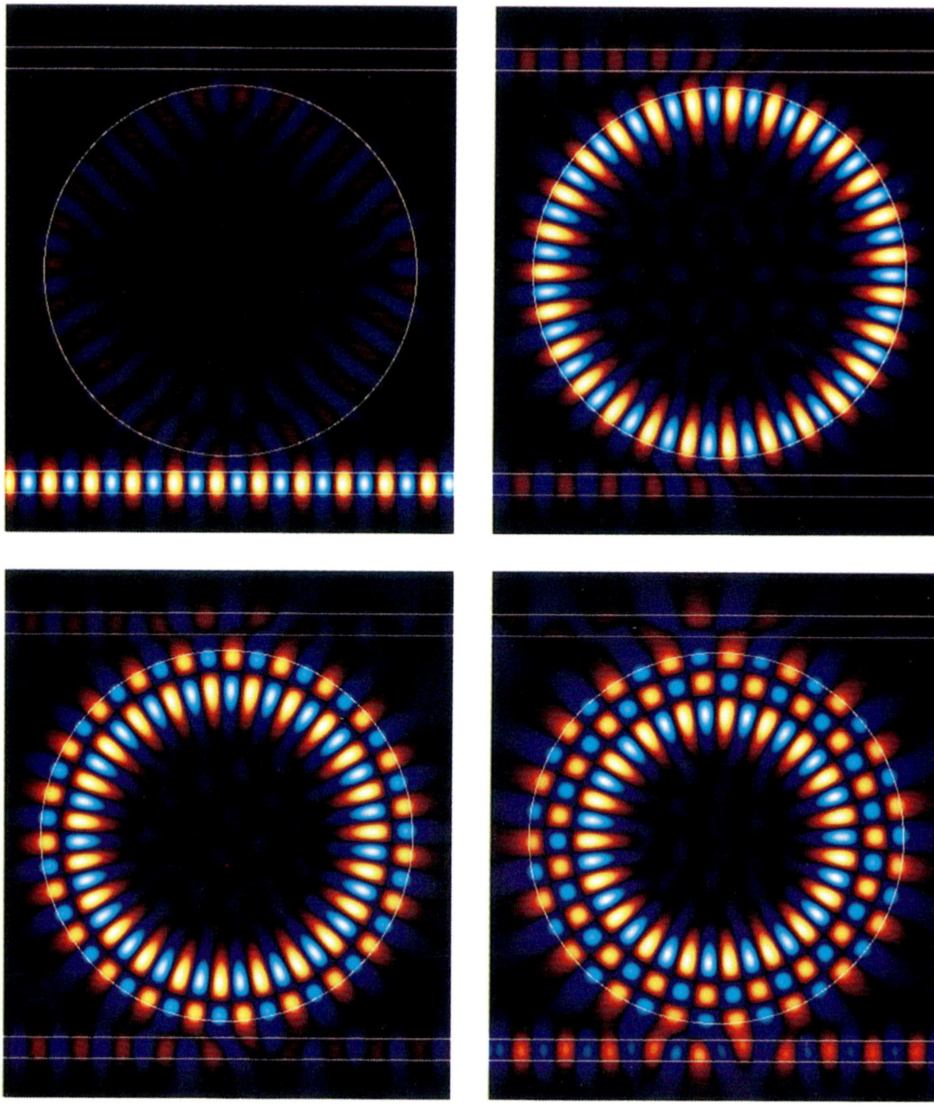

Plate 6. Visualizations of the FDTD-computed sinusoidal steady-state electric field pattern in a 5.0-μm-diameter microcavity disk resonator coupled to straight 0.3-μm-wide waveguides for single-frequency excitations in the lower waveguide, left side. Top left: off-resonance signal (193.4 THz); top right: on-resonance signal, first-order radial mode (189.2 THz); bottom left: second-order radial mode resonance (191.3 THz); bottom right: third-order radial mode resonance (187.8 THz). (See also Fig. 10.31.)

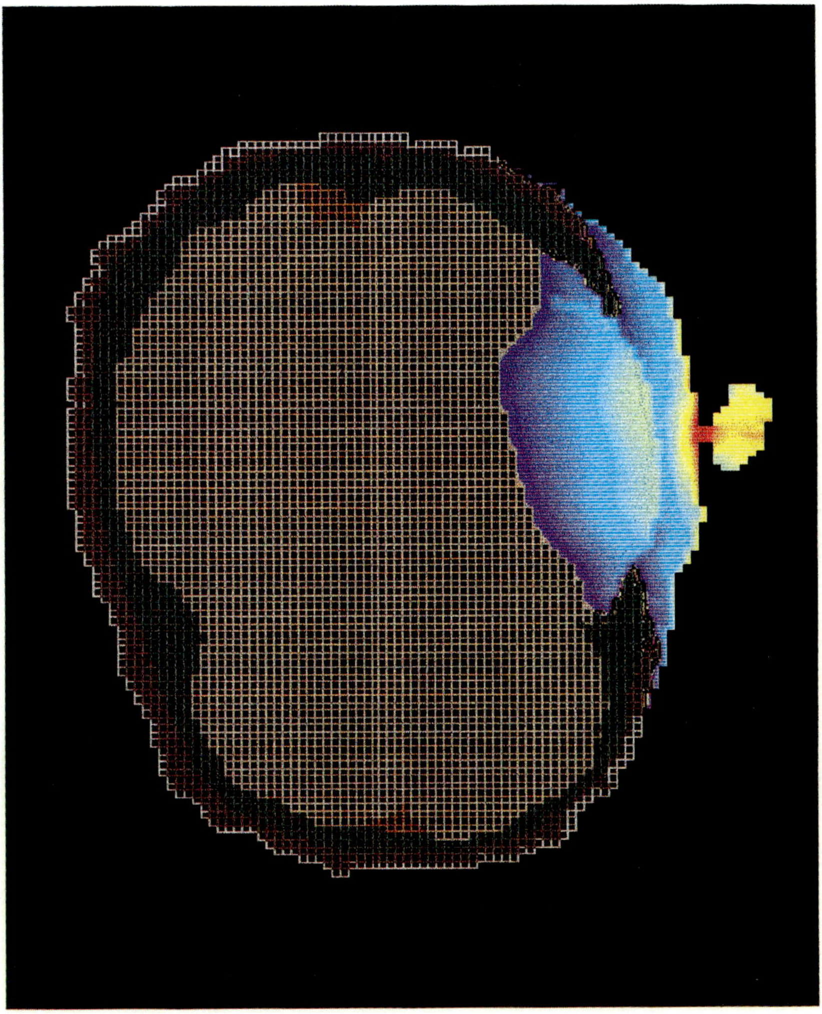

Plate 7. Visualization of the FDTD-computed SAR distribution in an ear-level plane (10 cm below the top of the head) for a telephone held vertically against a tilted-head model. Radiated frequency = 1900 MHz, one-quarter wavelength antenna, 125 mW radiated power. (See also Fig. 11.14.)

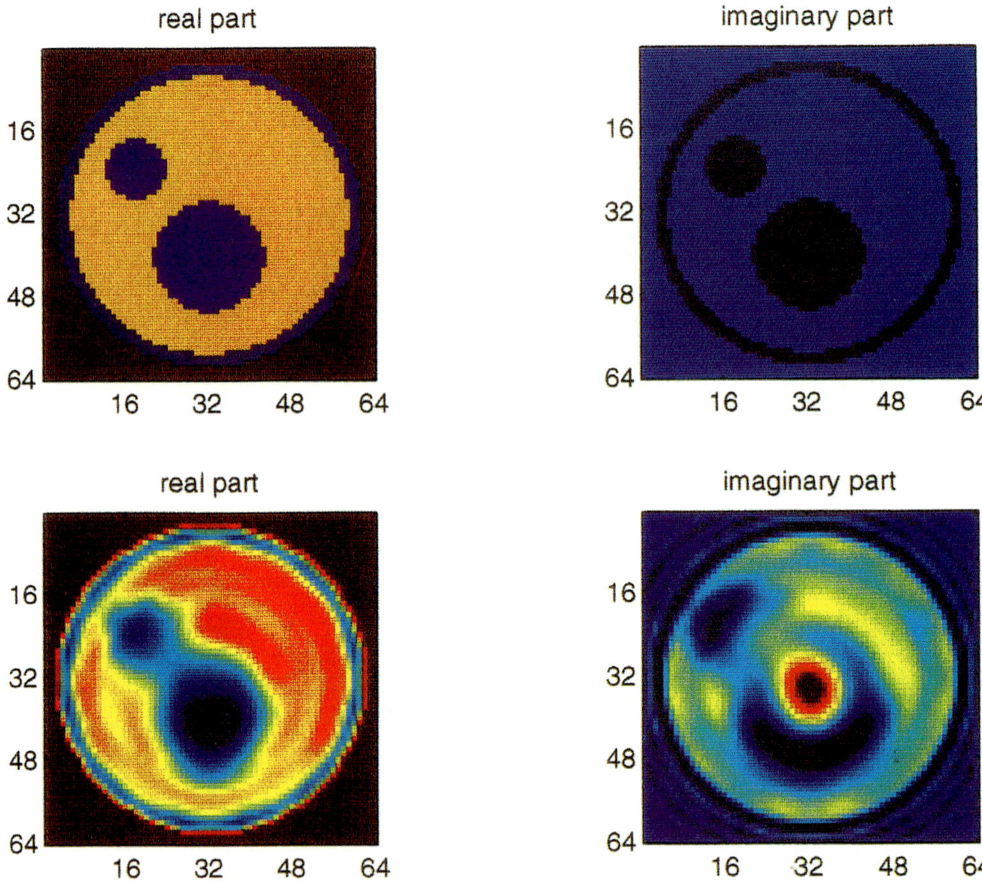

Plate 8. Reconstruction of the real and imaginary parts of the permittivity for an object resembling a human arm using the frequency-hopping approach where the frequencies used are 0.5, 0.9, 1.65, and 3 GHz. The reconstructed area is 10.2×10.2 wavelengths. Top: original dielectric image; bottom: reconstructed dielectric image. (See also Fig. 12.10.)

and substitute into (6.22). Here, the superscript s denotes the scattered field, which can be at an arbitrary direction with respect to the angle of incidence. The result is

$$\begin{bmatrix} \left(\frac{j\omega}{c}\cos^2\phi - jk_y^s \sin\phi\right) & jk_x^s \sin\phi & -jk_y^s \varepsilon_r \\ 0 & \frac{j\omega\mu_r}{c} & jk_x^s \\ -jk_y^s \mu_r & jk_x^s \mu_r & \left(\frac{j\omega}{c}\cos^2\phi - jk_y^s \sin\phi\right) \end{bmatrix} \begin{bmatrix} Q_x \\ Q_y \\ P_z \end{bmatrix} = 0 \qquad (6.24)$$

The dispersion relation is found by setting the determinant of the matrix in (6.24) equal to zero. After some manipulation, the dispersion relation for free space simplifies to

$$\left(k_x^s\right)^2 + \left(k_y^s + \frac{\omega \sin\phi}{c}\right)^2 = \left(\frac{\omega}{c}\right)^2 \qquad (6.25)$$

If we define the angle α, such that $k_x^s = (\omega / v_p)\cos\alpha$ and $k_y^s = (\omega / v_p)\sin\alpha$, (6.25) becomes

$$\cos^2\phi \left(\frac{v_p}{c}\right)^2 - 2\sin\alpha\sin\phi\left(\frac{v_p}{c}\right) - 1 = 0 \qquad (6.26)$$

where we have introduced the phase velocity, v_p, as a function of α. The dispersion is characterized by finding $v_p(\alpha)$. The solution to (6.26) is simply

$$\frac{v_p}{c} = \frac{\sin\alpha\sin\phi + \sqrt{(\sin\alpha\sin\phi)^2 + \cos^2\phi}}{\cos^2\phi} \qquad (6.27)$$

A comment about notation is necessary to avoid confusion at this point. The angle ϕ is the angle at which an incident plane wave impinges on the periodic structure, while the angle α represents a direction of propagation in the structure. Note that the solution to a particular problem involves a single ϕ, but there are many possible values of α for a given ϕ. Any scattering in the structure can create propagation at different angles.

Plots of the phase velocity for different angles of incidence are shown in Fig. 6.14. The phase velocities along the principal directions are given by

$$\frac{v_p}{c} = \frac{1}{\cos\phi} \qquad \alpha = 0^\circ \qquad (6.28a)$$

$$\frac{v_p}{c} = \frac{1}{1 - \sin\phi} \qquad \alpha = 90^\circ \qquad (6.28b)$$

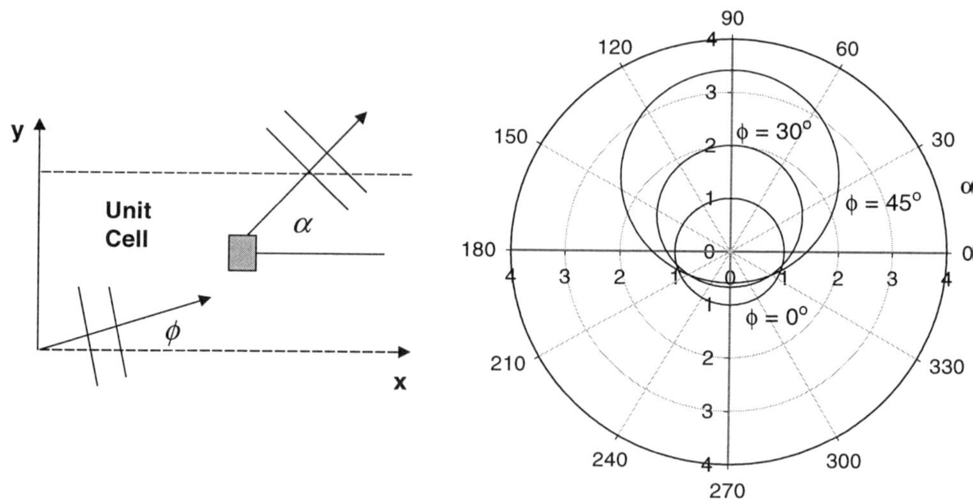

Fig. 6.14 Normalized phase velocity, v_p/c, as a function of the propagation angle, α, for three angles of incidence, ϕ.

$$\frac{v_p}{c} = \frac{1}{1 + \sin\phi} \qquad \alpha = -90^\circ \tag{6.28c}$$

The phase velocities for the continuous case (or equivalently, the wavelength) will be used to normalize the numerical results presented later in this section.

The minimum phase velocity determines the spatial discretization required to achieve an accurate solution. This minimum velocity occurs for $\alpha = -90^\circ$ and is given by (6.28c). This equation shows that, as the angle of incidence increases, a smaller cell size is required to maintain the same effective number of cells per wavelength. In general, the cell size needs to be reduced by as much as a factor of two as the angle of incidence approaches 90°.

A comment on the numerical stability of the system of (6.22) is also needed. We have investigated numerous techniques for discretizing these equations, and stability has always been a major concern. As a result, for each of the techniques that are discussed, a stability analysis will be a main focus. A simple estimate of the stability can be made from the maximum velocity in the grid, given by (6.28b). Using this value, we might expect the stability limit to be given by

$$\frac{c\Delta t}{\Delta x} \le \frac{1 - \sin\phi}{\sqrt{D}} \tag{6.29}$$

where D is the dimensionality of the problem. Stability analyses presented in later sections for various discretization schemes will show that this limit is conservative.

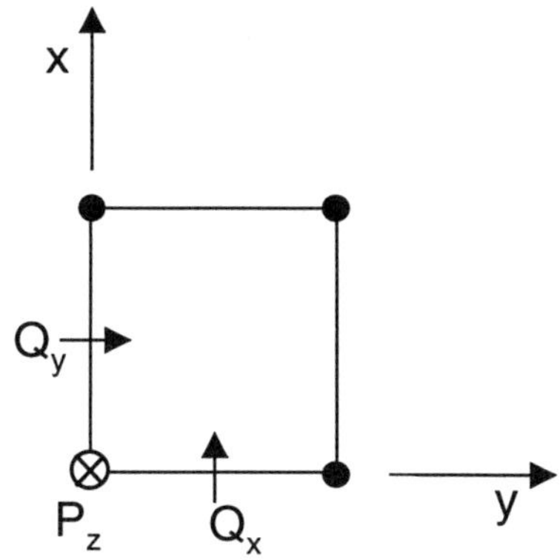

Fig. 6.15 Basic unit cell used in the discretization process.

6.4.1 Simple Discretization

Fig. 6.15 shows the location of the fields within the basic FDTD cell used in the discretization procedure for (6.22a) and (6.22c). The first approach we might consider to discretize these equations is to use the previous values in the spatial derivatives of Q_x and Q_y in (6.22a) and P_z in (6.22c). For example, with this approach, the update equation for P_z at grid location $(\bar{i}, \bar{j})$ is given by

$$\begin{aligned} P_z\Big|_{\bar{i},\bar{j}}^{n+1} &= P_z\Big|_{\bar{i},\bar{j}}^{n} - \frac{T_y A}{2}\sin\phi\left(P_z\Big|_{\bar{i},\bar{j}+1}^{n} - P_z\Big|_{\bar{i},\bar{j}-1}^{n}\right) \\ &+ \mu_r a T_y A\left(Q_y\Big|_{\bar{i},\bar{j}}^{n+1/2} - Q_y\Big|_{\bar{i}-1,\bar{j}}^{n+1/2}\right) - \mu_r T_y A\left(Q_x\Big|_{\bar{i},\bar{j}}^{n+1/2} - Q_x\Big|_{\bar{i},\bar{j}-1}^{n+1/2}\right) \end{aligned} \tag{6.30}$$

where the field component notation $Q_x\Big|_{\bar{i},\bar{j}}^{n} = Q_x(\bar{i}\Delta x, \bar{j}\Delta y, n\Delta t)$ is used, and

$$T_y = \frac{c\,\Delta t}{\Delta y} \tag{6.31a}$$

$$A = \frac{1}{\varepsilon_r \mu_r - \sin^2\phi} \tag{6.31b}$$

$$a = \frac{\Delta y}{\Delta x} \tag{6.31c}$$

The other equations are discretized in a similar fashion.

We next consider the numerical stability of this discretization by implementing a von Neumann local stability analysis. First, we assume a plane wave propagating through the grid at an arbitrary angle. The transformed fields are represented by

$$Q_x\Big|_{\bar{i},\bar{j}}^{n+1/2} = Q_x\Big|_{0,0}^{1/2} s^n \exp\left[-j\left(\bar{i}\,k_x^s\Delta x + \bar{j}k_y^s\Delta y\right)\right] \tag{6.32a}$$

$$Q_y\Big|_{\bar{i},\bar{j}}^{n+1/2} = Q_y\Big|_{0,0}^{1/2} s^n \exp\left[-j\left(\bar{i}\,k_x^s\Delta x + \bar{j}k_y^s\Delta y\right)\right] \tag{6.32b}$$

$$P_z\Big|_{\bar{i},\bar{j}}^{n} = P_z\Big|_{0,0}^{0} s^n \exp\left[-j\left(\bar{i}\,k_x^s\Delta x + \bar{j}k_y^s\Delta y\right)\right] \tag{6.32c}$$

Using this notation, the discretized version of the system of (6.22) for cell $(\bar{i},\,\bar{j})$ is

$$(s-1)Q_x^{n-1/2} = 2jT_yA\left(\varepsilon_r S_y P_z^n - aS_xC_y\sin\phi\, Q_y^{n-1/2} + S_yC_y\sin\phi\, Q_x^{n-1/2}\right) \tag{6.33a}$$

$$(s-1)Q_y^{n-1/2} = -2jT_y\frac{a}{\mu_r}S_x P_z^n \tag{6.33b}$$

$$(s-1)P_z^n = 2jT_yA\left(S_yC_y\sin\phi\, P_z^{n-1} - \mu_r aS_xQ_y^{n-1/2}s + \mu_r S_yQ_x^{n-1/2}s\right) \tag{6.33c}$$

where

$$S_x = \sin\left(k_x^s\Delta x/2\right) \tag{6.34a}$$

$$S_y = \sin\left(k_y^s\Delta y/2\right) \tag{6.34b}$$

$$C_y = \cos\left(k_y^s\Delta y/2\right) \tag{6.34c}$$

A nontrivial solution to the system of (6.33) requires that

$$\begin{vmatrix} \left(\dfrac{s-1}{2jT_y} - AS_yC_y\sin\phi\right) & aAS_xC_y\sin\phi & -\varepsilon_r AS_y \\ 0 & \dfrac{s-1}{2jT_y} & \dfrac{aS_x}{\mu_r} \\ -\mu_r AS_y s & \mu_r aAS_x s & \left(\dfrac{s-1}{2jT_y} - AS_yC_y\sin\phi\right) \end{vmatrix} = 0 \tag{6.35}$$

Equation (6.35) is a cubic polynomial in s. In order for the discretization to be stable for a given Courant number, T_y, the magnitude of the largest root of the polynomial must be less than or equal to 1 for the values of S_x and S_y that maximize the root. (The possible values of S_x and S_y are limited to lie between -1 and 1.) One solution to (6.35) is $s = 1$, while the other two solutions are given by

$$\begin{aligned} s \;=\; & 1 + 2jAT_y S_y C_y \sin\phi \;-\; 2T_y^2\left(Aa^2 S_x^2 + \mu_r\varepsilon_r A^2 S_y^2\right) \\ & \pm\; 2j\sqrt{AT_y^2\left(a^2 S_x^2 + \mu_r\varepsilon_r S_y^2\right)\left[1 + 2jAT_y S_y C_y \sin\phi - T_y^2\left(Aa^2 S_x^2 + \mu_r\varepsilon_r A^2 S_y^2\right)\right]} \end{aligned} \tag{6.36}$$

Numerically maximizing (6.36) shows that $|s| > 1$ for nonzero T_y. Thus, the method is unstable. Clearly, other schemes for discretization are needed.

6.4.2 Dual-Time Technique

The second approach that we will use to discretize the system of (6.22) is to collocate values of P_z, Q_x, and Q_y in time. In this case, the update equation for P_z is given by

$$\begin{aligned} P_z\Big|_{\bar{i},\bar{j}}^{n+1} \;=\; & P_z\Big|_{\bar{i},\bar{j}}^{n-1} \;-\; \frac{T_y A}{2}\sin\phi\left(P_z\Big|_{\bar{i},\bar{j}+1}^{n} \;-\; P_z\Big|_{\bar{i},\bar{j}-1}^{n}\right) \\ & +\; \mu_r a T_y A\left(Q_y\Big|_{\bar{i},\bar{j}}^{n} - Q_y\Big|_{\bar{i}-1,\bar{j}}^{n}\right) \;-\; \mu_r T_y A\left(Q_x\Big|_{\bar{i},\bar{j}}^{n} - Q_x\Big|_{\bar{i},\bar{j}-1}^{n}\right) \end{aligned} \tag{6.37}$$

Similar update equations exist for Q_x and Q_y. Two items are important to note about this technique. First, the time derivative extends over two time steps. The value of P_z at time step, $n+1$, is found from the value of P_z at $n-1$ and the values of P_z, Q_x, and Q_y at n. Thus, the space derivatives are computed using the field values at the intermediate time step. Second, unlike the traditional leapfrog approach, values for all variables are maintained at the same time levels and updated every time step. Hence, we refer to this method as the dual-time technique.

Numerical Stability Analysis

To derive the numerical stability of the dual-time technique, we again apply the von Neumann stability analysis. Using fields of the form (6.32), the update equations are

$$\left(s - \frac{1}{s}\right)Q_x^n \;=\; 2jT_y A\left(\varepsilon_r S_y P_z^n \;-\; aS_x C_y \sin\phi\, Q_y^n \;+\; S_y C_y \sin\phi\, Q_x^n\right) \tag{6.38a}$$

$$\left(s - \frac{1}{s}\right)Q_y^n \;=\; -2jT_y \frac{a}{\mu_r} S_x P_z^n \tag{6.38b}$$

$$\left(s - \frac{1}{s}\right) P_z^n = 2jT_y A\left(S_y C_y \sin\phi\, P_z^n - \mu_r a S_x Q_y^n + \mu_r S_y Q_x^n\right) \tag{6.38c}$$

The stability relation for this case is written in the form of the determinant

$$\begin{vmatrix} \left(AS_y C_y \sin\phi - \Lambda\right) & -aAS_x C_y \sin\phi & \varepsilon_r AS_y \\ 0 & -\Lambda & -aS_x / \mu_r \\ \mu_r AS_y & -\mu_r aAS_x & \left(AS_y C_y \sin\phi - \Lambda\right) \end{vmatrix} = 0 \tag{6.39a}$$

where

$$\Lambda = \frac{1}{2jT_y}\left(s - \frac{1}{s}\right) \tag{6.39b}$$

are recognized as the eigenvalues of the matrix formed by the right-hand side of the system of (6.38). In this case, the stability relation is somewhat simplified because it is formulated in terms of the eigenvalues of a 3×3 matrix rather than a sixth-order polynomial in s. In terms of the eigenvalues, the stability requirement is seen from the relationship between Λ and s:

$$s^2 - 2jT_y \Lambda s - 1 = 0 \tag{6.40a}$$

so that

$$s = j\Lambda T_y \pm \sqrt{1 - \left(\Lambda T_y\right)^2} \tag{6.40b}$$

Thus, if $\max\left(\left|\Lambda T_y\right|\right) \leq 1$, then $|s| = 1$ and the technique should be stable. This means that the time step can be made small enough to ensure that the local stability criterion is satisfied. The solutions to (6.39) are

$$\Lambda = 0 \tag{6.41a}$$

$$\Lambda = AS_y C_y \sin\phi \pm \sqrt{\varepsilon_r \mu_r A^2 S_y^2 + aAS_x^2} \tag{6.41b}$$

Maximizing (6.41b) over S_y and S_x allows the stability relation to be quantified. For free space and square grid cells, the result is of the form

$$T_y \leq \frac{\cos^2\phi}{\sin\phi\, F(\phi)\sqrt{1 - F^2(\phi)} + \sqrt{\cos^2\phi + F(\phi)}} \tag{6.42}$$

where $F(\phi)$ is a complicated trigonometric function of the incident angle.

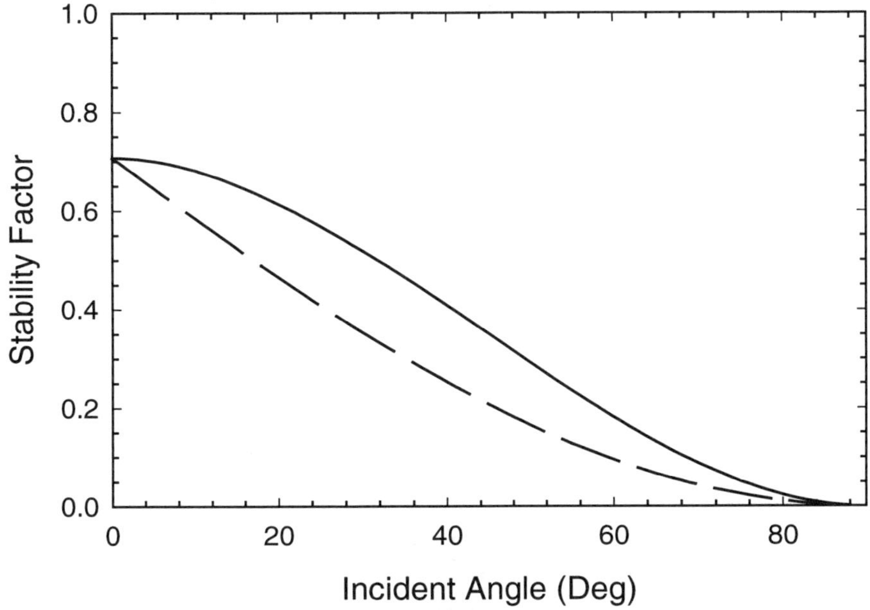

Fig. 6.16 Stability limit for the dual-time technique (solid line). Also shown is the stability limit derived from the maximum velocity in the grid (dashed line).

Fig. 6.16 shows a plot of (6.42) as a function of ϕ for the case of free space and square cells. Note that the maximum allowable time step decreases as ϕ increases, approaching zero as ϕ approaches 90°, indicating that run times for large angles may become impractical. However, this local analysis indicates that the technique should be stable as long as the time step is chosen to be less than what Fig. 6.16 indicates. For comparison, also shown in the figure is a heuristic stability criterion derived from the fact that the fastest velocity in the grid is given by $c/(1-\sin\phi)$. Thus, a simple, conservative stability limit is

$$T_y \leq \frac{1 - \sin\phi}{\sqrt{2}} \tag{6.43}$$

This stability criterion is slightly more restrictive than the one derived from the detailed analysis of the actual discretization used.

Numerical Dispersion Analysis

Another important aspect of a discretization technique is its numerical grid dispersion. Simply speaking, numerical wave solutions in the grid propagate with a slightly different phase velocity than the corresponding exact solutions of the underlying wave equation. Too much dispersion and the numerical solution is not accurate.

The dispersion analysis begins with the following expression obtained from (6.41b):

$$\left(\Lambda \cos^2\phi - S_y C_y \sin\phi\right)^2 = S_x^2 \cos^2\phi + S_y^2 \tag{6.44}$$

Substituting for Λ, S_x, S_y and C_y, and using $s = \exp(j\omega\Delta t/2)$, we obtain the dispersion relation

$$\begin{aligned}&\left[\frac{1}{c\Delta t}\sin\left(\frac{\omega\Delta t}{2}\right)\cos^2\phi - \frac{1}{\Delta y}\sin\left(\frac{\tilde{k}_y^s\Delta y}{2}\right)\cos\left(\frac{\tilde{k}_y^s\Delta y}{2}\right)\sin\phi\right]^2 \\ &= \frac{1}{(\Delta x)^2}\sin^2\left(\frac{\tilde{k}_x^s\Delta x}{2}\right)\cos^2\phi + \frac{1}{(\Delta y)^2}\sin^2\left(\frac{\tilde{k}_y^s\Delta y}{2}\right)\end{aligned} \tag{6.45}$$

where $\tilde{k}_x^s$ and $\tilde{k}_y^s$ are the numerical wavenumbers in the x and y directions, respectively, and the total numerical wavenumber is $\tilde{k}^2 = (\tilde{k}_x^s)^2 + (\tilde{k}_y^s)^2$. The numerical dispersion can now be characterized by comparing the numerical wavelength, $\tilde{\lambda} = 2\pi/\tilde{k}$, to the true wavelength in the medium as a function of the angle of propagation in the grid. Note that this angle is determined by $\arctan(\tilde{k}_y^s/\tilde{k}_x^s)$ and is independent of ϕ, the angle of incidence of the incident field. Using

$$\tilde{k}_x^s = \frac{2\pi}{\tilde{\lambda}}\cos\alpha, \qquad \tilde{k}_y^s = \frac{2\pi}{\tilde{\lambda}}\sin\alpha \tag{6.46}$$

then (6.45) becomes

$$\begin{aligned}&\left[\frac{1}{c\Delta t}\sin\left(\frac{\omega\Delta t}{2}\right)\cos^2\phi - \frac{1}{\Delta y}\sin\left(\frac{\pi\Delta y\sin\alpha}{\tilde{\lambda}}\right)\cos\left(\frac{\pi\Delta y\sin\alpha}{\tilde{\lambda}}\right)\sin\phi\right]^2 \\ &= \frac{1}{(\Delta x)^2}\sin^2\left(\frac{\pi\Delta x\cos\alpha}{\tilde{\lambda}}\right)\cos^2\phi + \frac{1}{(\Delta y)^2}\sin^2\left(\frac{\pi\Delta y\sin\alpha}{\tilde{\lambda}}\right)\end{aligned} \tag{6.47}$$

We note that, in the limit of small cell sizes and time steps, the numerical dispersion relation reduces to (6.26), the continuous case.

The procedure used for evaluating the dispersion is to select a value for α and numerically solve (6.47) for $\tilde{\lambda}$. This numerical wavelength is then compared with the actual wavelength, and a dispersion error per wavelength is computed using

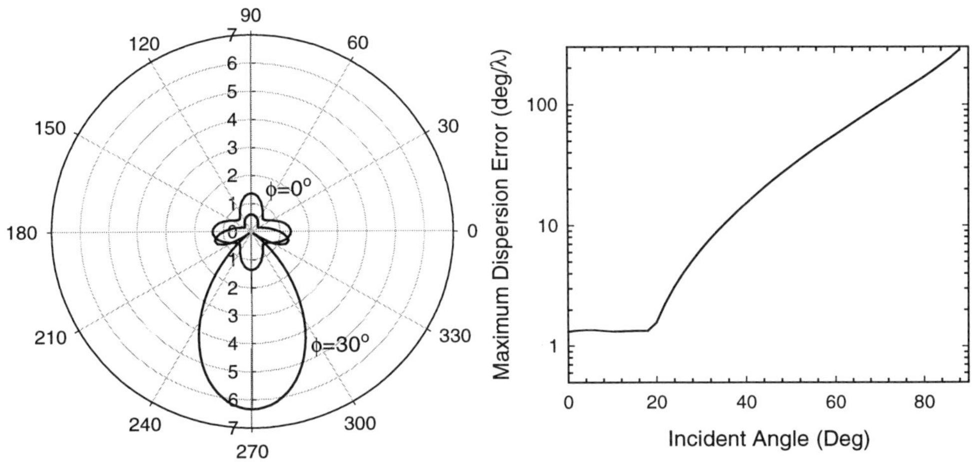

Fig. 6.17 Dispersion error for the dual-time technique. The dispersion error as a function of scattering direction is shown on the left for ϕ =0°, 30°. The maximum dispersion error as a function of the angle of incidence is shown on the right.

$$\Psi_{\text{error}} = \left(\frac{\lambda(\alpha)}{\tilde{\lambda}(\alpha)} - 1 \right) 360^\circ \tag{6.48}$$

Sample results for the numerical dispersion error are shown in Fig. 6.17. On the left side of this figure, the $\phi = 0$ curve is the typical butterfly plot for the numerical dispersion of the standard FDTD algorithm. For small incident angles, the shape of the dispersion error changes, but the size of the error is not significantly larger than the normal-incidence dispersion. However, for incident angles greater than about 20°, noticeable amounts of dispersion error arise. A plot of the maximum dispersion error as a function of the incident angle is shown on the right in Fig. 6.17 for the time step set at one-half of the stability limit. Note the logarithmic scale for the error. The sharp increase in dispersion error near 20° is a cause for concern with this technique.

Global Stability Analysis

To further analyze the dual-time technique, a global stability analysis is performed. This involves expressing the FDTD updates in terms of a matrix multiplied by the current field values. For the dual-time technique this update is written as

$$[x]^{n+1} = [x]^n + [M]\,[x]^{n+1/2} \tag{6.49}$$

where $[x]$ is a column vector of all the field values in the grid, and $[M]$ is a matrix representing how the old field values are combined to find the new ones. Global stability places certain restrictions on the properties of $[M]$. Equation (6.49) is written in a simpler first-order form:

$$[y]^{n+1} = \begin{bmatrix} M & I \\ I & 0 \end{bmatrix} [y]^n = [M'][y]^n \tag{6.50a}$$

where

$$[y]^n = \begin{bmatrix} x^n \\ x^{n-1/2} \end{bmatrix} \tag{6.50b}$$

The eigenvalues of $[M']$ determine whether the algorithm will be stable. If the magnitude of any eigenvalue is greater than one, then the algorithm eventually exhibits instability. (The stability condition can also be expressed in terms of the eigenvalues of $[M]$, but we will not discuss that relationship here [20].)

Global stability analyses of the dual-time technique indicate that it is unstable for any nonzero time step for angles greater than 20° when the grid contains a single dielectric cube or when the excitation is a point source. Thus, even though the local analysis indicates numerical stability, this is not a sufficient condition to ensure overall stability. These results are consistent with the outcome of our numerical experiments. Thus, this approach has very limited appeal.

6.4.3 Multiple-Spatial-Grid Approach

Another technique that has been used to discretize the transformed field equations involves multiple spatial grids [21]. For the simple two-dimensional case we have considered so far, only one additional spatial grid is prescribed. Fig. 6.18 shows the two grids needed for this case (compare with Fig. 6.15 for the single grid).

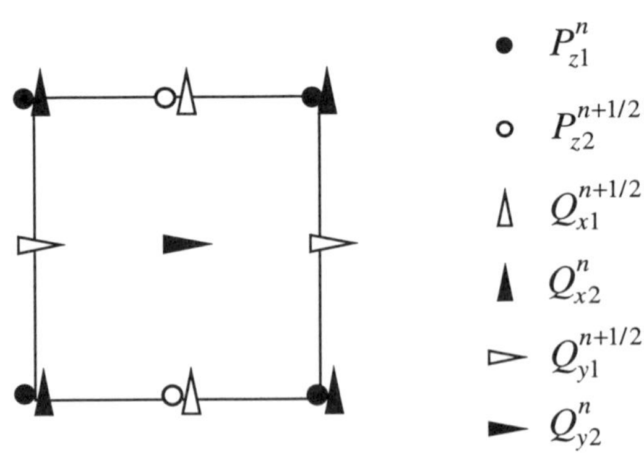

Fig. 6.18 Schematic of the two spatial grids required for the multiple-spatial-grid approach in two dimensions.

In Fig. 6.18, solid black symbols denote field components at time levels n, $n+1$, and so forth, while white symbols denote field components at time levels $n-1/2$, $n+1/2$, and so forth. The additional subscript indicates in which grid the field component resides. For example, P_{z1} is the P_z component in grid 1. Overall, this figure depicts the second grid as simply a shifted version of the first grid in both space and time. Spatially, the second grid is shifted one-half cell in the direction of periodicity (the y direction in this case). Temporally, it is shifted one-half time step from the first grid. (The more general three-dimensional case involves four spatial grids; see [21] for details.) This arrangement permits centered differences in space and time to be applied to the system of (6.22), with the space derivatives involving quantities located at most one-half cell away.

Numerical Stability Analysis

We begin with a stability analysis of the multiple-spatial-grid technique applied to the system of (6.22). The equations to update the white-symbol field components of Fig. 6.18 to time level $n+1/2$ are

$$\frac{Q_{x1}\big|_{\bar{i},\bar{j}}^{n+1/2} - Q_{x1}\big|_{\bar{i},\bar{j}}^{n-1/2}}{T_y} = -\varepsilon_r A\left(P_{z1}\big|_{\bar{i},\bar{j}}^{n} - P_{z1}\big|_{\bar{i},\bar{j}-1}^{n}\right) + aB\left(Q_{y2}\big|_{\bar{i},\bar{j}-1}^{n} - Q_{y2}\big|_{\bar{i}-1,\bar{j}-1}^{n}\right) - B\left(Q_{x2}\big|_{\bar{i},\bar{j}}^{n} - Q_{x2}\big|_{\bar{i},\bar{j}-1}^{n}\right) \tag{6.51a}$$

$$\frac{Q_{y1}\big|_{\bar{i},\bar{j}}^{n+1/2} - Q_{y1}\big|_{\bar{i},\bar{j}}^{n-1/2}}{T_y} = \frac{a}{\mu_r}\left(P_{z1}\big|_{\bar{i}+1,\bar{j}}^{n} - P_{z1}\big|_{\bar{i},\bar{j}}^{n}\right) \tag{6.51b}$$

$$\frac{P_{z2}\big|_{\bar{i},\bar{j}}^{n+1/2} - P_{z2}\big|_{\bar{i},\bar{j}}^{n-1/2}}{T_y} = -B\left(P_{z1}\big|_{\bar{i},\bar{j}}^{n} - P_{z1}\big|_{\bar{i},\bar{j}-1}^{n}\right) + \mu_r aA\left(Q_{y2}\big|_{\bar{i},\bar{j}-1}^{n} - Q_{y2}\big|_{\bar{i}-1,\bar{j}-1}^{n}\right) - \mu_r A\left(Q_{x2}\big|_{\bar{i},\bar{j}}^{n} - Q_{x2}\big|_{\bar{i},\bar{j}-1}^{n}\right) \tag{6.51c}$$

The equations to update the black-symbol field components to time level $n+1$ are

$$\frac{Q_{x2}\big|_{\bar{i},\bar{j}}^{n+1} - Q_{x2}\big|_{\bar{i},\bar{j}}^{n}}{T_y} = -\varepsilon_r A\left(P_{z2}\big|_{\bar{i},\bar{j}}^{n+1/2} - P_{z2}\big|_{\bar{i},\bar{j}-1}^{n+1/2}\right) + aB\left(Q_{y1}\big|_{\bar{i},\bar{j}-1}^{n+1/2} - Q_{y1}\big|_{\bar{i}-1,\bar{j}-1}^{n+1/2}\right) - B\left(Q_{x1}\big|_{\bar{i},\bar{j}}^{n+1/2} - Q_{x1}\big|_{\bar{i},\bar{j}-1}^{n+1/2}\right) \tag{6.52a}$$

$$\frac{Q_{y2}\big|_{\bar{i},\bar{j}}^{n+1} - Q_{y2}\big|_{\bar{i},\bar{j}}^{n}}{T_y} = \frac{a}{\mu_r}\left(P_{z2}\big|_{\bar{i}+1,\bar{j}}^{n+1/2} - P_{z2}\big|_{\bar{i},\bar{j}}^{n+1/2}\right) \tag{6.52b}$$

$$\begin{aligned}\frac{P_{z1}\big|_{\bar{i},\bar{j}}^{n+1} - P_{z1}\big|_{\bar{i},\bar{j}}^{n}}{T_y} &= -B\left(P_{z2}\big|_{\bar{i},\bar{j}}^{n+1/2} - P_{z2}\big|_{\bar{i},\bar{j}-1}^{n+1/2}\right) \\ &+ \mu_r aA\left(Q_{y1}\big|_{\bar{i},\bar{j}-1}^{n+1/2} - Q_{y1}\big|_{\bar{i}-1,\bar{j}-1}^{n+1/2}\right) - \mu_r A\left(Q_{x1}\big|_{\bar{i},\bar{j}}^{n+1/2} - Q_{x1}\big|_{\bar{i},\bar{j}-1}^{n+1/2}\right)\end{aligned} \tag{6.52c}$$

where $B = A\sin\phi$, $A = 1/(\mu_r\varepsilon_r - \sin^2\phi)$, and $a = \Delta y/\Delta x$.

Assuming a generalized plane wave propagating in the grid as specified in (6.32), the stability matrix determinant expression is given by

$$\begin{vmatrix} \Lambda & 0 & 0 & -BS_y & aBS_x & -\varepsilon_r AS_y \\ 0 & \Lambda & 0 & 0 & 0 & aS_x/\mu_r \\ 0 & 0 & \Lambda & -\mu_r AS_y & \mu_r aAS_x & -BS_y \\ -BS_y & aBS_x & -\varepsilon_r AS_y & \Lambda & 0 & 0 \\ 0 & 0 & aS_x/\mu_r & 0 & \Lambda & 0 \\ -\mu_r AS_y & \mu_r aAS_x & -BS_y & 0 & 0 & \Lambda \end{vmatrix} = 0 \tag{6.53}$$

Solving (6.53) for the nonzero values of Λ yields

$$\Lambda = \pm BS_y \pm \sqrt{aAS_x^2 + \mu_r\varepsilon_r A^2 S_y^2} \tag{6.54}$$

Since S_x and S_y lie between ± 1, the stability limit for free space is found to be

$$T_y \le \frac{1}{\Lambda_{\max}} = \frac{\cos^2\phi}{\sin\phi + \sqrt{1 + a\cos^2\phi}} \tag{6.55}$$

Fig. 6.19 is a plot of the stability limit as a function of the angle of incidence. For comparison, stability criterion (6.43) derived from the fastest velocity in the grid is also shown. The multiple-grid stability limit is less restrictive by a factor of 1.22 at 40° and by a factor of 1.4 at 88°.

Numerical Dispersion Analysis

The numerical dispersion for the multiple-grid technique is readily found from (6.54). For free space, the dispersion equation is of the form

$$\left(\Lambda\cos^2\phi - S_y\sin\phi\right)^2 = aS_x^2\cos^2\phi + S_y^2 \tag{6.56}$$

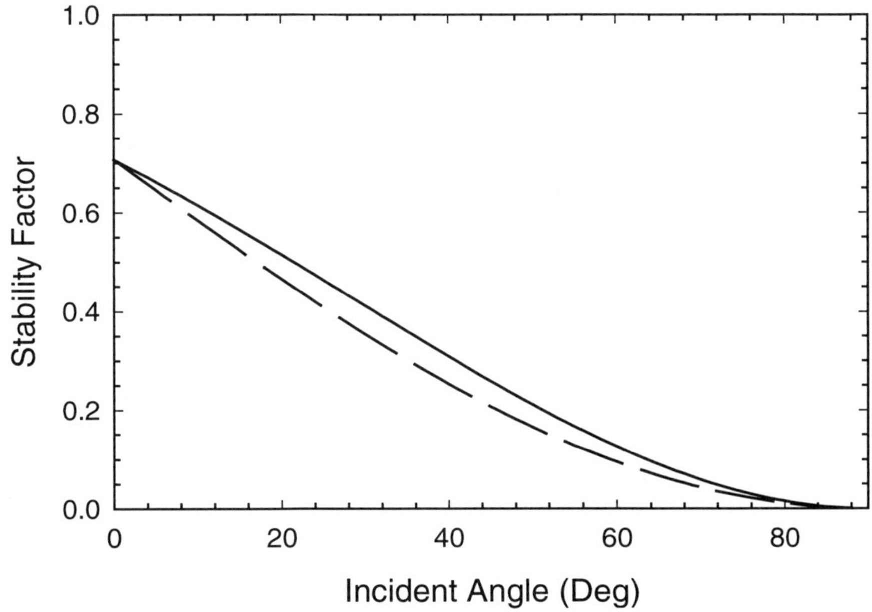

Fig. 6.19 Stability limit for the two-dimensional multiple-spatial-grid technique (solid line). Also shown is the conservative stability limit derived from the maximum velocity in the grid (dashed line).

where

$$\Lambda = \frac{1}{T_y}\sin\left(\frac{\omega\Delta t}{2}\right) = \frac{1}{T_y}\sin\left(\frac{\pi\Delta y T_y}{\lambda_0}\right) \tag{6.57a}$$

$$S_x = \sin\left(\frac{\pi\Delta x\cos\alpha}{\tilde{\lambda}}\right), \qquad S_y = \sin\left(\frac{\pi\Delta y\sin\alpha}{\tilde{\lambda}}\right) \tag{6.57b}$$

The numerical dispersion is characterized by finding $\tilde{\lambda}(\alpha)$ that satisfies (6.56) for a given angle of incidence, ϕ, and discretization parameters. Fig. 6.20 shows results for Ψ_{error} defined in (6.48) for 0° and 30° angles of incidence. In each case, the discretization is at 20 cells per wavelength with a time step of one-half the stability limit. To characterize the dispersion for all angles, we look at the maximum dispersion error as a function of the angle of incidence. This result is shown on the right in Fig. 6.20. The dispersion error increases with the angle of incidence. However, the magnitude is much more reasonable than with the dual-time technique.

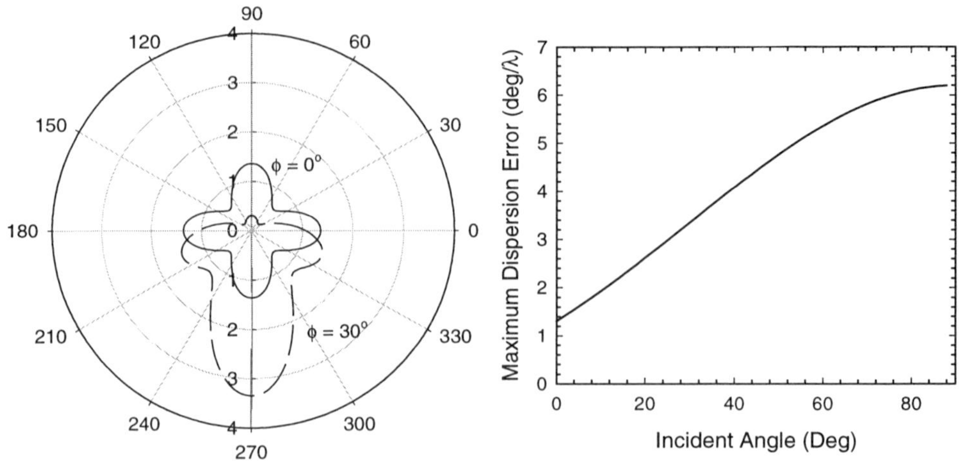

Fig. 6.20 Dispersion error per wavelength for the multiple-grid technique. The dispersion error as a function of scattering direction is shown on the left for $\phi = 0°, 30°$. The maximum dispersion error as a function of the angle of incidence is shown on the right.

Lossy Materials

Introducing lossy materials into the grid makes the update equations more complex. We start with the transformed field equations with both magnetic and electric loss:

$$\frac{\mu_r}{c}\frac{\partial Q_x}{\partial t} + \frac{\sigma^*}{\eta_0}Q_x = -\frac{\partial P_z}{\partial y} + \left\{\frac{\sin\phi}{c}\frac{\partial P_z}{\partial t}\right\} \tag{6.58a}$$

$$\frac{\mu_r}{c}\frac{\partial Q_y}{\partial t} + \frac{\sigma^*}{\eta_0}Q_y = \frac{\partial P_z}{\partial x} \tag{6.58b}$$

$$\frac{\varepsilon_r}{c}\frac{\partial P_z}{\partial t} + \sigma\eta_0 P_z = \frac{\partial Q_y}{\partial x} - \frac{\partial Q_x}{\partial y} + \left\{\frac{\sin\phi}{c}\frac{\partial Q_x}{\partial t}\right\} \tag{6.58c}$$

where σ and σ^* are the electric and magnetic conductivities, respectively.

The time derivatives on the right-hand sides of (6.58a) and (6.58c) are eliminated by substituting (6.58c) into (6.58a) and vice versa. This yields

$$\begin{aligned}\left(\frac{\varepsilon_r\mu_r - \sin^2\phi}{c}\right)\frac{\partial Q_x}{\partial t} + \frac{\varepsilon_r\sigma^*}{\eta_0}Q_x &= \\ -\varepsilon_r\frac{\partial P_z}{\partial y} + \left(\frac{\partial Q_y}{\partial x} - \frac{\partial Q_x}{\partial y}\right)&\sin\phi - \sigma\eta_0\sin\phi\, P_z\end{aligned} \tag{6.59a}$$

$$\frac{\mu_r}{c}\frac{\partial Q_y}{\partial t} + \frac{\sigma^*}{\eta_0}Q_y = \frac{\partial P_z}{\partial x} \tag{6.59b}$$

$$\left(\frac{\varepsilon_r \mu_r - \sin^2\phi}{c}\right)\frac{\partial P_z}{\partial t} + \mu_r \sigma \eta_0 P_z = - \frac{\partial P_z}{\partial y}\sin\phi + \mu_r\left(\frac{\partial Q_y}{\partial x} - \frac{\partial Q_x}{\partial y}\right) - \frac{\sigma^* \sin\phi}{\eta_0}Q_x \tag{6.59c}$$

In using the multiple spatial grid technique to discretize the system of (6.59), there are several options on what to do with the loss terms on the left-hand sides. Consider the space-time diagram in Fig. 6.21.

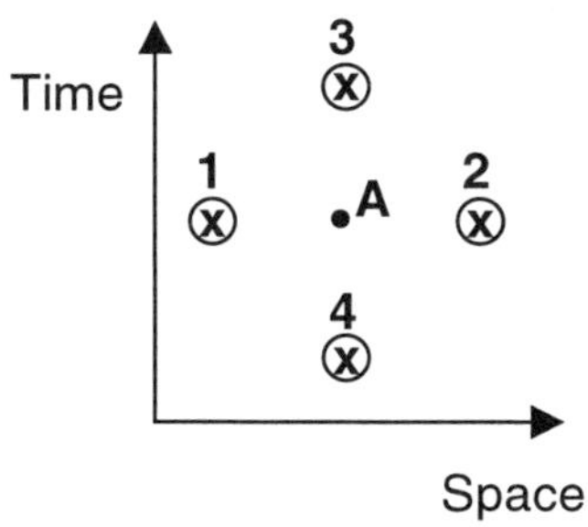

Fig. 6.21 Space-time diagram of the fields in the multiple-grid technique. The field is needed at point A, and must be estimated from the fields at points 1, 2, 3, and 4.

To maintain a centered approach, we need to have the field at the point A, but the field is only known at points 1 through 4. Points 1 and 2 are at the correct time, but not at the desired space point, while points 3 and 4 are correct spatially but not temporally. We may choose to perform a spatial average of points 1 and 2, a temporal average of points 3 and 4, or some combination of both. A general average using all four points is written as

$$\frac{\varepsilon_r \sigma^*}{\eta_0}Q_x \rightarrow \left(\frac{1-2\beta}{2}\right)\left[Q_x(1) + Q_x(2)\right] + \beta\left[Q_x(3) + Q_x(4)\right] \tag{6.60}$$

The first term on the right-hand side of (6.60) is a spatial average of the fields in the second grid, while the second term is a temporal average of the fields in the same grid. The parameter β determines the mixture of the two types of averages; $\beta = 0$ implements spatial averaging only, while $\beta = 0.5$ implements temporal averaging only. The choice of β impacts the stability of the algorithm. It is found that $\beta \geq 0.5$ is a good choice from a stability standpoint. In our modeling efforts, we use a value of 0.5, resulting in temporal averaging only.

Lossy Screen Example

To illustrate the multiple-spatial-grid technique, the transmission through the lossy screen discussed earlier is calculated. Fig. 6.22 shows the results for an angle of incidence of 30°, along with the results from the frequency-domain mode-matching technique. There is excellent agreement between the two techniques.

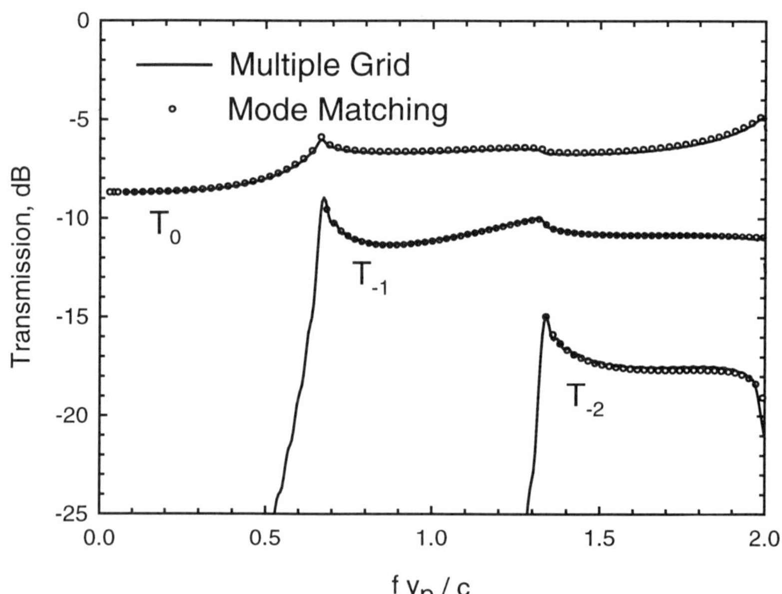

Fig. 6.22 Comparison of calculated transmission through the lossy screen using the multiple-spatial-grid technique and mode matching.

The multiple-grid technique appears to be a viable means to handle the transformed field equations. Numerical stability problems are abated at the cost of introducing additional field components. Compared to the traditional three-dimensional Yee algorithm that has 6 field components per unit cell, the multiple-grid technique has 24 per cell. Using Berenger's PML ABC as in [21], there are additional variables introduced that bring the total to 40 per cell. However, because of the periodic boundary conditions, this large number of variables per cell does not cause a significant problem. (The transverse directions usually contain far fewer cells than the longitudinal direction.)

Usually running time, not memory, is the limiting factor if data at many angles is desired.

A potentially troubling issue with this technique is that the multiple spatial grids can be decoupled from each other [22]. That is, it is possible to excite only the fields associated with one of the grids, while the fields in the other grids remain zero. Also, certain types of scattering objects affect the fields in one grid only. For example, a thin wire along the edge of a cell scatters into only one of the grids. Other scattering objects impact the separate grids differently. For objects covering many cells, the level of error caused by the grid decoupling should be small, but it is something to bear in mind when using this technique.

6.4.4 Split-Field Approach

The final method to discretize the transformed field equations is called the split-field approach [23 – 25]. We first discuss the two-dimensional version to introduce the technique, leaving the three-dimensional case for a separate section.

Consider the system of transformed field equations of (6.21) before any substitutions are made to eliminate the additional time-derivative terms in brackets on the right-hand sides. To eliminate these terms, we define a new set of split variables as follows:

$$Q_x = Q_{xa} + \frac{\sin\phi}{\mu_r} P_z \tag{6.61a}$$

$$P_z = P_{za} + \frac{\sin\phi}{\varepsilon_r} Q_x \tag{6.61b}$$

Substituting (6.61a) into the left-hand side of (6.21a), and (6.61b) into the left-hand side of (6.21c), the resulting equations are

$$\frac{\mu_r}{c}\frac{\partial Q_{xa}}{\partial t} = -\frac{\partial P_z}{\partial y} \tag{6.62a}$$

$$\frac{\mu_r}{c}\frac{\partial Q_y}{\partial t} = \frac{\partial P_z}{\partial x} \tag{6.62b}$$

$$\frac{\varepsilon_r}{c}\frac{\partial P_{za}}{\partial t} = \frac{\partial Q_y}{\partial x} - \frac{\partial Q_x}{\partial y} \tag{6.62c}$$

Equations (6.62a) and (6.62c) show that the new variables, P_{za} and Q_{xa}, defined by (6.61) satisfy the standard Maxwell's equations, with the caveat that the curl terms on the right involve the total fields. Once Q_{xa} and P_{za} are known, the total fields can be found using (6.61). We note also that (6.61) require that the field components be collocated in time.

The systems of equations to be solved are (6.62) and then (6.61). System (6.62) allows updating the "a" portions of the split fields in terms of the total fields. However, system (6.61) still involves the total fields on the right-hand sides, and must be manipulated to eliminate this dependence. (The total fields are not known for the updates yet.) There are three ways to accomplish this, with each way having unique numerical stability and dispersion characteristics. The three possibilities are now summarized:

1. Substitute (6.61b) into (6.61a) to eliminate P_z to form a new equation to use with (6.61b). The two equations to use are thus

$$Q_x\left(1 - \frac{\sin^2\phi}{\mu_r\,\varepsilon_r}\right) = Q_{xa} + \frac{\sin\phi}{\mu_r}P_{za} \tag{6.63a}$$

$$\varepsilon_r P_z = \varepsilon_r P_{za} + Q_x \sin\phi \tag{6.63b}$$

Equation (6.63a) determines Q_x at time step, n, from Q_{xa} and P_{za} at the same time step. This value for Q_x can then be used to determine P_z using (6.63b). Spatial averaging is needed for the P_{za} term in (6.63a) and the Q_x term in (6.63b) since the fields are not collocated in the grid.

2. Substitute (6.61a) into (6.61b) to form an equation for P_z in terms of Q_{xa} and P_{za}. In this case the two equations are

$$P_z\left(1 - \frac{\sin^2\phi}{\mu_r\,\varepsilon_r}\right) = P_{za} + \frac{\sin\phi}{\varepsilon_r}Q_{xa} \tag{6.64a}$$

$$\mu_r Q_x = \mu_r Q_{xa} + P_z \sin\phi \tag{6.64b}$$

The comments about spatial averaging made above apply in this case as well. Both approaches have similar, but not identical, dispersion characteristics. However, the first approach has a slightly more relaxed stability limit.

3. Eliminate the total fields from the right-hand sides by substituting (6.61b) into (6.61a) and (6.61a) into (6.61b). The result is simply (6.63a) and (6.64a). While this approach has appeal due to the symmetry with which the field components appear, the dispersion error associated with it is significant. In fact, the stability and dispersion characteristics are identical to that of the dual-time approach given by (6.41b). For this reason, either of the first two approaches should be used rather than this one.

In discretized form, the system of (6.62) becomes

$$Q_{xa}\Big|_{\bar{i},\bar{j}}^{n+1/2} = Q_{xa}\Big|_{\bar{i},\bar{j}}^{n-1/2} - \frac{T_y}{\mu_r}\left(P_z\Big|_{\bar{i},\bar{j}}^{n} - P_z\Big|_{\bar{i},\bar{j}-1}^{n}\right) \tag{6.65a}$$

$$Q_y\Big|_{\bar{i},\bar{j}}^{n+1/2} = Q_y\Big|_{\bar{i},\bar{j}}^{n-1/2} + \frac{aT_y}{\mu_r}\left(P_z\Big|_{\bar{i},\bar{j}}^{n} - P_z\Big|_{\bar{i}-1,\bar{j}}^{n}\right) \tag{6.65b}$$

$$P_{za}\Big|_{\bar{i},\bar{j}}^{n+1/2} = P_{za}\Big|_{\bar{i},\bar{j}}^{n-1/2} + \frac{aT_y}{\varepsilon_r}\left(Q_y\Big|_{\bar{i},\bar{j}}^{n} - Q_y\Big|_{\bar{i}-1,\bar{j}}^{n}\right) - \frac{T_y}{\varepsilon_r}\left(Q_x\Big|_{\bar{i},\bar{j}}^{n} - Q_x\Big|_{\bar{i},\bar{j}-1}^{n}\right) \tag{6.65c}$$

The system of (6.63) is discretized as

$$Q_x\Big|_{\bar{i},\bar{j}}^{n+1/2} = \mu_r\varepsilon_r A\, Q_{xa}\Big|_{\bar{i},\bar{j}}^{n+1/2} + \frac{\varepsilon_r A\sin\phi}{2}\left(P_{za}\Big|_{\bar{i},\bar{j}+1}^{n+1/2} + P_{za}\Big|_{\bar{i},\bar{j}}^{n+1/2}\right) \tag{6.66a}$$

$$P_z\Big|_{\bar{i},\bar{j}}^{n+1/2} = P_{za}\Big|_{\bar{i},\bar{j}}^{n+1/2} + \frac{\sin\phi}{2\varepsilon_r}\left(Q_x\Big|_{\bar{i},\bar{j}}^{n+1/2} + Q_x\Big|_{\bar{i},\bar{j}-1}^{n+1/2}\right) \tag{6.66b}$$

where $T_y = c\Delta t/\Delta y$, $A = 1/(\varepsilon_r\mu_r - \sin^2\phi)$, and $a = \Delta y/\Delta x$. These two systems of equations are the basic update relations for the split-field technique in the two-dimensional case under consideration.

Numerical Stability Analysis

As with the other techniques, it is important to understand the numerical stability and dispersion characteristics. Assuming a generalized plane wave propagating in the grid, the determinant expression for the stability matrix of the combined systems of (6.65) and (6.66) is given by

$$\begin{vmatrix} \Lambda & 0 & 0 & 0 & -S_y/\mu_r \\ 0 & \Lambda & 0 & 0 & aS_x/\mu_r \\ 0 & aS_x/\varepsilon_r & \Lambda & -S_y/\varepsilon_r & 0 \\ \mu_r\varepsilon_r A & 0 & \varepsilon_r A C_y\sin\phi & -1 & 0 \\ 0 & 0 & 1 & C_y\sin\phi/\varepsilon_r & -1 \end{vmatrix} = 0 \tag{6.67a}$$

where

$$\Lambda = \frac{1}{2jT_y}\left(s - \frac{1}{s}\right) \tag{6.67b}$$

and the order of the field components in the column vector is $[Q_{xa}, Q_y, P_{za}, Q_x, P_z]^T$. Solving (6.67) for Λ yields

$$\Lambda = AS_yC_y\sin\phi \pm \sqrt{\left(AS_y^2 + \frac{a^2}{\mu_r\,\varepsilon_r}S_x^2\right)\left(1 + AC_y^2\sin^2\phi\right)} \tag{6.68}$$

Stability requires that the largest value of $|\Lambda T_y|$ be no greater than one for any value of S_x and S_y between ± 1. It is easy to see that the maximum Λ will occur for $S_x = \pm 1$, but it is less evident which value of S_y should be used since both C_y and S_y are functions of k_y^s. After considerable manipulation, the stability limit for square cells and free space is derived:

$$T_y \le \frac{\cos^2\phi}{\sqrt{1+\cos^2\phi}} \tag{6.69}$$

A plot of this stability limit is shown in Fig. 6.23. The split-field technique has a more relaxed stability requirement than the other transformed field techniques.

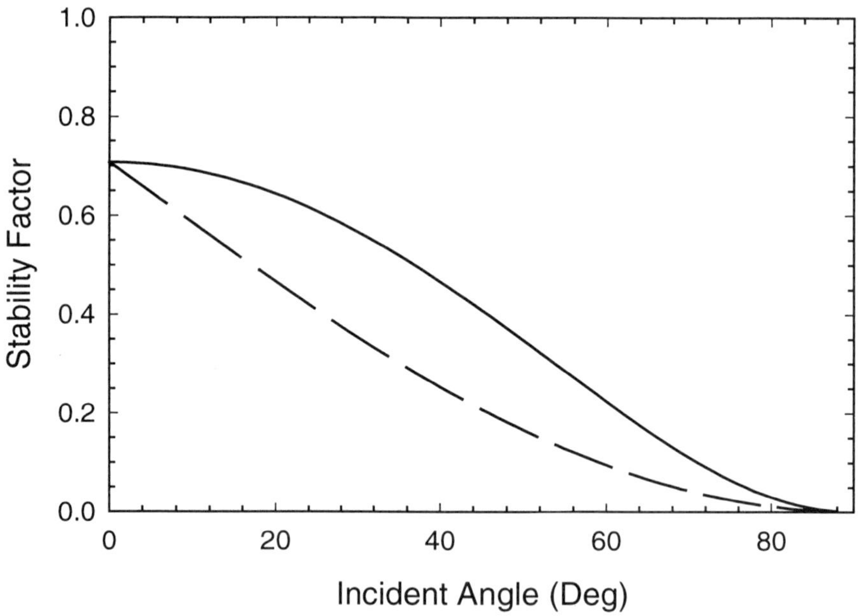

Fig. 6.23 Stability limit of the split-field technique (solid line). Also shown is the conservative stability limit derived from the maximum velocity in the grid (dashed line).

The numerical dispersion of the split-field technique is found by rewriting (6.68) in free space as

$$\left(\Lambda\cos^2\phi - S_y C_y \sin\phi\right)^2 = \left(S_y^2 + a^2 S_x^2 \cos^2\phi\right)\left(\cos^2\phi + C_y^2 \sin^2\phi\right) \tag{6.70}$$

where

$$\Lambda = \frac{1}{T_y}\sin\left(\frac{\omega\Delta t}{2}\right) = \frac{1}{T_y}\sin\left(\frac{\pi\Delta y T_y}{\lambda_0}\right) \tag{6.71a}$$

$$S_x = \sin\left(\frac{\pi\Delta x\cos\alpha}{\tilde{\lambda}}\right), \qquad S_y = \sin\left(\frac{\pi\Delta y\sin\alpha}{\tilde{\lambda}}\right) \tag{6.71b}$$

$$C_y = \cos\left(\frac{\pi\Delta y\sin\alpha}{\tilde{\lambda}}\right) \tag{6.71c}$$

Equation (6.70) is solved to determine the numerical wavelength as a function of the scattered direction α for a given angle of incidence and discretization parameters.

Fig. 6.24 shows the calculated dispersion error in degrees per wavelength for 0° and 30° angles of incidence. In each case, the discretization is at 20 cells per wavelength, and a time step of one-half the stability limit is used. To characterize the dispersion for all angles, we look at the maximum dispersion error as a function of the angle of incidence. This result is shown on the right in Fig. 6.24.

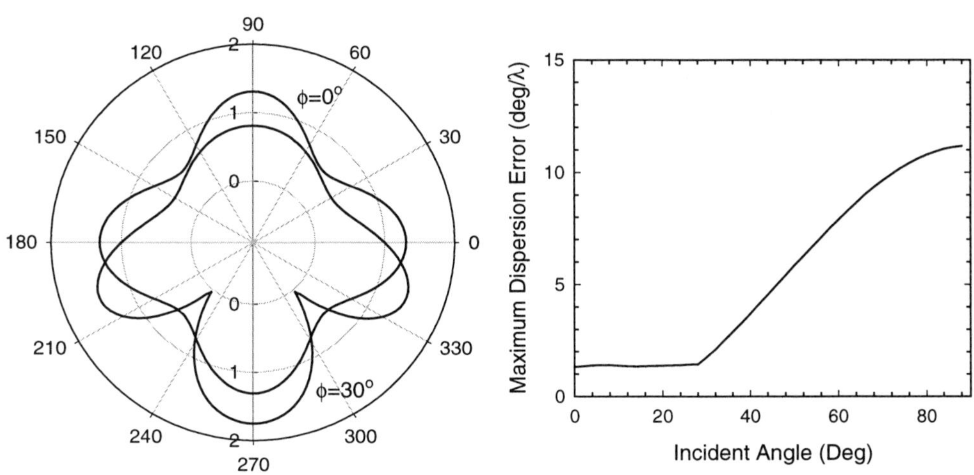

Fig. 6.24 Dispersion error per wavelength for the split-field technique.

As with all of the techniques discussed, the dispersion error increases with the angle of incidence. Although the magnitude of the dispersion error is larger than for the multiple-grid technique at angles larger than 40°, the magnitude is much less than with the dual-time technique.

Lossy Materials

The equations for the split-field technique must be modified to allow the inclusion of lossy materials in the grid. In fact, the introduction of lossy materials requires additional stability considerations. Whereas modeling lossy materials using the Yee cell with the standard FDTD equations does not impact the stability criterion, the usual approach fails when applied to the split-field equations.

The transformed equations with both electric and magnetic loss are given by the system of (6.58). Substituting (6.61a) and (6.61b) into the left-hand sides of (6.58a) and (6.58c), respectively, yields

$$\frac{\mu_r}{c}\frac{\partial Q_{xa}}{\partial t} + \frac{\sigma^*}{\eta_0}Q_{xa} = -\frac{\partial P_z}{\partial y} - \frac{\sigma^* \sin\phi}{\eta_0 \mu_r}P_z \tag{6.72a}$$

$$\frac{\mu_r}{c}\frac{\partial Q_y}{\partial t} + \frac{\sigma^*}{\eta_0}Q_y = \frac{\partial P_z}{\partial x} \tag{6.72b}$$

$$\frac{\varepsilon_r}{c}\frac{\partial P_{za}}{\partial t} + \sigma\eta_0 P_{za} = \frac{\partial Q_y}{\partial x} - \frac{\partial Q_x}{\partial y} + \frac{\sigma\eta_0 \sin\phi}{\varepsilon_r}Q_x \tag{6.72c}$$

The system of (6.72) along with the system of (6.63) are the equations to be solved. The usual approach in FDTD to discretize a field update with loss (for example, (6.72a)) is to time-average the loss term. With that approach, (6.72a) becomes

$$\begin{aligned}\frac{\mu_r}{c\Delta t}\left(Q_{xa}\Big|_{\bar{i},\bar{j}}^{n+1} - Q_{xa}\Big|_{\bar{i},\bar{j}}^{n}\right) + \frac{\sigma^*}{2\eta_0}\left(Q_{xa}\Big|_{\bar{i},\bar{j}}^{n+1} + Q_{xa}\Big|_{\bar{i},\bar{j}}^{n}\right) = \\ -\frac{1}{\Delta y}\left(P_z\Big|_{\bar{i},\bar{j}+1}^{n+1/2} - P_z\Big|_{\bar{i},\bar{j}}^{n+1/2}\right) - \frac{\sigma^* \sin\phi}{2\eta_0\mu_r}\left(P_z\Big|_{\bar{i},\bar{j}+1}^{n+1/2} + P_z\Big|_{\bar{i},\bar{j}}^{n+1/2}\right)\end{aligned} \tag{6.73}$$

Unfortunately, with this formalism, the approach is unstable for angles larger than about 30°. A more general approach for the time average is to use [20, 26]

$$\frac{\sigma^*}{\eta_0}Q_{xa} \rightarrow \frac{\sigma^*}{\eta_0}\left[\beta Q_{xa}\Big|_{\bar{i},\bar{j}}^{n+1} + (1-2\beta)Q_{xa}\Big|_{\bar{i},\bar{j}}^{n+1/2} + \beta Q_{xa}\Big|_{\bar{i},\bar{j}}^{n}\right] \tag{6.74}$$

By allowing β to be a function of the angle of incidence, the technique can be made stable. Specifically, there is a minimum value of β required for stability at each angle of incidence. However, the minimum β for stability does not have the largest allowable time step and is therefore not optimal. The choice for β which gives the largest allowable time step for stability is empirically determined to be

$$\beta_{\text{opt}} = \frac{0.5}{\cos^2 \phi} \tag{6.75}$$

With this choice of β, the stability criterion is well approximated by (6.69), the stability limit for the lossless case.

Lossy Screen Example

To illustrate the applicability of the split-field technique to lossy materials, the transmission through the standard lossy screen is calculated. Fig. 6.25 shows the results for an angle of incidence of 30°, along with the results from the frequency-domain mode-matching technique. There is excellent agreement between the two methods.

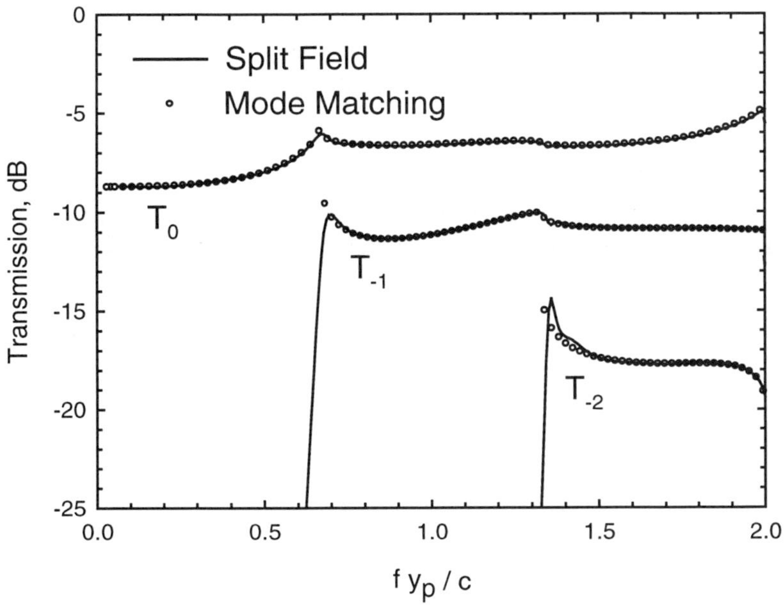

Fig. 6.25 Comparison of transmission through the lossy screen calculated using the split-field and mode-matching techniques.

6.4.5 Three-Dimensional Split-Field Technique

The split-field technique can be easily generalized to the three-dimensional case. This section contains a discussion of the technique as applied to a general, periodic structure that may contain lossy, anisotropic materials. Fig. 6.26 shows the geometry that is considered. The periodic boundaries are the y and z sides, while absorbing boundaries are used to truncate the lattice in the x direction.

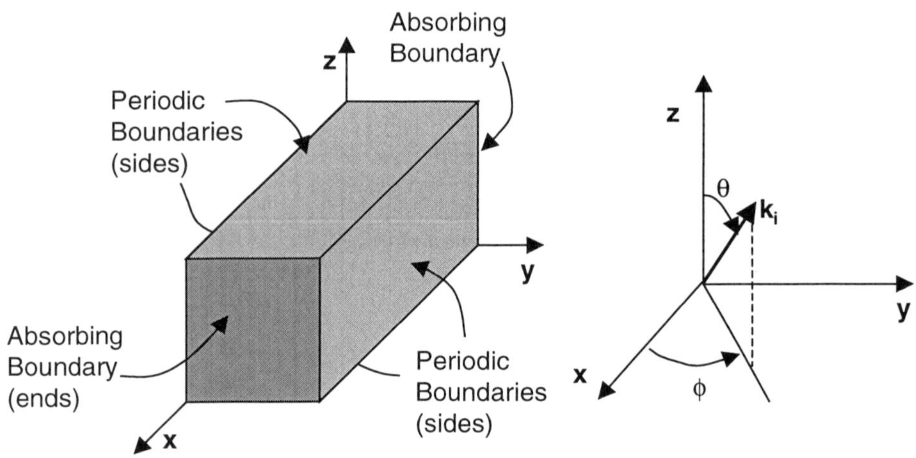

Fig. 6.26 Geometry used in the three-dimensional analysis.

We start with the Maxwell's equations for the transformed field variables, given by

$$\frac{\varepsilon_{xr}}{c}\frac{\partial P_x}{\partial t} + \sigma_x \eta_0 P_x = \frac{\partial Q_z}{\partial y} - \frac{\partial Q_y}{\partial z} - \frac{\bar{k}_y}{c}\frac{\partial Q_z}{\partial t} + \frac{\bar{k}_z}{c}\frac{\partial Q_y}{\partial t} \tag{6.76a}$$

$$\frac{\varepsilon_{yr}}{c}\frac{\partial P_y}{\partial t} + \sigma_y \eta_0 P_y = -\frac{\partial Q_z}{\partial x} + \frac{\partial Q_x}{\partial z} - \frac{\bar{k}_z}{c}\frac{\partial Q_x}{\partial t} \tag{6.76b}$$

$$\frac{\varepsilon_{zr}}{c}\frac{\partial P_z}{\partial t} + \sigma_z \eta_0 P_z = \frac{\partial Q_y}{\partial x} - \frac{\partial Q_x}{\partial y} + \frac{\bar{k}_y}{c}\frac{\partial Q_x}{\partial t} \tag{6.76c}$$

and

$$\frac{\mu_{xr}}{c}\frac{\partial Q_x}{\partial t} + \frac{\sigma_x^*}{\eta_0} Q_x = -\frac{\partial P_z}{\partial y} + \frac{\partial P_y}{\partial z} + \frac{\bar{k}_y}{c}\frac{\partial P_z}{\partial t} - \frac{\bar{k}_z}{c}\frac{\partial P_y}{\partial t} \tag{6.77a}$$

$$\frac{\mu_{yr}}{c}\frac{\partial Q_y}{\partial t} + \frac{\sigma_y^*}{\eta_0}Q_y = \frac{\partial P_z}{\partial x} - \frac{\partial P_x}{\partial z} + \frac{\bar{k}_z}{c}\frac{\partial P_x}{\partial t} \tag{6.77b}$$

$$\frac{\mu_{zr}}{c}\frac{\partial Q_z}{\partial t} + \frac{\sigma_z^*}{\eta_0}Q_z = -\frac{\partial P_y}{\partial x} + \frac{\partial P_x}{\partial y} - \frac{\bar{k}_y}{c}\frac{\partial P_x}{\partial t} \tag{6.77c}$$

where we have introduced the notation

$$\hat{k}_i = \sin\theta\cos\phi\,\hat{x} + \sin\theta\sin\phi\,\hat{y} + \cos\theta\,\hat{z} \tag{6.78a}$$

$$\bar{k}_x = \hat{k}_i\cdot\hat{x}\,, \qquad \bar{k}_y = \hat{k}_i\cdot\hat{y}\,, \qquad \bar{k}_z = \hat{k}_i\cdot\hat{z} \tag{6.78b}$$

Defining new variables as in the two-dimensional case, the extra time derivatives on the right-hand sides can be eliminated. Because the angle of incidence is arbitrary, all of the field components are split. The new "a" variables are given by

$$P_x = P_{xa} + \frac{\bar{k}_z}{\varepsilon_{xr}}Q_y - \frac{\bar{k}_y}{\varepsilon_{xr}}Q_z\,, \qquad Q_x = Q_{xa} - \frac{\bar{k}_z}{\mu_{xr}}P_y + \frac{\bar{k}_y}{\mu_{xr}}P_z \tag{6.79a}$$

$$P_y = P_{ya} - \frac{\bar{k}_z}{\varepsilon_{yr}}Q_x\,, \qquad Q_y = Q_{ya} + \frac{\bar{k}_z}{\mu_{yr}}P_x \tag{6.79b}$$

$$P_z = P_{za} + \frac{\bar{k}_y}{\varepsilon_{zr}}Q_x\,, \qquad Q_z = Q_{za} - \frac{\bar{k}_y}{\mu_{zr}}P_x \tag{6.79c}$$

Substituting these variables into the systems (6.76) and (6.77) results in the equations that are discretized. The first six equations are updates for the "a" portions of the fields. These update equations utilize only the total field values, and are given by

$$\frac{\varepsilon_{xr}}{c}\frac{\partial P_{xa}}{\partial t} + \sigma_x\eta_0 P_{xa} = \frac{\partial Q_z}{\partial y} - \frac{\partial Q_y}{\partial z} + \frac{\sigma_x\eta_0\bar{k}_y}{c}Q_z - \frac{\sigma_x\eta_0\bar{k}_z}{c}Q_y \tag{6.80a}$$

$$\frac{\varepsilon_{yr}}{c}\frac{\partial P_{ya}}{\partial t} + \sigma_y\eta_0 P_{ya} = -\frac{\partial Q_z}{\partial x} + \frac{\partial Q_x}{\partial z} + \frac{\sigma_y\eta_0\bar{k}_z}{c}Q_x \tag{6.80b}$$

$$\frac{\varepsilon_{zr}}{c}\frac{\partial P_{za}}{\partial t} + \sigma_z\eta_0 P_{za} = \frac{\partial Q_y}{\partial x} - \frac{\partial Q_x}{\partial y} - \frac{\sigma_z\eta_0\bar{k}_y}{c}Q_x \tag{6.80c}$$

and

$$\frac{\mu_{xr}}{c}\frac{\partial Q_{xa}}{\partial t} + \frac{\sigma_x^*}{\eta_0} Q_{xa} = -\frac{\partial P_z}{\partial y} + \frac{\partial P_y}{\partial z} - \frac{\sigma_x^* \bar{k}_y}{\eta_0 c} P_z + \frac{\sigma_x^* \bar{k}_z}{\eta_0 c} P_y \tag{6.81a}$$

$$\frac{\mu_{yr}}{c}\frac{\partial Q_{ya}}{\partial t} + \frac{\sigma_y^*}{\eta_0} Q_{ya} = \frac{\partial P_z}{\partial x} - \frac{\partial P_x}{\partial z} - \frac{\sigma_y^* \bar{k}_z}{\eta_0 c} P_x \tag{6.81b}$$

$$\frac{\mu_{zr}}{c}\frac{\partial Q_{za}}{\partial t} + \frac{\sigma_z^*}{\eta_0} Q_{za} = -\frac{\partial P_y}{\partial x} + \frac{\partial P_x}{\partial y} + \frac{\sigma_z^* \bar{k}_y}{\eta_0 c} P_x \tag{6.81c}$$

Spatial averaging of the nonderivative terms on the right-hand sides is needed to retain the centered nature of the method.

Once the "a" portions are updated, the new values of the total fields P_x and Q_x are found by algebraically manipulating the system (6.79) to obtain

$$\left(1 - \frac{\bar{k}_y^2}{\mu_{zr}\varepsilon_{xr}} - \frac{\bar{k}_z^2}{\mu_{yr}\varepsilon_{xr}}\right) P_x = -\frac{\bar{k}_y}{\varepsilon_{xr}} Q_{za} + \frac{\bar{k}_z}{\varepsilon_{xr}} Q_{ya} + P_{xa} \tag{6.82a}$$

$$\left(1 - \frac{\bar{k}_y^2}{\mu_{xr}\varepsilon_{zr}} - \frac{\bar{k}_z^2}{\mu_{xr}\varepsilon_{yr}}\right) Q_x = \frac{\bar{k}_y}{\mu_{xr}} P_{za} - \frac{\bar{k}_z}{\mu_{xr}} P_{ya} + Q_{xa} \tag{6.82b}$$

As before in the two-dimensional case, these relations are equalities and not standard update expressions. Spatial averaging is also needed for these equations. We have chosen the x components of the fields to update first. Other combinations are also possible. However, given the behavior observed in the two-dimensional case, we expect stability problems with some of them. We have found that computing the field components normal to the absorbing boundary first gives good results.

Finally, the remaining total fields are computing using

$$P_y = P_{ya} - \frac{\bar{k}_z}{\varepsilon_{yr}} Q_x \tag{6.83a}$$

$$P_z = P_{za} + \frac{\bar{k}_y}{\varepsilon_{zr}} Q_x \tag{6.83b}$$

and

$$Q_y = Q_{ya} + \frac{\bar{k}_z}{\mu_{yr}} P_x \tag{6.84a}$$

$$Q_z = Q_{za} - \frac{\bar{k}_y}{\mu_{zr}} P_x \tag{6.84b}$$

The foregoing equations illustrate that the split-field technique is a three-step update procedure. First, the “a” portions of the fields are updated using the total fields. Second, the normal components are found. Third and last, the remaining tangential components are determined. Each of the equations is spatially discretized with the usual Yee cell as shown in Fig. 6.27. However, spatial and temporal averaging is needed to properly deal with some of the terms. We include the discretized version of one equation from each group to illustrate the procedure.

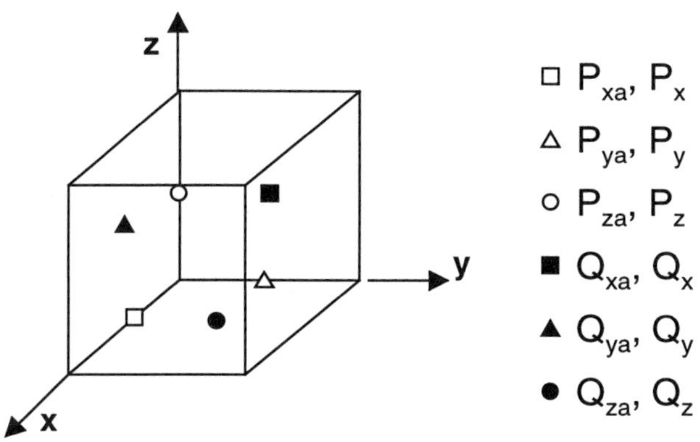

Fig. 6.27 Three-dimensional Yee cell used in the discretization procedure.

The discretized version of (6.80a) is given by

$$\begin{aligned}
P_{xa}\Big|_{\bar{i},\bar{j},\bar{k}}^{n+1} &= \left[\frac{-\xi_x(1-2\beta)}{1+\beta\xi_x}\right] P_{xa}\Big|_{\bar{i},\bar{j},\bar{k}}^{n+1/2} + \left(\frac{1-\beta\xi_x}{1+\beta\xi_x}\right) P_{xa}\Big|_{\bar{i},\bar{j},\bar{k}}^{n} \\
&+ \left[\frac{T_y}{\varepsilon_{xr}(1+\beta\xi_x)}\right]\left(Q_z\Big|_{\bar{i},\bar{j},\bar{k}}^{n+1/2} - Q_z\Big|_{\bar{i},\bar{j}-1,\bar{k}}^{n+1/2}\right) \\
&+ \left[\frac{-T_z}{\varepsilon_{xr}(1+\beta\xi_x)}\right]\left(Q_y\Big|_{\bar{i},\bar{j},\bar{k}}^{n+1/2} - Q_y\Big|_{\bar{i},\bar{j},\bar{k}-1}^{n+1/2}\right) \\
&+ \left[\frac{\bar{k}_y\xi_x}{2\varepsilon_{xr}(1+\beta\xi_x)}\right]\left(Q_z\Big|_{\bar{i},\bar{j},\bar{k}}^{n+1/2} + Q_z\Big|_{\bar{i},\bar{j}-1,\bar{k}}^{n+1/2}\right) \\
&+ \left[\frac{-\bar{k}_z\xi_x}{2\varepsilon_{xr}(1+\beta\xi_x)}\right]\left(Q_y\Big|_{\bar{i},\bar{j},\bar{k}}^{n+1/2} + Q_y\Big|_{\bar{i},\bar{j},\bar{k}-1}^{n+1/2}\right)
\end{aligned} \tag{6.85}$$

where $\xi_x = \sigma_x \Delta t / \varepsilon_{xr} \varepsilon_0$, $T_y = c\Delta t / \Delta y$, and $T_z = c\Delta t / \Delta z$. The modified temporal averaging of the loss term is evident by the presence of β. In the three-dimensional case, an optimal choice for β is found to be

$$\beta_{\text{opt}} = \frac{0.5}{\left(\bar{k}_x\right)^2} = \frac{0.5}{\sin^2\theta \cos^2\phi} \tag{6.86}$$

Using this value for β results in a maximum time step that is the same as for the lossless case. The lossless stability case is discussed later in this section.

The discretized version of (6.82a) is straightforward and is given by

$$\begin{aligned} P_x\Big|_{\bar{i},\bar{j},\bar{k}}^{n+1} &= \left(1 - \frac{\bar{k}_y^2}{\mu_{zr}\varepsilon_{xr}} - \frac{\bar{k}_z^2}{\mu_{yr}\varepsilon_{xr}}\right)^{-1} P_{xa}\Big|_{\bar{i},\bar{j},\bar{k}}^{n+1} \\ &- \left(1 - \frac{\bar{k}_y^2}{\mu_{zr}\varepsilon_{xr}} - \frac{\bar{k}_z^2}{\mu_{yr}\varepsilon_{xr}}\right)^{-1} \left(\frac{\bar{k}_y}{2\varepsilon_{xr}}\right) \left(Q_{za}\Big|_{\bar{i},\bar{j},\bar{k}}^{n+1} + Q_{za}\Big|_{\bar{i},\bar{j}-1,\bar{k}}^{n+1}\right) \\ &+ \left(1 - \frac{\bar{k}_y^2}{\mu_{zr}\varepsilon_{xr}} - \frac{\bar{k}_z^2}{\mu_{yr}\varepsilon_{xr}}\right)^{-1} \left(\frac{\bar{k}_z}{2\varepsilon_{xr}}\right) \left(Q_{ya}\Big|_{\bar{i},\bar{j},\bar{k}}^{n+1} + Q_{ya}\Big|_{\bar{i},\bar{j},\bar{k}-1}^{n+1}\right) \end{aligned} \tag{6.87}$$

Here, spatial averaging is used to preserve the centered nature of the algorithm. Finally, the discretized version of (6.83a) is given by

$$P_y\Big|_{\bar{i},\bar{j},\bar{k}}^{n+1} = P_{ya}\Big|_{\bar{i},\bar{j},\bar{k}}^{n+1} - \frac{\bar{k}_z}{2\varepsilon_{yr}}\left(Q_x\Big|_{\bar{i},\bar{j},\bar{k}}^{n+1} + Q_x\Big|_{\bar{i},\bar{j},\bar{k}-1}^{n+1}\right) \tag{6.88}$$

These equations give a full prescription of the split-field technique in three-dimensions in the main portion of the lattice. The only remaining issues are the stability relation and how to truncate the lattice with an absorbing boundary condition.

Numerical Stability Analysis

The stability relation for the three-dimensional split-field technique is derived by implementing the von Neumann analysis, as illustrated for the two-dimensional cases. This results in a 12×12 stability matrix for the general case with loss. Setting the determinant of this matrix to zero results in a twelfth-order polynomial that must be solved to determine the stability limit. Upon performing this evaluation, we determine that using the optimum value of β from (6.86) results in a stability limit that is essentially the same as for the lossless case.

The roots of the characteristic equation for the lossless case are determined analytically with the largest one given by

$$\Lambda = \frac{1}{\bar{k}_x^2}\left\{\bar{k}_y S_y C_y + \bar{k}_z S_z C_z + \sqrt{\begin{array}{c}\left(\bar{k}_y S_y C_y + \bar{k}_z S_z C_z\right)^2 + \\ \bar{k}_x^2\left[S_y^2 + S_z^2 + S_x^2\left(\bar{k}_x^2 + \bar{k}_y^2 C_y^2 + \bar{k}_z^2 C_z^2\right)\right]\end{array}}\right\} \tag{6.89}$$

The stability limit is determined by maximizing (6.89) over all possible values of S_x, S_y and S_z (which are limited to lie between ± 1). However, since C_y and C_z are not independent of S_y and S_z, respectively, the values of S_y and S_z which maximize (6.89) are not obvious. An analytical solution with either $k_y = 0$ or $k_z = 0$ has been derived for the general case of non-square cells [25], but the result is quite complicated. The result is somewhat simpler for square cells, and we present an analytical solution for the special cases of propagation along a coordinate axis ($k_y = 0$ or $k_z = 0$) or along the direction $k_y = k_z$. These two cases bound the stability limit for a general direction. The simplest and most stringent result occurs for $k_y = k_z$. Here, we have

$$T_y \le \frac{\bar{k}_x^2}{\sqrt{2 + \bar{k}_x^2}} \tag{6.90}$$

Along a coordinate axis ($k_z = 0$ in this case), the stability limit is

$$T_y \le \frac{\bar{k}_x^2}{\bar{k}_y F_2 \sqrt{1 - F_2^2} + \sqrt{\left(\bar{k}_x^2 + \dfrac{\bar{k}_y^4}{F_1}\right)\left(2 + F_2^2 \bar{k}_x^2\right)}} \tag{6.91}$$

where

$$F_1 = 8\bar{k}_y^2 + \bar{k}_x^6 + 2\bar{k}_x^2 \bar{k}_y^2 \tag{6.92a}$$

$$F_2^2 = \frac{\left(F_1 - \bar{k}_x^2 \bar{k}_y^2\right) + \sqrt{\left(F_1 - \bar{k}_x^2 \bar{k}_y^2\right)^2 - 8F_1 \bar{k}_y^2}}{2F_1} \tag{6.92b}$$

Fig. 6.28 shows plots of the stability limits for these two cases. To present the data, these plots use polar angles with respect to the *x*-axis:

$$\bar{k}_x = \cos\theta_x\,, \qquad \bar{k}_y = \sin\theta_x \cos\phi_{yz}\,, \qquad \bar{k}_z = \sin\theta_x \sin\phi_{yz} \tag{6.93}$$

With this notation, (6.90) is for $\phi_{yz} = 45°$, while (6.91) is for $\phi_{yz} = 0°$. The curves are similar, and it appears that just using the lower curve would be suitable. However, for large steering angles where the stability limit requires small time steps, (6.90) is considerably lower, about 25% lower by 76°. To reduce the long run times at these angles, it is desirable to use the largest possible time step. Thus, it is beneficial to know the behavior of the stability limit as a function of ϕ_{yz} for a fixed θ_x.

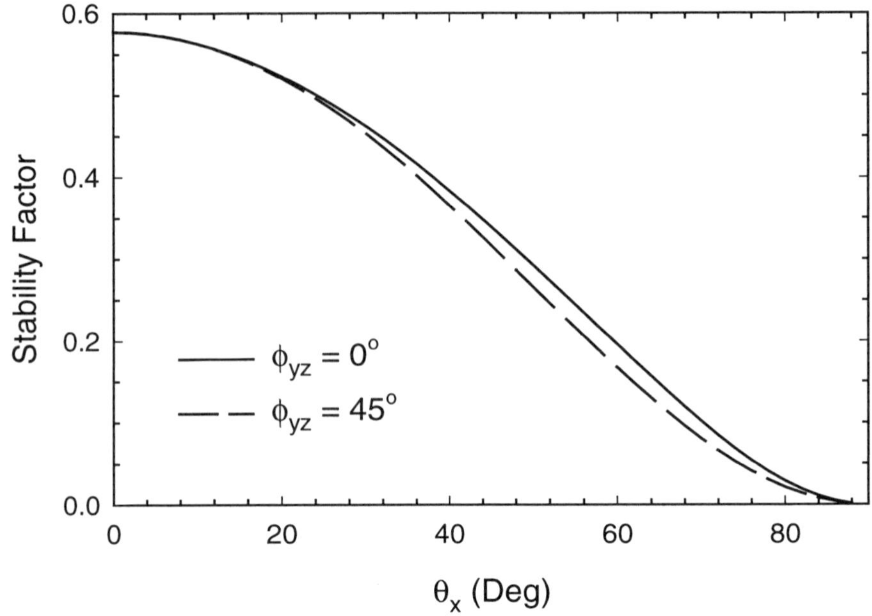

Fig. 6.28 Stability limit for the three-dimensional split-field technique for two special cases.

The behavior of the stability limit is investigated by numerically maximizing (6.89) as a function of ϕ_{yz} for different values of θ_x. Fig. 6.29 shows results for $\theta_x = 45^\circ$ and $\theta_x = 80^\circ$.

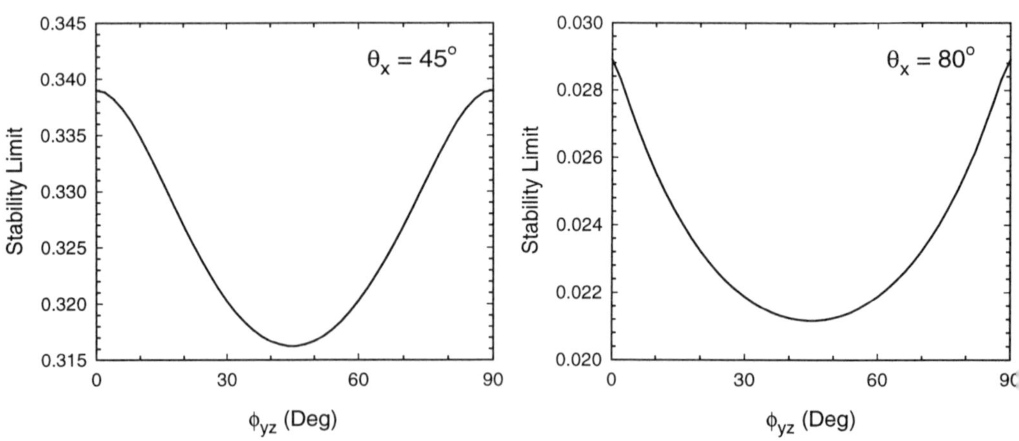

Fig. 6.29 Stability limit of the three-dimensional split-field technique for two sets of incidence angles.

Fig. 6.29 shows that the variation of the stability limit with ϕ_{yz} depends on θ_x. One approach to determining an approximate stability limit for a general angle of incidence is to fit a quadratic function of ϕ_{yz} through the minimum and maximum points given by (6.90) and (6.91), respectively. We find that using 98% of this approximate limit is adequate to ensure that the time step is below the true stability limit.

PML ABC

An important part of any FDTD simulation is the type of ABC used to truncate the grid. Since 1994, the PML concept has proven to be very effective for this purpose (see Chapter 5). In this section, we provide the details of how the anisotropic-material PML formulation can be applied to the split-field technique.

Since we only have absorbing boundaries at the ends of the lattice in a periodic problem, the anisotropic PML is a material having the properties

$$\frac{\varepsilon}{\varepsilon_0} = \frac{\mu}{\mu_0} = \begin{bmatrix} 1/s_x & 0 & 0 \\ 0 & s_x & 0 \\ 0 & 0 & s_x \end{bmatrix} \tag{6.94}$$

where $s_x = \kappa + \chi c / j\omega$ characterizes the PML material. The attenuation in the PML is determined by χ, and the usual tapering of χ should be used to minimize the numerical reflection from the free-space/PML interface and the steps in the discrete profile. With these material properties, the transformed equations are given in the frequency domain by

$$\frac{j\omega}{s_x c} P_x = \frac{\partial Q_z}{\partial y} - \frac{\partial Q_y}{\partial z} - \frac{j\omega \bar{k}_y}{c} Q_z + \frac{j\omega \bar{k}_z}{c} Q_y \tag{6.95a}$$

$$\frac{j\omega s_x}{c} P_y = -\frac{\partial Q_z}{\partial x} + \frac{\partial Q_x}{\partial z} - \frac{j\omega \bar{k}_z}{c} Q_x \tag{6.95b}$$

$$\frac{j\omega s_x}{c} P_z = \frac{\partial Q_y}{\partial x} - \frac{\partial Q_x}{\partial y} + \frac{j\omega \bar{k}_y}{c} Q_x \tag{6.95c}$$

and

$$\frac{j\omega}{s_x c} Q_x = -\frac{\partial P_z}{\partial y} + \frac{\partial P_y}{\partial z} + \frac{j\omega \bar{k}_y}{c} P_z - \frac{j\omega \bar{k}_z}{c} P_y \tag{6.96a}$$

$$\frac{j\omega s_x}{c} Q_y = \frac{\partial P_z}{\partial x} - \frac{\partial P_x}{\partial z} + \frac{j\omega \bar{k}_z}{c} P_x \tag{6.96b}$$

$$\frac{j\omega s_x}{c}Q_z = -\frac{\partial P_y}{\partial x} + \frac{\partial P_x}{\partial y} - \frac{j\omega \bar{k}_y}{c}P_x \tag{6.96c}$$

Because of the presence of s_x in the denominators of (6.95a) and (6.96a), we change variables to $\hat{P}_x = P_x / s_x$ and $\hat{Q}_x = Q_x / s_x$ [27]. Then, (6.95) and (6.96) become

$$\frac{j\omega}{c}\hat{P}_x = \frac{\partial Q_z}{\partial y} - \frac{\partial Q_y}{\partial z} - \frac{j\omega \bar{k}_y}{c}Q_z + \frac{j\omega \bar{k}_z}{c}Q_y \tag{6.97a}$$

$$\frac{j\omega s_x}{c}P_y = -\frac{\partial Q_z}{\partial x} + s_x\frac{\partial \hat{Q}_x}{\partial z} - \frac{j\omega s_x \bar{k}_z}{c}\hat{Q}_x \tag{6.97b}$$

$$\frac{j\omega s_x}{c}P_z = \frac{\partial Q_y}{\partial x} - s_x\frac{\partial \hat{Q}_x}{\partial y} + \frac{j\omega s_x \bar{k}_y}{c}\hat{Q}_x \tag{6.97c}$$

and

$$\frac{j\omega}{c}\hat{Q}_x = -\frac{\partial P_z}{\partial y} + \frac{\partial P_y}{\partial z} + \frac{j\omega \bar{k}_y}{c}P_z - \frac{j\omega \bar{k}_z}{c}P_y \tag{6.98a}$$

$$\frac{j\omega s_x}{c}Q_y = \frac{\partial P_z}{\partial x} - s_x\frac{\partial \hat{P}_x}{\partial z} + \frac{j\omega s_x \bar{k}_z}{c}\hat{P}_x \tag{6.98b}$$

$$\frac{j\omega s_x}{c}Q_z = -\frac{\partial P_y}{\partial x} + s_x\frac{\partial \hat{P}_x}{\partial y} - \frac{j\omega s_x \bar{k}_y}{c}\hat{P}_x \tag{6.98c}$$

Defining new variables as was done in the three-dimensional non-PML case, the extra time derivatives on the right-hand side are eliminated. These new variables are slightly different than before, with the tangential components now split into "a" and "b" parts:

$$\hat{P}_x = P_{xa} + \bar{k}_z Q_y - \bar{k}_y Q_z \tag{6.99a}$$

$$P_y = P_{ya} + c\int_{-\infty}^{t} \frac{\partial \hat{Q}_x}{\partial z}dt - \bar{k}_z \hat{Q}_x\,, \qquad P_{yb} = c\int_{-\infty}^{t} \frac{\partial \hat{Q}_x}{\partial z}dt \tag{6.99b}$$

$$P_z = P_{za} - c\int_{-\infty}^{t} \frac{\partial \hat{Q}_x}{\partial y}dt + \bar{k}_y \hat{Q}_x\,, \qquad P_{zb} = -c\int_{-\infty}^{t} \frac{\partial \hat{Q}_x}{\partial y}dt \tag{6.99c}$$

and

$$\hat{Q}_x = Q_{xa} - \bar{k}_z P_y + \bar{k}_y P_z \tag{6.100a}$$

$$Q_y = Q_{ya} - c\int_{-\infty}^{t} \frac{\partial \hat{P}_x}{\partial z} dt + \bar{k}_z \hat{P}_x , \qquad Q_{yb} = -c\int_{-\infty}^{t} \frac{\partial \hat{P}_x}{\partial z} dt \tag{6.100b}$$

$$Q_z = Q_{za} + c\int_{-\infty}^{t} \frac{\partial \hat{P}_x}{\partial y} dt - \bar{k}_y \hat{P}_x , \qquad Q_{zb} = c\int_{-\infty}^{t} \frac{\partial \hat{P}_x}{\partial y} dt \tag{6.100c}$$

Substituting these into the left-hand sides of (6.97) and (6.98) yields

$$\frac{1}{c}\frac{\partial P_{xa}}{\partial t} = \frac{\partial Q_z}{\partial y} - \frac{\partial Q_y}{\partial z} \tag{6.101a}$$

$$\frac{\kappa}{c}\frac{\partial P_{ya}}{\partial t} + \chi P_{ya} = -\frac{\partial Q_z}{\partial x} , \qquad \frac{1}{c}\frac{\partial P_{yb}}{\partial t} = \frac{\partial \hat{Q}_x}{\partial z} \tag{6.101b}$$

$$\frac{\kappa}{c}\frac{\partial P_{za}}{\partial t} + \chi P_{za} = \frac{\partial Q_y}{\partial x} , \qquad \frac{1}{c}\frac{\partial P_{zb}}{\partial t} = -\frac{\partial \hat{Q}_x}{\partial y} \tag{6.101c}$$

and

$$\frac{1}{c}\frac{\partial Q_{xa}}{\partial t} = -\frac{\partial P_z}{\partial y} + \frac{\partial P_y}{\partial z} \tag{6.102a}$$

$$\frac{\kappa}{c}\frac{\partial Q_{ya}}{\partial t} + \chi Q_{ya} = \frac{\partial P_z}{\partial x} , \qquad \frac{1}{c}\frac{\partial Q_{yb}}{\partial t} = -\frac{\partial \hat{P}_x}{\partial z} \tag{6.102b}$$

$$\frac{\kappa}{c}\frac{\partial Q_{za}}{\partial t} + \chi Q_{za} = -\frac{\partial P_y}{\partial x} , \qquad \frac{1}{c}\frac{\partial Q_{zb}}{\partial t} = \frac{\partial \hat{P}_x}{\partial y} \tag{6.102c}$$

The systems (6.101) and (6.102) show how the "a" and "b" portions of the fields are updated in terms of the total-field values. This is similar to the split field in the non-PML regions, but the additional splitting eliminates the nonderivative terms from the right-hand sides. As a result, no spatial averaging is needed in these equations. Also, the loss terms (terms multiplied by χ) are handled in the standard FDTD fashion. This is equivalent to using $\beta_{\text{PML}} = 0.5$ in the modified temporal averaging in (6.74).

Once the "a" and "b" portions of the fields are known, then the components normal to the PML boundary are found using the relations

$$\left(1-\bar{k}_y^2-\bar{k}_z^2\right)\hat{P}_x = P_{xa} - \bar{k}_y\left(Q_{za}+Q_{zb}\right) + \bar{k}_z\left(Q_{ya}+Q_{yb}\right) \tag{6.103a}$$

$$\left(1-\bar{k}_y^2-\bar{k}_z^2\right)\hat{Q}_x = Q_{xa} + \bar{k}_y\left(P_{za}+P_{zb}\right) - \bar{k}_z\left(P_{ya}+P_{yb}\right) \tag{6.103b}$$

Finally, the remaining field components are found from

$$P_y = P_{ya} + P_{yb} - \bar{k}_z\hat{Q}_x\,, \qquad P_z = P_{za} + P_{zb} + \bar{k}_y\hat{Q}_x \tag{6.104}$$

$$Q_y = Q_{ya} + Q_{yb} + \bar{k}_z\hat{P}_x\,, \qquad Q_z = Q_{za} + Q_{zb} - \bar{k}_y\hat{P}_x \tag{6.105}$$

Note that it is not necessary to find P_x from $\hat{P}_x$ because the fields within the PML are irrelevant to the solution in the non-PML region. The only requirement is that the reflection from the PML region be sufficiently small.

Although this procedure is more complex than for the non-PML regions, it is necessary to perform the additional splitting of the tangential components in the PML region to make the technique stable. In developing this method, both a local stability analysis and simulations involving many time steps have been used to verify that the method is stable.

6.5 APPLICATION OF THE PERIODIC FDTD TECHNIQUES

We have applied the split-field FDTD technique to analyze several types of periodic structures [25]. In this section, three particular applications are discussed: photonic bandgap structures, frequency-selective surfaces, and antenna arrays.

6.5.1 Photonic Bandgap Structures

We first discuss periodic FDTD modeling of a three-dimensional PBG structure: the "woodpile" geometry shown in Fig. 6.30. (See [10] for a discussion of the periodic FDTD modeling of other types of PBG structures.) The woodpile PBG structure is formed by stacking dielectric rods [28]. Each dielectric rod has a square cross section 1/8 inch (3.175 mm) on a side and is characterized by the relative permittivity $\varepsilon_r = 8$. The spacing between adjacent rods in each layer is $L = 0.4375$ inch (6.11 mm), and the rod orientation is alternated from layer to layer. Every fourth layer is located directly beneath each other.

Fig. 6.31 shows the transmission through three periods of this woodpile structure for a normally incident plane wave. The results agree very well with the calculations of a frequency-domain mode-matching code. Also shown in this figure are experimental data obtained using a focused-beam measurement system. The measured and calculated results are in relatively good agreement.

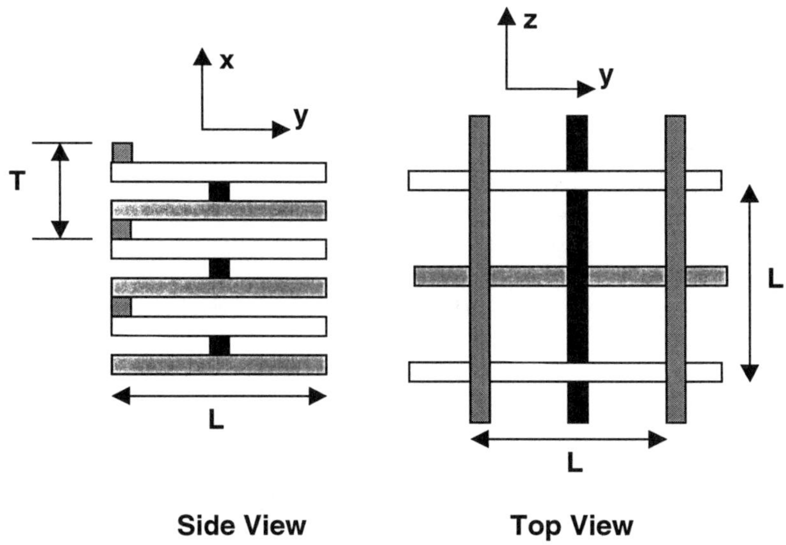

Fig. 6.30 Side and top views of the "woodpile" PBG structure comprised of dielectric rods of square cross section $W = T/4 = 1/8$ in $= 3.175$ mm on a side with $\varepsilon_r = 8$. The rod spacing is $L = 0.4375$ in $= 11.11$ mm, and the periodic cell is $L \times L$.

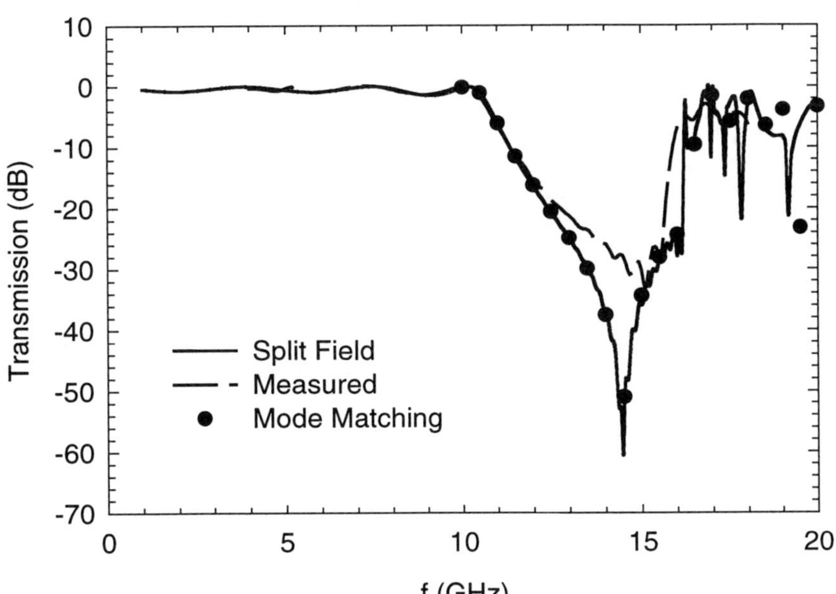

Fig. 6.31 Comparison of periodic FDTD, frequency-domain mode-matching, and measured results for transmission through the woodpile PBG of Fig. 6.30 for the normal-incidence case.

Characteristic of a PBG structure, there is a bandgap frequency range in which electromagnetic propagation is not allowed. In the bandgap, the transmission through the structure decreases sharply. Since the structure is lossless, the magnitude of its reflection coefficient in the bandgap is very close to unity, and the PBG behaves much like a metal reflector. However, the phase of the reflection coefficient is usually other than 180°.

As the angle of incidence increases, the bandgap tends to shift somewhat in frequency. Fig. 6.32 shows the transmission versus frequency of the woodpile PBG structure for an incident angle of 30°. The shape of the bandgap is different in this case than in Fig. 6.31, and the bandgap is shifted to lower frequencies. Again there is good agreement between the split-field FDTD technique and the mode-matching method.

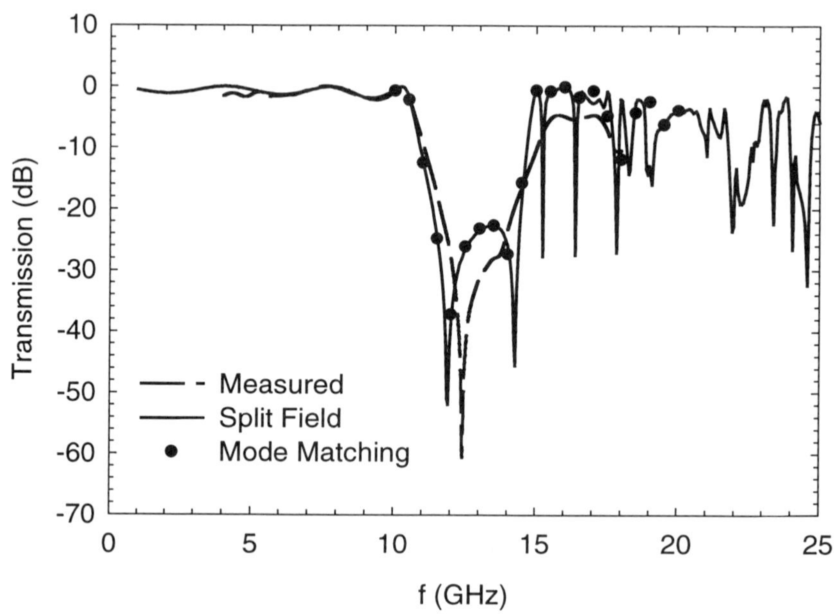

Fig. 6.32 Comparison of periodic FDTD, frequency-domain mode-matching, and measured results for transmission through the woodpile PBG of Fig. 6.30 for an incident angle of 30°.

6.5.2 Frequency-Selective Surfaces

The second application of the split-field FDTD technique is to an FSS. Fig. 6.33 shows the top view of one cell of the doubly periodic FSS [14, 29]. Thick conductors having square cross sections form the loops. Data for the percentage of reflected power are computed using the split-field technique and are compared with the MoM and sine-cosine FDTD results reported in [14, 29]. Fig. 6.34 shows the results for a plane wave at an incident angle of 0°, and Fig. 6.35 shows the results for a TM plane wave (H_y) at an

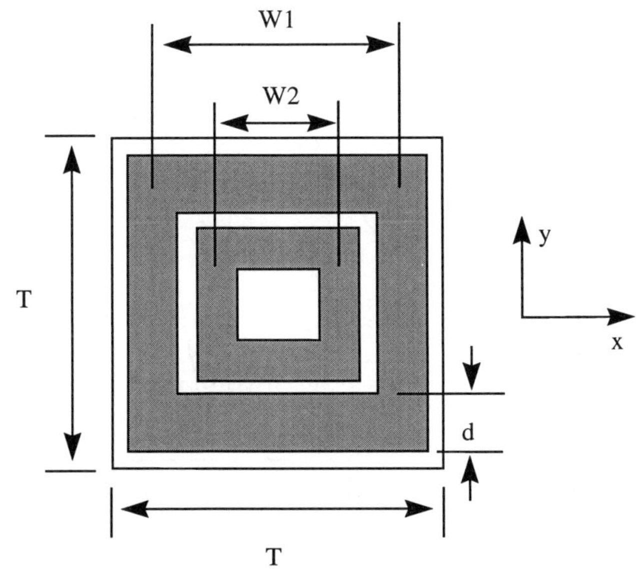

Fig. 6.33 Top view of a double-concentric-loop FSS that is doubly periodic in x and y. The conductors forming the loops have a thickness of T/16, where T is the periodic cell dimension. Also, $d = T/16$, $W_1/T = 0.875$, and $W_2/T = 0.6875$, where W_1 and W_2 are measured from the centers of the conductors.

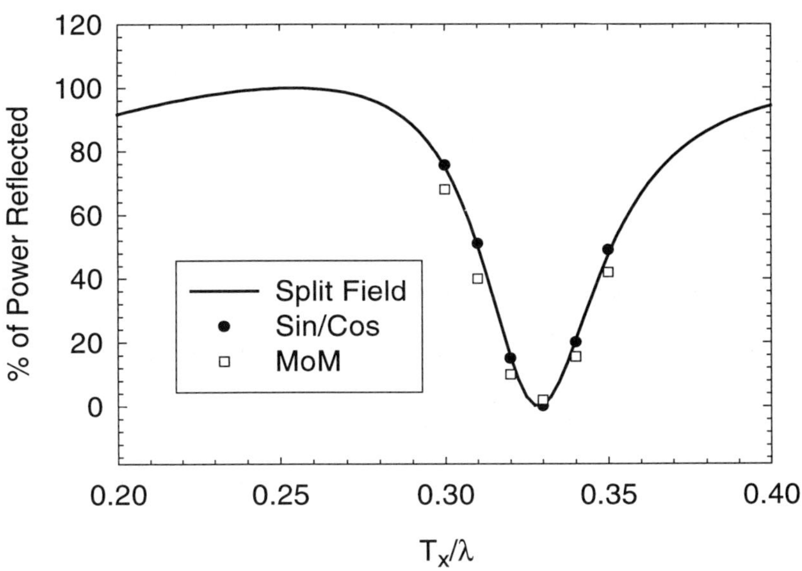

Fig. 6.34 Percentage reflected power for FSS of Fig. 6.33 for a plane wave at an incident angle of 0°.

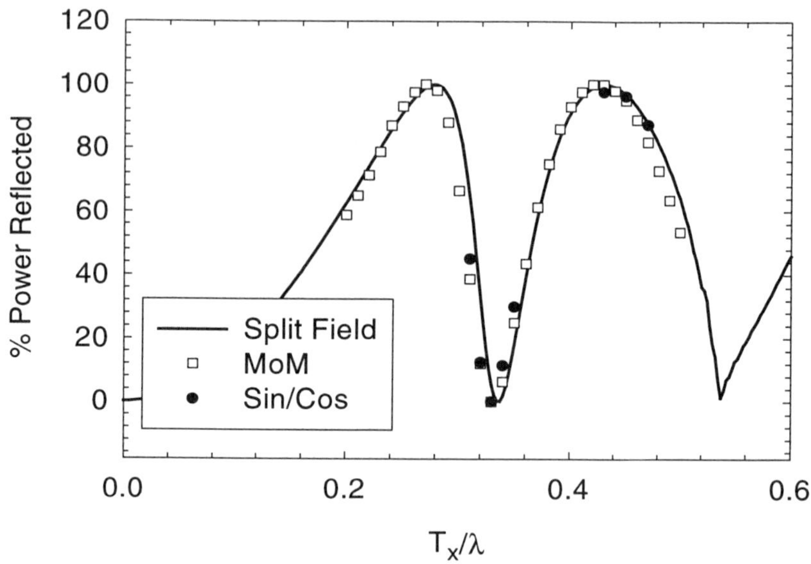

Fig. 6.35 Percentage of power reflected for the FSS of Fig. 6.33 for a plane wave at an incident angle of 60°.

incident angle of 60°. The split-field results agree best with the sine-cosine results for the 0° incidence case. For the 60° case, the split-field results are slightly shifted from the other data, but in general, good agreement is retained.

We note that the split-field approach is much more efficient than either the MoM or sine-cosine approaches. In either Figs. 6.34 or 6.35, *all* of the split-field data are obtained from a single computer run, whereas each symbol represents a single run for the other methods.

6.5.3 Antenna Arrays

The third and final class of application discussed is the analysis of a large antenna array. A single element in the array is modeled along with a simple transmission line feed structure, and periodic boundary conditions are used to model the effect of the element being in a large array.

Fig. 6.36 illustrates the geometry chosen for the simulation, an array of dipole antennas. The dimensions of the problem match those given in [30] where a MoM code is used to analyze the structure. Although the FDTD calculation uses a pulsed excitation and thus contains broadband data, we can compare its results only to the specific frequency used in [30]. At this frequency, the dipole length is $L = 0.39\lambda$, including a 0.01λ gap between the arms, while the dipole width is 0.01λ. The size of a unit cell is 0.5λ in both the y and z directions. Beneath the dipole array is a 0.19λ-thick dielectric

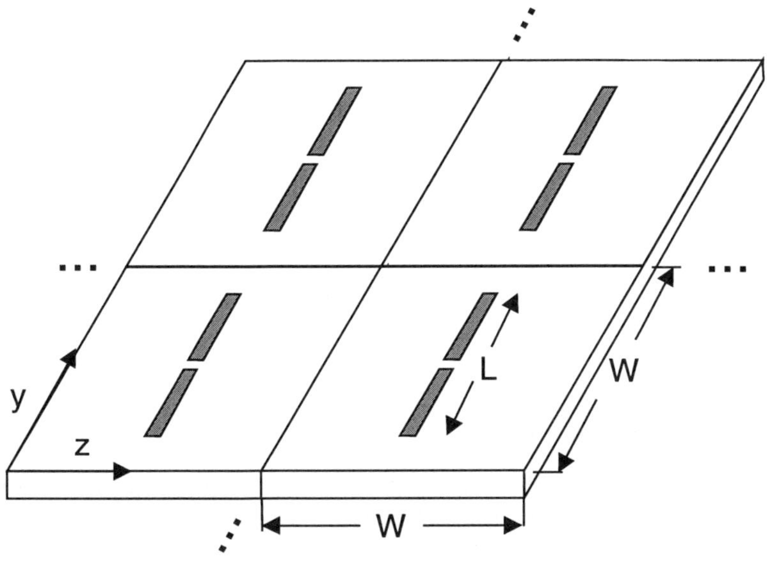

Fig. 6.36 Four unit cells of the infinite dipole array. The dipole spacing W is 0.5λ and the dipole length L is 0.39λ with a gap of 0.01λ. The substrate is 0.19λ thick with $\varepsilon_r = 2.55$ and is backed by a metal layer.

substrate ($\varepsilon_r = 2.55$) above a ground plane. When discretized, the fundamental unit cell consists of $50 \times 50 \times 20$ FDTD cells with an additional 20 free-space cells above the antenna, for a total grid (excluding the PML region) of $50 \times 50 \times 40$ cells.

Each dipole element is assumed to be excited at its terminals using a balanced 50Ω transmission line. The element's impedance is calculated from the reflected voltage wave on the transmission line. (See Chapter 7 for details on the simple transmission line feed.) We note that, because of the periodic boundary conditions, all of the antenna elements are driven simultaneously and phased to radiate in the scan direction. Thus, the reflected voltage wave on the feeding transmission line includes not only the reflection from the antenna terminals, which represents the isolated antenna impedance, but also any signal received because the other elements in the array are radiating as well. Therefore, the impedance calculated from the reflected voltage is the so-called "active impedance" of the element when the array is fully excited.

To compare the results of the split-field FDTD technique to those in [30], we use the computed antenna impedance in the transmission line circuit shown in Fig. 6.37. A series reactance is used to match the antenna impedance for a scan angle of 0° to the transmission line feed. Since the antenna impedance varies with scan angle, the match is no longer perfect as the array is steered away from 0°. With this matching arrangement, the reflection coefficient is given by

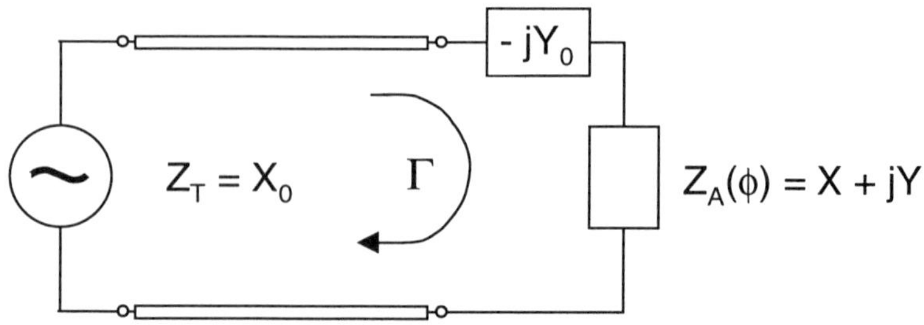

Fig. 6.37 Transmission line circuit used to determine the reflection coefficient from the antenna impedance. The conjugate reactance is used to match the antenna impedance at $\phi = 0°$.

$$\Gamma = \frac{Z_A(\phi) - Z_A(0)}{Z_A(\phi) + Z_A^*(0)} \tag{6.106}$$

where $Z_A(0) = X_0 + jY_0$ is the active impedance of the dipole element at 0° and $Z_A(\phi)$ is the active impedance when the array is scanned to the angle ϕ.

The magnitude of the reflection coefficient computed using (6.106) is shown in Fig. 6.38, along with the results from [30]. The agreement between the two techniques

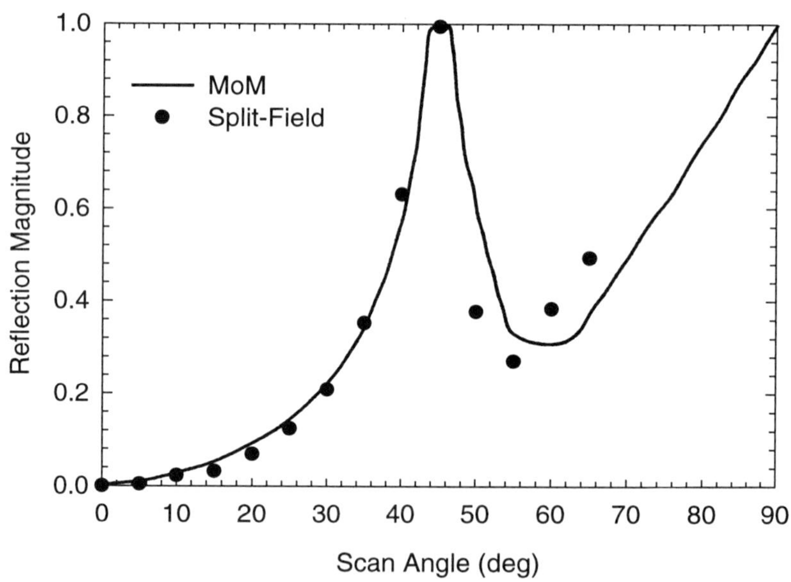

Fig. 6.38 Magnitude of the reflection coefficient of the infinite dipole array of Fig. 6.36 as a function of the scan angle. The results from the split-field code are compared with the MoM results from [30].

is good, especially below 45°, considering that there are slight differences in the physical models used. Note that the magnitude of the reflection coefficient approaches unity near 45°. This is the array *scan-blindness* phenomenon that occurs because of the excitation of surface waves in the dielectric substrate.

The radiated far field was also computed with the split-field FDTD code. Since the element spacing is $\lambda/2$ at the comparison frequency, only the lowest-order Floquet mode is computed. This result is then used to determine the gain of the element for comparison to the results given in [30]. The conversion from far field to gain is given by

$$G(\phi) = T_0(\phi)\cos\phi \tag{6.107}$$

Fig. 6.39 compares the split-field results for the gain of a single dipole element embedded in the infinite array of Fig. 6.36 with the MoM results from [30]. Again, there is good agreement between the two computational approaches.

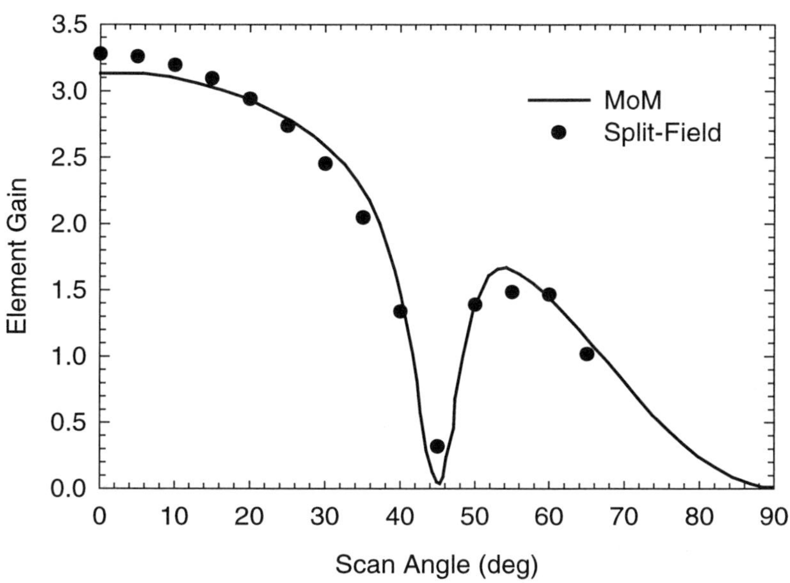

Fig. 6.39 Comparison of the split-field results for the gain of a single dipole element embedded in the infinite array of Fig. 6.36 with the MoM results from [30].

In Fig. 6.39, the scan blindness shows up as a reduction in the active gain of the element. The angle at which the scan blindness occurs is a function of the operating frequency. Since the FDTD results cover a broad frequency range, they can be used to track the angle at which scan blindness occurs.

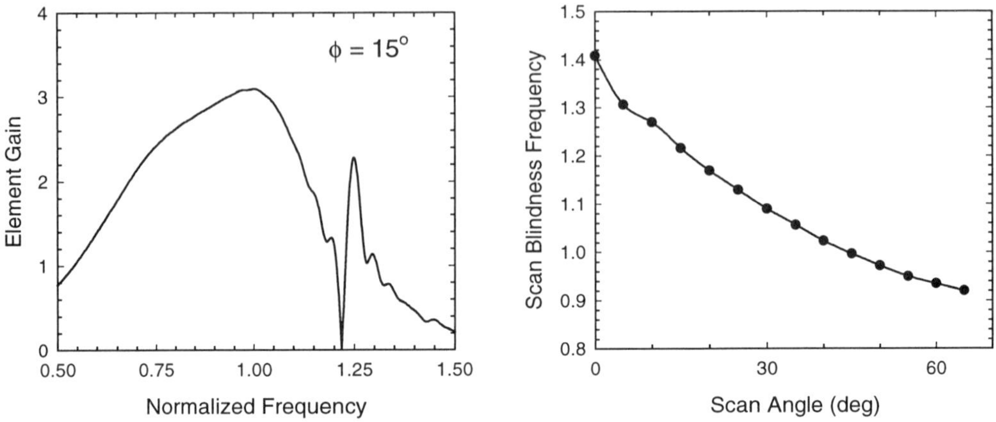

Fig. 6.40 Left: element gain as a function of normalized frequency when the array of Fig. 6.36 is scanned to 15°. The sharp minimum in the gain near 1.2 indicates scan blindness. Right: variation of the normalized scan blindness frequency with scan angle.

The phenomenon of scan blindness is further examined in Fig. 6.40. The plot on the left in this figure shows the element gain as a function of frequency when the array is steered to 15°. Here, a normalized frequency of 1.0 corresponds to the λ used earlier in the description of the array geometry of Fig. 6.36. The sharp dip near 1.2 indicates the frequency at which scan blindness occurs for this angle. The plot on the right side of Fig. 6.40 shows the variation of the scan-blindness frequency with scan angle. Overall, we comment that scan blindness can be a strong limitation on the useable scan range of broadband antenna arrays.

6.6 SUMMARY AND CONCLUSIONS

The FDTD method has strong appeal for modeling periodic structures. However, there are fundamental difficulties in applying periodic boundary conditions in the time domain. Several techniques have been proposed for this problem. Some methods approximate the problem by using multiple unit cells, while others use a single frequency to allow the periodic boundary condition to be applied in the frequency domain.

The most promising approaches involve applying a field transformation to eliminate the time delay across the FDTD grid. Most of this chapter is devoted to discussing techniques for discretizing Maxwell's equations when such a field transformation is applied. Because of our experiences in coding and applying the different techniques, the issue of numerical stability receives significant attention in this chapter. Of all the techniques investigated, two stand out as the best: the multiple-spatial-grid technique and the split-field technique. A computer program utilizing either of these approaches provides the researcher with a valuable tool for analyzing the electromagnetic wave behavior of periodic structures.

ACKNOWLEDGMENTS

The authors specifically acknowledge the significant contributions of Dr. Paul Harms, Dr. Alan Roden, and Dr. Eric Kuster to this chapter. Dr. Harms and Dr. Roden performed most of the initial development of the periodic-FDTD computer programs that we are using, and were the originators of the split-field concept. Dr. Kuster developed the mode-matching computer program and participated in many fruitful discussions. Without their contributions, the work discussed in this chapter would not have been possible. The authors also thank Prof. Stephen Gedney for his many helpful discussions on numerical stability.

REFERENCES

[1] Cangellaris, A. C., M. Gribbons, and G. Sohos, "A hybrid spectral / FDTD method for the electromagnetic analysis of guided waves in periodic structures," *IEEE Microwave and Guided Wave Letters*, Vol. 3, 1993, pp. 375–377.

[2] Celuch-Marcysiak, M., and W. K. Gwarek, "Spatially looped algorithms for time-domain analysis of periodic structures," *IEEE Trans. Microwave Theory and Techniques*, Vol. 43, 1995, pp. 860–865.

[3] Okoniewski, M., "Vector wave equation 2-D FDTD method for guided wave problems," *IEEE Microwave and Guided Wave Letters*, Vol. 3, 1993, pp. 307–309.

[4] Kuster, E. J., private communication.

[5] Tsay, W. J., and D. M. Pozar, "Application of the FDTD technique to periodic problems in scattering and radiation," *IEEE Microwave and Guided Wave Letters*, Vol. 3, 1993, pp. 250–252.

[6] Navarro, E. A., B. Gimeno, and J. L. Cruz, "Modeling of periodic structures using the finite difference time domain method combined with the Floquet theorem," *Electronics Letters*, Vol. 29, 1993, pp. 446–447.

[7] Alexanian, N. J. Kolias, R. C. Compton, and R. A. York, "Three-dimensional FDTD analysis of quasi-optical arrays using Floquet boundary conditions and Berenger's PML," *IEEE Microwave and Guided Wave Letters*, Vol. 6, 1996, pp. 138–140.

[8] Chan, H., S. H. Lou, L. Tsang, and J. A. Kong, "Electromagnetic scattering of waves by random rough surface: a finite-difference time-domain approach," *Microwave and Optical Tech. Letters*, Vol. 4, 1991, pp. 355–359.

[9] Kelly, P. K., J. G. Maloney, B. L. Shirley, and R. L. Moore, "Photonic bandgap structures of finite thickness: theory and experiment," *Proc. 1994 IEEE Antennas and Propagation Society International Symposium*, Seattle, WA, June 1994, Vol. 2, pp. 718–721.

[10] Kesler, M. P., J. G. Maloney, and G. Smith, "Antenna design with the use of photonic bandgap materials as all-dielectric planar reflectors," *Microwave and Optical Technology Letters*, Vol. 11, 1996, pp. 169–174.

[11] Ren, J., O. P. Gandhi, L. R. Walker, J. Fraschilla, and C. R. Boerman, "Floquet-based FDTD analysis of two-dimensional phased array antennas," *IEEE Microwave and Guided Wave Letters*, Vol. 4, 1994, pp. 109–112.

[12] Lee, K. H., and S. R. Laxpati, "FDTD analysis of an infinite array of microstrip patches," *Proc. 1996 IEEE Antennas and Propagation Society International Symposium*, Baltimore, MD, July 1996, Vol. 2, pp. 1284–1287.

[13] Ko, W. L., and R. Mittra, "Implementation of Floquet boundary condition in FDTD for FSS analysis," *Proc. 1993 IEEE Antennas and Propagation Society International Symposium*, Ann Arbor, MI, June 28-July 2, 1993, Vol. 1, pp. 14–17.

[14] Harms, P., R. Mittra, and W. Ko, "Implementation of the periodic boundary condition in the finite-difference time-domain algorithm for FSS structures," *IEEE Trans. Antennas and Propagation*, Vol. 42, 1994, pp. 1317–1324.

[15] Prescott, D. T., and N. V. Shuley, "Extensions to the FDTD method for the analysis of infinitely periodic arrays," *IEEE Microwave and Guided Wave Letters*, Vol. 4, 1994, pp. 352–354.

[16] Davidson, B., A. G. Smith, and J. J. van Tonder, "The analysis, measurement and design of frequency selective surfaces," *10th International Conference on Antennas and Propagation*, April 1997, IEE conference publication no. 436, pp. 156–159.

[17] Noda, T., T. Kanetani, and K. Uchidat, "FVTD analysis of metallic grating," *IEICE Transactions on Electronics*, Vol. E79-C, 1996, pp. 1772–1775.

[18] Marek, J. R., and J. MacGillivray, "A method for reducing run-times of out-of-core FDTD problems," *Proc. 9th Annual Review of Progress in Applied Computational Electromagnetics*, Monterey, CA, March 1993, pp. 344–351.

[19] Veysoglu, M. E., R. T. Shin, and J. A. Kong, "A finite-difference time-domain analysis of wave scattering from periodic surfaces: oblique incidence case," *J. Electromagnetic Waves and Applications*, Vol. 7, 1993, pp. 1595–1607.

[20] Gedney, S. D., and U. Navsariwala, "An unconditionally stable finite-element time-domain solution of the vector wave equation," *IEEE Microwave and Guided Wave Letters*, Vol. 5, 1995, pp. 332–334.

[21] Kao, Y-C., and R. G. Atkins, "A finite-difference time-domain approach for frequency selective surfaces at oblique incidence," *Proc. 1996 IEEE Antennas and Propagation Society International Symposium*, Baltimore, MD, July 1996, pp. 1432–1435.

[22] Ray, S. L., "Grid decoupling in finite element solutions of Maxwell's equations," *IEEE Trans. Antennas and Propagation*, Vol. 40, 1992, pp. 443–445.

[23] Roden, J. A., *Electromagnetic Analysis of Complex Structures Using the FDTD Technique in General Curvilinear Coordinates*, Ph.D. Thesis, University of Kentucky, Lexington, KY, 1997.

[24] Roden, J. A., S. D. Gedney, M. P. Kesler, J. G. Maloney, P. H. Harms, "Time-domain analysis of periodic structures at oblique incidence: Orthogonal and nonorthogonal FDTD implementations," *IEEE Trans. Microwave Theory and Techniques*, Vol. 46, 1998, pp. 420–427.

[25] Harms, P. H., J. A. Roden, J. G. Maloney, M. P. Kesler, E. J. Kuster, and S. D. Gedney, "Numerical analysis of periodic structures using the split-field algorithm," *Proc. 13th Annual Review of Progress in Applied Computational Electromagnetics*, Monterey, CA, March 1997, pp. 104–111.

[26] Zienkiewicz, O. C., "A new look at the Newmark, HouBolt, and other time-stepping formulas: a weighted residual approach," *Earthquake Engineering and Structural Dynamics*, Vol. 5, 1977, pp. 413–418.

[27] Roden, J. A., and S. D. Gedney, "Efficient implementation of the uniaxial-based PML media in three-dimensional nonorthogonal coordinates with the use of the FDTD technique," *Microwave and Optical Technology Letters*, Vol. 14, 1997, pp. 71–75.

[28] Ho, K. M., C. T. Chan, C. M. Soukoulis, R. Biswas, and M. Sigalis, "Photonic band gaps in three dimensions: new layer-by-layer periodic structures," *Solid State Communications*, Vol. 89, 1994, pp. 413–416.

[29] Webb, D. B., *Design Techniques for Bandpass Frequency Selective Surfaces*, M.S. thesis, University of Illinois, Urbana-Champaign, IL, 1992.

[30] Pozar, D. M., "Analysis of finite phased arrays of printed dipoles," *IEEE Trans. Antennas and Propagation*, Vol. 33, 1985, pp. 1045–1053.

Chapter 7

Modeling of Antennas

James G. Maloney and Glenn S. Smith

7.1 INTRODUCTION

The FDTD method was initially used primarily to model the scattering of electromagnetic waves from objects. Later, after techniques were developed for including sources within the computation lattice, the method was used to model radiating structures. The first antennas analyzed with the FDTD method were simple structures such as rotationally symmetric, perfectly conducting, cylindrical and conical monopoles in free space [1]. The excellent agreement between the FDTD results and measured data for these simple antennas showed the potential of the method for modeling antennas having realistic complexity.

Since these early investigations, the complexity of the antennas modeled with FDTD has steadily increased. A partial list of the different types of antennas analyzed includes loaded linear elements [2, 3] and arrays of linear elements [4 – 6], waveguide and TEM horns [7 – 12], planar designs (microstrip patches and spirals) [13 – 16], dielectric resonators and lenses [17 – 19], and active antennas [20]. A more complete accounting is in the detailed annotated review in Chapter 1, Sections 1.10 and 1.14.

Today, the FDTD method is routinely used to analyze practical antennas transmitting or receiving energy in the presence of realistic material structures. The ability of this method to accurately determine wideband near-field and far-field antenna characteristics in the presence of complex materials (inhomogeneous, anisotropic, and frequency dispersive) allows antenna designers great flexibility, and opens possibilities for novel antenna concepts.

This chapter provides a detailed discussion of FDTD modeling theory and procedures for antennas. It concludes with a case study that indicates the accuracy of the method, and a review of selected recent applications that point toward future modeling possibilities.

7.2 FORMULATION OF THE ANTENNA PROBLEM

7.2.1 Transmitting Antenna

Fig. 7.1(a) shows the basic elements for the FDTD analysis of a transmitting antenna. This figure is for a cross section through the computational volume. The antenna is placed near the center of the volume and is fed by a transmission line or waveguide. As we discuss later, the details of the transmission line can be completely modeled within the FDTD lattice, or the line may be simply modeled as a one-dimensional FDTD grid attached to the antenna at the feed point. The computational volume is surrounded by an ABC to reduce the reflection of the radiated energy at the boundaries of the volume. At this time, the PML ABC discussed in Chapter 5 is the most effective for this purpose.

The FDTD method provides the electromagnetic field for all space lattice points within the computational volume. However, for many antenna applications, we want to know the electromagnetic field at a large distance from the antenna—the radiated or far-zone field. This field can be obtained by applying a *near-field to far-field* (NFFF) transformation. To perform this transformation, a virtual surface is placed around the antenna, as indicated by the dashed line in Fig. 7.1(a). The field on this surface is obtained for the time period of interest. This field is then transformed to obtain the far field using a version of Huygens' principle for electromagnetic fields. One analytic form for this principle is

$$\vec{E}^r(\bar{r},t) = \frac{\mu_0}{4\pi r} \oint\!\!\!\oint_S \left\{ \hat{r}\times\hat{r}\times\frac{\partial}{\partial t'}\left[\hat{n}\times\vec{H}(\bar{r}',t')\right] - \frac{1}{\eta_0}\hat{r}\times\frac{\partial}{\partial t'}\left[\hat{n}\times\vec{E}(\bar{r}',t')\right]\right\}_{t'=t_r} dS' \qquad (7.1)$$

where the retarded time is

$$t_r = t - \frac{r - \hat{r}\cdot\bar{r}'}{c} \qquad (7.2)$$

Here, $\bar{r}'$ locates the point on the surface, S, with the outward-pointing, unit normal vector, $\hat{n}$; and $\bar{r}$ locates the point in the far zone where the radiated field, $\vec{E}^r$, is to be calculated.

In some situations, the near field is needed at points so far from the antenna that it is impractical to extend the computational volume to include these points. A *near-field to near-field* (NFNF) transformation may be used to obtain these results. For this case, the FDTD analysis is performed for a volume such as that shown in Fig. 7.1(a), and the field

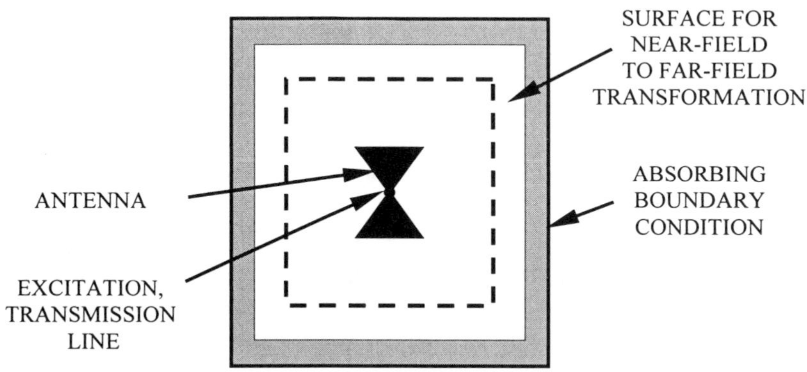

(a) Transmitting antenna case.

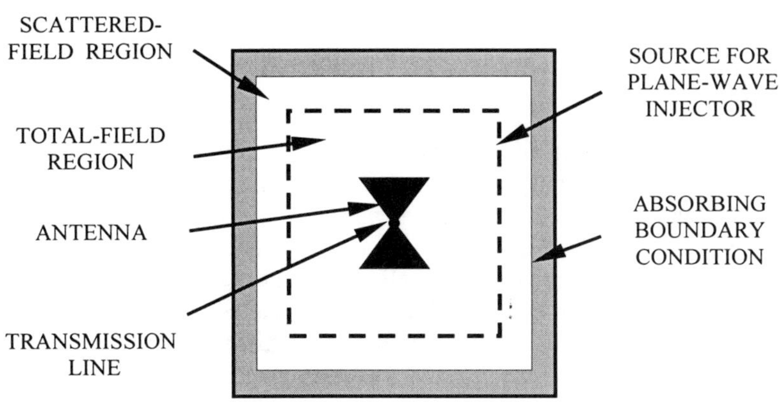

(b) Receiving antenna case.

Fig. 7.1 Basic elements in the FDTD analysis of antennas.

on the surface of the volume is transformed to obtain the near field outside the volume. Details of the NFNF transformation are found in [21] and [22].

Note that in certain situations (such as for a microstrip), more than one material region can intersect the transformation surface. The modifications to the NFFF transformation required for this case are discussed in [23].

7.2.2 Receiving Antenna

Fig. 7.1(b) shows the basic elements for the FDTD analysis of a receiving antenna. Again, the figure is for a cross section through the computational volume. The response of the antenna, usually the signal in the attached transmission line, is determined when the antenna is irradiated by an incident plane wave with the electromagnetic field, $(\vec{E}_{\text{inc}}(\bar{r}, t),\ \vec{H}_{\text{inc}}(\bar{r}, t))$.

In the FDTD formulation, a source is placed on the virtual surface of a volume containing the antenna; the dashed line in Fig. 7.1(b) indicates this virtual surface. The source produces the incident field within the volume and a null field outside the volume. An analytic form for this source specifies electric and magnetic surface currents with the values

$$\vec{J}_s(\bar{r}, t) = -\hat{n} \times \vec{H}_{\text{inc}}(\bar{r}, t) \tag{7.3a}$$

$$\vec{M}_s(\bar{r}, t) = \hat{n} \times \vec{E}_{\text{inc}}(\bar{r}, t) \tag{7.3b}$$

We note that the electromagnetic field within this volume is the total field (the incident field plus the scattered field), while outside the volume it is only the scattered field.

7.2.3 Symmetry

The symmetry of the antenna and the symmetry of the excitation are sometimes used to reduce the size of the computational volume. This is illustrated for a transmitting antenna in Figs. 7.2(a) and 7.2(b). Here, the antenna is a bow tie located in the y-z plane. It is formed from a thin PEC and fed at its center. First, this antenna is replaced by the monopole configuration. That is, the top half of the antenna is placed over a PEC image plane (the x-y plane), as shown in Fig. 7.2(b). Next, the x-z plane and the y-z plane are made PMCs to exploit the remaining symmetry. For this example, the use of symmetry reduces the computational volume and the accompanying computer storage and running time by a factor of 8.

Fig. 7.2(c) shows the same antenna as a receiver for a plane wave incident from an arbitrary direction. In this case, none of the symmetry of the antenna can be exploited to reduce the size of the computational volume. When the incident plane wave possesses some of the same symmetry as the antenna, a reduction in the computational volume is

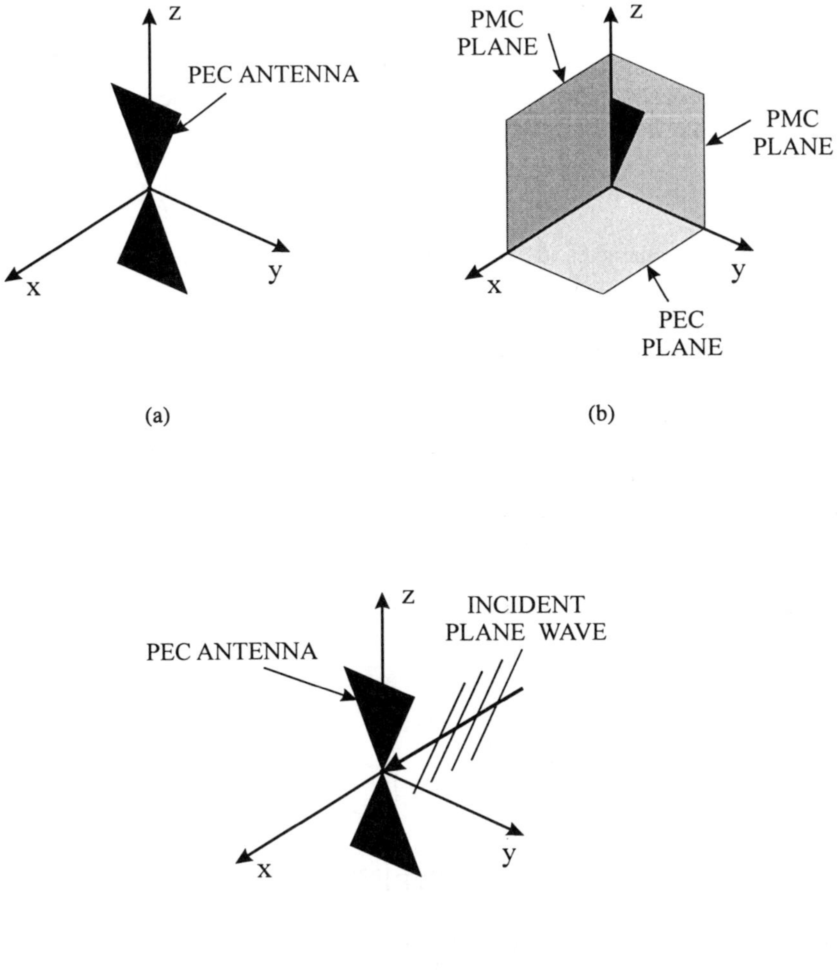

Fig. 7.2 Potential use of symmetry in the modeling of antennas. (a) Bow-tie antenna example. (b) For a transmitting bow-tie antenna, symmetry can be used to introduce PEC and PMC planes and reduce the computational volume by a factor of 8. (c) Receiving antenna with plane wave incident from an arbitrary direction.

possible. For example, when the plane wave is incident from the x direction with an electric field that is in the z direction, the PEC x-y plane and PMC x-z plane shown in Fig. 7.2(b) are used to decrease the computational volume by a factor of 4.

7.2.4 Excitation

The excitation for the antenna depends upon the particular application. For example, when we are determining the locations on an antenna at which reflections or radiation originate, a smooth, narrow pulse such as the Gaussian shown in Fig. 7.3(a) may be an appropriate excitation. For the transmitting antenna, the incident voltage in the feeding transmission line would then be

$$V_{\text{inc}}(t) = V_0 \exp\left[-(t/\tau_p)^2/2\right] \tag{7.4}$$

where τ_p is the characteristic time. Fig. 7.3(c) shows the Fourier transform of this signal, which is given by

$$V_{\text{inc}}(\omega) = \sqrt{2\pi}\,\tau_p\,V_0 \exp\left[-(\omega\tau_p)^2/2\right] \tag{7.5}$$

When the antenna is to be analyzed over a band of frequencies, it is efficient to excite the antenna with a pulsed signal and then use the Fourier transform to obtain the desired frequency-domain response. The Gaussian pulse can be used for this purpose; however, its spectrum has significant low-frequency content, as seen in Fig. 7.3(c). For some problems, this can produce an unacceptably long settling time for the solution. An alternative is to use the differentiated Gaussian pulse shown in Fig. 7.3(b). For the transmitting antenna, the incident voltage in the feeding transmission line would then be

$$V_{\text{inc}}(t) = -V_0\left(\frac{t}{\tau_p}\right)\exp\left\{-\left[(t/\tau_p)^2-1\right]/2\right\} \tag{7.6}$$

with the Fourier transform

$$V_{\text{inc}}(\omega) = -j\omega\sqrt{2\pi}\,\tau_p^2\,V_0 \exp\left\{-\left[(\omega\tau_p)^2-1\right]/2\right\} \tag{7.7}$$

shown in Fig. 7.3(c). This spectrum peaks at the frequency, $\omega_p = 1/\tau_p$; and the 10%-amplitude (-20 dB power) bandwidth extends from about $0.06\,\omega_p$ to $2.8\,\omega_p$.

When the antenna is to be analyzed at a single frequency, a ramped sinusoid is a good choice. An example is shown in Fig. 7.4, where the amplitude of the sinusoid is ramped from 0 to 1 over three periods. We apply this excitation and then wait a length of time sufficient for the electromagnetic field quantities of interest to reach the sinusoidal steady-state.

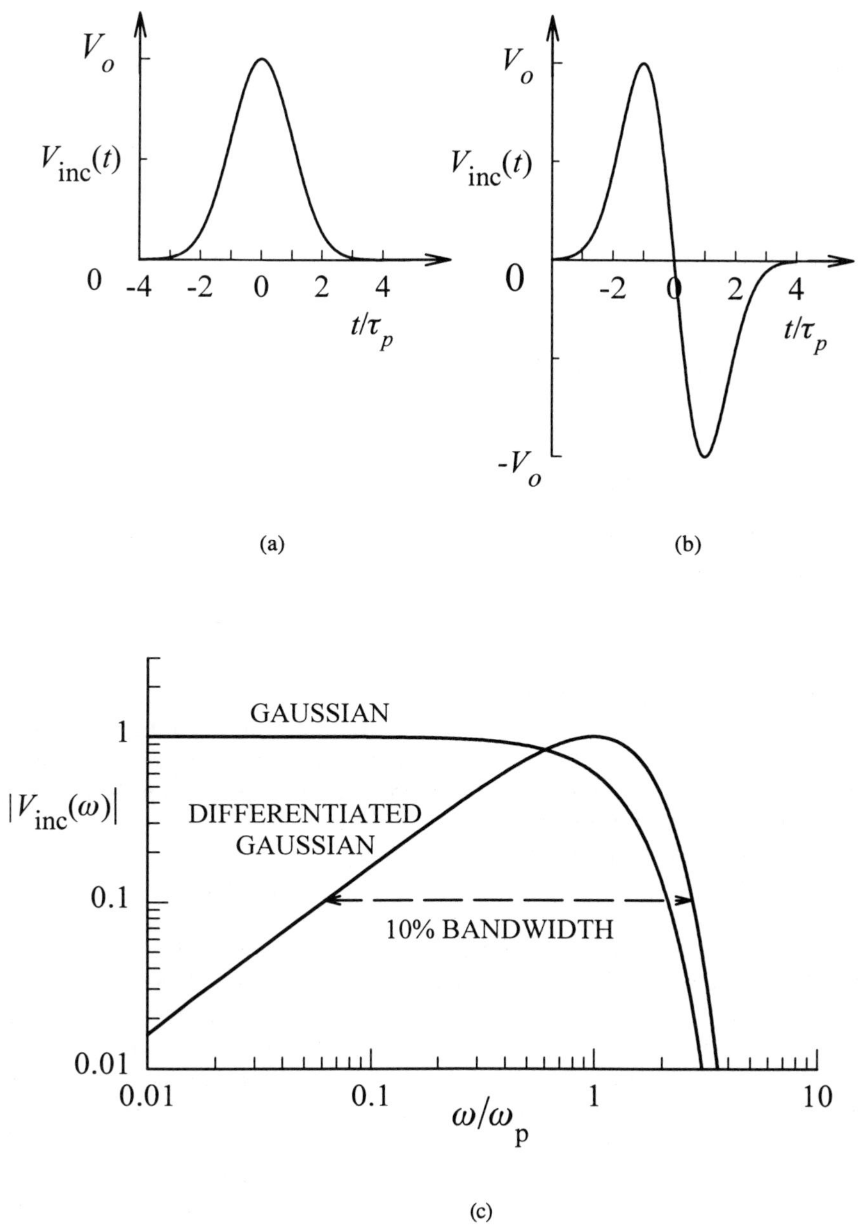

Fig. 7.3 (a) Gaussian pulse excitation; (b) differentiated Gaussian pulse excitation; (c) spectra for the Gaussian pulse and the differentiated Gaussian pulse.

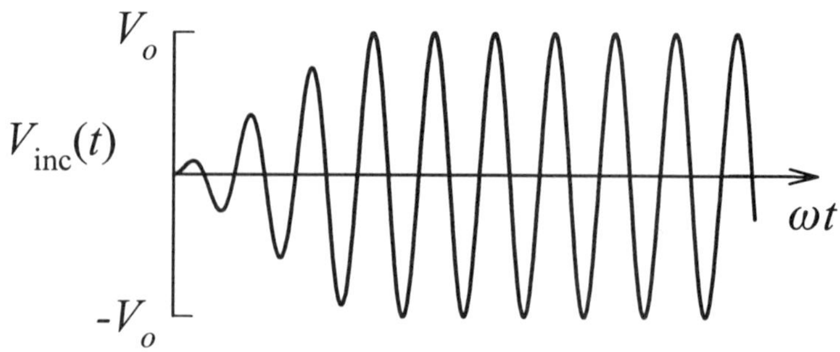

Fig. 7.4 Ramped sinusoidal excitation.

7.3 ANTENNA FEED MODELS

In the previous section, the general formulation of the antenna problem for use with the FDTD method was described. The need for key elements such as antenna feed models and NFFF transformations was introduced. This section and those that follow discuss the details of the implementation of these elements in the FDTD method. We begin with the antenna feed models.

7.3.1 Detailed Modeling of the Feed

The antenna is connected at its terminals to a transmission line or waveguide into which a signal is either introduced (transmitting) or extracted (receiving). When necessary, the full details of the transmission line and transmission line/antenna junction are included in the FDTD model. This may be desirable, for example, when designing an antenna for very low internal reflections.

We now review an example of such a detailed antenna feed model [24]. Fig. 7.5(a) is a schematic drawing showing the junction between a coaxial line and a metallic conical antenna. Using an FDTD analysis, the details of the junction (distance, *d*, and the size of the matching ring) were adjusted to make the peak of the reflected voltage from this junction -46 dB relative to the peak of the incident voltage.

The antenna shown in Fig. 7.5(a) is rotationally symmetric, so the surfaces of the FDTD cells can readily be made to conform to the surfaces of the conductors if a rotationally symmetric FDTD algorithm is used. For more complicated antennas that do not possess such symmetry, a staircased approximation to the geometry of the antenna and transmission line can be used. An example is shown in Fig. 7.5(b); this is the FDTD model for the junction between a bow-tie antenna and a coaxial line [25].

The dimensions of the transmission line and antenna feed region are usually smaller than the other key antenna dimensions. Thus, the FDTD cell size required to model these features is often much smaller than that needed to model only the antenna. If this small cell size is used for the entire grid, the computer storage and associated run time may be

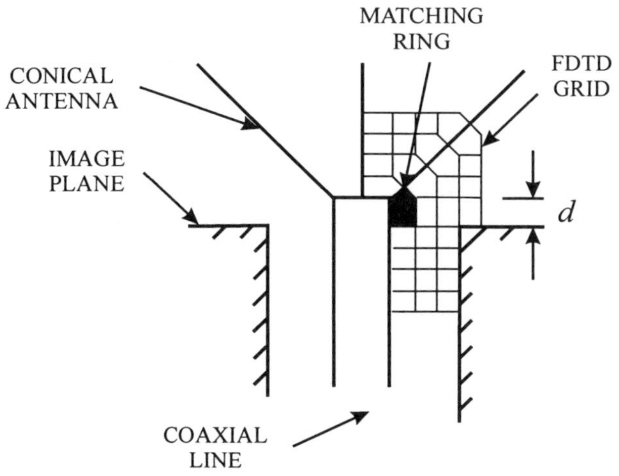

(a) Rotationally symmetric conical monopole antenna fed from coaxial line. *Source*: Maloney and Smith, *IEEE Trans. Antennas and Propagation*, 1993, pp. 940–947, © 1993 IEEE.

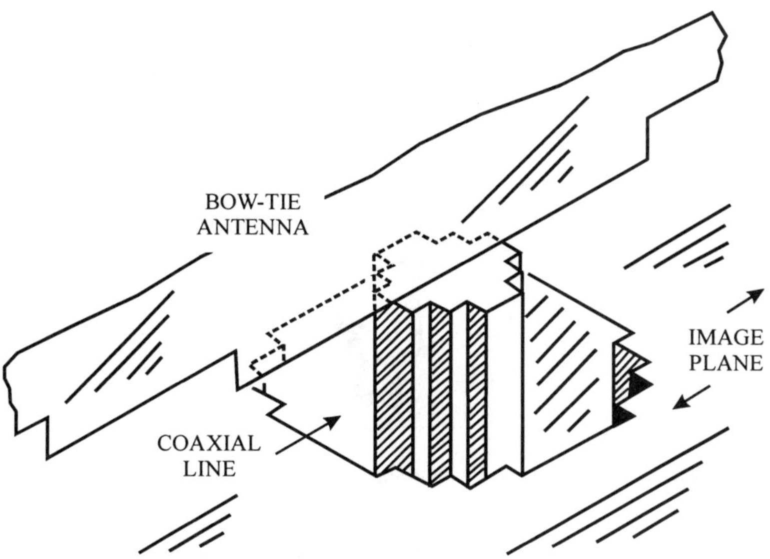

(b) Staircasing of geometry for bow-tie monopole antenna fed from coaxial line. *Source*: Shlager et al., *IEEE Trans. Antennas and Propagation*, 1994, pp. 975–982, © 1994 IEEE.

Fig. 7.5 Details of two antenna feeds directly modeled using FDTD.

very large. Alternatives are to use a ballooned mesh with the cell size increasing away from the terminals of the antenna, or to use a simplified, approximate model for the feedline and antenna junction. Different approaches to the latter are examined below.

7.3.2 Simple Gap Feed Model for a Monopole Antenna

For purposes of illustration, we use the monopole antenna shown in Fig. 7.6 in the following examples. The antenna of Fig. 7.6 is a perfectly conducting, cylindrical wire of radius a and length h driven through an image plane by a coaxial transmission line of characteristic impedance Z_0.

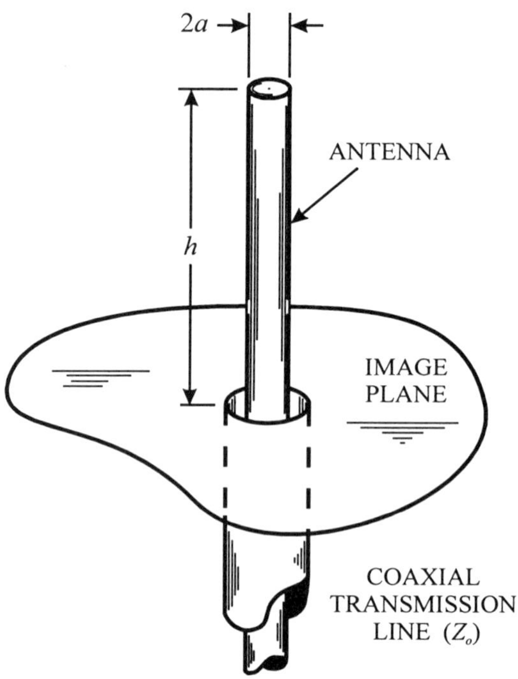

Fig. 7.6 Cylindrical monopole antenna fed through an image plane from a coaxial line.

Fig. 7.7 is a schematic drawing for the simplest method used for feeding this antenna. In the feed region, a gap of length Δz exists between the wire and the image plane. The lower end of the gap is on the image plane at the point $x = i_a \Delta x$, $y = j_a \Delta y$, $z = k_a \Delta z$, for which we use the notation (i_a, j_a, k_a). In the wire, the E_z component of the electric field is set to zero, and on the image plane the E_x and E_y components are set to zero.

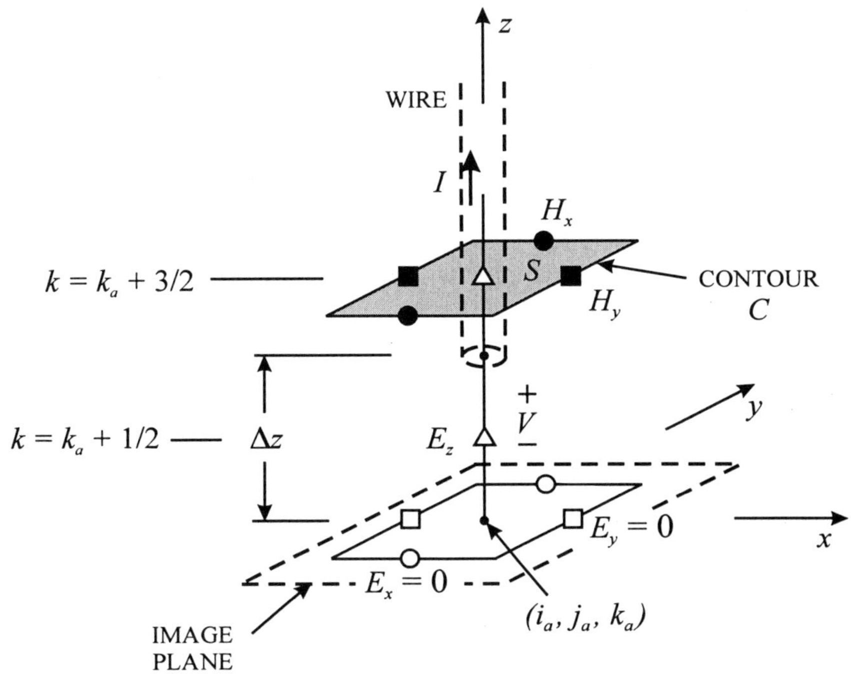

Fig. 7.7 Simple model for feeding the monopole antenna.

Transmitting Antenna

For a transmitting antenna, E_z in the gap is related to the specified voltage *V*:

$$E_z\Big|^n_{i_a,j_a,k_a+1/2} = -\frac{V(n\Delta t)}{\Delta z} \tag{7.8}$$

The current in the wire at the feedpoint is obtained by applying Ampere's law [26] to the surface *S* with the bounding contour *C* centered on the wire at $(i_a, j_a, k_a+3/2)$:

$$\oint_C \vec{H}\cdot d\vec{l} = \iint_S \vec{J}\cdot d\vec{S} + \varepsilon_0 \iint_S \frac{\partial \vec{E}}{\partial t}\cdot d\vec{S} \tag{7.9}$$

This gives

$$\begin{aligned} I\Big|^{n+1/2} = {} & \Delta x\left(H_x\Big|^{n+1/2}_{i_a,j_a-1/2,k_a+3/2} - H_x\Big|^{n+1/2}_{i_a,j_a+1/2,k_a+3/2}\right) \\ & + \Delta y\left(H_y\Big|^{n+1/2}_{i_a+1/2,j_a,k_a+3/2} - H_y\Big|^{n+1/2}_{i_a-1/2,j_a,k_a+3/2}\right) \end{aligned} \tag{7.10}$$

The Fourier transforms of the applied voltage and this current

$$V(t) \leftrightarrow V(\omega), \qquad I(t) \leftrightarrow I(\omega) \tag{7.11}$$

are used to compute the input impedance of the antenna:

$$Z(\omega) = V(\omega) / I(\omega) \tag{7.12}$$

For applications that involve pulse excitation, we are often interested in the reflected voltage, $V_{\text{ref}}(t)$, in the feeding transmission line for a specified incident voltage, $V_{\text{inc}}(t)$. This is obtained from the reflection coefficient

$$\Gamma(\omega) = \frac{Z(\omega) - Z_0}{Z(\omega) + Z_0} \tag{7.13}$$

and the Fourier transform of the incident voltage

$$V_{\text{inc}}(t) \leftrightarrow V_{\text{inc}}(\omega) \tag{7.14}$$

as follows:

$$V_{\text{ref}}(t) \leftrightarrow V_{\text{ref}}(\omega) = \Gamma(\omega)\, V_{\text{inc}}(\omega) \tag{7.15}$$

Receiving Antenna

For a receiving antenna, the open-circuit voltage is obtained from the E_z field in the gap:

$$V_{\text{oc}}\Big|^n = -\Delta z\, E_z\Big|^n_{i_a, j_a, k_a+1/2} \tag{7.16}$$

The voltage into a section of transmission line matched at the lower end is

$$V_{\text{rec}}(t) \leftrightarrow V_{\text{rec}}(\omega) \tag{7.17}$$

where

$$V_{\text{rec}}(\omega) = \left[\frac{Z_0}{Z(\omega) + Z_0}\right] V_{\text{oc}}(\omega) \tag{7.18a}$$

$$V_{\text{oc}}(t) \leftrightarrow V_{\text{oc}}(\omega) \tag{7.18b}$$

A review of the above procedures shows that they are quite awkward to use when only quantities in the time domain are of interest, such as the reflected voltage in the feeding transmission line of a transmitting antenna (7.15), or the voltage in the matched transmission line of a receiving antenna (7.17). This is due to the requirement for the

Fourier transformation and inverse Fourier transformation of several quantities. The FDTD method is inherently a time-domain approach, so it is advantageous, when possible, to carry out all computations in this domain. This can be accomplished by using the improved, simple feed model described below.

7.3.3 Improved Simple Feed Model

In this model, the one-dimensional FDTD grid shown in Fig. 7.8 is used to represent the transmission line attached to the antenna [27]. The index k' locates a point within this grid. The FDTD lattice used for the antenna is three-dimensional, and the set of indices (i, j, k) locates a point within this lattice. The top end of the transmission line is at point k'_{top} in the one-dimensional grid, and the feedpoint for the antenna (the center of the aperture in the coaxial line) is at point (i_a, j_a, k_a) in the three-dimensional lattice. The spatial step within the one-dimensional grid is Δz, the same as the spatial step for the z coordinate in the three-dimensional lattice.

The update equations for the current and voltage in the transmission line are

$$I\Big|_{k'+1/2}^{n+1/2} = I\Big|_{k'+1/2}^{n-1/2} - \left(\frac{1}{Z_0}\right)\left(\frac{v\,\Delta t}{\Delta z}\right)\left[V\Big|_{k'+1}^{n} - V\Big|_{k'}^{n}\right] \tag{7.19a}$$

$$V\Big|_{k'}^{n+1} = V\Big|_{k'}^{n} - Z_0\left(\frac{v\,\Delta t}{\Delta z}\right)\left[I\Big|_{k'+1/2}^{n+1/2} - I\Big|_{k'-1/2}^{n+1/2}\right] \tag{7.19b}$$

where v is the phase velocity within the transmission line. An ABC is used to match the transmission line at its bottom end, $k' = 0$.

The incident wave is introduced into the transmission line at the location k'_{source} by a "one-way" injector. This is an algorithm feature that launches a desired signal in the forward direction $(+z)$ while launching only a negligible, undesired signal in the backward direction $(-z)$. The update equations at the injector are

$$\begin{aligned} I\Big|_{k'_{\text{source}}+1/2}^{n+1/2} &= I\Big|_{k'_{\text{source}}+1/2}^{n-1/2} - \left(\frac{1}{Z_0}\right)\left(\frac{v\,\Delta t}{\Delta z}\right)\left[V\Big|_{k'_{\text{source}}+1}^{n} - V\Big|_{k'_{\text{source}}}^{n}\right] \\ &\quad + \left(\frac{1}{Z_0}\right)\left(\frac{v\,\Delta t}{\Delta z}\right) V_{\text{inc}}\Big|_{k'_{\text{source}}}^{n} \end{aligned} \tag{7.20a}$$

$$\begin{aligned} V\Big|_{k'_{\text{source}}}^{n+1} &= V\Big|_{k'_{\text{source}}}^{n} - Z_0\left(\frac{v\,\Delta t}{\Delta z}\right)\left[I\Big|_{k'_{\text{source}}+1/2}^{n+1/2} - I\Big|_{k'_{\text{source}}-1/2}^{n+1/2}\right] \\ &\quad + \left(\frac{v\,\Delta t}{\Delta z}\right) V_{\text{inc}}\Big|_{k'_{\text{source}}+1/2}^{n+1/2} \end{aligned} \tag{7.20b}$$

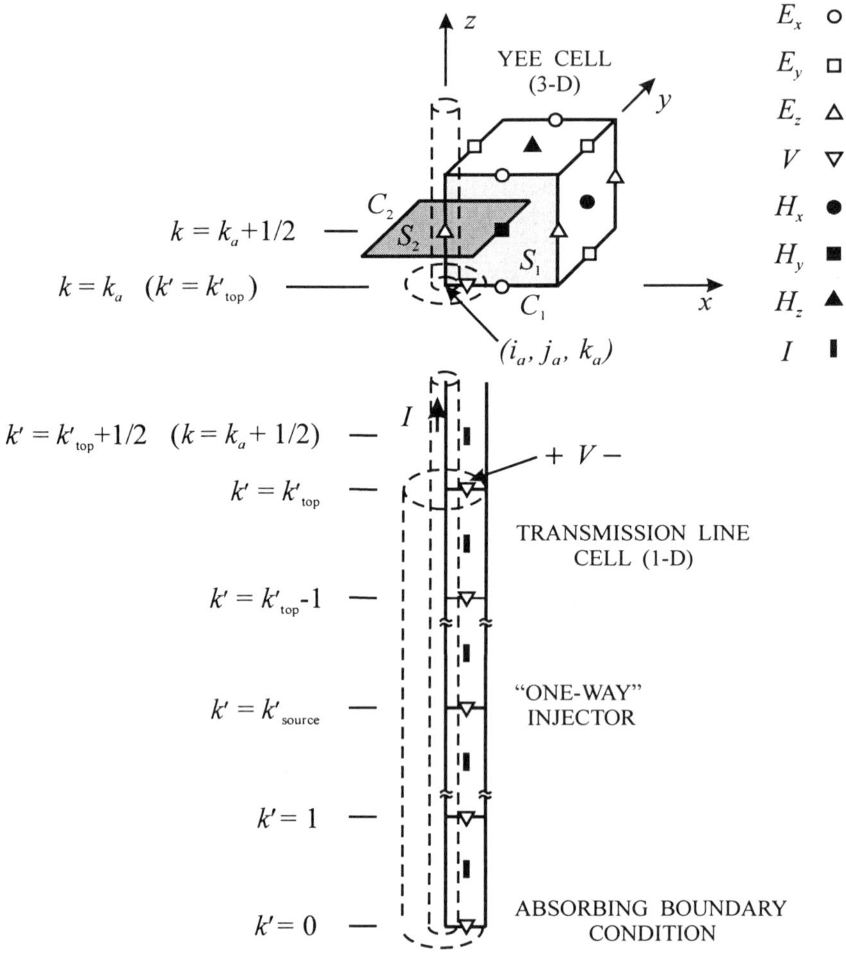

Fig. 7.8 Improved simple model for feeding the monopole antenna from a coaxial line, showing details of the one-dimensional FDTD grid for the transmission line and the three-dimensional lattice at the antenna feedpoint.

Fig. 7.9 illustrates the operation of the one-way injector [28]. This figure shows the computed voltage waveforms at a position above the injector (the forward signal) and below the injector (the backward signal). The forward signal is seen to be a 1V Gaussian pulse as given by (7.4), while the backward signal has a peak amplitude of less than 4 μV. The construction for this one-way injector is the same as that for the plane-wave injector referred to as "zero field on the right" in Section 7.5.2.

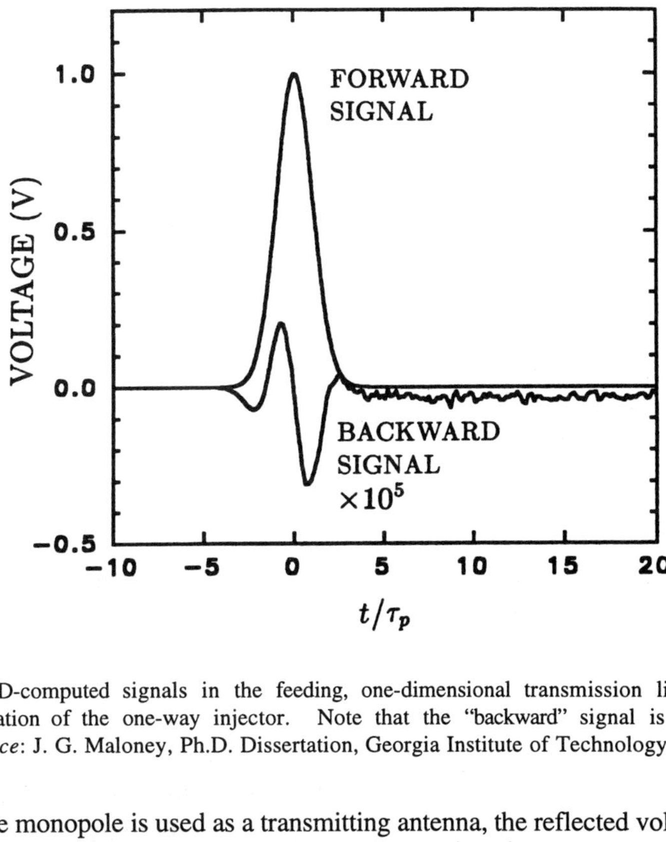

Fig. 7.9 FDTD-computed signals in the feeding, one-dimensional transmission line showing the operation of the one-way injector. Note that the "backward" signal is scaled by 10^5. *Source*: J. G. Maloney, Ph.D. Dissertation, Georgia Institute of Technology, 1992.

When the monopole is used as a transmitting antenna, the reflected voltage wave V_{ref} is the only wave present at points below the injector, $k' < k'_{\text{source}}$. When the monopole is used as a receiving antenna, the injector voltage is set to zero, $V_{\text{inc}} = 0$, and the voltage at any point in the matched transmission line is the received voltage wave, V_{rec}.

In the FDTD analysis, the one-dimensional transmission-line grid and the three-dimensional antenna lattice are attached in the following manner. The location $k' = k'_{\text{top}}$ in the one-dimensional grid is viewed as being coincident with the location (i_a, j_a, k_a) in the three-dimensional lattice. An update equation for H_y is then obtained by applying Faraday's law [26]:

$$\oint_{C_1} \vec{E}\cdot d\vec{l} \;=\; -\mu_0 \iint_{S_1} \frac{\partial \vec{H}}{\partial t}\cdot d\vec{S} \tag{7.21}$$

to the surface S_1 with the bounding contour C_1 in Fig. 7.8. This gives

$$\begin{aligned} V\big|_{k'_{\text{top}}}^{n} + \Delta z\, E_z\big|_{i_a+1,j_a,k_a+1/2}^{n} + \int_{\Delta x}^{a} \frac{\Delta x}{2x} E_x\big|_{i_a+1/2,j_a,k_a+1}^{n}\, dx \;=\; \\ \frac{\mu_0\,\Delta z}{\Delta t}\int_{a}^{\Delta x} \frac{\Delta x}{2x}\left(H_y\big|_{i_a+1/2,j_a,k_a+1/2}^{n+1/2} - H_y\big|_{i_a+1/2,j_a,k_a+1/2}^{n-1/2}\right) dx \end{aligned} \tag{7.22}$$

or

$$\begin{aligned} H_y\big|_{i_a+1/2,j_a,k_a+1/2}^{n+1/2} &= H_y\big|_{i_a+1/2,j_a,k_a+1/2}^{n-1/2} - \frac{\Delta t}{\mu_0 \Delta z}\cdot\left[E_x\big|_{i_a+1/2,j_a,k_a+1}^{n} - \frac{2}{\ln(\Delta x/a)}\cdot\frac{V\big|_{k'_{\text{top}}}^{n}}{\Delta x}\right] \\ &\quad + \frac{\Delta t}{\mu_0\,\Delta x}\cdot\frac{2}{\ln(\Delta x/a)}\cdot E_z\big|_{i_a+1,j_a,k_a+1/2}^{n} \end{aligned} \tag{7.23}$$

In this calculation, we assume that the E_x and H_y components of the field are zero within the wire, and that close to the wire they behave as $1/\rho$, where ρ is the radial distance from the center of the wire:

$$E_x(x) \;=\; \left(\frac{\Delta x}{2x}\right) E_x(\Delta x/2)\,, \qquad H_y(x) = \left(\frac{\Delta x}{2x}\right) H_y(\Delta x/2) \tag{7.24}$$

Update equations similar to (7.23) are obtained for the other three components of the magnetic field that surround the wire:

$$H_y\big|_{i_a-1/2,j_a,k_a+1/2}^{n+1/2}\,, \qquad H_x\big|_{i_a,j_a+1/2,k_a+1/2}^{n+1/2}\,, \qquad H_x\big|_{i_a,j_a-1/2,k_a+1/2}^{n+1/2}$$

The location $k' = k'_{\text{top}} + 1/2$ in the one-dimensional grid can be viewed as being coincident with the location $(i_a, j_a, k_a + 1/2)$ in the three-dimensional lattice. The update equation for the current $I\big|_{k'_{\text{top}}+1/2}^{n+1/2}$ in the transmission line is then calculated by applying Ampere's law (7.9) to the surface S_2 with the bounding contour C_2 in the three-dimensional lattice, as shown in Fig. 7.8:

$$\begin{aligned} I\big|_{k'_{\text{top}}+1/2}^{n+1/2} &= \Delta x\left(H_x\big|_{i_a,j_a-1/2,k_a+1/2}^{n+1/2} - H_x\big|_{i_a,j_a+1/2,k_a+1/2}^{n+1/2}\right) \\ &\quad + \Delta y\left(H_y\big|_{i_a+1/2,j_a,k_a+1/2}^{n+1/2} - H_y\big|_{i_a-1/2,j_a,k_a+1/2}^{n+1/2}\right) \end{aligned} \tag{7.25}$$

Equations (7.23) and (7.25) couple the transmission line to the antenna and vice versa.

At other points along the monopole, $k > k_a$, the monopole radius is easily accounted for in the update equations for the magnetic field [29]. For example, we use

$$\begin{aligned} H_y\Big|_{i_a+1/2,\,j_a,\,k+1/2}^{n+1/2} &= H_y\Big|_{i_a+1/2,\,j_a,\,k+1/2}^{n-1/2} - \frac{\Delta t}{\mu_0\,\Delta z}\cdot\left(E_x\Big|_{i_a+1/2,\,j_a,\,k+1}^{n} - E_x\Big|_{i_a+1/2,\,j_a,\,k}^{n}\right) \\ &+ \frac{\Delta t}{\mu_0\,\Delta x}\cdot\frac{2}{\ln(\Delta x/a)}\cdot E_z\Big|_{i_a+1,\,j_a,\,k+1/2}^{n} \end{aligned} \tag{7.26}$$

This equation is included to complete the description of the monopole antenna. It is not part of the feed model.

Fig. 7.10 compares FDTD results with measured data for the reflected voltage waveform in the transmission line feeding the transmitting monopole antenna. Here, time is normalized as t/τ_a, where $\tau_a = h/c$ is the time for light to travel the length of the monopole. Other parameters are $a/h = 0.0304$ and $Z_0 = 50\Omega$, and the excitation is a 1V Gaussian pulse characterized by $\tau_p/\tau_a = 0.161$. Very good agreement is seen.

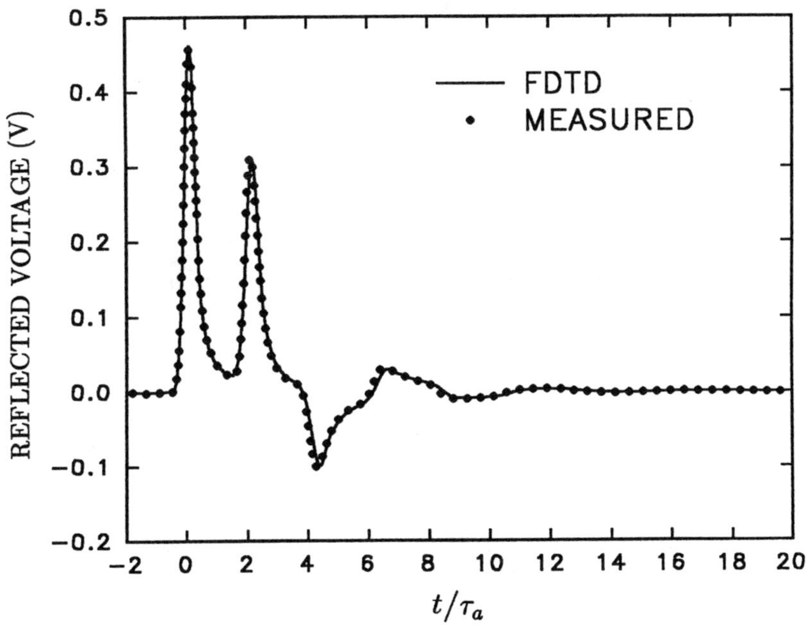

Fig. 7.10 Comparison of FDTD results with measured data for the reflected voltage in a coaxial transmission line feeding a cylindrical monopole antenna. The improved simple feed is used in the FDTD model. The excitation is a 1V Gaussian pulse with $\tau_p/\tau_a = 0.161$. *Source*: Maloney et al., *IEEE Trans. Antennas and Propagation*, vol. 42, 1994, pp. 289–292, © 1994 IEEE.

In the discussion presented above, the improved simple feed model is applied to a wire antenna connected to a coaxial transmission line (an unbalanced feed). This model is also adaptable to wire antennas connected to a two-wire transmission line (a balanced feed). The details of this application are presented in [30] and [31]. Other approaches to implementing the antenna feed are in [32] and [33].

7.4 NEAR-FIELD TO FAR-FIELD TRANSFORMATIONS

The FDTD method is inherently a near-field technique. It accurately computes the electromagnetic field within a computational volume surrounding the antenna as a function of time. Usually, it is impractical for the computational volume to extend from the antenna to the far-zone. Therefore, one of the NFFF transformations is needed to compute far-field antenna parameters such as radiation patterns and gain. These transformations are now discussed using the NFFF-transformation surface shown in Fig. 7.11.

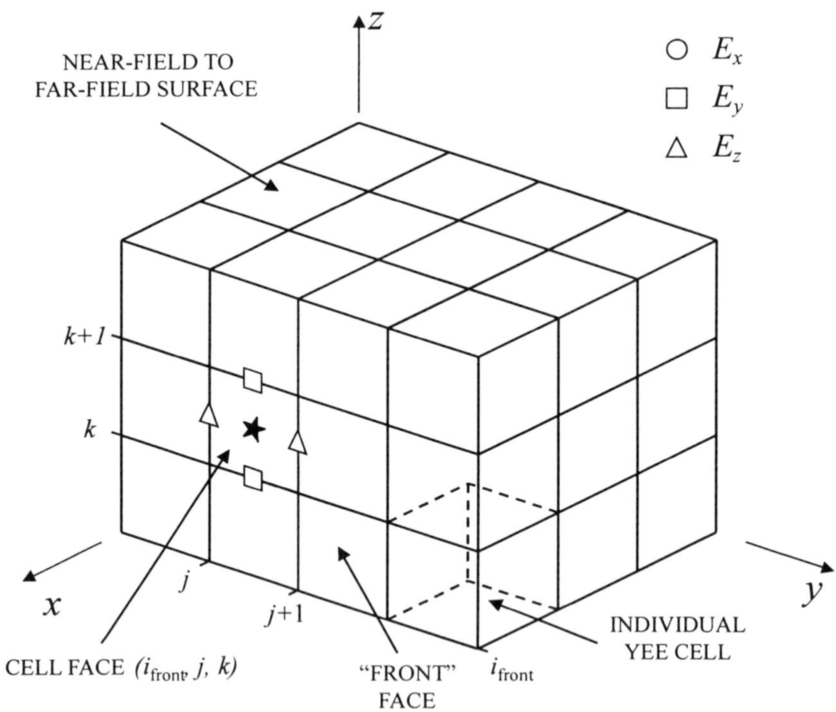

Fig. 7.11 NFFF-transformation surface constructed from the faces of individual Yee cells.

Because the standard FDTD method is based on an ordered, rectilinear gridding of the solution space, it is natural to consider a rectangular volume for the evaluation of a NFFF transformation. This volume is placed in the computational space so that its surface (the transformation surface) is between the exterior of the antenna and the beginning of the ABC, as shown in Fig. 7.1(a). In practice, the rectangular faces of the transformation surface are chosen to coincide with the faces of individual Yee cells, as shown in Fig. 7.11. Note that the six faces of the transformation surface are naturally subdivided into a rectangular array of patches by the faces of the individual Yee cells.

The far-field transformation surface must enclose the whole antenna. When PEC and PMC image planes are used to reduce the computational volume (as illustrated in Fig. 7.2(b)), image theory must be used to construct the field on the remainder of the transformation surface from the field on the portions of the surface inside the computational volume. Fig. 7.12(a) illustrates this case for the bow-tie transmitting antenna discussed earlier. Here, field components on the three faces inside the computational volume are used with image theory to compute the field on the rest of the transformation surface. For convenience, the relationships between the field components at PEC and PMC image planes are summarized in Fig. 7.12(b).

Two types of NFFF transformations are routinely used in FDTD antenna analysis: time-domain NFFF (TD-NFFF); and frequency-domain NFFF (FD-NFFF). The time-domain transformation computes the transient radiation response in a specific far-field direction. Upon Fourier transformation, this yields the broadband gain of the antenna in that direction. The frequency-domain transformation computes the complete far-zone radiation pattern at single frequencies. These two NFFF transformations are discussed below.

7.4.1 Time-Domain Near-Field to Far-Field Transformation

The TD-NFFF transformation is based on a discretized form of (7.1). To illustrate the derivation of the numerical TD-NFFF, consider the front face (planar surface parallel to the $y-z$ plane with the largest value of x) of the NFFF transformation surface shown in Fig. 7.11. The surface integral in the NFFF transformation (7.1) can be approximated by summations of the fields at the centers of the surface patches. Because of the spatially staggered nature of the field components in the Yee cell, the field at the center of each surface patch must be computed by averaging the nearest field components.

For example, consider the surface patch on the front face denoted by a star in Fig. 7.11. The tangential components of the electric field at the center of the patch (i_{front}, j, k) are found from the average of two neighboring values:

$$E_y\Big|_{i_{\text{front}},j+1/2,k+1/2}^{n+1} = 0.5\cdot\left(E_y\Big|_{i_{\text{front}},j+1/2,k}^{n+1} + E_y\Big|_{i_{\text{front}},j+1/2,k+1}^{n+1}\right) \tag{7.27a}$$

$$E_z\Big|_{i_{\text{front}},j+1/2,k+1/2}^{n+1} = 0.5\cdot\left(E_z\Big|_{i_{\text{front}},j,k+1/2}^{n+1} + E_z\Big|_{i_{\text{front}},j+1,k+1/2}^{n+1}\right) \tag{7.27b}$$

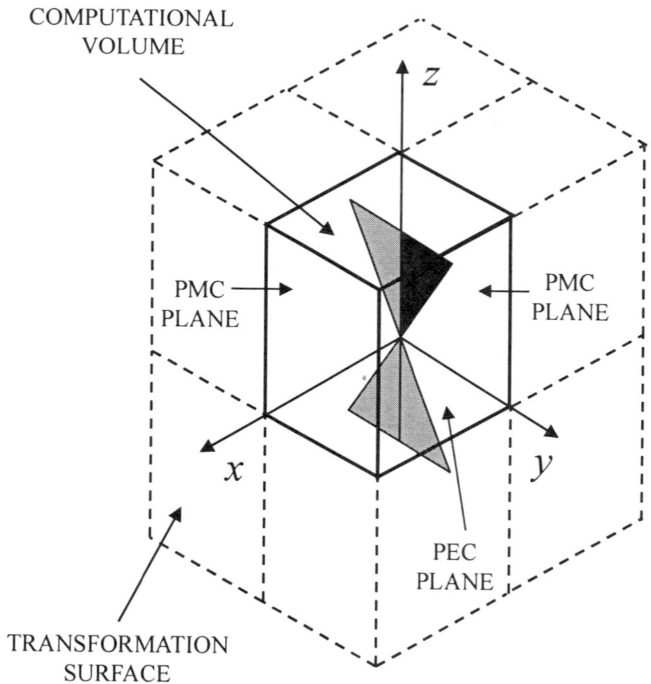

(a)

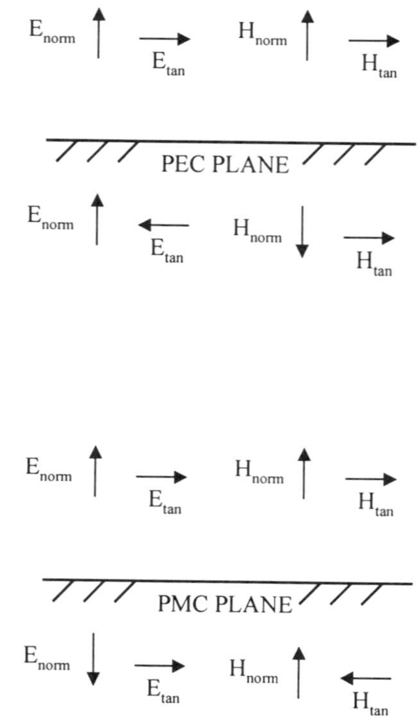

(b)

Fig. 7.12 (a) NFFF transformation surface developed from the surface of a smaller computational volume that exploits the symmetry of the antenna; (b) relationships for field components at the PEC and PMC image planes.

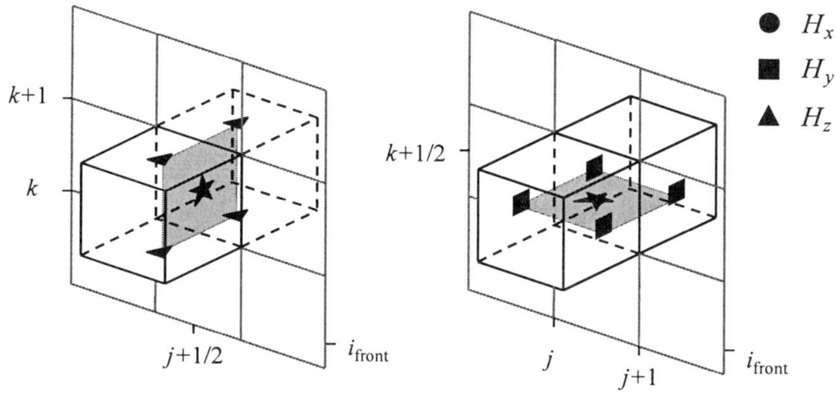

Fig. 7.13 Detail of spatial interpolation for the magnetic field.

Moreover, as illustrated in Fig. 7.13, the tangential components of the magnetic field at the center of the patch are found from the average of the four neighboring values:

$$H_y\Big|_{i_{\text{front}},\,j+1/2,\,k+1/2}^{n+1/2} = 0.25 \cdot \begin{pmatrix} H_y\Big|_{i_{\text{front}}+1/2,\,j+1,\,k+1/2}^{n+1/2} + H_y\Big|_{i_{\text{front}}+1/2,\,j,\,k+1/2}^{n+1/2} + \\ H_y\Big|_{i_{\text{front}}-1/2,\,j+1,\,k+1/2}^{n+1/2} + H_y\Big|_{i_{\text{front}}-1/2,\,j,\,k+1/2}^{n+1/2} \end{pmatrix} \tag{7.28a}$$

$$H_z\Big|_{i_{\text{front}},\,j+1/2,\,k+1/2}^{n+1/2} = 0.25 \cdot \begin{pmatrix} H_z\Big|_{i_{\text{front}}+1/2,\,j+1/2,\,k+1}^{n+1/2} + H_z\Big|_{i_{\text{front}}+1/2,\,j+1/2,\,k}^{n+1/2} + \\ H_z\Big|_{i_{\text{front}}-1/2,\,j+1/2,\,k+1}^{n+1/2} + H_z\Big|_{i_{\text{front}}-1/2,\,j+1/2,\,k}^{n+1/2} \end{pmatrix} \tag{7.28b}$$

The outward-pointing, unit vector normal to the front surface is $\hat{n} = \hat{x}$, and the vector products in (7.1) are

$$\hat{r} \times \hat{r} \times \left(\hat{n} \times \vec{H}\right) = \left(H_y \sin\theta + H_z \cos\theta \sin\phi\right)\hat{\theta} + \left(H_z \cos\phi\right)\hat{\phi} \tag{7.29a}$$

$$\hat{r} \times \left(\hat{n} \times \vec{E}\right) = \left(E_z \cos\phi\right)\hat{\theta} - \left(E_y \sin\theta + E_z \cos\theta \sin\phi\right)\hat{\phi} \tag{7.29b}$$

We note that the field in the far zone contains only $\hat{\theta}$ and $\hat{\phi}$ components, as expected. The contribution from the front face is then

$$E_\theta^r(\vec{r}, t) = \frac{\mu_0 \, \Delta y \, \Delta z}{4\pi r} \sum_{\substack{\text{Front} \\ \text{Face}}} \left(\sin\theta \, \frac{\partial H_y}{\partial t'} + \cos\theta \sin\phi \, \frac{\partial H_z}{\partial t'} - \frac{\cos\phi}{\eta_0} \frac{\partial E_z}{\partial t'} \right)_{t'=t_r} \tag{7.30a}$$

$$E_\phi^r(\vec{r}, t) = \frac{\mu_0 \Delta y \Delta z}{4\pi r} \sum_{\substack{\text{Front}\\\text{Face}}} \left(\cos\phi \frac{\partial H_z}{\partial t'} + \frac{\sin\theta}{\eta_0} \frac{\partial E_y}{\partial t'} + \frac{\cos\theta \sin\phi}{\eta_0} \frac{\partial E_z}{\partial t'} \right)_{t'=t_r} \tag{7.30b}$$

where the retarded time (7.2) changes from patch to patch. The contributions from the other five faces of the transformation surface are similar.

The postprocessing approach for implementing (7.30) involves storing at every time step the spatially interpolated tangential field components for every surface patch, as obtained from (7.27) and (7.28). To estimate the computer storage required for this operation, consider a cubic computational volume with $N \times N \times N$ cells and an iteration of $M = 10N$ time steps. When the transformation surface is 13 cells inside the boundaries of the computational volume, the computer storage required for the field components on the transformation surface is $24(N-26)^2 M = 240N(N-26)^2$. For large N, the required storage is about $240N^3$ real numbers. For comparison, the required computer storage for the full FDTD simulation alone is about $6N^3$ real numbers. Thus, the amount of additional computer storage needed for the postprocessing approach to the TD-NFFF transformation is about *40 times larger* than that for the full FDTD simulation alone. This amount of additional computational storage is prohibitive.

A computationally viable alternative is the concurrent-processing approach wherein (7.30) is implemented "on-the-fly" in step with the FDTD simulation. This approach incrementally computes the transient far field. It only requires storage of the interpolated fields on the transformation surface at the current time step, and a few linear arrays per far-field observation direction. The required storage for computation of the far field in P directions is now $24(N-26)^2 + 4MP$, which for large N is about $24N^2$ real numbers. We note that the storage required for the TD-NFFF transformation is now much less than that required for the full FDTD simulation alone: $24N^2$ versus $6N^3$. The details of the concurrent-processing approach are discussed below.

In the concurrent-processing approach, (7.30a) is rewritten as

$$r\,E_\theta^r(\vec{r}, t) = \frac{\mu_0}{4\pi} \left(\frac{\partial}{\partial t} \psi_\theta^H + \frac{1}{\eta_0} \frac{\partial}{\partial t} \psi_\theta^E \right) \tag{7.31}$$

with the new variables $\psi_\theta^H(t)$ and $\psi_\theta^E(t)$ given by

$$\psi_\theta^H(t) = \Delta y \Delta z \sum_{\substack{\text{front}\\\text{face}}} \left[\sin\theta \; H_y(t - t_{\text{front}}^{\text{delay}}) + \cos\theta \sin\phi \; H_z(t - t_{\text{front}}^{\text{delay}}) \right] \tag{7.32a}$$

$$\psi_\theta^E(t) = \Delta y \Delta z \sum_{\substack{\text{front}\\\text{face}}} \left[-\cos\phi \; E_z(t - t_{\text{front}}^{\text{delay}}) \right] \tag{7.32b}$$

where $t_{\text{front}}^{\text{delay}}$ for a front surface patch is given by

$$t_{\text{front}}^{\text{delay}}(j,k) = \frac{1}{c}\left[\begin{array}{c} r_{\text{offset}} - i_{\text{front}}\,\Delta x \sin\theta \cos\phi - \\ (j+0.5)\Delta y \sin\theta \sin\phi - (k+0.5)\Delta z \cos\theta \end{array}\right] \tag{7.32c}$$

and r_{offset} is an arbitrary constant that is chosen large enough to insure that $t_{\text{delay}} > 0$ for every surface patch. We note that a set of four ψ arrays must be computed per far-field observation direction.

During the FDTD time-stepping, $\psi_\theta^H(t)$ and $\psi_\theta^E(t)$ are computed incrementally at each time step. Consider the contribution to (7.32b) from a single surface patch on the front face (i_{front}, j, k). Because the field components in the FDTD simulation are only known at discrete time levels, the time retardation requires temporal interpolation. Applying linear interpolation, we have

$$\psi_\theta^E\Big|^n = -\cos\phi\,\Delta y\,\Delta z \left\{\begin{array}{l} \left[1 - q_{\text{front}}^E(j,k)\right] E_z\Big|_{i_{\text{front}},j+1/2,k+1/2}^{n-n_{\text{front}}^E(j,k)} \\ + \left[q_{\text{front}}^E(j,k)\right] E_z\Big|_{i_{\text{front}},j+1/2,k+1/2}^{n-n_{\text{front}}^E(j,k)+1} \end{array}\right\} \tag{7.33}$$

where

$$n_{\text{front}}^E(j,k) = \text{int}\left[t_{\text{front}}^{\text{delay}}(j,k)/\Delta t\right] \tag{7.34a}$$

$$q_{\text{front}}^E(j,k) = \left[t_{\text{front}}^{\text{delay}}(j,k)/\Delta t\right] - \text{int}\left[t_{\text{front}}^{\text{delay}}(j,k)/\Delta t\right] \tag{7.34b}$$

We note that each field component on the transformation surface affects at most two future time bins. Hence, in the concurrent-processing approach, (7.33) is rewritten as

$$\underbrace{\psi_\theta^E\Big|^{n+n_{\text{front}}^E}}_{\text{new value}} = \underbrace{\psi_\theta^E\Big|^{n+n_{\text{front}}^E}}_{\text{previous value}} - \Delta y\,\Delta z \cos\theta\left[q_{\text{front}}^E(j,k)\right] E_z\Big|_{i_{\text{front}},j+1/2,k+1/2}^{n} \tag{7.35a}$$

$$\underbrace{\psi_\theta^E\Big|^{n+n_{\text{front}}^E+1}}_{\text{new value}} = \underbrace{\psi_\theta^E\Big|^{n+n_{\text{front}}^E+1}}_{\text{previous value}} - \Delta y\,\Delta z \cos\theta\left[1 - q_{\text{front}}^E(j,k)\right] E_z\Big|_{i_{\text{front}},j+1/2,k+1/2}^{n} \tag{7.35b}$$

Note that (7.35a) and (7.35b) are update relations for each surface patch; they are not equalities. The incremental computation of ψ_θ^H is similar to that for ψ_θ^E.

At the conclusion of the FDTD time stepping, the far field is simply computed from a discretized form of (7.31), that is,

$$r\,E_\theta^r\Big|_{\bar{r}}^n = \frac{\mu_0}{4\pi\,\Delta t}\left[\left(\psi_\theta^H\Big|^{n+1} - \psi_\theta^H\Big|^n\right) + \frac{1}{\eta_0}\left(\psi_\theta^E\Big|^{n+1} - \psi_\theta^E\Big|^n\right)\right] \tag{7.36}$$

Equation (7.36) and a similar equation for $rE_\phi^r\Big|_{\bar{r}}^{n}$ give the far field in the time domain. These results are used to construct time-domain patterns for pulse-excited antennas (plots of the time-varying radiated field in various directions). They are also useful for calculating the antenna gain (a frequency-domain parameter) over a wide bandwidth. Since the radiated field contains both $\hat{\theta}$ and $\hat{\phi}$ components, it is convenient to define separate gains for each of these components:

$$G_\theta(\omega) = \frac{4\pi\left|rE_\theta^r(\omega)\right|^2/2\eta_0}{P_{\text{ant}}(\omega)} \tag{7.37a}$$

$$G_\phi(\omega) = \frac{4\pi\left|rE_\phi^r(\omega)\right|^2/2\eta_0}{P_{\text{ant}}(\omega)} \tag{7.37b}$$

where $P_{\text{ant}}(\omega)$ is the net power (incident minus reflected in a single mode) that enters the transmission line or waveguide feeding the antenna. The transient radiated field computed from (7.36) can be processed with the FFT to obtain the frequency-domain quantities in (7.37); for example, $rE_\theta^r(\bar{r},\omega) \leftrightarrow rE_\theta^r(\bar{r},t)$. $P_{\text{ant}}(\omega)$ is computed from

$$P_{\text{ant}}(\omega) = \left[1-\left|\Gamma(\omega)\right|^2\right]\left[\left|V_{\text{inc}}(\omega)\right|^2/2Z_0\right] \tag{7.38}$$

where $V_{\text{inc}}(\omega) \leftrightarrow V_{\text{inc}}(t)$, $V_{\text{refl}}(\omega) \leftrightarrow V_{\text{refl}}(t)$, and $\Gamma(\omega) = V_{\text{refl}}(\omega)/V_{\text{inc}}(\omega)$.

The TD-NFFF transformation is seen to be an efficient approach for computing the wideband, frequency-domain gain at a number of discrete directions. However, the TD-NFFF transformation is an inefficient approach for computing frequency-domain antenna patterns because of the fine angular spacing often required. The more efficient approach for computing frequency-domain antenna patterns is the FD-NFFF transformation, which is discussed next.

7.4.2 Frequency-Domain Near-Field to Far-Field Transformation

The FD-NFFF transformation is based on the frequency-domain counterpart of (7.1):

$$\vec{E}^r(\bar{r},\omega) = \frac{j\omega\mu_0}{4\pi r}\oiint_S \left\{\hat{r}\times\hat{r}\times[\hat{n}\times\vec{H}(\bar{r}',\omega)] - \frac{1}{\eta_0}\hat{r}\times[\hat{n}\times\vec{E}(\bar{r}',\omega)]\right\} \times \exp\left[-\frac{j\omega(r-\hat{r}\cdot\bar{r}')}{c}\right]dS' \tag{7.39}$$

The same surface used for the TD-NFFF transformation is used for the FD-NFFF transformation. Again, the time-dependent fields at the center of each surface patch are computed by averaging nearest-neighbor's values as in (7.27) and (7.28). The frequency-domain fields are determined by Fourier transformation of the time-dependent fields: $\vec{H}(\bar{r}, \omega) \leftrightarrow \vec{H}(\bar{r}, t)$ and $\vec{E}(\bar{r}, \omega) \leftrightarrow \vec{E}(\bar{r}, t)$.

To illustrate the derivation of the FD-NFFF transformation, again consider the front face (planar surface parallel to the $y-z$ plane with largest value of x) of the transformation surface in Fig. 7.11. The surface integral in (7.39) is approximated by a summation of the frequency-domain fields at the centers of the surface patches. The contribution from the front face is then

$$r E_\theta^r(\bar{r}, \omega) = \frac{j\omega\mu_0 \Delta y \Delta z}{4\pi} \sum_{\substack{\text{front}\\ \text{face}}} \left(\sin\theta\, H_y + \cos\theta \sin\phi\, H_z - \frac{\cos\phi}{\eta_0} E_z \right) \times \exp\left(j\omega t_{\text{front}}^{\text{delay}}\right) \tag{7.40a}$$

$$r E_\phi^r(\bar{r}, \omega) = \frac{j\omega\mu_0 \Delta y \Delta z}{4\pi} \sum_{\substack{\text{front}\\ \text{face}}} \left(\cos\phi\, H_z + \frac{\sin\theta}{\eta_0} E_y + \frac{\cos\theta \sin\phi}{\eta_0} E_z \right) \times \exp\left(j\omega t_{\text{front}}^{\text{delay}}\right) \tag{7.40b}$$

where $t_{\text{front}}^{\text{delay}}$ is given by (7.32c).

As discussed for the TD-NFFF transformation, the storage of the full time history of the tangential field components at every surface patch requires a prohibitively large amount of storage ($\approx 240\,N^3$ real numbers). Hence, it is impractical to compute the frequency-domain fields using Fourier transformation at the conclusion of the FDTD time stepping. The computationally attractive approach is to compute the Fourier transformation on-the-fly in step with the FDTD simulation using a running-sum implementation of the DFT. The DFT is given by

$$E_z(\bar{r}, \omega) = \sum_{n=1}^{N} E_z\big|_{\bar{r}}^{n} \exp(-j\omega n\Delta t) \tag{7.41}$$

In the running sum implementation, the update relation for time step, n, is simply

$$\underbrace{E_z(\bar{r}, \omega)}_{\text{new}} = \underbrace{E_z(\bar{r}, \omega)}_{\text{previous}} + E_z\big|_{\bar{r}}^{n} \exp(-j\omega n\Delta t) \tag{7.42}$$

Four complex-valued running sums (one per tangential field component) are required per surface patch per transformation frequency. For Q transformation frequencies, the required storage is about $24\,N^2 Q$ complex numbers.

In practice, the running-sum DFTs are computed during the execution of the FDTD code and stored for postprocessing. During postprocessing, the radiated fields (E_θ^r, E_ϕ^r) are computed using (7.40) with the stored frequency-domain field components (for the front face: E_y, E_z, H_y, H_z). The antenna patterns (gain versus angle) are then computed using (7.37) and (7.38). This approach gives the antenna analyst the flexibility to compute various far-field radiation patterns without rerunning the full FDTD simulation.

In practice, the FDTD analysis of transmitting antennas uses a short-pulse excitation such as the Gaussian pulse (7.4) or the differentiated Gaussian pulse (7.6), and both the TD-NFFF and FD-NFFF transformations are used. The FD-NFFF transformation is inefficient for computing broadband gain, because the field components on the transformation surface have to be computed for each frequency. Therefore, the TD-NFFF transformation with an FFT is used for this purpose. The frequency-domain surface fields computed during the FD-NFFF transformation are used during postprocessing to compute radiation patterns at selected frequencies.

7.5 PLANE-WAVE INJECTOR

The FDTD modeling of antennas in reception uses a plane-wave injector to create the incident electromagnetic plane wave. Here, the plane wave is produced by electric and magnetic current sources placed tangentially on a virtual surface surrounding the antenna. As shown in Fig. 7.1(b), the injector surface is located between the ABC and the antenna.

Ideally, the tangential electric and magnetic current sources are placed on the same surface. However, in the standard FDTD method, these sources cannot be coplanar because of the staggered nature of the Yee cell. Hence, a straightforward application of (7.3) leads to non-ideal behavior. That is, the field inside the computation volume only approximates the desired plane wave, and the field outside is small but not zero.

7.5.1 Effect of an Incremental Displacement of the Surface Currents

Consider the creation of a plane wave $[\vec{E}_{\text{inc}} = f(t + x/c)\vec{E}_0,\ \vec{H}_{\text{inc}} = -\hat{x} \times \vec{E}_{\text{inc}}/\eta_0]$ propagating in the $-\hat{x}$ direction by a set of planar current sources. The electric and magnetic surface currents are defined by

$$\vec{J}_s = \frac{1}{\eta_0} f(t)\, \hat{x} \times \hat{x} \times \vec{E}_0 \tag{7.43a}$$

$$\vec{M}_s = f(t)\, \hat{x} \times \vec{E}_0 \tag{7.43b}$$

In the ideal case, these currents are collocated, as shown in Fig. 7.14(a). The superposition of the radiation from these surface currents produces zero field on the right and the desired plane wave propagating toward the left.

Now consider the case shown in Fig. 7.14(b) where the surface currents are displaced by the small distance δ_x. The radiated fields are given by

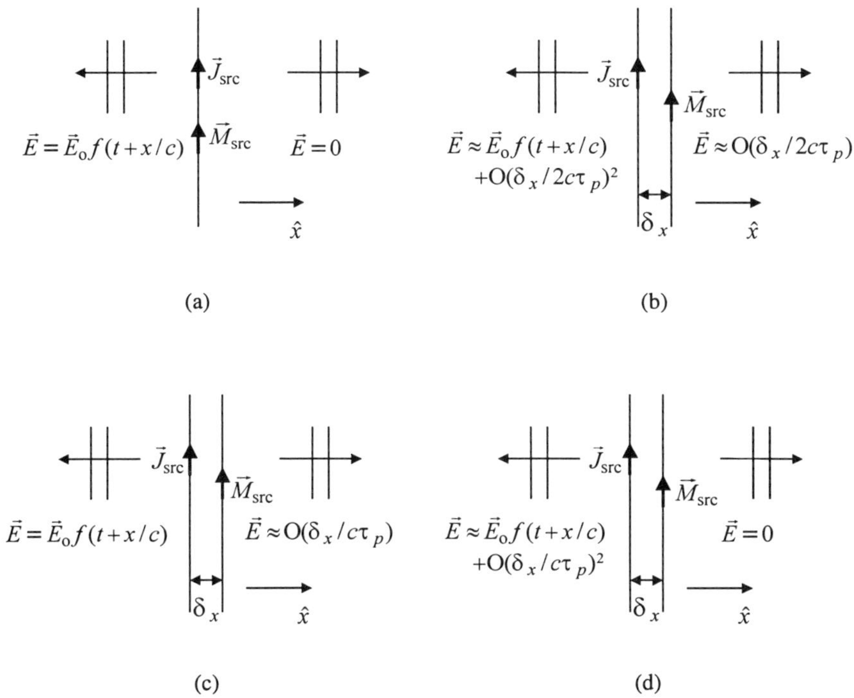

Fig. 7.14 Fields produced to the right and left of plane sheets of uniform electric and magnetic current.

$$\vec{E}_{\text{right}}(t - x/c) \approx -\left(\frac{\delta_x}{2c\tau_p}\right)\left[\tau_p f'(t + \delta_x/2c - x/c)\right]\vec{E}_0 + O\left(\frac{\delta_x}{2c\tau_p}\right)^3 \qquad (7.44a)$$

$$\vec{E}_{\text{left}}(t + x/c) \approx f\left(t - \delta_x/2c + x/c\right)\vec{E}_0 + \frac{1}{8}\left(\frac{\delta_x}{c\tau_p}\right)^2\left[\tau_p^2 f''(t - \delta_x/2c + x/c)\right]\vec{E}_0 + O\left(\frac{\delta_x}{2c\tau_p}\right)^4 \qquad (7.44b)$$

where τ_p is a characteristic time associated with $f(t)$, such as τ_p for the Gaussian pulse of (7.4), and $c\tau_p >> \delta x$. The field radiated to the right is no longer zero; it is now a small replica of the derivative of the source signal. The field radiated to the left is no longer the desired field; it is the desired signal with a slight time shift and an additional small replica of the second derivative of the source signal.

7.5.2 Effect of an Incremental Time Shift

A modification can be introduced that improves this situation. Here, the source currents are time shifted to produce either the desired signal propagating to the left or zero signal on the right, but not both conditions simultaneously. These two cases are illustrated in Figs. 7.14(c) and 7.14(d), respectively. The currents generating the exact field on the left (Fig. 7.14(c)) are

$$\vec{J}_s = \frac{1}{\eta_0} f(t)\, \hat{x} \times \hat{x} \times \vec{E}_0 \tag{7.45a}$$

$$\vec{M}_s = f(t + \delta_x / c)\, \hat{x} \times \vec{E}_0 \tag{7.45b}$$

with the resulting radiated fields

$$\vec{E}_{\text{right}}(t - x/c) \approx \left(\frac{\delta_x}{c\tau_p}\right)\left[\tau_p f'(t + \delta_x/c - x/c)\right]\vec{E}_0 + O\left(\frac{\delta_x}{c\tau_p}\right)^3 \tag{7.46a}$$

$$\vec{E}_{\text{left}}(t + x/c) = f(t + x/c)\, \vec{E}_0 \tag{7.46b}$$

For this case, the field to the left is the desired field, while the field to the right is again a small replica of the derivative of the source signal. While small, the amplitude of the signal propagating to the right is doubled; that is, all of the error is shifted onto the unwanted signal propagating to the right.

The currents for zero field on the right (Fig. 7.14(d)) are given by

$$\vec{J}_s = \frac{1}{\eta_0} f(t + \delta_x / c)\, \hat{x} \times \hat{x} \times \vec{E}_0 \tag{7.47a}$$

$$\vec{M}_s = f(t)\, \hat{x} \times \vec{E}_0 \tag{7.47b}$$

with the resulting radiated fields

$$\vec{E}_{\text{right}}(t - x/c) = 0 \tag{7.48a}$$

$$\vec{E}_{\text{left}}(t + x/c) \approx f(t + x/c)\, \vec{E}_0 + \frac{1}{2}\left(\frac{\delta_x}{c\tau_p}\right)^2\left[\tau_p^2 f''(t + x/c)\right]\vec{E}_0 + O\left(\frac{\delta_x}{c\tau_p}\right)^4 \tag{7.48b}$$

For this case, the field to the right is identically zero, while the field to the left is the desired signal plus an additional small replica of the second derivative of the source signal. All the error is shifted onto the signal propagating to the left: in (7.48b), the amplitude of the error is increased by a factor of 4 over that in (7.44b).

Traditionally, the FDTD method uses the zero-external-field case (Fig. 7.14(d)) for the plane-wave injector. We note that this is the preferred situation for computation of the scattered field, for example, when calculating the RCS. For modeling the receiving antenna, however, the exact-signal case (Fig. 7.14(c)) is more appropriate. Overall, for an all-purpose computer code, it makes sense to use only one form of the plane-wave injector for modeling both scattering and antenna reception. The configuration with zero external field (Fig. 7.14(d)) is generally used. This causes a small error, $O(\delta_x/c\tau_p)^2$, in the injected field.

7.5.3 Field Component Update Equations

Consider the update equations for field components near the front face of the plane-wave injector surface. This is a planar surface parallel to the $y-z$ plane with the largest value of x. Fig. 7.15 shows a cross section of the FDTD lattice on a slice of constant z. We note that the source currents are on planes perpendicular to this cross section. The field components involved in the update equations that are affected by the source currents on the front face are indicated by symbols. To derive the FDTD update equations for the field components near the plane-wave injector surface, consider Maxwell's equations with current-source terms:

$$\varepsilon_0 \frac{\partial \vec{E}}{\partial t} + \vec{J}_{\text{src}} = \nabla \times \vec{H} \tag{7.49a}$$

$$\mu_0 \frac{\partial \vec{H}}{\partial t} + \vec{M}_{\text{src}} = -\nabla \times \vec{E} \tag{7.49b}$$

where $\vec{J}_{\text{src}}$ and $\vec{M}_{\text{src}}$ are impressed volume current densities. The FDTD update equations for the field components affected by the electric current sources on the front surface of the plane-wave injector are

$$\begin{aligned} E_y\Big|_{i_{\text{front}}^{\text{src}},\,j+1/2,\,k}^{n+1} &= E_y\Big|_{i_{\text{front}}^{\text{src}},\,j+1/2,\,k}^{n} - \frac{\Delta t}{\varepsilon_0} J_{\text{src},y}\Big|_{i_{\text{front}}^{\text{src}},\,j+1/2,\,k}^{n+1/2} \\ &\quad - \frac{\Delta t}{\varepsilon_0 \Delta x}\left(H_z\Big|_{i_{\text{front}}^{\text{src}}+1/2,\,j+1/2,\,k}^{n+1/2} - H_z\Big|_{i_{\text{front}}^{\text{src}}-1/2,\,j+1/2,\,k}^{n+1/2} \right) \\ &\quad + \frac{\Delta t}{\varepsilon_0 \Delta z}\left(H_x\Big|_{i_{\text{front}}^{\text{src}},\,j+1/2,\,k+1/2}^{n+1/2} - H_x\Big|_{i_{\text{front}}^{\text{src}},\,j+1/2,\,k-1/2}^{n+1/2} \right) \end{aligned} \tag{7.50a}$$

(a)

(b)

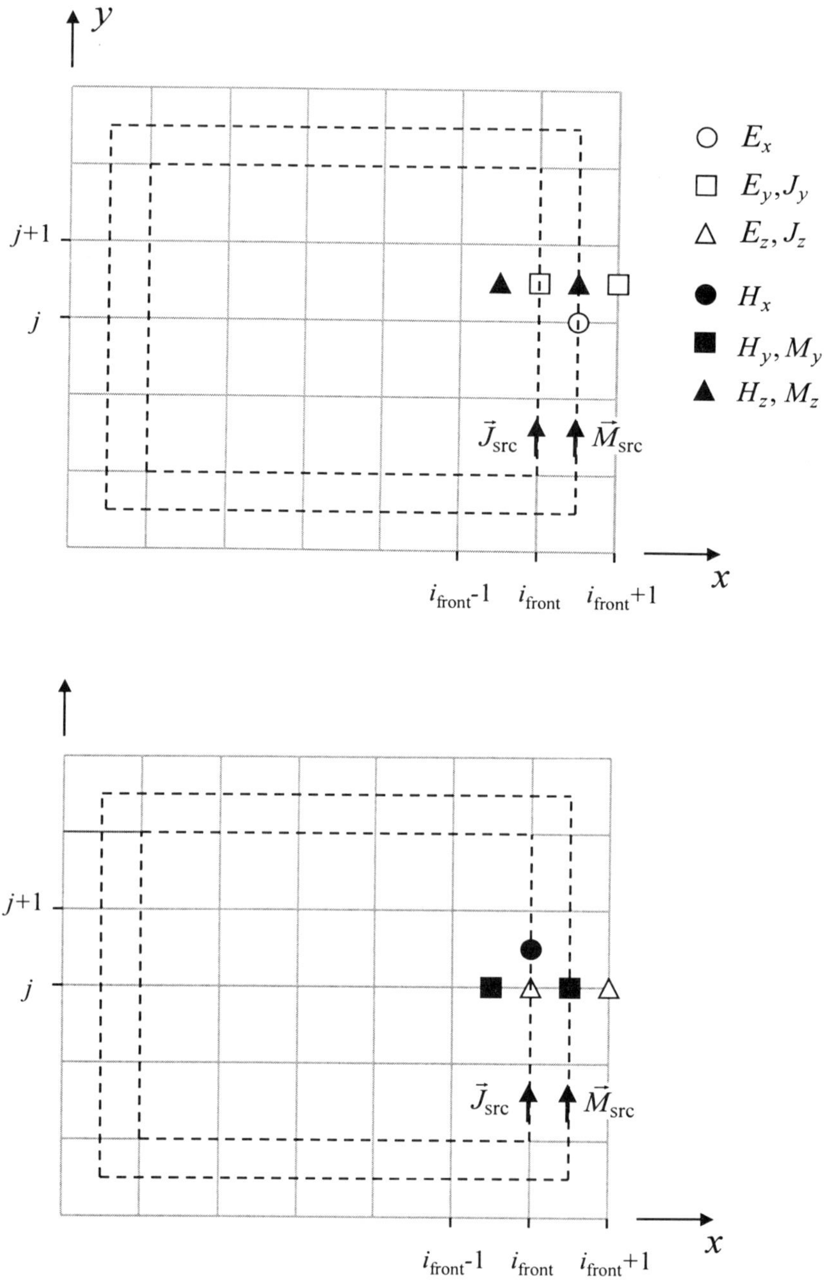

Fig. 7.15 Cross section of the FDTD lattice on a slice of constant z. The source currents are on a plane perpendicular to this cross section. (a) k, $k+1$ level; (b) $k \pm 1/2$ level.

$$E_z\Big|^{n+1}_{i^{\text{src}}_{\text{front}},j,k+1/2} = E_z\Big|^{n}_{i^{\text{src}}_{\text{front}},j,k+1/2} - \frac{\Delta t}{\varepsilon_0} J_{\text{src},z}\Big|^{n+1/2}_{i^{\text{src}}_{\text{front}},j,k+1/2}$$

$$+ \frac{\Delta t}{\varepsilon_0\,\Delta x}\left(H_y\Big|^{n+1/2}_{i^{\text{src}}_{\text{front}}+1/2,j,k+1/2} - H_y\Big|^{n+1/2}_{i^{\text{src}}_{\text{front}}-1/2,j,k+1/2} \right) \tag{7.50b}$$

$$- \frac{\Delta t}{\varepsilon_0\,\Delta y}\left(H_x\Big|^{n+1/2}_{i^{\text{src}}_{\text{front}},j+1/2,k+1/2} - H_x\Big|^{n+1/2}_{i^{\text{src}}_{\text{front}},j-1/2,k+1/2} \right)$$

and the update equations for field components affected by the magnetic current sources are

$$H_y\Big|^{n+1/2}_{i^{\text{src}}_{\text{front}}+1/2,j,k+1/2} = H_y\Big|^{n-1/2}_{i^{\text{src}}_{\text{front}}+1/2,j,k+1/2} - \frac{\Delta t}{\mu_0} M_{\text{src},y}\Big|^{n}_{i^{\text{src}}_{\text{front}}+1/2,j,k+1/2}$$

$$+ \frac{\Delta t}{\mu_0\,\Delta x}\left(E_z\Big|^{n}_{i^{\text{src}}_{\text{front}}+1,j,k+1/2} - E_z\Big|^{n}_{i^{\text{src}}_{\text{front}},j,k+1/2} \right) \tag{7.51a}$$

$$- \frac{\Delta t}{\mu_0\,\Delta z}\left(E_x\Big|^{n}_{i^{\text{src}}_{\text{front}}+1/2,j,k+1} - E_x\Big|^{n}_{i^{\text{src}}_{\text{front}}+1/2,j,k} \right)$$

$$H_z\Big|^{n+1/2}_{i^{\text{src}}_{\text{front}}+1/2,j+1/2,k} = H_z\Big|^{n-1/2}_{i^{\text{src}}_{\text{front}}+1/2,j+1/2,k} - \frac{\Delta t}{\mu_0} M_{\text{src},z}\Big|^{n}_{i^{\text{src}}_{\text{front}}+1/2,j+1/2,k}$$

$$- \frac{\Delta t}{\mu_0\,\Delta x}\left(E_y\Big|^{n}_{i^{\text{src}}_{\text{front}}+1,j+1/2,k} - E_y\Big|^{n}_{i^{\text{src}}_{\text{front}},j+1/2,k} \right) \tag{7.51b}$$

$$+ \frac{\Delta t}{\mu_0\,\Delta y}\left(E_x\Big|^{n}_{i^{\text{src}}_{\text{front}}+1/2,j+1,k} - E_x\Big|^{n}_{i^{\text{src}}_{\text{front}}+1/2,j,k} \right)$$

In (7.50) and (7.51), the discretized volume density of electric current for the case of zero external field is

$$J_{\text{src},y}\Big|^{n+1/2}_{i^{\text{src}}_{\text{front}},j+1/2,k} = \frac{1}{\Delta x} H_{\text{inc},z}\Big|^{n+1/2}_{i^{\text{src}}_{\text{front}}+1/2,j+1/2,k} \tag{7.52a}$$

$$J_{\text{src},z}\Big|^{n+1/2}_{i^{\text{src}}_{\text{front}},j,k+1/2} = -\frac{1}{\Delta x} H_{\text{inc},y}\Big|^{n+1/2}_{i^{\text{src}}_{\text{front}}+1/2,j,k+1/2} \tag{7.52b}$$

The discretized volume density of magnetic current for the case of zero external field is

$$M_{\mathrm{src},y}\Big|^{n}_{i^{\mathrm{src}}_{\mathrm{front}}+1/2,\,j,\,k+1/2} = -\frac{1}{\Delta x}\,E_{\mathrm{inc},z}\Big|^{n}_{i^{\mathrm{src}}_{\mathrm{front}},\,j,\,k+1/2} \tag{7.53a}$$

$$M_{\mathrm{src},z}\Big|^{n}_{i^{\mathrm{src}}_{\mathrm{front}}+1/2,\,j+1/2,\,k} = \frac{1}{\Delta x}\,E_{\mathrm{inc},y}\Big|^{n}_{i^{\mathrm{src}}_{\mathrm{front}},\,j+1/2,\,k} \tag{7.53b}$$

where H_{inc} and E_{inc} are components of the specified incident field. We note that these volume current densities are the surface current densities of (7.3) divided by Δx. This converts the surface current density to an equivalent volume current density. The other five surfaces are similarly treated.

The traditional derivation of the plane-wave injector is referred to as the total-field / scattered-field formulation [34]. A virtual surface that lies between the electric current source and the magnetic current source denotes the boundary between the total-field region and the scattered-field region, as shown in Fig. 7.16.

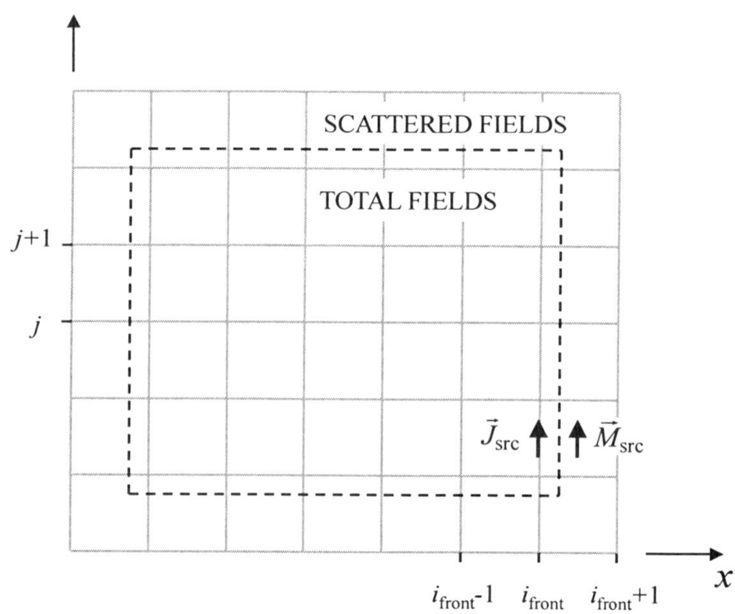

Fig. 7.16 Cross section of the FDTD lattice on a slice of constant z showing the total-field / scattered-field zoning. The source currents are on a plane perpendicular to this cross section.

Inside this boundary, all field components are considered total-field quantities; outside this boundary, all field components are considered scattered-field quantities. The relationship between the total field and the scattered field is

$$\text{total field} = \text{scattered field} + \text{incident field} \tag{7.54}$$

The total-field/scattered-field formulation is equivalent to the formulation discussed above, i.e., (7.50) to (7.53). For example, substituting (7.52a) into (7.50a) yields

$$\begin{aligned}
\underbrace{E_y\Big|^{n+1}_{i^{\text{src}}_{\text{front}},\,j+1/2,\,k}}_{\text{total field}} = {} & \underbrace{E_y\Big|^{n}_{i^{\text{src}}_{\text{front}},\,j+1/2,\,k}}_{\text{total field}} - \frac{\Delta t}{\varepsilon_0\,\Delta x}\underbrace{H_{\text{inc},z}\Big|^{n+1/2}_{i^{\text{src}}_{\text{front}}+1/2,\,j+1/2,\,k}}_{\text{incident field}} \\
& - \frac{\Delta t}{\varepsilon_0\,\Delta x}\left(\underbrace{H_z\Big|^{n+1/2}_{i^{\text{src}}_{\text{front}}+1/2,\,j+1/2,\,k}}_{\text{scattered field}} - \underbrace{H_z\Big|^{n+1/2}_{i^{\text{src}}_{\text{front}}-1/2,\,j+1/2,\,k}}_{\text{total field}}\right) \\
& + \frac{\Delta t}{\varepsilon_0\,\Delta z}\left(\underbrace{H_x\Big|^{n+1/2}_{i^{\text{src}}_{\text{front}},\,j+1/2,\,k+1/2}}_{\text{total field}} - \underbrace{H_x\Big|^{n+1/2}_{i^{\text{src}}_{\text{front}},\,j+1/2,\,k-1/2}}_{\text{total field}}\right)
\end{aligned} \tag{7.55}$$

where the total-field or scattered-field identification is indicated below each term. In (7.55), we note that the source term is simply the incident H_z field evaluated at the same space and time location as the lone scattered H_z-field term on the right-hand side. Conceptually, the incident magnetic field (source) is combining with the scattered magnetic field term to make (7.55) into a self-consistent total-field equation, as discussed in [34].

In a similar manner, substituting (7.53b) into (7.51b) yields

$$\begin{aligned}
\underbrace{H_z\Big|^{n+1/2}_{i^{\text{src}}_{\text{front}}+1/2,\,j+1/2,\,k}}_{\text{scattered field}} = {} & \underbrace{H_z\Big|^{n-1/2}_{i^{\text{src}}_{\text{front}}+1/2,\,j+1/2,\,k}}_{\text{scattered field}} - \frac{\Delta t}{\mu_0\,\Delta x}\underbrace{E_{\text{inc},y}\Big|^{n}_{i^{\text{src}}_{\text{front}},\,j+1/2,\,k}}_{\text{incident field}} \\
& - \frac{\Delta t}{\mu_0\,\Delta x}\left(\underbrace{E_y\Big|^{n}_{i^{\text{src}}_{\text{front}}+1,\,j+1/2,\,k}}_{\text{scattered field}} - \underbrace{E_y\Big|^{n}_{i^{\text{src}}_{\text{front}},\,j+1/2,\,k}}_{\text{total field}}\right) \\
& + \frac{\Delta t}{\varepsilon_0\,\Delta z}\left(\underbrace{E_x\Big|^{n}_{i^{\text{src}}_{\text{front}}+1/2,\,j+1,\,k}}_{\text{scattered field}} - \underbrace{E_x\Big|^{n}_{i^{\text{src}}_{\text{front}}+1/2,\,j,\,k}}_{\text{scattered field}}\right)
\end{aligned} \tag{7.56}$$

In (7.56), we note that the source term is simply the incident E_y field evaluated at the same space and time location as the lone total E_y-field term on the right-hand side. Conceptually, the incident electric field (source) is combining with the total electric field term to make (7.56) into a self-consistent scattered-field equation.

We see that the traditional plane-wave source condition for FDTD based on the total-field / scattered-field formulation is equivalent to the zero-external-field plane-wave injector case of Fig. 7.14(d). In practice, the external field is never exactly zero because the discretization of the differential equations causes a phenomenon known as grid dispersion. This introduces a small error that negates the perfect cancellation in the external region.

7.6 CASE STUDY

Now that the elements for the FDTD analysis of antennas have been discussed and their numerical implementation described, they will be applied to a particular antenna. The purpose of this example is twofold: (1) to show how the elements are combined in a practical situation, and (2) to show the accuracy that can be expected from the method when properly applied.

The antenna selected for this example is the Flann Microwave Instruments Ltd. Model 1624-20, an X-band (8.2 – 12.4 GHz) metal pyramidal horn fed by rectangular WR-90 waveguide. Antennas like this are used in many microwave applications and may serve as gain standards (standard-gain horns).

Fig. 7.17 shows the FDTD model for this horn antenna. The small inset drawings at the left of the figure define the lengths and angles that characterize the original antenna geometry. The numerical values of these parameters are: $a = 109.5$ mm, $b = 78.5$ mm, $D = 228.4$ mm, $l_w = 50.8$, $\alpha = 10.74^\circ$, and $\beta = 8.508^\circ$.

The FDTD model uses a uniform lattice composed of $519 \times 116 \times 183$ cubic Yee cells having $\Delta x = \Delta y = \Delta z = 0.635$ mm ($\approx \lambda_0 / 38$ at 12.4 GHz). The even symmetry of the horn geometry is used to define the x-z plane at $y = 0$ as a PMC. Looking closely at the perspective view of the model in Fig. 7.17, we see the faces of the individual cells that model the horn. In fact, the cells are drawn twice their actual size for clarity. (The linear dimensions of a cell face are twice actual size.) This high resolution permits direct modeling of the 1.35-mm wall thickness of the metallic components (assumed to be PECs). A 10-cell PML ABC is used on all sides of the lattice. It is spaced 20 cells from the closest surface of the antenna, except directly in front of the aperture where the spacing is 40 cells.

The antenna is assumed to be fed from a coaxial transmission line having the characteristic impedance $Z_0 = 50\Omega$. The center conductor of the coaxial line (the probe) has a radius $a_{\text{probe}} \approx 0.2$ mm and extends into the waveguide a distance $h_{\text{probe}} = 6.35$ mm. The probe is centered laterally in the waveguide and is spaced at the distance $l_{\text{probe}} = 5.08$ mm from the waveguide's back wall. This minimizes the reflection coefficient for the coaxial line / waveguide transition at midband, 10 GHz. In the FDTD analysis, the improved simple feed model of Section 7.3.3 is used for the transmission line and probe. For the transmitting case, the antenna is excited with the differentiated Gaussian pulse of (7.6) with $\tau_p = 1.59 \times 10^{-11}$ sec.

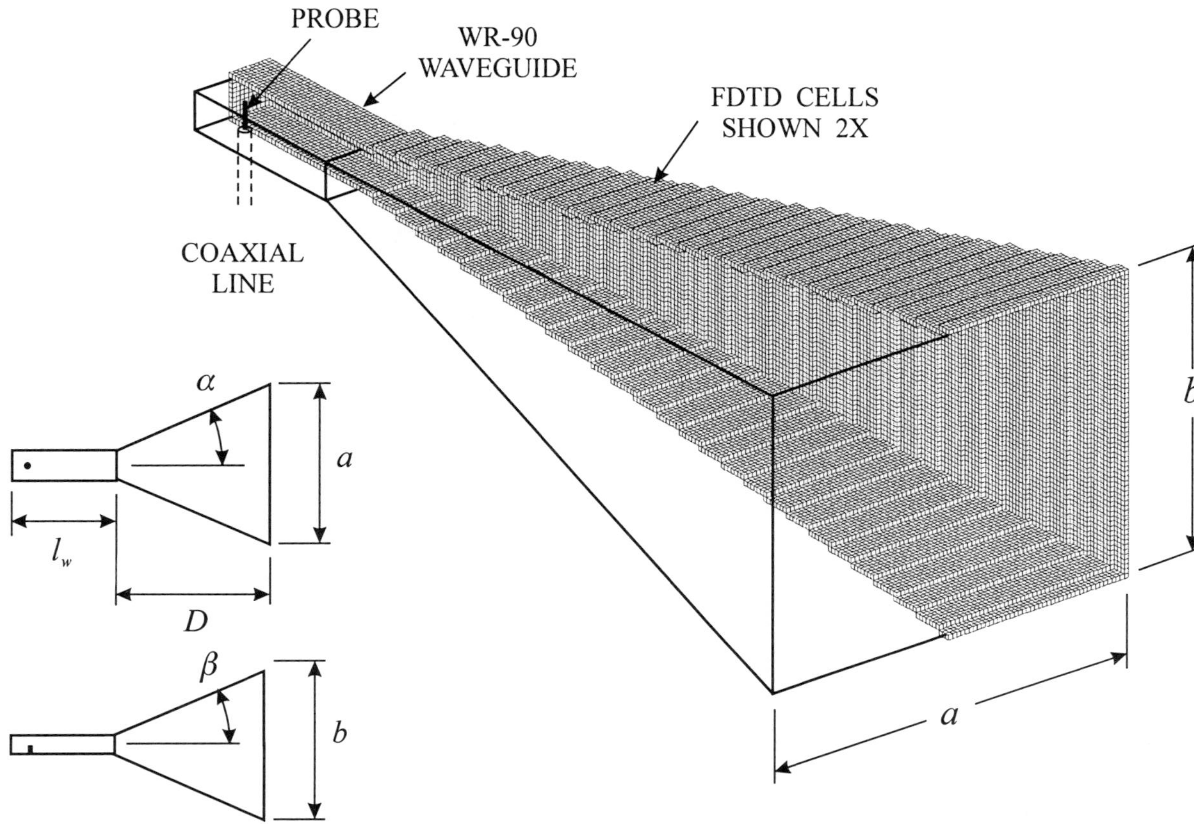

Fig. 7.17 Details of the geometry for the FDTD simulation of the Flann Model 1624-20 pyramidal horn antenna.

Fig. 7.18 compares FDTD calculations and measurements [35] of the far-field E-plane and H-plane radiation patterns of the Flann pyramidal horn antenna of Fig. 7.17 at 10 GHz.[1] Fig. 7.19(a) compares the FDTD-calculated and measured boresight gain as a function of frequency. The FDTD results are in excellent agreement with the measurements.

An FDTD analysis was also performed for the receiving horn antenna for the case of a z-polarized plane wave incident in the boresight direction. The time waveform of the impinging signal is the same differentiated Gaussian pulse used for the transmitting case. The effective area for the horn is derived from the FDTD-computed received voltage in the coaxial transmission line:

$$A_e(\omega) \;=\; \frac{\eta_0 \left| Z(\omega)/Z_0 + 1 \right|^2 \left| V_{\text{rec}}(\omega) \right|^2}{4\, R(\omega) \left| \vec{E}^i(\omega) \right|^2} \tag{7.57}$$

where $Z(\omega) = R(\omega) + jX(\omega)$ is the input impedance of the transmitting antenna. Fig. 7.19(b) compares the FDTD results for A_e obtained using (7.57) with the effective area, $A_e = \lambda_0^2\, G/4\pi$, derived from the measured gain, G. (Note that these results are shown in decibels relative to one square meter.) Again, the agreement is excellent. Overall, these data show that, when properly used, the FDTD analysis produces equally good results for the transmitting and receiving cases. Further, Fig. 7.19 demonstrates reciprocity for the FDTD results for transmitting and receiving antennas in the frequency domain. A demonstration of reciprocity for the time domain is in [36].

The FDTD method inherently provides information about the dynamics of the electromagnetic field within the entire computational volume over the entire period of the simulation. Only a small fraction of this information is used when investigating conventional antenna-performance metrics such as radiation pattern, gain, and effective area. Mapping or visualizing the computed space-time evolution of the electromagnetic field can provide useful additional information helping to improve our understanding of the radiation process for an antenna.

Fig. 7.20 illustrates how space-time visualization is used to help develop an understanding of how an antenna radiates. Here, the magnitude of the electric field on the vertical symmetry plane of the horn antenna of Fig. 7.17 is plotted on a gray scale for two different times. The incident voltage in the coaxial line feeding the horn is a sinusoid that is amplitude modulated by a Gaussian pulse:

$$V_{\text{inc}}(t) \;=\; V_0 \exp\left[-(t/\tau_p)^2/2 \right] \sin(\omega_0 t) \tag{7.58}$$

where $f_0 = \omega_0/2\pi = 10.0$ GHz and $\tau_p = 7.96 \times 10^{-11}$ s.

---

[1] Measurements for this horn antenna were made at the National Physical Laboratory, Teddington, Middlesex, U.K. and kindly provided by Dr. David Gentle.

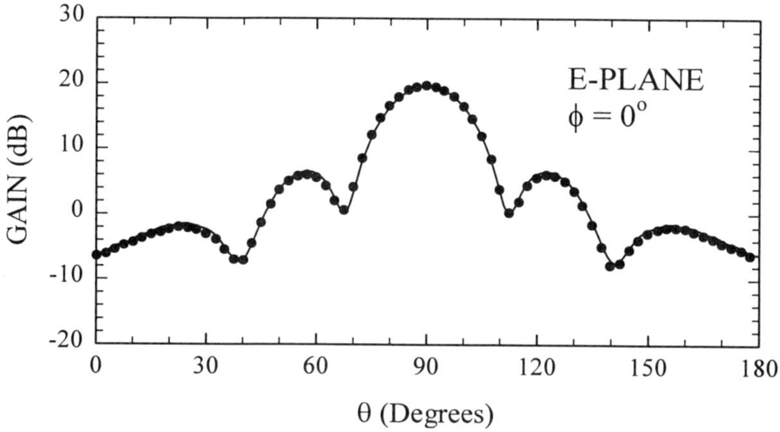

(a) *E*-plane radiation pattern.

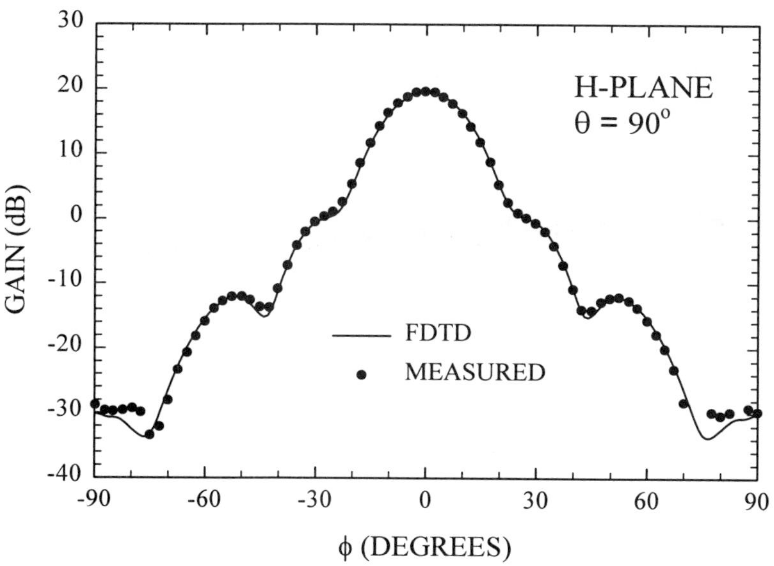

(b) *H*-plane radiation pattern.

Fig. 7.18 Comparison of FDTD-calculated and measured far-field patterns for the Flann pyramidal horn antenna of Fig. 7.17 at 10 GHz.

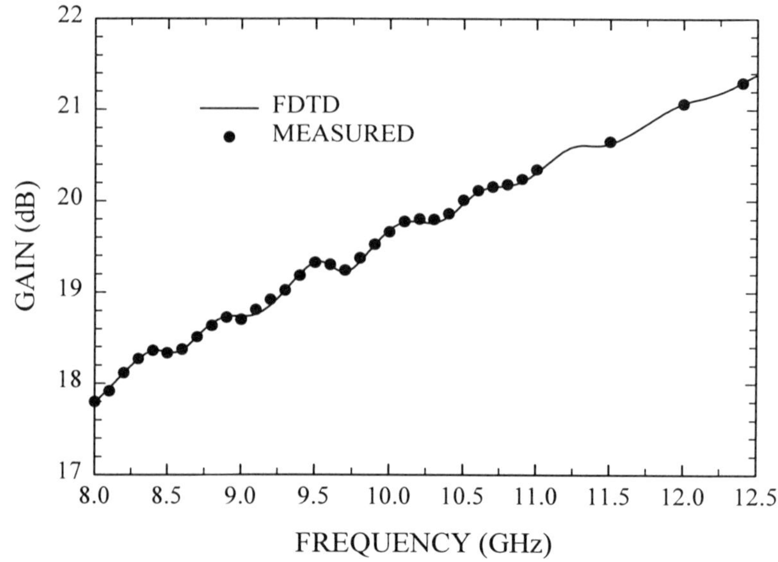

(a) Boresight gain versus frequency.

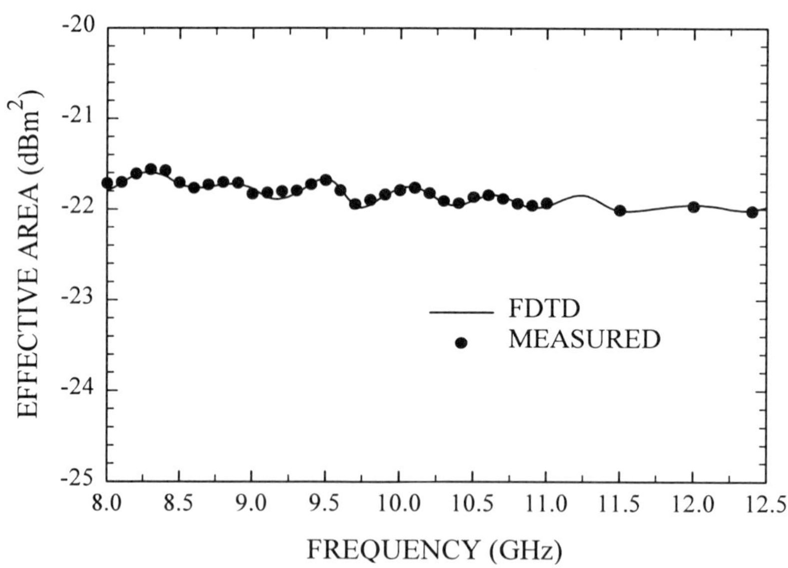

(b) Boresight effective area versus frequency.

Fig. 7.19 Comparison of FDTD-calculated and measured results for the Flann pyramidal horn antenna.

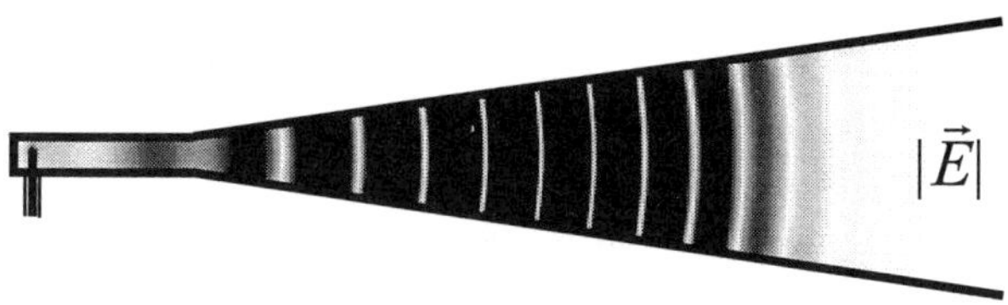

(a) The pulse has entered the horn from the waveguide.

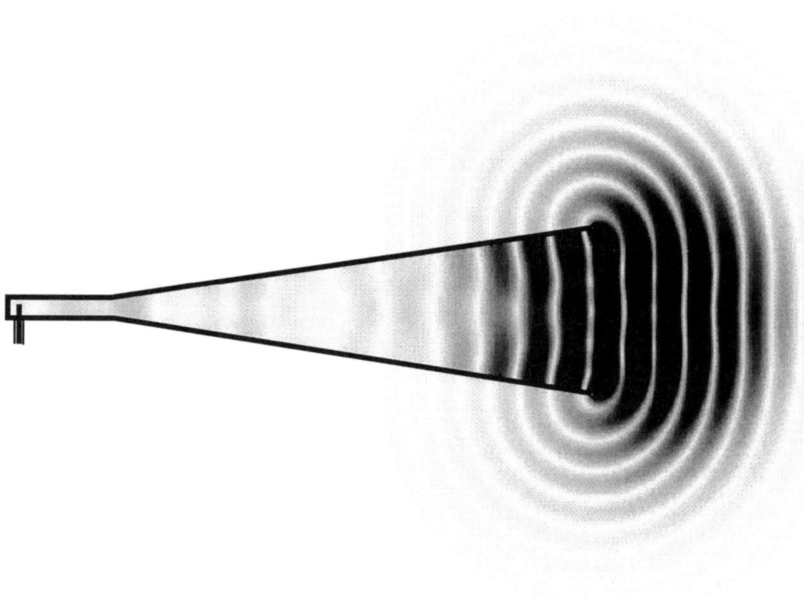

(b) The pulse has reached the aperture of the horn.

Fig. 7.20 Gray-scale visualization showing the magnitude of the FDTD-calculated electric field on the vertical symmetry plane of the Flann pyramidal horn antenna at two early times.

In Fig. 7.20(a), the pulse has entered the horn from the waveguide but has not yet reached the aperture. The spacing between the white lines (lines of null field) roughly corresponds to one-half of the guide wavelength. We observe that this spacing decreases on going from the throat of the horn toward the aperture. This observation is reasonable in that the wavelength in the rectangular waveguide is about 1.3 times the free-space wavelength, whereas at the aperture of the horn it is much closer to the free-space wavelength.

In Fig. 7.20(b), the pulse has reached the aperture. The white lines in the horn near the aperture are distorted; there is a small segment that is concave to the right. This is caused by the reflection from the aperture that is traveling back toward the throat of the horn. Directly in front of the aperture, the radiated wave is roughly planar; it becomes spherical at greater distances.

7.7 SELECTED RECENT APPLICATIONS

This section reviews several recent applications of FDTD modeling in the analysis and design of practical antennas used in complex environments. These include: (1) a reflector antenna constructed using a dielectric photonic-bandgap structure instead of a metal sheet, (2) a ground-penetrating radar for detection of buried all-dielectric land mines, (3) analysis of antenna-radome interactions, (4) modeling the effects of the human body on wireless personal communication devices, and (5) analysis and design of antennas for diagnostic and therapeutic biomedical applications.

7.7.1 Use of Photonic-Bandgap Materials

Consider first the usage of photonic-bandgap (PBG) materials in antenna design. PBG materials are periodic structures formed from dielectrics. They are an electromagnetic analog of electronic crystals, and are characterized by frequency bands (bandgaps) in which electromagnetic waves cannot propagate in any direction within the material. At frequencies within the bandgap, a plane wave incident on a half-space of PBG material is totally reflected. In this regard, the material acts like a good conductor [37].

Fig. 7.21 depicts the FDTD model of a PBG "crystal" used as an all-dielectric reflector for a simple dipole antenna. The crystal is a block of dielectric having the relative permittivity $\varepsilon_r = 13$ and penetrated with air-filled cylindrical voids at the sites of a body-centered cubic lattice. For clarity, only a small segment of the periodic reflector structure is shown in the figure. In fact, a dielectric block 21 periods square by 3 periods thick is used for the full reflector. Fig. 7.22(a) shows the FDTD-calculated three-dimensional pattern of this antenna structure for an excitation frequency within the bandgap of the PBG crystal. The antenna pattern is maximum for azimuth angles pointing away from the reflector, just as if it were metal. This is clearly seen in Fig. 7.22(b), which compares the FDTD-computed and measured antenna patterns in the horizontal plane ($\theta = 90^\circ$ in Fig. 7.22(a)). Very good agreement is noted.

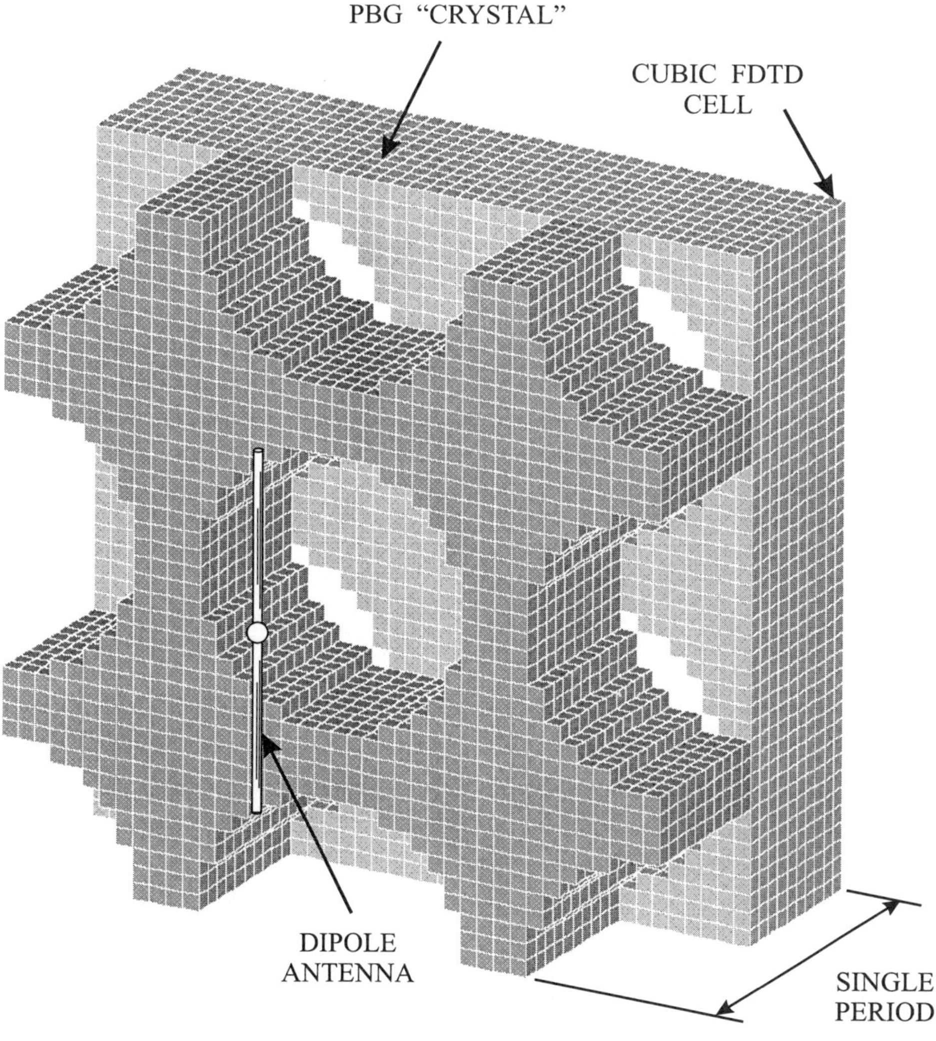

Fig. 7.21 FDTD model for a dipole antenna with an all-dielectric reflector formed from PBG material. Only a small portion of the reflector is shown. The actual reflector is 21 periods square and 3 periods thick.

(a)

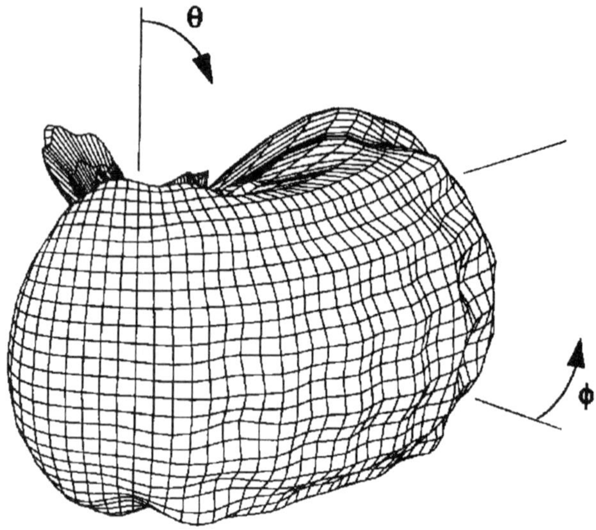

(b)

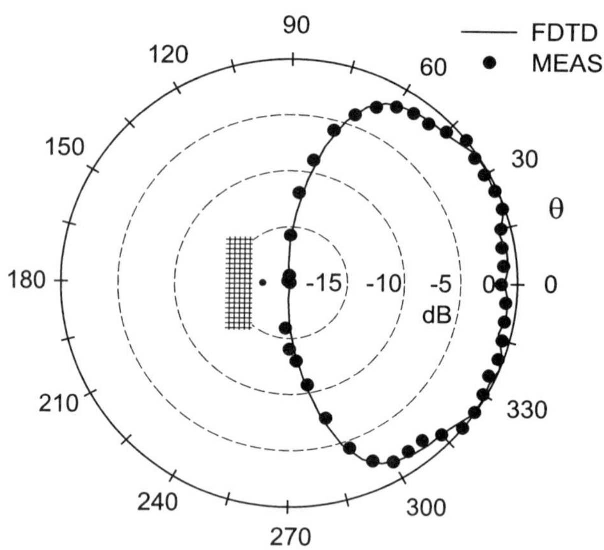

Fig. 7.22 FDTD-calculated radiation pattern of the reflector antenna structure of Fig. 7.21: (a) three-dimensional pattern for an excitation frequency within the bandgap of the PBG crystal; (b) comparison of FDTD-calculated and measured field patterns in the horizontal plane ($\theta = 90^\circ$) of (a).

7.7.2 Ground-Penetrating Radar

Fig. 7.23(a) shows a ground-penetrating radar (GPR) designed for the detection of buried, all-dielectric land mines. This arrangement of antennas for signal transmission and reception is designated as the separated-aperture sensor [31]. This sensor consists of two parallel dipole antennas, each contained in a metallic reflector. The dipoles are fed in the manner shown in Fig. 7.23(b), with the length of the dipoles and the position of the shorting screws on the tuning stubs adjusted to match the antennas at a specified frequency ($\approx$ 790 MHz). The coupling, $|S_{21}|$, between the two antennas is measured as the detector is moved over the surface of the ground, and an increase in the coupling indicates the presence of a buried anomaly such as a mine.

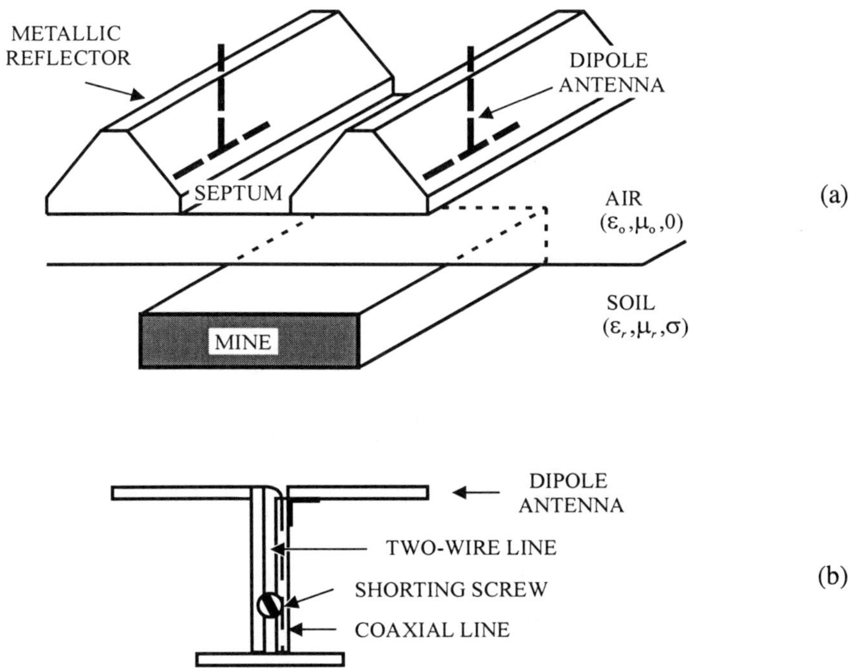

Fig. 7.23 Separated-aperture mine-detection sensor: (a) antenna arrangement; (b) details of the feed for the dipole antennas of (a). *Source*: J. M. Bourgeois, Ph.D. Dissertation, Georgia Institute of Technology, Atlanta, GA, 1997.

The complete system including two antennas with tuning stubs and reflectors, the half-space of soil, and a buried mine, has been modeled using the FDTD method [31]. Fig. 7.24 shows typical results from this study using gray-scale plots to visualize the magnitude of the electric field on a plane through the center of the detector.

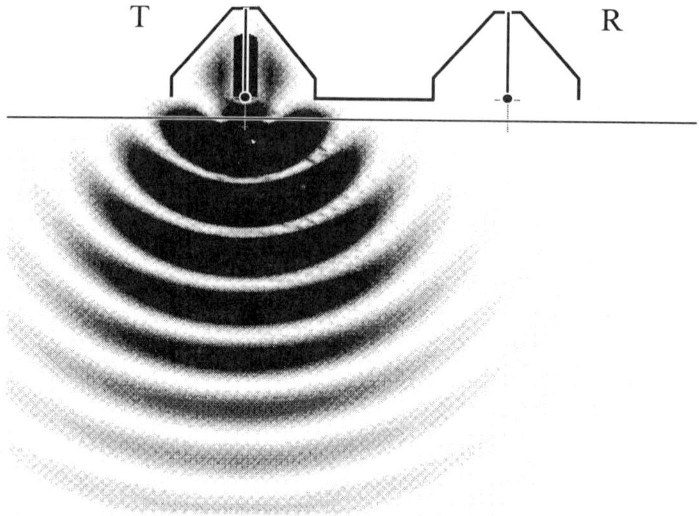

(a) Sensor over empty ground.

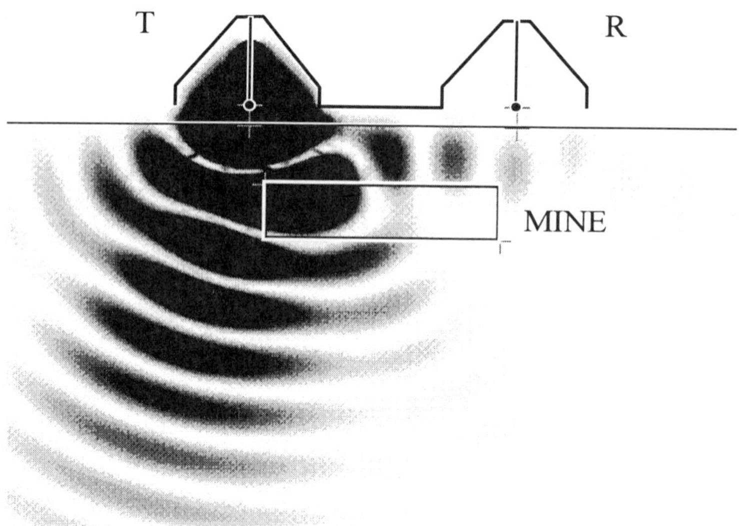

(b) Sensor centered over an all-dielectric land mine.

Fig. 7.24 Gray-scale visualizations showing the magnitude of the electric field on the transverse symmetry plane of the separated-aperture sensor of Fig. 7.23(a). *Source*: J. M. Bourgeois, Ph.D. Dissertation, Georgia Institute of Technology, Atlanta, GA, 1997.

In Fig. 7.24(a), the detector is over empty ground. In Fig. 7.24(b), the detector is over ground containing an all-dielectric mine. The increase in the coupling between the transmitting antenna, T, and the receiving antenna, R, caused by the presence of the mine is clearly shown.

7.7.3 Antenna-Radome Interaction

We now consider FDTD modeling of the antenna-radome interaction problem for a radar-guided missile. This is a long-standing defense-technology problem. Here, complex electromagnetic wave interactions between the missile's antenna and radome can cause errors in the perceived angular location of a target, thereby degrading the ability of the missile to guide itself to the target.

While the materials used in the construction of a missile radome are chosen to be as electromagnetically transparent as possible, the shape of the radome is usually dictated by aerodynamic considerations. Traditionally, the radome and its internal antenna are designed separately and their interaction is ignored. However, the current trend toward miniaturization and high-precision guidance is making this approach less valid. The FDTD method is well suited for use in analyzing both the antenna and the surrounding radome in the same computational model, thereby capturing the electromagnetic wave physics of the antenna-radome interaction.

Referring to Figs. 7.17 and 7.25(a), we consider an X-band pyramidal horn antenna having the characteristic dimensions $a = 7.30$ cm, $b = 5.32$ cm, and $D = 8$ cm. The missile is modeled as a 17.8-cm-diameter, hollow, circular, PEC tube of wall thickness 7.5 mm. Its 33.0-cm-long nose section (the radome) is an ogive body of revolution formed from a lossless dielectric of permittivity $\varepsilon_r = 4$. The nose cone is capped with a 2-cm-long metal tip of thickness 3.75 mm.

We shall consider the case of boresight illumination of the missile by an incident plane-wave pulse consisting of 6 sinusoidal cycles of a 10-GHz carrier. For this case, one plane of symmetry exists. Thus, the computational volume and run time can be reduced by a factor of 50% using a PMC image plane. The resulting FDTD model uses a uniform space lattice composed of $449 \times 150 \times 300$ cubic Yee cells, each spanning 1.27 mm ($\approx \lambda_0/24$ at 10 GHz).

Figs. 7.25(a) to 7.25(d) are gray-scale visualizations showing the magnitude of the electric field within and near the nose cone of the missile at four instants in time. These visualizations provide valuable insights into the complex interactions of the missile body, radome, and internal horn antenna, and are now briefly discussed in sequence.

In Fig. 7.25(a), the incident plane-wave pulse, which is propagating from right to left, is clearly displayed as its leading edge just contacts the metal tip of the radome. At the instant shown in Fig. 7.25(b), the plane wave has entered the radome. A quasi-spherical, radially propagating, scattered field is observed due to the action of the metal tip and the radome surface. The propagation delay due to the reduction in the speed of light inside the radome's dielectric wall is evident.

(a)

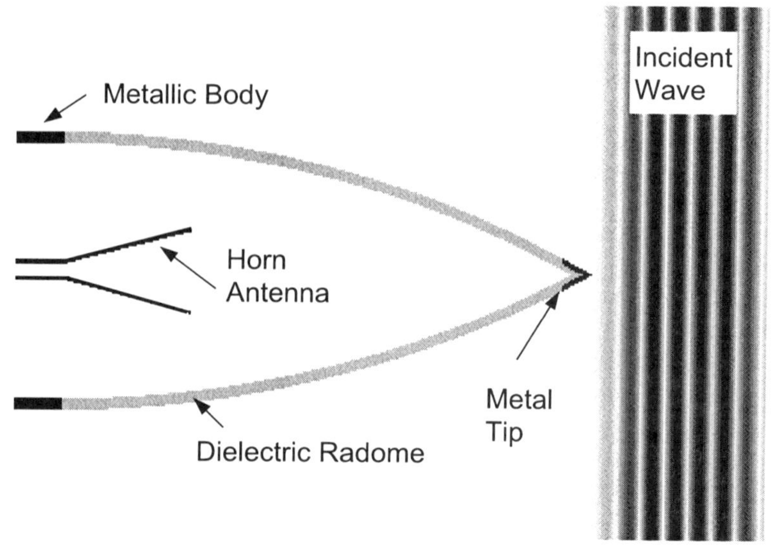

(b)

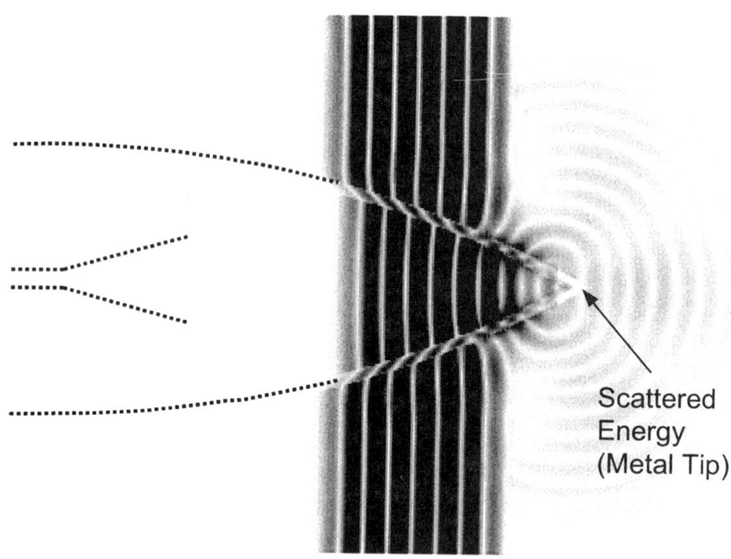

Fig. 7.25 Gray-scale visualizations showing the magnitude of the electric field within and near the radome of a missile at four instants in time. The plane-wave illumination propagates from right to left. For optimum display dynamic range, the gray scales are adjusted by a factor of two in (b) and (c), and by a factor of four in (d).

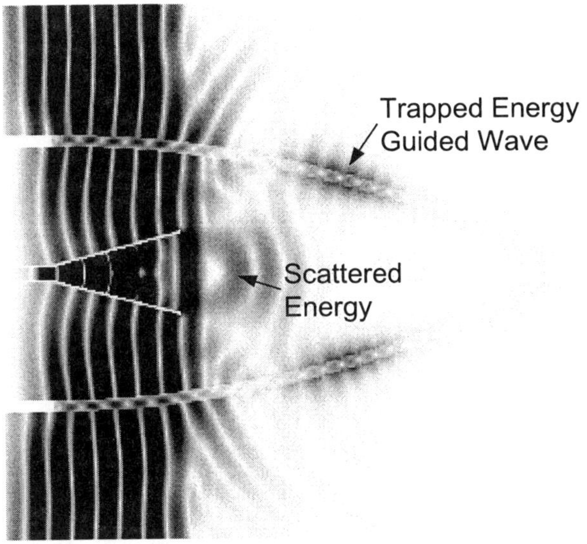

(c)

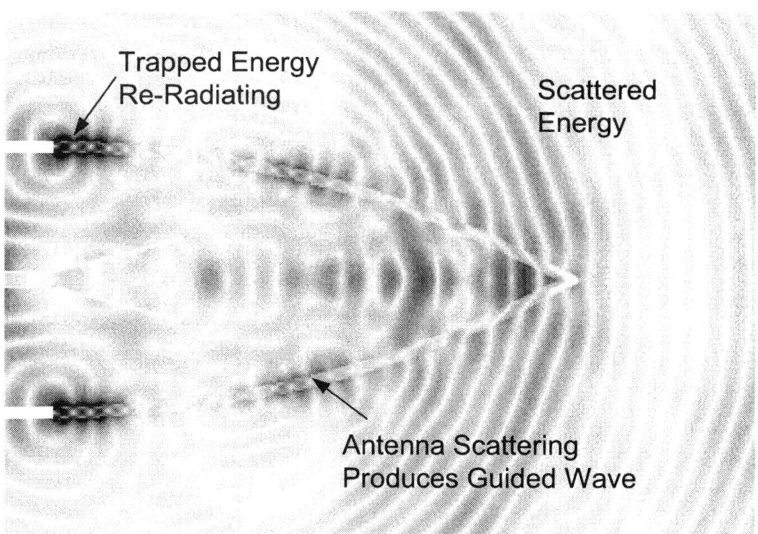

(d)

Fig. 7.25 *(cont.)*

In Fig. 7.25(c), the incident wave has encountered the horn antenna. A portion of the energy incident upon the horn antenna is scattered (structural-mode scattering). The waveguide attached to the horn is terminated in a matched load, and hence the antenna-mode scattering is zero. We note the packet of energy that has been trapped inside the wall of the dielectric radome. The strong evanescent fields near the metal tip and the curvature of the radome allow the excitation of this trapped, guided wave.

At the instant shown in Fig. 7.25(d), the illumination has passed the nose section of the missile. The trapped, guided mode identified in Fig. 7.25(c) has encountered the junction between the dielectric radome and the PEC missile body. The trapped energy is now reflected and re-radiated. A new guided wave is being formed in the dielectric radome due to structural scattering from the horn antenna.

Color visualizations of this radome problem for the case of an oblique incident wave propagating at 15° from boresight are provided in the color-plate section.

7.7.4 Effects of the Human Body on Personal Communications Devices

The 1990s have seen an explosive growth in the number and variety of personal wireless communication devices available to the consumer. Initial designs resulted in relatively large cellular telephones for cars and similar mobile use. With market forces pushing for miniaturization and low cost, all aspects of cellular phones have been re-engineered. The size and design of the antenna has been a major factor. Because the FDTD method allows easy and accurate modeling of essentially arbitrary configurations of inhomogeneous media, it is very attractive for this application.

Early work concentrated on the effects of the phone (modeled as a rectangular box) on the pattern of a top-mounted monopole [33]. More recently, researchers have studied various configurations and placements of planar inverted-F antennas (PIFAs) on the phone [38 – 40].

Because the operation of antennas can be significantly impacted by material loading (either dielectric or magnetic), it is important to consider the effects of the human hand and head on the performance of the antenna. Early work modeled the head as a sphere and the hand as a rectangular block of lossy dielectric [41]. More recently, researchers have used high-resolution models of the human head derived from computer tomography and magnetic resonance imaging to study the effects of the head on the antenna pattern [42]. The details of one such study are discussed in Chapter 11, Section 11.10.2.

In addition to the potential effects of the human on the antenna, we must consider the effects of the electromagnetic radiation on the human. Early work modeled the cellular phone as a dipole and examined the close coupling to a realistic model of the human head at 900 MHz and at 1.9 GHz [43]. Other researchers have considered more elaborate cellphone models and other frequency bands [42, 44 – 47]. The details of one such study are discussed in Chapter 11, Section 11.10.1.

7.7.5 Antennas for Biomedical Applications

Often in biomedical applications, the goal of the antenna system is to radiate into the human body. The applications fall into two broad categories: microwave hyperthermia and diagnostic imaging. Again, the ability of FDTD to easily model essentially arbitrary configurations of inhomogeneous, lossy materials makes the method very attractive.

The goal in microwave hyperthermia is to configure a set of antennas around the human body so that they irradiate and focus electromagnetic energy onto unwanted tumor cells within the body. The electromagnetic energy is converted into heat through the lossy nature of the tissue. This heat either directly damages the unwanted cells or renders them more vulnerable to damage by a concurrent program of chemotherapy or ionizing radiotherapy. For example, in [48] the temperature distribution patterns for arrays of microwave antennas located in the proximity of inhomogeneous tissues were studied.

Microwave imaging techniques employing large arrays of sensing antennas and sophisticated inversion algorithms (as applied to the human body and similar inhomogeneous structures) are discussed in considerable detail in Chapter 12, Sections 12.6 to 12.10. Recently, researchers have begun investigating the use of a novel, pulsed, confocal microwave approach to detect early-stage breast cancer [49] using a relatively small antenna array. In the early stage of this disease, X-ray and ultrasound detection approaches suffer due to poor physical-properties contrast between malignant and normal tissues. The high dielectric contrast observed at microwave frequencies makes this modality a promising alternative. Initial two- and three-dimensional FDTD modeling has shown that malignant tumors as small as 2 mm in diameter can be detected in the presence of background clutter generated by the inhomogeneity of the surrounding normal tissue.

7.8 SUMMARY AND CONCLUSIONS

This chapter summarized the primary elements involved in FDTD modeling of transmitting and receiving antennas, namely:

- Formulation of the antenna problem, including the use of symmetry;
- Antenna feed models;
- Near-field to far-field transformations in the time domain and frequency domain;
- The plane-wave injector for the receiving antenna;
- A detailed case study (with experimental validation) involving a pyramidal horn used as a transmitting and receiving antenna;
- Selected recent applications.

A key conclusion is that, when properly implemented, FDTD models of antennas produce results for radiation patterns, receiving apertures, and driving-point impedances that agree very well with carefully performed experiments. FDTD has a powerful ability to provide (in a straightforward manner) wideband results for complex antenna structures comprised of, or adjacent to, essentially arbitrary configurations of inhomogeneous, lossy, dispersive materials. This robustness allows the use of the method to confidently test proposed novel antenna designs on the computer before they are built.

REFERENCES

[1] Maloney, J. G., G. S. Smith, and W. R. Scott, Jr., "Accurate computation of the radiation from simple antennas using the finite-difference time-domain method," *IEEE Trans. Antennas and Propagation*, Vol. 38, 1990, pp. 1059–1068.

[2] Maloney, J. G., and G. S. Smith, "A study of transient radiation from the Wu-King resistive monopole — FDTD analysis and experimental measurements," *IEEE Trans. Antennas and Propagation*, Vol. 41, 1993, pp. 668–676.

[3] Montoya, T. P., and G. S. Smith, "A study of pulse radiation from several broad-band loaded monopoles," *IEEE Trans. Antennas and Propagation*, Vol. 44, 1996, pp. 1172–1182.

[4] Boonzaaier, J. J., and C. W. Pistorius, "Thin-wire Yagi antenna radiation patterns using the finite-difference time-domain method," *Microwave and Optical Technology Lett.*, Vol. 4, 1991, pp. 311–313.

[5] Luebbers, R., and K. Kunz, "Finite difference time domain calculations of antenna mutual coupling," *IEEE Trans. Electromagnetic Compatibility*, Vol. 34, 1992, pp. 357–359.

[6] Kashiwa, T., S. Tanaka, and I. Fukai, "Time-domain analysis of Yagi-Uda antennas using the FDTD method," *Electronics and Communications in Japan, Part 1: Communications*, Vol. 77, 1994, pp. 96–105.

[7] Katz, D. S., M. J. Piket-May, A. Taflove, and K. R. Umashankar, "FDTD analysis of electromagnetic wave radiation from systems containing horn antennas," *IEEE Trans. Antennas and Propagation*, Vol. 39, 1991, pp. 1203–1212.

[8] Tirkas, P. A., and C. A. Balanis, "Contour path FDTD methods for analysis of pyramidal horns with composite inner *E*-plane walls," *IEEE Trans. Antennas and Propagation*, Vol. 42, 1994, pp. 1476–1483.

[9] Thiele, E., and A. Taflove, "FDTD analysis of Vivaldi flared horn antennas and arrays," *IEEE Trans. Antennas and Propagation*, Vol. 42, 1994, pp. 633–641.

[10] Shlager, K. L., G. S. Smith, and J. G. Maloney, "TEM horn antenna for pulse radiation: An improved design," *Microwave and Optical Technology Lett.*, Vol. 12, 1996, pp. 86–90.

[11] Shlager, K. L., G. S. Smith, and J. G. Maloney, "Accurate analysis of TEM horn antennas for pulse radiation," *IEEE Trans. Electromagnetic Compatibility*, Vol. 38, 1996, pp. 414–423.

[12] Reig, C., E. A. Navarro, and V. Such, "FDTD analysis of *E*-sectoral horn antennas for broad-band applications," *IEEE Trans. Antennas and Propagation*, Vol. 45, 1997, pp. 1484–1487.

[13] Reineix, A., and B. Jecko, "Analysis of microstrip patch antennas using finite difference time domain method," *IEEE Trans. Antennas and Propagation*, Vol. 37, 1989, pp. 1361–1369.

[14] Wu, C., K.-L. Wu, Z.-Q. Bi, and J. Litva, "Accurate characterization of planar printed antennas using finite-difference time-domain method," *IEEE Trans. Antennas and Propagation*, Vol. 40, 1992, pp. 526–533.

[15] Reineix, A., and B. Jecko, "A time domain theoretical method for the analysis of microstrip antennas composed by slots," *Annales des Telecommunications*, Vol. 48, 1993, pp. 29–34.

[16] Penney, C. W., and R. J. Luebbers, "Input impedance, radiation pattern, and radar cross-section of spiral antennas using FDTD," *IEEE Trans. Antennas and Propagation*, Vol. 42, 1994, pp. 1328–1332.

[17] Shum, S. M., and K. M. Luk, "FDTD analysis of probe-fed cylindrical dielectric resonator antenna operating in fundamental broadside mode," *Electronics Lett.*, Vol. 31, 1995, pp. 1210–1212.

[18] Shum, S. M., and K. M. Luk, "Stacked annular ring dielectric resonator antenna excited by axi-symmetric coaxial probe," *IEEE Trans. Antennas and Propagation*, Vol. 43, 1995, pp. 889–892.

[19] Toland, B., C. C. Liu, and P. G. Ingerson, "Design and analysis of arbitrarily shaped dielectric antennas," *Microwave J.*, Vol. 40, 1997, pp. 278–286.

[20] Thomas, V. A., K. M. Ling, M. E. Jones, B. Toland, J. Lin, and T. Itoh, "FDTD analysis of an active antenna," *IEEE Microwave and Guided Wave Lett.*, Vol. 4, 1994, pp. 296–298.

[21] Shlager, K. L., and G. S. Smith, "Near-field to near-field transformation for use with FDTD method and its application to pulsed antenna problems," *Electronic Letters*, Vol. 30, 1994, pp. 1262–1264.

[22] Shlager, K. L., and G. S. Smith, "Comparison of two near-field to near-field transformations applied to pulsed antenna problems," *Electronic Letters*, Vol. 30, 1995, pp. 936–938.

[23] Reineix, A., J. Paillol, and B. Jecko, "FDTD method applied to the study of radar cross section of microstrip patch antennas," *Annales des Telecommunications*, Vol. 48, 1993, pp. 589–593.

[24] Maloney, J. G., and G. S. Smith, "Optimization of a conical antenna for pulse radiation: an efficient design using resistive loading," *IEEE Trans. Antennas and Propagation*, Vol. 41, 1993, pp. 940–947.

[25] Shlager, K. L., G. S. Smith, and J. G. Maloney, "Optimization of bow-tie antennas for pulse radiation," *IEEE Trans. Antennas and Propagation*, Vol. 42, 1994, pp. 975–982.

[26] Smith, G. S., *An Introduction to Classical Electromagnetic Radiation*, New York: Cambridge University Press, 1997.

[27] Maloney, J. G., K. L. Shlager, and G. S. Smith, "A simple FDTD model for transient excitation of antennas by transmission lines," *IEEE Trans. Antennas and Propagation*, Vol. 42, 1994, pp. 289–292.

[28] Maloney, J. G., *Analysis and Synthesis of Transient Antennas Using the Finite-Difference Time-Domain (FDTD) Method*, Ph.D. Dissertation, Georgia Institute of Technology, Atlanta, GA, January 1992.

[29] Umashankar, K. R., A. Taflove, and B. Beker, "Calculation and experimental validation of induced currents on coupled wires in an arbitrary shaped cavity," *IEEE Trans. Antennas and Propagation*, Vol. 35, 1987, pp. 1248–1257.

[30] Bourgeois, J. M., and G. S. Smith, "A fully three-dimensional simulation of a ground-penetrating radar: FDTD theory compared with experiment," *IEEE Trans. Geoscience and Remote Sensing*, Vol. 34, 1996, pp. 36–44.

[31] Bourgeois, J. M., *A Complete Three-Dimensional Electromagnetic Simulation of Ground-Penetrating Radars Using the Finite-Difference Time-Domain Method*, Ph.D. Dissertation, Georgia Institute of Technology, Atlanta, GA, 1997.

[32] Tirkas, P. A., and C. A. Balanis, "Finite-difference time-domain method for antenna radiation," *IEEE Trans. Antennas and Propagation*, Vol. 40, 1992, pp. 334–340.

[33] Luebbers, R., L. Chen, T. Uno, and S. Adachi, "FDTD calculation of radiation patterns, impedance, and gain for a monopole antenna on a conducting box," *IEEE Trans. Antennas and*

Propagation, Vol. 40, 1992, pp. 1577–1583.

[34] Taflove, A., *Computational Electrodynamics: The Finite-Difference Time-Domain Method*, Norwood, MA: Artech House, 1995.

[35] Nye, J. F., "Theory of the directional pattern of a pyramidal horn," *IEEE Trans. Antennas and Propagation*, Vol. 45, 1997, pp. 1357–1365.

[36] Maloney, J. G., and G. S. Smith, "The reception of short pulses by antennas: FDTD results and reciprocity," in *Ultra-Wideband, Short-Pulse Electromagnetics 2*, L. Carin and L.B. Felsen, eds., Plenum Press, NY, 1995.

[37] Kesler, M. P., J. G. Maloney, B. L. Shirley, and G. S. Smith, "Antenna design with the use of photonic bandgap materials as all-dielectric reflectors," *Microwave and Optical Technology Letters*, Vol. 11, 1996, pp. 169–174.

[38] Jensen, M. A., and Y. Rahmat-Samii, "Performance analysis of antennas for hand-held transceivers using FDTD," *IEEE Trans. Antennas and Propagation*, Vol. 42, 1994, pp. 1106–1113.

[39] Liu, Z. D., P. S. Hall, and D. Wake, "Dual-frequency planar inverted-F antenna," *IEEE Trans. Antennas and Propagation*, Vol. 45, 1997, pp. 1451–1458.

[40] Rowell, C. R., and R. D. Murch, "A capacitively loaded PIFA for compact mobile telephone handsets," *IEEE Trans. Antennas and Propagation*, Vol. 45, 1997, pp. 837–842.

[41] Toftgaard, J., S. N. Hornsleth, and J. B. Andersen, "Effects on portable antennas of the presence of a person," *IEEE Trans. Antennas and Propagation*, Vol. 41, 1993, pp. 739–746.

[42] Okoniewski, M., and M. A. Stuchly, "A study of the handset antenna and human body interaction," *IEEE Trans. Microwave Theory and Techniques*, Vol. 44, 1996, pp. 1855–1864.

[43] Dimbylow, "FDTD calculations of SAR for a dipole closely coupled to the head at 900 MHz and 1.9 GHz," *Phys. Med. Biol.*, Vol. 38, 1993, pp. 361–368.

[44] Martens, L., J. De Moerloose, D. De Zutter, J. De Poorter, and C. De Wagter, "Calculation of the electromagnetic fields induced in the head of an operator of a cordless telephone," *Radio Science*, Vol. 30, 1995, pp. 283–290.

[45] Watanabe, S., M. Taki, T. Nojima, and O. Fujiwara, "Characteristics of the SAR distributions in a head exposed to electromagnetic fields radiated by a hand-held portable radio," *IEEE Trans. Microwave Theory and Techniques*, Vol. 44, 1996, pp. 1874–1883.

[46] Gandhi, O. P., and J.-Y. Chen, "Electromagnetic absorption in the human head from experimental 6-GHz handheld transceivers," *IEEE Trans. Electromagnetic Compatibility*, Vol. 37, 1995, pp. 547–558.

[47] Gandhi, O. P., G. Lazzi, and C. M. Furse, "Electromagnetic absorption in the human head and neck for mobile telephones at 835 and 1900 MHz," *IEEE Trans. Microwave Theory and Techniques*, Vol. 44, 1996, pp. 1884–1897.

[48] Cherry, P. C., and M. F. Iskander, "Calculations of heating patterns of an array of microwave interstitial antennas," *IEEE Trans. Biomedical Engineering*, Vol. 40, 1993, pp. 771–779.

[49] Hagness, S. C., A. Taflove, and J. E. Bridges, "Two-dimensional FDTD analysis of a pulsed microwave confocal system for breast cancer detection: fixed-focus and antenna-array sensors," *IEEE Trans. Biomedical Engineering*, in press.

Chapter 8

High-Speed Electronic Circuits with Active and Nonlinear Components

Bijan Houshmand, Tatsuo Itoh, and Melinda Piket-May

8.1 INTRODUCTION

High-speed electronic circuits are traditionally grouped into two classes: analog microwave circuits and digital logic circuits.

1. Microwave circuits typically process bandpass signals at center frequencies above 1 GHz. Common circuit structures include waveguides, striplines, microstrips, and their transitions; directional couplers and circulators; filters and matching networks; and individual transistors and their feedback paths. The small-signal performance of a microwave circuit is usually characterized by its *S* parameters. The large-signal performance is typically characterized by specifications for gain compression and generation of harmonic and intermodulation products. Designers of microwave circuits account for electromagnetic wave effects as a matter of course since the circuit operation is based upon these effects.
2. Digital circuits typically process low-pass pulses having clock rates below 0.5 GHz. Typical circuit structures include densely packed, multiple planes of metal traces providing flow paths for the signals, dc power feeds, and ground returns. Via pins provide electrical connections between the planes. Digital designers may not always account for electromagnetic wave effects since the circuit operation is not based upon these effects. In fact, electromagnetic phenomena for such circuits are considered to be parasites.

However, the distinction between the design and packaging of these two classes of high-speed electronic circuits is blurring. Microwave circuits are becoming very complex systems composed of densely spaced elements, discontinuity structures, and passive and active devices. On the digital-circuit side, the rise of everyday clock speeds to 300 to 600 MHz implies low-pass signal bandwidths up to about 1.5 to 3 GHz, the low end of the microwave range. (In fact, clock speeds at the cutting edge of digital design are as high as 2.5 GHz, with bandwidths up to 12.5 GHz, as discussed in Section 8.11.) Electromagnetic wave effects that until now were in the domain of the microwave engineer are becoming a limiting factor to digital computer performance. At the chip level or the level of a device comprised of several interconnected chips, these effects are very complex, usually not well understood, and not trivial to analyze either theoretically, numerically, or experimentally. For instance, hard-won experience has shown that:

- Coupling between vias can distort signals;
- Mismatches between vias and signal lines can lead to ground bounce;
- Holes and other discontinuities in ground planes can increase coupling between the circuit-board layers;
- Metal traces with or without bends are likely to have reactive impedance components that can degrade system performance at high clock speeds;
- Signals can couple (crosstalk) from one parallel trace to another;
- Manufacturing tolerances can cause a range of coupling, crosstalk, and impedance parameters;
- Electromagnetic interference and compatibility problems can arise relative to other circuits and systems.

A key bottom line here is the *signal integrity*, a measure of how good the signal quality is at some digital circuit output relative to the specific design goals. Common signal-integrity specifications include rise time, threshold, and ringing.

Overall, the design of complex high-speed circuits, especially those having nonlinear active devices, encounters substantial problems in dealing with electromagnetic coupling and radiation effects between different circuit elements. Until recently, the focus of electromagnetic simulators has been the analysis of passive structures [1, 2]. Design tools have been based on an approach wherein the circuits are first divided into small elements, and then the characteristics of each element are cascaded to obtain the overall system performance [3, 4]. Here, coupling of the cascaded parts is ignored or approximated. Software packages such as Touchstone™ and MDS™ use equivalent-circuit models to represent the passive part of the circuits, and the complete circuit including the lumped elements is then analyzed using a circuit simulator. The High-Frequency Structure Simulator (HFSS™) provides a complete electromagnetic solution based on the finite-element method [5].

Recent progress in the FDTD method (including the development of the PML ABC, information-extraction techniques, and the efficient use of parallel and vector computer architectures) has made it possible to analyze realistic high-speed circuits in their complete form using full-vector computations [6, 7]. FDTD algorithms now allow the accurate calculation of S-parameters, radiation patterns, and coupling coefficients over a very broad frequency band with one run. In addition, the visualization of the time-domain data provides practical insights into the operation of a particular structure.

Sections 8.2 to 8.8 present the extension of the FDTD algorithm to model the reactive, nonlinear, and gain components of a high-speed circuit. Both Norton's and Thevenin's equivalent circuits are presented for the FDTD lattice. These provide flexible and straightforward options to connect the distributed, full-wave physics of Maxwell's equations to the behavior of potentially complex, lumped-parameter electrical circuits. Sample computed results are summarized that show very good accuracy for the extended FDTD algorithm, and the ability to model cases where the available circuit simulators cannot be applied. Examples include nonlinear cases involving the generation of harmonics and intermodulation products in large-signal microwave circuits.

Sections 8.9 to 8.11 present two specific examples of applications of FDTD to high-speed digital design: calculating the inductance of a complex power-distribution system; and modeling a 2.5-GHz clock-distribution network for an advanced multichip module (MCM). The modeling procedures and the interpretations of the results are discussed in sufficient detail to provide guidance for readers attempting similar work. These applications are typical concerns of present-day design engineers working with densely packaged high-speed digital circuits.

8.2 BASIC FDTD ALGORITHM

The FDTD algorithm is derived directly from Maxwell's equations in the time domain. The components of the electric and magnetic fields in Cartesian coordinates with the constitutive parameters (ε, μ) and an electric current vector source are given by

$$\nabla \times \vec{E} = -\mu \frac{\partial \vec{H}}{\partial t} \tag{8.1}$$

$$\nabla \times \vec{H} = \varepsilon \frac{\partial \vec{E}}{\partial t} + \vec{J} \tag{8.2}$$

In Cartesian coordinates, (8.1) and (8.2) are written as the component equations

$$-\mu \frac{\partial H_x}{\partial t} = \frac{\partial E_z}{\partial y} - \frac{\partial E_y}{\partial z} \tag{8.3a}$$

$$-\mu \frac{\partial H_y}{\partial t} = \frac{\partial E_x}{\partial z} - \frac{\partial E_z}{\partial x} \tag{8.3b}$$

$$-\mu\frac{\partial H_z}{\partial t} = \frac{\partial E_y}{\partial x} - \frac{\partial E_x}{\partial y} \tag{8.3c}$$

$$\varepsilon\frac{\partial E_x}{\partial t} = \frac{\partial H_z}{\partial y} - \frac{\partial H_y}{\partial z} - J_x \tag{8.4a}$$

$$\varepsilon\frac{\partial E_y}{\partial t} = \frac{\partial H_x}{\partial z} - \frac{\partial H_z}{\partial x} - J_y \tag{8.4b}$$

$$\varepsilon\frac{\partial E_z}{\partial t} = \frac{\partial H_y}{\partial x} - \frac{\partial H_x}{\partial y} - J_z \tag{8.4c}$$

The central-differencing scheme proposed by Yee results in second-order accuracy for the field components with respect to the discretization variables. The resulting difference equations for the electric and magnetic field components are

$$\begin{aligned} H_x\Big|_{i,j,k}^{n+1/2} &= H_x\Big|_{i,j,k}^{n-1/2} + \frac{\Delta t}{\mu_{i,j,k}\,\Delta z}\left(E_y\Big|_{i,j,k+1/2}^{n} - E_y\Big|_{i,j,k-1/2}^{n}\right) \\ &\quad - \frac{\Delta t}{\mu_{i,j,k}\,\Delta y}\left(E_z\Big|_{i,j+1/2,k}^{n} - E_z\Big|_{i,j-1/2,k}^{n}\right) \end{aligned} \tag{8.5a}$$

$$\begin{aligned} H_y\Big|_{i,j,k}^{n+1/2} &= H_y\Big|_{i,j,k}^{n-1/2} + \frac{\Delta t}{\mu_{i,j,k}\,\Delta x}\left(E_z\Big|_{i+1/2,j,k}^{n} - E_z\Big|_{i-1/2,j,k}^{n}\right) \\ &\quad - \frac{\Delta t}{\mu_{i,j,k}\,\Delta z}\left(E_x\Big|_{i,j,k+1/2}^{n} - E_x\Big|_{i,j,k-1/2}^{n}\right) \end{aligned} \tag{8.5b}$$

$$\begin{aligned} H_z\Big|_{i,j,k}^{n+1/2} &= H_z\Big|_{i,j,k}^{n-1/2} + \frac{\Delta t}{\mu_{i,j,k}\,\Delta y}\left(E_x\Big|_{i,j+1/2,k}^{n} - E_x\Big|_{i,j-1/2,k}^{n}\right) \\ &\quad - \frac{\Delta t}{\mu_{i,j,k}\,\Delta x}\left(E_y\Big|_{i+1/2,j,k}^{n} - E_y\Big|_{i-1/2,j,k}^{n}\right) \end{aligned} \tag{8.5c}$$

$$\begin{aligned} E_x\Big|_{i,j,k}^{n+1} &= E_x\Big|_{i,j,k}^{n} + \frac{\Delta t}{\varepsilon_{i,j,k}\,\Delta y}\left(H_z\Big|_{i,j+1/2,k}^{n+1/2} - H_z\Big|_{i,j-1/2,k}^{n+1/2}\right) \\ &\quad - \frac{\Delta t}{\varepsilon_{i,j,k}\,\Delta z}\left(H_y\Big|_{i,j,k+1/2}^{n+1/2} - H_y\Big|_{i,j,k-1/2}^{n+1/2}\right) - \frac{\Delta t}{\varepsilon_{i,j,k}} J_x\Big|_{i,j,k}^{n+1/2} \end{aligned} \tag{8.6a}$$

$$E_y\Big|_{i,j,k}^{n+1} = E_y\Big|_{i,j,k}^{n} + \frac{\Delta t}{\varepsilon_{i,j,k}\,\Delta z}\left(H_x\Big|_{i,j,k+1/2}^{n+1/2} - H_x\Big|_{i,j,k-1/2}^{n+1/2}\right) - \frac{\Delta t}{\varepsilon_{i,j,k}\,\Delta x}\left(H_z\Big|_{i+1/2,j,k}^{n+1/2} - H_z\Big|_{i-1/2,j,k}^{n+1/2}\right) - \frac{\Delta t}{\varepsilon_{i,j,k}} J_y\Big|_{i,j,k}^{n+1/2} \tag{8.6b}$$

$$E_z\Big|_{i,j,k}^{n+1} = E_z\Big|_{i,j,k}^{n} + \frac{\Delta t}{\varepsilon_{i,j,k}\,\Delta x}\left(H_y\Big|_{i+1/2,j,k}^{n+1/2} - H_y\Big|_{i-1/2,j,k}^{n+1/2}\right) - \frac{\Delta t}{\varepsilon_{i,j,k}\,\Delta y}\left(H_x\Big|_{i,j+1/2,k}^{n+1/2} - H_x\Big|_{i,j-1/2,k}^{n+1/2}\right) - \frac{\Delta t}{\varepsilon_{i,j,k}} J_z\Big|_{i,j,k}^{n+1/2} \tag{8.6c}$$

where Δx, Δy, and Δz are the spatial discretizations, and the indices (i, j, k) correspond to the x, y, and z directions, respectively. The index, n, corresponds to the temporal discretization, Δt. We observe that the electric and magnetic fields are displaced by a half-step both in the time and space indices.

8.3 INCLUSION OF A LUMPED-CIRCUIT ELEMENT

The inclusion of a lumped-circuit element within a single cell of the FDTD space lattice is implemented by solving the equations that govern the element's voltage-current characteristic along with Maxwell's equations. The coupling of the circuit-element equations and Maxwell's equations is through the independent current term in (8.4a) to (8.4c) for Maxwell's differential equations, and through (8.6a) to (8.6c) for the corresponding set of difference equations.

One approach to implement this extended capability is to explicitly include the governing equations of a circuit element into (8.6a) to (8.6c). A complete treatment of this approach and illustrative examples are presented in [8 – 10].

Another approach is to solve Maxwell's equations and the device equations simultaneously. For this approach, care must be taken to properly combine the values of current derived from the device equations into the discrete version of Maxwell's equations. For the FDTD algorithm, currents and voltages are computed with a half-step separation in time according to (8.5) and (8.6). For the device equations, however, the current and voltage values are computed at the same time. In general, the current value is interpolated to the half-time step for coupling to the FDTD algorithm. The interpolation scheme can be as simple as time averaging. This method provides a systematic approach for inclusion of lumped elements in the FDTD algorithm, but it can be numerically unstable [10 – 12].

A more versatile method is to combine the device terminal characteristics with Maxwell's equations. This implicit method was proposed in [13] and was adopted in [14]. The following sections will discuss aspects of all of these approaches.

8.4 NORTON EQUIVALENT CIRCUIT "LOOKING INTO" THE FDTD SPACE LATTICE

The basic idea underlying this approach is expressed very simply. We imagine ourselves to be a two-terminal lumped-circuit device embedded within the FDTD space lattice. We want to conveniently and self-consistently interface the lumped-circuit physics of the device (represented by a system of ordinary differential equations in current and voltage) with the distributed-electromagnetic-field physics of the space lattice (represented by Maxwell's partial differential equations). Therefore, we ask the question: What is the Norton equivalent circuit "looking into" the space lattice from the perspective of the circuit device? That is, we wish to represent the physics of the *entirety* of the FDTD space lattice containing the device by a conventional Norton equivalent circuit. This Norton circuit is then combined with the circuit of the device of interest, and the resulting overall circuit is solved to incorporate the properties of the device into the FDTD field solution.

8.4.1 Theory

Consider a circuit device positioned at the site of an electric-field component, E_w, in the FDTD space lattice (either E_x, E_y, or E_z). From the Norton equivalent-circuit standpoint, the excitation effect of the complete FDTD lattice provided to the circuit device can be distilled down to the action of the four H-field components immediately surrounding (looping around) E_w. From Ampere's law, the total Norton current, $I_{\mathrm{N}}(t)$, is obtained from these field components by implementing the closed contour integral

$$I_{\mathrm{N}}(t) \;=\; \oint_{\Gamma} \vec{H}(t) \cdot d\vec{\Gamma} \tag{8.7}$$

where the contour, Γ, bounds the space-cell cross section, A, where the device is embedded. For example, the following expressions implement (8.7) for the three possible Cartesian orientations of I_N within a cell:

$$I_x\Big|_{i,j,k}^{n+1/2} = \left(H_y\Big|_{i,j,k-1/2}^{n+1/2} - H_y\Big|_{i,j,k+1/2}^{n+1/2}\right)\Delta y + \left(H_z\Big|_{i,j+1/2,k}^{n+1/2} - H_z\Big|_{i,j-1/2,k}^{n+1/2}\right)\Delta z \tag{8.8a}$$

$$I_y\Big|_{i,j,k}^{n+1/2} = \left(H_x\Big|_{i,j,k+1/2}^{n+1/2} - H_x\Big|_{i,j,k-1/2}^{n+1/2}\right)\Delta x + \left(H_z\Big|_{i-1/2,j,k}^{n+1/2} - H_z\Big|_{i+1/2,j,k}^{n+1/2}\right)\Delta z \tag{8.8b}$$

$$I_z\Big|_{i,j,k}^{n+1/2} = \left(H_x\Big|_{i,j-1/2,k}^{n+1/2} - H_x\Big|_{i,j+1/2,k}^{n+1/2}\right)\Delta x + \left(H_y\Big|_{i+1/2,j,k}^{n+1/2} - H_y\Big|_{i-1/2,j,k}^{n+1/2}\right)\Delta y \tag{8.8c}$$

We note that $I_{\mathrm{N}}(t)$ sources two currents flowing in parallel through the cell: the displacement current, $I_{\text{displacement}}$; and the device current, I_{device}:

$$I_N(t) = I_{displacement}(t) + I_{device}(t) \tag{8.9}$$

The displacement current is related to the cell electric-field component, E_w, that is oriented perpendicular to the cell cross section, A, and spans the device terminals. This relation is given by

$$I_{displacement}(t) = \frac{d}{dt}\left(\iint_A \vec{D}(t) \cdot d\vec{A}\right) \cong \varepsilon A \frac{dE_w(t)}{dt} \tag{8.10}$$

In the FDTD algorithm, the electric field is constant with position within the space cell. Therefore, it can be related to the voltage across the cell, and thus the voltage, V_{device}, across the device terminals as

$$E_w(t) = \frac{V_{device}(t)}{\Delta w} \tag{8.11}$$

where Δw is the dimension of the lattice cell in the w direction spanning the device terminals. Substituting (8.11) into (8.10), we have

$$I_{displacement}(t) = \varepsilon A \frac{d}{dt}\left[\frac{V_{device}(t)}{\Delta w}\right] = \frac{\varepsilon A}{\Delta w}\frac{dV_{device}(t)}{dt} \equiv C\frac{dV_{device}(t)}{dt} \tag{8.12}$$

where $C = \varepsilon A / \Delta w$ is recognized as the equivalent capacitance of the FDTD lattice cell. Finally, substituting (8.12) into (8.9), we obtain

$$I_N(t) = C\frac{dV_{device}(t)}{dt} + I_{device}\left[V_{device}(t)\right] \tag{8.13}$$

where $I_{device}(V_{device})$ is the known voltage-current characteristic of the device as observed at its terminals. The lattice-cell capacitance, C, is seen to provide the Norton equivalent admittance in our model. In fact, we designate this capacitance as C_N.

We note that (8.13) is an ordinary differential equation for $V_{device}(t)$ that can be integrated in time given knowledge of the initial condition, $V_{device}(t_0)$; the driving function, $I_N(t)$; and the device's current-voltage characteristic, $I_{device}(V_{device})$. Starting at time t_0 and integrating to time $t_0 + \Delta t$, this leads to

$$V_{device}(t_0 + \Delta t) = V_{device}(t_0) + \frac{1}{C_N}\int_{t_0}^{t_0+\Delta t}\left\{I_N(t) - I_{device}\left[V_{device}(t)\right]\right\}dt \tag{8.14}$$

Equation (8.14) can be implemented using standard numerical integration techniques such as the predictor-corrector method and the Runge-Kutta method even if $I_{device}(V_{device})$ is nonlinear. If desired, (8.14) can be performed by connecting to an external software package such as SPICE that is structured specifically to analyze complex, potentially nonlinear circuits [13]. In this case, there would be no need to have specific knowledge about $I_{\text{device}}(V_{\text{device}})$, as this function would arise naturally from the interconnection of the linear and nonlinear elements placed into the SPICE model.

Equations (8.7) and (8.14) provide a numerically robust linkage between the lattice-based FDTD electromagnetic field calculation and the lumped-circuit-device solver. For the regions where devices are not present, the electric and magnetic fields are updated by the conventional Yee algorithm. For a lattice cell where a circuit device having a known terminal *I-V* characteristic is located at E_w, time-stepping proceeds as follows. Starting at time $t_0 + \Delta t/2$ with an assumed known magnetic field distribution:

1. I_{N} is computed at time $t_0 + \Delta t/2$ using (8.7) applied to the magnetic fields circulating around the cell (in the manner of (8.8)).
2. Equation (8.14) is integrated numerically from t_0 to $t_0 + \Delta t$ using the value of $I_{\text{N}}(t_0 + \Delta t/2)$ obtained in step 1 and the known *I-V* characteristic of the device. This yields $V_{\text{device}}(t_0 + \Delta t)$, and thus $E_w(t_0 + \Delta t)$ from (8.11).
3. The magnetic fields everywhere in the lattice can now be updated to time $t_0 + 3\Delta t/2$ and the computation cycle of steps 1 and 2 repeated.

In the procedure discussed above, the terminals of the lumped circuit device are confined to one FDTD cell. However, the contour and surface integrations of the fields of (8.7) and (8.10) can be expanded to several cells in order to account for the physical size of the circuit device being greater than one space cell. Fig. 8.1 shows the arrangement of the device currents for inclusion into the FDTD algorithm. Here, the FDTD cells that include current sources are connected to vias in order to establish a voltage reference. In this interpretation, an appropriate electric-field integration surface is used to specify the displacement current, and an appropriate magnetic-field integration contour is used to obtain the Norton current. For a multiport lumped-element network that is connected to a number of FDTD cells, Fig. 8.2 shows the interaction of the device model with the FDTD cells through the equivalent source method where (8.7) and (8.14) are used at each of the device ports.

8.4.2 Illustration for a Resistor

The method of Section 8.4.1 has been successfully used to incorporate various linear and nonlinear circuit devices in FDTD [13 – 17]. In the following example, illustrated in Fig. 8.3, the device of interest is a resistor, R, used to terminate a microstrip line. This simple linear device permits an exact solution of the integral of (8.14) and expansion of the resulting expression in a Taylor series.

Fig. 8.4(a) is the circuit block diagram of the FDTD cell that includes the resistor, R. This device is assumed to be located at the space lattice point (i, j, k) and oriented in the z direction. From (8.13) we have

$$I_z(t) = C\frac{dV_{\text{R}}(t)}{dt} + \frac{V_{\text{R}}}{R} \qquad (8.15)$$

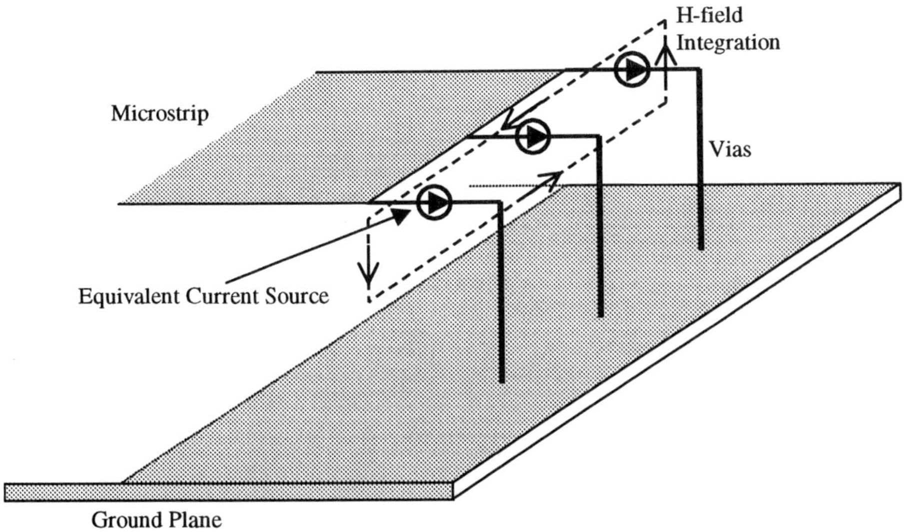

(a) Current flow is set in the horizontal direction.

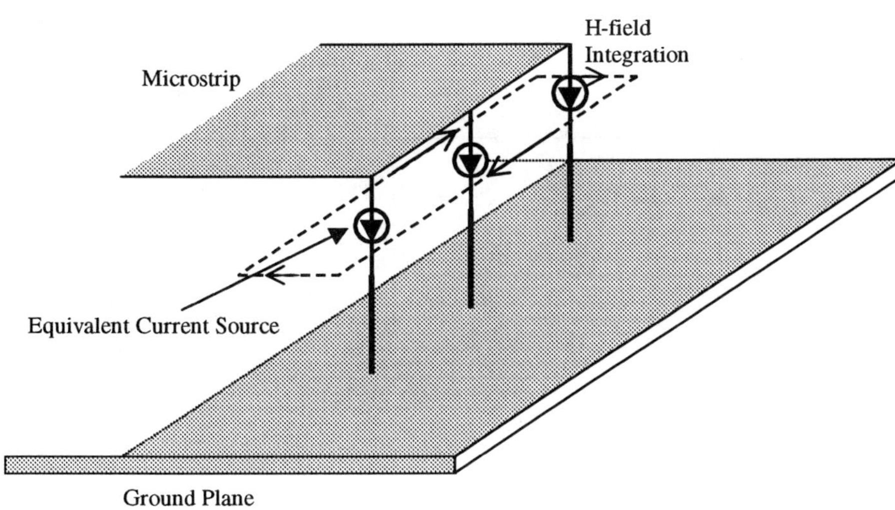

(b) Current flow is in the vertical direction.

Fig. 8.1 Arrangement of current sources for the equivalent current source formulation.

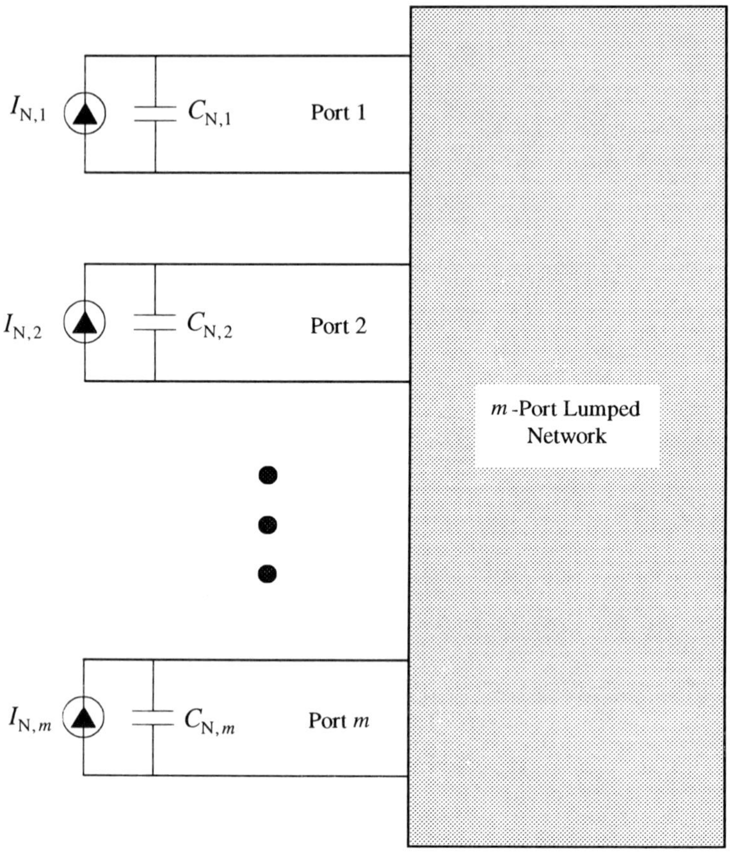

Fig. 8.2 Equivalent current-source representation of the extended FDTD algorithm with a multiport lumped element.

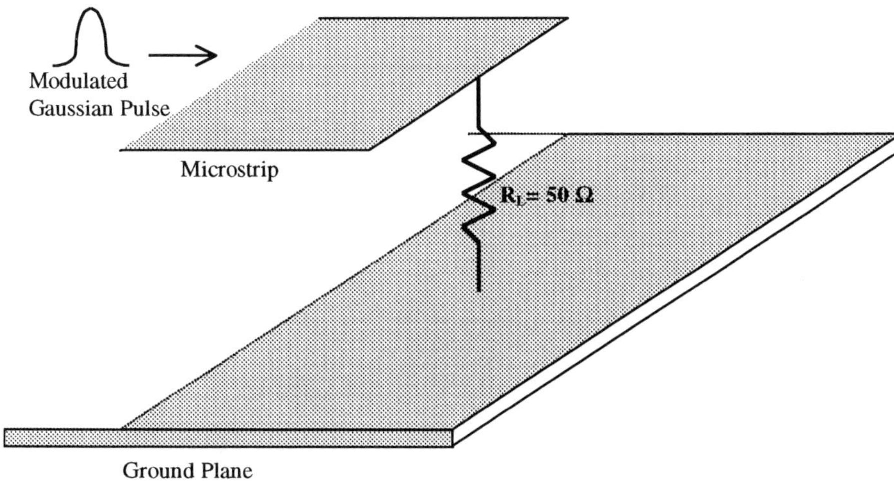

Fig. 8.3 Microstrip line terminated with a 50Ω lumped resistor. The FDTD lattice cells that include the lumped element are modified to include the device terminal characteristics.

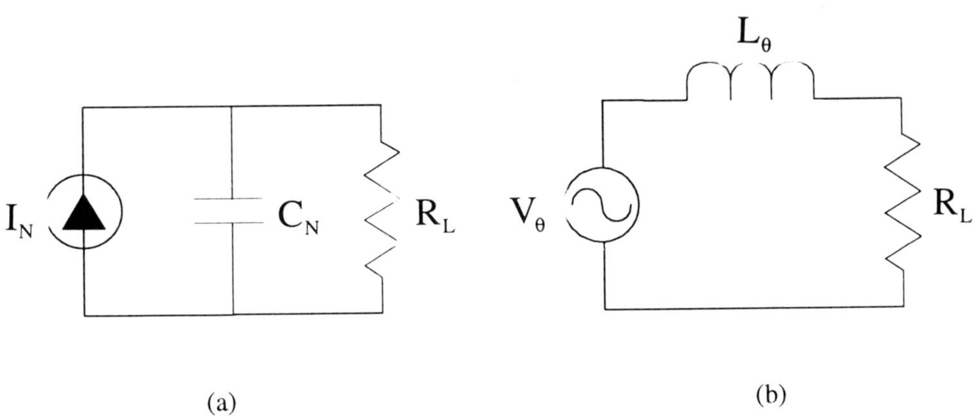

Fig. 8.4 Equivalent-circuit models for the FDTD cells that include lumped elements: (a) equivalent current-source model; (b) equivalent voltage-source model.

where V_R denotes the voltage across the resistor. Starting with (8.14), we obtain the analytical solution for the resistor voltage advanced by one time step:

$$V_R(t_0 + \Delta t) = V_R(t_0)e^{-(\Delta t/RC)} + \frac{1}{C}\int_{t_0}^{t_0+\Delta t} I_z(\tau)e^{-[(t-\tau)/RC]}\,d\tau \tag{8.16}$$

Using the average value of the current during the integration interval, (8.16) simplifies to

$$V_R(t_0 + \Delta t) = V_R(t_0)e^{-(\Delta t/RC)} + I_{z,\text{avg}}\,R\left(e^{\Delta t/RC} - 1\right) \tag{8.17}$$

Implementing a Taylor series expansion for the exponentials and retaining only the first-order terms yields

$$V_R(t_0 + \Delta t) = V_R(t_0)\left(1 - \frac{\Delta t}{RC}\right) + I_{z,\text{avg}}\left(\frac{\Delta t}{C}\right) \tag{8.18}$$

Then, dividing by the lattice-cell height, Δz, provides the updated electric field at the resistor's location:

$$E_R(t_0 + \Delta t) = \left(1 - \frac{\Delta t}{RC}\right)E_R(t_0) + \left(\frac{\Delta t}{C\Delta z}\right)I_{z,\text{avg}} \tag{8.19a}$$

This is expressed more conveniently in the usual shorthand space-time notation as

$$E_z\Big|_{i,j,k}^{n+1} = \left(1 - \frac{\Delta t}{RC}\right)E_z\Big|_{i,j,k}^{n} + \left(\frac{\Delta t}{C\Delta z}\right)I_{z,\text{avg}} \tag{8.19b}$$

Now, identifying $I_{z,\text{avg}}$ with (8.8c) and substituting for the cell capacitance, $C = \varepsilon_{i,j,k}\,\Delta x\,\Delta y/\Delta z$, in the second term, (8.19b) becomes

$$\begin{aligned} E_z\Big|_{i,j,k}^{n+1} &= \left(1 - \frac{\Delta t}{RC}\right)E_z\Big|_{i,j,k}^{n} \\ &\quad + \frac{\Delta t}{\varepsilon_{i,j,k}}\left(\frac{H_x\Big|_{i,j-1/2,k}^{n+1/2} - H_x\Big|_{i,j+1/2,k}^{n+1/2}}{\Delta y} + \frac{H_y\Big|_{i+1/2,j,k}^{n+1/2} - H_y\Big|_{i-1/2,j,k}^{n+1/2}}{\Delta x}\right) \end{aligned} \tag{8.20a}$$

Using the notation of [6] this is written more compactly as

$$E_z\Big|_{i,j,k}^{n+1} = \left(1 - \frac{\Delta t}{RC}\right)E_z\Big|_{i,j,k}^{n} + \frac{\Delta t}{\varepsilon_{i,j,k}}\nabla\times H_{i,j,k}^{n+1/2} \tag{8.20b}$$

Equation (8.20) is the required time-stepping expression at FDTD lattice point (i, j, k) to provide the physics of a resistor, R, connected across the microstrip at that point.

8.4.3 Equivalence to an Alternate Explicit Representation

We now consider an alternate approach for modeling the resistor of Section 8.4.2 in the FDTD lattice wherein the device characteristic is used directly in the FDTD computation. This approach is suitable primarily for linear circuit devices.

We begin by modifying (8.6c) for the lattice cell of cross-section area, $A = \Delta x \, \Delta y$, which includes the z-directed resistor:

$$\begin{aligned} E_z\Big|_{i,j,k}^{n+1} = E_z\Big|_{i,j,k}^{n} &+ \frac{\Delta t}{\varepsilon_{i,j,k}\,\Delta x}\left(H_y\Big|_{i+1/2,j,k}^{n+1/2} - H_y\Big|_{i-1/2,j,k}^{n+1/2}\right) \\ &- \frac{\Delta t}{\varepsilon_{i,j,k}\,\Delta y}\left(H_x\Big|_{i,j+1/2,k}^{n+1/2} - H_x\Big|_{i,j-1/2,k}^{n+1/2}\right) - \frac{\Delta t}{\varepsilon_{i,j,k}}\left(\frac{E_z\Big|_{i,j,k}^{n+1/2}\Delta z}{RA}\right) \end{aligned} \tag{8.21}$$

In (8.21), the last term represents the current density, J_R, due to the presence of the resistor. To evaluate this term at time step $n+1/2$, the arithmetic average of the electric-field values at time steps n and $n+1$ is used:

$$J_R\Big|_{i,j,k}^{n+1/2} = \frac{E_z\Big|_{i,j,k}^{n+1/2}\Delta z}{RA} = \frac{\left(E_z\Big|_{i,j,k}^{n} + E_z\Big|_{i,j,k}^{n+1}\right)\Delta z}{2RA} \tag{8.22}$$

This is the well known "semi-implicit" approximation for the electric-field value at half-time steps [6]. Using the shorthand notation of (8.20b), the update relation for the electric field at the location of the lumped resistor becomes

$$E_z\Big|_{i,j,k}^{n+1} = \left(\frac{1 - \dfrac{\Delta t}{2RC}}{1 + \dfrac{\Delta t}{2RC}}\right) E_z\Big|_{i,j,k}^{n} + \left(\frac{\dfrac{\Delta t}{\varepsilon_{i,j,k}}}{1 + \dfrac{\Delta t}{2RC}}\right) \nabla \times H_{i,j,k}^{n+1/2} \tag{8.23}$$

Equation (8.23) can be further simplified by observing the relative magnitude of the time step, Δt, and the cell time constant, RC. For many FDTD computations, $\Delta t << RC$. Using the approximations

$$\left.\frac{1-\delta}{1+\delta} \cong 1 - 2\delta, \qquad \frac{1}{1+\delta} \cong 1 - \delta \quad \right\} \; \delta << 1 \tag{8.24}$$

we finally have for (8.23)

$$E_z\Big|_{i,j,k}^{n+1} = \left(1 - \frac{\Delta t}{RC}\right) E_z\Big|_{i,j,k}^{n} + \frac{\Delta t}{\varepsilon_{i,j,k}}\left(1 - \frac{\Delta t}{2RC}\right) \nabla \times H_{i,j,k}^{n+1/2} \tag{8.25}$$

The update equations (8.25) and (8.20b) differ only due to the presence of the second-order term $(\Delta t)^2$ in (8.25). It is important to note that while both methods provide similar accuracy for the computed field, the general method resulting in (8.20b) can readily accommodate complex, nonlinear device models.

8.5 THEVENIN EQUIVALENT CIRCUIT "LOOKING INTO" THE FDTD SPACE LATTICE

By analogy (and duality) relative to the discussion of Section 8.4, the basic idea underlying this approach is expressed as follows. We imagine ourselves to be a two-terminal lumped-circuit device embedded within the FDTD space lattice. We want to conveniently and self-consistently interface the lumped-circuit physics of the device with the distributed-electromagnetic-field physics of the space lattice, but in a manner that exploits information regarding the device voltage as a function of the device current. Therefore, we ask the question: What is the Thevenin equivalent circuit "looking into" the space lattice from the perspective of the circuit device? That is, we wish to represent the physics of the *entirety* of the FDTD space lattice containing the device by a conventional Thevenin equivalent circuit. This Thevenin circuit is then combined with the circuit of the device of interest, and the resulting overall circuit is solved to incorporate the properties of the device into the FDTD field solution.

Consider a circuit device positioned at the site of an electric-field component, E_w, in the FDTD space lattice (either E_x, E_y, or E_z). From the Thevenin equivalent-circuit standpoint, the excitation effect of the complete FDTD lattice provided to the circuit device can be distilled down to the action of the four, single-cell E-field loops that have E_w along their common edge. The total Thevenin voltage, $V_\theta(t)$, is the sum of the loop-voltage contributions:

$$V_\theta(t) = L_\theta \sum_{i=1}^{4} \frac{V_{\text{loop},i}(t)}{L_i} \tag{8.26}$$

where

$$V_{\text{loop},i}(t) = -\oint_{\Gamma_i} \vec{E} \cdot d\vec{\Gamma}_i \qquad i = 1,\ 2,\ 3,\ 4 \tag{8.27a}$$

Here, Γ_i is the ith cell-face electric-field contour having E_w along one of its edges. (There are four such contours.) Further,

$$L_i = \frac{\mu A_i}{\Delta w} \tag{8.27b}$$

is the lattice-cell inductance associated with the ith contour, where A_i is the cross-section area of the ith cell immediately adjacent to the circuit device, and Δw is the dimension spanning the device terminals. Finally, we identify

$$L_\theta = \left(\sum_{i=1}^{4} \frac{1}{L_i} \right)^{-1} \tag{8.27c}$$

as the Thevenin-equivalent source inductance in our model. Note that this formulation permits unequal cross-section areas of the lattice cells immediately adjacent to the circuit device.

Having defined the Thevenin-equivalent voltage source and source inductance of the FDTD space lattice seen by the circuit device, the remainder of the formulation is most easily understood as a duality transformation of (8.13):

$$V_\theta(t) = L_\theta \frac{dI_{\text{device}}(t)}{dt} + V_{\text{device}}\left[I_{\text{device}}(t)\right] \tag{8.28}$$

We note that (8.28) is an ordinary differential equation for $I_{\text{device}}(t)$ that is integrated in time given knowledge of the initial condition, $I_{\text{device}}(t_0)$; the driving function, $V_\theta(t)$; and the device's voltage-current characteristic, $V_{\text{device}}(I_{\text{device}})$. Starting at time $t_0 - \Delta t/2$ and integrating to time $t_0 + \Delta t/2$, this leads to

$$I_{\text{device}}(t_0 + \Delta t/2) = I_{\text{device}}(t_0 - \Delta t/2) + \frac{1}{L_\theta} \int_{t_0 - \Delta t/2}^{t_0 + \Delta t/2} \left\{ V_\theta(t) - V_{\text{device}}\left[I_{\text{device}}(t)\right] \right\} dt \tag{8.29}$$

Equation (8.29) can be implemented using standard numerical integration techniques such as the predictor-corrector method and the Runge-Kutta method even if $V_{\text{device}}(I_{\text{device}})$ is nonlinear. If desired, (8.29) can be performed by connecting to an external software package such as SPICE that is structured specifically to analyze complex, potentially nonlinear circuits [13]. In this case, there would be no need to have specific knowledge about $V_{\text{device}}(I_{\text{device}})$, as this function would arise naturally from the interconnection of the linear and nonlinear elements placed into the SPICE model.

Equations (8.26) and (8.29) provide a numerically robust linkage between the lattice-based FDTD electromagnetic field calculation and the lumped-circuit-device solver. For the regions where devices are not present, the electric and magnetic fields are updated by the conventional Yee algorithm. For a lattice cell where a circuit device having a known terminal *V-I* characteristic is located at E_w, time stepping proceeds as follows. Starting at time t_0 with an assumed known electric field distribution:

1. V_θ is computed at time t_0 using (8.26) applied to the electric fields circulating around the four cell-face loops having the device as a common edge.

2. Equation (8.29) is integrated numerically from $t_0 - \Delta t/2$ to $t_0 + \Delta t/2$ using the value of $V_\theta(t_0)$ obtained in step 1 and the known *V-I* characteristic of the device. This yields $I_{\text{device}}(t_0 + \Delta t/2)$, and hence from Ampere's law, the $\int \vec{H} \cdot d\vec{\Gamma}$ circulation at time $t_0 + \Delta t/2$ around the device. (The cell displacement current, equivalently the cell capacitance, *C*, is by definition zero in the Thevenin equivalent circuit model. Thus, the device current can be equated directly to the magnetic field circulation around it.)
3. E_w can now be updated from t_0 to $t_0 + \Delta t$ using the $\int \vec{H} \cdot d\vec{\Gamma}$ circulation obtained in step 2. The computation cycle of steps 1 and 2 can now be repeated.

For linear circuit devices, $V_{\text{device}}(I_{\text{device}})$ can be readily derived from $I_{\text{device}}(V_{\text{device}})$. For example, Fig. 8.4 shows block diagrams for both the Norton's and Thevenin's equivalent formulations. However, for nonlinear devices, this transformation is not trivial. Both representations can be useful depending upon the available device model. Fig. 8.5 shows the placement of the equivalent voltage sources for circuit devices in the horizontal and vertical directions in the space lattice. Note that vertical vias are used to provide a voltage reference. Fig. 8.6 shows the Thevenin equivalent circuit technique used to model a multiport lumped element.

8.6 STATE-VARIABLE REPRESENTATION OF THE TERMINAL CHARACTERISTICS OF A LUMPED-CIRCUIT ELEMENT

In general, the terminal characteristics of a lumped-circuit element are represented by a system of nonlinear differential equations:

$$\overline{\mathbf{A}}(\mathbf{X}) \cdot \frac{d\mathbf{X}}{dt} = \overline{\mathbf{B}}(\mathbf{X}) \cdot \mathbf{X} + \mathbf{F}(\mathbf{X}) \tag{8.30}$$

Here, vector **X** denotes the continuous circuit state variables (for example, capacitor voltages and inductor currents); matrices $\overline{\mathbf{A}}$ and $\overline{\mathbf{B}}$ are derived from the circuit elements; and the forcing term, **F**, is comprised of the total voltages or total currents for the equivalent-voltage-source and equivalent-current-source formulations, respectively.

We observe that (8.30) can be integrated by a number of available softwares, including the popular SPICE. For example, [13] demonstrates a three-dimensional FDTD simulation of a nonlinear UHF amplifier with input and output impedance-matching networks. Here, the Norton equivalent-circuit technique of Section 8.4 permits a direct connection of the FDTD electromagnetic field model to the SPICE circuit solver. This tactic permits taking advantage of SPICE's built-in algorithms for the integration of the large systems of ordinary differential equations that arise in circuit analysis, as well as SPICE's versatile device library.

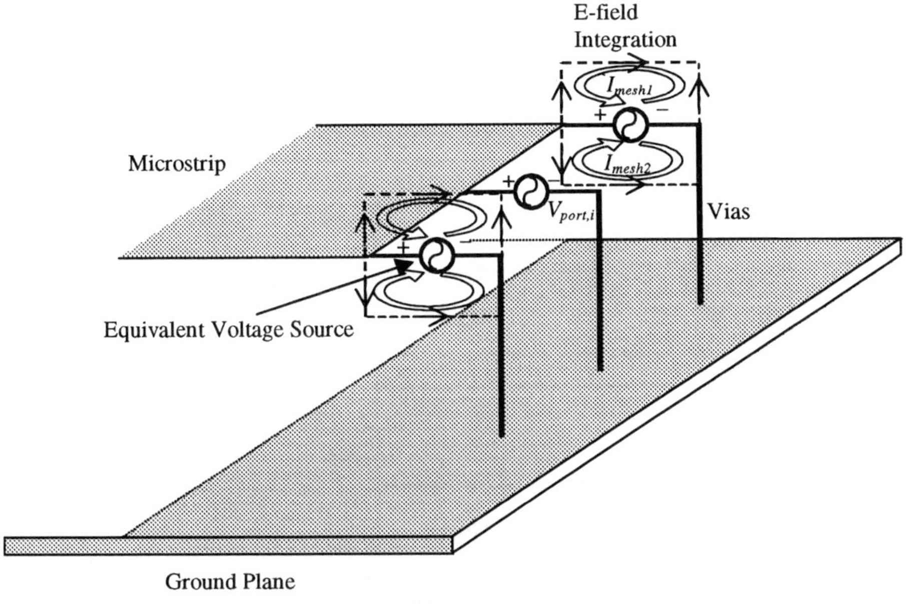

(a) Voltage source is set in the horizontal direction.

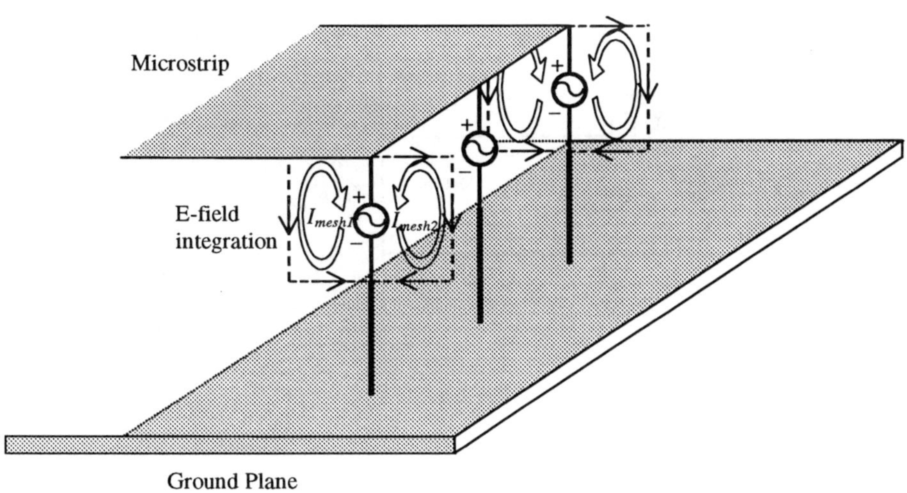

(b) Voltage source is set in the vertical direction.

Fig. 8.5 Arrangement of voltage sources for the equivalent voltage-source formulation.

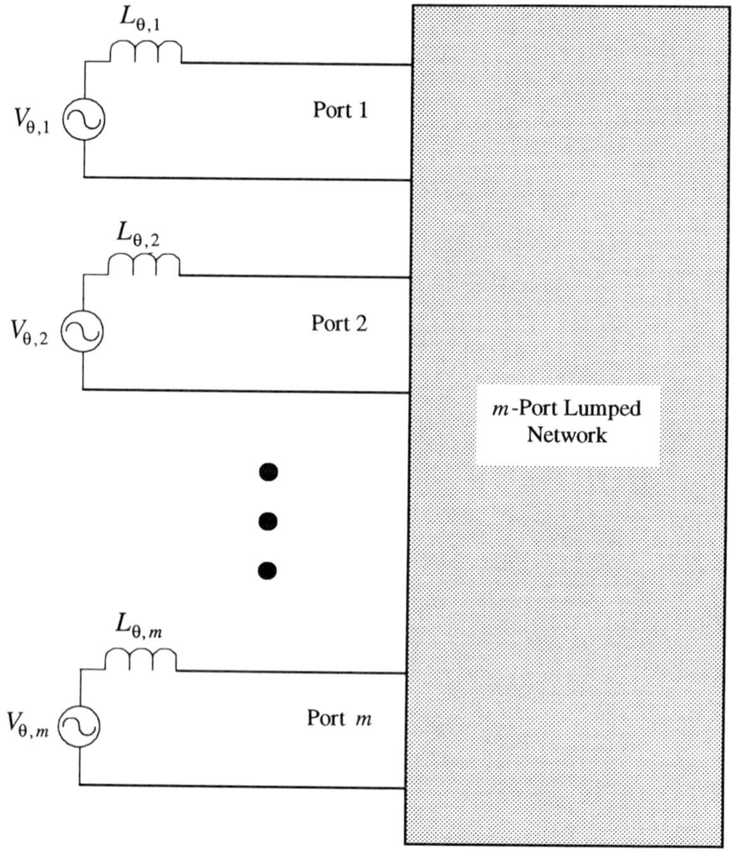

Fig. 8.6 Equivalent voltage-source representation of the FDTD algorithm for a multiport lumped element.

8.6.1 Discrete-State-Variable Technique

Another approach to numerically solve (8.30) is by using a technique wherein the original continuous state variables are approximated by discrete state variables. For example, in conjunction with a forward-difference solver, (8.30) yields

$$\mathbf{H}(\mathbf{X}^{n+1}) = \overline{\mathbf{A}}(\mathbf{X}^{n+1})\cdot\left(\frac{\mathbf{X}^{n+1} - \mathbf{X}^{n}}{\Delta t}\right) - \overline{\mathbf{B}}(\mathbf{X}^{n+1})\cdot\mathbf{X}^{n+1} - \mathbf{F}(\mathbf{X}^{n+1}) = \mathbf{0} \qquad (8.31)$$

where the index, n, denotes the nth time step. The state variable at time $(n+1)\Delta t$ is computed by searching for the root of the function, $\mathbf{H}$. An iterative search for $\mathbf{X}^{n+1}$ is performed by the Newton-Raphson method [18 – 20], a multidimensional root-finding technique. Given the initial trial (guess) for $\mathbf{X}^{n+1}$:

$$\mathbf{X}^{n+1}_{(\text{icount} = 0)} = \mathbf{X}^{n} \qquad (8.32a)$$

the following iterative process is conducted until it converges to the final value of $\mathbf{X}^{n+1}$:

$$\mathbf{X}^{n+1}_{(\text{icount}+1)} = \mathbf{X}^{n+1}_{(\text{icount})} - \left\{\overline{\mathbf{J}}\left[\mathbf{X}^{n+1}_{(\text{icount})}\right]\right\}^{-1}\cdot\mathbf{H}\left[\mathbf{X}^{n+1}_{(\text{icount})}\right] \qquad (8.32b)$$

where $\overline{\mathbf{J}}$ is the Jacobian matrix having the elements

$$\mathrm{J}_{pq} = \frac{\partial H_p}{\partial x_q} \qquad (8.33)$$

For a pure circuit-analysis problem, it is important to choose Δt small enough such that the Jacobian matrix is not singular. However, for modeling most microwave circuits in the context of the FDTD electromagnetic field solver, this is not a problem. This is because the Courant stability condition for FDTD mandates a time step in the picosecond range for such problems, while the intrinsic time constants of microwave circuits are much larger (in the nanosecond range). Therefore, the discrete-state-variable circuit solver does not require a finer sampling in time than the usual FDTD algorithm.

8.6.2 Application to a Passive, Linear, Single-Port Device

Consider a passive, single-port, lumped-circuit device consisting of the parallel combination of a resistor, R, an inductor, L, and a capacitor, C, all contained within a single FDTD lattice cell. This device, shown in schematic-diagram form in Fig. 8.7, is used to model a transmission-line termination. Here, the state vector is composed of the lattice cell (capacitor) voltage and the inductor current. The matrices denoting the state equation are given by

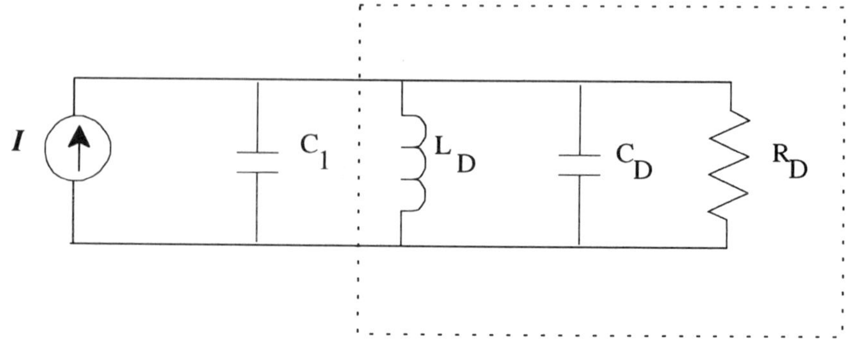

Fig. 8.7 Parallel RLC circuit consisting of passive lumped elements.

$$\overline{\mathbf{A}} = \begin{bmatrix} C_{\text{cell}} + C & 0 \\ 0 & L \end{bmatrix} \tag{8.34a}$$

$$\overline{\mathbf{B}} = \begin{bmatrix} -1/R & -1 \\ 1 & 0 \end{bmatrix} \tag{8.34b}$$

$$\mathbf{F} = \begin{bmatrix} 1 \\ 0 \end{bmatrix} \tag{8.34c}$$

where C_{cell} is the cell capacitance. Typically, $C >> C_{\text{cell}}$ and the resonance frequency is determined by the device inductance and capacitance.

It is interesting to consider the case where C and R are set to zero. For the explicit formulation of [9], the device current appears as the sum of the past values of the cell voltage. With the above approach, the state equation is solved at each FDTD time step using the cell voltage and the inductor current of the previous time step.

8.6.3 Application to a Passive, Linear, Two-Port Device

The next example illustrates the combining of the FDTD algorithm with a passive, linear, two-port device model. The two-port element is an RLC network used to model the interconnect transition between two transmission lines, as shown in Fig. 8.8. Here, the resistance, R, represents the radiation loss. The matrices denoting the state equation are given by

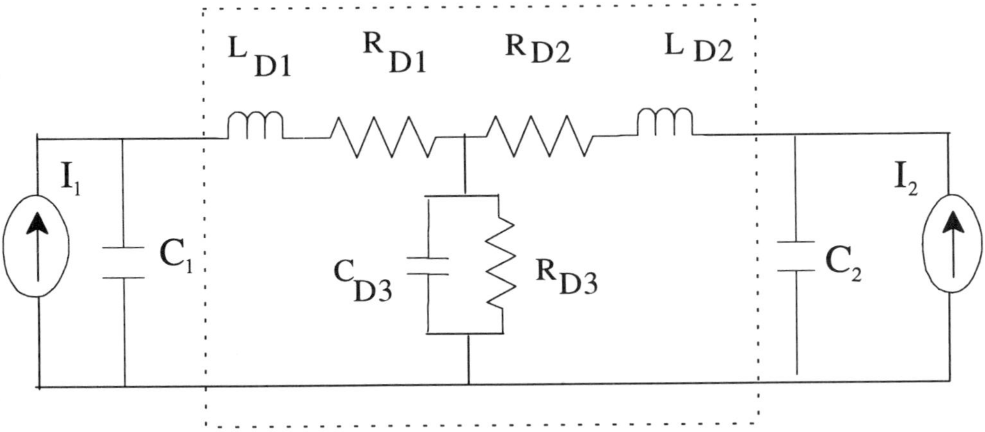

Fig. 8.8 Two-port RLC network consisting of passive lumped elements.

$$\overline{\mathbf{A}} = \begin{bmatrix} C_1 & 0 & 0 & 0 & 0 \\ 0 & L_{D1} & 0 & 0 & 0 \\ 0 & 0 & C_{D3} & 0 & 0 \\ 0 & 0 & 0 & C_2 & 0 \\ 0 & 0 & 0 & 0 & L_{D2} \end{bmatrix} \tag{8.35a}$$

$$\overline{\mathbf{B}} = \begin{bmatrix} 0 & -1 & 0 & 0 & 0 \\ 1 & -R_{D1} & -1 & 0 & 0 \\ 0 & 1 & -1/R_{D3} & 0 & 1 \\ 0 & 0 & 0 & 0 & -1 \\ 0 & 0 & -1 & 1 & -R_{D2} \end{bmatrix} \tag{8.35b}$$

$$\mathbf{F} = \begin{bmatrix} 1 & 0 \\ 0 & 0 \\ 0 & 0 \\ 0 & 1 \\ 0 & 0 \end{bmatrix} \tag{8.35c}$$

8.7 MODELING A NONLINEAR TWO-TERMINAL ELEMENT: THE SCHOTTKY DIODE

Microwave circuits utilize diodes for mixing and harmonic generation. The relation between the diode terminal current, I_D, and terminal voltage, V_D, is given by

$$I_D = I_{sat}\left[\exp(qV_D/\eta kT) - 1\right] \tag{8.36}$$

where I_{sat} is the saturation current; q is the electronic charge; η is the ideality factor; k is Boltzmann's constant; and T is the absolute temperature [21].

Consider the Hewlett-Packard HP HSMS-8202 Schottky diode used in a singly balanced mixer. The equivalent circuit of the diode is shown in Fig. 8.9, where L_{p1}, L_{p2}, and C_p represent the parasitic inductances and capacitance of the diode package.

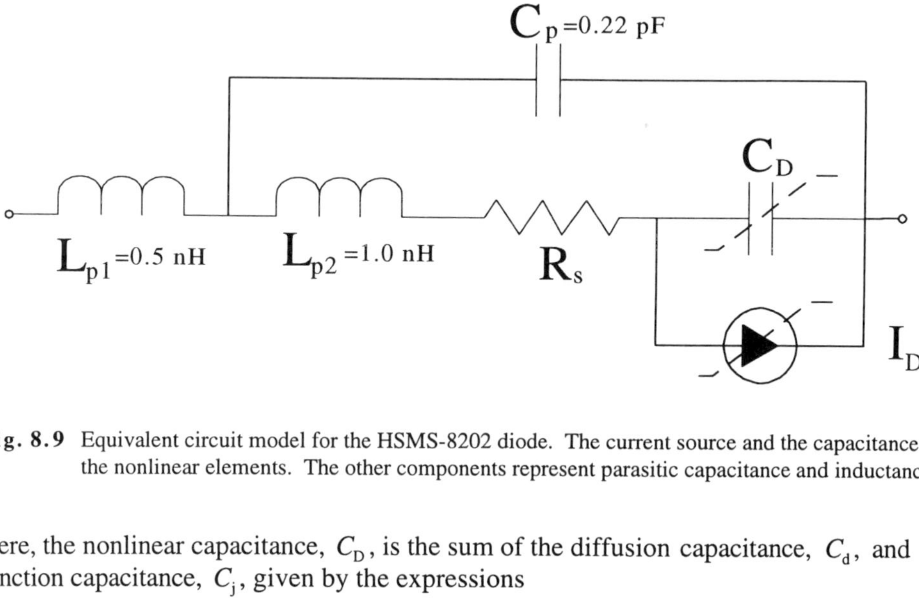

Fig. 8.9 Equivalent circuit model for the HSMS-8202 diode. The current source and the capacitance are the nonlinear elements. The other components represent parasitic capacitance and inductance.

Here, the nonlinear capacitance, C_D, is the sum of the diffusion capacitance, C_d, and the junction capacitance, C_j, given by the expressions

$$C_d = \tau_D \frac{dI_D}{dV_D} \tag{8.37}$$

$$C_j = \begin{cases} C_j(0)\left(1 - \dfrac{V_D}{\phi_0}\right)^{-m} & V_D < \mathrm{FC} * \phi_0 \\ \dfrac{C_j(0)}{\mathrm{F}_2}\left(\mathrm{F}_3 + \dfrac{mV_D}{\phi_0}\right) & V_D \geq \mathrm{FC} * \phi_0 \end{cases} \tag{8.38}$$

where

$$F_2 = (1 - FC)^{1+m} \quad , \qquad F_3 = 1 - FC(1+m) \tag{8.39}$$

Fig. 8.10 illustrates the microwave circuit layout of the singly balanced mixer, which is composed of two diodes and a 90° hybrid coupler. Inductors L_{g1} and L_{g2} provide the intermediate-frequency (IF) ground for the diodes. The combination of L_f, C_f, and the open-circuited microstrip provides a low-pass filter for the IF output signal.

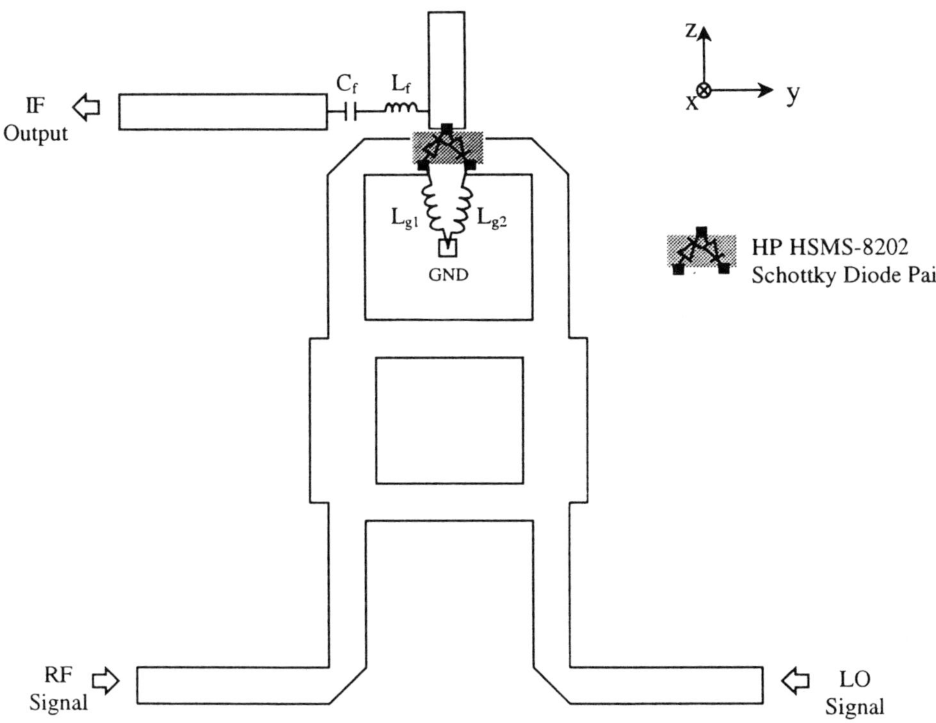

Fig. 8.10 Microwave circuit layout of the singly balanced diode mixer.

For the FDTD simulation of this circuit, all lumped elements are replaced by their equivalent-circuit models. In order to combine the circuit simulation of the lumped elements along with the full-wave electromagnetic simulation of the structure, the Norton equivalent-circuit formulation of Section 8.4 or the Thevenin equivalent-circuit formulation of Section 8.5 is used. Figs. 8.11(a) and 8.11(b) show, respectively, the equivalent-current-source and equivalent-voltage-source circuit models for the embedding of the mixer diodes within the FDTD space lattice.

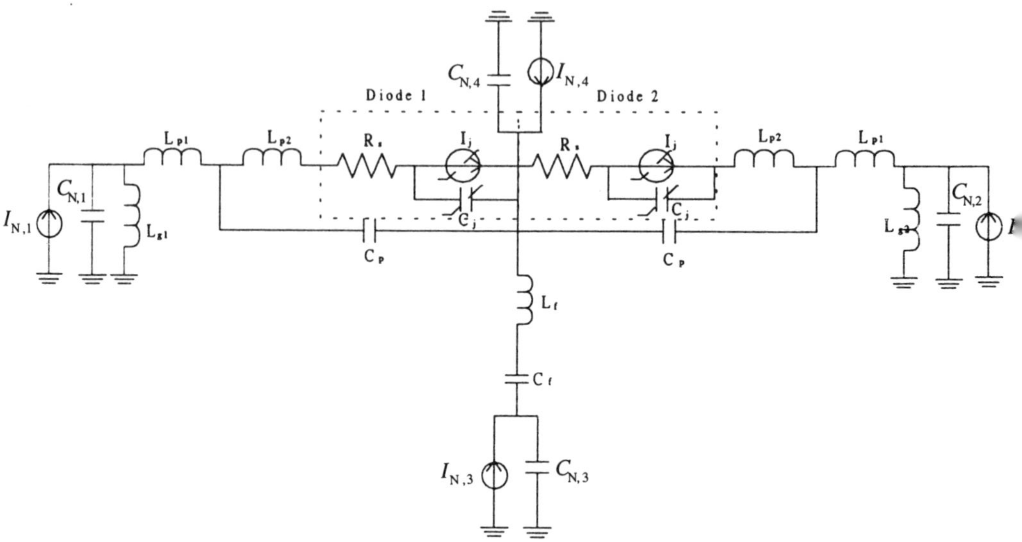

(a) Equivalent-current-source model.

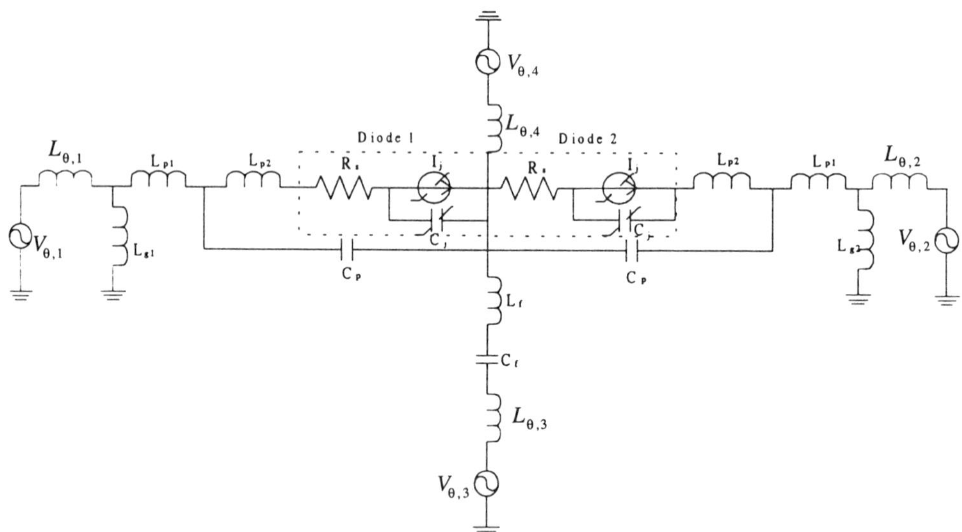

(b) Equivalent-voltage-source model.

Fig. 8.11 Equivalent-circuit models for the embedding of the mixer diodes within the FDTD lattice.

FDTD calculations for this mixer involve simulating the injection of a 10-GHz local-oscillator (LO) signal and a 10.25-GHz RF signal into the mixer's input ports, and observing the computed response at the IF output port. Fig. 8.12 shows the FDTD-computed mixer output, which is the 250-MHz RF–LO difference-frequency signal modulated by higher-order harmonics. This simulation agrees well with the MDS-Libra analysis [16].

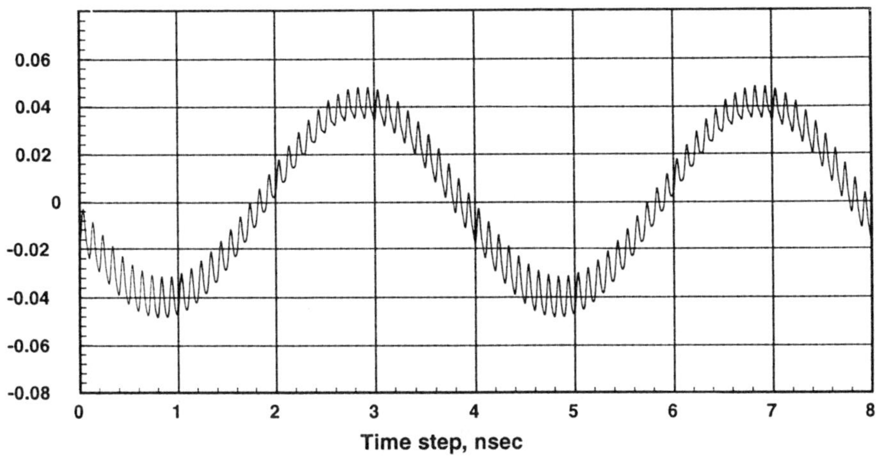

Fig. 8.12 Time-domain response of the balanced mixer of Figs. 8.10 and 8.11 consisting of a 250-MHz difference-frequency signal modulated by higher-order harmonics.

8.8 MODELING A NONLINEAR THREE-TERMINAL ELEMENT: THE MESFET TRANSISTOR

Transistors are active devices that can operate in linear (small-signal) or nonlinear (large-signal) regimes. This section summarizes the FDTD analysis of the nonlinear-equivalent-circuit model of a metal-semiconductor (MESFET) transistor used as a two-port network in a common-source amplifier. Fig. 8.13(a) shows the mounting of the packaged MESFET in a microstrip circuit. The MESFET is connected to the ground plane through vias at its source terminal.

8.8.1 Large-Signal Model

Following [17], Fig. 8.13(b) illustrates the large-signal model of the MESFET used in the FDTD simulation. In this figure, the MESFET model is enclosed by a dashed box, while the Thevenin equivalent circuits outside of this box represent the action of the FDTD space lattice located at the gate and drain terminals of the transistor (see Section 8.5). The internal nodes G', D', and S' represent the intrinsic part of the MESFET.

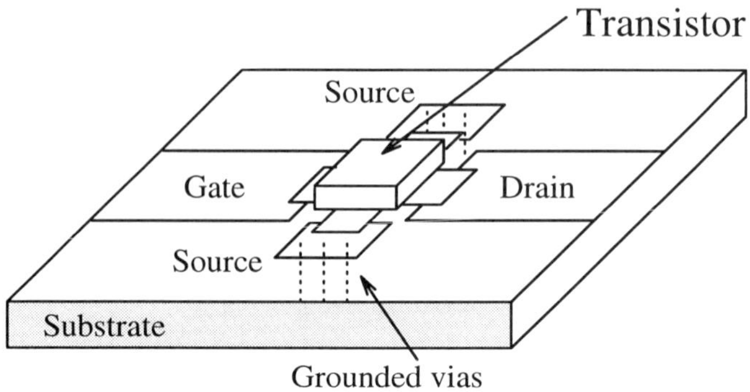

(a) Mounting of the packaged MESFET in a microstrip circuit. The transistor is connected to the ground plane through vias at its source terminal.

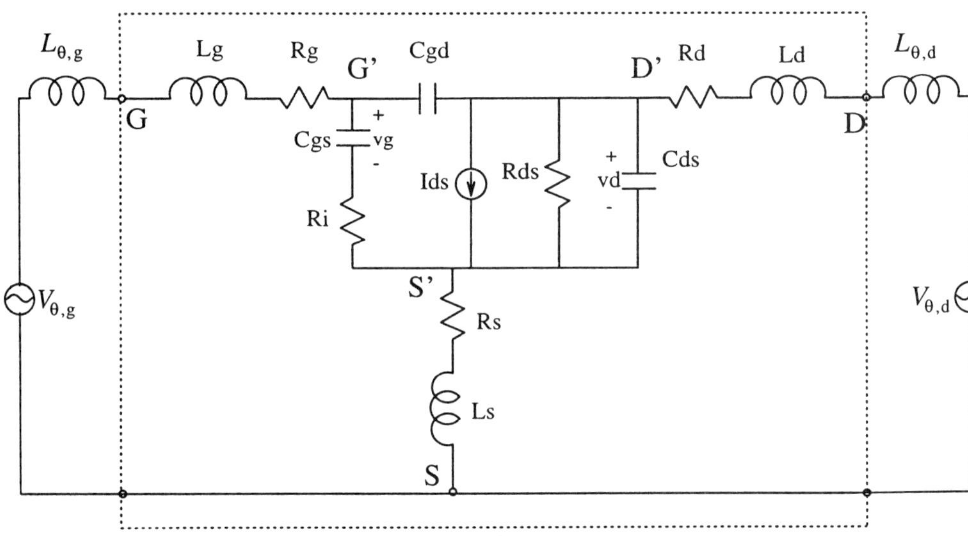

Rg = 0.5Ω
Rd = 0.5Ω

Rs = 0.7 Ω
Ri = 1.0 Ω

Lg = 0.05 nH
Ld = 0.05 nH
Ls = 0.1 nH

Cdg = 0.2 pF
Cds = 0.6 pF

(b) Large-signal model of the MESFET integrated with the Thevenin equivalent circuits for the FDTD lattice at its gate and drain terminals.

Fig. 8.13 MESFET transistor used in the FDTD model of a common-source amplifier. *Source*: Kuo et al., *IEEE Trans. Microwave Theory and Techniques*, 1997, pp. 819–826, © 1997 IEEE.

The MESFET circuit model of Fig. 8.13(b) contains two nonlinear elements: the gate-source capacitor, C_{gs}, and the drain current source, I_{ds}. Governed by the PN-junction capacitance model, the gate-source capacitor is expressed as

$$C_{gs}(v_g) = \frac{C_{gso}}{\sqrt{1 - v_g/\phi_{bi}}} \tag{8.40}$$

where $C_{gso} = 3$ pf and $\phi_{bi} = 0.7$ V for this transistor. The drain current source, I_{ds}, describing dc characteristics, relates to v_g and v_d as

$$I_{ds}(v_g, v_d) = \left[A_0 + A_1 v_{G'S'} + A_2 \left(v_{G'S'}\right)^2 + A_3 \left(v_{G'S'}\right)^3\right] \tanh(\alpha v_d) \tag{8.41}$$

where $A_0 = 0.5304$, $A_1 = 0.2595$, $A_2 = -0.0542$, $A_3 = -0.0305$, and $\alpha = 1.0$ for this transistor. The dc characteristics are plotted in Fig. 8.14.

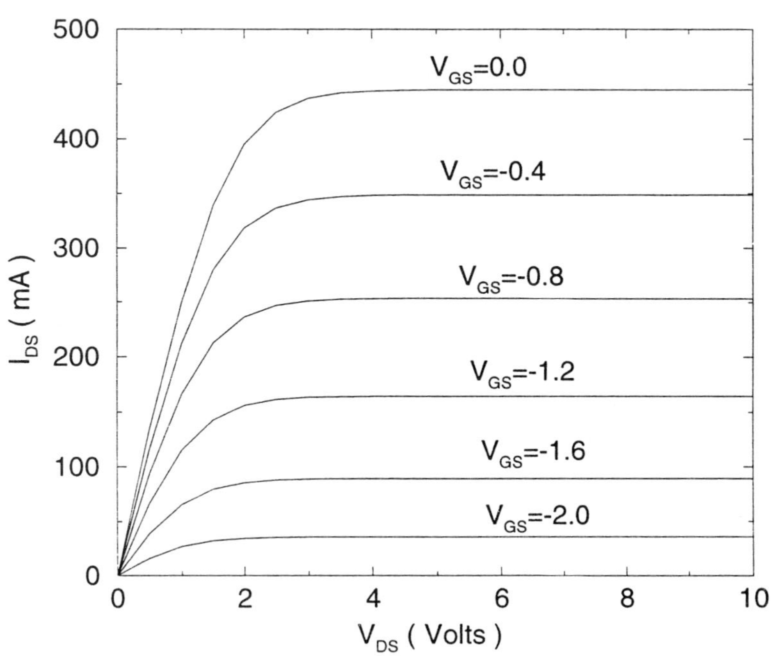

Fig. 8.14 DC characteristics of the MESFET transistor. *Source*: Kuo et al., *IEEE Trans. Microwave Theory and Techniques*, 1997, pp. 819–826, © 1997 IEEE.

After choosing the state variables as $[v_g, v_{G'D'}, v_d, i_{Lg}, i_{Ld}]^T$, it is straightforward to derive the state equation from the nodal equation of the circuit. This derivation yields the following for the $\overline{\mathbf{A}}$ matrix:

$$\overline{\mathbf{A}} = \begin{bmatrix} C_{gs} & 0 & 0 & 0 & 0 \\ C_{gs} & C_{gd} & 0 & 0 & 0 \\ 0 & C_{gd} & -C_{ds} & 0 & 0 \\ 0 & 0 & 0 & A_{44} & A_{45} \\ 0 & 0 & 0 & A_{54} & A_{55} \end{bmatrix} \tag{8.42}$$

wherc

$$A_{44} = G_g\left[\left(\frac{G_g}{G} - 1\right)L'_g - \left(\frac{G_s}{G}\right)L_s\right] \tag{8.43a}$$

$$A_{45} = G_g\left[\left(\frac{G_d}{G}\right)L'_d - \left(\frac{G_s}{G}\right)L_s\right] \tag{8.43b}$$

$$A_{54} = G_d\left[\left(\frac{G_g}{G}\right)L'_g - \left(\frac{G_s}{G}\right)L_s\right] \tag{8.43c}$$

$$A_{55} = G_d\left[\left(\frac{G_d}{G} - 1\right)L'_d - \left(\frac{G_s}{G}\right)L_s\right] \tag{8.43d}$$

$$L'_g = L_g + L_{\theta,g}\,, \qquad L'_d = L_d + L_{\theta,d}\,, \qquad G = G_g + G_d + G_s \tag{8.43e}$$

and $G_g = 1/R_g$, $G_d = 1/R_d$, and $G_s = 1/R_s$. The $\overline{\mathbf{B}}$ matrix for the nonlinear state equation is given by

$$\overline{\mathbf{B}} = \begin{bmatrix} -G_{gs} & G_{gs} & G_{gs} & 0 & 0 \\ 0 & 0 & 0 & 1 & 0 \\ 0 & 0 & 0 & 0 & -1 \\ 0 & B_{42} & B_{43} & 1 & 0 \\ 0 & B_{52} & B_{53} & 0 & 1 \end{bmatrix} \tag{8.44}$$

where

$$B_{42} = G_g\left(1 - \frac{G_g}{G}\right), \qquad B_{43} = \frac{G_g\,G_s}{G} \tag{8.45a}$$

$$B_{52} = -\frac{G_d\,G_g}{G}, \qquad B_{53} = \frac{G_d\,G_s}{G} \tag{8.45b}$$

Note that $\overline{\mathbf{B}}$ is independent of the state variables. In addition, the $\mathbf{F}$ vector for the nonlinear state equation is given by

$$\mathbf{F} = \begin{bmatrix} 0 \\ 0 \\ I_{\text{ds}} \\ F_4 \\ F_5 \end{bmatrix} \tag{8.46}$$

where

$$F_4 = G_{\text{g}}\left(1 - \frac{G_{\text{g}}}{\text{G}}\right) V_{\theta,g} - \frac{G_{\text{g}}\, G_{\text{d}}}{\text{G}} V_{\theta,d} \tag{8.47a}$$

$$F_5 = G_{\text{d}}\left(1 - \frac{G_{\text{d}}}{\text{G}}\right) V_{\theta,d} - \frac{G_{\text{g}}\, G_{\text{d}}}{\text{G}} V_{\theta,g} \tag{8.47b}$$

From (8.32) and (8.33), the iterative process for $\mathbf{X}^{n+1}$ is given by

$$\mathbf{X}^{n+1}_{(\text{icount}+1)} = \mathbf{X}^{n+1}_{(\text{icount})} - \left[\frac{1}{\Delta t}\left(\frac{\partial \overline{\mathbf{A}}}{\partial v_{\text{g}}} + \overline{\mathbf{A}} \right) - \overline{\mathbf{B}} - \frac{\partial \mathbf{F}}{\partial \mathbf{X}^{n+1}_{(\text{icount})}} \right]^{-1} \cdot \mathbf{F}\left[\mathbf{X}^{n+1}_{(\text{icount})}\right] \tag{8.48}$$

Finally, the device voltages are evaluated by

$$V_{\text{g}} = -V_{\theta,\text{g}} - L_{\theta,\text{g}} \frac{di_{\text{Lg}}}{dt} \tag{8.49a}$$

$$V_{\text{d}} = -V_{\theta,\text{d}} - L_{\theta,\text{d}} \frac{di_{\text{Ld}}}{dt} \tag{8.49b}$$

8.8.2 Simulation Results

The system under consideration is the 6-GHz microwave amplifier shown in Fig. 8.15(a), which contains the MESFET described in Section 8.8.1. The size of the MESFET (about 2.0 mm) is much smaller than the guided wavelength at 6 GHz. A PEC box is used to model the packaging structure. There are two rectangular holes at the input / output ports for feeding dc power from the sources, V_{gg} and V_{dd}, each with a source impedance of 50Ω. The biasing conditions chosen are $V_{\text{GS}} = -0.81$ V and $V_{\text{DS}} = 6.4$ V.

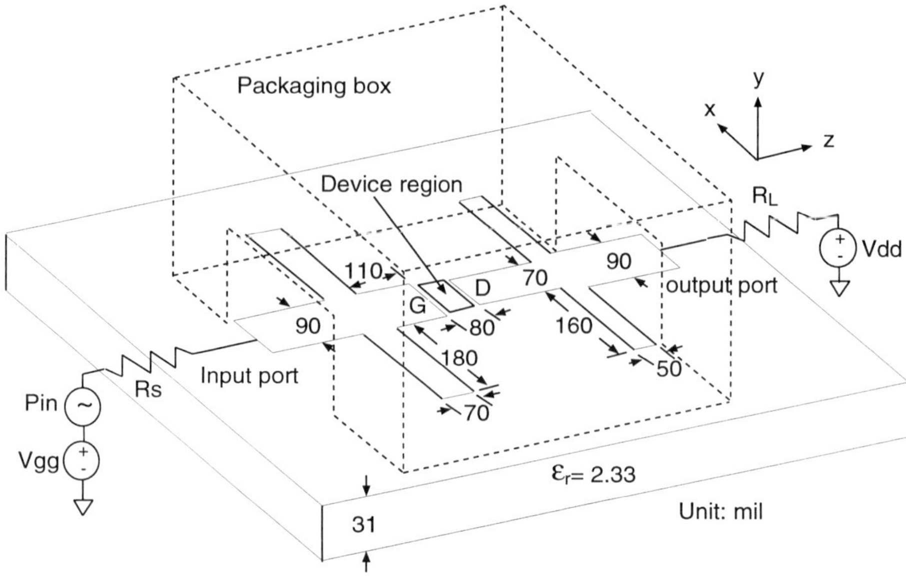

(a) Structure and dimensions of the amplifier.

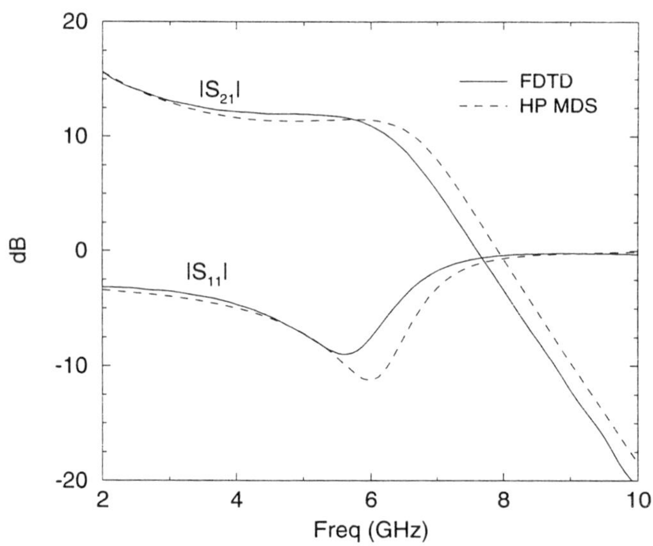

(b) Small-signal analysis of the amplifier without the packaging box.

Fig. 8.15 Common-source MESFET amplifier modeled using FDTD. *Source*: Kuo et al., *IEEE Trans. Microwave Theory and Techniques*, 1997, pp. 819–826, © 1997 IEEE.

The FDTD computation domain is a uniform space lattice of dimensions $74 \times 40 \times 128$ cells with $\Delta x = \Delta z = 0.254$ mm (10 mil) and $\Delta y = 0.197$ mm (7.75 mil). Higdon's second-order ABC is applied at the lattice truncation to absorb outgoing waves.

Analysis of the Circuit Without the Packaging Structure

The FDTD model is first applied to obtain the small-signal response of the amplifier without the packaging box [17]. Here, a small-amplitude Gaussian pulse modulating a 6-GHz carrier is used as the ac input signal to the nonlinear large-signal FDTD MESFET circuit model shown in Fig. 8.13. The signal amplitude is small enough to allow the circuit to operate in the linear region. Fig. 8.15(b) compares the FDTD-computed S parameters with results obtained using the Hewlett-Packard MDS circuit simulator. In this figure, the dip of S_{11} (the impedance-matched condition) in the FDTD simulation is at 5.58 GHz, 7% below that of the HP MDS simulation. This difference may result from a different modeling of the vias at the source port in the two simulations. The frequency of the S_{11} dip is sensitive to the effective inductance of the vias, which causes a series-feedback effect. In the HP MDS simulation, the frequency of the S_{11} dip decreases as the inductance of the vias decreases, approaching the FDTD value.

We now review results of the nonlinear large-signal case. Fig. 8.16(a) compares the FDTD- and MDS-calculated output power at 6 GHz (the fundamental frequency), 12 GHz (the second harmonic), and 18 GHz (the third harmonic) as a function of the input power for a single-tone excitation [17]. Very good agreement is observed.

Next, we review the calculation of intermodulation due to a two-tone, equal-input-power excitation at 3 and 6 GHz. The output power appears only at the mixing frequencies, or intermodulation products, which arise as linear combinations of 3 and 6 GHz. Fig. 8.16(b) compares the FDTD- and MDS-calculated intermodulation products as a function of the input power [17]. Again, good agreement is observed.

Analysis of the Circuit With the Packaging Structure

The FDTD simulation of the microwave amplifier permits a detailed evaluation of the effects of its packaging simply by properly locating the package geometry and materials in the space lattice. Such an evaluation is beyond the capability of many existing microwave circuit simulators, but can be readily accomplished using the extended FDTD method discussed in this chapter. Physically, the packaging structure forms a partially dielectric-filled cavity. Excited by the circuit, the cavity stores energy due to its natural resonances and the stored energy is inevitably coupled back to the circuit. For a nonlinear active circuit, this feedback makes the stability circles shift and may result in oscillation.

Consider the example of this MESFET circuit with a packaging box of dimensions $39.6 \times 4.7 \times 31.8$ mm ($1560 \times 186 \times 1250$ mil). The first resonant frequency of this box is found to be at 5.72 GHz by FDTD presimulation of a packaged uniform

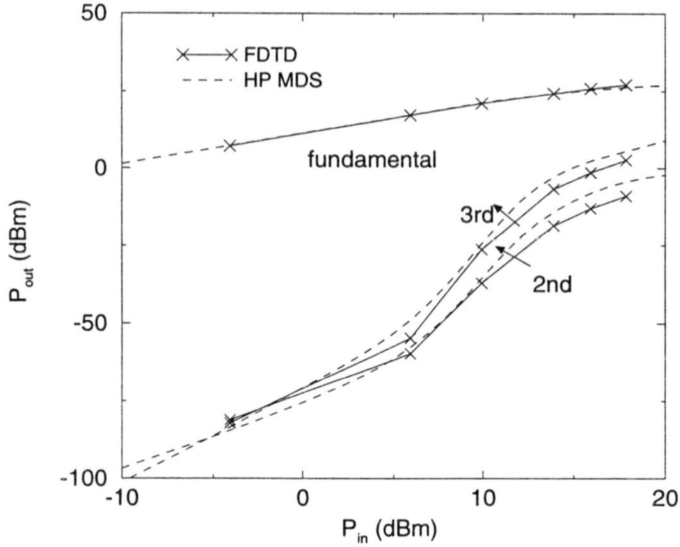

(a) Output power at 6-GHz fundamental frequency and 12- and 18-GHz harmonics.

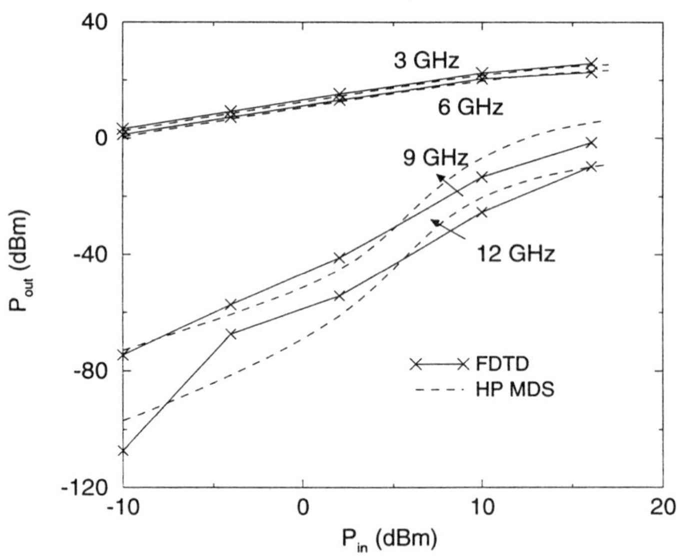

(b) Output power of 9- and 12-GHz intermodulation products for inputs at 3 and 6 GHz.

Fig. 8.16 Comparison of FDTD and HP-MDS results for the nonlinear large-signal excitation of the MESFET amplifier. *Source*: Kuo et al., *IEEE Trans. Microwave Theory and Techniques*, 1997, pp. 819–826, © 1997 IEEE.

microstrip line. Fig. 8.17 shows FDTD calculations indicating that this packaging interacts strongly with the circuit and causes oscillation at the resonant frequency [17]. Here, a low-level input Gaussian pulse modulating a 6-GHz carrier (close to the resonant frequency) generates large output power at the resonant frequency. This indicates the presence of essentially unlimited gain at the resonant frequency, and thus the condition of amplifier instability.

In order to avoid oscillation, the dimensions of the packaging structure are usually chosen such that the resonant frequency is raised far above the frequency range of interest. We now consider a second packaging example where the dimensions of the enclosing box are reduced to $16.3 \times 4.7 \times 17.5$ mm ($640 \times 186 \times 690$ mil) [17]. This elevates the FDTD-calculated first resonant frequency to 11.79 GHz, rendering the packaged circuit electrically stable. Fig. 8.18 shows the effect of the modified packaging upon the S parameters. In this case, the frequency of the FDTD-calculated match point (the S_{11} dip) remains at 5.58 GHz, but the output magnitude drops from -8.98 to -10.07 dB, while the amplifier gain at 6 GHz rises from 10.95 to 11.76 dB. Fig. 8.19 illustrates the effect of the packaging on the output power at the fundamental, second-harmonic, and third-harmonic frequencies for a large-signal, single-tone input at 6 GHz. The resonant frequency of the package is close to that of the second harmonic at 12 GHz. Therefore, package energy is coupled strongly to the second harmonic, and its output-power curve varies more significantly than the others.

8.9 INTRODUCTION TO MODELING IN HIGH-SPEED DIGITAL DESIGN

The contemporary computer marketplace demands reliable, high-performance systems at competitive prices with the shortest possible times from product conception to delivery. Ideally, using robust modeling techniques, even the most complex new systems would be rapidly delivered to the market at lowest cost and in working order without requiring debugging of hardware prototypes. Problems uncovered during modeling would be designed out of a system by establishing a reliable set of "ground rules" before any hardware is built. For high-speed digital systems, reliable ground rules require an intimate and accurate understanding of the electrical performance of key packaging, interconnection, and power-distribution components. Ground rules ensure that the intended performance and reliability of the system is achieved on the first pass of the design. The reader is directed to [22] and [23] for baseline information about high-speed digital circuit design, including typical ground rules.

We note that modeling ideally complements experience, especially in selecting the optimum model and in interpreting its results. For instance, we must decide on the order of accuracy of a simulation, its field-sampling density, its potential need to deal with dispersive materials and/or the conductor skin effect, and the choice of the ABC. These options provide significant trades-off between the solution's accuracy, programming complexity (and thus, the model-development time), and running time.

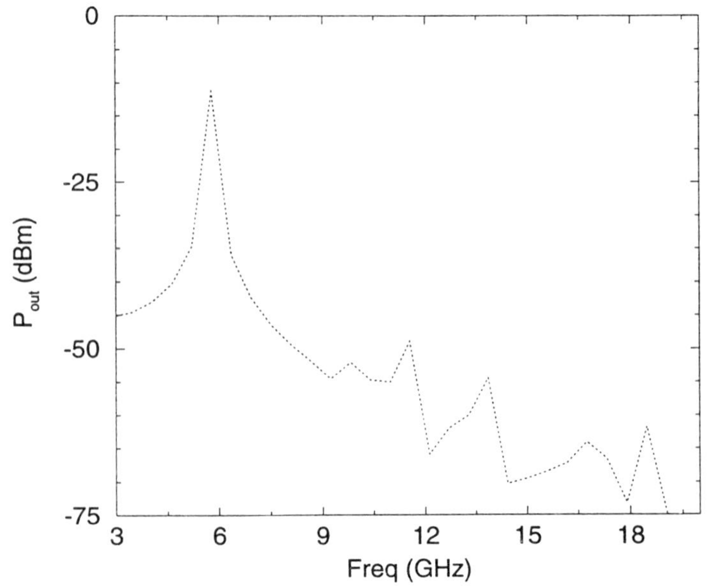

Fig. 8.17 The spectrum of the output power for a small-signal exciting the initial packaged amplifier showing the circuit oscillating after being placed in the packaging structure.

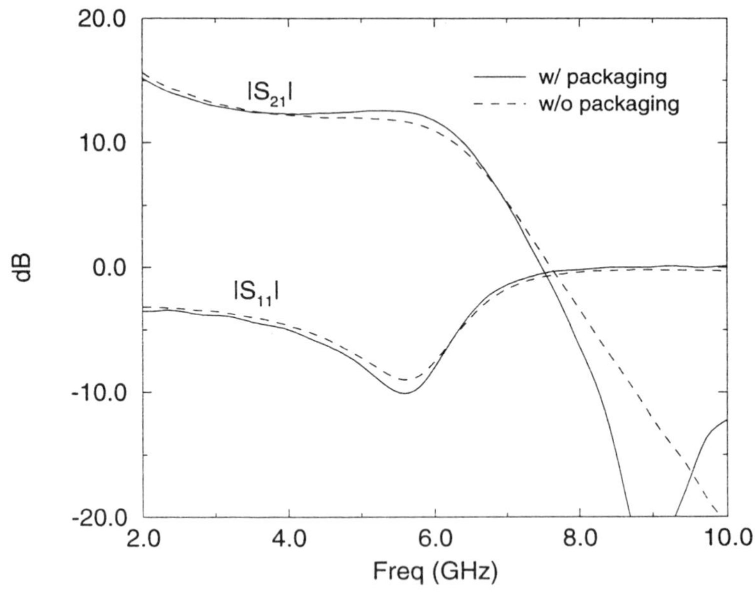

Fig. 8.18 Effect of modified packaging upon *S* parameters. *Source for Figs. 8.17 and 8.18*: Kuo et al. *IEEE Trans. Microwave Theory and Techniques*, 1997, pp. 819–826, © 1997 IEEE.

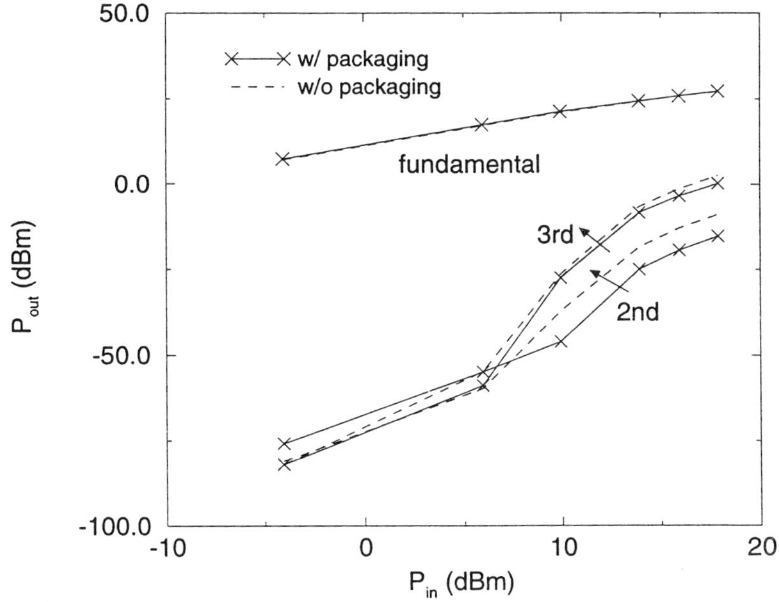

Fig. 8.19 Effect of the amplifier packaging on the output power at the fundamental, second-harmonic, and third-harmonic frequencies for a large-signal, single-tone input at 6 GHz. *Source*: Kuo et al., *IEEE Trans. Microwave Theory and Techniques*, 1997, pp. 819–826, © 1997 IEEE.

Further, some informed judgment is needed as to when the modeling technique is adequately validated, allowing subsequent results to be trusted. Even validation by comparing to measured data is subject to experience, since deembedding the required numbers from the measurements can prove to be an art. Overall, while no simulation can take into account every system parameter, the qualitative and quantitative data gained from a good simulation does lead to a more solid design. This is especially true when the basic numerical tool has been proven robust, that is, successfully used by many different people over many years for many different problems.

FDTD is precisely such a robust means to model high-speed digital circuits. It allows accurate characterization of three-dimensional structures in complex electronic systems operating well in excess of 100 MHz. It also provides a level of modeling complexity that is suitable for advanced electronic components such as MCMs, as well as more traditional components such as printed circuit boards, integrated circuit packages, and connectors. Not only is FDTD able to address the complex electromagnetic wave interactions within individual devices and components, FDTD can also model the coupling of energy into electronic devices from external sources (for example, due to noise or jamming signals).

In the following two sections, FDTD is used to examine power-distribution and clock-signal-distribution systems in complex digital systems. The characterization requirements for these problems include an intuitive and visual understanding of the electrodynamics along with a quantitative description of the *L*, *C*, *R*, and *Z* parameters.

8.10 INDUCTANCE OF COMPLEX POWER-DISTRIBUTION SYSTEMS

Determining the equivalent-circuit lumped inductance of a three-dimensional power-distribution system is of critical importance to successful package-level and system-level high-speed design. Physical and electrical design trades-off can be understood and optimal performance achieved if this inductance is known before commitment to hardware. This section discusses FDTD modeling of the lumped inductance of a complex printed-circuit-board power-distribution structure. The example shown demonstrates the versatility of the method for arbitrary structures of this type.

8.10.1 Method Description

Chapter 13 of [6] discusses the calculation of the signal-path inductance directly from the FDTD field data based upon the fundamental flux-current definition. That is, the inductance, L, is taken as the ratio of the magnetic flux, Φ, to the current, I, in the trace. Φ is calculated using a numerical "hockey net" to capture the magnetic flux appearing between the signal path and the return-current path. This technique works well for signal paths where the return current flows along a reasonably linear route adjacent to the signal trace. However, by design, low-inductance power-distribution systems usually have highly interdigitated geometries with globally distributed power and return paths. The inductance of these three-dimensional structures results from a complex interaction of the electromagnetic fields not amenable to the flux-current definition.

Consider an alternate approach [24] more suitable for modeling the inductance of an arbitrary power-distribution system. Fig. 8.20 illustrates this approach schematically.

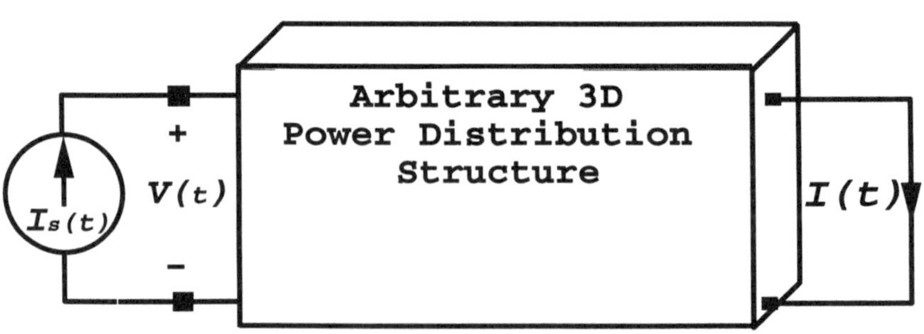

Fig. 8.20 Black-box model of inductance simulation setup.

Let a current pulse excite the input of the power-distribution structure. Short-circuiting the output simulates the operation of a load that is drawing significant current (for example, a CMOS integrated circuit experiencing a current spike). In this state, the conduction current causes magnetic flux to develop throughout the distribution structure, thereby emphasizing its inductive characteristic. A voltage, $V(t)$, develops across the input of the distribution structure in response to the current, $I(t)$, flowing through the effective lumped inductance of the structure according to the fundamental relation $V = L\,dI/dt$. Data for both $V(t)$ and dI/dt, shown in generic form in Fig. 8.21, are derived from the full-wave FDTD-calculated fields as described in [6].

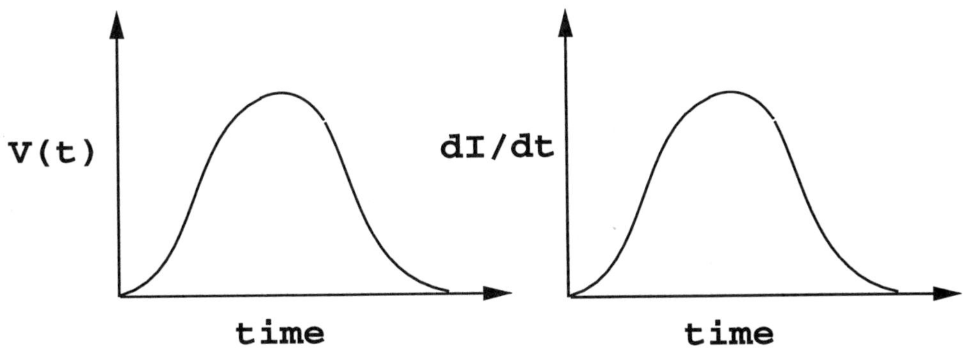

Fig. 8.21 Generic voltage and dI/dt curves used to calculate system inductance.

This allows calculation of the lumped inductance of the structure using the relation $L = V/(dI/dt)$. The presence of an inductive element can be inferred from the shape of $V(t)$ by observing its similarity to the shape of the corresponding dI/dt curve. In Fig. 8.21, the fact that these two waveforms have identical shapes indicates that they are related to each other by a multiplicative constant. This factor is the structure's inductance.

This approach permits the calculation of inductance to be free from any special considerations of the internal geometry of the power-distribution structure. In fact, the full-wave solution of Maxwell's equations by FDTD already accounts for the physics of the structure's internal and external geometry to the resolution limit of the space lattice. Thus, this approach can be used systematically to calculate the inductance of arbitrarily shaped, three-dimensional structures. In addition, the full-wave FDTD solver frees the inductance calculation from any potentially limiting TEM assumptions or approximations. This provides an inherently three-dimensional environment for the required simulation and parameter extraction.

Of key importance to the correct application of this method is the choice of the exciting-current waveform. There are two directly competing conditions for this waveform:

1. The leading edge of the exciting wave should rise *quickly* enough to yield a voltage response that corresponds mainly to the inductive component of the structure.
2. The leading edge of the exciting wave should rise *slowly* enough so that, upon propagation, this edge extends spatially over the entire structure to be characterized for inductance. This provides a comparable dynamic condition over each part of the structure even though these parts may be distributed in space, and permits the structure to be characterized as a single lumped element.

A Gaussian exciting current pulse can be used to meet these conditions. Validation studies using FDTD modeling of canonical microstrip and coaxial geometries [24] indicate that Gaussian leading-edge rise times that are 5 to 10 times longer than the propagation delay across the structure yield reliable inductance data over a 20% to 80% span of the rising edge of the FDTD-calculated response. We also find it useful to average the observed inductance over a time-windowed portion of the response. This averaging window can extend up to the 20% to 80% span. These findings are further supported by analogous circuit-based studies using SPICE to model the current-pulse excitation of an inductor [24].

8.10.2 Example: Multiplane Meshed Printed-Circuit Board

A significant problem in modeling and designing high-speed digital circuits has been understanding how to extract system-level parameters from complex power-distribution systems. Fig. 8.22 shows one such structure, a meshed printed-circuit-board (PCB) system having three power planes with current sources at the edges, five ground planes, and nine interdigitated vias connecting the power and ground planes.

This model is based on a portion of an actual PCB power-distribution structure that lies under a 2499-pin, 30-mm MCM [25]. The MCM has a pad pitch of 0.5 mm and is built with copper/polyimide technology similar to the memory-element MCM published in [26]. The modeled portion of the PCB covers a planar area of 3×3 mm and a via height of 1.02 mm. When the planes are meshed due to the signal-pin antipads, approximately 50% of the metal is removed. The meshed planes force the current to diverge around the antipad holes. Consequently, magnetic flux develops through the holes and the total inductance of the system is increased. In applying the inductance-modeling technique discussed in Section 8.10.1, the voltage response is measured between a power plane and a ground plane at the edge of the PCB structure. A via short to a ground plane on top of the structure (simulating a real load) provides a measurement of the input current that causes a voltage to develop across the structure.

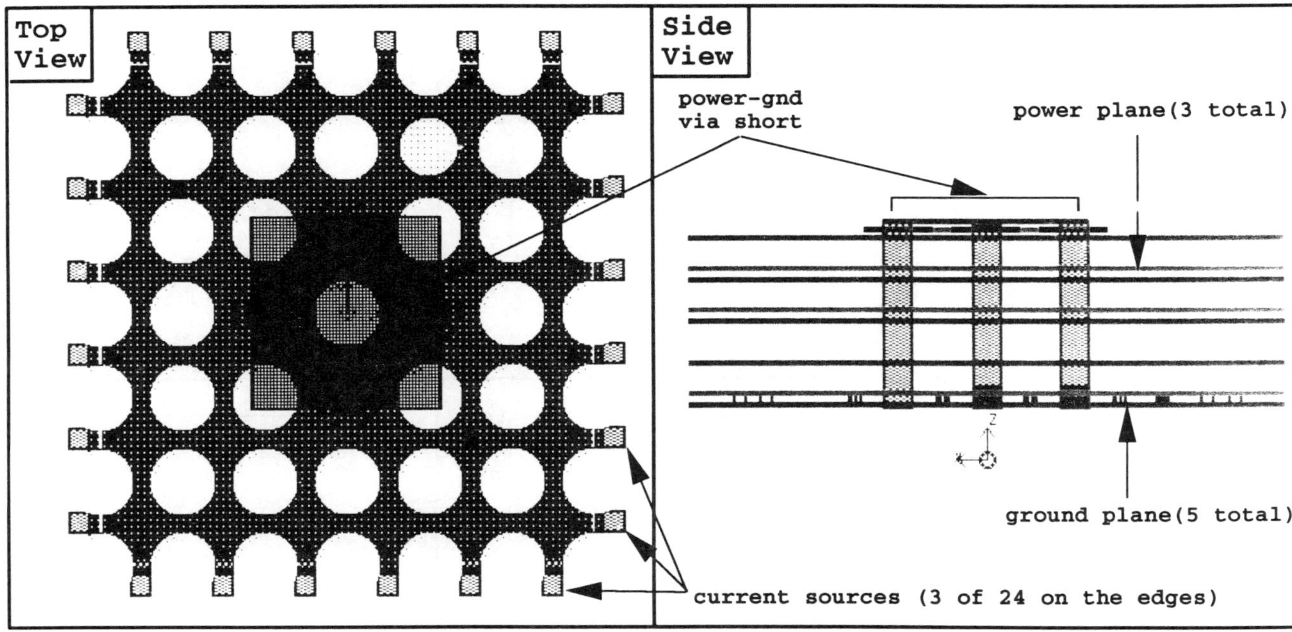

Fig. 8.22 Example of a complex power-distribution system: a meshed printed-circuit-board system having three power planes with current sources at the edges, five ground planes, and nine interdigitated vias connecting the power and ground planes.

In the past, inductance calculations for this type of structure were done in a piecewise manner; that is, by examining the effects of the power / ground planes on the vias and vice versa. However, by applying the method of Section 8.10.1 (based strictly on the structure's stimulus and response), it is possible to represent the entire power-distribution structure as a lumped inductor with one straightforward calculation.

Fig. 8.23 depicts the FDTD-computed voltage and current data for the structure of Fig. 8.22, along with the results of the inductance calculation. Note that the shapes of the *dI/dt* stimulus and the voltage response are similar, indicating the presence of an inductive element linking the two variables. The final plot in the figure is the structure's inductance as given by $L = V/(dI/dt)$. This shows an 83.8 pH average value over the 20% to 80% time window. This period, in which L shows little variation, corresponds to the state where the leading edge of the exciting pulse extends spatially over the entire structure, as discussed earlier. Because the complete PCB under the MCM repeats this 3×3 mm subsection eight times in both lateral directions, the composite inductance of the power distribution system feeding the MCM is approximately 10 pH.

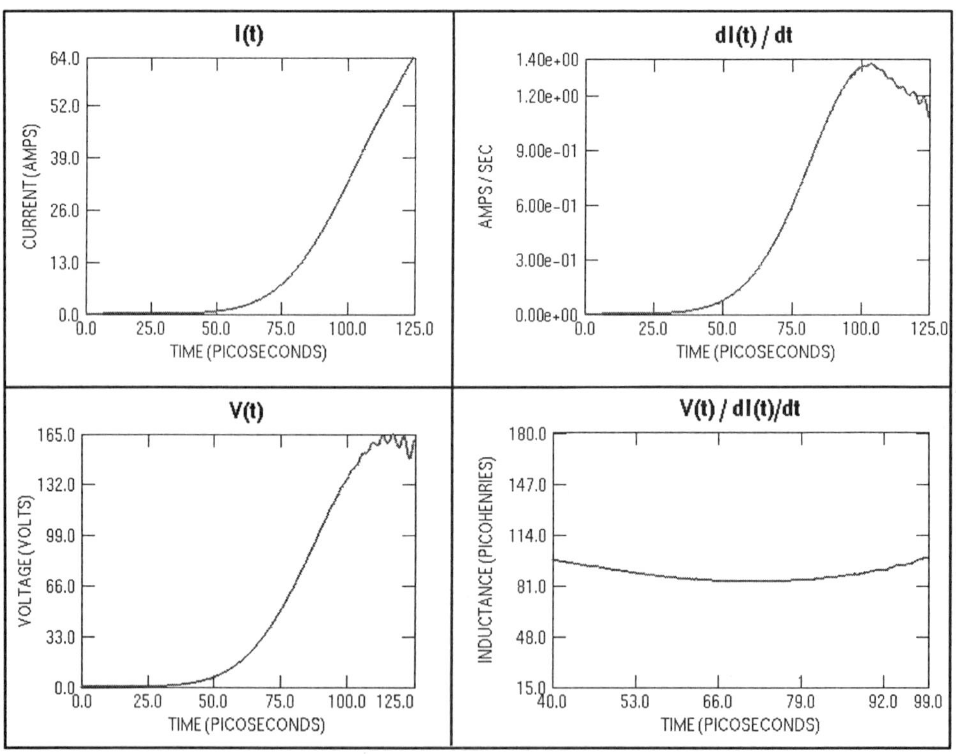

Fig. 8.23 FDTD data used to calculate the inductance of the multiplane power-distribution system of Fig. 8.22.

In the absence of measured inductance data for this structure, we obtained a qualitative feel for the usefulness of this procedure by comparing results for the inductance of the meshed PCB model to a PCB structure without meshed planes. Without the meshed holes present in the planes, we expect the inductance to decrease along with the magnetic flux.

Fig. 8.24 presents a comparison between the modeled inductance results for the meshed-plane and unmeshed-plane PCBs. Removing the signal-via antipad mesh holes from the planes drops the overall structure inductance 8.6% from 83.8 pH to 76.6 pH. Future work will involve experimental validations conducted by packaging test integrated circuits having programmable *dI/dt* generators into the MCM. Upon graphing the measured voltage drop as a function of increasing *dI/dt*, the slope of the graph will be the power-distribution inductance.

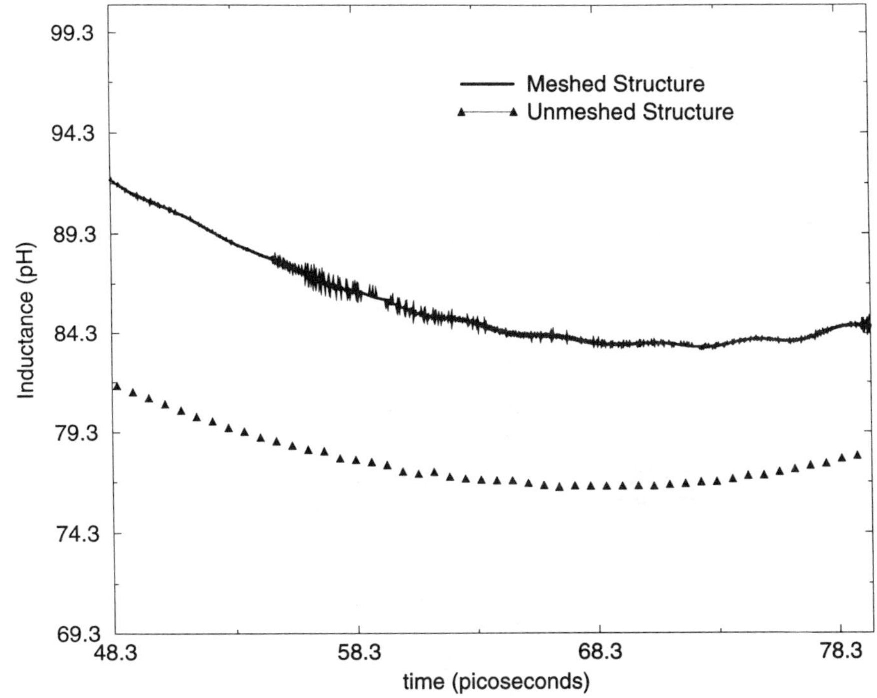

Fig. 8.24 FDTD-calculated meshed-structure versus unmeshed-structure inductances.

This approach appears to be useful for general optimization of the physical configuration of a high-speed digital system to permit conformance with inductance ground rules. For example, in addition to optimizing the geometry of the meshed planes, this method can optimize the height of the vias that transport the signal from the power and ground planes of the PCB into the MCM. An extension of this approach involves developing and validating a dual method for lumped-capacitance modeling and optimization. Overall, this method could be integral in the development of design guidelines for modern digital structures similar to the one in Fig. 8.22.

8.11 CLOCK-DISTRIBUTION NETWORK WITHIN A 2.5-GHZ MULTICHIP MODULE

This section describes an illustrative example of the use of FDTD modeling to design a 2.5-GHz clock-distribution network within an MCM. The clock network provides the timing for data signals traveling through a crossbar switch composed of superconducting Josephson-junction (J-J) logic devices. The design requirements along with the design changes made as a result of the modeling are now described.

8.11.1 Background

Many current high-performance digital systems make use of MCM technologies. MCMs can significantly outperform conventional PCBs in their electrical and thermal properties, reliability, size, and weight. Applications include telecommunications [27], avionics [28], and high-performance computing [29].

Interconnections within a typical MCM include package pins, wire bonds, bump bonds, circuit traces, and vias. The common method of modeling these interconnections is to construct equivalent-circuit models based on RLC parameters, and then compare and modify these models until a reasonable match with measured results is obtained [30]. There are several methods used to estimate the distributed elements including closed-form formulas [31] and two-dimensional software packages such as Linecalc or PISCES [32].

Accurate modeling of the clock distribution within high-speed MCMs is important. The clock signal provides the crucial timing mechanism required for all synchronized circuits. Particular attention must be paid to the shape of the clock-signal waveform as well as the relative timing, or skew, between the gates being synchronized [33, 34].

8.11.2 MCM Model and Clock Requirements

The application presented here involves the distribution of a 2.5-GHz clock within an MCM wherein the VLSI chips contain superconducting J-J logic. A complicating factor is the low impedance of this logic, which poses significant problems in terms of impedance mismatches within the clock-distribution network. (The input impedance of each J-J chip is only 8Ω.) Because of the high clock rate and impedance mismatch, a microwave-engineering approach is used to design candidate clock-distribution networks.

Fig. 8.25 is a simplified diagram of the 13 layers of the MCM. The clock-distribution network is located on the second tungsten metallization layer, W2.

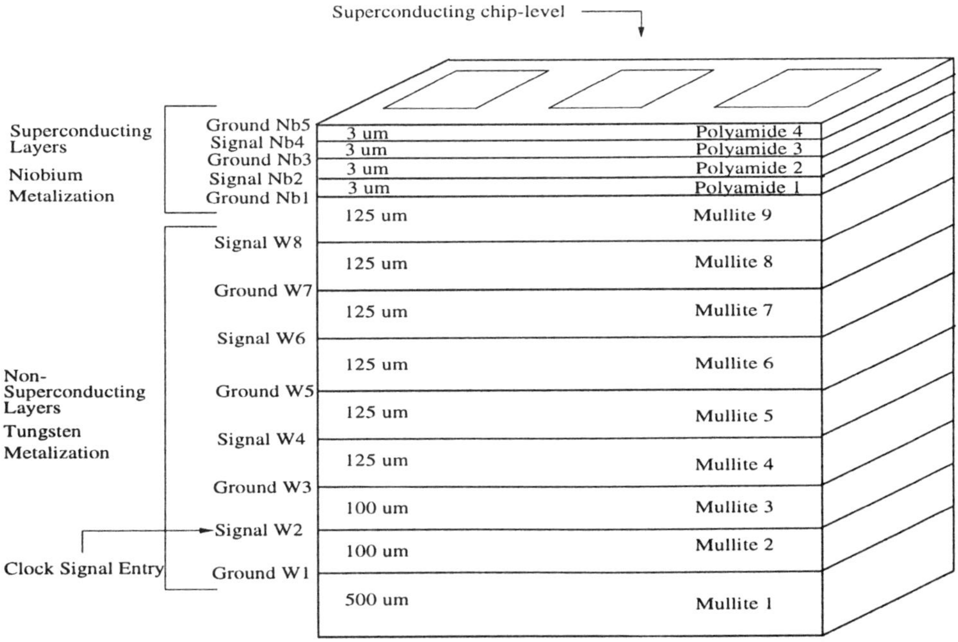

Fig. 8.25 Simplified diagram of the layers of the MCM structure that houses the 2.5-GHz clock-distribution network for the superconducting J-J chips.

The 2.5-GHz clock signal enters the MCM on a 50Ω ribbon cable, and then vias to the W2 clock layer for distribution to several J-J chips located at the top of the MCM.

Several formidable electromagnetic issues must be addressed in the design of the clock-distribution network for this MCM:

1. Multi-gigahertz broadband performance is required to preserve the shape of the clock pulses;
2. Impedance transformation from 50Ω to several 8Ω loads in parallel is needed;
3. The relative delay between the clock signals at successive output ports must be maintained;
4. A compact design is required (10×20 mm available footprint).

Early in the design process, it became apparent that a broadband impedance transformation of this type could not be achieved using a clock-distribution network that fits within the available footprint. The design goal was then to preserve an acceptable level of signal integrity at the clock inputs of the J-J chips. A key requirement is that the absolute amplitude of each clock signal must be the same for each chip. A second need is for a 6-ps time delay of the clock signal between the inputs of successive chips. This corresponds to propagation delays within the switching network as data signals pass between processors and memories. In addition, any reflection of the incident clock signal on the distribution network must be circulated and dissipated outside of the MCM.

8.11.3 Simulation Approach and Clock-Distribution Models

We now review FDTD results for the time-domain waveforms in the proposed clock-distribution networks. In each simulation, the input and output ports of the network are impedance matched through the use of lumped resistors. The excitation applied to the input port is a 1V, 2.5-GHz digital signal having a 55-ps pulse rise time (10% to 90%). The voltages at all ports of the circuit are observed at each time step and converted into the frequency domain by applying the DFT. This yields the *S* parameters for the excitation of the input port.

The initial clock-distribution network analyzed is the 17-port fanout structure shown in Fig. 8.26. This structure is fed by a 50Ω trace that distributes its signal to sixteen 8Ω lines in parallel via a triangular impedance transformer. A characteristic impedance of 8Ω is used for the lines at the output of this structure to match the measured 8Ω input impedance of the J-J chips. The output lines are very closely spaced at 0.1 mm, and vary in length by 0.716 mm due to timing requirements mandated by the 6-ps delay between the chips within the switching network. The substrate material is 0.3-mm-thick Mullite having $\varepsilon_r = 6.4$.

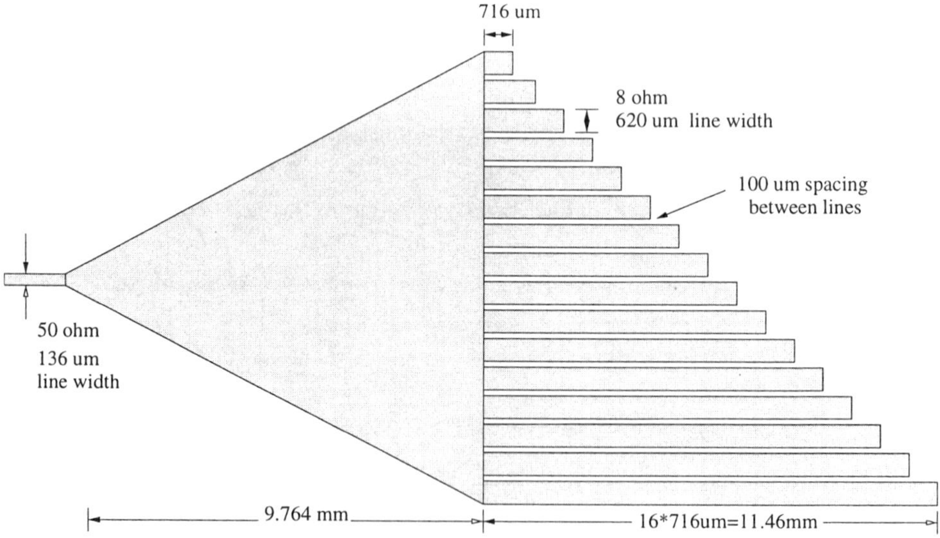

Fig. 8.26 The 17-port fanout clock-distribution network.

The performance of this fanout (see the next section) led designers to reduce the number of 8Ω loads from sixteen to only four. This mitigates the impedance mismatch between the 50Ω input and the multiple 8Ω loads, and reduces the overall size of the circuitry. The reduced number of outputs is made possible by allowing each output signal from the clock-distribution network to feed multiple chip inputs in parallel. As a result, the overall skew between the four output signals increases to 12 ps.

Subsequently, a set of serial-feed designs was investigated. These have the advantage of being smaller than the original fanout geometry. Such a compact design allows for the use of more clocks, which improves the performance and reliability of the switching network. Fig. 8.27 shows the serial-feed design that performs closest to the system requirements. This consists of a 50Ω input line that transitions abruptly to an 8Ω line. The large reflection that occurs as a result of this mismatch is circulated out of the system. The transmitted signal is distributed to four vias separated longitudinally by 1.4 mm to provide the necessary 12-ps delay. Each via carries the clock signal up two layers onto an 8Ω line that matches the input impedance of the J-J chip. The substrate used is 0.35-mm thick Mullite having $\varepsilon_r = 6.4$.

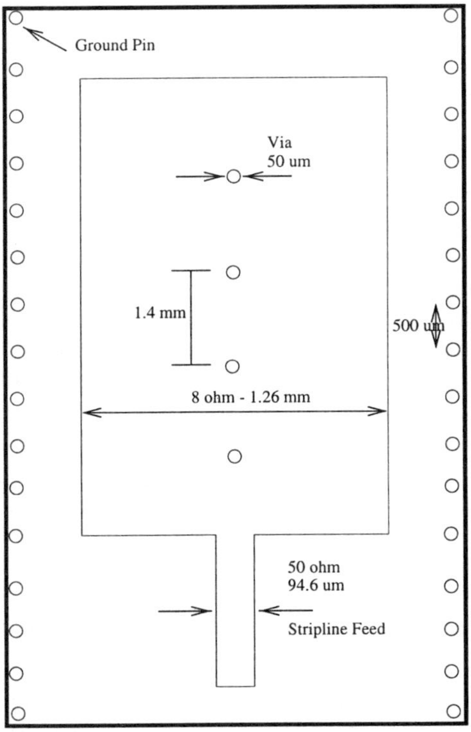

Fig. 8.27 Plan view of the simulation model for the serial-feed design.

Fig. 8.28 illustrates a new class of clock-distribution structures currently being modeled, the sectoral power divider [35]. Such structures were originally conceived for use as compact, broadband (1 – 15 GHz) microwave power dividers. This bandwidth implies good performance past the fifth harmonic of the 2.5-GHz clock signal of interest.

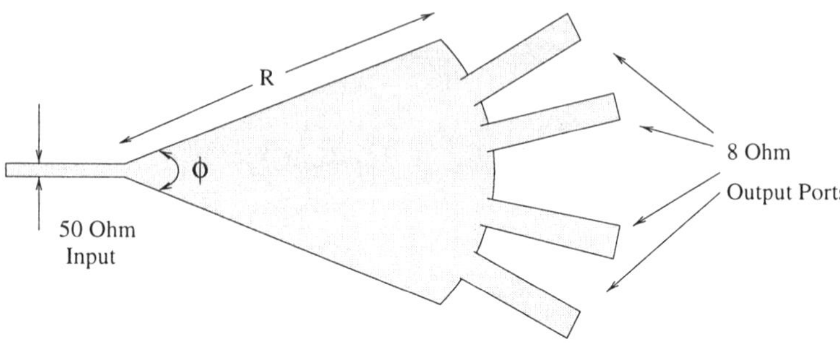

Fig. 8.28 Plan view of the simulation model for the sectoral clock-distribution structure.

The sectoral geometry of Fig. 8.28 ideally generates an equal-amplitude and equal-time-delay signal distribution among the output ports. Required incremental time delays are obtained by using variable lengths of lines between these ports and the vias. The design is optimized by selecting the angle, ϕ, and the radius, R, of the sectoral structure. ϕ is chosen to yield sufficient pitch between the output lines for proper isolation. For the MCM application discussed here, $\phi = 72^\circ$ yields a pitch of about three substrate thicknesses. Setting R to approximately one-fourth wavelength at the clock frequency provides impedance transformation between the 50Ω input line and the 8Ω output lines. For a Dupont 901 substrate ($\varepsilon_r = 5.2$) of thickness 0.3 mm, $R = 13.15$ mm at 2.5 GHz. The input line is 0.125 mm wide to provide a 50Ω input impedance, and the output vias are placed directly on the sector and extend through the upper ground layer to terminated 8Ω output lines.

8.11.4 Simulation Results

Fig. 8.29 shows the FDTD-calculated waveforms for the 17-port fanout structure of Fig. 8.26. The clock input signal is shown as the solid trace in the top graph of this figure. The bottom graph depicts the FDTD-computed output signal waveforms at ports 2, 9, and 17, representing the shortest, median, and longest signal paths, respectively.

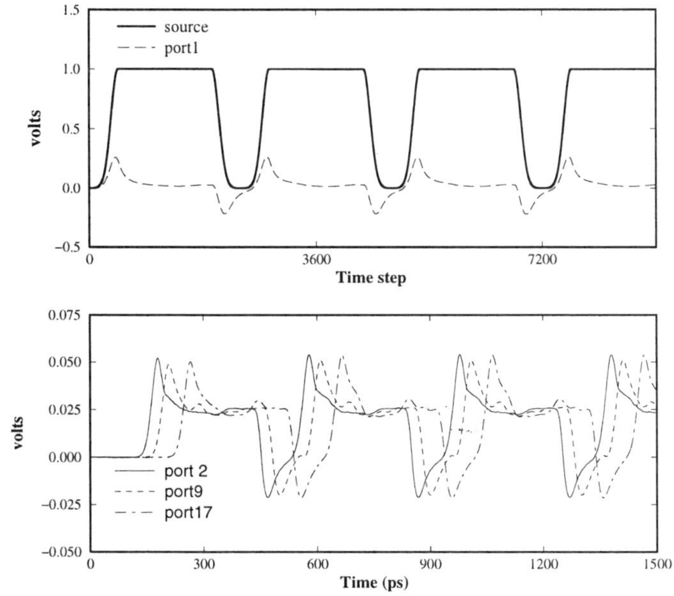

Fig. 8.29 FDTD-calculated waveforms for the 17-port fanout structure of Fig. 8.26.

Fig. 8.29 shows the presence of a 100% overshoot relative to the steady-state value of 25 mV at the leading edge of the signal at all three ports. This overshoot is attributed to two factors: (1) an imperfect impedance transformation through the triangular section from the 50Ω source to the 0.5Ω effective load (sixteen 8Ω loads in parallel); and (2) crosstalk between the tightly coupled output lines. These problems rule out the use of this fanout structure for the final clock-distribution design.

The serial feed structure of Fig. 8.27 has also been modeled using FDTD, and its S parameters computed for frequencies ranging from the first through the seventh harmonics of the clock (2.5 – 17.5 GHz). Modeling results show that power transmission levels from the input to the output ports vary substantially with frequency and from one output port to another, especially for the higher harmonics. Significantly degraded output-pulse rise times as well as leading-edge overshoots are observed. In fact, the degradation of the rise times is such that the relative delay of 12 ps between clock signals no longer exists, and the signals appear to trigger simultaneously. The physical basis of these problems appears to arise from the longitudinal alignment of the low-impedance loads. This causes progressive impedance mismatches to occur along the feed. Coupled with the fact that there is no isolation between the output ports, these mismatches result in multiple reflections that severely degrade both skew and signal integrity. Overall, the negatives uncovered by the modeling rule out the use of the serial feed despite its small footprint.

Finally, FDTD modeling shows that the transmission characteristics of the sectoral structure of Fig. 8.28 are less broadband than reported for the planar structure of [35]. This decrease in bandwidth is attributed to impedance mismatches introduced by the inductive reactance of the vias. For the case under consideration, the vias present a reactance of 8.6Ω to the 55-ps rising edge of the clock, a value that is significant in comparison with the load impedance. Since the circuit is electrically long at higher harmonics of the clock frequency, ringing can persist over the length of a clock period. Although the clock signal may be acceptable in terms of overshoot (<20%), it does not stabilize. Therefore, the viability of the sectoral structure for use as the clock-distribution network remains to be determined.

8.11.5 Systems Perspective

The primary design criterion for the 2.5-GHz clock-distribution network in the MCM discussed here is that it must maintain skew and signal integrity while transforming from a single 50Ω input to multiple 8Ω outputs feeding the superconducting J-J chips. If the clock-distribution network can be placed on one of the upper thin-film layers of the MCM, its design becomes simplified due to significant shrinkage of the circuitry and shortening of the vias. In fact, a potential design indicates that nearly ideal performance in terms of signal integrity and skew can be obtained [36]. However, the two thin-film signal layers are primarily used for the high-density 2Ω and 8Ω data nets. Because of space constraints, it is preferable that no additional circuitry is placed on these layers.

Therefore, it is more desirable from a systems perspective to place the clock-distribution network on one of the lower ceramic layers of the MCM. This complicates the design considerably. In terms of the circuit layout, the large widths of 8Ω lines on this layer increase the circuit's footprint substantially. In terms of signal integrity, the inductive reactance of the vias introduces large reflections for the very fast edge rates considered. Three clock designs were studied for use on the ceramic layer. The fanout design was first examined with the goal of distributing its signal to sixteen 8Ω outputs. Though the fanout design was unsuccessful, a modified version of this structure, the sectoral design, shows promise in satisfying the necessary system requirements. The serial-feed design, motivated by its small size, performed poorly and failed to preserve signal integrity and skew.

Studies of clock distribution for this MCM are ongoing. The results presented here show the important role that simulation can play in the design process. To aid in the final design, a test MCM has been fabricated to allow measurements of the characteristics of the various candidate distributions discussed above. These data will also be used to validate and, where necessary, improve the simulation models. Comparison of the simulated and measured results will surely lead to many more interesting conclusions.

8.12 SUMMARY AND CONCLUSIONS

This chapter summarized several key aspects of the formulation and application of FDTD for modeling contemporary high-speed digital and microwave circuits, including those having active and nonlinear components. Topics reviewed included:

- Norton and Thevenin equivalent circuits "looking into" the FDTD space lattice for the purpose of seamlessly coupling FDTD full-wave electromagnetic simulations to circuit models of linear and nonlinear components and subsystems;
- State-variable representation of the terminal characteristics of lumped-circuit elements;
- Modeling the Schottky diode as a nonlinear two-terminal element with application to the singly balanced microwave mixer;
- Modeling the MESFET transistor as a nonlinear three-terminal element with application to the common-source microwave amplifier to calculate small-signal *S* parameters, large-signal harmonic and intermodulation response, and the potentially destabilizing effects of the packaging;
- Modeling the inductance of complex power-distribution systems in high-speed digital-circuit packages with application to a multi-plane meshed printed-circuit board;
- Modeling the clock-distribution network within a multichip module with application to a crossbar switch employing 2.5-GHz Josephson-junction logic.

We conclude that recent progress in the FDTD method has made it possible to analyze many high-speed digital and microwave circuits of engineering interest in their complete form using full-wave electromagnetic field computations and accounting for relevant circuit nonlinearities. A self-consistent interface with SPICE and related circuit solvers is feasible. In the future, this interfacing of FDTD and SPICE may help to unify the electromagnetics and circuit-analysis disciplines within electrical engineering. Overall, FDTD can provide ultrawideband modeling of nonlinear devices and circuits with a single simulation run. The method is straightforward, robust, and physically appealing. Given these attributes, FDTD modeling of microwave and digital circuits will likely become widely used in industry as the capabilities of workstations and personal computers improve from year to year.

ACKNOWLEDGMENTS

Dr. Houshmand and Prof. Itoh are pleased to acknowledge the original contributions made by Brent Toland, Tian Huang, Chien-nan Kou, Min Chen, and K. P. Ma in the extended FDTD algorithm. Prof. Piket-May is pleased to recognize the numerous contributions of Andy Byers (University of Colorado) and Roger Gravrok (Sequent) for their work on the distributed power system, Paul Vichot (Applied Physics Lab) for his work on the clock distribution system, and Cray Research for its computing support.

REFERENCES

[1] Maas, S. A., *Microwave Mixers*, 2nd ed., Norwood, MA: Artech House, 1993.

[2] Pozar, D. M., *Microwave Engineering*, New York: Addison-Wesley, 1990.

[3] Cendes, Z. C., "EM simulators = CAE tools," *IEEE Spectrum*, Vol. 27, 1990, pp. 73–79.

[4] Horng, T. S., "Extending the three-dimensional spectral-domain approach to hybrid microwave integrated circuits with passive and active lumped elements," in *1994 IEEE MTT-S Int. Microwave Symp. Dig.*, San Diego, CA, May 1994, pp. 709–712.

[5] Guillouard, K., M. F. Wong , V. F. Hanna, and J. Citerne, "A new global finite-element analysis of microwave circuits including lumped elements," in *1996 IEEE MTT-S Int. Microwave Symp. Dig.*, San Francisco, CA, June 1996, pp. 355–358.

[6] Taflove, A., *Computational Electrodynamics: The Finite-Difference Time-Domain Method*, Norwood, MA: Artech House, 1995.

[7] Itoh, T., and B. Houshmand, eds., *Time-Domain Methods for Microwave Circuits*, New York: IEEE Press, 1998.

[8] Sui, W., D. A. Christensen, and C. H. Durney, "Extending the two-dimensional FDTD method to hybrid electromagnetic systems with active and passive lumped elements," *IEEE Trans. Microwave Theory and Techniques*, Vol. 40, 1992, pp. 724–730.

[9] Kunz, K., and R. Luebbers, *The Finite-Difference Time-Domain Method for Electromagnetics*, Boca Raton, FL: CRC Press, 1993.

[10] Toland, B., B. Houshmand, and T. Itoh, "Modeling of nonlinear active regions with the FDTD method," *IEEE Microwave and Guided Wave Lett.*, Vol. 3, 1993, pp. 333–335.

[11] Toland, B., J. Lin, B. Houshmand, and T. Itoh, "FDTD analysis of an active antenna," *IEEE Microwave and Guided Wave Lett.*, Vol. 3, 1993, pp. 423–425.

[12] Piket-May, M., A. Taflove, and J. Baron, "FDTD modeling of digital signal propagation in 3-D circuits with passive and active loads," *IEEE Trans. Microwave Theory and Techniques*, Vol. 42, 1994, pp. 1514–1523.

[13] Thomas, V. A., M. E. Jones, M. Piket-May, A. Taflove, and E. Harrigan, "The use of SPICE lumped circuits as sub-grid models for FDTD analysis," *IEEE Microwave and Guided Wave Lett.*, Vol. 4, 1994, pp. 141–143.

[14] Kuo, C.-N., V. A. Thomas, S. T. Chew, B. Houshmand, and T. Itoh, "Small-signal analysis of active circuits using FDTD algorithm," *IEEE Microwave and Guided Wave Lett.*, Vol. 5, 1995, pp. 216–218.

[15] Kuo, C.-N., R.-B Wu, B. Houshmand, and T. Itoh, "Modeling of microwave active devices using the FDTD analysis based on the voltage-source approach," *IEEE Microwave and Guided Wave Lett.*, Vol. 6, 1996, pp. 199–201.

[16] Chen, M., S. T. Chew, C.-N. Kuo, and T. Itoh, "The extended FDTD method incorporating equivalent-source technique for full-wave analysis of linear and nonlinear microwave circuits," in *1997 IEEE MTT-S Int. Microwave Symp. Dig.*, June 1997, pp. 67–70.

[17] Kuo, C.-N., B. Houshmand, and T. Itoh, "Full-wave analysis of packaged microwave circuits with active and nonlinear devices: An FDTD approach," *IEEE Trans. Microwave Theory and Techniques*, Vol. 45, 1997, pp. 819–826.

[18] Choma, J., *Electrical Networks: Theory and Analysis*, New York: Wiley, 1985, Chap. 6.

[19] Press, W. H., B. Flannery, S. Teukolsky, and W. T. Vetterling, *Numerical Recipes, The Art of Scientific Computing*, Cambridge, U.K.: Cambridge University Press, 1990.

[20] Ortega, J., and W. Rheinboldt, *Iterative Solution of Nonlinear Equations in Several Variables*, New York: Academic Press, 1970.

[21] Massobrio, G., and P. Antognetti, *Semiconductor Device Modeling with SPICE*, 2nd ed., New York: McGraw-Hill.

[22] Johnson, H. W., and M. Graham, *High Speed Digital Design: A Handbook of Black Magic*, New York: Prentice Hall, 1993.

[23] Poon, *Computer Circuits Electrical Design*, New York: Prentice Hall, 1995.

[24] Gravrok, R., Byers, A., and M. Piket-May, "Numerical modeling of inductance for a distributed system," *Proc. IEEE 6th Conf. on Electronics Packaging*, Oct. 27–29, 1997, pp. 83–88.

[25] Gravrok, R., D. Scheid, J. Ficke, K. VanGoor, and V. Rao, "Advanced MCM-Ds break the I/O bottleneck with embedded area-array integrated circuits," *Proc. ISHM MCM Conference*, Denver, CO, April 17, 1996.

[26] Ficke, J., K. Hokanson, C. Berry, and R. Gravrok, "High-bandwidth, low-latency 3-D memory module," *Int. J. of Microcircuits and Electronic Packaging*, Vol. 19, 1996, pp. 369–374.

[27] Licciardi, Peratti, et al., "A multichip module solution for high performance ATM switching," Tech. Rep. 4, CSELT, Torino, Italy, Aug. 1996.

[28] Hagge, "State of the art multichip modules in avionics," *IEEE Trans. Components, Hybrids, Manuf. Technol.*, p. 25, 1992.

[29] Tummala, Potts, and Ahmed, "Packaging technology for IBM's latest mainframe computers," *Proc. 41st Electronic Components and Technology Conf.*, 1991, pp. 682–688.

[30] Sercu and Martens, "A new algorithm for experimental circuit modeling of interconnection structures based on causality," *Comp.*, Vol. 19, 1996, pp. 289–295.

[31] You and Soma, "Analysis and simulation of multiconductor transmission lines for high speed interconnect and package design," *Comp.*, Vol. 13, 1990, pp. 839–846.

[32] Nayak, Hwang, and Turlik, "Calculation of electrical parameters of a thin film multichip package," *Comp.*, Vol. 12, 1989, pp. 303–309.

[33] Yu and Chao, "Signal distortion and EM radiation of a clock distribution net on a multichip module," *Proc. IEEE Int. Symp. on Electromagnetic Compatibility*, 1991, pp. 416–417.

[34] Friedman and Menon, *Theory and Design of Switching Circuits*, Woodland Hills, CA: Computer Science Press, Digital System Design Series, 1975.

[35] Gupta and Abouzahra, eds., *Analysis and Design of Planar Microwave Components*, New York: IEEE Press, 1994.

[36] Vichot, Mix, Schoenborn, Dunn, and Piket-May, "Numerical modeling of a clock distribution network for a superconducting MCM," *Proc. IEEE Symp. on Components, Manufacturing, and Packaging Tech.*, Feb. 1998.

Chapter 9

Physics-Based Modeling of Millimeter-Wave Devices

Samir M. El-Ghazaly

9.1 INTRODUCTION

Semiconductor device simulation is a powerful tool for understanding, developing, and designing optimal devices. Over the last three decades, physics-based device models were developed based on quasi-static electric forces derived from Poisson's equation. Recently, the importance of using a complete representation of the electromagnetic fields inside high-frequency semiconductor devices has become more evident [1–8]. New models and simulation tools that combine both electromagnetic waves and semiconductor device characteristics are currently being developed.

The development of an accurate physics-based semiconductor device simulation is a complex procedure that involves three steps:

1. Identification of the distinct physical phenomena relevant to the device operation (physical modeling);
2. Development of the mathematical formulation that satisfactorily describes the physical phenomena (mathematical modeling);
3. Development of the numerical algorithms and computer programs to solve the set of equations developed in step 2 with the required accuracy.

Throughout this process, approximations and simplifications may be introduced. The validity and the applicability of the computed results are primarily limited by the approximations introduced in steps 1 and 2. Step 3 can generally be implemented in a straightforward manner since both numerical techniques and modern computers are well developed.

This chapter discusses recent progress in coupling dynamic electromagnetic wave solutions with charge-transport models to achieve effective physics-based simulations of devices operating at millimeter-wave frequencies. Sufficient details regarding the physical phenomena, mathematical formulations, numerical algorithms, and results are provided to permit an assessment of the state of the field.

9.2 SPECIAL CONSIDERATIONS FOR HIGH-FREQUENCY DEVICE SIMULATIONS

RF circuits and devices perform basic operations such as the generation, amplification, modulation, frequency conversion, and switching of electrical signals. For a semiconductor device operating in the millimeter-wave frequency range, these operations can be viewed as the interaction between electromagnetic waves and the electron flow within the device. The electron flow acts as a means of converting energy gained from a direct current (dc) source to a high-frequency signal (i.e., the desired electromagnetic wave).

At low alternating-current (ac) or radio frequencies, this conversion process is normally analyzed using an equivalent-circuit approach, discussed in Chapter 8. For historical reasons, this approach was extended to the microwave and millimeter-wave ranges. For example, an amplifier is classically analyzed as follows:

- The dc bias is obtained in a quasi-static manner. In this step, the physical characteristics of the electron flow (or the solid-state device controlling the flow) are introduced. An ac equivalent circuit, or model, is then developed based on the dc solution.
- The amplifier gain characteristics are developed based on the equivalent circuit, and an amplifier gain model is developed. This model is then integrated with the ac or RF models of the other system elements.

While the details of the actual analysis may differ from one implementation of the equivalent-circuit method to another, the basic assumption underlying each implementation is the same. Namely, there is postulated to be a separation of the physics of electron flow from the physics of electromagnetic wave propagation. Thus, the gain-producing mechanism and its model are developed without taking electromagnetic waves into consideration. While such an approach is acceptable at RF and many microwave frequencies, it is not clear that it can always be applied at higher frequencies, especially in the millimeter-wave range. Some of the reasons are:

- Critical device or circuit dimensions could be comparable to the electromagnetic wavelength;
- The electron transit time could be comparable to the electromagnetic oscillation period;

- Relaxation times governing electron-transport phenomena could be comparable to the electromagnetic oscillation period;
- For large-signal conditions, the ac amplitude is a significant fraction of the dc bias. Then, the ac electric and magnetic fields interact nonlinearly with the electron flow and significantly perturb it relative to the quasi-static case.

A fundamental thesis of this chapter is that accurate design of millimeter-wave devices requires technologists to accurately account for these possibilities. Numerical simulations that combine the physics of the internal device fields due to the dc bias and the electromagnetic wave must be developed in pursuit of this goal. Examination of the large-signal nonlinear problem discussed above reinforces this conclusion, as this problem inherently involves a multifrequency device-circuit interaction. Frequency conversion due to the nonlinearity can produce significant harmonic content that affects system performance. Thus, understanding the operation of a 100-GHz mixer, for example, requires accurate modeling over at least the first several harmonics to frequencies approaching 1 THz. Further, a recent trend has been to move away from circuit models of the broadband behavior of the linear components of the system toward full-wave electromagnetic models. The problem of coupling a "no-wave" nonlinear device model to such "full-wave" circuit models needs thorough investigation.

As another example, consider the field-effect transistor (FET). A wide variety of physics-based models have been developed for this device [9 – 25]. These are grouped into two main categories: particle-based models and fluid-based models. The first category is represented by Monte Carlo techniques [9 – 13]. The second category includes analytical models [14, 15], drift-diffusion based simulators [16, 17], and hydrodynamic simulators [18 – 25]. These models involve different approximations of the electron transport phenomena in FETs. However, they are all based on quasi-static electric-field solutions derived from Poisson's equation, and none of them includes a full-wave dynamic-field electromagnetic model.

Given this basis, the calculated characteristics of FETs should be labeled "quasi-static." Such a characterization may be acceptable at relatively low frequencies. However, the validity of quasi-static thinking declines as the operating frequency increases. As frequencies approach the millimeter-wave range, the FET width becomes comparable to the wavelength. In this regime, the transistor can no longer be treated as a dimensionless point or a lumped element. [26, 27]. High-frequency aspects including distributed effects, propagation delays, electron transit time, parasitic elements, and discontinuity effects become important. In addition, efficient coupling of the electromagnetic energy from the device to the circuit (the matching problem) is not trivial. Analyzing matching using a simple circuit concept that neglects the distributed effects of the feeding and output lines can lead to poor results. Matching becomes even more complicated when the characteristic impedances of the transmission lines in use are either undefined or not uniquely defined if the lines are not operated in a pure TEM mode. Overall, we need to go back to the physical modeling step and thoroughly investigate the new aspects of device operation.

9.3 THE ACTIVE TRANSMISSION LINE MODEL

Electromagnetic wave phenomena in semiconductor devices can be modeled using the active transmission line approach. This approach is useful for analyzing distributed effects on semiconductor devices, primarily FETs. It can be segmented into three fundamental steps [26]:

1. Development of a small-signal equivalent-circuit model for the semiconductor device;
2. Inclusion of the equivalent circuit developed in step 1 in a transmission-line model representing the waveguiding structure in which the device is embedded;
3. Development and solution of a system of linear equations for the propagation characteristics of the composite active transmission line.

The small-signal equivalent-circuit is normally obtained by perturbing the steady-state dc solution. It must be linearized to avoid any complications in step 3. In developing the transmission line model, the quasi-TEM approach is employed to achieve a resistance-inductance-capacitance (RLC) model. Since these RLC elements are independent of frequency, possible dispersion characteristics of the guiding structure are not easily incorporated. This can be justified for some FETs where the gap between the electrodes is in the order of few microns [26–29]. For other devices having larger gaps or operating at higher frequencies, this can lead to significant errors, especially when the model is used over a wide frequency band. We may try to include the dispersion effects in the model by using frequency-dependent RLC elements and empirically derived formulas. Though the use of empirical formulas is acceptable in computer-aided design, this is not the case in theoretical investigations undertaken to assure device potential. This also leads to nonlinear effects that complicate step 3 .

In summary, the active transmission line approach is good for circuit-oriented applications. However, it is not useful for providing deep physical insight into device operation, or analyzing the interactions between the various physical phenomena.

9.4 EFFECTS OF THE AC FIELDS ON THE DEVICE RESPONSE

The question we ask here is: Do large-amplitude ac fields have significant effects on the device response, or are the effects of such fields randomized through the various scattering mechanisms? This issue is investigated in this section.

From the viewpoint of this study, semiconductor device models are classified into three main categories according to two ratios. The first is the ratio between the electron's transit time inside the device, τ_t, to the longest relaxation time of the physical phenomena of interest inside the device, τ_r. The second is the ratio between the device width, W, and the wavelength, λ. When $\tau_t >> \tau_r$, a quasi-static semiconductor model can be used. This means that the state of the electron gas is completely determined by the local dc electric field. For this case, Poisson's equation is an adequate representation of the electromagnetic field physics within the device except when W approaches λ and distributed field effects have to be considered. When $\tau_t \approx \tau_r$, a dynamic semiconductor transport model should be used. When $\tau_t << \tau_r$ or when a characteristic length of the system approaches that of the wave packet, quantum mechanical models should be used.

The suitability of the electromagnetic model can similarly be judged. When the electromagnetic oscillation period, T, is much larger than τ_r (i.e., $\omega\tau_r << 1$), a quasi-static model can be used. By a quasi-static model, we mean that the ac device response is based on dc current / voltage (I/V) characteristics, which are in turn based on Poisson's equation. When $\omega\tau_r \approx 1$, a dynamic electromagnetic model should be used. By a dynamic model, we mean that the effect of the large ac signal should be considered while developing the device response. One should recognize that this is a separate issue from the wave propagation effects that are judged by considering the ratio W/λ.

Let us consider a modern high-frequency FET as an example of how to decide which model is needed. Here, we recognize that scattering by longitudinal optical phonons is the dominant scattering mechanism in MODFETs, and thus a relaxation time in the order of 1 ps is expected. For an operating frequency of 100 GHz, $\omega\tau_r \approx 0.6$. Hence, a dynamic electromagnetic model is needed in addition to the dynamic semiconductor transport model.

9.5 FET SIMULATIONS BASED ON ELECTROMAGNETIC AND ELECTRON-TRANSPORT MODELS

Having reviewed general types of semiconductor device models and their ranges of applicability, we next consider a comprehensive modeling approach based on both electron transport and electromagnetic wave interactions. With this approach, the moving carriers within the device become a source of electromagnetic fields. In turn, these fields influence the carrier transport. This coupling is believed to have an important effect on the operation of millimeter-wave electronic devices. Previously, El-Ghazaly et al. [1] used a combination of direct solutions of Maxwell's equations and Monte Carlo models of photocarrier transport to study the behavior of photoconductive switches. However, for transistor modeling, several tens of picoseconds of simulation time are needed to obtain both the small-signal and the large-signal responses. Therefore, an alternative to the computationally intensive Monte Carlo approach is preferred.

This section describes the analysis of electron / electromagnetic wave interactions in FETs having submicron gates. The analysis is a time-domain simulation that couples a three-dimensional, full-wave solution of Maxwell's equations to an active-device model based on the moments of Boltzmann's transport equation. The coupling between the two models is established by using the fields obtained from the solution of Maxwell's equations in the active-device model to calculate the current densities inside the device. These current densities are then used to update the electromagnetic fields. The problem is solved in the time domain because the carrier-transport processes are highly nonlinear.

The physical model used in this work allows a flexible description of the device along with the appropriate representation of the simulation parameters. It is capable of representing short-gate effects such as nonisothermal transport and nonstationary electron dynamics, as well as effects related to electromagnetic wave propagation. The application of this model to electronic structures such as MODFETs having various material profiles and boundaries is straightforward. It should be noted, however, that as for all models, the accuracy of the results is significantly affected by the precision of the available device-geometry and material data.

9.5.1 Active Device Model

The active device model is based on the moments of Boltzmann's transport equation obtained by integration over momentum space. The integration results in a strongly coupled and highly nonlinear set of partial differential equations called the conservation equations [30–33]. These equations provide a self-consistent time-dependent solution for carrier density, carrier energy, and carrier momentum and are given by:

Current continuity

$$\frac{\partial n}{\partial t} + \nabla \cdot (n\vec{v}) = 0 \tag{9.1}$$

Energy conservation

$$\frac{\partial (nw)}{\partial t} + \nabla \cdot (n\vec{v}w) = qn\vec{v} \cdot (\vec{E} + \vec{v} \times \vec{B}) - \nabla \cdot (nkT\vec{v}) - \frac{n(w - w_0)}{\tau_w} \tag{9.2}$$

Momentum conservation

$$\frac{\partial (np_x)}{\partial t} + \nabla \cdot (n\vec{v}\,p_x) = qn\left[E_x + (\vec{v} \times \vec{B})_x\right] - \nabla(nkT) - \frac{np_x}{\tau_p} \tag{9.3}$$

where n is the electron density, t is time, $\vec{v}$ is the electron velocity, w is the electron energy, q is the electronic charge, k is Boltzmann's constant, T is the electron temperature, w_0 is the equilibrium thermal energy, p_x is the x-component of electron momentum, and τ_w and τ_p are the energy and momentum relaxation times, respectively. Similar equations are written for the momentum in the other directions. The electronic current-density distribution, $\vec{J}$, inside the active device at any time t is given by:

$$\vec{J}(t) = -q\,n(t)\,\vec{v}(t) \tag{9.4}$$

Considering the scale of the problem, we choose to solve the hydrodynamic equations in their complete form. Unlike most reported simulations, we do not simplify the momentum-conservation equation by neglecting the time and spatial dependencies of the electron. This simplification is equivalent to assuming that the electron momentum is able to adjust itself to a change in the electric field within a very short time. While this assumption is justified for long-gate devices because of the negligible effects of the overshoot phenomenon, it leads to inaccurate estimations of the internal charge distributions and microwave characteristics for submicron-gate devices [34]. The present work avoids this problem.

Another simplifying assumption that is eliminated here is the constant-effective-mass approximation. The variations in effective mass with respect to time, space, and electron energy are all included in the models reported here. The transport parameters in the energy and momentum equations are energy dependent and obtained from steady-state Monte Carlo simulations. The low-field electron mobility, m_{e_0}, is taken as

$$m_{e_0} = \frac{8000}{1 + \sqrt{N_d / 1 \times 10^{17}}} \tag{9.5}$$

where N_d is the doping density.

9.5.2 Electromagnetic Model

The physics of electromagnetic wave propagation can be completely characterized by solving Maxwell's equations. These are first-order, linear, coupled partial-differential equations relating the electric and magnetic field vectors, current densities, and charge densities at any point in space at any time. They must be, however, supplemented by the constitutive relations of the medium. Assuming a uniform, linear, isotropic medium for both the electric and magnetic fields, Maxwell's equations are given as

$$\nabla \times \vec{E} = -\mu \frac{\partial \vec{H}}{\partial t} \tag{9.6}$$

$$\nabla \times \vec{H} = \varepsilon \frac{\partial \vec{E}}{\partial t} + \vec{J} \tag{9.7}$$

where $\vec{E}$ is the electric field, $\vec{H}$ is the magnetic field, $\vec{J}$ is the conduction current density, and ε and μ are, respectively, the permittivity and the permeability of the medium. The conductivity of the medium is obtained using the active device model.

9.5.3 Coupling the Two Models

The electromagnetic field model simulates the evolution of electric and magnetic fields due to moving free charges. On the other hand, the hydrodynamic model calculates the response of the charges to an applied field. Hence, coupling the two models results in the simulation of the overall high-frequency characteristics of the semiconductor device. The link between the two models is established by properly transforming the physical parameters (fields and current densities) from one model to the other. The simulation starts by obtaining the steady-state dc solution using Poisson's equation and the semiconductor device model. This solution is then used in the ac analysis as initial values. Then, the ac excitation is applied. Maxwell's equations are solved for the electric and magnetic fields, which are then used in the semiconductor model to find the current density. This process is then iterated. The coupling procedure and the mesh arrangements for the various quantities are discussed in detail in the following sections.

Initializations

The steady-state dc solutions for the electric field, current density, carrier density, carrier velocity, carrier energy, and transport parameters are obtained from the semiconductor model by solving Poisson's equation and the continuity, energy balance, and momentum balance equations. These dc solutions serve as the initial values inside the active device for the time-domain model. Since the dc solution is performed in two dimensions (the x-y plane) and the ac analysis is performed in three dimensions, all of the sections of the semiconductor device along the third dimension (the z direction) are initialized with the same dc values for the above quantities.

To complete the initialization, the magnetic field is specified using the following approach. At steady-state, Maxwell's equations are given by

$$\nabla \times \vec{E} = 0 \tag{9.8}$$

$$\nabla \times \vec{H}_{\mathrm{dc}} = \vec{J}_{\mathrm{dc}} \tag{9.9}$$

where $\vec{H}_{\mathrm{dc}}$ is the dc magnetic field, and $\vec{J}_{\mathrm{dc}}$ is the dc current density. Equation (9.9) suggests that it is not necessary to find the dc distribution of the magnetic field since the dc current density carries the information about the dc magnetic field distribution. This approach can be implemented as one of the measures taken to reduce the computation time when the effects of $\vec{H}_{\mathrm{dc}}$ are not of interest in the simulation.

Time-Domain Solution

After completing the above initializations, the ac excitation is applied. The mesh for the electromagnetic fields is extended along the z axis to each side of the semiconductor device. The extended regions are assumed to be comprised of a passive transmission line. At the device input, the passive transmission line section allows the correct mode of the excitation wave to develop before it is fed to the active device. At the device output, the transmission line section is used for absorbing the electromagnetic wave.

The time-domain distributions of the electric and magnetic fields are obtained using Maxwell's equations. Using (9.9), the curl of $\vec{H}$ is written as:

$$\nabla \times \vec{H} = \nabla \times \vec{H}_{\text{ac}} + \vec{J}_{\text{dc}} \tag{9.10}$$

The electric field is then specified as:

$$\frac{\partial \vec{E}}{\partial t} = \frac{1}{\varepsilon}(\nabla \times H_{\text{ac}} + \vec{J}_{\text{dc}} - \vec{J}_{\text{tot}}) \tag{9.11}$$

where $\vec{J}_{\text{tot}}$ is the total current density obtained from the active device. The magnetic field is given by:

$$\frac{\partial \vec{H}}{\partial t} = -\frac{1}{\mu}\nabla \times \vec{E} \tag{9.12}$$

Equations (9.11) and (9.12) give the electric and magnetic field distributions at each time step. These new values are used by the semiconductor model to update the current density at the same time step. These current densities are fed back to the Maxwell's equations again to calculate the electric and magnetic fields in the following time step. This process of updating the fields and the current densities marches in time in response to the applied excitation, and continues for the time needed to study the device behavior. The flowchart in Fig. 9.1(a) outlines the calculation steps and the coupling procedure between the semiconductor and the electromagnetic models. The flowchart in Fig. 9.1(b) describes the sequence of operations in the active device model.

9.6 IMPLEMENTATION

Both finite-difference and finite-element schemes were used in the semiconductor device simulations. Although the finite-element method provides a flexible way of modeling complex geometries and requires fewer nodes than the finite-difference scheme, it requires a matrix re-ordering algorithm to deal with the generated complex matrix structure. This process reduces the speed of the solution. On the other hand, the finite-difference method lends itself naturally to the simple rectangular geometries generally considered for semiconductor device simulations. Also, finite-difference schemes are easier to formulate and considerable information is available on their stability and convergence properties [35]. In this work, several finite-difference techniques such as the upwind and Lax methods are used in conjunction with the basic finite-difference formulation to achieve stable and accurate solutions. The space increments are adjusted to satisfy both Debye length criteria for the semiconductor simulation and wavelength-resolution criteria for the electromagnetic simulation to achieve a well-resolved and stable solution. The same space increments are used in the two models and are identical in all three dimensions.

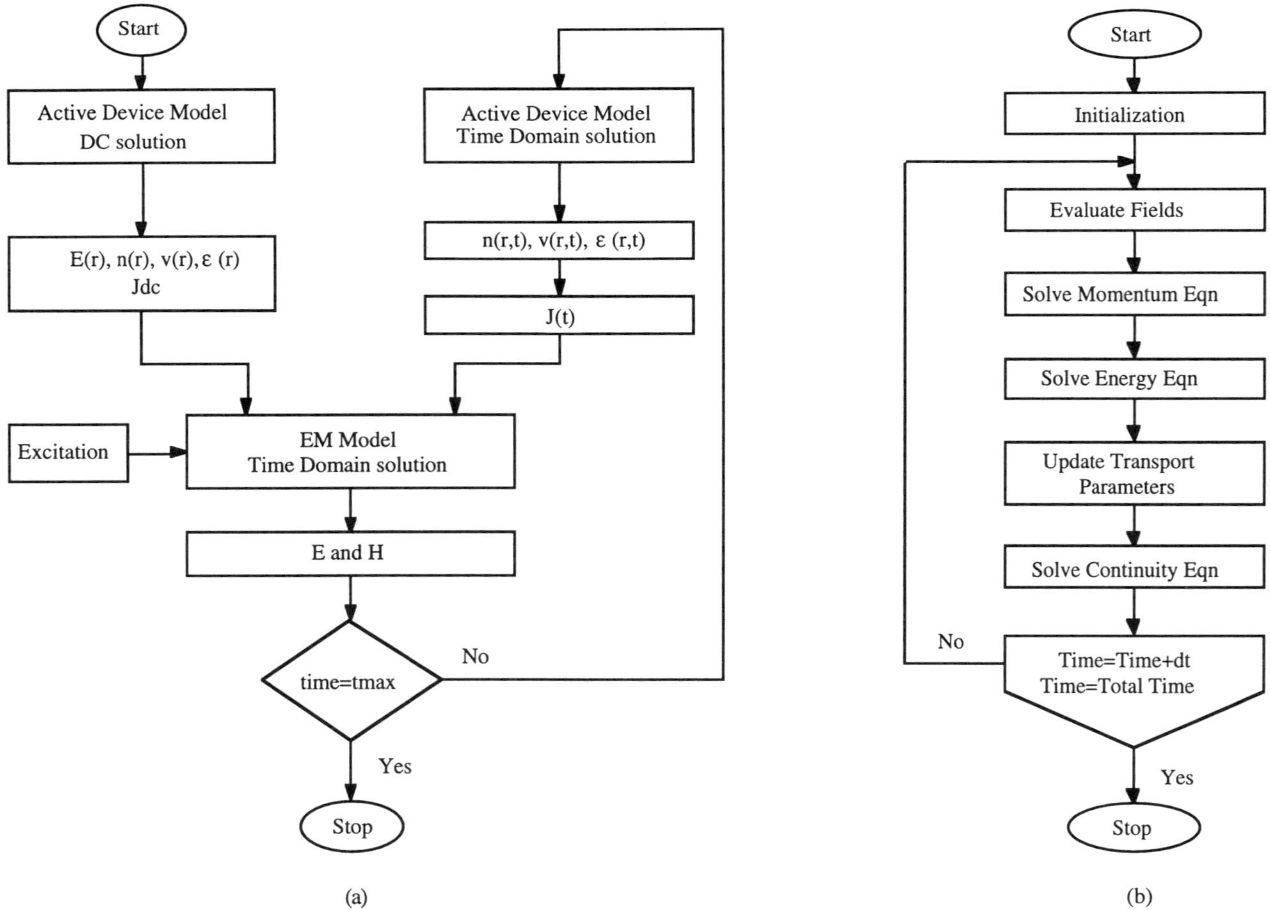

Fig. 9.1 Calculation flowcharts for the combined electromagnetic and solid-state (CESS) simulator: (a) calculation steps and coupling procedure between the semiconductor and the electromagnetic models; (b) sequence of operations in the active device model.

9.6.1 The Simulated Structure

The two-dimensional structure used in the charge-transport simulation of a millimeter-wave FET is shown in Fig. 9.2. The device consists of three basic layers: the active layer, the buffer layer, and the substrate. The active layer is a heavily doped n-type region where high-mobility electrons are generated. Due to the fact that the fields and currents inside the device are confined to a small region, and due to the large difference in the dimensions of the three layers, only a small region of the device needs to be simulated. This region encloses the gate electrode and small portions of the source and drain electrodes, and extends well into the buffer layer. Normally, the source is connected to ground. The dc operating point of the device is defined by applying dc voltages at the gate and drain. The metallic gate creates a Schottky-diode contact with the semiconductor that suppresses the gate current when it is reverse-biased. On the other hand, the drain and source contacts are ohmic, presenting virtually zero resistance to the carriers entering or leaving the device.

The three-dimensional structure used in the full-wave electromagnetic simulation is shown in Fig. 9.3. In this figure, the semiconductor part is considered to be a lossy dielectric having a nonlinear conductivity that is dependent upon time, space, and the electromagnetic field. This nonlinear conductivity is obtained from the solution of the active device model. The electrodes shown in Fig. 9.3 represent the source, gate, and drain terminals. The upper space above the transistor is air.

9.6.2 Solution of the Hydrodynamic Equations

Equations (9.1) to (9.3) are coupled, highly nonlinear, partial differential equations. To decouple these equations in time, a staggered-mesh finite-difference scheme is used. The scalar variables (potential, carrier density, carrier energy, and transport parameters) are defined at the basic nodes, whereas the vector quantities (carrier velocity, electric field, and current density) are defined at complementary nodes located between adjacent basic nodes. This arrangement separates the components of a vector variable and is convenient for decoupling the variables and approximating the space derivatives using finite differences. The solution is obtained in a self-consistent evaluation of the three equations in conjunction with Maxwell's equations.

We find that the order in which these equations are solved is critical to the stability of the solution. As shown in the flowchart of Fig. 9.1(b), the change in carrier momentum is computed first. The carrier density and energy equations are then solved using the updated distribution of carrier velocity. The reason is that the relaxation time of the electron momentum is about one order of magnitude shorter than the dielectric and energy relaxation times. If the order of the solution is reversed, large changes in momentum are introduced at every following time step leading to potential numerical instabilities in the solution of the velocity equation.

All of the equations are represented by explicit finite-difference schemes. Since the time step is determined according to the stability condition of the E and H field updates

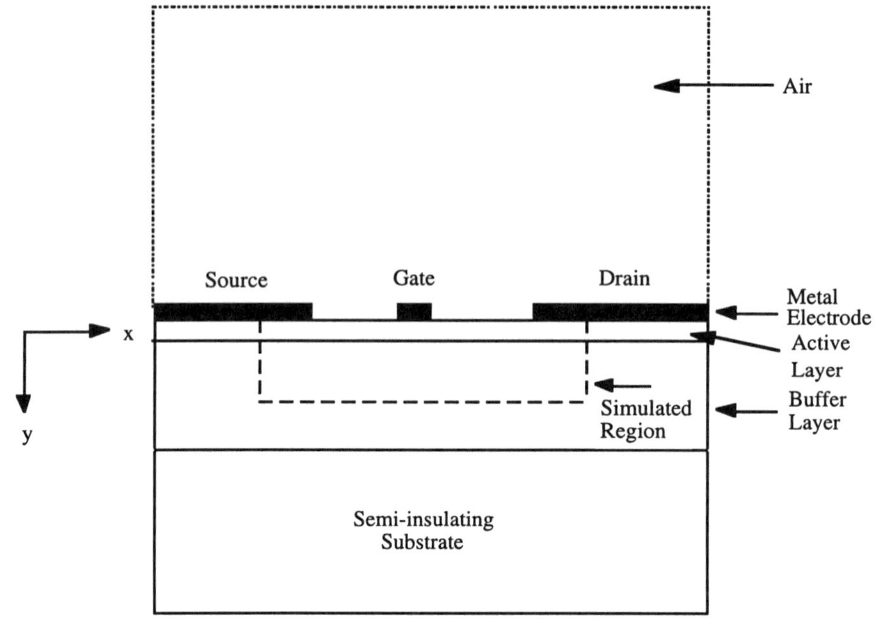

Fig. 9.2 Two-dimensional structure used in the charge-transport simulation of a millimeter-wave FET.

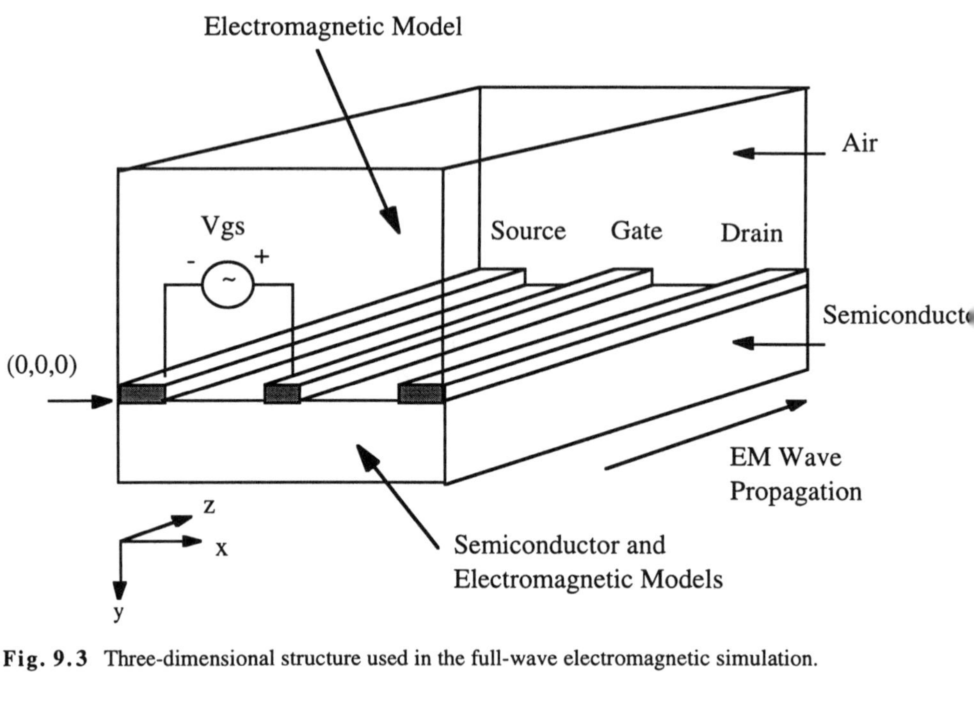

Fig. 9.3 Three-dimensional structure used in the full-wave electromagnetic simulation.

($\Delta t \approx 10^{-17}$ s) [36], there is no gain in using implicit or semi-implicit finite-difference formulations. The best accuracy and stability in solving the continuity and energy equations is obtained using the upwind scheme where the direction of electron velocity dictates the correct differencing. Because all of the inertia terms in the momentum equation are taken into account, extra attention is given to its solution method. The velocity-gradient terms in the momentum equation make a direct finite-difference solution sensitive to small fluctuations and prone to instability. To overcome this problem, a scheme that utilizes average values over two time and space steps is used. This scheme, called the Lax method [37], furnishes a stable solution within the required accuracy.

9.6.3 Solution of Maxwell's Equations

The solution of Maxwell's equations (9.11) and (9.12) is obtained using a standard three-dimensional FDTD Yee mesh [38] employing second-order differencing. The electrodes are taken as perfect conductors, implying zero tangential electric fields at their surfaces.

ABCs are used to simulate the infinite space surrounding the structure. The choice of ABC is key to the numerical stability of the finite-difference scheme. It has been observed that, for long-time simulations or for absorbing boundaries located where the field strength is high, instability is likely to occur if the absorption is not adequate [39]. This phenomenon is primarily observed when using first-order ABCs. A discussion of some of the most common analytical second-order ABCs is provided in [40], where it is concluded that Higdon's ABC [41] gives the best overall efficiency for waveguiding structures. Higdon's ABC is based on the decomposition of an arbitrary wave into a sum of plane waves having different angles of incidence θ_i. For second-order Higdon ABCs, two velocities are chosen according to the relation $v_i = c/\cos\theta_i$, where c is the propagation velocity of the wave inside the medium. In the present work, one of the velocities is related to the effective dielectric constant of the medium which corresponds to a 0° angle of incidence [42]. For the other velocity, however, numerical experiments were performed to obtain the optimum absorption for the given geometry and simulation parameters. It has been observed that the wave reflections generated by Higdon's ABC are an order of magnitude less than those from the first-order ABCs.

Another FDTD model has been developed based on the PML ABC concept [43]. It is used for generating results on transistor discontinuities to be presented later in this chapter.

9.6.4 Parallel-Processing Implementation

Parallel implementation of these numerical techniques is necessary due to the inherent computational intensity of the model. This intensity is a result of the limitations on the space and time increments and the long simulation time needed to obtain steady-state results. In the present work, the simulations were performed on the MasPar massively parallel computer. With proper mapping of the variable arrays on the computer's processing elements and with efficient FORTRAN 90 coding, the overall scheme provides a relatively fast method for obtaining device response compared to a similar serial code on a vector supercomputer.

9.7 RESULTS FOR A MILLIMETER-WAVE MESFET

The device considered in this simulation is a 0.2-μm gate-length MESFET having the transverse geometry of Fig. 9.2. The MESFET's active layer is assumed to be 0.1 μm thick and doped to 2×10^{17} cm^{-3}. An abrupt transition is assumed between this layer and the semi-insulating buffer. The device operates at 300°K and is characterized by the following additional parameters:

Schottky barrier height = 0.8 V	Drain and source contacts = 0.5 μm
Gate-source separation = 0.4 μm	Gate-drain separation = 0.5 μm
Device thickness = 0.4 μm	Device width (W) = 250 μm
Device length (L) = 2.1 μm	DC gate-source voltage (V_{gs}) = -0.5 V
DC drain-source voltage (V_{ds}) = 5 V	AC gate-source voltage = 0.1 V

A uniform $80 \times 26 \times 31$ Cartesian space lattice is used to model the three-dimensional MESFET geometry of Fig. 9.3 in the full-wave FDTD electromagnetic simulation.

9.7.1 DC Steady-State Results

The dc steady-state distributions are obtained by solving the active device model only. Here, the state of the MESFET is represented by its spatial distribution of potential, carrier density, and carrier energy.

Fig. 9.4(a) depicts the computed equipotential lines inside the MESFET in steps of 0.5 V. This figure shows that most of the applied voltage is dropped in the channel under the gate. Fig. 9.4(b) plots contours of the carrier concentration, n/N_d, in steps of 0.1. This figure shows the depletion region under the gate where the field is at a maximum value. The corresponding energy contour lines are shown in Fig. 9.4(c) for energies from 0.1 to 0.55 eV. These indicate a high-energy region between the gate and the drain. Here, the electric field is strong, especially at the gate edge where the energy exceeds 0.55 eV.

This two-dimensional model is used to estimate the device gain at the bias points as shown in Fig. 9.5. From this graph, the 3-dB bandwidth is found to be about 90 GHz, and the maximum frequency of oscillation is 170 GHz. Therefore, the operating frequency is chosen to be 80 GHz for the ac analysis. Note that no parasitic elements are included in this estimate. Such elements lower the cutoff frequency.

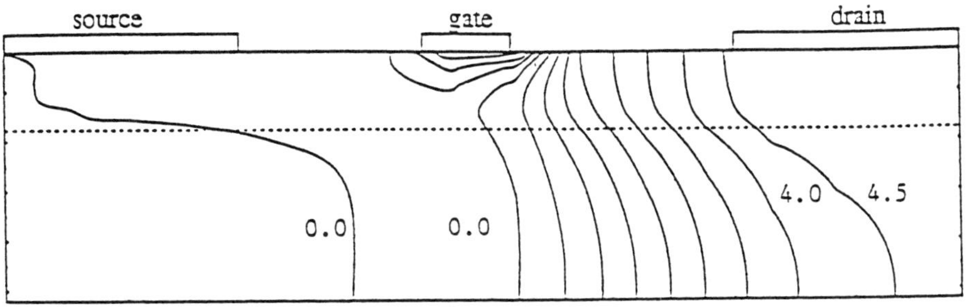

(a) Equipotential lines in steps of 0.5 V.

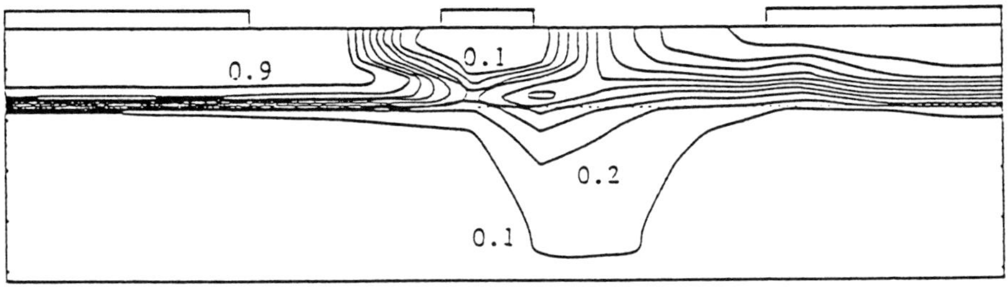

(b) Carrier concentration, n/N_d, in steps of 0.1.

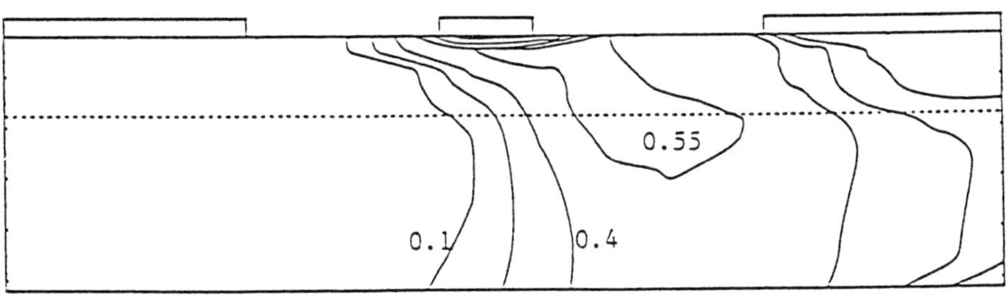

(c) Carrier energy in steps of 0.15 eV.

Fig. 9.4 Contour plots for the dc distributions within the two-dimensional MESFET model of Fig. 9.2.

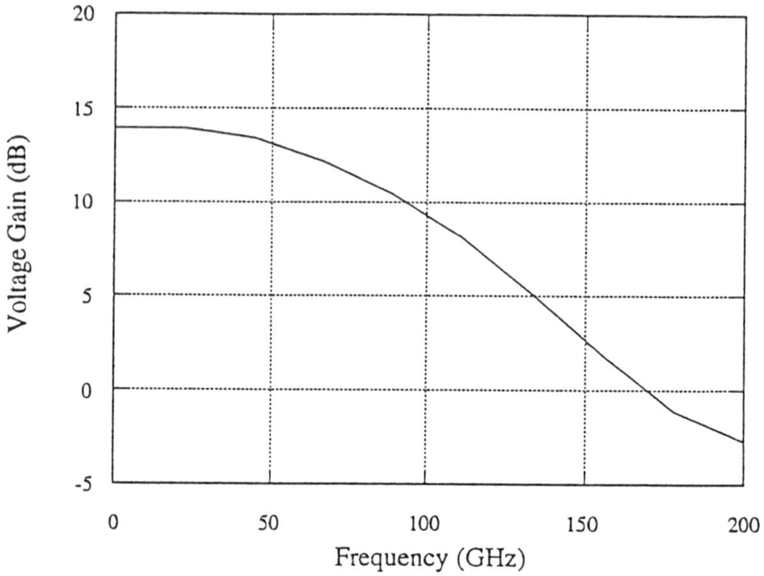

Fig. 9.5 Frequency response of the MESFET obtained from the two-dimensional model.

9.7.2 Electromagnetic Wave Results

For the electromagnetic-wave analysis, a sinusoidal excitation is applied between the gate and the source electrodes. The total gate voltage becomes:

$$V_{gs}(t) = V_{gso} + \Delta v_{gs} \sin \omega t \tag{9.13}$$

where V_{gso} and Δv_{gs} are the dc and the peak ac voltages, respectively. As shown in Fig. 9.3, this excitation is applied as a TEM wave source at $z = 0$. At each time step, this source corresponds to the solution of Laplace's equation for the ac signal. The full-wave electromagnetic model is then marched in time for several cycles of the exciting wave (several tens of picoseconds) to reach the steady-state ac solution.

Fig. 9.6 depicts the evolution of the input wave (the voltage signal between the gate and source electrodes) at several points along the z-axis. The input wave decreases in magnitude as it propagates along the device, an effect primarily due to energy loss to the conducting electrons. As shown in Fig. 9.7, the role of the active device in attenuating the input wave is assessed by comparing two waves: one propagating on the active device, and the other propagating on a passive structure made of semi-insulating GaAs with the same transmission line configuration. The active device significantly attenuates the input wave as it propagates in the z direction, while the reduction in the input wave amplitude in the passive device is insignificant.

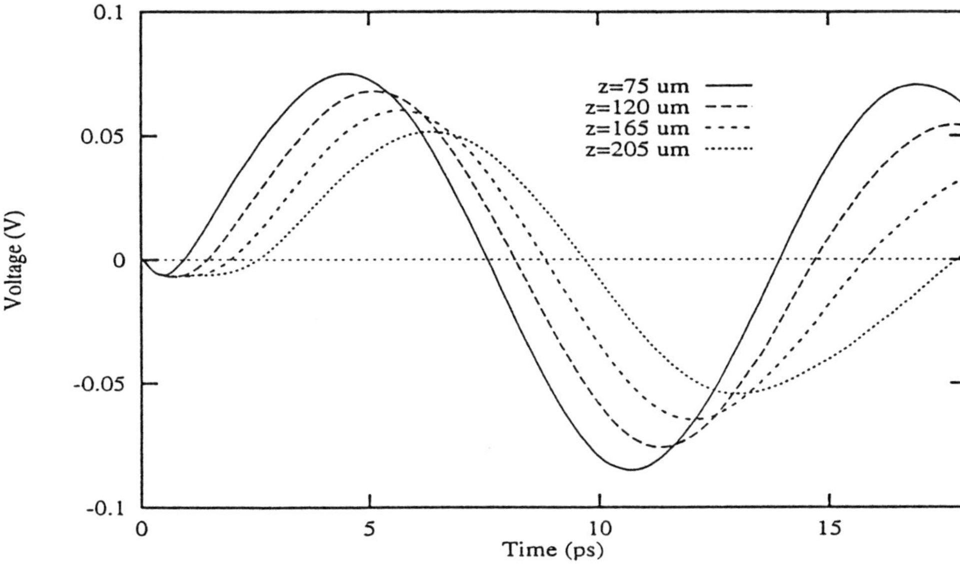

Fig. 9.6 Calculated input voltage waveform at several points in the *z* direction along the MESFET.

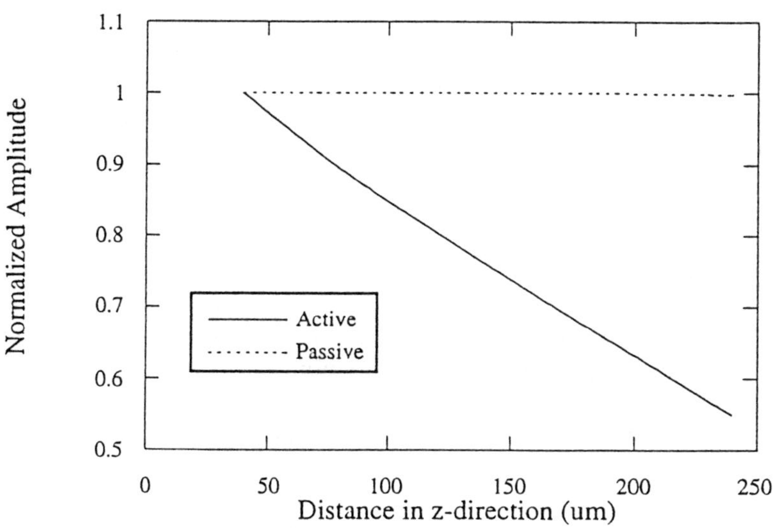

Fig. 9.7 Comparative decays of the input voltage wave along the active MESFET and along a passive GaAs structure having the same transmission line configuration.

Additional understanding of electron/electromagnetic wave interactions in the MESFET is gained by examining the calculated input-output waveforms at $z = 75$ μm shown in Fig. 9.8. Upon factoring in the normal sign inversion between the ac gate input and the amplified ac drain output, we see that the output wave lags the input wave by about 5 ps. This delay is due to the device's switching time caused by finite-velocity electron transit from the source to the drain. A simple time-distance calculation verifies this observation. The total phase shift between the input and output waves is more than 180° since it accounts for the sign inversion of the amplifier as well as the finite device switching time.

Fig. 9.9 shows the calculated output voltage waveforms at different points along the z direction. These waveforms are shown at the distances $z = 75$, 120, 165, and 205 μm from the excitation plane. Two important observations are made from this figure:

1. Early in the simulation, the electronic effect is not present and the output wave amplitude decreases along the device. Again, this is due to the finite device switching time. Later, as more electromagnetic energy is propagated along the device, the wave energy builds up and the wave amplitude increases.
2. At a particular z location, the voltage amplification is optimum. For the case considered in this simulation, the device width where the optimum gain is attained is about 120 μm. Beyond this point, the voltage gain decreases.

To understand the phenomenon of observation 2, we must consider the mismatch of the phase velocities of the input and output waves caused by their different electrode geometries. Phase-velocity mismatch can lead to phase cancellation and a consequent reduction of the device gain. This phenomenon is one of the most interesting aspects of electromagnetic wave propagation along the electrodes of an active device. It is discussed in more detail in [26].

One of the advantages of the approach presented here is the ability to analyze the device's internal parameters and primary physical variables during the various phases of its operation. For example, electromagnetic wave effects on the carrier density and the carrier energy within the conducting channel near the drain side of the gate are shown in Figs. 9.10 and 9.11, respectively. In Fig. 9.10, the change in the electron density lags the input wave by at least 3.5 ps. Moreover, the electron density variations are in phase with the input wave. Note that the lag time for the carrier density is shorter than the device switching time, since the latter is the accumulation of the lag times of all quantities including carrier density, momentum, and energy. Fig. 9.11 shows that the electron energy takes a slightly longer time than the carrier density to respond to the input wave (6 to 8 ps). The variations in energy are apparently in phase with the output wave rather than the input wave.

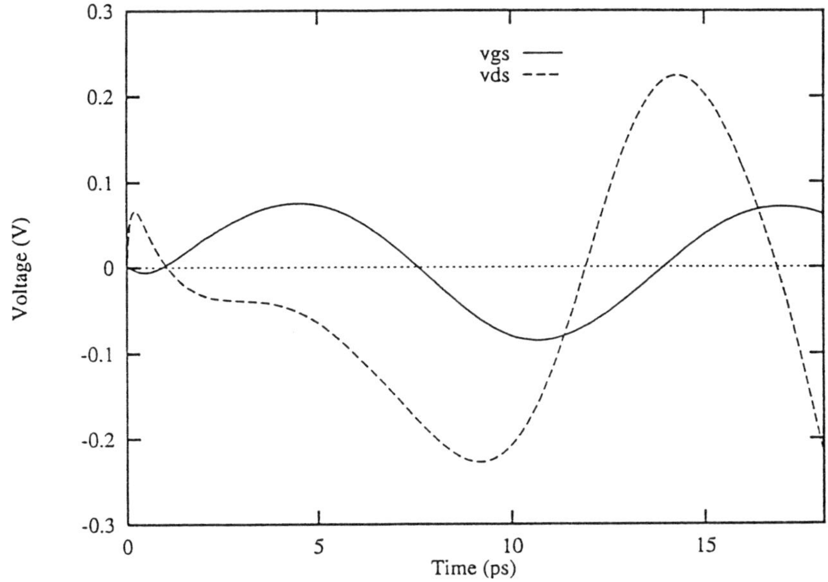

Fig. 9.8 Gate (input) and drain (output) voltage waveforms of the MESFET at $z = 75\ \mu m$.

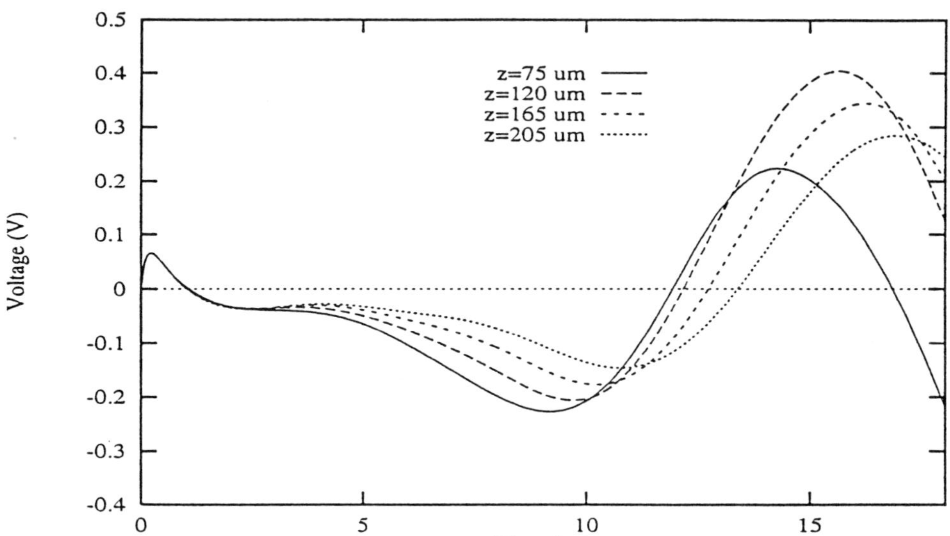

Fig. 9.9 Output voltage waveforms at several points along the z direction.

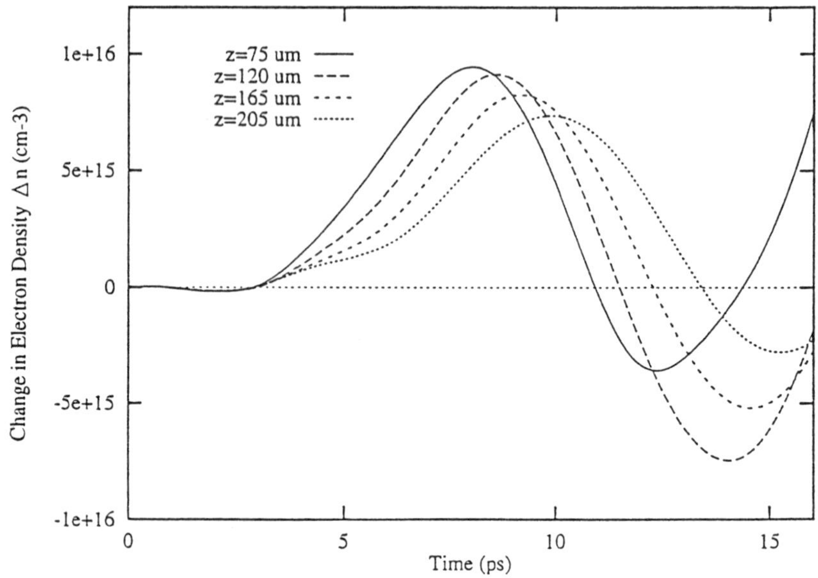

Fig. 9.10 Time waveforms of the change in electron density in the channel near the drain side of the gate at several points along the z direction.

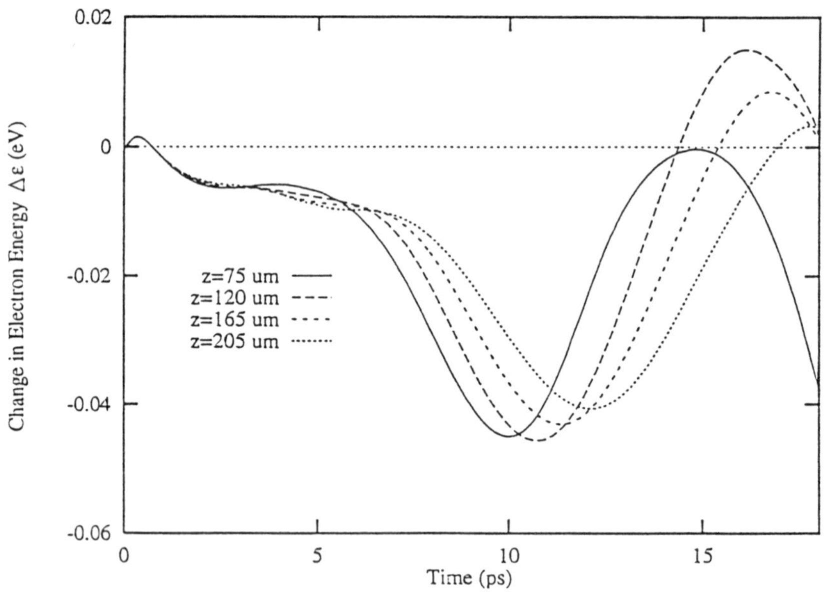

Fig. 9.11 Time waveforms of the change in electron energy in the channel near the drain side of the gate at several points along the z direction.

9.7.3 Frequency-Domain Characteristics

Fourier analysis can be performed on the time-domain output signal of the MESFET to obtain its amplitude and phase response as a function of frequency. Fig. 9.12 shows the relative amplitude response of the MESFET output calculated in this manner at four points along the z direction near the wave-excitation plane. We see that the amplitude response is maximum at the fundamental frequency, $f_0 = 80$ GHz, and has smaller relative peaks at dc and at multiples of f_0. Further, in this observation regime close to the wave-excitation plane, the amplitude response increases approximately linearly with distance along the z direction.

The relative phase response of the MESFET output is shown in Fig. 9.13 at two points along the z direction near the wave-excitation plane. We see that the phase response is not quite a linear function of frequency, with the deviation from linearity more pronounced at higher frequencies. This effect is due to dispersive electromagnetic wave propagation along the MESFET structure, which is an inhomogeneous transmission line. The wave dispersion is further enhanced by the presence of electrons that provide additional slowing of the wave. Using the phase data given in this figure, the phase velocity of the output wave can be calculated. For example, at 80 GHz this velocity is $v_p \cong 1.01 \times 10^8$ m/s.

9.8 DISCONTINUITIES IN SEMICONDUCTOR STRUCTURES AND DEVICES

Several physical configurations are implemented in conventional high-frequency GaAs transistors. Especially at high operating frequencies, the layout of the physical device affects its performance. Typical configurations for a MESFET or HMET include the interdigitated and the four-finger geometries shown respectively in Figs. 9.14(a) and 9.14(b). It is clear that the electrodes of the active device form a guided-wave structure. At high frequencies, these electrodes must be treated as transmission lines. The transmission lines can be regarded as slot lines or coplanar waveguides that form a discontinuity at the input and output ports of the device. Therefore, the input and output impedance must be carefully analyzed. A common practice in analyzing discontinuity effects in active devices is to model the passive structure separately from the behavior of the active device [44]. This approach does not take into account the interaction between the electromagnetic wave propagating along the device electrodes and the electron transport in the active device, which impacts the overall device characteristics.

It is difficult to determine the discontinuity effects from measurement. First, a disconnection and reconnection may be necessary to switch a tuner to the measuring device. Second, there usually is some physical distance between the FET and the tuner. In many cases, this situation prevents matching the FET, and consequently, the true performance of the FET might be unknown. The model described in the previous section is an excellent tool for studying this problem.

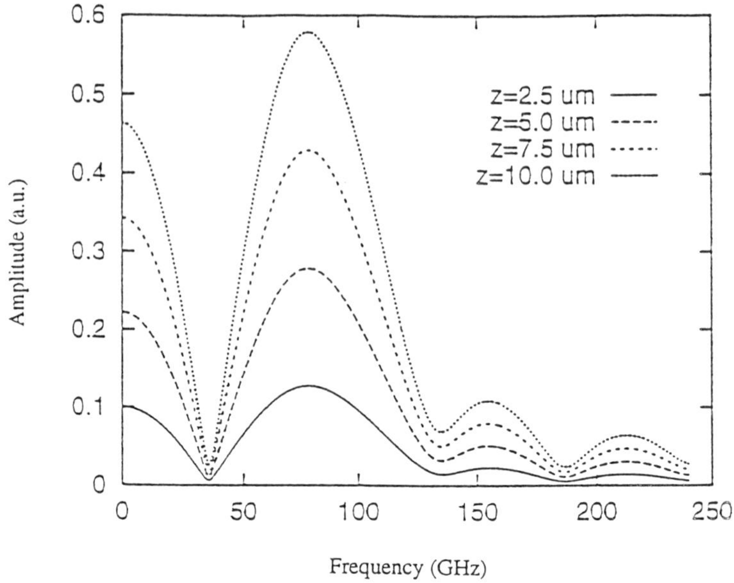

Fig. 9.12 Relative amplitude response of the MESFET output versus frequency at four points along the z direction near the wave-excitation plane.

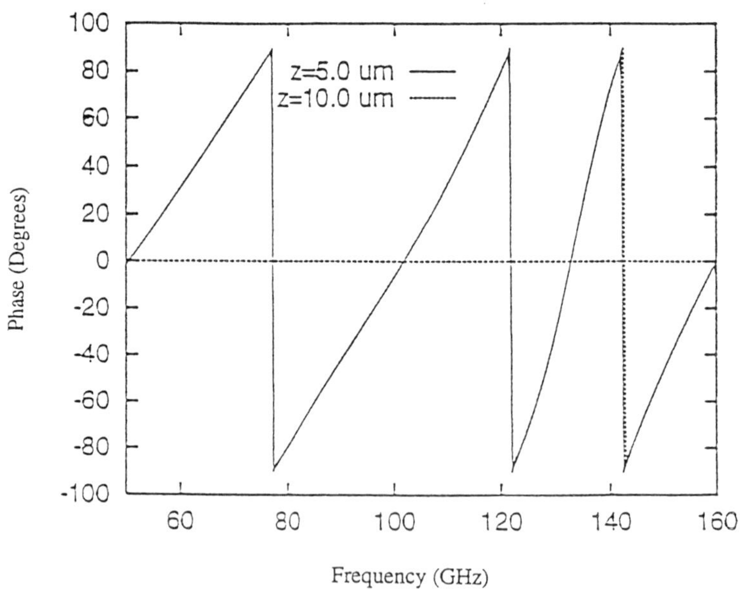

Fig. 9.13 Relative phase response of the MESFET output versus frequency at two points along the z direction near the wave-excitation plane.

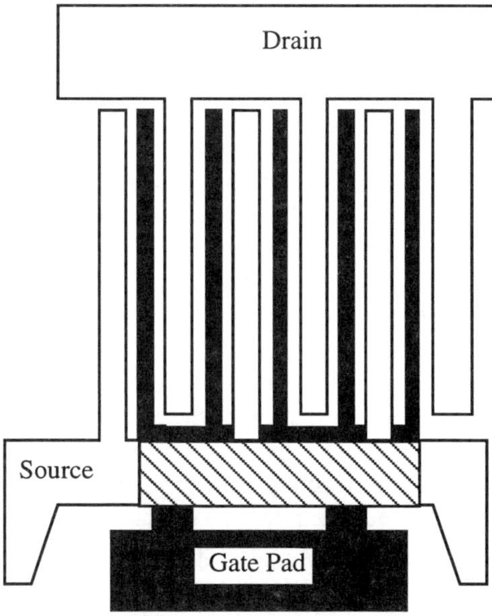

(a) Interdigitated topology used for power FETs.

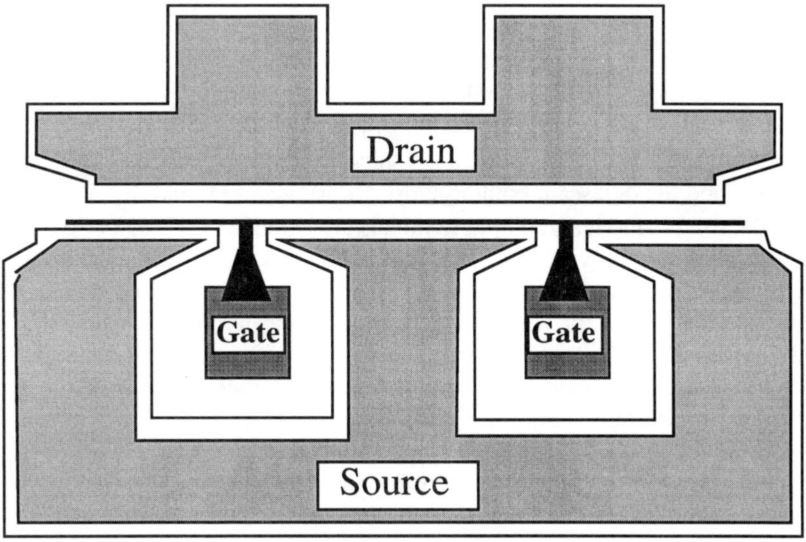

(b) Four-finger FET configuration.

Fig. 9.14 Typical electrode configurations for a high-frequency MESFET.

9.8.1 Discontinuities and Equivalent Circuits

A transmission-line discontinuity is a variation from a straight and longitudinally homogeneous circuit. The change is generally abrupt, which means that the dimensions of the discontinuity are usually small compared with the wavelength. At the discontinuity, the flux lines of the electric and magnetic fields and currents are distorted compared with those of a longitudinally homogeneous line. Part of the propagating energy is radiated and can excite undesired resonances in a conducting enclosure. Another part of the energy is transferred into propagating surface-wave modes that can lead to parasitic coupling between the various parts of the circuit.

To account for these effects, transmission-line discontinuities can be analyzed as individual one-port, two-port, or n-port elements characterized by frequency-dependent S parameters. At low frequencies, discontinuities are often approximated by equivalent circuits comprised of lumped LC elements that are independent of the transmission line. This allows a fair consideration of reactive field components that can impact the operation of the circuit [45]. It has been demonstrated that accounting for transmission-line discontinuities in the computer-aided design of circuits leads to a remarkable improvement in modeling accuracy [46]. However, the discontinuities associated with active circuits should not be analyzed separately because the energy scattered from any such discontinuity is coupled to the active device. This significantly affects the device performance.

9.8.2 Transistor Model

The common-source MESFET amplifier modeled in this section involves a full-wave, three-dimensional, physics-based simulation [47] very similar to that discussed in Section 9.7. Here, the MESFET gate length is 0.2 μm; the device width is 150 μm; the active-layer thickness is 0.1 μm; the channel and substrate doping is 2×10^{17} and 1×10^{14} cm^{-3}, respectively; and the quiescent point is chosen to be $V_{ds} = 2\,V$ and $V_{gs} = -1V$.

The MESFET's input and output terminals are extended for a distance of 30 μm and are fully matched by applying the PML ABC [43] to truncate the electromagnetic field computational domain at its front and back. The left, right, top, and bottom truncations of the domain are placed far from the active device and terminated using the first-order Mur ABC. Since evanescent electromagnetic fields are excited at each discontinuity, the reference planes are placed far enough away so that these decaying fields have negligible amplitude there.

As in Section 9.7, Maxwell's equations are coupled with the complete hydrodynamic system wherein the electron density, energy, and velocity are obtained from the continuity, energy-conservation, and momentum-balance equations, respectively. Here, however, the amplifier circuit is assumed to be linear since the applied signal (a wideband Rayleigh pulse) has an amplitude of only 0.05 V. Thus, scattered-wave amplitudes are linearly related to the incident-wave amplitude.

9.8.3 Evaluation of S Parameters

The common-source MESFET amplifier can be regarded as a two-port network where the input port is defined to be between the gate and the source, and the output port is between the drain and the source. The current definition is used to calculate the S parameters. Here, the currents are calculated from the magnetic field by applying Ampere's law at the input and output terminals. The definition of the current is more reliable in this case than voltage since the voltage can be a path-dependent quantity. The S parameters are calculated using the following expression:

$$S_{ij} = \left(I_j\sqrt{Z_j}\right)/\left(I_i\sqrt{Z_i}\right) \tag{9.14}$$

where Z is the characteristic impedance at the input and output ports. For the MESFET model, the S parameters are calculated in the frequency range from 5 to 150 GHz.

9.8.4 Calculated Results and Discussion

Four types of connections to the ports of the common-source MESFET amplifier have been modeled:

1. Coplanar waveguides at both the input and output ports (CP/CP);
2. Coplanar waveguide at the input port, slot line at the output (CP/SL);
3. Slot line at the input port, coplanar waveguide at the output (SL/CP);
4. Slot lines at both the input and output ports (SL/SL).

Note that the slot-line connection presents an electromagnetic field discontinuity relative to the natural coplanar waveguide geometry of the transistor electrodes.

Fig. 9.15(a) compares $|S_{11}|$ for the different cases studied. The SL/SL and CP/SL cases have $|S_{11}| > 0$ dB, indicating circuit instability. For this case, the capacitive nature of the discontinuity at the output port of the transistor provides capacitance coupling between the gate and the drain. The resulting positive feedback drives the amplifier to instability. Although $|S_{11}| < 0$ dB for the SL/CP case, it is relatively large compared with the CP/CP case, as expected. For the CP/CP case, there are no electromagnetic discontinuities at either the input or output ports of the active device, and the amplifier is electrically stable.

Fig. 9.15(b) compares $|S_{21}|$ for the different cases studied. $|S_{21}|$ is largest for the SL/SL configuration, and smallest for the CP/CP case. This behavior is explained by the multiple wave reflections within the MESFET caused by discontinuities at its input and output ports for the SL/SL case. The wave reverberations along the transistor electrodes interact with the electron transport inside the active device to yield more gain (at the expense of instability). Note that in this example, $|S_{21}|$ for the CP/CP case is quite small since the transistor geometry is not optimized.

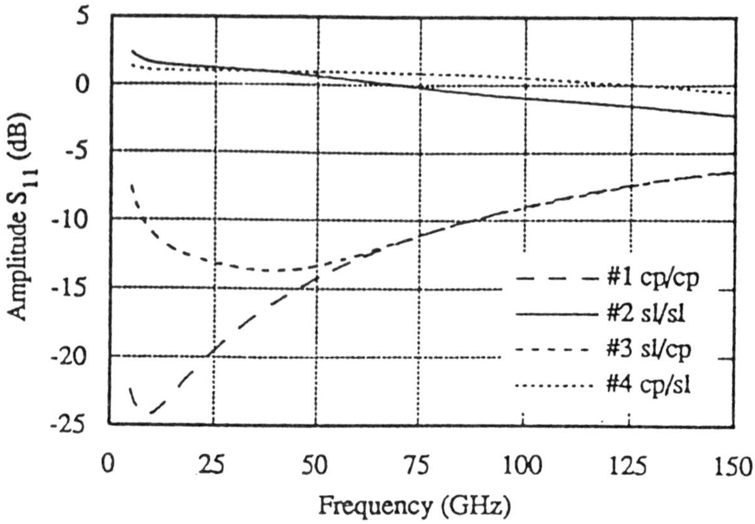

(a) $|S_{11}|$

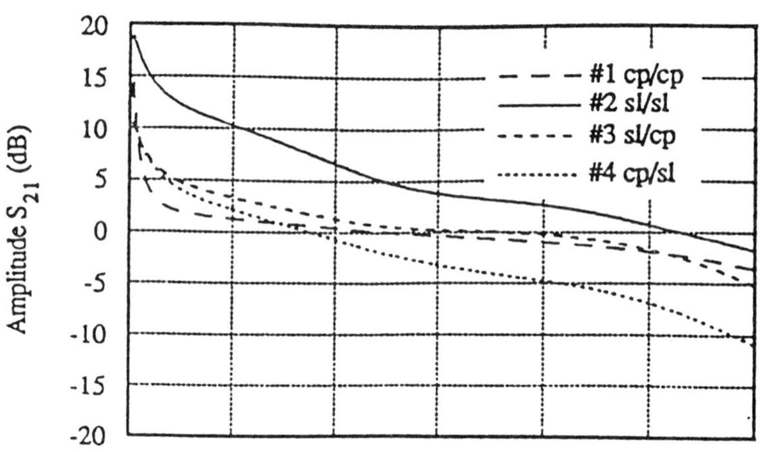

(b) $|S_{21}|$

Fig. 9.15 Comparison of the calculated S parameters for the four types of connections to the ports of the common-source MESFET amplifier.

9.9 RESULTS OF THE PHYSICS-BASED MODEL VERSUS QUASI-STATIC AND CIRCUIT MODELS

As mentioned in Sections 9.1 and 9.2, full-wave physics-based models, quasi-static models, and circuit-based models have been developed for incorporating electromagnetic wave effects in semiconductor device simulations. This section provides comparative data generated by these approaches.

Fig. 9.16 compares simulation results for a MODFET and a MESFET obtained using the combined electromagnetic and solid-state (CESS) simulator discussed in the previous sections with results from a quasi-static model. In the CESS, Maxwell's equations are solved to obtain the time-dependent electric and the magnetic fields. In the quasi-static model, Poisson's equation is solved for the electric field.

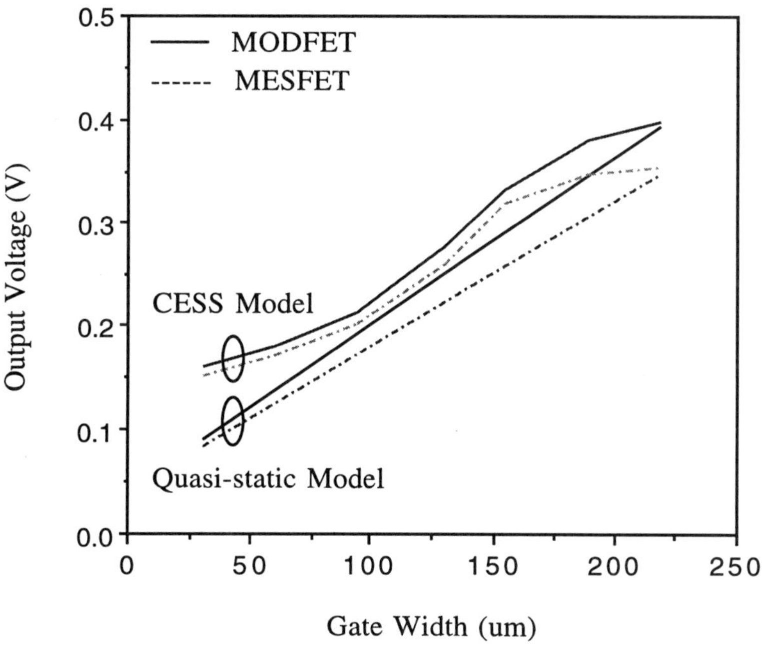

Fig. 9.16 Output voltage versus device width for a MODFET and MESFET obtained from the CESS simulator and the quasi-static model.

From Fig. 9.16, it is evident from both models that the MODFET has higher output voltage than the MESFET. The amplitude of the output voltage wave monotonically increases along the device width in the quasi-static model. However, in the CESS model, the output voltage behavior is more complicated due to the interaction of the device with the electromagnetic wave. Depending upon the device width, the electromagnetic wave can receive energy from the device or can give back some of its energy to the device. The latter phenomenon is due to phase cancellation caused by phase-velocity mismatches between the waves on the different electrodes of the device. This important and complex behavior is absent when the output is calculated using a quasi-static model. Thus, the full-wave electromagnetics model provides the opportunity to choose the optimized device width for maximum output voltage.

A comparison between a circuit-based model and a full-wave model is provided by simulating a MESFET structure similar to that presented in [48]. Here, the gate-source spacing is 1.5 μm, the gate length is 0.5 μm, the gate width is 1 mm, the active-layer thickness = 0.2 μm, and the active layer doping is 10^{-17} cm^{-3}. The amplifier gain characteristics resulting from the two simulation techniques are shown in Fig. 9.17.

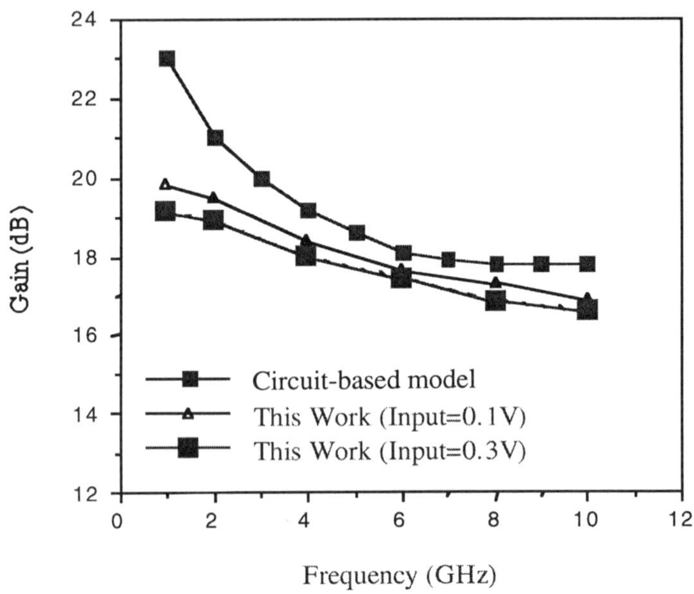

Fig. 9.17 Gain of a MESFET obtained from the full-wave simulator and the circuit-based model.

While the results of Fig. 9.17 exhibit the same trend, there are quantitative differences. The circuit-based approach is strictly a small-signal analysis, while the full-wave model retains the large-signal behavior of the device. To demonstrate this, the input signal level was taken to be both 0.1V and 0.3V for the full-wave model. From Fig. 9.17, the full-wave model shows that the gain decreases as the signal amplitude increases, as expected due to considerations of device saturation.

The strength of the full-wave approach is demonstrated not only by predicting the gain decrease at higher input levels, but also by calculating the harmonics generated by the device nonlinearity. The harmonic content of the output is analyzed by taking its Fourier transform, as shown in the next section. To the best of our knowledge, the full-wave technique is the only direct approach that can perform such a study.

9.10 MODELING DEVICE-CIRCUIT INTERACTIONS: A GLOBAL APPROACH

Continuing demand for processing and transmitting information at ever-faster rates drives electronic systems to operate at higher frequencies or higher clock speeds. At the same time, circuit complexity and density is increasing, resulting in closely spaced active and passive devices and interconnects. As a result, circuit performance can be degraded due to crosstalk caused by electromagnetic coupling, surface waves, and radiation effects.

In such cases, the circuit-modeling issue becomes quite pressing. In fact, it is becoming clear that circuit design should be based on a global model that takes such electromagnetic wave effects into consideration [49]. At the device level, recent developments in semiconductor technology are leading to an increasing demand for detailed modeling. As device dimensions drop to well below 1 μm, researchers are compelled to develop techniques for device analysis, design, and optimization that account for the key physical phenomena at such distance scales. These phenomena include hot-electron and quantum-mechanical effects and the velocity-overshoot phenomenon.

The hybridization technique in microwave and millimeter-wave circuits was previously used in coupling lumped-circuit elements and a distribution network of transmission lines [50 – 54]. Hybrid systems with active and passive lumped elements have been modeled using FDTD [55, 56] and the TLM method [57, 58]. Three-terminal active devices were incorporated into FDTD either by placing appropriate two-terminal lumped elements in adjacent space cells [56, 59], or by locally connecting FDTD to a SPICE circuit solver to implement a complete device in a single FDTD cell [60]. Subsequently, microwave amplifiers were simulated by using equivalent voltage-source and current-source formulations for a transistor [61, 62]. (See also Chapter 8.) Finally, the characterization of microwave circuits containing various lumped components was performed using a global finite-element electromagnetic analysis in [63].

We note that, in [55 – 63], the physical devices were not simulated; lumped elements were used instead. Especially at millimeter-wave frequencies, this process of replacing the active device with equivalent sources or equivalent circuits may not represent all aspects of the device nonlinearity and the wave-device interactions. The alternative approach is to simulate the circuit by coupling the physical equations representing the semiconductor devices with the electromagnetic fields using an accurate full-wave solver. However, full-wave field solvers such as FDTD require significant computer memory and running time for most problems of engineering interest.

The problem of large computer memory and running-time requirements can be mitigated by breaking a circuit into several subcircuits, solving the subcircuits individually, and reconnecting them together. This method was successfully applied to steady-state networks [64, 65] and to time-domain solutions using TLM [66, 67]. Computer resource requirements have also been significantly reduced by using Prony's method and system-identification techniques [68 – 73]. The global modeling technique discussed below is based upon an idea that is similar in its fundamental approach to that of the subcircuit reconnection concept, except that here the modeling of the semiconductor devices uses the much more rigorous, physics-based CESS theory discussed in the previous sections.

The global modeling technique for millimeter-wave devices is still a new concept. In this technique, the circuit of interest is divided into multiple regions. A full-wave analysis of each region is performed separately. Each active device is individually simulated by combining its electron-transport model with the full-wave electromagnetic model. Each passive device is simulated using the full-wave electromagnetic model. The physical characteristics of the entire multiple-region circuit are calculated by incorporating a simple but rigorous hybridization technique that couples information from the adjacent regions.

Although global modeling is simple in concept, its implementation is not easy. For example, potential problem areas include maintenance of numerical stability and accuracy, coping with a wide range of device dimensions, and properly modeling the wide variety of physical and electromagnetic aspects of semiconductor devices and circuits. In a modeling example explored in the next section (a millimeter-wave amplifier), the input and the output matching networks are physically large compared to the transistor. However, the grid mesh size for proper modeling of the transistor is constrained by the Debye length of the semiconductor. In turn, this constrains the time step, Δt, of the full-wave electromagnetic solver to be on the order of 10^{-17} s to ensure numerical stability. If the input and the output matching networks were to be simulated using these criteria, the computer storage and running time would render the problem virtually impossible using present computers.

To avoid these problems, the millimeter-wave amplifier with associated matching networks is divided into three regions. A full-wave analysis of each region is performed individually and properly coupled to the next stage with all of the required information from the preceding stage. This technique enables the use of a large space step, and hence, a large time step, in modeling the matching networks. The required computer memory and simulation time is greatly reduced compared to a potential alternative approach involving usage of a spatially nonuniform mesh for the whole amplifier. As we will see, the global model can calculate both the small-signal and the large-signal responses of the amplifier (including its nonlinearity and distortion), as well as the wideband frequency content of the output spectrum. These characteristics are perhaps more vivid in this global model than in any other simulator. The amplifier analysis, the multiregion coupling procedure, and sample results are summarized next.

9.11 GLOBAL MODELING OF A MILLIMETER-WAVE AMPLIFIER

As a step towards global modeling of millimeter-wave circuits, a complete amplifier implemented in a coplanar waveguide configuration has been simulated. Consider first the active MESFET device. Fig. 9.18(a) is a perspective view of one-half of the MESFET geometry. The other half has mirror-image symmetry about the center line of the gate (at the bottom of the figure). Fig. 9.18(b) is the plan view of the MESFET corresponding to Fig. 9.18(a). This figure shows the details of the device dimensions.

The MESFET's channel length and aspect ratio were selected to achieve the required transconductance and the cut-off frequency. Upon performing the dc analysis for the MESFET, its cut-off frequency was estimated as 77 GHz. A 40-GHz operating frequency was chosen to be well within the device bandwidth. The S parameters were computed for the MESFET at the selected bias point, and the equivalent circuit admittance matrix of the device was obtained from the simulated S parameters.

Fig. 9.19 is a plan view of the complete amplifier including its associated matching networks. Figs. 9.20(a) and 9.20(b) show geometric details of the input and output matching networks contained in Fig. 9.19. The dimensions of these networks' center and outer conductors were selected to achieve a characteristic impedance of about 50Ω and to permit an exact physical connection at the gate and the drain sides of the MESFET. The line and stub lengths of the matching networks were designed from the MESFET's calculated equivalent circuit admittance matrix.

9.11.1 Computational Linkage of the Matching Networks to the MESFET

The millimeter-wave amplifier is divided into three regions. The simulation of each region is performed separately and properly coupled to the next stage with all the required information of the preceding stage.

First, an impulse, $I_1(t)$, is applied to the input matching network at point A as shown in Fig. 9.20(a). Using FDTD, the transmitted impulse response, $R_1(t)$, is calculated at point B. Then, $I_1(t)$ is applied at point B and the reflected impulse response, $\Gamma_1(t)$, is calculated.

Second, an arbitrary waveform, $V_1(t)$, is applied at point A in Fig. 9.20(a). The resulting response, $V_2(t)$, at point B is obtained by convolving $V_1(t)$ with $R_1(t)$. $V_2(t)$ is now applied to the MESFET at point C as shown in Fig. 9.18(a). The CESS simulator is used to propagate this signal within the transistor. $V_{R1}(t)$, the reflected wave from the MESFET arising from the CESS model, is identified and convolved with $\Gamma_1(t)$. The resulting signal is added to $V_2(t)$. Thus at point C of Fig. 9.18(a), we have

$$V_C(t) = V_2(t) + \sum \Gamma_1(\tau)\, V_{R1}(t-\tau) \tag{9.15}$$

Third, an impulse, $I_2(t)$, is applied to the output matching network at point E as shown in Fig. 9.20(b). Using FDTD, the reflected impulse response, $\Gamma_2(t)$, is calculated at point E and the transmitted impulse response, $R_2(t)$, is calculated at point F.

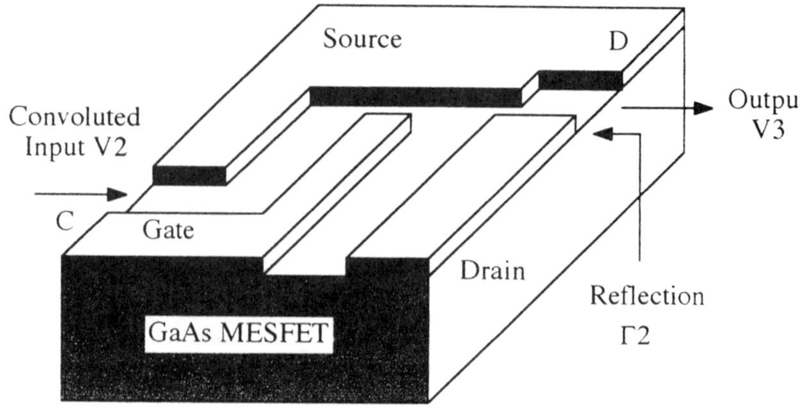

(a) Perspective view.

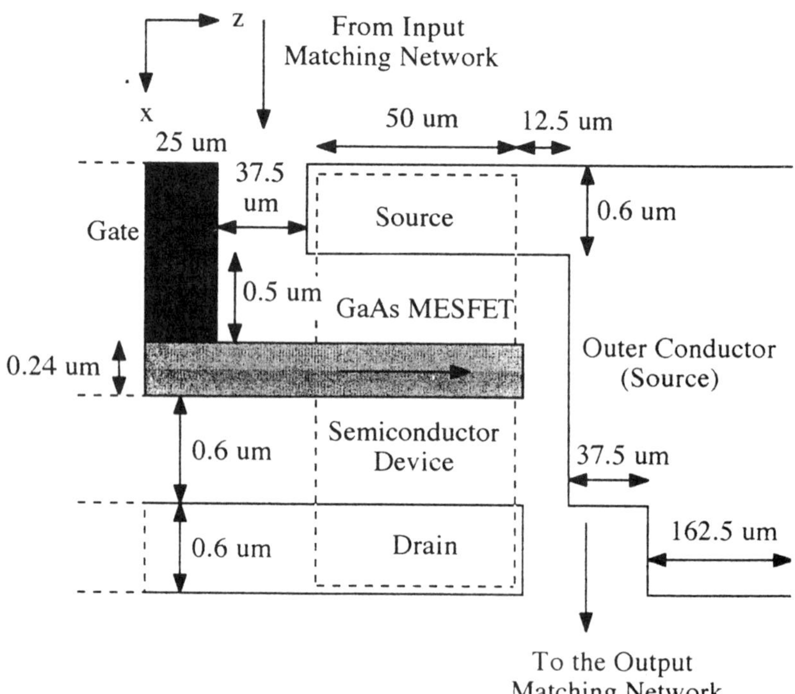

(b) Plan view.

Fig. 9.18 MESFET used for the millimeter-wave amplifier example. *Source*: Imtiaz and El-Ghazaly, *IEEE Trans. Microwave Theory and Techniques*, 1997, pp. 2208–2216, © 1997 IEEE.

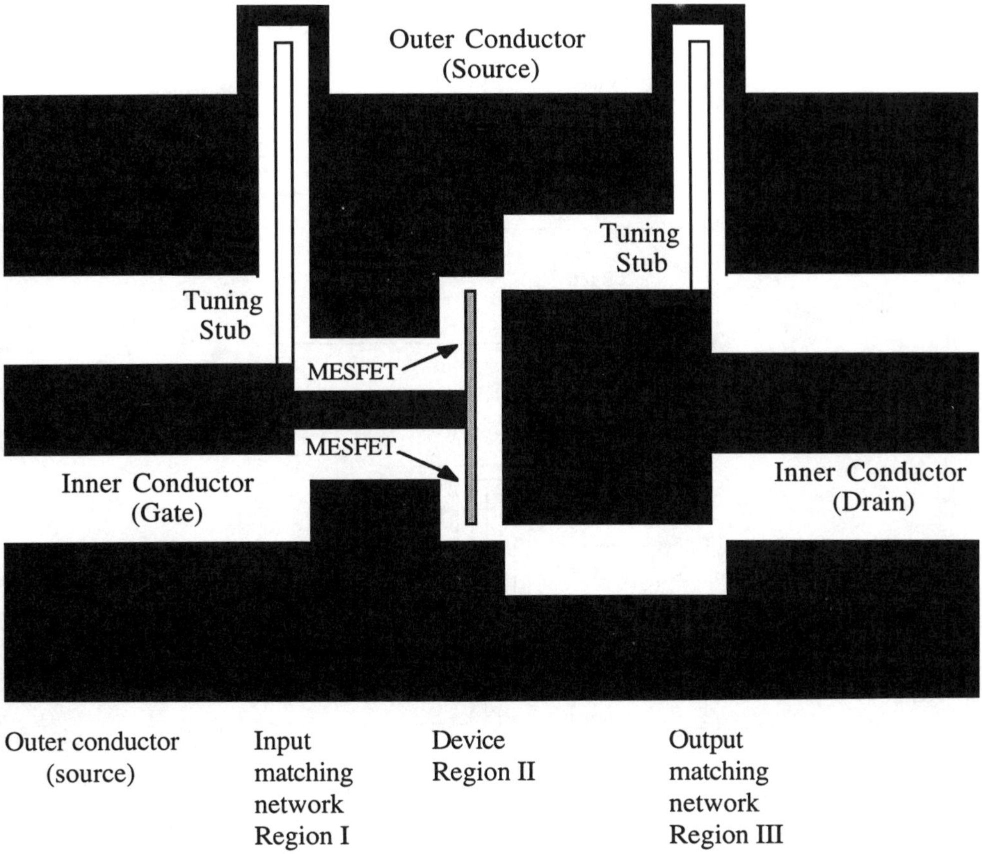

Fig. 9.19 Plan view of the complete millimeter-wave amplifier. *Source*: Imtiaz and El-Ghazaly, *IEEE Trans. Microwave Theory and Techniques*, 1997, pp. 2208-2216, © 1997 IEEE.

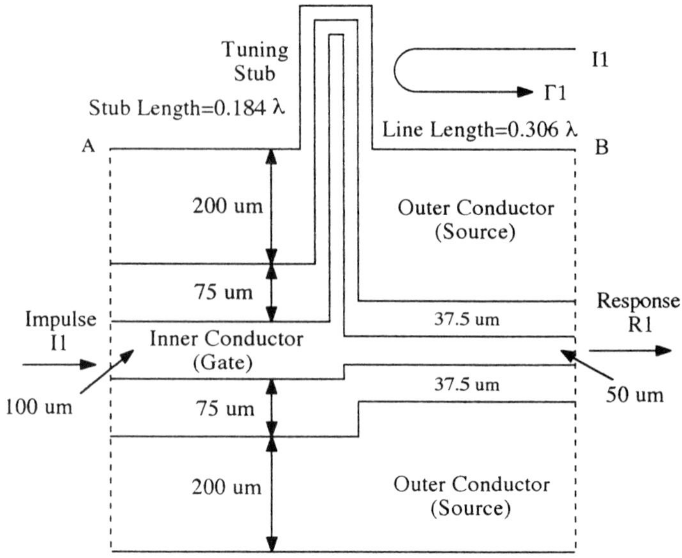

(a) Input matching network.

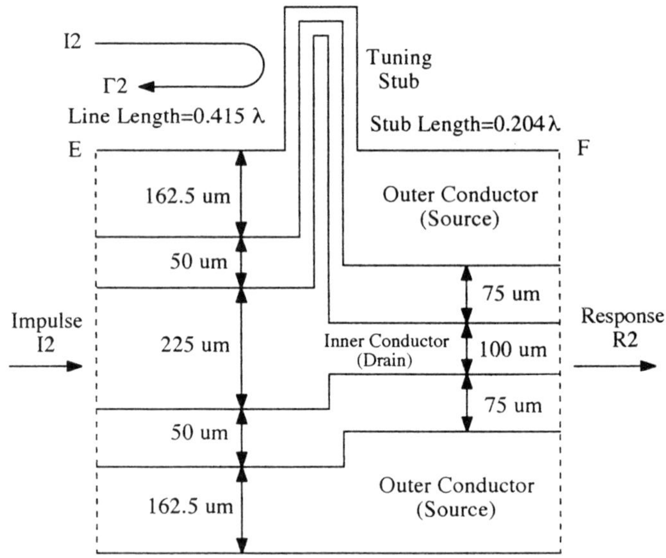

(b) Output matching network.

Fig. 9.20 Details of input / output matching networks of Fig. 9.19. *Source*: Imtiaz and El-Ghazaly, *IEEE Trans. Microwave Theory and Techniques*, 1997, pp. 2208–2216, © 1997 IEEE.

Fourth, $\Gamma_2(t)$ is convolved with the output of the MESFET. This is applied as a secondary source at point D of Fig. 9.18(a). The output wave, $V_3(t)$, is collected between the drain and the source at point D.

Fifth and last, the output at point F of Fig. 9.20(b) is calculated by convolving $V_3(t)$ with $R_2(t)$.

9.11.2 Amplifier Optimization Procedure

The S parameters for the complete amplifier are initially calculated without any optimization of the MESFET. Then, to obtain better gain characteristics, the MESFET is optimized in terms of its channel length, aspect ratio, and the active-layer doping density. Next, the line and stub lengths for the input and output matching networks are recalculated for the optimized transistor, and the S parameters for the complete amplifier recalculated. Finally, the large-signal response of the amplifier is obtained, including gain compression and harmonic generation.

9.11.3 Validation of the Convolutional Description of the Matching Networks

The convolutional description of the input matching network is tested (with the MESFET assumed to be inactive) by applying a 2.5-ps square pulse of magnitude 0.1V at the input. The output of this network is calculated both directly using FDTD and by convolving the square input pulse with the network's impulse response, $R_1(t)$. ($R_1(t)$ is precalculated by applying a 0.15-ps Gaussian pulse at the input. This narrow pulse serves as a quasi-Dirac delta function.) Fig. 9.21(a) compares the output voltages for the continuous FDTD simulation (solid line) with that obtained using convolution (dashed line). The two output voltages are seen to be in good agreement.

Fig. 9.21(b) shows the transfer functions for this case. Here, the solid line denotes the transfer function for the continuous simulation, and the two dashed lines represent the results obtained using quasi-Dirac delta functions of Gaussian shape and pulse widths 0.15 and 0.3 ps. These results show that the transfer function derived using the shorter, 0.15-ps Gaussian pulse corresponds better to the continuous simulation results (with a global difference of only 1.2%) than does the transfer function derived using the 0.3-ps Gaussian pulse (a global difference of 5%).

Fig. 9.22 compares the output voltages for the continuous simulation and the convolutional calculation for the case where the convolution of the impulse response is performed in both the input and output matching networks. Again, the MESFET is assumed to be inactive. Here, the overall response to the input 2.5-ps square pulse is obtained by convolving it with $R_1(t)$, and then taking the outcome of this convolution and convolving it with $R_2(t)$. (Both $R_1(t)$ and $R_2(t)$ are precalculated using a 0.15-ps Gaussian pulse which serves as a quasi-Dirac delta function.) The output voltages are in reasonable agreement with each other.

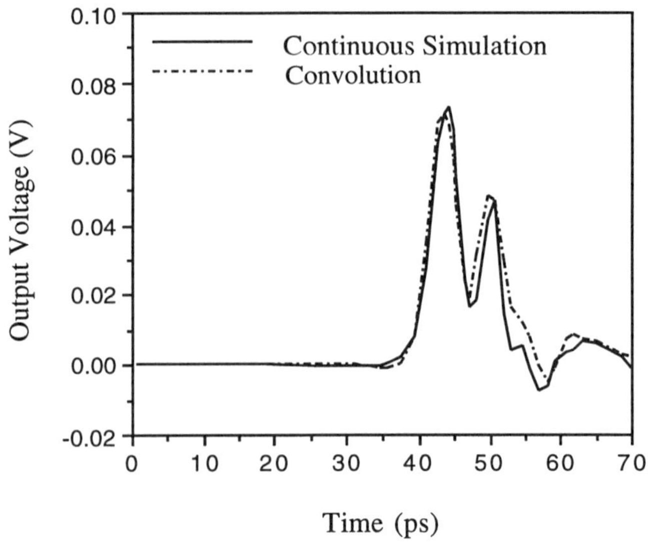

(a) Voltage at the output of the input matching network due to a 0.1-V, 2.5-ps square input pulse.

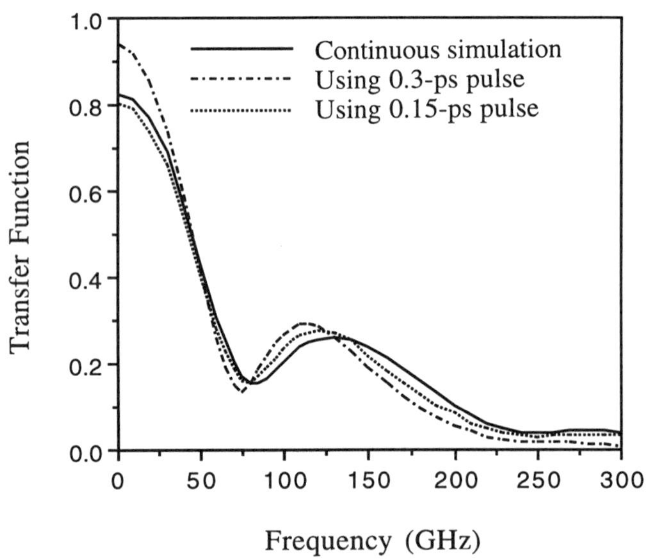

(b) Transfer function of the input matching network for 0.15-ps and 0.3-ps Gaussian input pulses

Fig. 9.21 Characterization of the input matching network of Fig. 9.19. *Source*: Imtiaz and El-Ghazaly, *IEEE Trans. Microwave Theory and Techniques*, 1997, pp. 2208-2216, © 1997 IEEE.

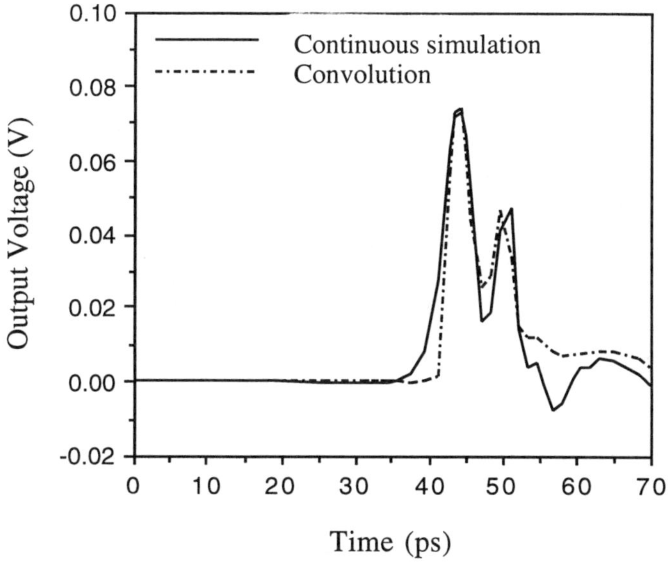

Fig. 9.22 Voltage at the output of both matching networks of Fig. 9.19 due to a 0.1-V, 2.5-ps square input pulse. *Source*: Imtiaz and El-Ghazaly, *IEEE Trans. Microwave Theory and Techniques*, 1997, pp. 2208-2216, © 1997 IEEE.

This technique of separating the linear electromagnetic field behavior of the matching networks from the nonlinear behavior of the MESFET requires simply a one-time calculation of the impulse response of each matching network. These impulse responses are stored and subsequently used to obtain the output due to an arbitrary input signal. In addition to significantly reducing the required computer memory and running time, this approach is directly applicable to the common engineering situation where the transistor parameters and biasing are changed during the design process, but the passive networks are left alone. Thus, the hybridization approach in global modeling provides good efficiency in simulating the amplifier in the practical engineering sense.

9.11.4 Impulsive Electric Field Propagation Within the MESFET

Fig. 9.23 shows snapshots of the distribution of E_z at the air-semiconductor interface of the MESFET of Fig. 9.18(a) for a narrow Gaussian pulse excitation at 2 ps and 6 ps. (Note that only one-half of the MESFET is modeled due to its even symmetry about the center line of the gate.) The x-axis and z-axis correspond to Fig. 9.18(b), and the grid numbers indicate the mesh points for $\Delta x = 32$ nm and $\Delta z = 8.13$ μm. At 2 ps, the input pulse is at its peak and the field is highest at the gate input. At 6 ps, the tail of the Gaussian pulse passes through the MESFET to the drain and considerable lateral wave propagation is evident in the middle portion of the device.

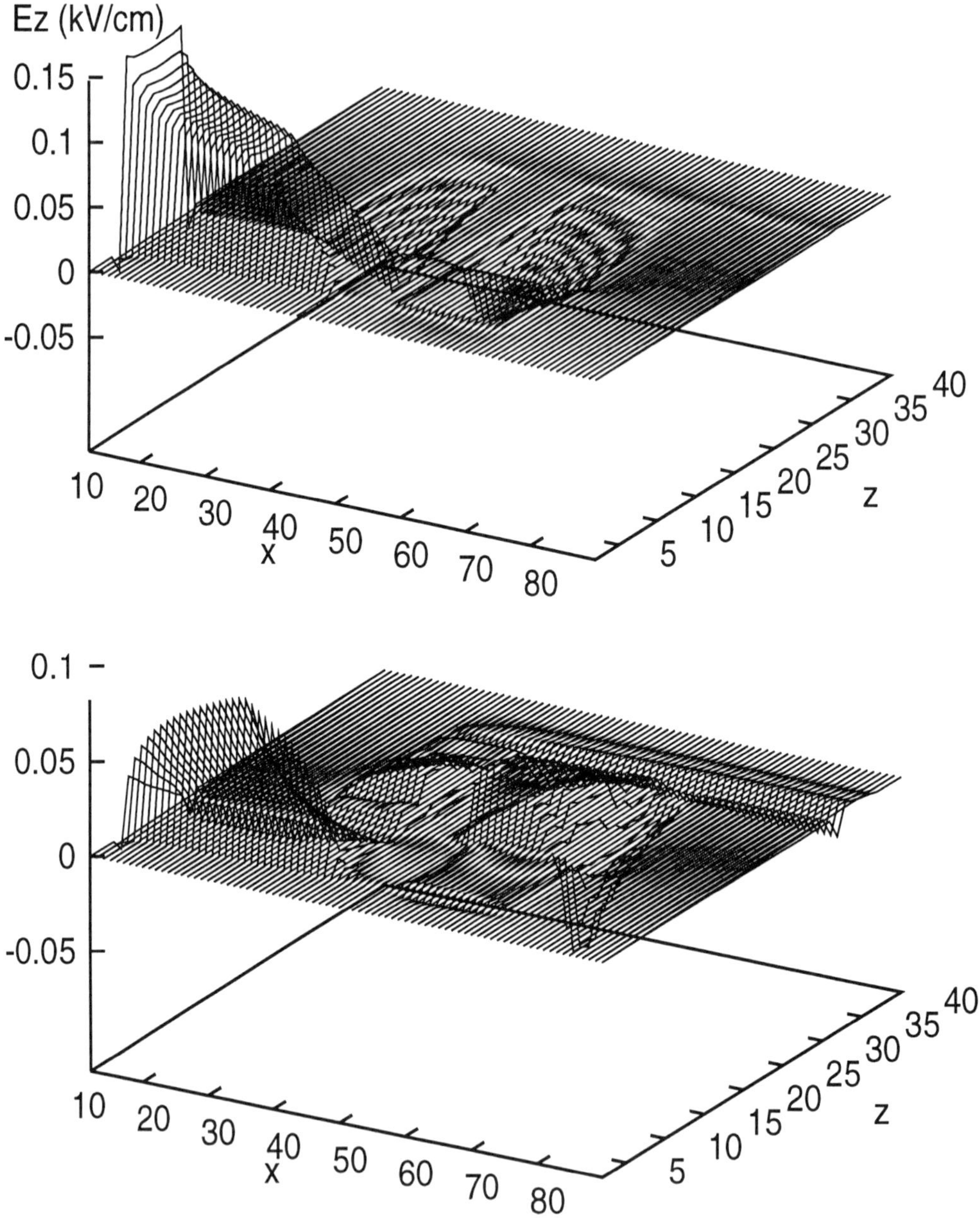

Fig. 9.23 Snapshots of the E_z field distribution within the MESFET of Fig. 9.18(a) due to a narrow Gaussian input pulse: top: at 2 ps; bottom: at 6 ps. *Source*: Imtiaz and El-Ghazaly, *IEEE Trans. Microwave Theory and Techniques*, 1997, pp. 2208–2216, © 1997 IEEE.

9.11.5 Small-Signal Gain for the Complete Amplifier, Initial Transistor Design

The initial design used for the GaAs MESFET in this amplifier assumes a gate length of 0.24 μm, an aspect ratio of 2.2, a GaAs substrate thickness of 0.5 μm, and an active layer doping of 2×10^{17} cm^{-3}. The dc drain bias is 3.0V and the gate bias is -0.4V.

The complete amplifier is simulated including the input and output matching networks designed for a 40-GHz center frequency. With a small 0.1V signal applied to the input, the amplifier's S parameters are calculated as a function of frequency. Fig. 9.24 shows the results for S_{21} obtained under simultaneously matched conditions. While S_{21} peaks slightly at the design frequency, its value is quite low because the transistor width is very small compared to any commercial amplifier. Increasing the transistor width and aspect ratio increases the value of S_{21}.

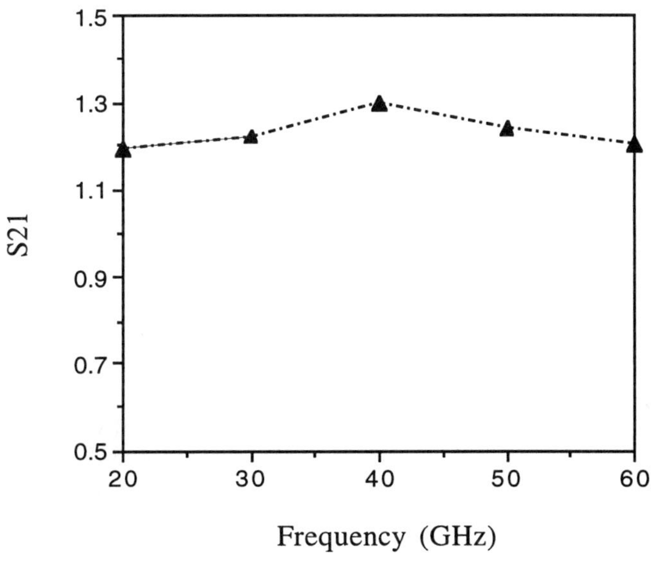

Fig. 9.24 Small-signal $|S_{21}|$ of the complete amplifier for the initial transistor design. *Source*: Imtiaz and El-Ghazaly, *IEEE Trans. Microwave Theory and Techniques*, 1997, pp. 2208–2216, © 1997 IEEE.

9.11.6 Amplifier Characteristics for the Optimized Transistor Design

Based on the results presented above, the MESFET is optimized in terms of its channel length, aspect ratio, and active-layer doping to increase the amplifier gain at 40 GHz while keeping the drain current the same. The optimized transistor has a gate length of 0.22 μm, an aspect ratio of 2.5, and an active-layer doping of 2.2×10^{17} cm^{-3}.

The complete amplifier is simulated by first applying a small 0.1V signal and then larger 0.3V and 1.0V signals. Each simulation is performed for signals at 20, 30, 40, 50, and 60 GHz. The *S* parameters of the amplifier are extracted from the simulation results by applying Fourier transforms to the calculated waveforms. This procedure allows separation of the response at the excitation frequency from the response at harmonics of the excitation frequency.

Fig. 9.25 depicts S_{21} as a function of frequency for the three input-amplitude cases. The large-signal *S* parameters are calculated at each excitation frequency. At the design frequency of 40 GHz, the gain is 7.38 dB for the 0.1V input signal, 6.44 dB for the 0.3V signal, and 5.15 dB for the 1.0 V signal. As expected, the gain at the fundamental frequency drops for large signals because an increasing amount of the output power is shifted to harmonic frequencies. We also note that, for small signals, the gain drops slightly as the input frequency shifts away from the 40-GHz design center. However, for large signals, the gain reduction away from 40 GHz is more pronounced. This, too, is a consequence of the increased harmonic distortion for the large-signal case.

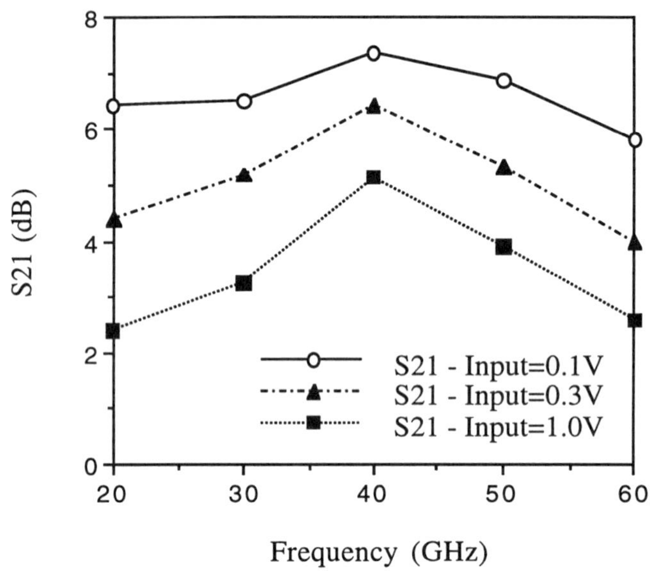

Fig. 9.25 Dependence of $|S_{21}|$ of the complete amplifier upon the input signal level for the optimized transistor design. *Source*: Imtiaz and El-Ghazaly, *IEEE Trans. Microwave Theory and Techniques*, 1997, pp. 2208–2216, © 1997 IEEE.

Fig. 9.26 depicts S_{12}, S_{11}, and S_{22} as a function of frequency for the 0.1V and 0.3V input-amplitude cases. As expected, S_{12} is very small and decreases with frequency. In addition, S_{12} decreases for the large-signal case. Also as expected, S_{11} and S_{22} are minimum at the design frequency. However, S_{11} and S_{22} increase slightly for large

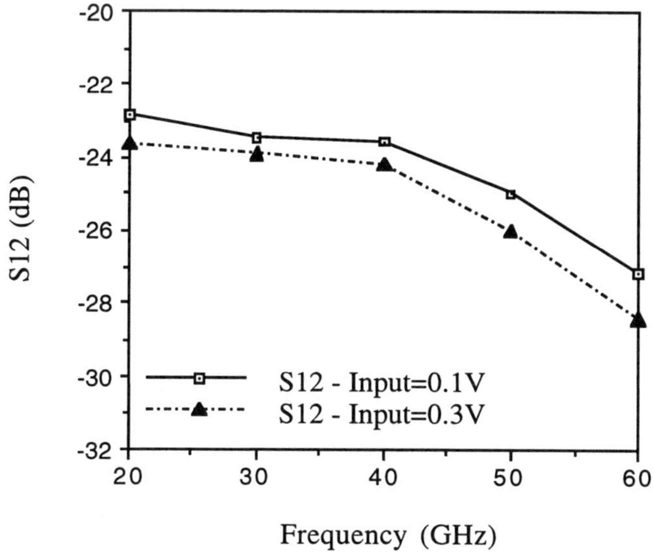

(a) $|S_{12}|$

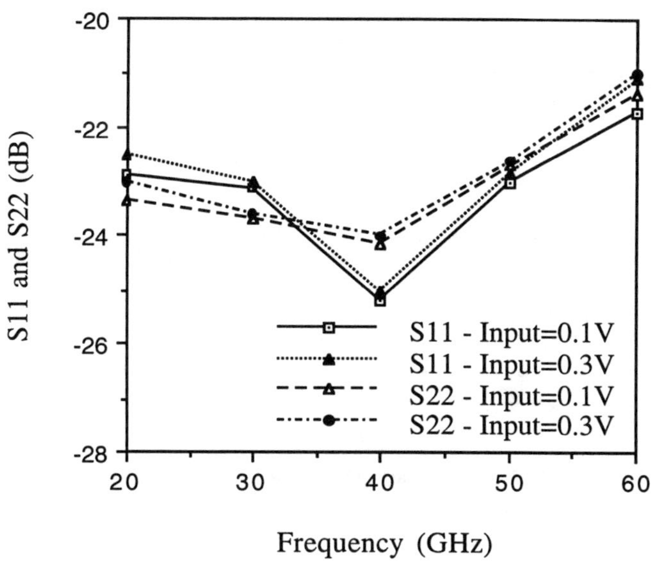

(b) $|S_{11}|$ and $|S_{22}|$

Fig. 9.26 Dependence of the *S* parameters of the complete amplifier upon the input signal level for the optimized transistor design. *Source*: Imtiaz and El-Ghazaly, *IEEE Trans. Microwave Theory and Techniques*, 1997, pp. 2208–2216, © 1997 IEEE.

signals. Further, the difference between the small-signal and large-signal results for S_{11} and S_{22} increases as the frequency deviates from the 40-GHz design center.

Once the large-signal response of the amplifier is obtained, it is useful to study the frequency content of its output and to identify the distortion products, which can be quite significant especially at the third and fifth harmonics of the input frequency. Fig. 9.27 shows the harmonic content of the output for 0.3V and 1.0V input signals at fundamental frequencies of 30, 40, and 50 GHz.

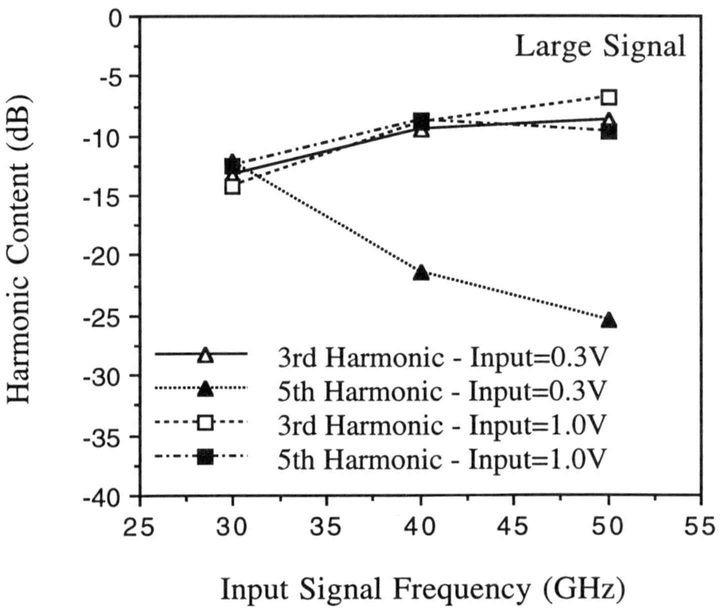

Fig. 9.27 Harmonic content of the output with respect to output at the fundamental frequency for the complete amplifier with the optimized transistor design. The signal input is 0.3V or 1.0V at 30, 40, or 50 GHz. *Source*: Imtiaz and El-Ghazaly, *IEEE Trans. Microwave Theory and Techniques*, 1997, pp. 2208–2216, © 1997 IEEE.

For the 0.3V input case, the third-harmonic relative output power levels are 4.8%, 11.6%, and 13% of the output power at the fundamental frequency for inputs at 30, 40, and 50 GHz, respectively For the 1.0V input case, the corresponding third-harmonic relative output power levels are 3.7%, 12.6%, and 21.2%.

For the 0.3V input case, the fifth-harmonic relative output power levels are 6%, 0.7%, and 0.3% of the fundamental power output at 30, 40, and 50 GHz, respectively. For the 1.0V input case, the corresponding fifth-harmonic relative output power levels are 5.5%, 13.3%, and 11%.

It is clear that a considerable amount of power is transferred to the harmonic components. The calculated large-signal (1.0V) efficiency of the amplifier at the fundamental is only 17.8%, 27.3%, and 20.5% at 30, 40, and 50 GHz, respectively. However, we note that, for the positive half-cycle of the 1.0V input signal, the gate of the transistor is forward-biased. This is partially responsible for the significant harmonic generation calculated for this case.

9.12 SUMMARY AND CONCLUSIONS

This chapter summarized means to self-consistently combine the modeling of full-wave electromagnetic fields with electron transport in GaAs, thereby permitting simulation of millimeter-wave transistors. The approach was then extended to model a complete millimeter-wave amplifier, including its matching sections. A strong feature of this technique is its ability to predict the gain and nonlinearity of transistors operating at many tens of gigahertz from first principles, and to quantify the resulting harmonic components generated at the amplifier output.

In the future, additional physical phenomena can be added to this model, including thermal and electromagnetic radiation effects. By incorporating all of these effects in the circuit simulation, the electrical engineer should be able to formulate complex circuit designs at millimeter-wave frequencies with a level of confidence similar to that presently enjoyed by circuit designers at much lower frequencies. Further, ideas for novel devices employing traveling-wave phenomena unique to this regime could be pursued in a rapid and inexpensive manner relative to the experimental option of fabricating and testing prototype devices and circuits.

ACKNOWLEDGMENT

The author thanks all of his colleagues whose results are used in this chapter, especially M. Megahed, S. M. Imtiaz, and M. Alsunaidi. Most of the results presented here have been generated from research supported by the U.S. Army Research Office under Contract DAAH04-95-1-0521.

REFERENCES

[1] El-Ghazaly, S., R. Joshi, and R. Grondin, "Electromagnetic and transport considerations in sub-picosecond photoconductive switch modeling," *IEEE Trans. Microwave Theory Tech.*, Vol. 38, 1990, pp. 629–637.

[2] Connolly, K., S. El-Ghazaly, R. Grondin, and R. Joshi, "Coupling Maxwell's equation time-domain solution with Monte Carlo technique to simulate ultrafast optically controlled switches," *1990 IEEE MTT-S Int. Microwave Symp. Dig.*, pp. 295–298.

[3] Lu, Y., R. Joshi, S. El-Ghazaly, and R. Grondin, "Time-domain finite difference and EMC study of hot-carrier transport in GaAs on a picosecond scale," *Solid State Electron Devices*, Vol. 32, 1989, pp. 1297–1301.

[4] El-Ghazaly, S., and T. Itoh, "Two-dimensional numerical simulation of short-gate-length GaAs MESFETs and application to the traveling Gunn domain phenomenon," *Int. J. Numerical Modeling*, Vol. 1, 1988, pp. 19–30.

[5] Taflove, A., "Re-inventing electromagnetics: supercomputing solution of Maxwell's equations via direct integration on space grids," *Computational Electromagnetics, Computer-Aided Design & Supercomputing Symposium*, Office of Research and Development, Central Intelligence Agency, Tyson's Corner, VA, July 1992.

[6] Taflove, A., *Computational Electrodynamics: The Finite-Difference Time-Domain Method*, Norwood, MA: Artech House, 1995.

[7] Sano, E., and T. Shibata, "Fullwave analysis of picosecond photoconductive switches," *IEEE J. Quantum Electron.*, Vol. 26, 1990, pp. 372–377.

[8] Harrigan, E., "The Cray Research — Los Alamos National Laboratory collaborative research and development agreement in computational electromagnetics," *Computational Electromagnetics, Computer-Aided Design & Supercomputing Symposium*, Office of Research and Development, Central Intelligence Agency, Tyson's Corner, VA, July 1992.

[9] Moglestue, C., *IEE Proceedings*, Vol. 130, Pt. 1, 1983, pp. 275–280.

[10] Pone, J.-F., R. Castagné, J.-P. Courat, and C. Arnodo, "Two-dimensional particle modeling of submicrometer gate GaAs FETs near pinchoff," *IEEE Trans. Electron Devices*, Vol. 29, 1982, pp. 1244–1255.

[11] Fauquembergure, R., M. Pernisek, and E. Constant, "Monte Carlo simulation of space-charge injection FET," *Electronics Letters*, Vol. 18, 1982, pp. 670–671.

[12] Ravaioli, U., and D. Ferry, "MODFET ensemble Monte Carlo model including the quasi-two-dimensional electron gas," *IEEE Trans. Electron Devices*, Vol. 33, 1986, pp. 677–681.

[13] Kuzuhara, M., T. Itoh, and K. Hess, "Ensemble Monte Carlo simulation of sub-0.1 μm gate length GaAs MESFETs," *Solid State Electronics*, Vol. 32, 1989, pp. 1857–1861.

[14] Chen, T.-H., and M. Shur, "Analytical models of ion-implanted GaAs FET's," *IEEE Trans. Electron Devices*, Vol. 32, 1987, pp. 18–28.

[15] Pucel, R., H. Haus, and H. Statz, "Noise characteristics of gallium arsenide field-effect transistors," *IEEE Trans. Electron Devices*, Vol. 21, 1974, pp. 549–562.

[16] Riser, M., "Large-scale numerical simulation in semiconductor device modeling," *Computer Methods in Applied Mathematics and Engineering*, Vol. 1, 1972, pp. 17–38.

[17] Yamaguchi, K., S. Asai, and H. Kodera, "Two-dimensional numerical analysis of stability criteria of GaAs FET's," *IEEE Trans. Electron Devices*, Vol. 23, 1976, pp. 1283–1290.

[18] Madjar, A., and F. Rosenbaum, "A large-signal model for the GaAs MESFET," *IEEE Trans. Microwave Theory Tech.*, Vol. 29, 1981, pp. 781–788.

[19] Curtice, W., and Y. Yun, "A temperature model for the GaAs MESFET," *IEEE Trans. Electron Devices*, Vol. 28, 1981, pp. 954–962.

[20] Shur, M., *GaAs Devices and Circuits*, New York: Plenum Press, 1987.

[21] Cook, R., and J. Frey, "Two-dimensional numerical simulation of energy transport effects in Si and GaAs MESFETs," *IEEE Trans. Electron Devices*, Vol. 29, 1982, pp. 970–977.

[22] El-Ghazaly, S., M. Lefebvre, G. Salmer, M. Ibrahim, and O. L. El-Sayed, " Two-dimensional FET simulation in non-stationary conditions," *Proc. 13th European Solid State Device Research Conf.*, Canterbury, England, 1983, p. 127.

[23] El-Sayed, O. L., S. El-Ghazaly, G. Salmer, and M. Lefebvre, "Performance analysis of sub-micron gate MESFETs," *Solid-State Electron.,* Vol. 30, 1987, pp. 643–654.

[24] Widiger, D., I. Kizilyalli, K. Hess, and J. Coleman, "Two-dimensional transient simulation of an idealized high electron mobility transistor," *IEEE Trans. Electron Devices*, Vol. 32, 1985, pp. 1092–1102.

[25] Snowden, C., and D. Loret, "Two-dimensional hot-electron models for short-gate-length GaAs MESFETs," *IEEE Trans. Electron Devices*, Vol. 34, 1987, pp. 212–223.

[26] El-Ghazaly, S., and T. Itoh, "Traveling-wave inverted-gate field-effect transistors: concept, analysis, and potential," *IEEE Trans. Microwave Theory Tech.,* Vol. 37, 1989, pp. 1027–1032.

[27] El-Ghazaly, S., and T. Itoh, "Inverted-gate GaAs field-effect transistors: novel high-frequency structures," *IEEE Trans. Electron Devices*, Vol. 35, 1988, pp. 810–817.

[28] Heinrich, W., "Distributed equivalent-circuit model for traveling-wave FET design," *IEEE Trans. Microwave Theory Tech.,* Vol. 35, 1987, pp. 487–491.

[29] Holden, A., D. Daniel, I. Davies, C. Oxley, and H. Rees, "Gallium arsenide traveling-wave field-effect transistors," *IEEE Trans. Electron Devices*, Vol. 32, 1985, pp. 61–66.

[30] Shawki, T., G. Salmer, and O. El-Sayed, "MODFET 2-D hydrodynamic energy modeling: optimization of subquarter-micron-gate structure," *IEEE Trans. Electron Devices*, Vol. 37, 1990, pp. 21–30.

[31] Blotekjaer, K., "Transport equations for electrons in two-valley semiconductors," *IEEE Trans. Electron Devices*, Vol. 17, 1970, pp. 38–47.

[32] Shur, M., "Influence of nonuniform field distribution on frequency limits of GaAs field-effect transistors," *Electron Lett.,* Vol. 12, 1976, pp. 615–616.

[33] Carnez, B., A. Cappy, A. Kaszynski, E. Constant, and G. Salmer, "Modeling of a sub-micrometer gate field-effect transistors including effects of nonstationary electron dynamics," *J. Appl. Phys.* Vol. 51, 1980, pp. 784–790.

[34] Feng, Y. K., and A. Hintz, "Simulation of sub-micrometer GaAs MESFETs using a full dynamic transport model," *IEEE Trans. Electron Devices*, Vol. 35, 1988, pp. 1419–1431.

[35] Reiser, M., "A two-dimensional numerical FET model for dc, ac, and large-signal analysis," *IEEE Trans. Electron Devices*, Vol. 20, 1973, pp. 35–45.

[36] Taflove, A., and M. E. Brodwin, "Numerical solution of steady-state electromagnetic scattering using the time-dependent Maxwell's equations," *IEEE Trans. Microwave Theory Tech.*, Vol. 23, 1975, pp. 623–630.

[37] Hoffman, K. A., *Computational Fluid Dynamics for Engineers*, 1989.

[38] Yee, K. S., "Numerical solution of initial boundary value problems involving Maxwell's equations in isotropic media," *IEEE Trans. Antennas Propagat.*, Vol. 14, 1966, pp. 302–307.

[39] Railton, C. J., E. M. Daniel, D. Paul, and J. P. McGeehan, "Optimized absorbing boundary conditions for the analysis of planar circuits using the finite-difference time-domain method," *IEEE Trans. Microwave Theory Tech.*, Vol. 41, 1993, pp. 290–297.

[40] Railton, C. J., and E. M. Daniel, "A comparison of the properties of radiating boundary conditions in the FDTD method for finite discretization and non-planar waves," *IEEE Trans. Antennas Propagat.*, Vol. 42, 1994, pp. 276–281.

[41] Higdon, R. L., "Numerical absorbing boundary conditions for the wave equation," *Math. Comp.*, Vol. 49, 1987, pp. 65–91.

[42] Balanis, C. A., *Advanced Engineering Electromagnetics*, New York: Wiley, 1989.

[43] Berenger, J.-P., "A perfectly matched layer for the absorption of electromagnetic waves," *J. Computational Physics*, Vol. 114, 1994, pp. 185–200.

[44] Akello, R. J., B. Easter, and I. M. Stephenson, "Effects of microstrip discontinuities on GaAs MESFET amplifier gain performance," *Electronics Lett.*, Vol. 13, 1977, pp. 160–162.

[45] Hoffmann, R. K., *Handbook of Microwave Integrated Circuits*, Norwood, MA: Artech House, 1987.

[46] Dydyk, M., "Master the T-junction and sharpen your MIC designs," *Microwaves*, 1977, pp. 119–121.

[47] Megahed, M., and S. El-Ghazaly, "Discontinuity effects on high-frequency transistors," *Proc. 1997 IEEE MTT-S Int. Microwave Symp.*, pp. 609–612.

[48] Hakim, M. S., and S. M. El-Ghazaly, "Analysis of wave propagation effects on microwave field effect transistors," *Proc. 1993 IEEE MTT-S Int. Microwave Symp.*, pp. 1303–1306.

[49] Imtiaz, S. M., and S. M. El-Ghazaly, "Global modeling of millimeter-wave circuits: electromagnetic simulation of amplifiers," *IEEE Trans. Microwave Theory Tech.*, Vol. 45, 1997, pp. 2208–2216.

[50] Kobayashi, K., Y. Nemoto, and R. Sato, "Kuroda's identity for mixed lumped and distributed circuits and their application to nonuniform transmission lines," *IEEE Trans. Microwave Theory Tech.*, Vol. 29, 1981, pp. 81–86.

[51] Cottee, R. M., and W. T. Joines, "Synthesis of lumped and distributed networks for impedance matching of complex loads," *IEEE Trans. Circuits Syst.*, Vol. 26, 1979, pp. 316–329.

[52] Endo, I., Y. Nemoto, and R. Sato, "Design of transformerless quasi-broad-band matching networks for lumped complex loads using nonuniform transmission lines," *IEEE Trans. Microwave Theory Tech.*, Vol. 36, 1988, pp. 629–634.

[53] Nemoto, Y., K. Kobayashi, and R. Sato, "Equivalent transformations for mixed-lumped Richards section and distributed transmission line," *IEEE Trans. Microwave Theory Tech.*, Vol. 36, 1988, pp. 635–641.

[54] Winklestein, D., et al., "Simulation of arbitrary transmission line networks with nonlinear terminations," *IEEE Trans. Circuits Syst.*, Vol. 38, 1991, pp. 418–422.

[55] Sui, W., D. A. Christensen, and C. H. Durney, "Extending the two-dimensional FDTD method to hybrid electromagnetic systems with active and passive lumped elements," *IEEE Trans. Microwave Theory Tech.*, Vol. 40, 1992, pp. 724–730.

[56] Piket-May, M., A. Taflove, and J. Baron, "FDTD modeling of digital signal propagation in 3-D circuits with passive and active loads," *IEEE Trans. Microwave Theory Tech.*, Vol. 42, 1994, pp. 1514–1523.

[57] Voelker, R. H., and R. J. Lomax, "A finite-difference transmission line matrix method incorporating a nonlinear device model," *IEEE Trans. Microwave Theory Tech.*, Vol. 38, 1990, pp. 302–312.

[58] Russer, P., P. P. M. So, and W. J. R. Hoefer, "Modeling of nonlinear active regions in TLM," *IEEE Microwave and Guided Wave Lett.*, Vol. 1, 1991, pp. 10–13.

[59] Tsuei, Y.-S., A. C. Cangellaris, and J. L. Prince, "Rigorous electromagnetic modeling of chip-to-package (first-level) interconnections," *IEEE Trans. Components, Hybrids, and Manufacturing Tech.*, Vol. 16, 1993, pp. 876–883.

[60] Thomas, V. A., M. E. Jones, M. Piket-May, A. Taflove, and E. Harrigan, "The use of SPICE lumped circuits as sub-grid models for FDTD analysis," *IEEE Microwave and Guided Wave Lett.*, Vol. 4, 1994, pp. 141–143.

[61] Kuo, C.-N., V. A. Thomas, S. T. Chew, B. Houshmand, and T. Itoh, "Small signal analysis of active circuits using FDTD algorithm," *IEEE Microwave and Guided Wave Lett.*, Vol. 5, 1995, pp. 216–218.

[62] Kuo, C.-N., R.-B. Wu, B. Houshmand, and T. Itoh, "Modeling of microwave active devices using the FDTD analysis based on the voltage-source approach," *IEEE Microwave and Guided Wave Lett.*, Vol. 6, 1996, pp. 199–201.

[63] Guillouard, K., M.-F. Wong, V. F. Hanna, and J. Citerne, "A new global finite element analysis of microwave circuits including lumped elements," *IEEE Trans. Microwave Theory Tech.*, Vol. 44, 1996, pp. 2587–2594.

[64] Brameller, A., M. N. John, and M. R. Scott, *Practical Diakoptics for Electrical Networks*, London, U. K.: Chapman and Hall, 1969.

[65] Brewitt-Taylor, C. R., and P. B. Johns, "On the construction and numerical solution of transmission-line and lumped network models of Maxwell's equations," *Int. J. Num. Meth. Engg.*, Vol. 15, 1980, pp. 13–30.

[66] Johns, P. B., and K. Akhtarzad, "The use of time-domain diakoptics in time-discrete models of fields," *Int. J. Num. Meth. Engg.*, Vol. 17, 1981, pp. 1–14.

[67] Hoefer, W. J. R., "The discrete time-domain Green's function or Johns matrix — A new powerful concept in transmission line modeling (TLM)," *Int. J. Num. Meth. Engg.*, Vol. 2, 1989, pp. 215–225.

[68] Dubard, J. L., D. Pompei, J. Le Roux, and A. Papiernik, "Characterization of microstrip antennas using the TLM simulation associated with a Prony-Pisarenko method," *Int. J. Num. Modeling: Electronic Networks, Devices and Fields*, Vol. 3, 1990, pp. 269–285.

[69] Ko, W., and R. Mittra, "A combination of FDTD and Prony methods for analyzing microwave integrated circuits," *IEEE Trans. Microwave Theory Tech.*, Vol. 39, 1991, pp. 2176–2181.

[70] Hammadi, S. M., and S. El-Ghazaly, "Extending FDTD/convolution techniques to 3-D microwave structures," *Proc. IEEE Antennas and Propagation Soc. Int. Symp.*, 1997, pp. 2180–2183.

[71] Hammadi, S. M., and S. El-Ghazaly, "Simulation of large electromagnetic structures with an application to quasi-optical amplifier;" *Proc. 1997 European Microwave Conf.*, Vol. II, pp. 870–876.

[72] Kuempel, W., and I. Wolff, "Digital signal processing of time-domain field simulation results using the system identification method," *IEEE MTT-S Int. Symp. Dig.*, Vol. 2, 1992, pp. 793–796.

[73] Houshmand, B., T. W. Huang, and T. Itoh, "Microwave structure characterization by a combination of FDTD and system identification methods," *IEEE Microwave and Guided Wave Lett.*, Vol. 3, 1993, pp. 262–264.

Chapter 10

Microcavity Resonators

Susan C. Hagness

10.1 INTRODUCTION

The FDTD method has been extensively applied to model and design complicated RF and microwave devices. Until recently, most optical devices had distance and / or time scales that made them inappropriate for FDTD analysis. However, for the emerging class of micrometer- and nanometer-scale integrated optical devices, FDTD modeling has the potential to play a useful and practical design role.

Recent advances in nanofabrication techniques permit optical resonators having physical dimensions of the order of the optical wavelength ("microcavities") to be seriously considered as components in high-density photonic integrated circuits. Such components include wideband tunable filters, ultrafast all-optical switches, low-threshold lasers, and highly efficient light-emitting diodes (for example, see [1 – 4]). Microcavity structures can be categorized into two groups based on the physical mechanism by which the cavity confines the optical mode [5].

1. Group 1: Confinement by multiple reflections, as in vertical-cavity surface-emitting lasers (VCSELs) and photonic crystals;
2. Group 2: Confinement by total internal reflection, as in whispering-gallery-mode disk and single-mode ring cavities.

Microcavity resonators offer several desirable properties such as a very high quality factor (Q), a large spacing between longitudinal mode resonances, and enhanced spontaneous-emission and stimulated-emission rates. Further, they are suitable for large-scale, high-density integration.

This chapter reviews the application of FDTD to the modeling of integrated optical microcavities. Section 10.2 discusses basic issues related to modeling optical phenomena including optical waveguide dispersion, material dispersion, and optical nonlinearities. Section 10.3 reviews an FDTD formulation suitable for modeling a saturable homogeneously broadened gain medium, and then focuses on the modeling of VCSELs. A comparison is made between uniform-gain and periodic-gain configurations within the microcavity. Sections 10.4 and 10.5 discuss integrated optical microcavities in which light propagates parallel to, rather than perpendicular to, the wafer substrate. Section 10.4 presents investigations of one-dimensional PBG structures, and considers the resonant modes that arise from the introduction of a defect in the photonic crystal. Finally, Section 10.5 provides FDTD simulation results for microcavity ring and disk resonators that are coupled to adjacent input and output waveguides via the evanescent field. Key design parameters and tradeoffs are discussed.

10.2 ISSUES RELATED TO FDTD MODELING OF OPTICAL PHENOMENA

The FDTD method permits accurate analysis of the electromagnetic radiation, scattering, diffraction, reflection, and polarization effects that are inherent in integrated optical microcavity resonators. This section discusses considerations for modeling highly dispersive optical waveguides in the context of FDTD Maxwell's equations solutions. The discussion also includes a brief review of algorithms that have been developed for FDTD modeling of the frequency-dependent linear and nonlinear properties of optical materials. A more in-depth treatment of optical gain media is presented in Section 10.3.

10.2.1 Optical Waveguides

The submicron width and large refractive index contrast of waveguides commonly found in micron-scale integrated optical devices leads to significant waveguide dispersion. The relatively strong dependence of the waveguide propagation constant upon frequency impacts several key design parameters for microcavity resonators composed of such waveguides, including their Q and the free spectral range (FSR), defined as the spacing in frequency or wavelength between adjacent resonances. Accurate FDTD modeling of highly dispersive waveguides requires careful consideration of two important issues:

- How to terminate waveguides that extend beyond the grid boundaries;
- How to source specific waveguide modes for the case of pulsed excitation.

The accuracy of the FDTD method for problems in linear optics was first demonstrated for the directional coupler [6]. Since then, Berenger's PML ABC [7] has been introduced. The improved performance of the PML ABC relative to earlier techniques has provided the increased computational dynamic range required for sensitive numerical simulations, such as the analysis of antireflection coatings [8] or resistively loaded antennas [9]. Of particular importance here is the demonstration by Reuter et al.

[10] that multimodal and dispersive waveguiding structures extending beyond the edges of the FDTD grid can be terminated accurately by the PML ABC with reflection coefficients below -75 dB for a wide range of group velocities.

Consider a dielectric slab waveguide and its TE optical modes. Sourcing the numerical analog of a propagating TE mode is accomplished by specifying an electric field distribution along a one-dimensional transverse cross section of the waveguide. For the case of sinusoidal excitation, the transverse field distribution of the desired guided mode is obtained by numerically solving a determinantal equation for the transverse propagation constant and decay constants. For example, given the refractive indices of the core and cladding layers, the dispersion relations for the various waveguide modes can be computed [11]. With these data, the longitudinal propagation constant, β, corresponding to the desired frequency of excitation and mode number is selected and used to calculate the transverse propagation constant and decay constants.

The case of a pulsed source, such as a Gaussian pulse modulating a carrier wave, poses more of a challenge. Even if the frequency spectrum of the pulse is narrow enough and the carrier frequency is low enough to allow only the fundamental mode to propagate, there still exists a spread of β values along the fundamental mode dispersion curve corresponding to the spread of frequencies of interest, as illustrated in Fig. 10.1.

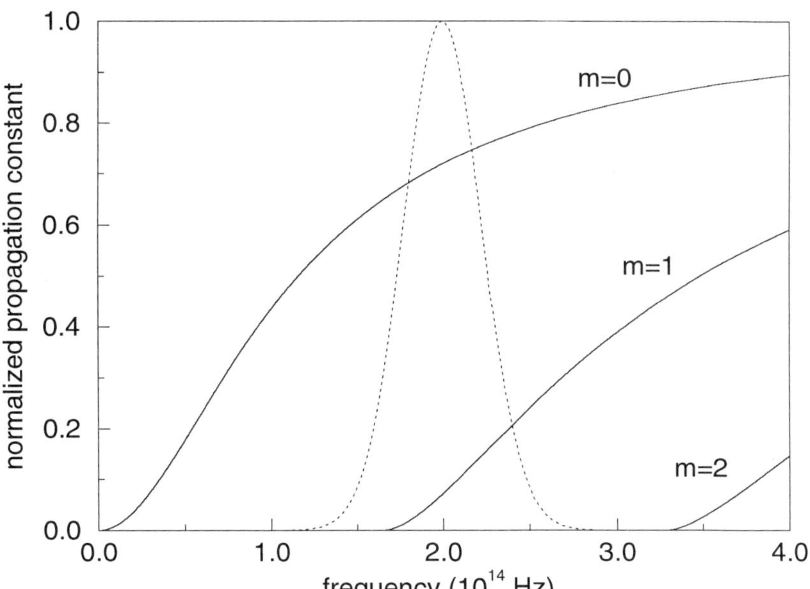

Fig. 10.1 Excitation spectrum (dotted line) superimposed on the waveguide dispersion relations (solid lines) for the first three TE modes of a symmetric strongly guiding dielectric slab waveguide. The index of refraction for the 0.3-μm-wide core is 3.2; for the cladding layers, $n = 1.0$.

In this case, the transverse mode profile for the pulse can be approximated using the value of β at the carrier frequency. The fact that the transverse distribution of the source is not exact for all frequencies in the pulse spectrum means that a small amount of energy is shed from the waveguide early in the FDTD simulation. The accuracy of simulations in which the waveguide is immediately surrounded by the PML ABC, as in the examples of Section 10.4, is not affected by this shed energy. However, Section 10.5 presents examples in which shed energy from the waveguide interacts with other structures in the grid, thereby introducing unwanted noise into the computation.

An effective solution to this problem is to use "bootstrapping," that is, to perform a preliminary FDTD simulation of a sufficiently long version of the waveguide, and store the time history of the transverse electric field distribution as the pulse passes an observation plane at the far end of the grid. The stored data are used as the waveguide source excitation in all subsequent simulations. The level of numerical noise introduced into the grid from this bootstrapped source falls well below the noise level arising from reflections at the PML ABC. Therefore, for all practical purposes, this method serves as a perfect pulsed source condition.

The accuracy of the FDTD method for modeling propagating waveguide modes is now demonstrated in a sample computation of the longitudinal propagation constant, β, and the group velocity, v_g, of a strongly guiding planar waveguide. Here, the waveguide consists of a 0.3-μm-wide core of high refractive index ($n = 3.2$) having low-refractive-index ($n = 1.0$) cladding on both sides. The PML ABC is used to terminate the waveguide in the grid. Since the only structure in this simulation is the straight waveguide, a bootstrapped source is not required and the inexact mode profile for the pulsed source of the waveguide is used. (Any energy shed from the waveguide is absorbed by the PML ABC.) The waveguide must simply be made long enough to ensure that the mode has settled down before recording any data.

A single 20-fs Gaussian pulse modulating a carrier of frequency $\nu_0 = 2.0 \times 10^{14}$ Hz is launched at one end of the grid using the fundamental ($m = 0$) mode profile computed for ν_0. Fig. 10.1 shows the source spectrum superimposed on the normalized waveguide dispersion curves for the first three TE modes. At ν_0, this waveguide can support only the $m = 0$ mode, which has spatially even symmetry about the longitudinal axis of the waveguide, and the $m = 1$ mode, which has odd symmetry. Because the numerical excitation has precisely even symmetry, only the $m = 0$ mode is generated by the impulsive source. At the far end of the grid, two fixed observation points located along the longitudinal axis of the core record electric-field data as the pulse passes. By taking the ratio of the DFTs of the two time histories, β is computed over the full bandwidth of the pulse.

Fig. 10.2(a) compares the FDTD-computed β as a function of frequency with the exact denormalized dispersion curve. Fig. 10.2(b) compares the corresponding FDTD-computed and exact variations of v_g with frequency. Here, v_g is computed from the numerical data for β using a second-order-accurate central difference approximation to the derivative. The error in β at ν_0 is less than 0.2% at a grid resolution of 13.6 nm. Similar accuracy is obtained for v_g.

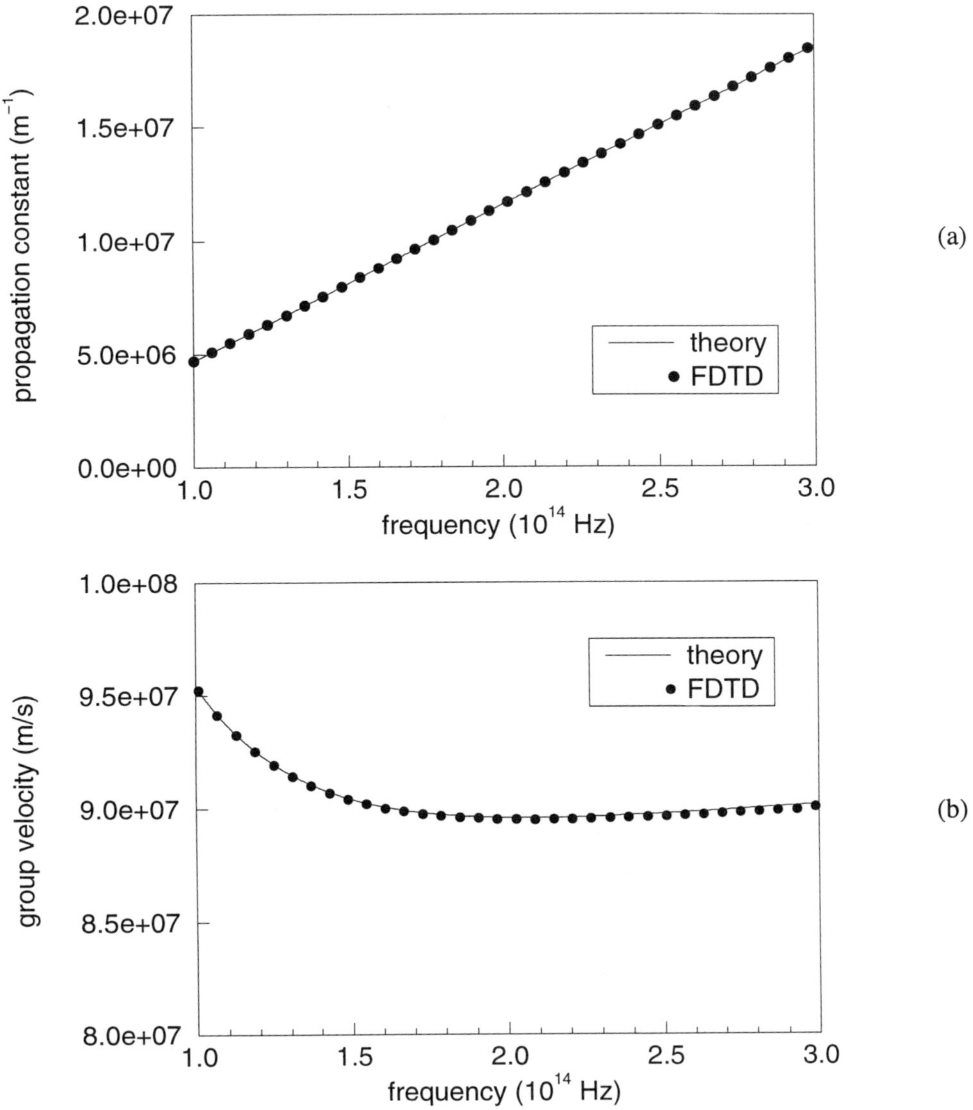

Fig. 10.2 Comparison of FDTD-computed and theoretical values for properties of the fundamental TE mode propagating in the strongly guiding dielectric-slab waveguide characterized in Fig. 10.1: (a) propagation constant, β; (b) group velocity, v_g.

10.2.2 Material Dispersion and Nonlinearities[1]

To fully treat the broadband behavior of many integrated optical devices, the effect of electronic material dispersion may need to be included in the FDTD calculations. As discussed in Chapter 1 (Section 1.5.1), a number of different methods have been proposed over the past several years to account for material dispersion with absorption. These include RC methods that implement a discrete recursive convolution of the dispersion relation [12, 13], ADE methods that discretize an auxiliary differential equation obtained from the relevant constitutive relation [14, 15], and Z-transform methods based on digital signal processing techniques [16].

Additional extensions to the ADE and Z-transform methods have enabled the modeling of $\chi^{(2)}$ and $\chi^{(3)}$ nonlinear optical materials [17 – 20]. With this advent, the range of modeling applications has been substantially expanded to include temporal and spatial soliton propagation [17, 19 – 22], self-focusing of optical beams [18], scattering from linear-nonlinear interfaces [23], pulse propagation through nonlinear corrugated waveguides [24], pulse-selective behavior in nonlinear Fabry-Perot cavities [25], and second-harmonic generation in nonlinear waveguides [19].

FDTD algorithms for modeling optical gain media on a macroscopic level have also been reported. Using their approach for modeling linear dispersion, Hawkins and Kallman selected parameters to provide gain rather than loss and modeled tilted waveguide traveling wave amplifiers [26]. Hagness et al. developed an ADE formulation to account for materials exhibiting frequency-dependent gain with saturation, and applied the technique to model VCSELs [27]. The details of this work are highlighted in the next section. Advances in modeling the interaction of light with optical materials on a microscopic level have enabled the modeling of ultrafast pulse interactions with two-level atoms [28]. This approach involves the solution of the semiclassical Maxwell-Bloch equations and has been used to model self-induced transparency.

10.3 VERTICAL-CAVITY SURFACE-EMITTING LASERS

This section begins with a review of an FDTD formulation that permits modeling frequency-dependent optical gain media, including the effect of gain saturation. Two validation studies illustrate the potential for high accuracy. The approach is applied to the class of microcavities (VCSELs) where light propagates perpendicular to the substrate. A comparison between uniform-gain and periodic-gain configurations within the laser cavity illustrates the potential for a substantial reduction in the lasing threshold for a VCSEL having a periodic-gain structure.

[1] The references listed in this section provide examples of algorithms and applications for optical devices. For additional citations, the reader is referred to Chapter 1, references 156 to 193 and 228 to 243.

10.3.1 FDTD Modeling of Optical Gain Media[2]

The ADE method described below incorporates the frequency-dependent linear and nonlinear properties of optical gain media through the constitutive relation for the current density, as reported in [27]. The idea of using a negative conductance for gain is an extension of the decades-old concept in microwave engineering and electronics. For example, the tunnel diode has a long history of use as a microwave negative-resistance amplifier. The algorithm assumes that the optical gain medium is homogeneously broadened wherein the atoms of the gain medium are indistinguishable and have the same atomic transition frequency, ω_0. The relaxation processes are included in a phenomenological manner through the time constant, T_2. This dipole relaxation time accounts for the fact that any phase coherence introduced into the system of atoms by the action of an electric field is lost in a finite time interval once the electric field is turned off. Therefore, the small-signal linear gain coefficient is governed in the frequency domain by a single Lorentzian profile having a width determined by T_2. The gain coefficient should also include the large-signal nonlinear effect of saturation which is due to the decrease of the population inversion with field intensity.

For simplicity, the algorithm is described in the context of a one-dimensional problem with electric and magnetic field components, E_z and H_y, propagating along the x direction through a nonmagnetic, isotropic medium. In this case, Maxwell's curl equations are

$$\frac{\partial H_y}{\partial t} = \frac{1}{\mu_0}\frac{\partial E_z}{\partial x} \tag{10.1}$$

$$J_z + \varepsilon_r \varepsilon_0 \frac{\partial E_z}{\partial t} = \frac{\partial H_y}{\partial x} \tag{10.2}$$

The frequency-dependent conductivity that relates the electric field and the current density is given by

$$\sigma(\omega) = \frac{J_z(\omega)}{E_z(\omega)} = \frac{1}{1 + I/I_s}\left[\frac{\sigma_0 / 2}{1 + j(\omega - \omega_0)T_2} + \frac{\sigma_0 / 2}{1 + j(\omega + \omega_0)T_2}\right] \tag{10.3}$$

where Hermitian symmetry is used for the Lorentzian. In (10.3), σ_0 is related to the peak value of the gain set by the pumping level and the resulting population inversion, and I_s is the saturation intensity. Since $\sigma(\omega)$ accounts for any gain (or loss) in the medium, the susceptibility can be reduced to a frequency-independent real constant such that $D_z = \varepsilon E_z$ where $\varepsilon = \varepsilon_r \varepsilon_0$.

---

[2]For a more detailed physical description of the physics of a homogeneously broadened gain medium, the reader is referred to a quantum electronics textbook (for example, see [29, 30]).

To see how this gives gain, consider the linear case when I is small compared with I_s. The right-hand side of (10.3) simplifies and is separated into real and imaginary parts:

$$\sigma(\omega) = \sigma_R(\omega) + j\sigma_I(\omega)$$
$$= \frac{\sigma_0\left[1+\left(\omega_0^2+\omega^2\right)T_2^2\right]}{\left[1+\left(\omega_0^2-\omega^2\right)T_2^2\right]^2+4\omega^2T_2^2} + \frac{j\sigma_0\,\omega T_2\left[-1+\left(\omega_0^2-\omega^2\right)T_2^2\right]}{\left[1+\left(\omega_0^2-\omega^2\right)T_2^2\right]^2+4\omega^2T_2^2} \tag{10.4}$$

Upon assuming that the propagating wave is of the form

$$E_z(x,t) = E_0\exp(-\alpha x)\exp\left[-j(\beta x-\omega t)\right] \tag{10.5}$$

and that the material is low-gain (or low-loss) such that

$$\frac{\sigma_R(\omega)}{\omega\,\varepsilon_0\,\varepsilon_{r,\text{eff}}(\omega)} << 1 \tag{10.6a}$$

where the effective relative permittivity is

$$\varepsilon_{r,\text{eff}}(\omega) = \varepsilon_r - \frac{\sigma_I(\omega)}{\omega\,\varepsilon_0} \tag{10.6b}$$

it can be shown that [27]

$$\alpha(\omega) \cong \frac{\sigma_R(\omega)}{2c\,\varepsilon_0\,n_{\text{eff}}} \tag{10.7a}$$

$$\beta(\omega) \cong n_{\text{eff}}\frac{\omega}{c} \tag{10.7b}$$

where $n_{\text{eff}} = \sqrt{\varepsilon_{r,\text{eff}}}$. From (10.7a), we conclude that if $\sigma_R(\omega)$ is negative, then α is negative, providing gain. According to (10.4), this is the case when the parameter σ_0 is chosen to be negative. Physically, this corresponds to a population inversion.

Taking the inverse Fourier transformation of (10.3) provides the following auxiliary differential equation that is solved simultaneously with (10.2):

$$(1+\omega_0^2T_2^2)J_z + 2T_2\frac{\partial J_z}{\partial t} + T_2^2\frac{\partial^2 J_z}{\partial t^2} = s\sigma_0 E_z + s\sigma_0 T_2\frac{\partial E_z}{\partial t} \tag{10.8}$$

Here $s \equiv (1+I/I_s)^{-1}$ is the saturation coefficient that contains feedback information of the latest peak electric field. The saturation intensity, I_s, is treated as a frequency-independent constant; the spatially dependent intensity, I, is treated as a feedback parameter in time. An auxiliary variable is defined in order to write (10.8) as two first-order differential equations:

$$F_z \equiv \frac{\partial J_z}{\partial t} \tag{10.9}$$

$$(1+\omega_0^2 T_2^2) J_z + 2T_2 F_z + T_2^2 \frac{\partial F_z}{\partial t} = s\sigma_0 E_z + s\sigma_0 T_2 \frac{\partial E_z}{\partial t} \tag{10.10}$$

The coupled differential equations (10.2), (10.9), and (10.10) are converted to finite-difference expressions using second-order central differences at time step $n+1/2$. Solving the resulting system for $F_z\big|_i^{n+1}$, $J_z\big|_i^{n+1}$, and $E_z\big|_i^{n+1}$ yields the following explicit update equations:

$$F_z^{n+1}(i) = A_1\left(H_y\big|_{i+1/2}^{n+1/2} - H_y\big|_{i-1/2}^{n+1/2}\right) + A_2 E_z\big|_i^n + A_3 J_z\big|_i^n + A_4 F_z\big|_i^n \tag{10.11}$$

$$J_z\big|_i^{n+1} = J_z\big|_i^n + \frac{\Delta t}{2}\left(F_z\big|_i^{n+1} + F_z\big|_i^n\right) \tag{10.12}$$

$$E_z\big|_i^{n+1} = E_z\big|_i^n - \frac{\Delta t}{2\varepsilon}\left(J_z\big|_i^{n+1} + J_z\big|_i^n\right) + \frac{\Delta t}{\varepsilon\Delta x}\left(H_y\big|_{i+1/2}^{n+1/2} - H_y\big|_{i-1/2}^{n+1/2}\right) \tag{10.13}$$

where

$$A_1 = \frac{4\Delta t\, s(i)\sigma_0(\Delta t + 2T_2)}{\beta\Delta x} \tag{10.14a}$$

$$A_2 = \frac{8\varepsilon\, s(i)\sigma_0\Delta t}{\beta} \tag{10.14b}$$

$$A_3 = -\frac{4\Delta t\left[2\varepsilon(1+\omega_0^2 T_2^2) + s(i)\sigma_0(\Delta t + 2T_2)\right]}{\beta} \tag{10.14c}$$

$$A_4 = -\frac{8\varepsilon T_2(\Delta t - T_2)}{\beta} - \frac{(\Delta t)^2\left[2\varepsilon(1+\omega_0^2 T_2^2) + s(i)\sigma_0(\Delta t + 2T_2)\right]}{\beta} \tag{10.14d}$$

and

$$\beta = 8\varepsilon T_2(\Delta t + T_2) + (\Delta t)^2\left[2\varepsilon(1+\omega_0^2 T_2^2) + s(i)\sigma_0(\Delta t + 2T_2)\right] \tag{10.15a}$$

$$s(i) = \left[1 + \frac{I(i)}{I_s}\right]^{-1}, \qquad I(i) = 0.5cn\varepsilon_0\left(E_z\big|_i^{\text{peak}}\right)^2 \tag{10.15b}$$

For a linear medium, there is no feedback; the saturation coefficient reduces to $s = 1$ because the intensity is negligible compared to the saturation intensity. For a nonlinear medium, the saturation coefficient is updated as follows. If the electric field at time step, n, is greater than the electric field at $n-1$ at the same location in space, then the saturation coefficient is updated using the latest value of the electric field. On the other hand, if the electric field at n has decreased from its previous value, then the saturation coefficient is not updated; hence $s(i)$ remains based on the latest peak electric field. In this manner, intensity feedback in the time domain retains as much as possible its frequency-domain meaning. Note that since the feedback is performed independently at each grid location, this model represents a spatially inhomogeneously broadened medium in which spatial hole burning can occur.

The following update expression for H_y is obtained at time step, n, in the standard manner from (10.1):

$$H_y\big|_{i+1/2}^{n+1/2} = H_y\big|_{i+1/2}^{n-1/2} + \frac{\Delta t}{\mu_0\,\Delta x}\left(E_z\big|_{i+1}^{n} - E_z\big|_i^{n}\right) \tag{10.16}$$

The complete FDTD time-stepping algorithm for a single-Lorentzian optical gain medium is given by (10.16), (10.11), (10.12), and (10.13). The computational model is a four-step recursive process that retains the fully explicit nature of the original frequency-independent FDTD algorithm and requires storage of fields only one step back in time. It is easily extended to arbitrary two- and three-dimensional problems with heterogeneous media. More complicated gain spectra can be approximated using a linear combination of Lorentzians.

10.3.2 Validation Studies

First, the accuracy of the algorithm is reviewed for a linear (small-signal) gain medium. As reported in [27], the electric field propagation factor, defined as the ratio of the Fourier transforms $E_z(x_2, \omega) / E_z(x_1, \omega)$, is computed for a linear GaAs gain medium having $\lambda_0 = 0.89$ μm, $T_2 = 0.07$ ps (gain spectrum bandwidth = 5 THz), and $n = 3.59$. A Gaussian pulse of carrier frequency, ω_0, and bandwidth greater than 5 THz is excited in a one-dimensional grid filled entirely with the gain medium. Data are recorded every time step at two observation points separated by a distance of $l = x_2 - x_1 = \lambda_0 / n$. By taking the ratio of the DFTs of the time-domain waveforms recorded at x_1 and x_2, the complex-valued propagation factor is calculated over the full bandwidth of the pulse. Fig. 10.3 compares with the exact values the FDTD-computed amplification factor $e^{-\alpha l}$ and phase βl. The latter is corrected by $n\omega l / c$ to give the phase due solely to the presence of the gain medium.

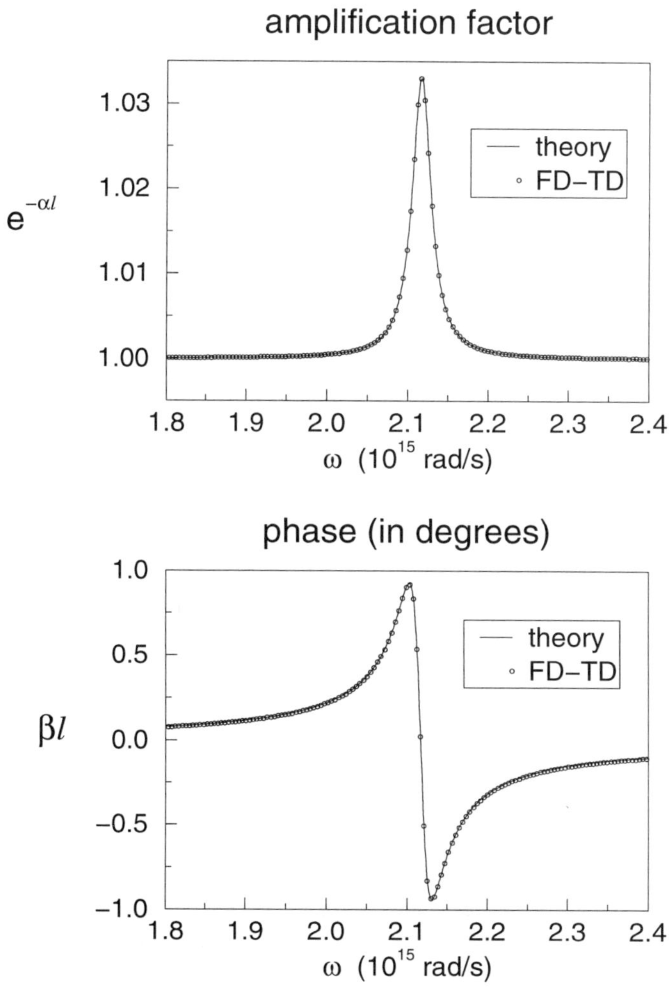

Fig. 10.3 Comparison of FDTD results and exact theory for the amplification factor and phase spectra of a pulse propagating a distance of one dielectric wavelength in a linear gain medium with $\sigma_0 = -5000$ S/m. *Source*: Hagness et al., *Radio Science*, 1996, pp. 931–941.

Fig. 10.3 shows that, for a very fine grid resolution of $\Delta x = \lambda_0 / 400n = 0.62$ nm, the deviation from the exact values over the complete frequency range is less than 2 parts per 10,000 in $e^{-\alpha l}$ and less than 0.006° in βl. This indicates the validity and potential for high accuracy of the FDTD model over extremely large instantaneous bandwidths.

Next, the accuracy of the algorithm is reviewed for a nonlinear (large-signal) gain medium. The model used for this study [27] consists of a one-dimensional Fabry-Perot cavity filled entirely with a GaAs gain medium ($\lambda_0 = 0.89$ μm, $T_2 = 0.07$ ps, $n = 3.59$, and $I_s = 65.2$ kW/cm^2) of length $l = 12.4$ μm. The mirrors at the ends of the laser cavity are formed by planar GaAs/air interfaces. Therefore, the reflectivity, R, of each end facet is independent of frequency. The mth resonance for this simple cavity is given by $\lambda_m = 2nl/m$. Fig. 10.4 shows an above-threshold unsaturated gain curve for $\sigma_0 = -7000$ S/m superimposed on the cavity resonances marked by the vertical lines. Three longitudinal modes lie in the region between 880 and 900 nm where the unsaturated gain exceeds the round-trip loss. Since this is a homogeneously broadened system, the longitudinal mode with the highest unsaturated gain clamps the gain curve at the loss line, yielding single-mode operation. In this example, the frequency of the lasing mode is designed to be the $m = 100$ resonance, which corresponds to λ_0, that is, the peak of the gain curve. For a cavity with no internal loss, the minimum material gain required to achieve lasing (the point at which the round-trip loss of the cavity equals the round-trip gain) is given by the gain threshold, $\alpha_{\text{th}} = (1/2l)\ln R$. Using (10.7a) to determine the threshold value of σ_0 yields $\sigma_{0,\text{th}} = -1760$ S/m.

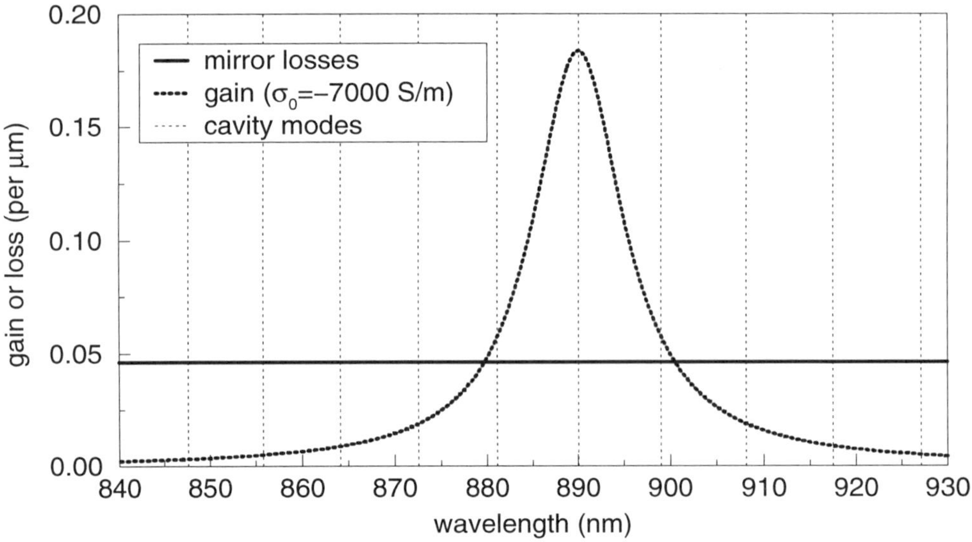

Fig. 10.4 Unsaturated gain and loss spectra for a Fabry-Perot microcavity laser. The longitudinal modes of the cavity are shown as vertical lines. *Adapted from*: Hagness et al., *Radio Science*, 1996, pp. 931–941.

Within a semiclassical framework, spontaneous emission can be included in Maxwell's equations as a noise current [31]. A pseudorandom number generator for a Gaussian random number sequence [32] with zero mean and a variance of 1.0×10^{-6}

(a standard deviation of $\sigma_x = 0.001$) is used to implement a noise current in the FDTD model of the laser cavity. The approximately Gaussian random number sequence is obtained by summing uniformly distributed random numbers; in this case, the generation of one Gaussian pseudorandom number, X, requires the generation of twelve uniformly random numbers. Fig. 10.5 shows a histogram of the first 10,000 outputs for X. Here, each bar gives the relative frequency of X falling within that range. The noise current is implemented in the FDTD algorithm at each time step as an additive virtual current source at an electric field component located within the laser cavity:

$$E_z\big|_i^{n+1} = E_z\big|_i^{n+1} + X \qquad (10.17)$$

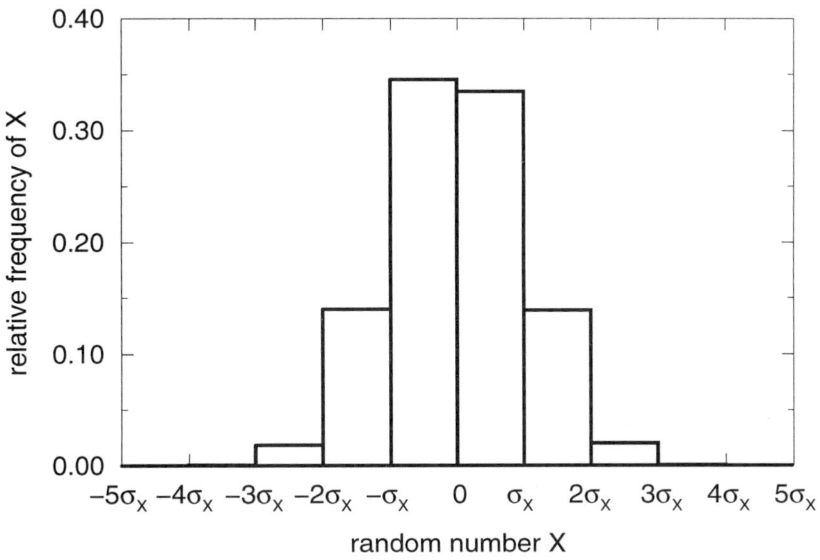

Fig. 10.5 Histogram of the first 10,000 outputs of a pseudorandom number generator for a Gaussian random number with mean = 0.0 and variance = σ_x^2.

Fig. 10.6(a) shows the FDTD-computed time evolution of the electric field outside the laser cavity for $\sigma_0 = -7000$ S/m. The electric field oscillations build up from the noise background and then level off to a steady-state amplitude as the gain saturates. Fig. 10.6(b) shows an expanded time scale of the sinusoidal steady-state region of Fig. 10.6(a), which illustrates that the FDTD model correctly predicts a lasing frequency of ω_0. This simulation is then repeated for smaller values of σ_0. For each run, the output intensity is obtained from the steady-state data. For gain levels above threshold, the output intensity is found to vary linearly with σ_0, a behavior that is expected for a homogeneously broadened system. Extrapolating the output intensity data towards zero yields $\sigma_{0,\text{th}} \cong -1780$ S/m, a value that corresponds closely to -1760 S/m calculated above. Thus, this study of a simple Fabry-Perot laser demonstrates the validity of using FDTD to provide accurate predictions of the gain threshold and lasing wavelength.

(a)

(b)

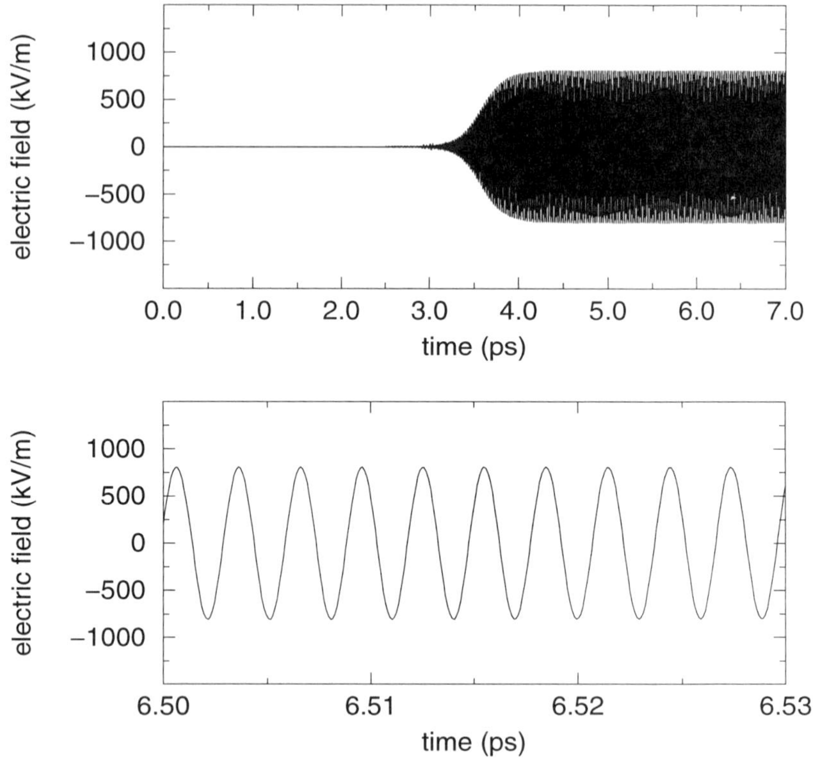

Fig. 10.6 (a) FDTD-computed time evolution of the electric field outside the Fabry-Perot laser cavity; (b) expanded time scale of the steady-state region showing a single-mode oscillation at ω_0. *Source*: Hagness et al., *Radio Science*, 1996, pp. 931–941.

10.3.3 Application to Vertical-Cavity Surface-Emitting Lasers

Vertical microcavity lasers based on simple Fabry-Perot cavities having high-reflectivity dielectric quarter-wave stacks for mirrors have attracted much attention over the last decade because of their desirable lasing characteristics: low power consumption, high output power, and single-mode operation. The low lasing threshold of VCSELs can be further improved by using a periodic-gain configuration within the cavity. By placing thin gain segments along the electric field standing-wave maxima, the longitudinal confinement factor is maximized, thereby reducing the material gain threshold [33].

This section reviews the use of the FDTD algorithm described in Section 10.3.1 to model VCSELs having uniform-gain and periodic-gain structures (UGS and PGS) [27]. The microcavity geometries, shown schematically in Fig. 10.7, are modeled as one-dimensional structures since the lateral dimensions are relatively large (> 5 μm) [33]. Each is assumed to be grown on a GaAs substrate ($n = 3.59$) with mirrors comprised of pairs of alternating layers of AlAs ($n_L = 2.971$) and $Al_{0.2}Ga_{0.8}As$ ($n_H = 3.452$) having a Bragg wavelength $\lambda_B = 0.87$ μm. The top mirror has nine pairs while the bottom mirror has 12.5 pairs; both mirrors are designed to have a peak reflectivity of 92.8%. The optical path length of each cavity is assumed to be $27\lambda_B/2$.

For the uniform-gain case, the cavity consists of a solid, active GaAs region ($n = 3.59$) of length, d, surrounded on each side by a passive GaAs region (also $n = 3.59$). The fill factor, d/L, is chosen to be 0.5. For the periodic-gain case, the cavity consists of thin active GaAs segments, each with a thickness, t, separated by passive GaAs segments. The spacing between the gain segments is equal to one-half the lasing wavelength, λ_{lasing}, and the total thickness of the gain segments is d. These design parameters permit comparison of the FDTD results with the analysis presented in [33].

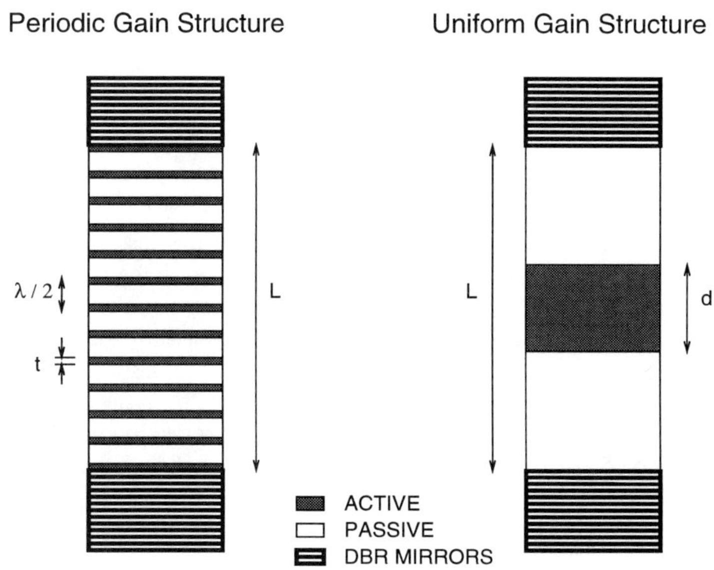

Fig. 10.7 Schematic of microcavity lasers having periodic (left) or uniform (right) gain structures within the cavity. *Source*: Hagness et al., *Radio Science*, 1996, pp. 931–941.

Following [33], for a cavity with no internal loss, the material gain required at threshold is given by

$$\alpha_{\text{th}} = \frac{1}{2\Gamma L} \ln R \tag{10.18}$$

where R is the geometric mean mirror reflectivity and Γ is the longitudinal confinement factor. For the UGS, Γ is simply the fill factor: $\Gamma_{\text{UGS}} = d/L$. Assuming a uniform standing wave pattern and ideal mirror reflectivities, the longitudinal confinement factor for the PGS is approximated as

$$\Gamma_{\text{PGS}} \cong \frac{d}{L}\left[1 + \frac{\sin(2\pi t/\lambda)}{2\pi t/\lambda}\right] \tag{10.19a}$$

where $\lambda = \lambda_{\text{lasing}}/n$. When $t = \lambda/2$, (10.19a) shows that Γ for the PGS reduces to that of the UGS. When $t \to 0$, that is, when the gain segments are extremely thin, Γ is maximized at twice that of the uniform-gain system. Therefore, according to (10.18), the gain threshold can be reduced by as much as 50%. Equation (10.19a) is expressed in terms of the fill factor as follows:

$$\Gamma_{\text{PGS}} \cong \frac{d}{L}\left[1 + \frac{\sin(\pi d/L)}{\pi d/L}\right] \tag{10.19b}$$

since $L = M\lambda/2$ and $d = Mt$, where M is the number of thin gain segments in the PGS. For this case ($d/L = 0.5$), the gain threshold for the PGS according to (10.18) and (10.19b) is approximately 39% less than the gain threshold for the UGS.

Passive Studies

First, the FDTD method without gain is used to determine the passive characteristics of the microcavity. Here, the UGS and PGS are identical because the refractive indices of the passive and active regions are assumed to be the same. For each mirror, the reflectivity spectrum is calculated with a single run using a short-pulse excitation and DFTs of the incident and reflected time-domain data. Fig. 10.8(a) shows the FDTD results for the bottom mirror compared with the spectra of the reflectivity (the squared amplitude of the reflection coefficient) and the phase of the reflection coefficient obtained using the matrix method [34]. At a grid resolution $\Delta x = \lambda_{\text{B}}/124 n_{\text{H}} \cong 2$ nm, the error at the reflectivity peak for the bottom mirror is less than 3 parts per 10,000. Similar accuracy is obtained for the top mirror, shown in Fig. 10.8(b).

Fig. 10.9 graphs the FDTD-computed reflectivity spectrum of the entire microcavity. Five cold-cavity resonances (at 821, 843, 870, 897, and 923 nm) fall within the broad stopband. Knowledge of the wavelength spacing between these resonances, here about 27 nm, permits determination of the effective cavity length. This is longer than the physical length of the cavity because of the field penetration into the mirrors. In this case, the effective cavity length is approximately $34\lambda_B/2$, compared to the physical length of $27\lambda_B/2$.

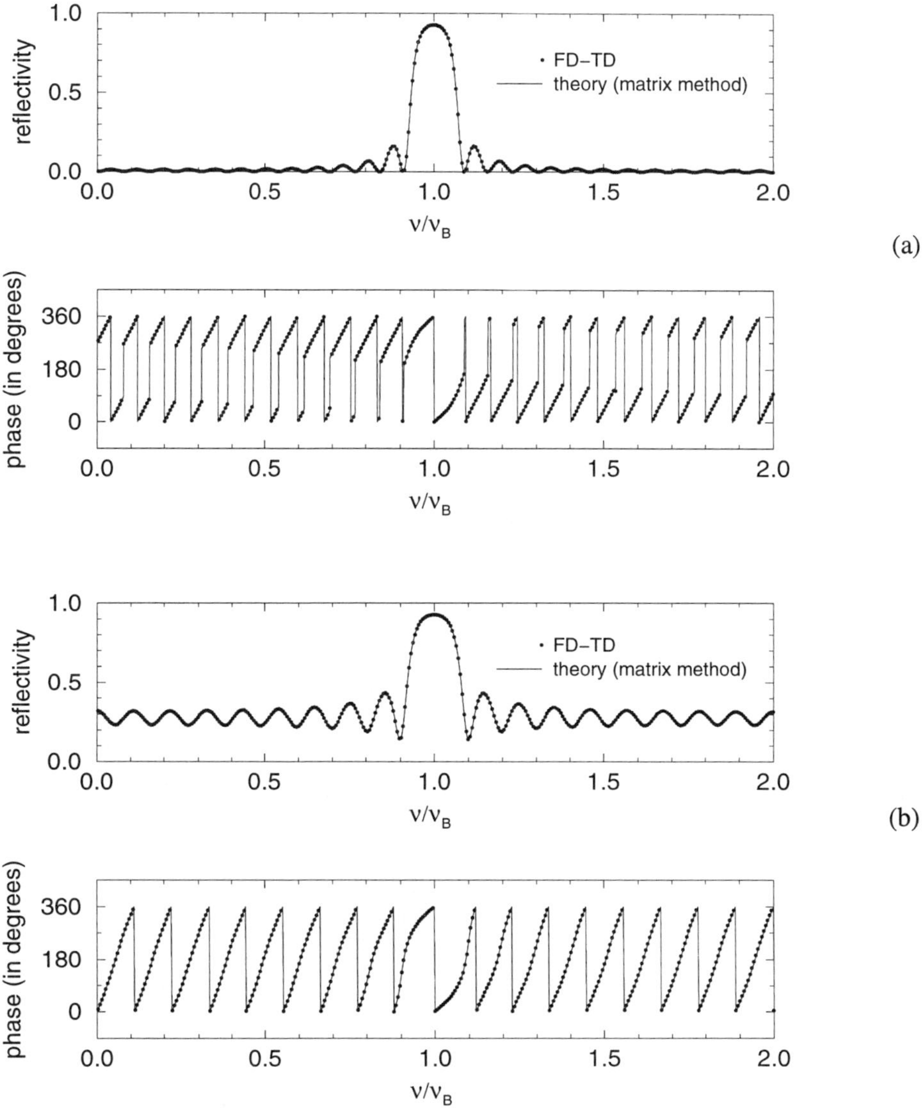

Fig. 10.8 FDTD-computed reflectivity and phase spectra for the vertical microcavities illustrated in Fig. 10.7: (a) bottom mirror alone; (b) top mirror alone.

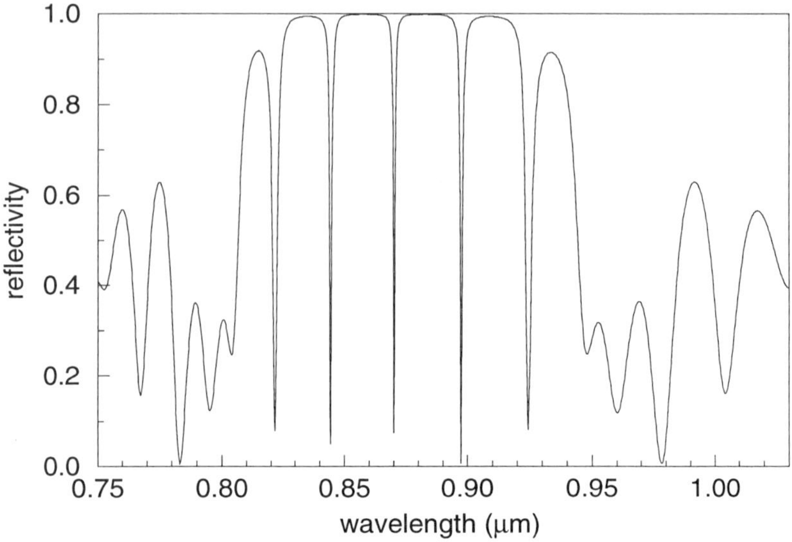

Fig. 10.9 FDTD-computed reflectivity spectrum of the entire passive vertical microcavities illustrated in Fig. 10.7. *Source*: Hagness et al., *Radio Science*, 1996, pp. 931–941.

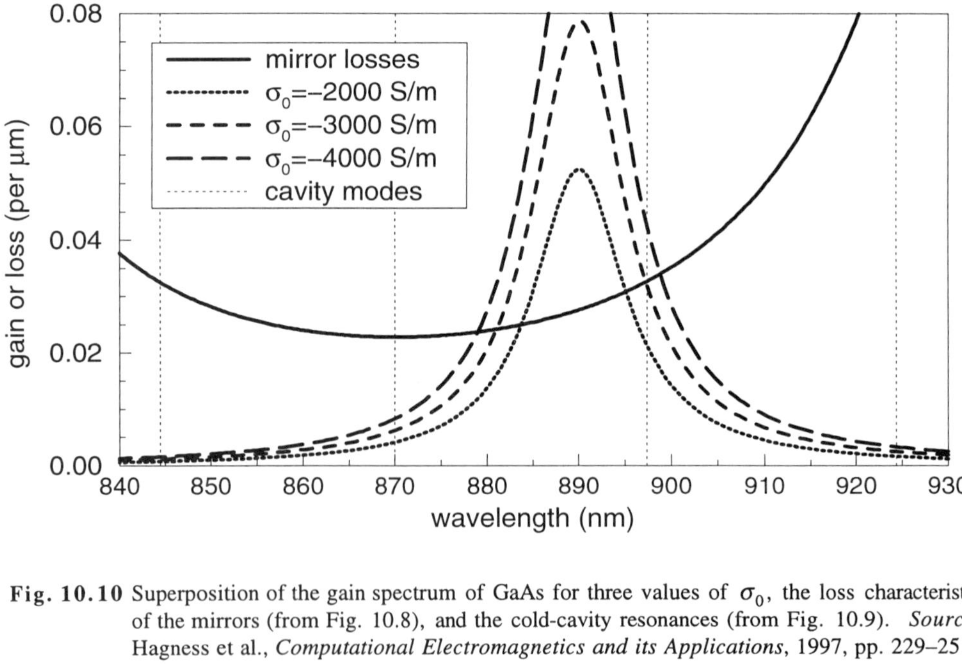

Fig. 10.10 Superposition of the gain spectrum of GaAs for three values of σ_0, the loss characteristic of the mirrors (from Fig. 10.8), and the cold-cavity resonances (from Fig. 10.9). *Source*: Hagness et al., *Computational Electromagnetics and its Applications*, 1997, pp. 229–251.

Fig. 10.10 graphs the gain spectrum of GaAs for three different values of σ_0, the frequency-dependent mirror loss characteristic calculated from Fig. 10.8, and the cold-cavity resonances obtained from Fig. 10.9 (shown as vertical dotted lines) [35]. The gain spectrum is centered at 890 nm while the mirror reflectivity spectrum is centered at the Bragg wavelength of 870 nm. At the 870-nm cavity resonance, the mirror loss exceeds the optical gain for each value of σ_0 shown, even though the mirror loss is at a minimum here. However, at the 897-nm cavity resonance, the optical gain exceeds the mirror loss for values of σ_0 more negative than approximately –3000 S/m. Therefore, since the 897-nm longitudinal mode experiences the greatest net gain, lasing should occur at this wavelength.

Following the methodology of Section 10.3.2, the FDTD algorithm with gain and a noise current is used to determine the lasing wavelength and gain threshold for the VCSELs of Fig. 10.7. Fig. 10.11 (top) shows the assumed spatial variation of the index of refraction for both the UGS and PGS microcavities. Here, the index varies periodically in the mirror regions between 2.97 and 3.45 and is independent of the specific gain structure. Fig. 10.11 (middle) plots σ_0 as a function of position within the microcavity to show the region of uniform gain in the UGS that is centered within the otherwise passive cavity-mirror structure. Fig. 10.11 (bottom) illustrates the periodic-gain configuration where σ_0 alternates between 0 (a passive segment) and –10,000 S/m (an active segment).

Fig. 10.12 plots the FDTD-computed envelope of the electric field within the uniform-gain VCSEL at the sinusoidal steady state. Here, the decaying fields on either side of the cavity illustrate the penetration of the mode into the passive mirrors. This simulation yields a lasing wavelength of just under 900 nm, as expected from the discussion of Fig. 10.10. The standing-wave maxima within the cavity are spaced periodically by an optical path length of $\lambda_{\text{lasing}}/2$. These maxima determine the locations of the thin gain segments within the periodic-gain VCSEL, as shown in Fig. 10.11 (bottom). For each lasing structure, the simulation is performed for several different values of σ_0 and the output intensity is recorded, as shown in Fig. 10.13.

A comparison of the FDTD-computed UGS and PGS data sets reveals the following advantages of the PGS: (1) a substantial reduction in the gain threshold for lasing; and (2) a higher output intensity for a given gain level. Extrapolation of the data of Fig. 10.13 towards zero output intensity yields a lasing gain threshold of $\sigma_{0,\text{th}} \cong -3260$ S/m for the case of the UGS, about as expected from the discussion of Fig. 10.10. For the case of the PGS, $\sigma_{0,\text{th}} \cong -1790$ S/m. Therefore, the FDTD simulations indicate that the lasing threshold for the PGS is approximately 45% less than the threshold for the UGS. This compares to a 38.5% reduction in the gain threshold predicted by the approximate analysis of (10.18) and (10.19). The FDTD calculation is a more accurate measure of the threshold reduction since it accounts for the fact that the mirrors are not perfect reflectors.

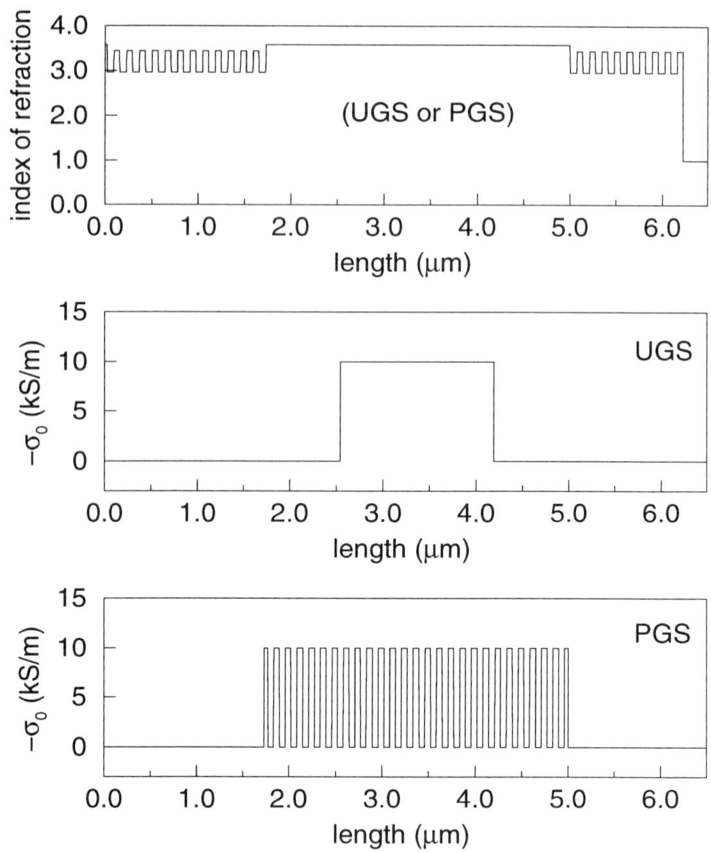

Fig. 10.11 Spatial variation of refractive index and optical gain within the VCSELs. *Adapted from*: Hagness et al., *Computational Electromagnetics and its Applications*, 1997, pp. 229–251.

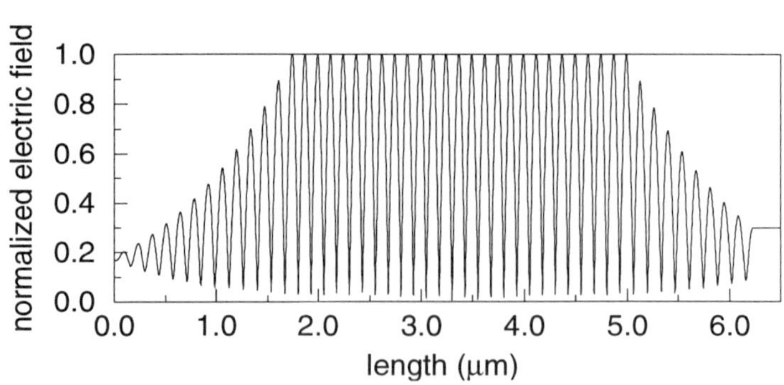

Fig. 10.12 Normalized steady-state electric field envelope within the uniform-gain VCSEL. *Adapted from*: Hagness et al., *Radio Science*, 1996, pp. 931–941.

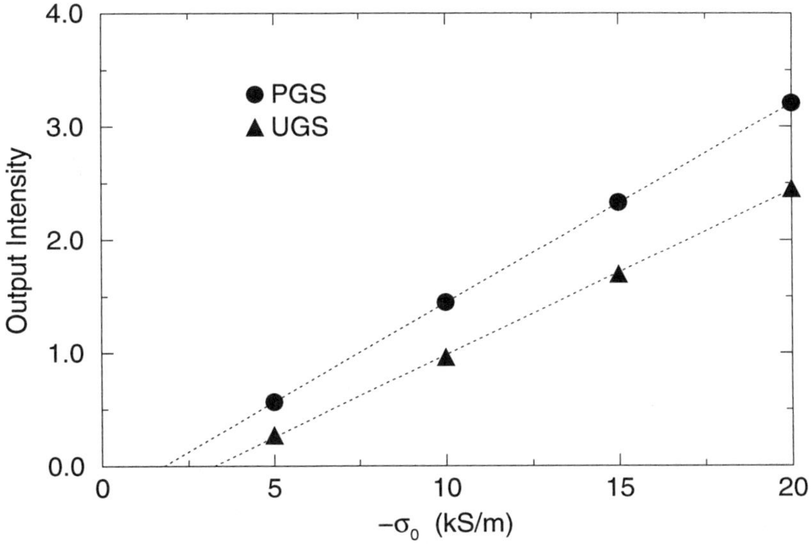

Fig. 10.13 FDTD-computed output intensity versus gain for UGS and PGS VCSELs. *Source*: Hagness et al., *Radio Science*, 1996, pp. 931–941.

10.4 MICROCAVITIES BASED ON PHOTONIC BANDGAP STRUCTURES

Photonic crystals [36] are artificial structures that have a periodic variation of the refractive index in one, two, or three dimensions. Analogous to the energy gap in pure semiconductor crystals in which electrons are forbidden, these PBG structures have a frequency stopband over which there is no transmission of electromagnetic waves [37]. Similar to a donor or acceptor state in a doped semiconductor, a small defect introduced into the photonic crystal creates a resonant mode at a frequency that lies inside the bandgap [38]. The defect in the periodic array behaves as a microcavity resonator.

A novel class of coplanar microcavities is based on a strongly guiding semiconductor waveguide that confines the light in the transverse dimensions, and a one-dimensional PBG structure that confines the light in the longitudinal dimension. Fig. 10.14(a) is a scanning electron microscope image of one such structure, an "air-bridge" [39] or "photonic-wire" [40] microcavity consisting of a series of etched air holes along a semiconductor waveguide suspended in air. The microcavity of Fig. 10.14(a) is comprised of a 0.45×0.79 μm InGaAsP / InGaAs waveguide with a defect introduced into a linear array of 40 etched holes. Fig. 10.14(b) is a schematic of a PBG reflector formed by a five-hole array, and a microcavity formed by a defect in an array of ten holes (or, equivalently, by two sets of reflectors at either end of the photonic wire).

(a)

(b)

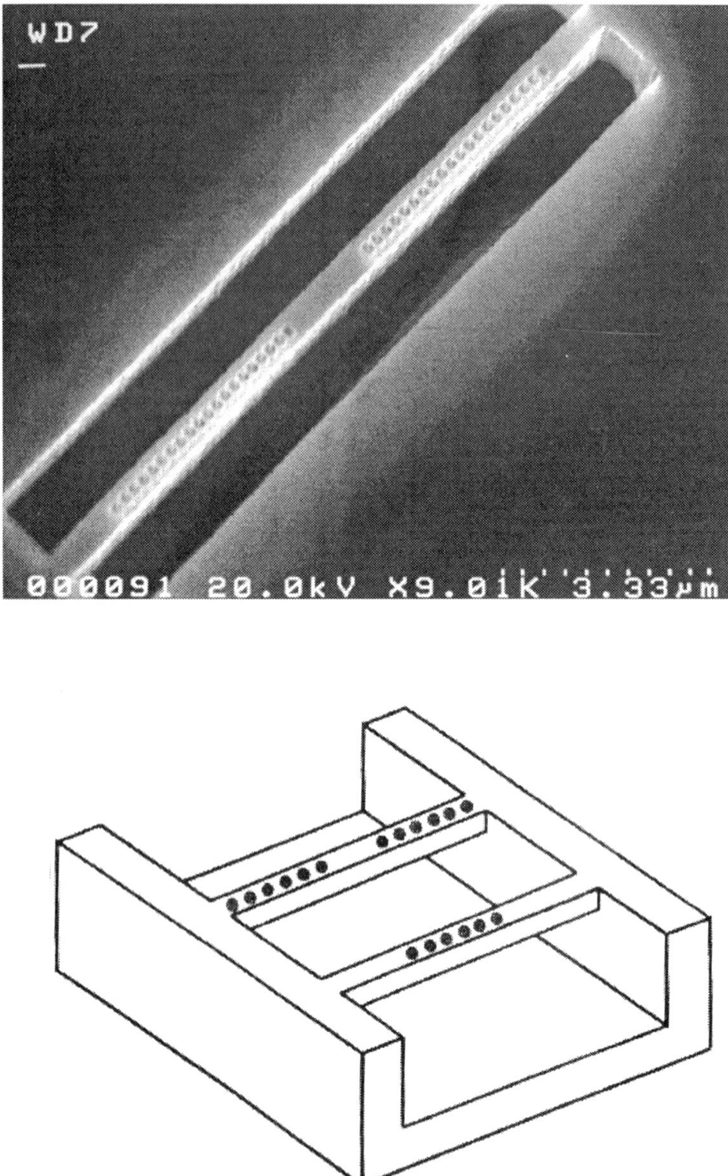

Fig. 10.14 (a) Scanning electron microscope image of an InGaAsP / InGaAs suspended photonic wire microcavity. Holes etched in the waveguide form one-dimensional PBG structures. (b) Schematic of a photonic wire mirror and the corresponding microcavity. *Adapted from*: Zhang et al., *IEEE Photonics Technology Letters*, 1996, pp. 491–493, © 1996 IEEE.

Numerical investigations of a variety of one-dimensional PBG structures, including waveguides with holes or notches and multilayer dielectric stacks, have been carried out using FDTD [27, 35, 40 – 43]. These simulations provide transmission and reflection characteristics, radiation losses, field patterns, and cavity Q.

This section discusses FDTD modeling of PBG reflectors and microcavities of the type shown in Fig. 10.14 comprised of 0.3-μm-wide, 0.79-μm-thick semiconductor waveguide ($n = 3.3$) surrounded by air ($n = 1$). To permit high resolution (12.5 nm), simulations are performed in two dimensions (the horizontal plane of the structures) for the field components E_z, H_x, and H_y. The vertical waveguide composition is accounted for by calculating the effective index of propagation in the three-dimensional waveguide and using it as the bulk material index in the two-dimensional model. All waveguide and grid boundaries are terminated using the PML ABC [7]. To obtain broadband spectral information from a single computer run, a 30-fs Gaussian pulse modulating a $\lambda = 1.5\ \mu\text{m}$ carrier is excited at one end of the waveguide using a fundamental-mode transverse source profile. Reflection and transmission spectra for the PBG structures are obtained from the FDTD data by taking the appropriate DFTs.

The center-frequency and width of the reflectivity spectrum of a single PBG structure are governed by the size, shape, number, and spacing of the holes. Fig. 10.15 shows the FDTD-computed reflectivity of PBG structures formed by arrays of four, five, six, and seven rectangular air holes (each $0.25 \times 0.1\ \mu\text{m}$). The center-to-center spatial periodicity of the holes along the waveguide is 0.325 μm.

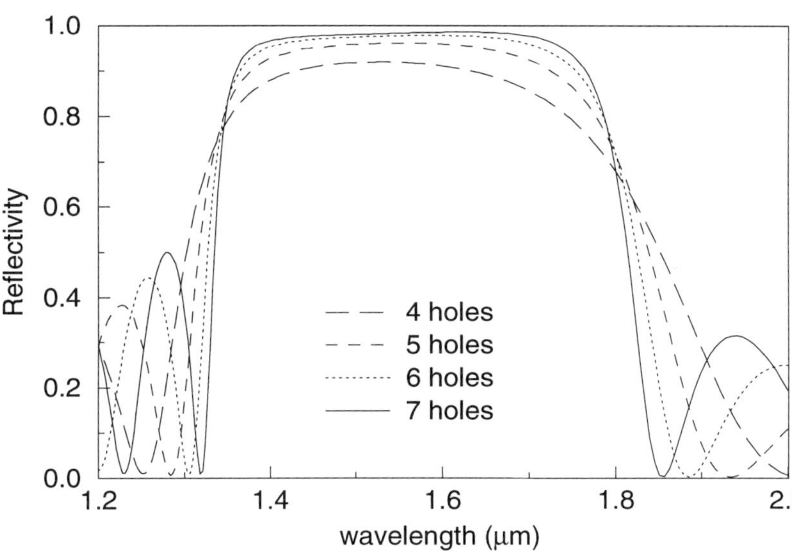

Fig. 10.15 FDTD-computed reflectivity spectra for PBG structures formed by arrays of rectangular air holes etched along a strongly guiding semiconductor waveguide.

Fig. 10.15 shows that the edges of the stopband sharpen and the peak reflectivity increases as the number of holes increases. With seven holes, the PBG structure gives a reflectivity bandwidth of approximately 500 nm (about 33% of the carrier wavelength) and a peak reflectivity that approaches 99%.

Fig. 10.16 illustrates the effect of the shape of the holes on the reflectivity spectrum. Here, each PBG structure consists of seven rectangular, elliptical, or circular holes with a spatial periodicity of 0.325 μm. Note that the top (rectangular-hole) PBG structure in Fig. 10.16(a) is the same as that considered in Fig. 10.15.

(a)

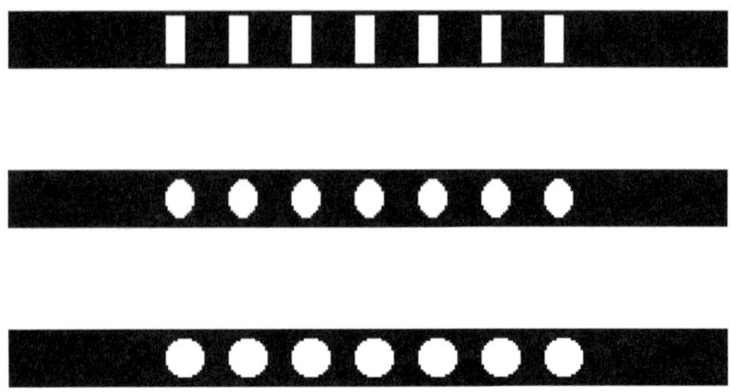

(b)

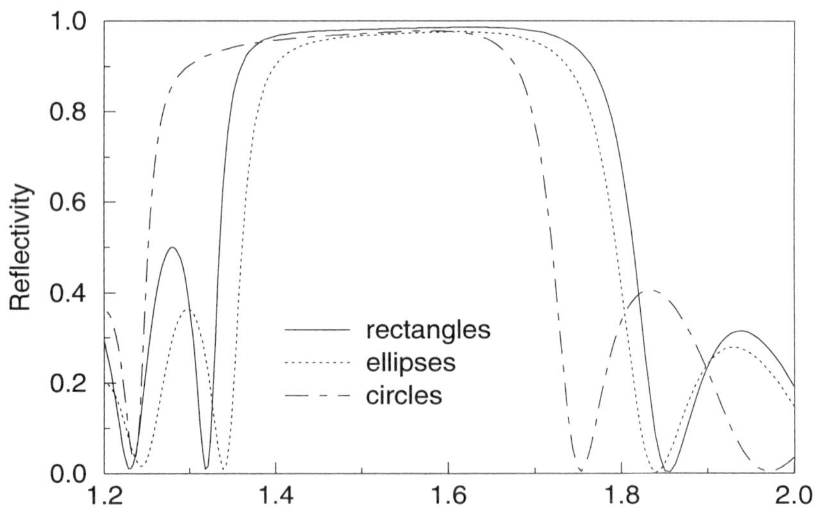

Fig. 10.16 (a) FDTD models of three PBG structures formed by arrays of seven rectangular, elliptical, or circular air holes with a spatial periodicity of 0.325 μm etched along a strongly guiding 0.3–μm-wide semiconductor waveguide; (b) reflectivity spectra corresponding to (a).

In Fig. 10.16(a), each elliptical hole spans 0.20 μm in the transverse direction and 0.15 μm in the longitudinal direction; the diameter of each circular hole is 0.20 μm. Fig. 10.16(b) shows that, while the reflectivity spectra of both the rectangular- and elliptical-hole PBG structures are centered near 1.55 μm, the bandwidth and peak reflectivity of the elliptical-hole PBG is somewhat reduced. The bandwidth of the circular-hole PBG structure is similar to that of the rectangular-hole PBG, but is shifted towards shorter wavelengths.

Fig. 10.17(a) shows the geometry of a microcavity formed by an array of ten rectangular holes of the type shown in Fig. 10.16(a), with a 0.6625-μm-long defect introduced in the center.

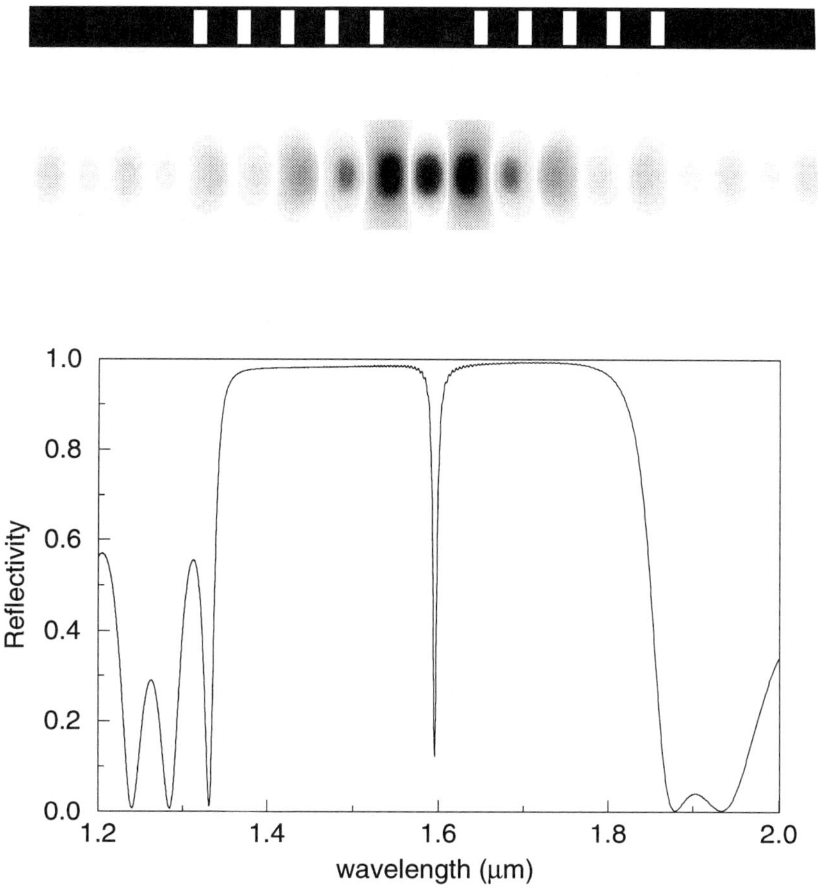

(a)

(b)

Fig. 10.17 (a) Geometry of a microcavity formed by introducing a defect in a PBG structure, and visualization of its electric field at $t = 1.67$ ps; (b) FDTD-computed reflectivity spectrum.

Fig. 10.17(a) also shows a gray-scale visualization of a snapshot of the electric field 1.67 ps after the launching of a short pulse at the left end of the waveguide. This observation time is long after the exciting pulse propagates off the grid to the right. The continued presence of significant field energy within the defect illustrates the resonant nature of the microcavity. Fig. 10.17(b) plots the reflectivity spectrum of the microcavity obtained from the time-domain data using a high-resolution DFT. A single resonant mode is seen within the stopband at $\lambda = 1.595$ μm. From the width and location of the resonance, the cavity Q is estimated to be about 300 [27]. By changing the length of the defect, the number and location of resonant modes within the stopband can be customized.

FDTD modeling has also been used to investigate the properties of two- and three-dimensional photonic crystals. For example, in [4], the FDTD method is used to compute the extraction efficiency (fraction of the emitted flux through the top and bottom surfaces) of dielectric slabs of two-dimensional photonic crystals. Results for a slab containing a triangular lattice of circular air holes and a slab containing a honeycomb lattice of short dielectric posts indicate that the extraction efficiency of light from the slab with the photonic crystal is enhanced relative to the case of a uniform slab. In [44], FDTD is used to compute the coupling efficiency and scattering losses in microcavities formed by local defects in two- and three-dimensional photonic crystals. As with the one-dimensional PBG structures, the properties of the resonant modes can be controlled by changing the nature and the size of the defect.

10.5 MICROCAVITY RING AND DISK RESONATORS

Optical ring and disk resonators are useful components for wavelength filtering, routing, switching, modulation, and multiplexing/demultiplexing applications [45–46]. The quality of a resonator for optical communications is measured by several properties including: the FSR, or spacing between adjacent resonances; the finesse, or ratio of the FSR to the width of the resonance; and the extinction ratio, or the ratio of the transmission at resonance to the off-resonance transmission. The ideal resonator for a wavelength-division multiplexing (WDM) system has a wide FSR and a high finesse to accommodate many channels, high on-resonance transmission to minimize insertion loss, and a large extinction ratio to minimize crosstalk. In particular, a ring or disk resonator having an FSR larger than 30 nm is desirable for accommodating the WDM channels within the erbium-amplifier communications window.

To achieve such a large FSR, the diameter of the ring or disk must be on the order of just a few micrometers. Semiconductor ring and disk resonators based on high-index-contrast, strongly guiding waveguides can be designed with diameters as small as 1 to 2 μm with negligible bending loss [45]. Current nanofabrication techniques permit the realization of such resonators coupled to submicron-width waveguides across submicron-width air gaps via evanescent waves [1, 47]. For example, Fig. 10.18 shows scanning

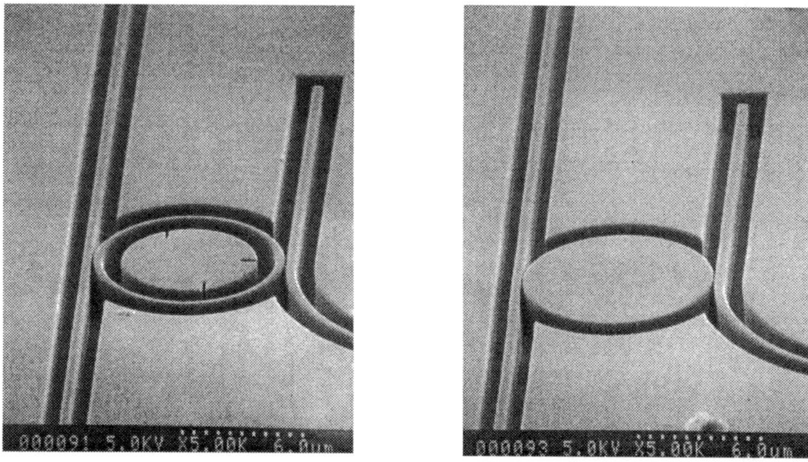

Fig. 10.18 Scanning electron microscope images of 10.5-μm-diameter AlGaAs / GaAs ring and disk resonators coupled to 0.5-μm-wide waveguides. The air gap between the adjacent waveguides and the ring (left) or disk (right) is only 100 nm. *Source*: Rafizadeh et al., *Optics Letters*, 1997, pp. 1244–1246.

electron microscope images of 10.5-μm-diameter AlGaAs / GaAs microcavity ring and disk resonators coupled to 0.5-μm-wide waveguides.

10.5.1 FDTD Modeling Considerations

This section reviews the use of the FDTD method for analysis of waveguide-coupled microcavity ring and disk resonators. In the models discussed here, the standard FDTD method is used to solve Maxwell's equations in two dimensions for the E_z, H_x, and H_y field components.

Fig. 10.19 shows a typical geometry of a microcavity ring or disk resonator. Adjacent waveguides, WG1 and WG2, serve as evanescent wave input and output couplers. Using the bootstrapping technique discussed in Section 10.2.1, a fundamental mode is sourced at the left end of WG1. Here, the excitation is a 20-fs Gaussian pulse modulating a 200-THz carrier. The field components in the pulsed waveguide mode are recorded as the mode propagates past cross sections, P1 and P2, in WG1, and P3 in the microring or microdisk. By taking DFTs of the time waveforms of the fields, calculating the Poynting power densities along the cross sections, and summing, the longitudinal power flux passing through each cross section is obtained as a function of frequency. In this manner, a single computer run provides a complete spectral characterization of the coupling efficiency (the flux through P3 normalized by the incident flux through P1) or the transmittance (the normalized flux through P2).

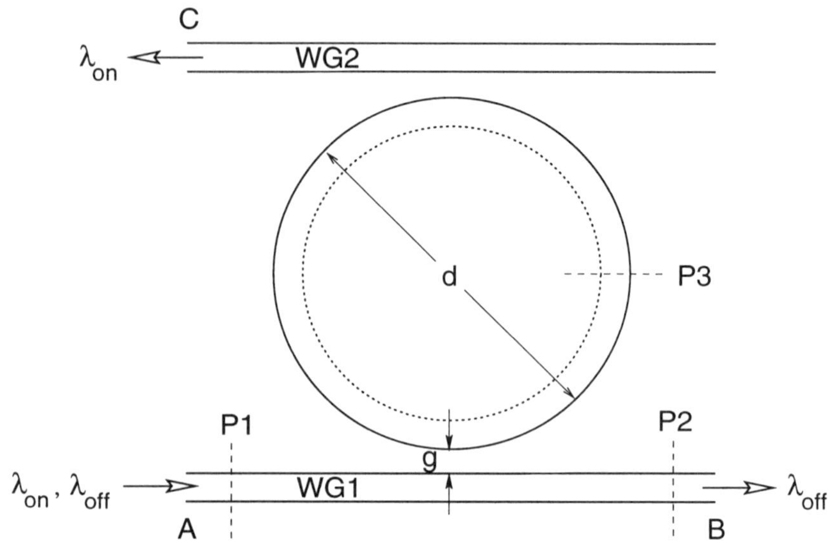

Fig. 10.19 Schematic of a microcavity ring or disk resonator coupled to two straight waveguides. An off-resonance signal entering port A remains in WG1 and exits at port B. An on-resonance signal is switched to WG2 and exits at port C. *Source*: Hagness et al. *J. Lightwave Technology*, 1997, pp. 2154–2165, © 1997 IEEE.

Where benchmark data are available, the accuracy level of this modeling approach is found to be excellent for the assumed standard grid resolution of 13.6 nm. For example, power balances in the propagating mode computed in this manner are found to hold within parts per ten thousand for cases where the optical radiation is negligible.

In the vertical direction, each waveguiding structure considered here consists of a 0.45-μm-thick GaAs guiding layer with AlGaAs cladding layers above and below. Strong lateral confinement is achieved in the semiconductor waveguide by etching down through the lower cladding layer. The structure is weakly guiding in the vertical direction to allow for the input coupling of light from an external fiber or planar-emitting device. The vertical waveguide thickness and material composition is accounted for by computing the effective refractive index, n_{eff}, for the fundamental mode at $\lambda = 1.5$ μm, and then using n_{eff} as the bulk material index of the core in the two-dimensional simulations. In the cases considered here, each waveguide core ($n = 3.2$) is bordered on each side by air ($n = 1$).

The etching process results in a surface roughness of the sidewalls of the waveguides measured to be on the order of 10 to 20 nm. This fabrication artifact causes two problems due to its tendency to scatter light propagating within the waveguiding structure: (1) radiation to the far field, and hence, a partial loss of the stored energy and a reduction of the resonator Q [48]; and (2) backreflection into the counterpropagating mode [49]. The 13.6-nm grid-cell size is approximately 1/40th of the optical wavelength within the waveguide core material. This gridding is fine enough to directly model the 10 to 20-nm waveguide surface roughness by implementing a standard, staircased approximation of the curving waveguide in the manner of the basic FDTD algorithm. (In fact, several variations of the staircase fit can be constructed to examine the optical effects of randomness in the fabrication process). Thus, the FDTD model naturally includes the scattering losses. This grid-cell size is also fine enough to permit highly detailed models of electromagnetic coupling across the air gaps between the coupling waveguides and the resonators, even down to the minimum distance that can be reliably fabricated with current technology (on the order of 100 nm). Radiation losses to the far field can also be caused by waveguide bends. However, because of the strong mode confinement, the bending losses are negligible in these resonators. For the passive devices considered here, the intrinsic absorption loss due to the materials is also negligible.

The importance of accurate modeling of the highly dispersive semiconductor waveguides is illustrated with the following equation for the FSR between longitudinal modes m and $m+1$:

$$\Delta\nu_{\text{FSR}} = \frac{c}{\pi}\left[\frac{m+1}{d_{\text{eff}}(\nu_{m+1})\, n_{\text{eff}}(\nu_{m+1})} - \frac{m}{d_{\text{eff}}(\nu_m)\, n_{\text{eff}}(\nu_m)}\right] \tag{10.20}$$

where d_{eff} is the effective diameter of the ring or disk. The quantity πd_{eff} corresponds to the circumferential path traveled by the peak of the guided mode; that is, d_{eff} is the distance between the two peaks of the transverse mode profile found along the diameter of the ring or disk. For semiconductor microcavity ring and disk resonators, $n_{\text{eff}}(\nu_{m+1}) \neq n_{\text{eff}}(\nu_m)$ because the waveguide dispersion is strong and the longitudinal mode spacing is so wide. Also, d_{eff} is smaller than the diameter, d, and is a function of frequency. Thus, (10.20) cannot be reduced to $\Delta\nu_{\text{FSR}} = c/\pi d n_{\text{eff}}$, as is commonly done for larger weakly guiding resonators. Similarly, the Q is affected by the relatively large dispersion [50].

10.5.2 Microcavity Ring Resonators Coupled to Straight Waveguides

In Fig. 10.19, the coupling between the waveguides and the microcavity ring or disk resonator is controlled by two key parameters: g, the size of the coupler air gap; and d, the outer diameter of the ring or disk. Accordingly, the coupling efficiency of this geometry can be investigated by varying g and d [47]. In the present simulations, the width of WG1, WG2, and the ring waveguide is 0.3 μm. The straight waveguides support only one symmetric and one antisymmetric mode at $\lambda = 1.5$ μm. The coupling coefficient, κ, defined as the percentage of power coupled into the ring from WG1, is computed using the procedure outlined in Section 10.5.1.

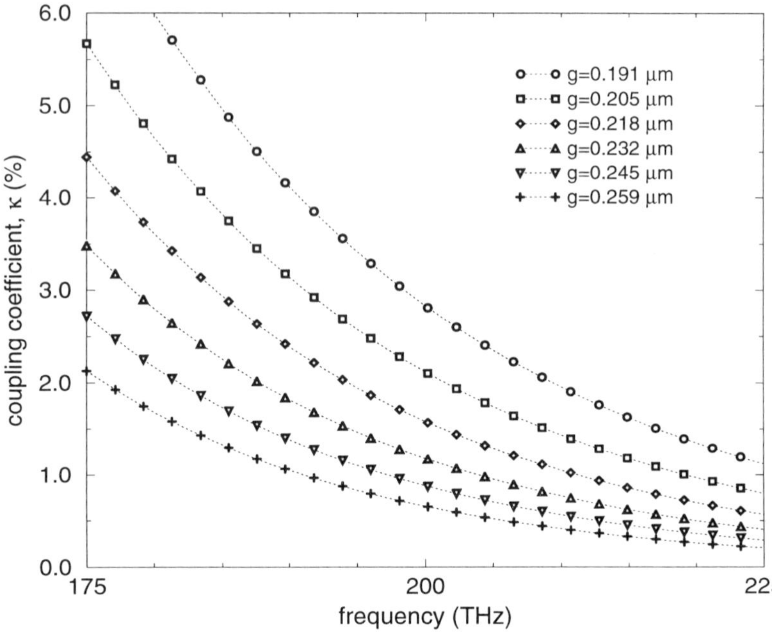

Fig. 10.20 FDTD-computed coupling coefficient, κ, as a function of frequency and gap size, g, for the geometry of Fig. 10.19, $d = 5.0\,\mu\text{m}$ ring resonator case. The electric field is vertically polarized, i.e., perpendicular to the plane of Fig. 10.19. *Source*: Hagness et al. *J. Lightwave Technology*, 1997, pp. 2154–2165, © 1997 IEEE.

Consider a 5.0-μm-diameter ring coupled to a straight, single-mode, input waveguide. Fig. 10.20 shows the FDTD-computed coupling coefficient, κ, for the vertical-polarization (V-pol) case where the electric field is perpendicular to the plane of the ring for g varying from 0.19 to 0.26 μm. Two well-known coupling phenomena are illustrated in this figure [47]:

1. For a fixed frequency, κ decreases as the physical gap size increases;
2. For a fixed physical gap size, κ decreases at higher frequencies where the effective gap size increases.

At $\lambda = 1.5$ μm ($\nu = 200$ THz), κ ranges between 0.7% and 2.8% for a change of only 68 nm in g. This compares favorably to the desired range for κ, typically 0.5% to 3% needed to achieve good transmission characteristics and large extinction ratios. (Convenient rules of thumb are that κ should be greater than the round-trip cavity loss and should be symmetric for both WG1 and WG2 relative to the ring or disk.) While it is feasible to fabricate the small gaps required to achieve the desired range of κ, the FDTD simulations indicate that κ is sensitive to g. Therefore, since it is difficult to fabricate identical gaps to very high precision, the coupling from the ring to WG1 and WG2 may be slightly asymmetric.

Fig. 10.21 compares the V-pol coupling coefficient curve of Fig. 10.20 obtained for $g = 0.232$ µm with analogous FDTD results for the same gap size computed for the horizontal-polarization (H-pol) case where the electric field is in the plane of the ring [47]. The coupling coefficient for the H-pol case at 1.5 µm is 1.9% compared with 1.2% for the V-pol case. These results show that the H-pol coupling exceeds the V-pol coupling for most of the spectrum of interest. In order to avoid high lateral coupling of light to the media surrounding the waveguides and the ring or disk, relatively wide boundary trenches must be etched, as seen in the scanning electron microscope images of Fig. 10.18. While the remainder of this chapter focuses on the V-pol case, note that with a sufficiently wide trench, the H-pol case is qualitatively the same.

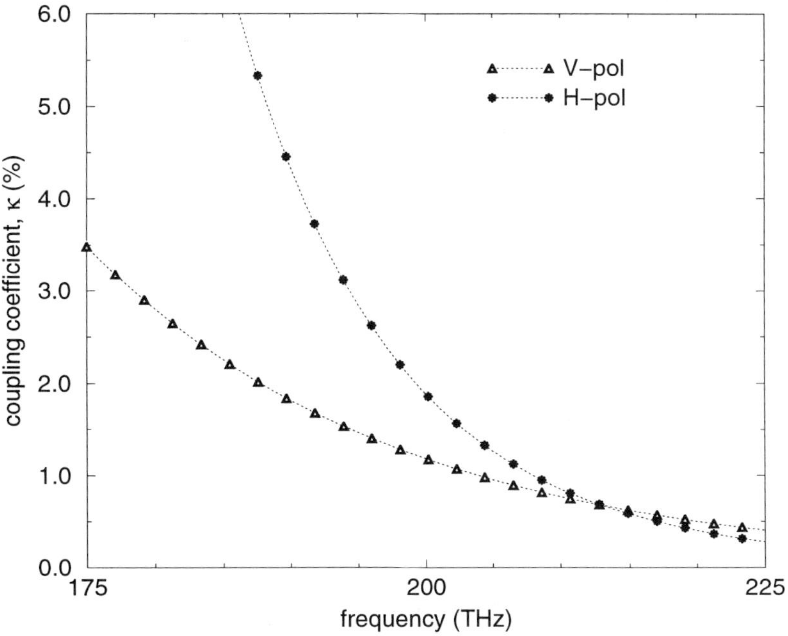

Fig. 10.21 Comparison of the FDTD-computed coupling coefficient κ for the V-pol and H-pol cases of Fig. 10.19 (ring) with $d = 5.0\,\mu$m and $g = 0.232$ µm. *Source*: Hagness et al., *J. Lightwave Technology*, 1997, pp. 2154–2165, © 1997 IEEE.

The V-pol FDTD results for a 10-µm-diameter ring resonator have also been obtained for the same range of gap sizes investigated in Fig. 10.20. The 10-µm ring has about twice the κ of the 5.0-μm ring for a given gap size and frequency; for example, ranging from 1.3% to 5.5% at $\lambda = 1.5$ μm. The interaction length for the larger ring is still short enough to avoid backcoupling from the ring to the waveguide.

10.5.3 Microcavity Ring Resonator Coupled to Curved Waveguides

In addition to varying the gap size and ring diameter, the waveguide-to-ring coupling can be manipulated by adding curvature to the input and output waveguides. This section reviews the case wherein the adjacent waveguide remains straight from the source plane to the point of smallest separation from the ring, and then is provided with a controlled double curvature to allow variability of the interaction length [47].

Fig. 10.22 illustrates the geometry used for the curved-waveguide models. At the point of smallest separation from the 5.0-μm-diameter ring, the gap width is 0.205 μm. The curved part of the input waveguide has a bend radius of ρ. For a given value of ρ, the curved waveguide section bends toward the ring for 1/8th of the ring's circumference, and then bends away from the ring in exactly the same manner.

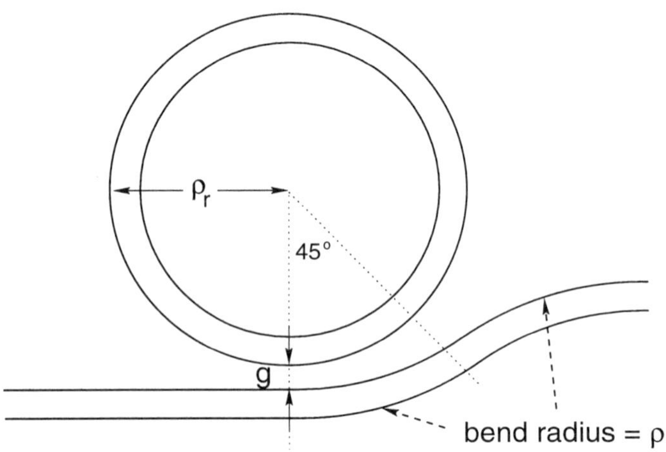

Fig. 10.22 Schematic diagram of a microcavity ring resonator coupled to a curved waveguide. The bend radius of the waveguide's outer curves is ρ. *Source*: Hagness et al., *J. Lightwave Technology*, 1997, pp. 2154–2165, © 1997 IEEE.

Fig. 10.23 graphs the FDTD-computed κ for this structure. The top curve ($\rho = \infty$) shows the previous case of the straight waveguide. In successive curves, the bend radius decreases slightly. The minimum possible value for ρ is $\rho_r + g$, where ρ_r is the radius of the ring. In this extreme, the gap between the ring and the waveguide is constant over a 45° angle. The FDTD results illustrate the following phenomena:

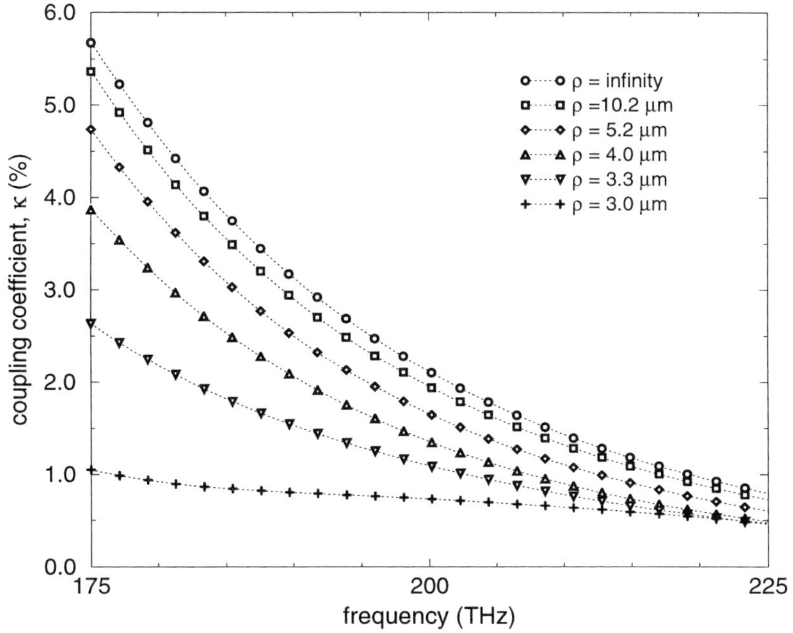

Fig. 10.23 FDTD-computed coupling coefficient, κ, as a function of frequency and waveguide bend radius, ρ, for a 5.0-μm-diameter ring (g = 0.205 μm, V-pol case). *Source*: Hagness et al., *J. Lightwave Technology,* 1997, pp. 2154–2165, © 1997 IEEE.

1. The coupling drops as the input waveguide bends more toward the ring, suggesting an increasing phase mismatch between the signals in the waveguide and in the ring;
2. For the smaller values of ρ, κ is less sensitive to frequency.

Therefore, curving the adjacent waveguide toward the ring increases the interaction length, but introduces an undesirable path length difference due to the differing bend radii of the ring and adjacent waveguide.

10.5.4 Elongated Ring Designs

As discussed in Section 10.5.2, the strong optical confinement of the waveguides requires very close proximity of the waveguide to the ring in order to achieve the desired level of coupling. An important step towards practical fabrication and commercialization of these devices is the exploration of designs that might alleviate the need for such narrow gaps. In particular, a wider air gap may allow reliable and repeatable fabrication of these resonators with reduced sensitivity to fabrication variations.

To achieve the desired coupling level with a wider gap, the interaction length over which the coupling takes place must be increased. In the previous section, it was shown that the coupling actually drops when the adjacent waveguide curves toward the ring. Here, changing the shape of the ring instead of the adjacent waveguide is investigated. For example, elongating the ring to form a "racetrack" configuration [51], as shown in Fig. 10.24, leads to an increased interaction length without introducing a phase mismatch.

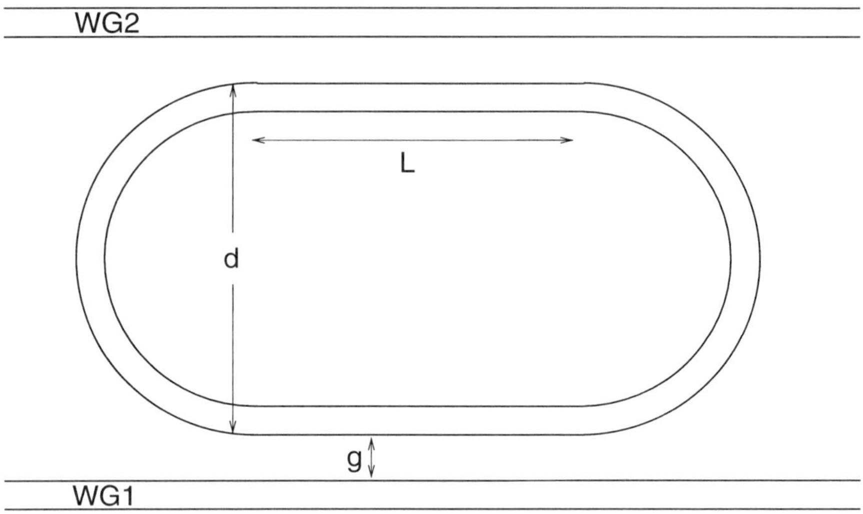

Fig. 10.24 Schematic diagram of an elongated "racetrack" microcavity ring resonator coupled to straight waveguides.

As an example, consider the resonator of Fig. 10.24 configured with $d = 5.0\ \mu$m semicircular sections and $L = 1.5\ \mu$m straight sections. Fig. 10.25 compares the FDTD-computed κ for this case with the previous 5.0-µm-diameter circular ring resonator as a function of the gap width, g [52]. The dotted curves represent κ at the wavelengths 1.5, 1.55, and 1.6 µm for the circular ring, and the solid curves represent κ at the same three wavelengths for the racetrack geometry. At a given wavelength, a specific coupling level is achieved at a larger gap width with the racetrack resonator in comparison to the circular ring resonator. These results illustrate the potential advantage of the racetrack structure in allowing wider air gaps while providing the desired coupling levels.

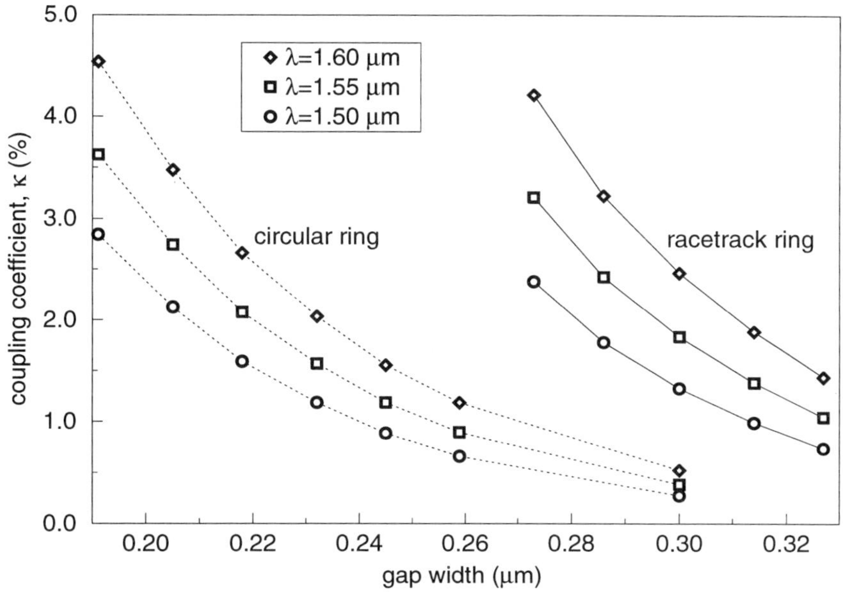

Fig. 10.25 FDTD-computed coupling coefficient, κ, as a function of wavelength and the gap width, g, for a 5.0-μm-diameter circular ring resonator (dotted lines) and a racetrack ring resonator with 5.0-μm-diameter semicircular sections and 1.5-μm-long straight sections (solid lines). The electric field is vertically polarized in this case. *Source*: Hagness et al., *Integrated Photonics Research*, OSA Technical Digest Series, in press.

10.5.5 Microcavity Ring Resonances

The focus of this section and the next is on the resonance behavior of microcavity rings and disks, as reported in [47]. The resonant nature of these structures is illustrated in Fig. 10.19. If the signal that enters port A is on-resonance with the ring or disk, then that signal couples into the cavity from WG1, couples out from the cavity into WG2, and exits the device at port C. Alternatively, a signal that is off-resonance with the cavity remains in WG1 and exits at port B. The output at port B is often called the channel-dropping transmittance; the output at port C is correspondingly called the channel-passing transmittance. The rings and disks support both a clockwise and a counterclockwise propagating mode. By sending a signal into port A, only the counterclockwise propagating mode is directly excited. Backward reflections in the coupling regions and backscattering due to sidewall roughness can excite the counterpropagating mode.

Referring to Fig. 10.19, consider a 5.0-μm-diameter ring resonator with a $g = 0.232\ \mu\text{m}$ gap between the ring and each adjacent straight waveguide. All straight and ring waveguides are 0.3 μm wide. Exciting WG1 in the same manner as in the coupling studies, the simulation is continued in time as the pulse initially couples into the ring and repeatedly travels around it. Fig. 10.26 is a visualization of snapshots in time of the FDTD-computed electric field (V-pol) as the pulse first couples into the ring and completes one round trip. For best display dynamic range, the gray scale in each snapshot is normalized to the maximum electric field in that snapshot.

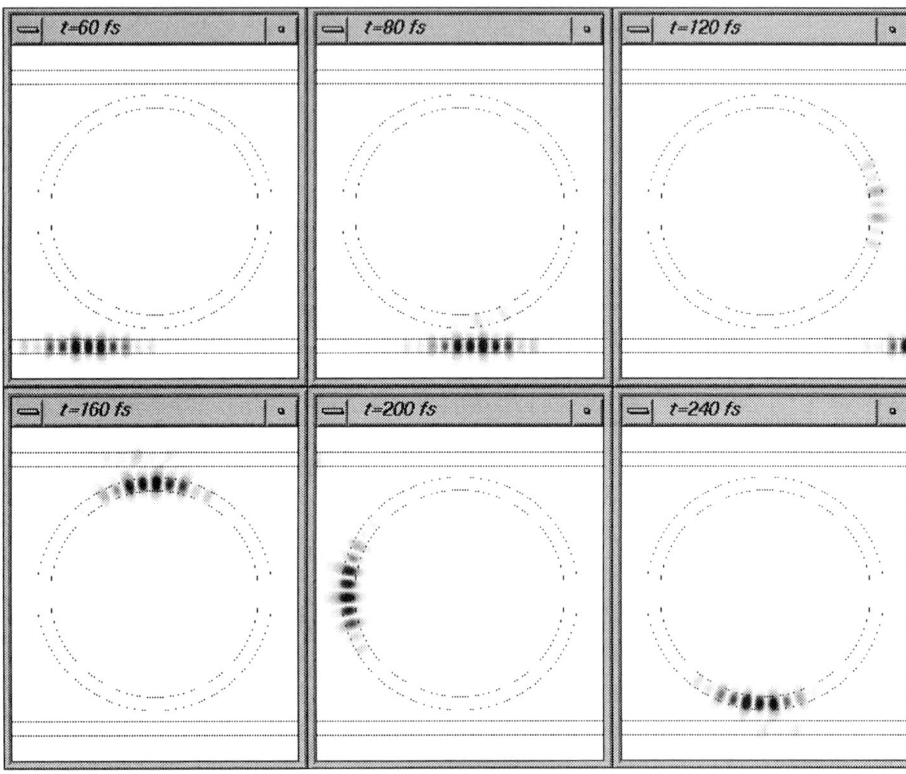

Fig. 10.26 Visualization of the FDTD-computed initial coupling and circulation of the exciting 20-fs optical carrier pulse around the 5.0-μm-diameter microcavity ring resonator. *Source*: Hagness et al., *J. Lightwave Technology*, 1997, pp. 2154–2165, © 1997 IEEE.

The coupling of energy out to the two adjacent waveguides at each pass results in a series of output pulses of diminishing amplitude at the output ports of WG1 and WG2. To determine the spectral properties of the transmittance at port B, the FDTD simulation is run until the energy in the resonator and the amplitude of the pulse train have sufficiently died down. This permits convergence of the DFT that is performed concurrently with the FDTD time stepping.

Fig. 10.27 graphs the calculated transmittance spectrum for this microcavity ring. The Q of the mth resonance is obtained directly from the power spectrum by forming the ratio of the center frequency, ν_m, to the width of the resonance, $\Delta\nu_m$, at the half-power points. The resonant frequencies and Q's calculated from the data in Fig. 10.27 are listed in Table 10.1. The FSR values, also listed in Table 10.1, decrease as the wavelength decreases. This behavior is a consequence of the relatively large waveguide dispersion, as seen in (10.20). The finesse, calculated from the FSR values and the widths of the resonances, falls in the range between 115 and 160.

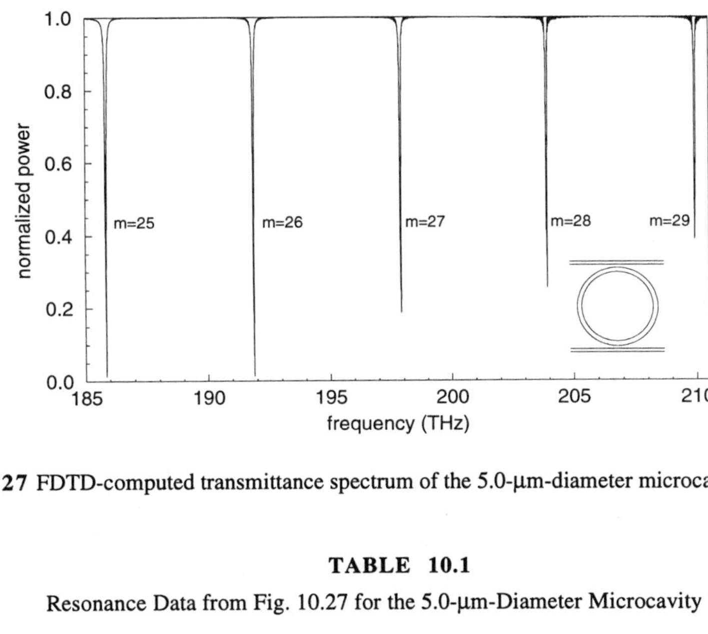

Fig. 10.27 FDTD-computed transmittance spectrum of the 5.0-μm-diameter microcavity ring resonator.

TABLE 10.1

Resonance Data from Fig. 10.27 for the 5.0-μm-Diameter Microcavity Ring

m	ν_m (THz)	λ_m (nm)	Q	FSR (nm)
25	185.85	1613.10	3700	
				50.66
26	191.88	1562.44	5000	
				47.56
27	197.90	1514.88	4000	
				44.73
28	203.92	1470.15	4000	
				42.17
29	209.94	1427.98	4000	

Impact of Sidewall Roughness

Based on only the coupling, which decreases at higher frequencies as seen in Fig. 10.20, the Q should increase with frequency. Such behavior is observed for the first two resonances listed in Table 10.1 ($m = 25, 26$), but then the apparent Q drops for the higher-frequency resonances. This drop is due to stronger scattering at higher frequencies where imperfections in the etched sidewalls (approximated in the FDTD model using staircasing) have more of an effect. In fact, on an expanded frequency scale, it is clear that each of the $m = 28$ and $m = 29$ resonances actually represents two subresonances. The sidewall roughness causes backscattering into a counterpropagating mode that splits the resonance peak [49]. In the case of the higher-frequency resonances of Fig. 10.27, the subresonances are not split far enough to be individually discernible, so the apparent Q of each overall resonance is lowered.

The drop in the apparent Q is verified by simulating a microcavity ring resonator that has rougher sidewalls by a factor of two. This involves modifying the staircased approximation of the sidewalls to mimic a twice-coarser grid in the vicinity of the walls, but actually retaining the fine 13.6-nm grid-cell size everywhere. In this case, all five resonances are split as a result of higher scattering losses across the 25-THz spectrum. Further, the two highest-frequency resonances are split far enough apart to individually resolve each subresonance. This study reveals that the Q increases with increasing mode number for each subresonance, as expected based on the strong dependence of coupling on frequency.

Rejection Ratio

The on/off ratio, or rejection ratio, is defined here as the ratio of power transmitted at a resonance frequency to the power not transmitted at that frequency. Consider the 30-nm communications window of erbium amplifiers that is centered around 1.55 μm. Focusing on this window, that is, around the $m = 26$ resonance (191.9 THz) in Fig. 10.27, the rejection ratio is approximately 72:1. By slightly decreasing the gap size, the rejection ratio is increased to more than 500:1. In general, this ratio can be optimized over any desired band spanning a couple of THz by adjusting the gap size to change the coupling with respect to the scattering losses in that wavelength range.

Extinction Ratio

Fig. 10.28 illustrates the potential for high extinction ratios. First, the sinusoidal steady-state electric field is computed for an excitation at a nonresonant frequency of 193.4 THz (1550 nm), shown in Fig. 10.28(a). In this case, nearly 100% of the signal remains in WG1. Second, the sinusoidal steady-state electric field is computed for an excitation at the $m = 26$ resonant frequency of 191.9 THz. As shown in Fig. 10.28(b), nearly 100% of the signal in this case is switched to WG2.

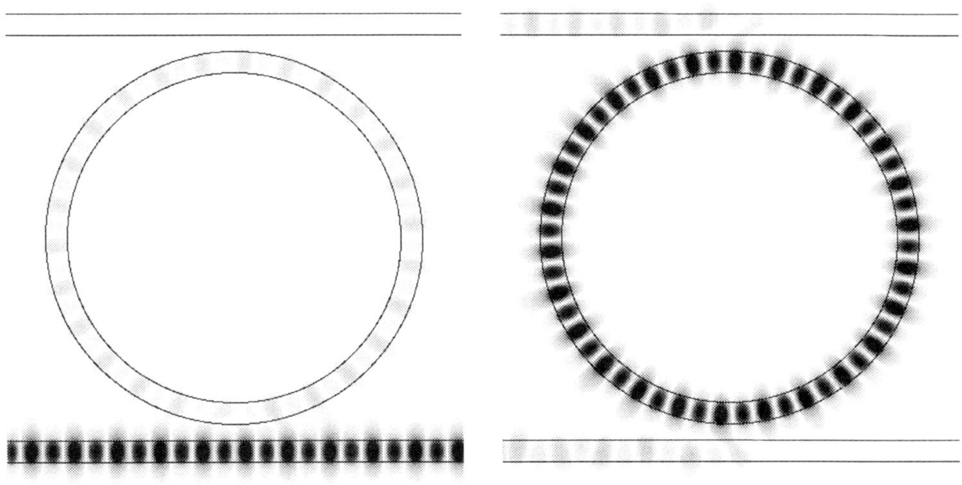

(a) (b)

Fig. 10.28 Visualizations of the FDTD-computed sinusoidal steady-state electric field in the 5.0-μm-diameter ring having the properties summarized in Fig. 10.26, Fig. 10.27, and Table 10.1. Here, a single-frequency excitation is applied at port A of WG1: (a) off-resonance signal at 193.4 THz; (b) on-resonance signal at 191.9 THz (the m = 26 resonance).

Effective Diameter

Since n_{eff} is a known function of frequency for the straight waveguides, an estimate of d_{eff} can be obtained using $mc/\pi\nu_m n_{\text{eff}}$ for the ring resonator. For the resonances in Fig. 10.27, d_{eff} increases with frequency from 4.717 to 4.720 μm. If the circumferential path of the mode were along the center of the waveguide, d_{eff} would be 4.7 μm (halfway between the outer-rim diameter of 5.0 μm and the inner-rim diameter of 4.4 μm). The estimate of d_{eff} is slightly larger, agreeing with the expectation that the peak of the transverse mode profile shifts toward the outer edge of the curved waveguide.

Overall Assessment

The 5.0-μm-diameter microcavity ring resonator demonstrates narrow channel wavelength selectivity (high finesse) and constant transmittance over FSR values as wide as 6.026 THz (50.66 nm). This structure exhibits excellent rejection and extinction ratios. Its Q is limited primarily by the degree of sidewall roughness, which is a function of the fabrication process and its quality control.

10.5.6 Microcavity Disk Resonances

Referring again to Fig. 10.19, consider a 5.0-μm-diameter disk resonator with a $g = 0.232\ \mu m$ gap between the disk and each adjacent straight 0.3-μm-wide waveguide. The disk geometry is identical to that of the 5.0-μm-diameter ring studied above except that there is no inner rim. Exciting WG1 in the same manner as in the coupling and ring-resonance studies, the simulation is continued in time as the pulse initially couples into the disk and repeatedly travels around it. Fig. 10.29 is a visualization of snapshots in time of the FDTD-computed electric field (V-pol) as the pulse first couples into the disk and completes one round trip. Again, for best display dynamic range, the gray scale in each snapshot is normalized to the maximum electric field in that snapshot.

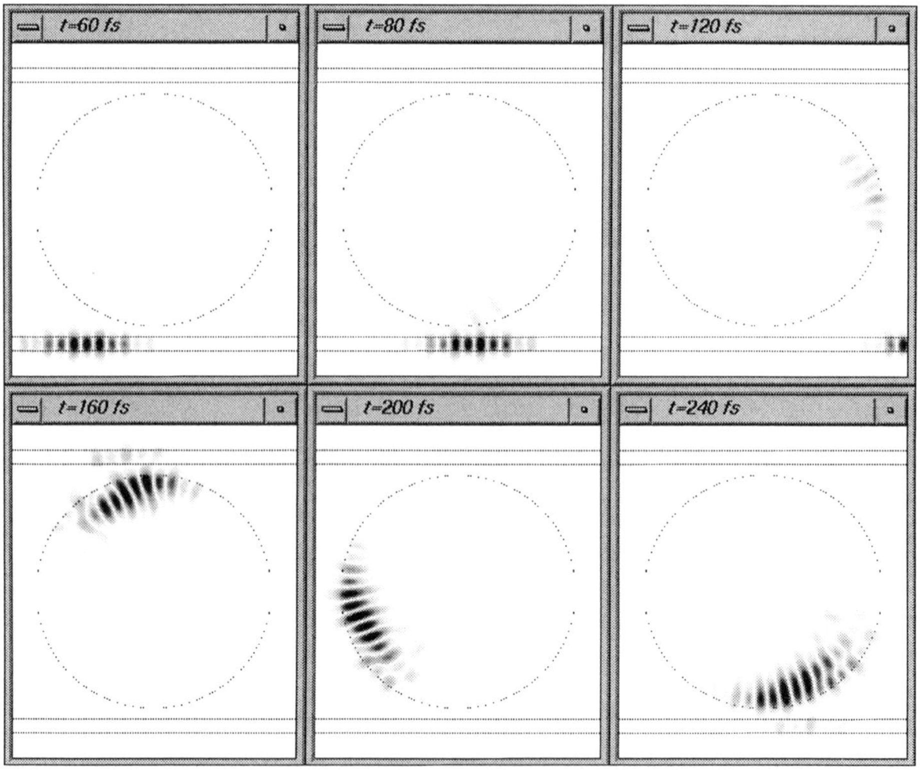

Fig. 10.29 Visualization of the FDTD-computed initial coupling and circulation of the exciting 20-fs optical carrier pulse around the 5.0-μm-diameter microcavity disk resonator. *Source*: Hagness et al., *J. Lightwave Technology*, 1997, pp. 2154–2165, © 1997 IEEE.

Fig. 10.29 can be compared directly with Fig. 10.26, the corresponding visualization for the ring resonator studied earlier. These figures illustrate that there is a difference in the group velocities between the ring and the disk. In the final frame of Fig. 10.26, the pulse coupled into the ring is just short of completing one round trip after 240 fs. The pulse also maintains its single-mode transverse profile. However, the same pulse has a greater group velocity in the disk and therefore propagates farther around the disk after 240 fs. Further, the pulse in the disk is also broadened and is starting to trace out the whispering-gallery modes.

Fig. 10.30 graphs the calculated transmittance spectrum for the microcavity disk. The three sets of resonances in this figure correspond to the first-order, second-order, and third-order radial whispering-gallery modes of the disk. As the radial order (the number of electric field intensity maxima in the radial direction) increases, the transmission characteristics worsen because the coupling between the disk and the adjacent waveguides is not optimal for the higher-order radial modes.

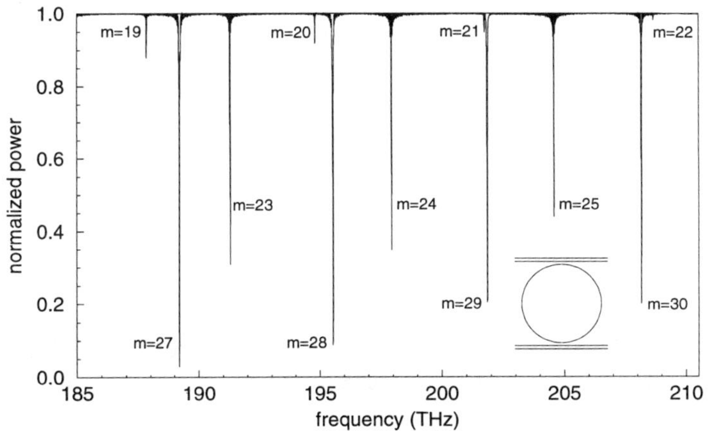

Fig. 10.30 FDTD-computed normalized transmittance spectrum at port B of the 5.0-μm-diameter microcavity disk resonator.

Table 10.2 lists the resonant frequencies, Q's, and FSR values calculated from the data in Fig. 10.30. In this table, q indicates the radial order of the mode and "NA" indicates that the resonance is too weak to obtain accurate data. In comparison with the resonance data for the ring, the first-order radial mode resonances for the disk exhibit several interesting properties:

1. The locations of the resonances are at lower frequencies because the effective index of the fundamental whispering-gallery modes in the solid disk is different from that of the narrow-waveguide ring.

TABLE 10.2

Resonance Data from Fig. 10.30 for the 5.0-μm-Diameter Microcavity Disk

q	m	ν_m (THz)	λ_m (nm)	Q	FSR (nm)
1	27	189.20	1584.51	7600	
					51.32
1	28	195.54	1533.19	7500	
					47.98
1	29	201.85	1485.21	7000	
					45.03
1	30	208.16	1440.18	9100	
2	23	191.29	1567.24	8700	
					52.69
2	24	197.94	1514.55	9900	
					49.12
2	25	204.58	1465.43	11,000	
3	19	187.83	1596.10	9900	
					57.09
3	20	194.80	1539.01	9700	
					52.96
3	21	201.74	1486.05	8800	
					NA
3	22	~208.65	NA	NA	

2. The FSR values are even larger that those for the ring because the lack of an inner rim allows the effective diameter to be smaller for the fundamental whispering-gallery modes of the disk than for the ring. This phenomenon is also evident in comparing the FSR of the first-order modes of the disk with the FSR of the higher-order modes. As the order of the radial mode increases, the peak of the mode moves toward the center of the disk. Therefore, d_{eff} decreases even further for the higher-order modes. Just as with the microcavity ring, the FSR decreases as the wavelength decreases within a given set of microdisk resonances.
3. The linewidths are narrower than for the ring. The higher fundamental-mode Q's of the disk resonator result from the reduced scattering loss arising from the lack of an inner rim. In an ideal disk, the low-order modes have the highest Q. However, since the lowest-order mode is confined most closely to the perimeter of the disk, it is most sensitive to sidewall roughness. Therefore, as seen in Table 10.2, the higher-order mode resonances have slightly higher Q's because of lower scattering loss.

To observe the whispering-gallery modes of the microdisk resonator, single-frequency FDTD simulations are performed at the resonance frequencies pinpointed by Fig. 10.30. Fig. 10.31(a) visualizes the sinusoidal steady-state E_z-field distribution for a continuous excitation at the nonresonant frequency, $\nu = 193.4$ THz ($\lambda = 1550.0$ nm). Calculations made from the FDTD-computed field data for this case show that 99.98% of the power in the incident signal at 193.4 THz remains in WG1.

Fig. 10.31(b) visualizes the steady-state E_z distribution for an excitation at $\nu = 189.2$ THz ($\lambda = 1584.5$ nm). According to Fig. 10.30, this frequency corresponds to the $m = 27$, $q = 1$ resonance. Indeed, the field pattern within the microdisk illustrates a first-order radial whispering-gallery mode. Fig. 10.31(b) also illustrates the effect of resonant field enhancement inside the microcavity. Since the gray scale here is normalized to the peak resonant fields, the fields in WG1 and WG2 appear to be weak. In fact, they are at the same level as the fields in WG1 in Fig. 10.31(a). For this on-resonance case, 99.79% of the incident power in WG1 switches to WG2.

Fig. 10.31(c) visualizes the steady-state E_z distribution for an excitation at $\nu = 191.3$ THz ($\lambda = 1567.2$ nm), the frequency of the $m = 23$, $q = 2$ resonance in Fig. 10.30. This graphic illustrates the second-order radial whispering-gallery mode, and shows the effect of the modal peak shifting away from the disk edge.

Finally, Fig. 10.31(d) shows the steady-state E_z distribution for an excitation at $\nu = 187.8$ THz ($\lambda = 1596.1$ nm), the frequency of the $m = 19$, $q = 3$ resonance in Fig. 10.30. This graphic illustrates the third-order radial whispering-gallery mode, and shows the progressive shifting of the modal peak away from the edge of the disk.

The visualizations of Fig. 10.31 (see also the color-plate section) show how each resonant whispering-gallery mode is confined to an annulus around the perimeter of the disk wherein the width of the annulus increases as the radial order increases. A comparison between Figs. 10.31(a) and 10.31(b) indicates the potential for high extinction ratios and low crosstalk between channels in these devices if the higher-order modes of the microdisk can be suppressed.

10.5.7 Suppression of Higher-Order Radial Whispering-Gallery Modes

As demonstrated above, microcavity disk resonators supporting whispering-gallery modes have less sidewall scattering loss and higher Q in comparison to single-mode microcavity rings. However, microdisks have multiple sets of resonances due to the presence of fundamental and higher-order radial modes. Suppression of the higher-order modes is necessary to use the disks as single-mode laser sources or as WDM devices having low crosstalk across a wide spectrum.

This section reviews two methods for suppressing the higher-order radial modes of microcavity disk resonators [47, 53]: (1) etching out the center of the disk; and (2) choosing the width of the adjacent waveguide to minimize the coupling between the waveguide mode and the higher-order modes of the disk. In each case, the resonances are manipulated by taking advantage of the spatial characteristics of the whispering-gallery modes of the disk.

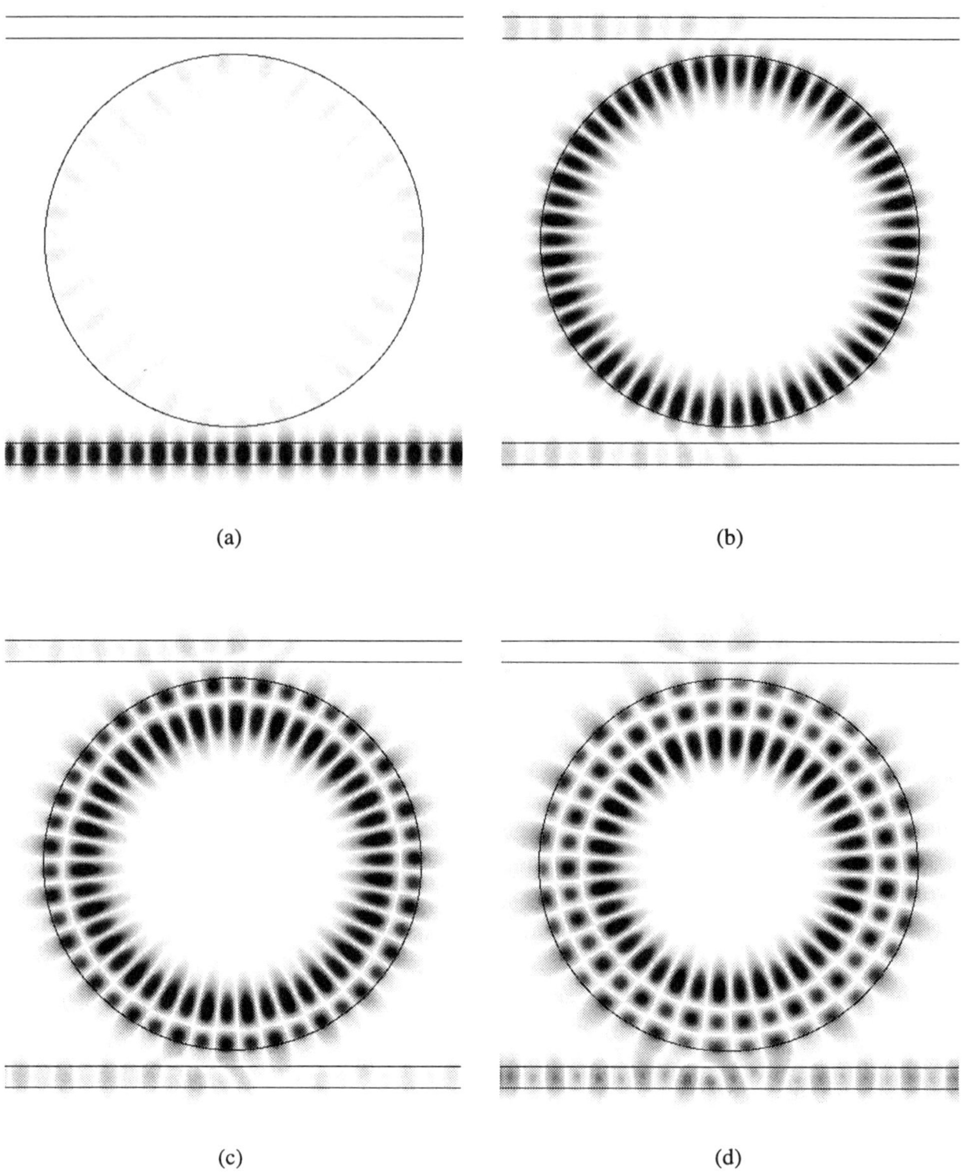

Fig. 10.31 Visualizations of the FDTD-computed sinusoidal steady-state electric field pattern in a 5.0-μm-diameter microcavity disk resonator coupled to straight 0.3-μm-wide waveguides for single-frequency excitations at port A of WG1: (a) off-resonance signal (193.4 THz); (b) on-resonance signal, first-order radial mode (189.2 THz); (c) second-order radial mode resonance (191.3 THz); and (d) third-order radial mode resonance (187.8 THz). *Source*: Hagness et al., *J. Lightwave Technology*, 1997, pp. 2154–2165, © 1997 IEEE.

Etching Out the Center of the Disk

The center of the disk can be etched out to create larger losses for the higher-order modes that propagate in that region. Here, it is instructive to consider a series of intermediate disk geometries that span the extremes of the solid disk and the narrow-waveguide ring. Only the cavity geometry is perturbed. The air gap between the 0.3-μm-wide coupling waveguides and the microcavity remains constant ($g = 0.232\ \mu\text{m}$).

Consider starting with the solid 5.0-μm-diameter microcavity disk resonator. For convenience, Fig. 10.32(a) repeats the FDTD-computed transmittance of this solid disk previously shown in Fig. 10.30.

Now, assume that the center of the disk is etched out to form a wide ring with an inner-rim diameter, d_i, of 3.0 μm, thereby yielding a ring-waveguide width, w, of 1.0 μm. Fig. 10.32(b) shows the FDTD-computed transmittance for this structure. The first-order and second-order whispering-gallery-mode resonances remain unchanged from that of the solid disk. The effect of creating an inner ring boundary is to shift the third-order resonances toward higher frequencies and to lower their Q. This begins to suppress the third-order radial modes that have fields concentrated toward the center of the disk.

In the next step, d_i is increased to 3.5 μm so that $w = 0.75\ \mu\text{m}$. Fig. 10.32(c) shows the computed transmittance for this case. Here, the third-order-mode resonances are completely suppressed and each second-order resonance is shifted upward in frequency. Only the first-order resonances remain unaffected.

Proceeding further, d_i is increased to 4.0 μm so that $w = 0.5\ \mu\text{m}$. Fig. 10.32(d) shows the computed transmittance for this case. There is now substantial suppression and upshifting of the second-order resonances. Further, the fundamental-mode resonances are now slightly upshifted and their Q's slightly lowered.

Finally, d_i is increased to 4.4 μm so that $w = 0.3\ \mu\text{m}$. This yields the geometry of the single-mode, narrow-waveguide ring previously studied in Section 10.5.5. Fig. 10.32(e) repeats for convenience the computed transmittance for this ring shown in Fig. 10.27.

In summary, as the etched-out inner diameter of the disk increases, each set of resonances monotonically shifts toward higher frequencies. This is seen by tracking the location of specific resonances through the sequence of Fig. 10.32(a–e). A smooth transition occurs between the multi-whispering-gallery-mode resonances of the solid 5.0-μm-diameter disk and the single-mode resonances of the 5.0-μm-diameter ring formed by 0.3-μm-wide waveguide.

Choosing the Width of the Adjacent Waveguide

Properly choosing the width of the adjacent waveguide minimizes coupling between the waveguide mode and the higher-order whispering-gallery modes of the disk. Consider the FDTD-computed transmittance spectra for the 5.0-μm-diameter microcavity disk resonator coupled to two straight adjacent waveguides with an air gap width of 0.2 μm.

(a)

(b)

(c)

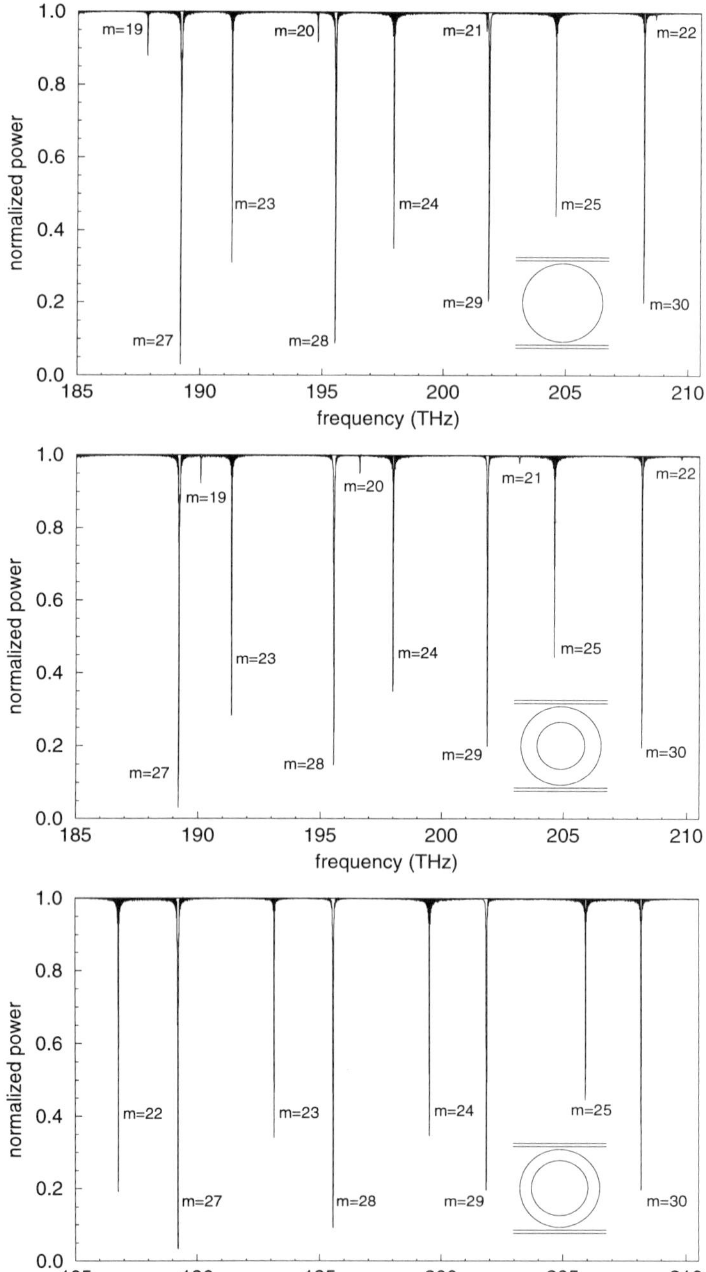

Fig. 10.32 Progression of the FDTD-computed transmittance of a 5.0-μm-diameter microcavity.

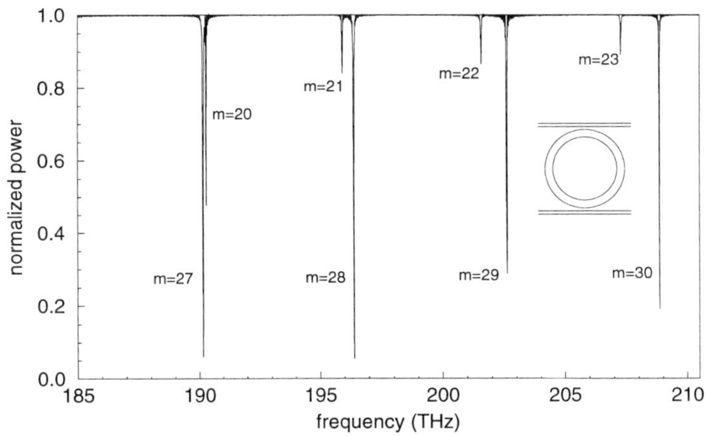

(d)

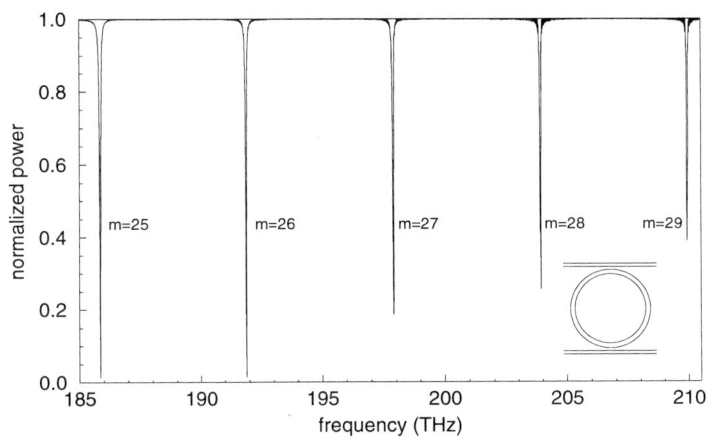

(e)

Fig. 10.32 *(cont.)* The resonator is coupled to straight 0.3-μm-wide waveguides. The inner-rim diameter, d_i, is increased in four steps: (a) solid disk ($d_i = 0$); (b) $d_i = 3.0$ μm, ring-waveguide width, $w = 1.0$ μm; (c) $d_i = 3.5$ μm, $w = 0.75$ μm; (d) $d_i = 4.0$ μm, $w = 0.5$ μm; (e) $d_i = 4.4$ μm, $w = 0.3$ μm. *Source*: Hagness et al., *J. Lightwave Technology*, 1997, pp. 2154–2165, © 1997 IEEE.

Fig. 10.33(a) shows transmittance results when the adjacent waveguides are 0.2 μm wide. Here, there are four sets of resonances corresponding to the first four radial whispering-gallery modes of the disk. (Each resonance is labeled with its radial number.)

In Fig. 10.33(b), the widths of the adjacent waveguides are increased to 0.3 μm. For this case, the fourth-order radial mode resonances disappear and the third-order radial mode resonances are weak.

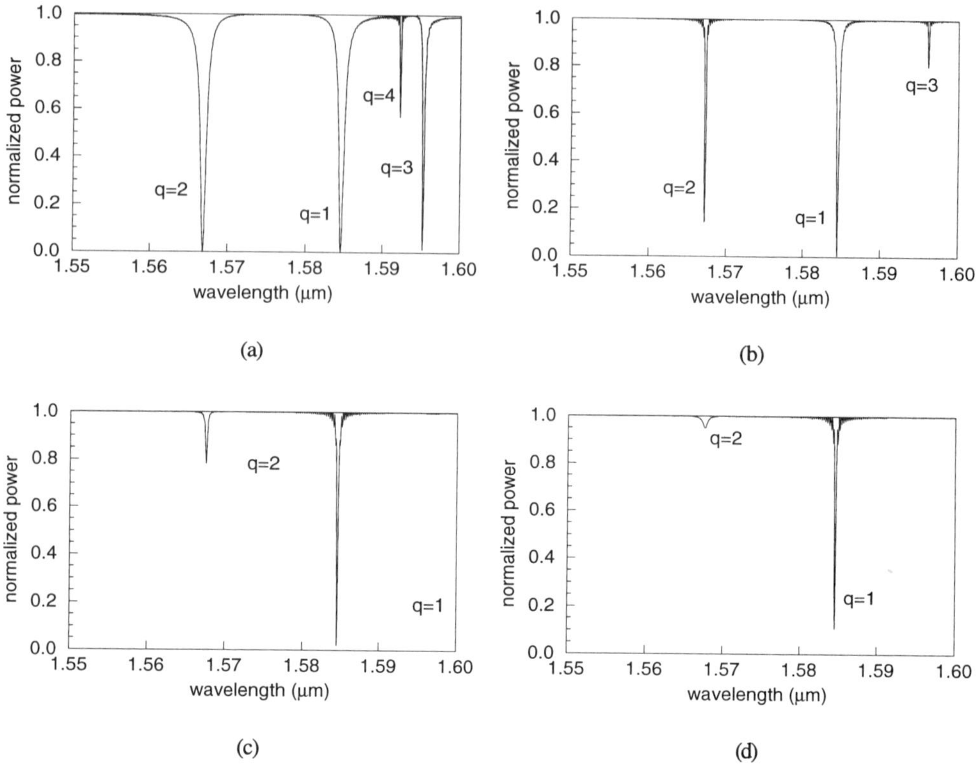

Fig. 10.33 FDTD-computed transmittance of a 5.0-μm-diameter microcavity disk resonator coupled to straight waveguides of the following widths: (a) 0.20 μm; (b) 0.30 μm; (c) 0.35 μm; (d) 0.38 μm. *Adapted from*: Hagness et al., *Proc. IEEE LEOS Annual Meeting*, 1997, pp. 160–161, © 1997 IEEE.

In Fig. 10.33(c), the widths of the adjacent waveguides are further increased to 0.35 μm. Now, the third-order resonances are eliminated, and the second-order-mode resonances are weak.

Finally, in Fig. 10.33(d), the widths of the adjacent waveguides are increased to 0.38 μm. Here, the second-order resonances are strongly suppressed, verging on extinction. As the waveguide width approaches this optimum value, the Q's of the fundamental modes increase. However, without changing the gap size, the minimum on-resonance transmission also increases. Once the optimum waveguide width is determined from the standpoint of suppressing unwanted higher-order modes, the gap size can be adjusted to give the best fundamental-mode transmission characteristics (namely, low on-resonance transmission and high Q).

10.5.8 Additional FDTD Modeling Studies

The FDTD method has also been used to investigate cascaded and series-coupled microcavity ring resonators with the goal of improving their filter characteristics [54]. A related class of devices based on the elevated microdisk (also called a "thumbtack" structure) has been modeled as well. For example, FDTD analyses of the whispering-galley modes of isolated microdisks on pedestals [55] or suspended in air [56] have been reported. FDTD simulations of isolated microdisks have also been used as baselines for comparison with analytical mode calculations [48].

10.6 SUMMARY AND CONCLUSIONS

This chapter first discussed basic issues related to modeling optical phenomena using FDTD, including highly dispersive optical waveguides, material dispersion, and optical nonlinearities. It then reviewed an FDTD formulation suitable for modeling a saturable homogeneously broadened gain medium, and applied this technique to analyze the operation of VCSELs. Specific applications to one-dimensional photonic bandgap structures were then reviewed. The chapter concluded with FDTD modeling results for high-Q, micron-sized, waveguide-coupled, ring and disk resonators. Key design parameters and tradeoffs identified as a result of these modeling studies were discussed in considerable detail.

A primary technology advance underlying all of these analyses is the advent of nanofabrication techniques. This technology is leading to the realization of optical resonators having physical dimensions of the order of the optical wavelength. These microcavities have been proposed for a rich variety of applications in high-density photonic integrated circuits. FDTD modeling has the potential to play a useful and practical design role for this emerging class of integrated optical devices and circuits. In fact, the usage of FDTD in micrometer- and nanometer-scale integrated optics should eventually be similar in scope to current FDTD applications in the engineering design of linear and nonlinear microwave circuits, as discussed in earlier chapters.

REFERENCES

[1] Rafizadeh, D., J. P. Zhang, S. C. Hagness, A. Taflove, K. A. Stair, S. T. Ho, and R. C. Tiberio, "Waveguide-coupled AlGaAs / GaAs microcavity ring and disk resonators with high finesse and 21.6-nm free spectral range," *Optics Lett.*, Vol. 22, 1997, pp. 1244–1246.

[2] Blom, F. C., D. R. van Dijk, H. J. W. M. Hoekstra, A. Driessen, and Th. J. A. Popma, "Experimental study of integrated-optics microcavity resonators: Toward an all-optical switching device," *Applied Physics Lett.*, Vol. 71, 1997, pp. 747–749.

[3] Yamamoto, Y., and R. Slusher, "Optical processes in microcavities," *Physics Today*, Vol. 46, 1993, pp. 66–73.

[4] Fan, S., P. R. Villeneuve, J. D. Joannopoulos, and E. F. Schubert, "High extraction efficiency of spontaneous emission from slabs of photonic crystals," *Physical Review Lett.*, Vol. 78, 1997, pp. 3294–3297.

[5] Baba, T., "Photonic crystals and microdisk cavities based on GaInAsP - InP systems," *IEEE J. Selected Topics in Quantum Electronics*, Vol. 3, 1997, pp. 808–830.

[6] Chu, S.-T., and S. K. Chaudhuri, "A finite-difference time-domain method for the design and analysis of guided-wave optical structures," *J. Lightwave Technology*, Vol. 7, 1989, pp. 2033–2038.

[7] Berenger, J.-P., "A perfectly matched layer for the absorption of electromagnetic waves," *J. Computational Physics*, Vol. 114, 1994, pp. 185–200.

[8] Yamauchi, J., M. Mita, S. Aoki, and H. Nakano, "Analysis of antireflection coatings using the FDTD method with the PML absorbing boundary condition," *IEEE Photonics Technology Lett.*, Vol. 8, 1996, pp. 239–241.

[9] Hagness, S. C., A. Taflove, and J. E. Bridges, "Wideband ultralow reverberation antenna for biological sensing," *Electronics Lett.*, Vol. 33, 1997, pp. 1594–1595.

[10] Reuter, C. E., R. M. Joseph, E. T. Thiele, D. S. Katz, and A. Taflove, "Ultrawideband absorbing boundary condition for termination of waveguiding structures in FDTD simulations," *IEEE Microwave and Guided Wave Lett.*, Vol. 4, 1994, pp. 344–346.

[11] Kogelnik, H., and V. Ramaswamy, "Scaling rules for thin-film optical waveguides," *Applied Optics*, Vol. 13, 1974, pp. 1857–1862.

[12] Luebbers, R. J., and F. Hunsberger, "FDTD for Nth-order dispersive media," *IEEE Trans. Antennas and Propagation*, Vol. 40, 1992, pp. 1297–1301.

[13] Kelley, D. F., and R. J. Luebbers, "Piecewise linear recursive convolution for dispersive media using FDTD," *IEEE Trans. Antennas and Propagation*, Vol. 44, 1996, pp. 792–797.

[14] Joseph, R. M., S. C. Hagness, and A. Taflove, "Direct time integration of Maxwell's equations in linear dispersive media with absorption for scattering and propagation of femtosecond electromagnetic pulses," *Optics Lett.*, Vol. 16, 1991, pp. 1412–1414.

[15] Korner, T. O., and W. Fichtner, "Auxiliary differential equation: efficient implementation in the finite-difference time-domain method," *Optics Lett.*, Vol. 22, 1997, 1586–1588.

[16] Sullivan, D. M., "Frequency-dependent FDTD methods using *Z* transforms," *IEEE Trans. Antennas and Propagation*, Vol. 40, 1992, pp. 1223–1230.

[17] Goorjian, P. M., A. Taflove, R. M. Joseph, and S. C. Hagness, "Computational modeling of femtosecond optical solitons from Maxwell's equations," *IEEE J. Quantum Electronics*, Vol. 28, 1992, pp. 2416–2422.

[18] Ziolkowski, R. W., and J. B. Judkins, "Full-wave vector Maxwell's equations modeling of the self-focusing of ultrashort pulses in a nonlinear Kerr medium exhibiting a finite response time," *J. Optical Society of America B*, Vol. 10, 1993, pp. 186–198.

[19] Joseph, R. M., and A. Taflove, "FDTD Maxwell's equations modeling for nonlinear electrodynamics and optics," *IEEE Trans. Antennas and Propagation*, Vol. 45, 1997, pp. 364–374.

[20] Sullivan, D. M., "Nonlinear FDTD formulations using *Z* transforms," *IEEE Trans. Microwave Theory and Techniques*, Vol. 43, 1995, pp. 676–692.

[21] Joseph, R. M., P. M. Goorjian, and A. Taflove, "Direct time integration of Maxwell's equations in two-dimensional dielectric waveguides for propagation and scattering of femtosecond electromagnetic solitons," *Optics Lett.*, Vol. 18, 1993, pp. 491–493.

[22] Joseph, R. M., and A. Taflove, "Spatial soliton deflection mechanism indicated by FDTD Maxwell's equations modeling," *IEEE Photonic Technology Lett.*, Vol. 6, 1994, pp. 1251–1254.

[23] Ziolkowski, R. W., and J. B. Judkins, "Applications of the nonlinear finite difference time domain (NL-FDTD) method to pulse propagation in nonlinear media: self-focusing and linear-nonlinear interfaces," *Radio Science*, Vol. 28, 1993, pp. 901–911.

[24] Ziolkowski, R. W., and J. B. Judkins, "Nonlinear finite-difference time-domain modeling of linear and nonlinear corrugated waveguides," *J. Optical Society of America B*, Vol. 11, 1994, pp. 1565–1575.

[25] Basinger, S. A., and D. J. Brady, "Finite-difference time-domain modeling of dispersive nonlinear Fabry-Perot cavities," *J. Optical Society of America B*, Vol. 11, 1994, pp. 1504–1511.

[26] Hawkins, R. J., and J. S. Kallman, "Lasing in tilted-waveguide semiconductor laser amplifiers," *Optical and Quantum Electronics*, Vol. 26, 1994, pp. S207–S217.

[27] Hagness, S. C., R. M. Joseph, and A. Taflove, "Subpicosecond electrodynamics of distributed Bragg reflector microlasers: results from finite difference time domain simulations," *Radio Science*, Vol. 31, 1996, pp. 931–941.

[28] Ziolkowski, R. W., J. M. Arnold, and D. M. Gogny, "Ultrafast pulse interactions with two-level atoms," *Physical Review A*, Vol. 52, 1995, pp. 3082–3094.

[29] Yariv, A., *Quantum Electronics*, 3rd ed., New York: Wiley, 1989.

[30] Siegman, A., *Lasers*, Mill Valley, CA: University Science Books, 1986.

[31] Agrawal, G. P., and N. K. Dutta, *Semiconductor Lasers*, 2nd ed., New York: Van Nostrand Reinhold, 1993.

[32] Stearns, D. S., and R. David, *Signal Processing Algorithms Using Fortran and C*, Englewood Cliffs, NJ: PTR Prentice Hall, 1993.

[33] Corzine, S. W., R. S. Geels, J. W. Scott, R. H. Yan, and L. A. Coldren, "Design of Fabry-Perot surface-emitting lasers with a periodic gain structure," *IEEE J. Quantum Electronics*, Vol. 25, 1989, pp. 1513–1524.

[34] Born, M., and E. Wolf, *Principles of Optics*, Oxford: Pergamon Press, 1980.

[35] Hagness, S. C., S. T. Ho, and A. Taflove, "Finite-difference time-domain (FDTD) computational electrodynamics simulations of microlaser cavities in one and two spatial dimensions," in *Computational Electromagnetics and Its Applications*, T. G. Campbell, R. A. Nicolaides, and M. D. Salas, eds., Dordrecht, Netherlands: Kluwer, 1997, pp. 229–251.

[36] Yablonovitch, "Photonic band-gap structures," *J. Optical Society of America B*, Vol. 10, 1993, pp. 283–295.

[37] Yablonovitch, E., "Inhibited spontaneous emission in solid-state physics and electronics," *Physical Review Lett.*, Vol. 58, 1987, pp. 2059–2062.

[38] Yablonovitch, E., T. J. Gmitter, R. D. Meade, A. M. Rappe, K. D. Brommer, and J. D. Joannopoulos, "Donor and acceptor modes in photonic band structure," *Physical Review Lett.*, Vol. 67, 1991, pp. 3380–3383.

[39] Villeneuve, P. R., S. Fan, J. D. Joannopoulos, K. Y. Lim, G. S. Petrich, L. A. Kolodziejski, and R. Reif, "Air-bridge microcavities," *Applied Physics Lett.*, Vol. 67, 1995, pp. 167–169.

[40] Zhang, J. P., D. Y. Chu, S. L. Wu, W. G. Bi, R. C. Tiberio, R. M. Joseph, A. Taflove, C. W. Tu, and S. T. Ho, "Nanofabrication of 1-D photonic bandgap structures along a photonic wire," *IEEE Photonic Technology Lett.*, Vol. 8, 1996, pp. 491–493.

[41] Ziolkowski, R. W., and T. Liang, "Design and characterization of a grating-assisted coupler enhanced by a photonic-band-gap structure for effective wavelength-division demultiplexing," *Optics Lett.*, Vol. 22, 1997, pp. 1033–1035.

[42] Villeneuve, P. R., D. S. Abrams, S. Fan, and J. D. Joannopoulos, "Single-mode waveguide microcavity for fast optical switching," *Optics Lett.*, Vol. 21, 1996, pp. 2017–2019.

[43] Chen, J. C., H. A. Haus, S. Fan, P. R. Villeneuve, and J. D. Joannopoulos, "Optical filters from photonic band gap air bridges," *J. Lightwave Technology*, Vol. 14, 1996, pp. 2575–2580.

[44] Villeneuve, P. R., S. Fan, and J. D. Joannopoulos, "Microcavities in photonic crystals: mode symmetry, tuneability, and coupling efficiency," *Physical Review B*, Vol. 54, 1996, pp. 7837–7842.

[45] Marcatili, E. A. J., "Bends in optical dielectric guides," *Bell System Technical J.*, Vol. 48, 1969, pp. 2103–2132.

[46] Oda, K., N. Takato, and H. Toba, "A wide-FSR waveguide double-ring resonator for optical FDM transmission systems," *J. Lightwave Technology*, Vol. 9, 1991, pp. 728–736.

[47] Hagness, S. C., D. Rafizadeh, S. T. Ho, and A. Taflove, "FDTD microcavity simulations: design and experimental realization of waveguide-coupled single-mode ring and whispering-gallery-mode disk resonators," *J. Lightwave Technology*, Vol. 15, 1997, pp. 2154–2165.

[48] Little, B. E., and S. T. Chu, "Estimating surface-roughness loss and output coupling in microdisk resonators," *Optics Lett.*, Vol. 21, 1996, pp. 1390–1392.

[49] Little, B. E., J.-P. Laine, and S. T. Chu, "Surface-roughness-induced contradirectional coupling in ring and disk resonators," *Optics Lett.*, Vol. 22, 1997, pp. 4–6.

[50] Slusher, R. E., A. F. J. Levi, U. Mohideen, S. L. McCall, S. J. Pearton, and R. A. Logan, "Threshold characteristics of semiconductor microdisk lasers," *Applied Physics Lett.*, Vol. 63, 1993, pp. 1310–1312.

[51] Griffel, G., and S. Arnold, "Synthesis of variable optical filters using meso-optical ring arrays," *IEEE LEOS Annual Meeting*, San Francisco, CA, Nov. 1997, paper WT4.

[52] Hagness, S. C., D. Rafizadeh, S. T. Ho, and A. Taflove, "FDTD analysis and comparison of circular and elongated ring designs for waveguide-coupled microcavity ring resonators," in *Integrated Photonics Research*, 1998 OSA Technical Digest Series (Optical Society of America, Washington DC), in press.

[53] Hagness, S. C., D. Rafizadeh, S. T. Ho, and A. Taflove, "Suppression of higher-order radial whispering gallery modes in waveguide-coupled microcavity disk resonators," in *Proc. IEEE LEOS Annual Meeting*, San Francisco, CA, Nov. 1997, pp. 160–161.

[54] Little, B. E., S. T. Chu, H. A. Haus, J. Foresi, and J.-P. Laine, "Microring resonator channel dropping filters," *J. Lightwave Technology*, Vol. 15, 1997, pp. 998–1005.

[55] Hawkins, R. J., N. K. Madsen, J. S. Kallman, M. D. Feit, C. C. Shang, B. W. Shore, and J. F. DeFord, "Full-wave simulation of the thumbtack laser," in *Integrated Photonics Research*, 1993 OSA Technical Digest Series (Optical Society of America, Washington DC), Vol. 10, pp. 116–119.

[56] Li, B. J., and P. L. Liu, "Numerical analysis of the whispering gallery modes by the finite-difference time-domain method," *IEEE J. Quantum Electronics*, Vol. 32, 1996, pp. 1583–1587.

Chapter 11

FDTD in Bioelectromagnetics: Safety Assessment and Medical Applications

Om P. Gandhi

11.1 INTRODUCTION

The FDTD method has been used extensively in the two major areas in bioelectromagnetics: the assessment of safety for exposure to electromagnetic fields; and the design of therapeutic exposure systems (for example, hyperthermia systems for cancer treatment). Almost all of the RF safety standards worldwide are set in terms of mass-normalized rates of electromagnetic energy absorption for tissues (specific absorption rates or SARs in W/kg) that must not be exceeded for exposure to electromagnetic fields [1, 2]. For radiated near-field or far-field exposure conditions, FDTD has been used to calculate SAR distributions for tissue-specified, anatomically based models of the human body from low power-line frequencies (60 Hz in North America and 50 Hz in Europe and Japan) to microwave frequencies. For low frequencies to 100 kHz, the SARs are generally negligible and the more pertinent parameters of biological relevance are the induced current densities. Safety standards at low frequencies have therefore been proposed in terms of induced current densities [3, 4], which may be predicted using numerical methods including FDTD [5, 6].

The challenge for therapeutic medical applications of electromagnetic energy is to design applicators that provide desirable distributions of absorbed energy. Here, a controlled and uniform heating of the diseased body-tissue region with minimal heating of adjacent healthy tissues is the goal. This goal is difficult to attain because of the great heterogeneity and geometric complexity of the body's tissues.

11.2 FDTD WITH UNIFORM AND EXPANDING GRID

The FDTD method is extremely versatile for bioelectromagnetic problems. FDTD has been used for modeling whole-body or partial-body exposures due to spatially uniform or nonuniform near or far fields that are sinusoidally varying or transient in nature. The latter includes the fields associated with an electromagnetic pulse (EMP) [5 – 12].

As is typical in FDTD, the time-dependent Maxwell's curl equations, given by

$$\nabla \times \vec{E} = -\mu \frac{\partial \vec{H}}{\partial t} \tag{11.1}$$

$$\nabla \times \vec{H} = \sigma \vec{E} + \varepsilon \frac{\partial \vec{E}}{\partial t} \tag{11.2}$$

are implemented for a lattice of subvolumes or Yee space cells that may be cubical or parallelepiped with different dimensions Δx, Δy, and Δz in the x, y, and z directions, respectively. The components of $\vec{E}$ and $\vec{H}$ are positioned about each of the cells at half-cell intervals and calculated alternately (leapfrogged) with half-time steps, $\Delta t/2$. The details of the method are given in several of the above-referenced publications and are therefore not repeated here.

In the FDTD method, it is necessary to represent not only the scatterer / absorber such as the human body or a part thereof, but also the electromagnetic sources, including their shapes, excitations, and so forth, if these sources are in the near-field region. On the other hand, the far-field sources are described by means of incident plane-wave fields specified on a three-dimensional source "box" located typically 6 to 10 cells away from the exposed body. The interaction space consisting of several hundred thousand to several million Yee cells is truncated by means of absorbing boundaries. The penetrating fields are tracked in time for all cells of the interaction space. The computation is considered to be completed when either the $\vec{E}$ and $\vec{H}$ fields have died off for an impulsive excitation or, for a sinusoidal excitation, when the sinusoidal steady-state is observed for all cells within the interaction space.

Even though a uniform FDTD grid has primarily been used for the bioelectromagnetic problems to date, an expanding-grid formulation has also been proposed [13]. This offers the advantage of modeling the tightly coupled region with a fine resolution (small cell size) while allowing cell sizes to increase as one moves further away from the region of primary interest. The expanding-grid algorithm allows different cell-to-cell expansion factors along the three coordinate axes, and can reduce by a factor of 4 to 10 the total number of cells needed to model a given volume as compared with a uniform grid formulation wherein the cells with the finest resolution are used throughout the volume. Some of the applications to date are for calculations of induced current and SAR distributions for a worker model exposed to an RF dielectric heater [14], for SAR distributions due to cellular telephones [15, 16], and for dosimetry of the human body at higher frequencies. For all of these applications, the coupled fields diminish rapidly into the exposed parts of the body. By using smaller cell sizes for the coupled region, we are able to determine the SARs with higher precision and better model the irradiating source (for example, the personal wireless device, its antenna, and the feed region).

A variation of the expanding-grid formulation is the use of subgridding [17]. Whereas a slowly expanding cell size with cell-to-cell expansion factors on the order of 1.1 to 1.25 are used in the expanding-grid formulation, a stepped expansion factor of two, and at times three, may be used in the subgridding method to represent the regions where the electromagnetic coupling is considerably lower.

11.3 FREQUENCY-DEPENDENT FDTD METHOD

As illustrated in Fig. 11.1, the electrical properties of biological tissues vary significantly with frequency. A frequency-dependent FDTD, or $(FD)^2TD$ method, is therefore needed to model the exposure of biological bodies to short pulses such as those associated with ultrawideband radar. Several general approaches have been used for the $(FD)^2TD$ method. One approach is to convert the complex permittivity from the frequency domain to the time domain and convolve this with the time-domain electric field to obtain time-domain fields within the dispersive material. The convolution can be updated recursively for some rational forms of complex permittivity, which removes the need to store the time history of the fields and makes the method feasible. This method has been applied to materials described by a first-order Debye relaxation equation [18 – 20], a second-order Lorentz equation with multiple poles [21], and to a gaseous plasma [22].

Another approach is to add a differential equation relating the electric flux density, $\vec{D}$, to the electric field, $\vec{E}$, and solve this new equation simultaneously with the standard FDTD time-stepping equations. This method has been applied to one-dimensional and two-dimensional examples with materials described by a first-order Debye equation or second-order single-pole Lorentz equation [23]. It has also been applied to a three-dimensional sphere and a homogeneous, two-thirds muscle-equivalent man model with properties described by a second-order Debye equation [24, 25]. In the following, we describe this differential equation approach that has been used for induced current and SAR calculations for a heterogeneous model of the human body exposed to ultrawideband electromagnetic pulses [26, 27].

The time-dependent Maxwell's curl equations used for the FDTD method are given in (11.1) and (11.2). Equation (11.2) can also be written as

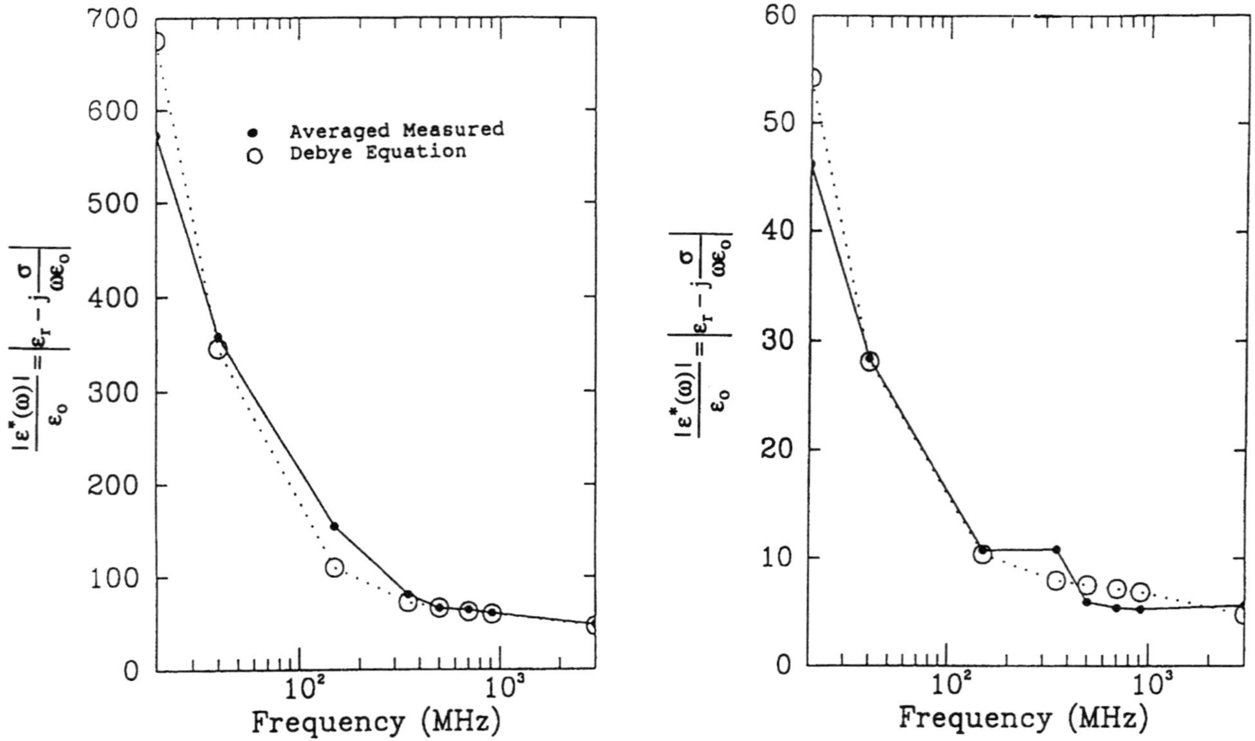

(a) (b)

Fig. 11.1 Fit of Debye equation with two relaxation constants (Equation 11.5) to measured tissue properties of: (a) muscle; (b) fat. *Source*: Gandhi and Furse, *IEEE Trans. Electromagnetic Compatibility*, 1997, pp. 174–180.

$$\nabla \times \vec{H} = \frac{\partial \vec{D}}{\partial t} \tag{11.3}$$

In the frequency domain, the electric flux density, $\vec{D}(\omega)$, is related to the electric field, $\vec{E}(\omega)$, by

$$\vec{D} = \varepsilon(\omega)\vec{E} \tag{11.4}$$

where $\varepsilon(\omega)$ is the complex permittivity of the local tissue. Since (11.1) and (11.3) are solved iteratively in the time domain, (11.4) must also be expressed in the time domain. This is done by choosing a rational function for $\varepsilon(\omega)$ such as the Debye equation with two relaxation constants:

$$\varepsilon(\omega) = \varepsilon_0\left(\varepsilon_\infty + \frac{\varepsilon_{s1} - \varepsilon_\infty}{1 + j\omega\tau_1} + \frac{\varepsilon_{s2} - \varepsilon_\infty}{1 + j\omega\tau_2}\right) \tag{11.5}$$

Rearranging (11.5) and substituting in (11.4) gives

$$\vec{D}(\omega) = \varepsilon(\omega)\vec{E}(\omega) = \varepsilon_0\left[\frac{\varepsilon_s + j\omega(\varepsilon_{s1}\tau_2 + \varepsilon_{s2}\tau_1) - \omega^2\tau_1\tau_2\varepsilon_\infty}{1 + j\omega(\tau_1 + \tau_2) - \omega^2\tau_1\tau_2}\right]\vec{E}(\omega) \tag{11.6}$$

where the dc (zero frequency) dielectric constant is given by

$$\varepsilon_s = \varepsilon_{s1} + \varepsilon_{s2} - \varepsilon_\infty \tag{11.7}$$

Assuming $\exp(j\omega t)$ time dependence, we can write (11.6) as a differential equation in the time domain

$$\tau_1\tau_2\frac{\partial^2\vec{D}}{\partial t^2} + (\tau_1 + \tau_2)\frac{\partial\vec{D}}{\partial t} + \vec{D} = \varepsilon_0\left[\varepsilon_s\vec{E} + (\varepsilon_{s1}\tau_2 + \varepsilon_{s2}\tau_1)\frac{\partial\vec{E}}{\partial t} + \varepsilon_\infty\tau_1\tau_2\frac{\partial^2\vec{E}}{\partial t^2}\right] \tag{11.8}$$

For the $(\text{FD})^2\text{TD}$ method, we need to solve (11.1) and (11.3) subject to (11.8). As in [24, 25], we write these equations in the difference form, apply (11.8) to find $\vec{E}$, apply (11.1) to find $\vec{H}$, and apply (11.3) to find $\vec{D}$ at each electric-field component location. The $\vec{E} \to \vec{H} \to \vec{D}$ cycle is then repeated until the pulse has died off.

11.4 VALIDATIONS FOR NEAR-FIELD BIOELECTROMAGNETIC SIMULATIONS

The FDTD method has been thoroughly tested against analytical solutions, measured data, and other numerical techniques for far-field exposures. However, until recently, it has not been subjected to such rigorous validations for near-field exposures.

Because of the increasing use of FDTD for SAR compliance testing of all new personal wireless devices mandated by the Federal Communications Commission [28], we recently started to focus on this issue. In this section, we describe four near-field test cases where excellent agreement is found between the results obtained using FDTD and analytical and/or other procedures [29].

11.4.1 Dipole Antenna Near a Layered Bone-Brain Half-Space

The first validation involves a 12-cm-long, 900-MHz dipole antenna located 2 cm away in air from a layered bone-brain half-space. The problem geometry is shown as the insert of Fig. 11.2. An exact analytical solution based on expansion of the dipole fields and enforcement of the boundary conditions at the planar material interfaces is given in [30]. For this specific case, the bone layer is 1-cm thick with $\varepsilon_r = 5.98$ and $\sigma = 0.099$ S/m. The underlying brain is infinitely thick with $\varepsilon_r = 50.3$ and $\sigma = 1.34$ S/m [30].

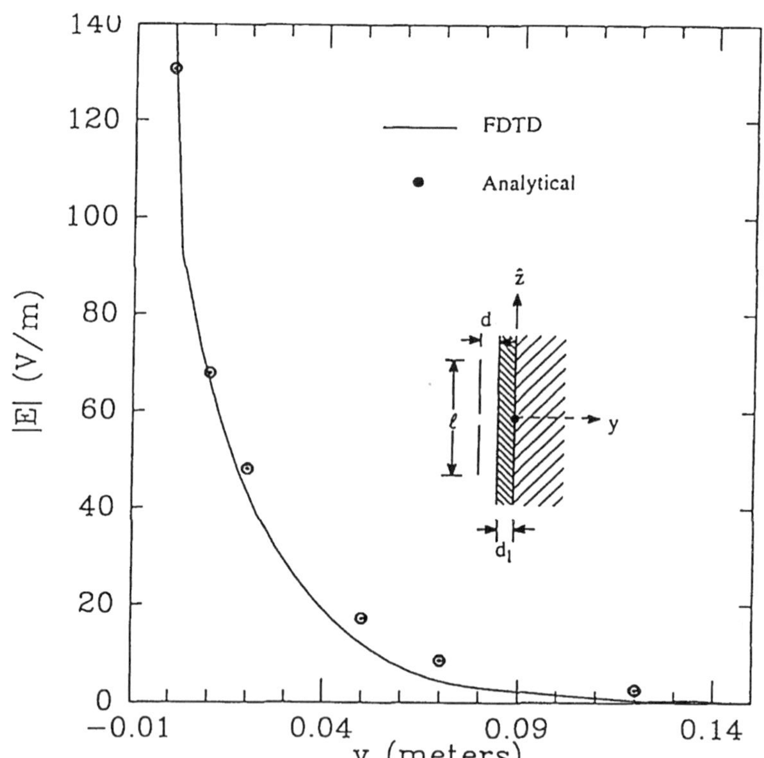

Fig. 11.2 Comparison of FDTD-computed and analytical results for the electric-field magnitude along the y axis penetrating into a layered bone-brain half-space excited by a 12-cm-long dipole antenna operating at 900 MHz.

The FDTD grid uses a uniform cell size of 5 mm in the x and y directions and 9.23 mm in the z direction. This gives an antenna length of 13 cells with the feedpoint in the center cell, and a distance of four free-space cells between the antenna and the first (bone) layer, which is two cells thick. The second (brain) layer, supposed to be infinitely thick, is modeled by a finite thickness of 24.5 cm (119 cells). This is sufficient to attenuate the fields to well below 1% of their peak values before the end of the brain layer is reached. The infinite nature of the layered half-space in the x and z directions is modeled by considering a sufficiently large computational volume such that the fields are very small at its edges. This volume is taken to be 12 cm (24 cells) in the x direction and 33.3 cm (36 cells) in the z direction. The simulation is run for 10 sinusoidal cycles, but convergence is observed at about 8 cycles.

Fig. 11.2 compares the magnitude of the FDTD-computed electric field along the y axis (penetrating into the half-space) with the analytical solution (Table 1 of [30]). The field values are normalized to an antenna feedpoint current of 100 mA rms. Agreement is seen to be very good.

11.4.2 Dipole Antenna Near an Acrylic Box Containing Brain-Equivalent Phantom Material

The second validation is for a half-wavelength dipole antenna near an acrylic box filled with brain-equivalent phantom material. Here, comparison is made to measured data.

The 17.4 cm-long dipole operating at 840 MHz is placed at several distances from the acrylic box with the antenna feedpoint centered in front of the box. Similar to [31], the box dimensions are $30 \times 15 \times 50$ cm in the x, y, and z directions, respectively.

The walls of the box are made of 6.35-mm-thick acrylic, assumed to have the electrical properties of polystyrene ($\varepsilon_r = 2.55$). Since the FDTD cell size used in this simulation is slightly larger than the wall thickness, the effective dielectric constant for the acrylic is calculated from the formula given in [32]

$$\varepsilon_{\text{eff}} = \frac{\varepsilon_r \Delta}{\varepsilon_r(\Delta - w) + w} \tag{11.9}$$

where Δ is the size of the FDTD cell, w is the thickness of the acrylic, and ε_r is the dielectric constant of polystyrene. In [32], this formula is derived by finding the average field at the air-acrylic interface and setting this equal to the field in the larger-sized FDTD cell, thereby giving the smaller effective dielectric constant of (11.9). The acrylic box is filled with brain-simulating material with $\varepsilon_r = 41.1$ and $\sigma = 1.06$ S/m.

SAR measurements are taken using a Narda Model 8021B E-field probe. All data are parameterized relative to the distance, d, between the dipole and the exterior surface of the acrylic box for $d = 11.0$, 16.0, 21.0, and 26.5 mm. FDTD simulations are run for 2500 time steps, and convergence is observed after about 1500 time steps. Values are normalized to a total radiated power of 500 mW.

Fig. 11.3 compares the FDTD-computed and the measured SAR distributions along the y axis (penetrating into the phantom). Fig. 11.4 provides a similar comparison along a line parallel to the z axis and 7 mm beneath the inside edge of the box wall within the phantom brain material. Excellent agreement is seen between the measured and computed data. This agreement demonstrates the accuracy of both the measurements and the FDTD simulations in this near-field region. Note that the physical dimensions of the E-field probe make it impossible to measure closer than 4 mm to the front of the box. However, this would be necessary to obtain the highest 1-gram-averaged SAR values since the peak fields (and hence the peak SARs) occur at the very front of the phantom nearest the antenna. The excellent agreement between the FDTD results and the measured data extrapolated to the front of the box supports the conclusion that FDTD can provide a good estimate of the peak SARs even in locations where direct measurements are not feasible.

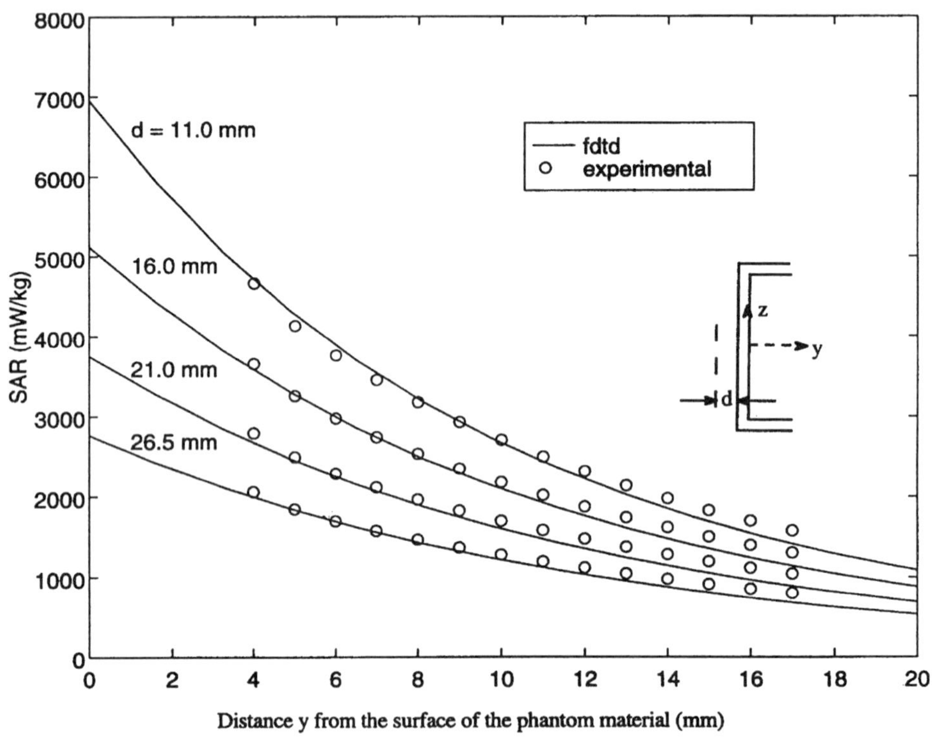

Fig. 11.3 Comparison of FDTD-computed and measured SAR distributions along the y axis penetrating into an acrylic box filled with brain-equivalent phantom material excited by a half-wavelength dipole antenna radiating 0.5 W at 840 MHz.

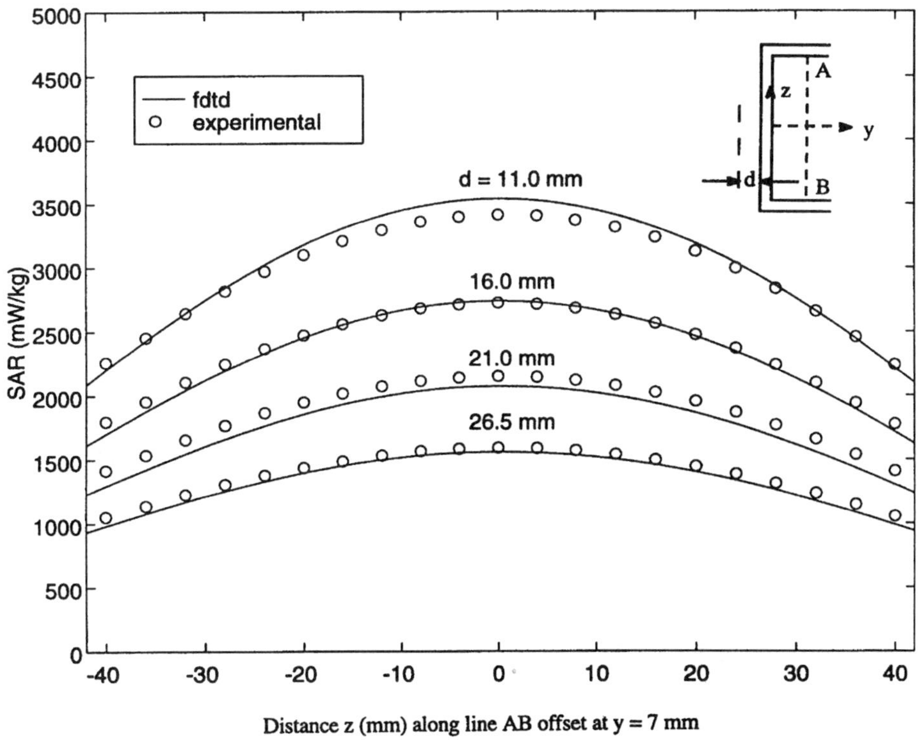

Fig. 11.4 FDTD-computed and measured SAR distributions corresponding to Fig. 11.3 within the acrylic box filled with brain-equivalent phantom material excited by a half-wavelength dipole antenna radiating 0.5 W at 840 MHz. Here, the comparison is along a line AB parallel to the z axis and 7-mm deep within the phantom brain material.

11.4.3 Infinitesimal Dipole Near a Brain-Equivalent Sphere

The above examples give good confidence in the ability of FDTD to correctly predict the fields and SAR distributions within planar phantoms as a function of the distance between the antenna and the phantom. However, these simulations do not require modeling of the curved surfaces associated with the high-resolution head models required for analysis of wireless communication devices.

The next test case demonstrates the ability of FDTD to correctly analyze fields within a sphere that does require modeling of its curved surface. Here, simple staircasing is used (the best fit to the sphere's surface by the rectangular unit cells of the lattice). Note that the smaller the sphere is relative to the cell size, the coarser is its staircased model.

For this case, an infinitesimal dipole excited at 900 MHz is located 1.5 cm from a 20-cm-diameter, homogeneous brain-equivalent sphere ($\varepsilon_r = 43$, $\sigma = 0.83$ S/m) [33]. The infinitesimal dipole is modeled as a single E_z feedpoint location. The FDTD cell size is $\Delta x = \Delta y = \Delta z = 5$ mm, which makes the sphere 40 cells in diameter. Care is taken in modeling the sphere so that the front edge contains five cells, rather than just one. This is found to be critical to obtain good FDTD agreement with analytical results.

Fig. 11.5 compares the FDTD-computed and exact normalized SAR along the y axis penetrating into the sphere from its front surface. The exact solution is obtained with a Bessel function expansion [33]. Excellent agreement is seen.

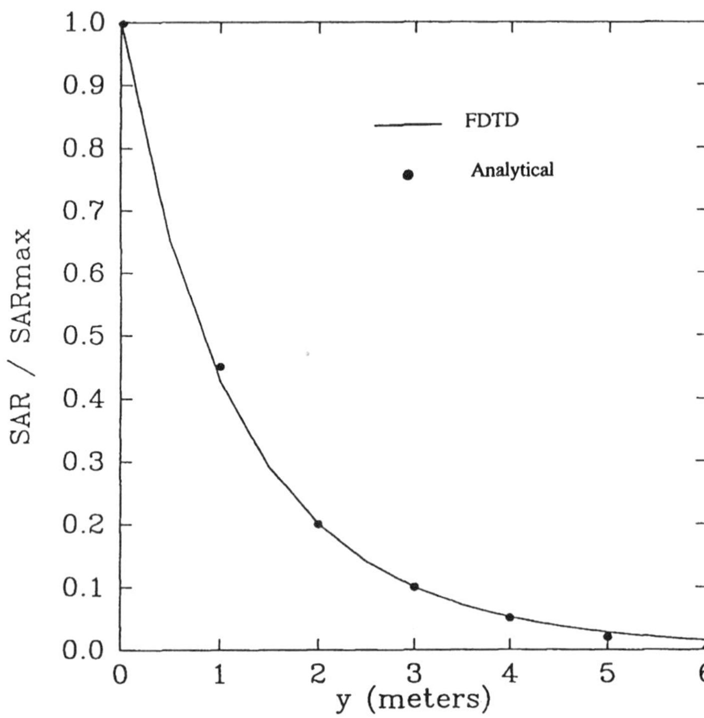

Fig. 11.5 Comparison of FDTD-computed and analytical results for the relative SAR distribution along the y axis penetrating into a 20-cm-diameter, homogeneous, brain-equivalent sphere excited by an infinitesimal dipole located 1.5 cm from the sphere and radiating at 900 MHz.

11.4.4 Half-Wavelength Dipole Near a Brain-Equivalent Sphere

The next test case involves an approximately half-wavelength dipole antenna ($\ell = 17.3$ cm) excited at 840 MHz that is located 2.5 cm from a 20-cm-diameter, homogeneous brain-equivalent sphere ($\varepsilon_r = 55$, $\sigma = 1.4$ S/m). The FDTD cell size is $\Delta x = \Delta y = 5$ mm and $\Delta z = 5.9655$ mm, and the simulation is run for 1500 time steps.

Fig. 11.6 compares the absolute SAR distributions along the y axis penetrating into the sphere from its front surface as computed by both FDTD and the multiple-multipole method (MMP) [31]. Excellent agreement is seen. The high level of agreement between the FDTD and analytical or MMP results observed in Figs. 11.5 and 11.6 gives confidence that the FDTD method can accurately model curved absorbing objects for near-field simulations.

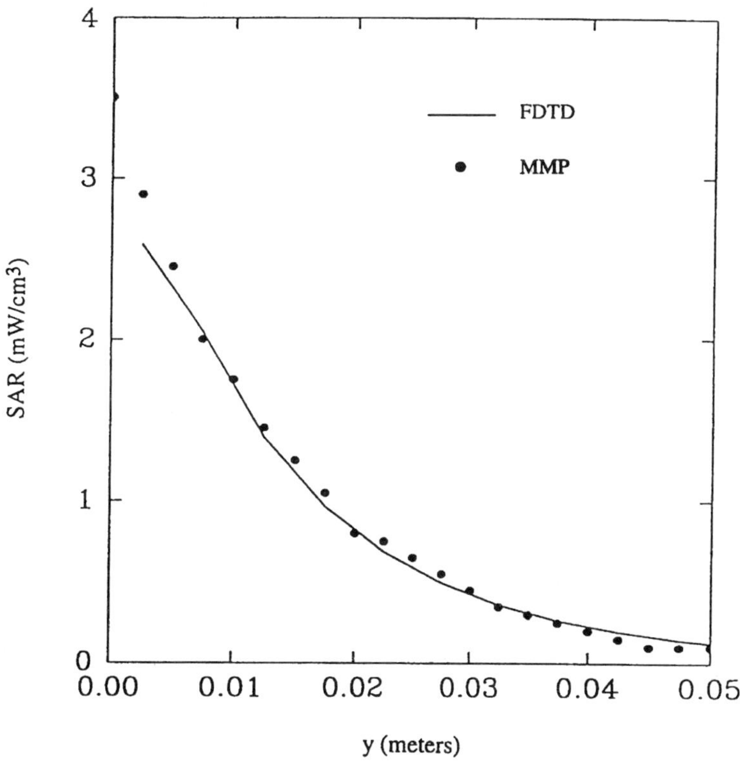

Fig. 11.6 Comparison of FDTD-computed and MMP results for the SAR distribution along the y axis penetrating into a 20-cm-diameter, homogeneous, brain-equivalent sphere excited by a half-wavelength dipole located 2.5 cm from the sphere and radiating at 840 MHz.

11.5 ANATOMICALLY BASED MODELS

Since the beginning of bioelectromagnetics, there has been a desire to obtain SAR distributions modeling the nonuniformity of the interacting fields as realistically as possible. It has also been known that the electromagnetic fields associated with several sources do not extend to more than a small fraction of the human body. Commonly used sources such as RF sealers, for example, create leakage fields that decrease rapidly in the vertical direction with fields dropping off to a small fraction of the maximum values within 10 to 20 cm in each direction along the vertical axis. For biomedical applications, too, RF and microwave applicators are generally quite small to apply the electromagnetic energy to the desired part of the body.

11.5.1 Analytical Models

The earliest analyses of absorbed electromagnetic energy in biological tissue models were performed by Schwan and Li [34] and subsequently by Johnson et al. [35]. These analyses involved planar layered structures. Although these models were useful at the time, they are incapable of taking into account shape-dependent physical phenomena. Subsequently, the planar models were followed by lossy homogeneous sphere models and lossy layered sphere models implemented by Kritikos and Schwan [36], Lin et al. [37], Joines and Spiegel [38], Weil [39], and others. An advantage of using spherical shapes is that the classical Mie series solution [40] can be written for the various layers of the sphere.

The sphere models revealed that SAR is very shape dependent. To help account for the effects of body shape and consequently wave polarization, Johnson, Durney, and their colleagues developed prolate spheroidal, ellipsoidal, and layered-cylindrical models. These were relatively simple to use and provided excellent information on whole-body-averaged SARs as a function of frequency for different polarizations of the illuminating wave [41–45].

11.5.2 Early Moment-Method Block Models

By the mid-1970s, it was becoming apparent that analytical models using either simple homogeneous or layered-inhomogeneity treatments of the body were incapable of providing the level of detail of SAR distributions needed to understand possible hazards and set standards. Therefore, the first fully numerical models of electromagnetic wave interactions with the human body were initiated.

Livesay and Chen [46] and Chen et al [47, 48] used the frequency-domain method of moments cast in a volumetric form to calculate SAR distributions in models of the human body comprised of cubic tissue cells. Subsequently, Hagmann et al. [49, 50] developed an improved moment-method body model. Here, 180 tissue cubes of different sizes were combined to form an overall size and shape closely resembling that of an average man, as given by biometric data. Also, different complex permittivities based on measured tissue properties for the various regions of the body were used. Later, to further improve the resolution for SAR calculations, somewhat smaller cell sizes were used, resulting in the human body being represented by 1132 cells [51].

11.5.3 Initial Human Body Cross-Section Models

With the advent of efficient numerical techniques such as the impedance method [52, 53] and FDTD [7, 9, 10] in the mid-1980s, it became possible to obtain SAR distributions within the human body or parts thereof being represented by several hundred-thousand cells, rather than the 180 to 1132 cells that had been possible using the method of moments. This level of detail permits anatomically based models of the human body to be implemented.

Early models of this type were reported in [9, 10, 53] using the sectional diagrams given by Eycleshymer and Schoemaker [54]. As detailed in [9 – 11], these models defined the predominant tissue type (one of 16 tissues such as muscle, fat, bone, blood, brain, etc.) with a resolution of 0.25×0.25 inch for each of the sectional diagrams. Since the sectional diagrams were available for somewhat variable separations, typically 2.3 to 2.7 cm, a new set of equispaced layers was defined at 0.635-cm (0.25-inch) intervals by interpolating the data onto these layers. The database at this stage consisted of fractional volumes occupied by the various tissues in each of the cubical subvolumes or cells of dimension 0.635 cm (0.25 inch) for each of the sides.

Because the number of such cells representing the human body (approximately 360,000) was not easy to handle with the readily accessible computers of the mid-1980s, two coarser models of cell dimensions 1.27 cm (0.5 inch) and 2.54 cm (1 inch) were developed by combining the data for groups of $2 \times 2 \times 2$ or $4 \times 4 \times 4$ of the original cells [55, 56]. Without changes in the anatomy, this process allowed some variations in the total height and weight of the models. To obtain the biometrically correct height and weight of 175.5 cm and 69.6 kg for the average man, it was necessary to slightly increase the cell size to 1.31 and 2.62 cm for the two models. These human body models of 45,024 and 5,628 cells were used for a number of applications involving whole-body or partial-body exposures due to spatially uniform or nonuniform (far-field or near-field) sinusoidally varying fields, or for electromagnetic pulses [5, 6, 11, 55, 56].

11.5.4 Today's Capabilities

With increasing computer speed and memory and improving efficiencies of the numerical codes, including use of parallelized FDTD codes [57], it is now possible to obtain SAR distributions for coupled parts of the body represented by 1 to 30 million cells. High-resolution models with cell sizes in the order of millimeters have been developed by a number of groups [5, 6, 32, 58 – 62]. These models are needed to extend dosimetric calculations to microwave frequencies where penetration depths are generally less than 1 or 2 cm. They are also needed for nonuniform, at times highly localized, electromagnetic fields where the exposed region is of the dimensions of a few centimeters. Examples of such sources at microwave frequencies are personal wireless communications devices and police radars; at power-line frequencies there are hair dryers and power drills. In the following section, we describe the development of one of the millimeter-resolution models of the human body.

11.6 MILLIMETER-RESOLUTION MODELS BASED ON MRI SCANS OF THE HUMAN BODY

For bioelectromagnetic calculations, several millimeter-resolution models of the human body have recently been developed which employ tissue-geometry data obtained from magnetic resonance imaging (MRI) scans of the human body [5, 6, 32, 57 – 62]. One of these models was developed from MRI scans of a male volunteer of height 176.4 cm and weight 64 kg [5, 6, 32, 57, 58]. These scans have a resolution of 3 mm along the height of the body and 1.875 mm in each cross section. While the height of the volunteer is average for an adult male, his weight is somewhat less than the average of 71 kg. This problem is addressed by assuming a $(71/64)^{1/2} \cdot 1.875 = 1.974$ mm cell size in each cross section. The volume of the model is thereby increased by 71/64, or 10.9%, increasing its weight by the same percentage to a new value of approximately 71 kg.

The MRI sections have been converted into images involving 31 tissue types whose electrical properties are prescribed at the frequency of interest. The tissue types are:

fat	pancreas	lung
muscle	heart	bladder
bone	blood	stomach
cartilage	eye	ligament
skin	eye humor	compact bone
brain	eye sclera	testicle
nerve	eye lens	spermatic cord
cerebrospinal fluid	ear	prostate gland
intestine	liver	erectile tissue
spleen	kidney	pituitary gland
		pineal gland

The last two glands have been separately identified because of the alleged effects of power-frequency fields upon the glands' hormonal secretions, particularly from the pineal gland.

To date, this model has been used to calculate electromagnetic wave absorption in the human head, neck, and shoulders for personal wireless devices at frequencies up to 6 GHz [32, 58]. It has also been used to calculate plane-wave exposures of the grounded and ungrounded man at frequencies up to 350 MHz [63]. These have identified the frequencies of maximum absorption (resonant frequencies) of the head and neck, 207 and 193 MHz, respectively, for the ungrounded and grounded man. The corresponding absorption cross sections are 3.27 and 2.62 times the shadow cross section [63].

Also recently developed is a second model of the human head and neck based on MRI scans of another male volunteer of height 176.5 cm and weight 79.0 kg. This model and the model of the head and neck from the whole-body model discussed above were used to compare the SAR distributions due to cellular telephones both at 835 and 1900 MHz [57].

11.7 CURRENTS INDUCED IN THE HUMAN BODY BY FIELDS OF POWER TRANSMISSION LINES

With several epidemiologic studies linking electromagnetic fields with higher rates of incidence of cancer, there is some public concern regarding the potential health effects of these fields. Some of the strongest fields are encountered under high-voltage power-transmission lines where vertically polarized electric fields up to 10 kV/m and horizontally polarized magnetic fields up to 50 μT have been measured.

11.7.1 Quasi-Static FDTD Model

As described in [12], FDTD can be used to calculate the fields and induced currents within the human body when exposed to pure electric, pure magnetic, or combined electric and magnetic fields at power-line frequencies. The method is so robust that it can accommodate the highly anisotropic conductivities of skeletal muscle and bone that exist at these low frequencies [64, 65]. In this section, we discuss the use of the full-wave FDTD technique (normally applied to high-frequency electromagnetic wave phenomena) to model what is really a quasi-static situation at power-line frequencies [66, 67].

It is clear that both sinusoidal and prescribed time-varying incident fields can be modeled with FDTD. For continuous sinusoidal excitation, the FDTD solution is normally considered to be completed when sinusoidal steady-state behavior for $\vec{E}$ and $\vec{H}$ is observed at each grid cell. At UHF and microwave frequencies, this typically requires time-stepping on the order of three to four oscillation periods for lossy biological bodies [68]. Since Δt is bounded for a given grid-cell size by considerations of numerical stability, and assuming millimeter-scale grid cells, many billions of time steps would be needed at power-line frequencies if the oscillation-period criterion were followed blindly. This would render the use of FDTD at power-line frequencies completely inappropriate.

However, it is precisely the quasi-static nature of the coupling at power-line frequencies that opens the door to FDTD modeling. Using a logic similar to [66, 67], the fields outside the body in this regime depend not on the internal tissue properties, but only on the shape of the body. This holds as long as the size of the body is a factor of 10 or more smaller than the wavelength and $|\sigma + j\omega\varepsilon| >> \omega\varepsilon_0$, where σ and ε are respectively the conductivity and permittivity of the tissues, $\omega = 2\pi f$ is the radian frequency, and ε_o is the permittivity of the free space outside the body. Under these conditions, the electric field in air is normal to the body surface, and the internal-tissue electric field is given from the boundary condition in terms of the outside field as

$$j\omega\varepsilon_0\,\hat{n}\cdot\vec{E}_{\text{air}} = (\sigma + j\omega\varepsilon)\,\hat{n}\cdot\vec{E}_{\text{tissue}} \tag{11.10}$$

We see that a much higher frequency, f', that still falls within the quasi-static regime can be used to excite the model, and the internal fields, $\vec{E}'$, calculated at that frequency can be scaled back to the power frequency, f, of interest (usually 60 Hz). From (11.10) we write the necessary scaling relationship:

$$\vec{E}_{\text{tissue}}(f) = \left(\frac{\omega}{\omega'}\right) \cdot \left[\frac{\sigma(f') + j\omega'\varepsilon(f')}{\sigma(f) + j\omega\varepsilon(f)}\right] \vec{E}'_{\text{tissue}}(f')$$

$$\cong \left[\frac{f\,\sigma(f')}{f'\,\sigma(f)}\right] \vec{E}'_{\text{tissue}}(f') \tag{11.11}$$

assuming that $\omega\varepsilon(f) << \sigma(f)$ and $\omega'\varepsilon(f') << \sigma(f')$.

Reference [12] reported the use of this method for power-frequency coupling calculations using a scaled frequency, f', of 5 to 10 MHz. The computation time was reduced by five orders of magnitude relative to that implied by the oscillation-period criterion at 60 Hz. Simply speaking, the frequency-scaled FDTD calculation had to be time-stepped through just 3 to 4 sinusoidal periods at 5 to 10 MHz, rather than 3 to 4 sinusoidal periods at the much lower frequency of 60 Hz. Reference [12] assumed that $\sigma(f') = \sigma(60)$, that is, the conductivities of the various tissues at the scaled frequency were taken to be those at 60 Hz. Further, the incident electric field was taken as $\vec{E}_{\text{inc}}(f') = 60\,\vec{E}_{\text{inc}}(f)/f'$ to obtain $\vec{E}_{\text{tissue}}(f)$ for $E_{\text{inc}}(f) = 10$ kV/m. The incident magnetic field $\vec{H}_{\text{inc}}(f')$ was similarly taken to be $60\,\vec{H}_{\text{inc}}(f)/f'$ to account for the fact that the induced current densities and internal electric fields are proportional to the frequency of the incident fields and would therefore be higher at the scaled frequency f'.

11.7.2 Calculated Results

Reference [12] first verified the accuracy of the FDTD scaling approach of (11.11) with models of homogeneous and layered spheres. These models allowed direct comparison of the results for the internal field distributions with the analytic Mie series [69].

Reference [12] also reported FDTD results for a 16-tissue, 1.31-cm-resolution, anatomically based model of the human body. These results were compared to available experimental data for a mannequin given by Deno [70], and to variations of the induced currents calculated along the height of the body by DiPlacido et al. [71]. Here, the assumed excitation was a vertical electric field such as that under a high-voltage power line. Table 11.1 lists the tissue conductivities used for the human model at 60 Hz. Two different values of muscle conductivities are taken in recognition of muscle anisotropy. The higher conductivity of 0.52 S/m is taken for the skeletal muscle, and an average value of 0.11 S/m is taken for the muscle in the interior of the body.

Fig. 11.7 graphs the FDTD-calculated body-layer currents for the anatomically based grounded and ungrounded human body models subjected to strong 60-Hz power-line fields. Curves (1), (3), and (11) are for a vertical electric field of 10 kV/m and a horizontal magnetic field of 26.5 μT. For curve (2), the electric field is retained while the magnetic field is set to zero. Also, for curves (1) and (2), $\sigma = 0.52$ S/m for skeletal muscle and 0.11 S/m for interior muscle. For curves (3) and (11), $\sigma = 0.11$ S/m for all muscle. Curves (1) and (3) can be directly compared to observe the effect of the muscle conductivity upon the layer current at each height within the grounded man model.

TABLE 11.1

Tissue Conductivities Used for FDTD Calculations at the Power Line Frequency of 60 Hz
Source: Gandhi and Chen, *Bioelectromagnetics Supplement*, 1992, pp. 43–60.

Tissue Type	σ (S/m)	Tissue Type	σ (S/m)
Air	0	Liver	0.13
Muscle	0.52, 0.11	Kidney	0.16
Fat, bone	0.04	Pancreas	0.11
Blood	0.60	Spleen	0.18
Intestine	0.11	Heart	0.11
Cartilage	0.04	Nerve, brain	0.12
Lung (33% tissue, 67% air)	0.04	Skin	0.11
		Eye	0.11

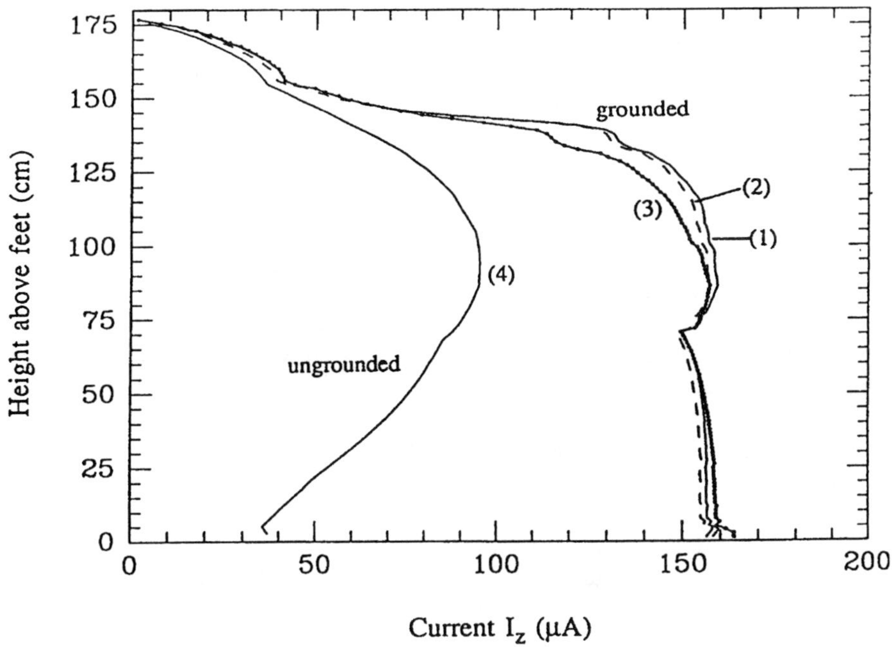

Fig. 11.7 FDTD-calculated body-layer currents for anatomically based grounded and ungrounded human body models subjected to strong 60-Hz power-line fields. Curves (1), (3), and (4) are for a vertical E field of 10 kV/m and a horizontal H field of 26.5 μT. For curve (2), E is retained while H is set to zero. Also, for curves (1) and (2), σ = 0.52 S/m for skeletal muscle and 0.11 S/m for interior muscle. For curves (3) and (4), σ = 0.11 S/m for all muscle.

Comparison of curves (1) and (2) in Fig. 11.7 shows that the vertical electric field of the power line is dominant in generating body current. In fact, the body-layer currents due to the electric field alone comprise 98% to 99% of the currents calculated for the combined electric and magnetic fields. Finally, the calculated foot currents of 155 to 160 μA for the grounded-man model are in excellent agreement with the 165 μA value projected from the measurements of Deno [70].

The anisotropic conductivity of the skeletal muscle has recently been assumed for FDTD calculations [64]. A 6-mm-resolution model of the human body was generated by combining $3 \times 3 \times 2$ cells of the $1.974 \times 1.974 \times 3.0$ mm MRI-based model described in Section 11.6. As expected, the calculated variation of the vertically flowing induced current is similar to that given in curves (1) and (2) of Fig. 11.7.

11.8 INDUCED CURRENTS / SARS FOR PLANE-WAVE EXPOSURE: CONDITIONS FOR WHOLE-BODY AND HEAD RESONANCE

For plane-wave exposures at 10 to 915 MHz, the FDTD method has been used to calculate local, layer-averaged, and whole-body-averaged SARs and induced currents for anatomically based man models with resolutions from 2 to 26.2 mm [55, 63, 68]. Fig. 11.8 graphs the computed cross-section layer-average RF current distributions for one such man model at 40 and 60 MHz, assuming a vertically polarized incident electric field of 1 V/m [55]. These frequencies are close to the whole-body resonances under the grounded and isolated conditions of the model, respectively. Fig. 11.9 graphs the corresponding layer-average SARs at these frequencies [55]. As expected, the highest SARs are calculated for the horizontal sections through the neck, knee, and ankle. The high neck SAR is due to its small cross section relative to the head and upper body on either side of the neck, which elevates the neck's layer-averaged current density. The high knee and ankle SARs are due to their high bone contents as well as their reduced cross-section areas relative to adjacent body parts.

The resonant frequency for whole-body exposure of a human to plane-wave illumination is about 40 MHz under grounded conditions, and 65 to 75 MHz under isolated conditions or situations where the ground is several tens of centimeters removed from the feet [55]. For grounded and isolated conditions, respectively, whole-body absorption cross sections up to 6 and 4.5 times the physical cross section of the body have been calculated and measured for models of the human body [55, 63, 68, 72].

Because of interest in possible neurological effects of RF radiation, the FDTD method has also been used to obtain SAR distributions and frequencies of maximum absorption for the head and neck for plane-wave exposure of grounded and isolated models of the human body [63, 68]. As discussed in Section 11.6, a high-resolution MRI-based model was constructed with a voxel size of $1.974 \times 1.974 \times 3.0$ mm. In addition, a coarse-resolution model with a voxel size of $5.922 \times 5.922 \times 6.0$ mm was constructed by combining $3 \times 3 \times 2$ cells of the high resolution model along the x, y, and z axes, respectively. The resonant frequencies for the head and neck are calculated to be 207 and 193 MHz for isolated and grounded conditions of the man model, respectively.

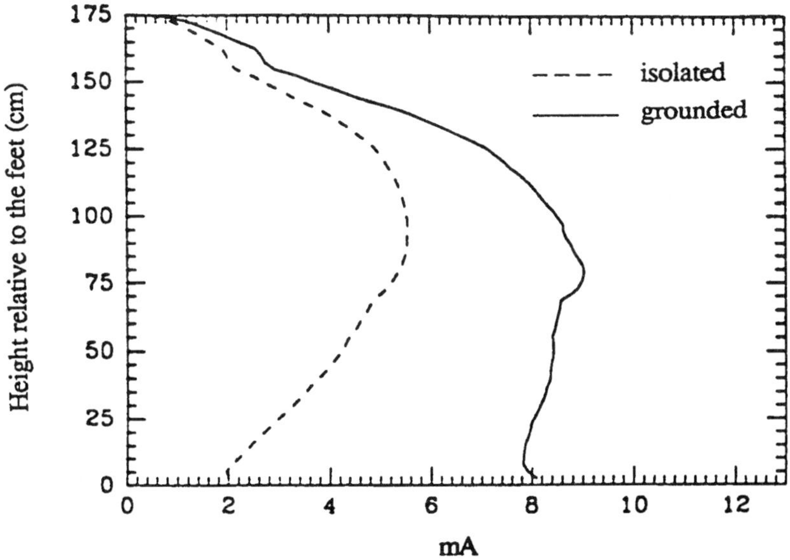

(a)

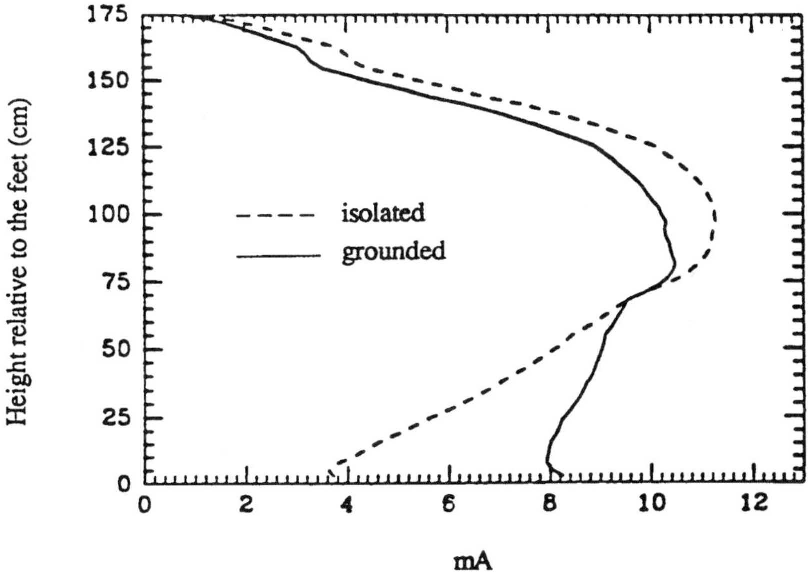

(b)

Fig. 11.8 FDTD-computed cross-section layer-average RF current distributions for the man model exposed to a vertically polarized plane wave with a 1 V/m electric field at: (a) 40 MHz; (b) 60 MHz. *Source*: Chen and Gandhi, *Health Physics*, 1989, pp. 89–98.

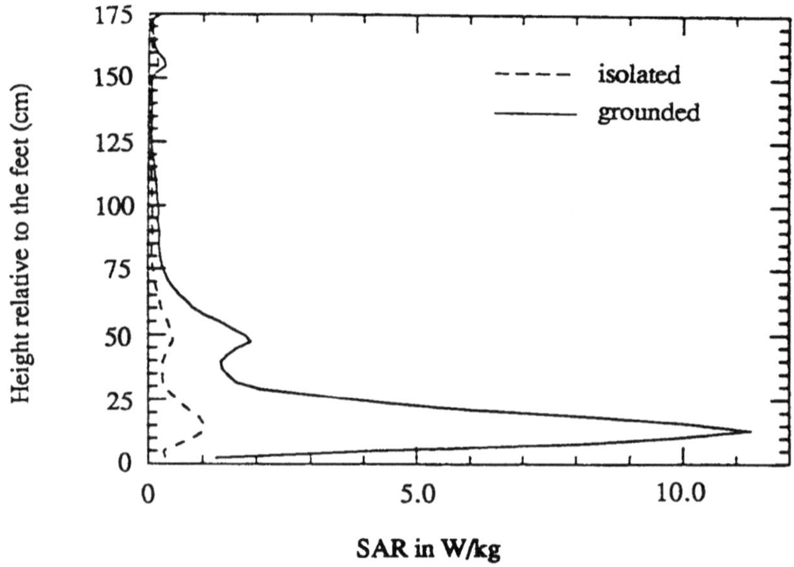

(a)

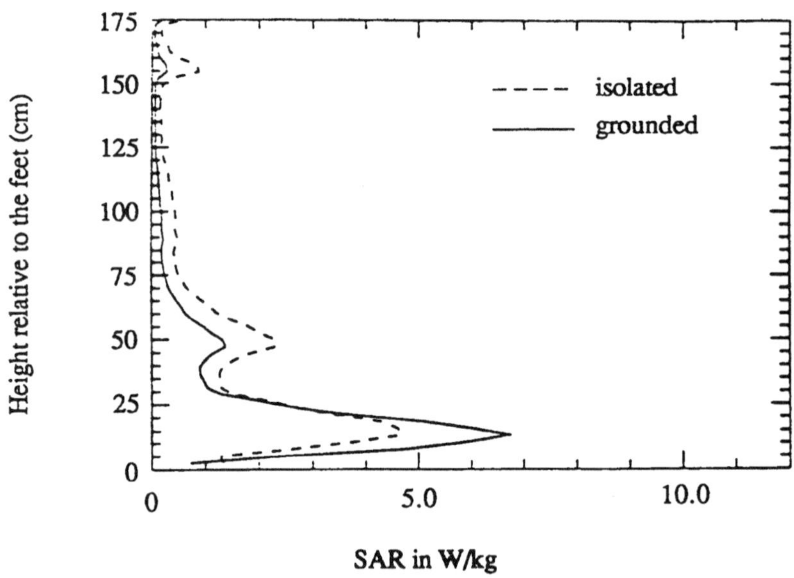

(b)

Fig. 11.9 FDTD-computed cross-section layer-average SARs corresponding to the current distributions of Fig. 11.8 for the man model exposed to a vertically polarized plane wave with a 1 V/m electric field at: (a) 40 MHz; (b) 60 MHz. *Source*: Chen and Gandhi, *Health Physics*, 1989, pp. 89–98.

The corresponding absorption cross sections are 3.27 and 2.62 times the physical (shadow) cross section for this region of the body [63].

11.9 COUPLING OF ULTRAWIDEBAND PULSES TO THE HUMAN BODY

11.9.1 Modeling of the Tissue Properties with the Debye Equation

For ultrawideband models using the $(FD)^2TD$ method of Section 11.3, the measured properties of various biological tissues of interest [73] have been fitted to (11.5), a Debye equation with two relaxation constants [24–27]. Tissues modeled in this manner include muscle, fat, bone, blood, intestine, cartilage, lung, kidney, pancreas, spleen, lung, heart, brain / nerve, skin, and eye. Optimum values for ε_{s1}, ε_{s2}, ε_{∞}, τ_1, and τ_2 in (11.5) were obtained by nonlinear least-squares matching to the measured data for fat and muscle [26, 27]. All other tissues have properties falling roughly between these two extremes.

Table 11.2 lists optimum values for ε_{s1}, ε_{s2}, and ε_{∞} for these tissues. These values were obtained by assuming that τ_1 and τ_2 are constant for all tissues of the model. Further, the values of τ_1 and τ_2 are assumed to equal the average of the optimized values for fat and muscle ($\tau_1 = 46.2$ ns, $\tau_2 = 0.091$ ns). These assumptions facilitate linear (volume) averaging of the ε values of each tissue in a given volume cell of the heterogeneous man model to obtain the effective overall ε value for that cell.

TABLE 11.2

Debye Constants for the Tissues Used in the Ultrawideband Pulse Exposure Model
Source: Gandhi and Furse, *IEEE Trans. Electromagnetic Compatibility*, 1997, pp. 174–180.

Tissue Type	ε_{s1}	ε_{s2}	ε_{∞}
Muscle	3950	59.1	40.0
Bone / cartilage	313	7.1	3.4
Blood	3560	66.4	35.0
Intestine	4720	66.1	39.0
Liver	2860	57.1	36.3
Kidney	3330	67.1	35.0
Pancreas / spleen	3790	73.9	10.0
Lung (33% tissue, 67% air)	1220	13.1	10.0
Heart	4310	54.6	38.5
Brain / nerve	2060	56.9	32.5
Skin	3400	55.6	23.0
Eye	2190	57.0	40.0

The measured tissue properties and those computed from the Debye equation subject to the assumptions underlying Table 11.2 are shown in Fig. 11.1 for fat and muscle. Similar comparisons were also obtained for the other tissue types.

11.9.2 Induced Current and Specific Absorption

The $(FD)^2TD$ method has been used to calculate the coupling of an ultrashort pulse to the heterogeneous model of the human body [26, 27]. From the calculated internal field distribution, the vertical (z-directed) current passing through the k_0th layer of the body model at time step, n, is given by

$$I_z\Big|^n_{\text{layer } k_0} = \Delta x\, \Delta y \sum_{i,j} \frac{\partial}{\partial t} D_z\Big|^n_{i,j,k_0} \tag{11.12}$$

where $\Delta x\, \Delta y$ is the grid-cell area in the k_0th layer, and the summation is carried out for all of the cells in that layer. The layer-averaged absorbed energy density or specific absorption, SA, and the total energy, W, absorbed by the whole body are given by

$$\text{SA}\Big|_{\text{layer } k_0} = \frac{\Delta t}{N_{k_0}} \sum_{i,j,n} \left(\frac{\vec{E}\Big|^n_{i,j,k_0}}{\rho_{i,j,k_0}} \cdot \frac{\partial}{\partial t} \vec{D}\Big|^n_{i,j,k_0} \right) \quad \text{joules/kg} \tag{11.13}$$

$$\text{W} = \Delta x\, \Delta y\, \Delta z\, \Delta t \sum_{i,j,k,n} \left(\vec{E}\Big|^n_{i,j,k} \cdot \frac{\partial}{\partial t} \vec{D}\Big|^n_{i,j,k} \right) \quad \text{joules} \tag{11.14}$$

where Δt is the time step used for the FDTD model, N_{k_0} is the number of grid cells in the k_0th layer of the body, and ρ_{i,j,k_0} is the mass density in kg/m^3 for cell (i, j, k_0).

11.9.3 Computed Example

Results have been obtained for the uniform plane-wave illumination of the whole body by the ultrawideband pulse waveform of Fig. 11.10(a) [27]. This wave has a peak amplitude of 1.1 V/m, a rise time of about 0.2 ns, and a total duration of about 7 to 8 ns. The corresponding Fourier spectrum of the pulse is shown in Fig. 11.10(b). Most of the energy in the pulse is concentrated in the 200 to 900 MHz band, with the peak of the energy being at about 500 MHz.

The incident wave is assumed to be vertically polarized since this is known to result in the strongest coupling to standing individuals. Two cases were considered. First, the man model was assumed to be wearing shoes and grounded. Second, the man model was without shoes and isolated from ground. For both cases, the $(FD)^2TD$ procedure was used to calculate the temporal variation of the total vertical current at each horizontal body cross section.

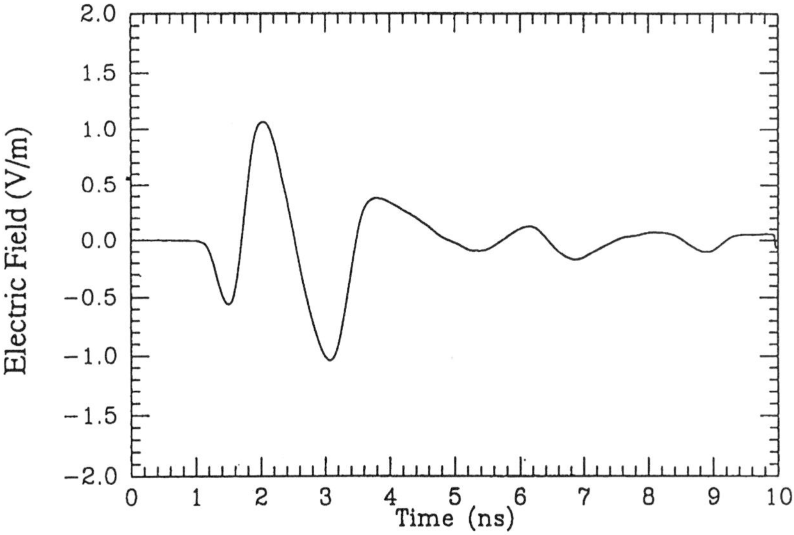

(a) Time waveform

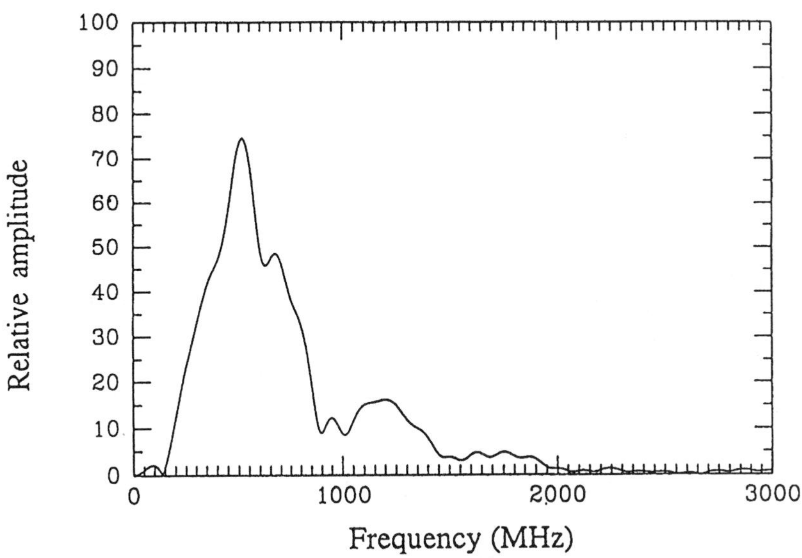

(b) Fourier spectrum

Fig. 11.10 Characteristics of the plane-wave electromagnetic pulse illuminating the whole-body model. *Source*: Gandhi and Furse, *IEEE Trans. Electromagnetic Compatibility*, 1997, pp. 174–180, © 1997 IEEE.

Figs. 11.11(a) and 11.11(b) graph the vertical-current waveforms at representative cross sections through the eyes and through the bladder [27]. Normalized to an incident-wave pulse amplitude of 1 V/m, the calculated peak currents for these cross sections are 1.1 and 3.2 mA, respectively. We note that there is little difference in the induced current caused by the grounding of the man model. This is due to the fact that most of the energy in the incident wave is at frequencies above 300 MHz, where the effect of the ground plane on the induced currents or the SARs is minimal [68].

Fig. 11.12 graphs the peak total current for each horizontal cross section of the body. The maximum peak current is 3.5 mA (equivalent to 3.2 mA for a 1 V/m illumination). This occurs at a height of 96.3 cm above the bottom of the feet. A very similar result was previously calculated for isolated and grounded models of the human body subjected to sinusoidal plane-wave exposures at 350 to 700 MHz. There, the highest induced currents were approximately 3.0 to 3.2 mA for a 1 V/m illumination, and were located at horizontal cross sections 85 to 100 cm above the bottom of the feet [68].

Using (11.13) and (11.14), the specific absorption and the total absorbed energy due to exposure to an ultrawideband pulse can be calculated. For the single pulse of Fig. 11.10, this procedure yields a total energy absorbed by the body of 2.0 pJ for the isolated case and 1.91 pJ for the shoe-wearing grounded case.

11.10 SAR DISTRIBUTIONS AND RADIATION PATTERNS FOR PERSONAL WIRELESS COMMUNICATIONS DEVICES

The FDTD method has been used to calculate SAR distributions and design antennas for personal wireless communications devices. Since these devices are used in close proximity to the human head, it is important that the handset/antenna systems be modeled as realistically as possible adjacent to anatomically based human-head models.

11.10.1 Specific Absorption Rate

References [32] and [58 – 62] report FDTD calculations of SAR distributions for a variety of wireless telephones operating at 835 and 1900 MHz using different antennas such as the monopole, the helix, the helix-monopole whip, and the stacked-patch antenna. In these models, the handset is assumed to be held against the ear at one or more angles such as 0°, 20°, 30°, and 45° to model realistic usage.

MRI-Based Anatomical Model, Generic Wireless Telephone

We now consider several aspects of modeling the human head for exposure to a generic wireless telephone. The $1.974 \times 1.974 \times 3.0$-mm-resolution, MRI-based, anatomical model described in Section 11.6 is used in this example. To avoid a staircased representation of the handset and the antenna, it is desirable to rotate the head model instead of the handset through the desired angle at which the telephone is to be used [32, 74].

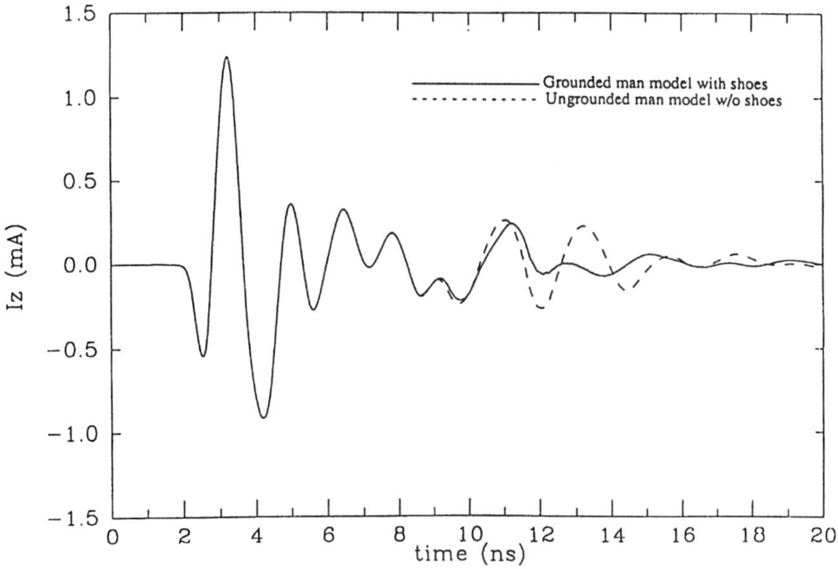

(a)

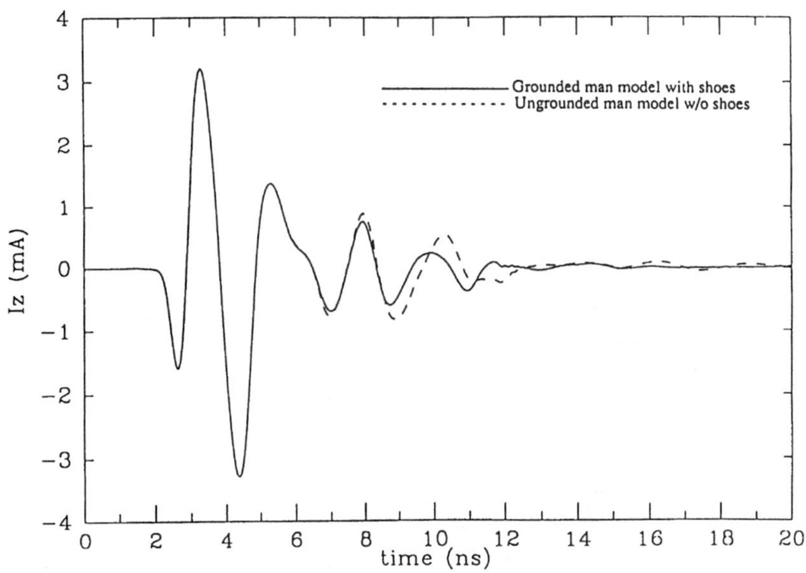

(b)

Fig. 11.11 FDTD-computed vertical-current waveforms at two horizontal sections of the whole-body model when illuminated by the electromagnetic pulse of Fig. 11.10: (a) section through the eyes (height above the bottom of the feet = 168.3 cm); (b) section through the bladder (height above the bottom of the feet = 91.0 cm). *Source*: Gandhi and Furse, *IEEE Trans. Electromagnetic Compatibility*, 1997, pp. 174–180, © 1997 IEEE.

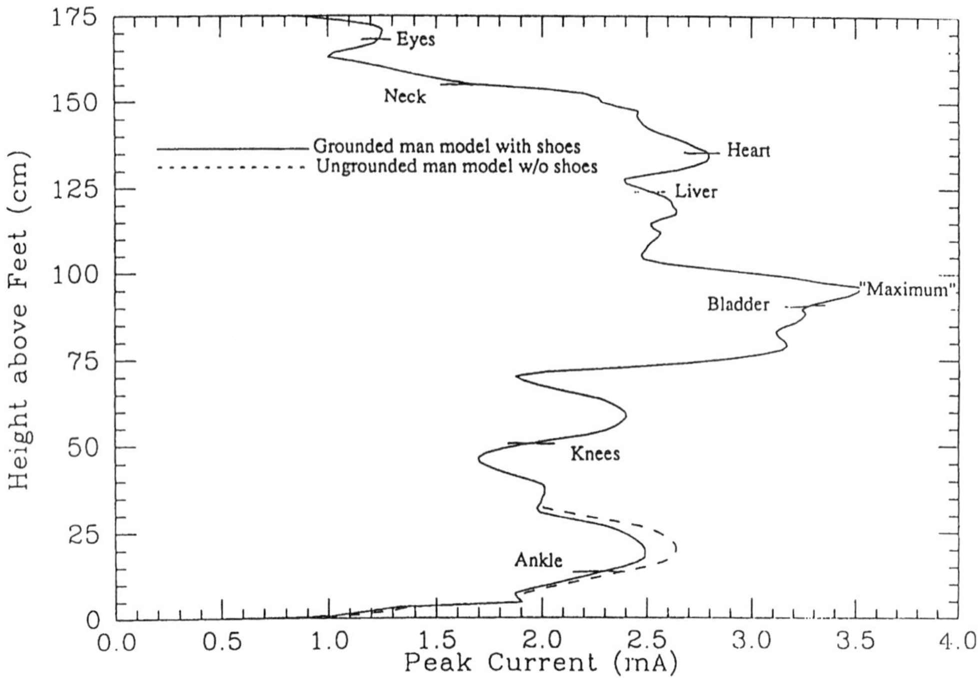

Fig. 11.12 FDTD-computed peak vertical current at each horizontal cross section of the whole-body model when illuminated by the electromagnetic pulse of Fig. 11.10. *Source*: Gandhi and Furse, *IEEE Trans. Electromagnetic Compatibility*, 1997, pp. 174–180, © 1997 IEEE.

Fig. 11.13 is a visualization of the head model tilted forward by 30°, which highlights the surface contour of the brain. Since the head model is tilted forward by 30°, the handset placed vertically against the ear is tantamount to holding the telephone at an angle of 30° relative to a vertical (erect) head.

The generic wireless telephone is modeled here as a plastic-covered, rectangular, metal box of dimensions $2.96 \times 5.76 \times 15.5$ cm. Its antenna is assumed to be a quarter-wavelength monopole mounted at the top of the box. This size and configuration is representative of many current wireless handsets.

Table 11.3 lists the FDTD-calculated 1-g and 10-g peak SARs in units of W/kg for various tilt angles of this handset relative to the head model. Note that the proper time-averaged power outputs of current wireless telephones are used for the calculations. The final row of Table 11.3 provides the 1-g and 10-g SARs for a 30° forward-tilted head with an additional azimuthal head rotation of 9° moving the face toward the mouthpiece of the telephone. This may provide a more realistic representation of the way that telephones are often held against the head [74].

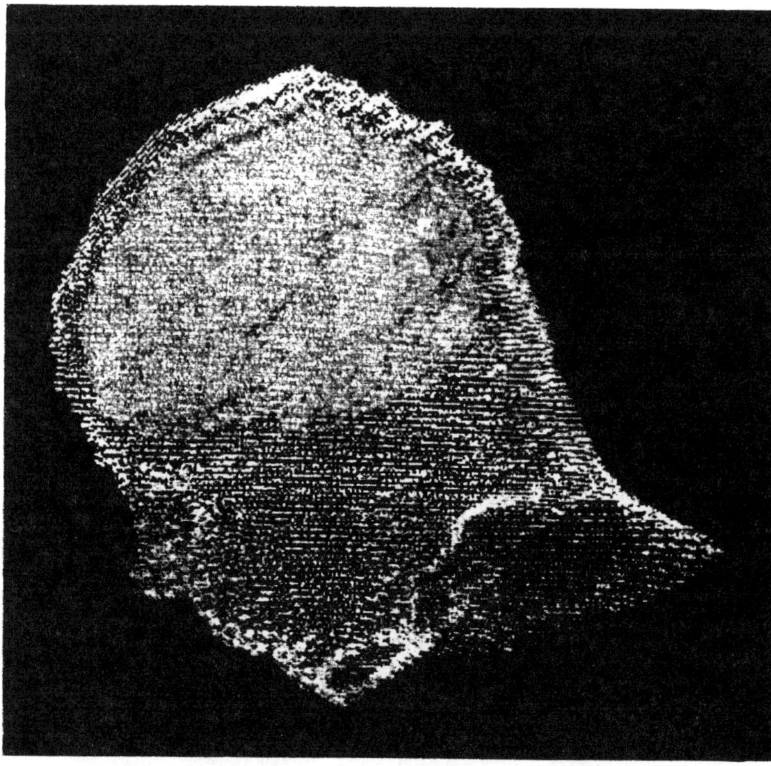

Fig. 11.13 Visualization of the anatomical head model tilted forward by 30°, which highlights the surface contour of the brain. A handset placed vertically against the ear is tantamount to holding the telephone at an angle of 30° relative to a vertical (erect) head.

TABLE 11.3

FDTD-Calculated 1-Gram-Average and 10-Gram-Average Peak SARs (in W/kg) for Various Tilt Angles of a Generic Wireless Telephone Relative to the MRI-Based Anatomical Model of the Head

Tilt angle	835 MHz (600 mW) 1-g SAR	835 MHz (600 mW) 10-g SAR	1900 MHz (125 mW) 1-g SAR	1900 MHz (125 mW) 10-g SAR
0°	2.93	1.41	1.11	0.59
20°	2.70	1.33	1.08	0.56
30°	2.44	1.21	1.08	0.57
45°	2.14	1.10	0.85	0.42
30° + 9° rot.	2.31	1.08	1.20	0.44

Fig. 11.14 visualizes the FDTD-computed SAR distribution in an ear-level plane (10 cm below the top of the head) for a telephone held vertically against the tilted-head model of Fig. 11.13. The telephone radiates 125 mW power at 1900 MHz from a one-quarter-wavelength antenna. We see that the SAR is fairly localized with almost all of the absorbed energy occurring within the ear and the side of the head adjacent to the handset. (See the color-plate section for a visualization of this SAR distribution with a calibrated color scale.)

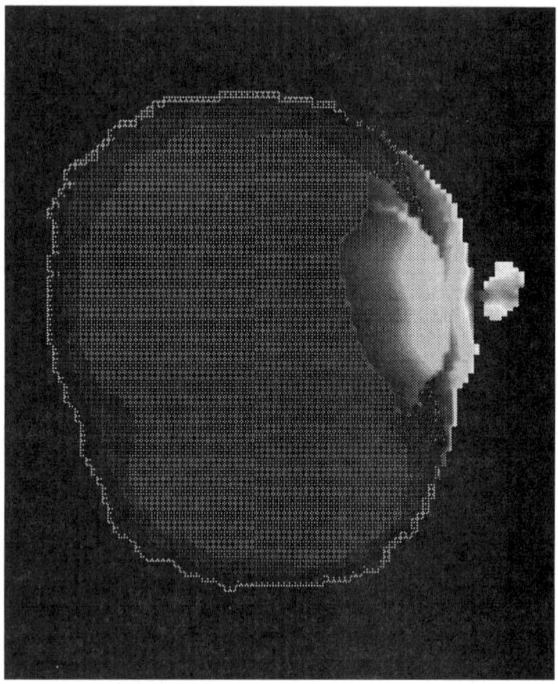

Fig. 11.14 Visualization of the FDTD-computed SAR distribution in an ear-level plane (10 cm below the top of the head) for a telephone held vertically against the tilted-head model of Fig. 11.13. Radiated frequency = 1900 MHz, one-quarter wavelength antenna, 125 mW radiated power. *Source*: Lazzi and Gandhi, *IEEE Trans. Electromagnetic Compatibility*, 1997, pp. 55–61, © 1997 IEEE.

In order to save computer memory, it is also possible to use spatially truncated models that use only the one-half or one-third portion of the head closest to the telephone. It is found that the SAR distribution obtained using these truncated models is virtually the same as that visualized in Fig. 11.14 for the model of the entire head [74].

Comparison with Measurements

Table 11.4 compares peak 1-g SAR values obtained using the FDTD method with measured data obtained with the Utah Experimental Phantom Model. This study involves 10 wireless telephones: five at 835 MHz; and five at the 1900-MHz personal communications services (PCS) frequencies. The telephones include research test samples from several manufacturers and use a variety of antennas. The experimental phantom head model is constructed with filler media simulating the average electrical properties of the skull, ear, brain, and eyes.

TABLE 11.4

Comparison of FDTD-Calculated and Measured Peak 1-g SARs
for 10 Wireless-Telephone Research Test Samples at 835 MHz and 1900 MHz

Telephone Type	Time-Averaged Radiated Power (mW)	FDTD-Calculated Peak 1-g SAR (W/kg)	Measured Peak 1-g SAR (W/kg)
Cellular telephones at 835 MHz			
Telephone A	600	3.90	4.02
Telephone B	600	4.55	5.41
Telephone C	600	3.52	4.48
Telephone D	600	2.80	3.21
Telephone E	600	0.53	0.54
PCS telephones at 1900 MHz			
Telephone F	125	1.47	1.48
Telephone G	125	0.15	0.13
Telephone H	125	0.81	0.65
Telephone I	125	1.56	1.32
Telephone J	99.3	1.25	1.41

Note: For all new personal wireless devices, the U. S. Federal Communications Commission requires that the peak SAR for any 1-g of tissue be less than 1.6 W/kg [28]. The corresponding safety guideline in the CENELEC European Prestandard [4] is a maximum of 2 W/kg for any 10 g of tissues.

The agreement between the measured and calculated 1-g SARs in Table 11.4 is good, generally within ±25%. This is noteworthy given that the MRI-based anatomical model of the head is used for the FDTD calculations, while a relatively simple four-tissue heterogeneous phantom is used for the measurements.

11.10.2 Far-Field Radiation Pattern

Since FDTD provides the electric and magnetic fields both external and internal to the head model, a near-to-far-field transformation can be used to calculate the radiation pattern of a wireless device operating close to the human head [59, 60, 75, 76].

Figs. 11.15(a) and 11.15(b) compare the FDTD-calculated and measured radiation patterns of a commercial 835-MHz wireless telephone using a helix antenna. Two cases are examined: the telephone in air (Fig. 11.15(a)); and the telephone held against the head model (Fig. 11.15(b)). The measured free-space radiation pattern for this telephone (without the human head) is also available from the manufacturer and is shown in Fig. 11.15(a). Following the procedure leading to the comparative SAR data of Table 11.4, the MRI-based anatomical model of the head and neck is used for the FDTD-calculated radiation patterns, while a relatively simple four-tissue heterogeneous model is used for the measurements. There is good agreement between the FDTD-calculated and the measured radiation patterns, with and without the human head. Similar agreement has been obtained for several other telephones [76].

11.11 BIOMEDICAL APPLICATIONS

11.11.1 Design of Annular Phased Arrays for Hyperthermia

Hyperthermia, the induction of tumor temperatures substantially greater than the normal physiologic range, is considered to be a useful form of cancer therapy when combined with chemotherapy or radiotherapy [77]. The use of hyperthermia in addition to radiotherapy in the treatment of superficially located tumors has been repeatedly shown to result in substantial improvement in tumor regression [77]. Annular phased arrays of multiple aperture antennas or dipoles placed symmetrically around the body have been designed to provide in-depth deposition of energy to the tumor site by controlling the phases of the fields radiated from each of the antennas [78]. The FDTD method has been used to obtain the optimum phases and magnitudes for the various radiating elements using either a generic model of the human body [79] or patient-specific models [80, 81].

11.11.2 SAR Distributions Due to RF Magnetic Fields of Magnetic Resonance Imaging

MRI is an increasingly important tool for medical diagnostic applications. Emerging MRI technologies are leading to use of higher static magnetic fields and associated higher radio frequencies up to 400 MHz [82–84]. Sixteen-rung birdcage coils especially designed for

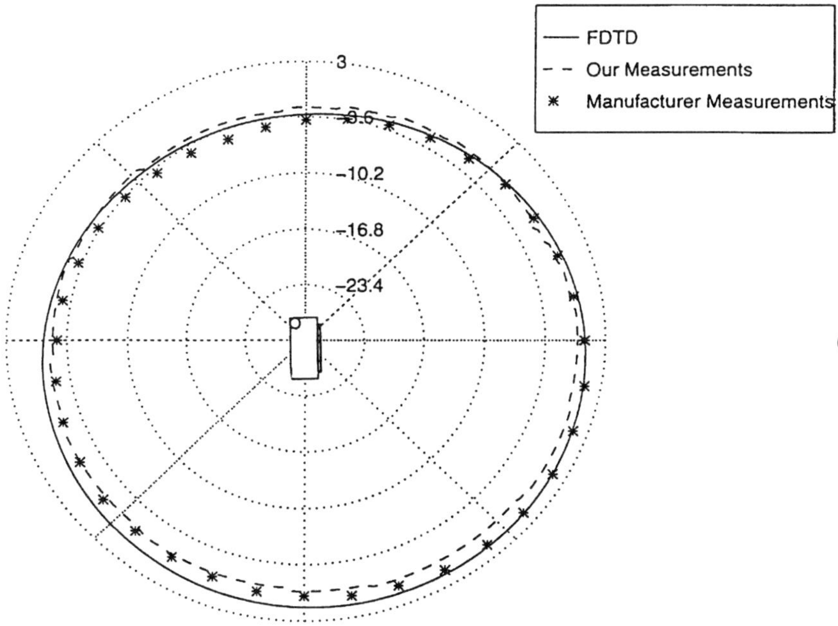

(a)

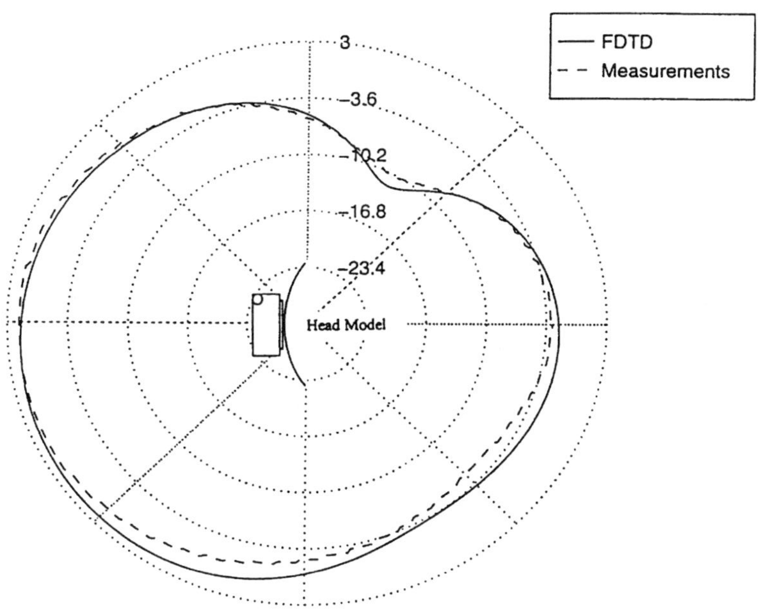

(b)

Fig. 11.15 Comparison of FDTD-computed and measured radiation patterns (in dBi) of a commercial 835-MHz wireless telephone using a 16-mm-long helical antenna: (a) in free space; (b) in the presence of the human head.

imaging of the head, and the larger coils for imaging of the torso, are used not only for today's MRI machines (64 MHz, 1.5T biasing magnetic field) [85], but are also planned for higher static-magnetic-field modalities [82 – 84].

The safety standards proposed by the U. S. Food and Drug Administration [86] and the National Radiological Protection Board of the United Kingdom [87] limit SARs that can be induced within the human body. The FDTD method has been used for this purpose to calculate the coupled electromagnetic fields in heterogeneous, anatomically based models of the human body for 16-rung birdcage coils having currents with progressive phase shifts of 22.5° between the adjacent rungs [85]. These phase shifts are used to generate circularly polarized RF magnetic fields within the interior volume of the coil occupied by the parts of the body to be imaged. SAR distributions within the body have been obtained for head and body coils operating at 64 MHz and also at higher frequencies up to 400 MHz [88, 89].

A related problem is to ensure the relative homogeneity of the RF magnetic fields within the body to meet the requirements of the imaging process. The FDTD method has been used for this purpose by calculating the distribution of internal RF magnetic fields [88 – 91].

11.12 SOME IN-PROGRESS AND LIKELY FUTURE APPLICATIONS

11.12.1 The Parallelized FDTD Code

A parallelized FDTD code offers the advantages of scalability of memory size and speed to permit SAR calculations at increasingly higher frequencies where smaller grid-cell sizes are needed to resolve the wavelength within body tissues. Parallelized versions of the FDTD code have been developed for use on large parallel-processing machines such as IBM SP-2 [57, 63]. Although the parallel FDTD code can be run on all of the 64 nodes of the machine, 8 and 16 nodes are routinely used for simulations involving the MRI-based anatomical model having $5.974 \times 5.974 \times 6.0$-mm resolution. Thirty-two nodes are used for the larger, finer-grained anatomical models having $1.974 \times 1.974 \times 3.0$-mm resolution of the entire human body, or $0.9375 \times 0.9375 \times 3.0$-mm resolution of the head and neck for mobile telephone simulations [57]. A general simulation strategy is to first apply the lower-resolution FDTD models to pinpoint resonant frequencies, for example, and then apply the higher-resolution models to accurately calculate the SAR distributions at these frequencies.

11.12.2 Use of CAD Files for Accurate Modeling of Sources

Numerical techniques such as the FDTD method have been accepted for demonstrating compliance of the SAR values with the limits suggested by the FCC [28]. The guideline for all new personal wireless devices is to demonstrate that the peak SAR for any 1-g of tissue taken in the form of a cube is less than 1.6 W/kg [28].

Current FDTD methodology employs highly realistic, anatomically based models of human body parts such as the head, neck, and hand as compared to measurement techniques which use simplistic, often homogeneous, tissue-phantom models. However, a perceived weakness of FDTD modeling practice to-date is the use of idealized, plastic-covered metal boxes to represent the wireless handsets [32, 58 – 62]. The exact shape of the handset should be accounted to properly model the RF source. Of course, this is not an inherent limitation of the FDTD method, but rather the fact that attention has not been focused on this issue.

In order to remove this perceived weakness, CAD files for mobile telephones obtained from their manufacturers are now beginning to be used for a realistic description of the handsets [57]. The details available in these CAD files can be used to prescribe exact shapes, volumes occupied by insulating buttons, LCD displays, and so forth. This formulation is expected to be used extensively in the future, especially as a tool for the design of antennas that couple minimally to the head and generate the desired far-field radiation patterns.

11.12.3 Optimized Designs of Medical Devices

Because of the capability for accurate modeling of tissue heterogeneities and shapes, numerical methods such as FDTD may increasingly be used in the future for optimized designs of medical devices such as MRIs, implantable and external cardiac defibrillators and pacemakers, cardiac ablation devices, electrotherapy devices for pain treatment, bone and wound healing, and muscle stimulation. As described in Section 11.11, FDTD has already been used in conjunction with realistic, anatomically based models of the human body to optimize the design of hyperthermia phased-array applicators and to determine the coupling of MRI RF magnetic fields. Many more applications are likely in the future for medical devices that have in the past been designed by using simplistic models and subsequently refined during clinical trials. Another area of likely application of numerical design is the emerging technology based on microbioelectromagnetics, where the interaction of electromagnetic fields at cellular and subcellular levels may be used for new and novel applications such as electroporation for drug delivery.

11.13 SUMMARY AND CONCLUSIONS

This chapter summarized the formulation and application of FDTD for bioelectromagnetics problems dealing with exposure safety assessment and medical applications. Topics reviewed included:

- FDTD with a uniform and expanding grid;
- Frequency-dependent FDTD method;
- Analytical and experimental validations for near-field simulations;
- Construction of anatomically based models of the human body;

- Quasi-static FDTD model for calculating currents induced in the body by 60-Hz power-line fields;
- Plane wave exposures leading to whole-body and head resonances;
- Coupling of ultrawideband pulses to the body;
- SAR distributions and radiation patterns for personal wireless communications devices;
- Biomedical applications;
- In-progress and potential future applications.

We conclude that the FDTD method is extremely versatile for modeling bioelectromagnetics systems. FDTD can model the complex geometrical and material details of such systems with sufficient accuracy to enable the understanding of the systems' engineering performance and probable conformance with safety standards. Further, there are emerging opportunities in FDTD modeling of bioelectromagnetics systems in a number of important diagnostic and therapeutic areas.

REFERENCES

[1] ANSI/IEEE C95.1-1992, *American National Standard—Safety Levels With Respect to Human Exposure to Radio Frequency Electromagnetic Fields, 3 kHz to 300 GHz*, New York: Institute of Electrical and Electronics Engineers, 1992.

[2] Repacholi, M. H., "Radiofrequency field exposure standards: current limits and relevant bioeffects data," Chap. 2 in *Biological Effects and Medical Applications of Electromagnetic Energy*, Gandhi, O. P., ed., Englewood Cliffs, NJ: Prentice-Hall, 1990.

[3] *Board Statement on Restrictions on Human Exposure to Static and Time Varying Electromagnetic Fields and Radiation, Vol. 4(5)*, Chilton, Didcot, Oxon, United Kingdom: National Radiological Protection Board, 1993.

[4] European Prestandard ENV 50166-1, *Human Exposure to Electromagnetic Fields, Low Frequencies (0 Hz to 10 kHz)*, Brussels, Belgium: CENELEC Central Secretariat, 1995.

[5] Gandhi, O. P., "Some numerical methods for dosimetry: Extremely low frequencies to microwave frequencies," *Radio Science*, Vol. 30, 1995, pp. 161–177.

[6] Lin, J. C., and O. P. Gandhi, "Computational methods for predicting field intensity," Chap. 9 in *Handbook of Biological Effects of Electromagnetic Fields*, Polk, C., and E. Postow, eds., Boca Raton, FL: CRC Press, 1995.

[7] Taflove, A., and M. E. Brodwin, "Computation of the electromagnetic fields and induced temperatures within a model of the microwave-irradiated human eye," *IEEE Trans. Microwave Theory and Techniques*, Vol. MTT-23, 1975, pp. 888–896.

[8] Spiegel, R. J., M. B. Fatmi, and K. S. Kunz, "Application of a finite-difference technique to the human radio-frequency dosimetry problem," *J. Microwave Power*, Vol. 20, 1985, pp. 241–254.

[9] Sullivan, D. M., D. T. Borup, and O. P. Gandhi, "Use of the finite-difference time-domain method in calculating EM absorption in human tissues," *IEEE Trans. Biomedical Engineering*, Vol. BME-34, 1987, pp. 148–157.

[10] Sullivan, D. M., O. P. Gandhi, and A. Taflove, "Use of the finite-difference time-domain method in calculating EM absorption in man models," *IEEE Trans. Biomedical Engineering*, Vol. BME-35, 1988, pp. 179–186.

[11] Chen, J. Y., and O. P. Gandhi, "Currents induced in an anatomically-based model of a human for exposure to vertically polarized electromagnetic pulses," *IEEE Trans. Microwave Theory and Techniques*, Vol. 39, 1991, pp. 31–39.

[12] Gandhi, O. P., and J. Y. Chen, "Numerical dosimetry at power-line frequencies using anatomically based models," *Bioelectromagnetics Supplement*, Vol. 1, 1992, pp. 43–60.

[13] Gao, B. Q., and O. P. Gandhi, "An expanding-grid algorithm for the finite-difference time-domain method," *IEEE Trans. Electromagnetic Compatibility*, Vol. 34, 1992, pp. 277–283.

[14] Gandhi, O. P., D. Wu, J. Y. Chen and D. L. Conover, "Induced current and SAR distributions for a worker model exposed to an RF dielectric heater under simulated workplace conditions," *Health Physics*, Vol. 72, 1997, pp. 236–242.

[15] Wiart, J., S. Chaillou, and S. Drago, "Calculation of the power absorbed by the head using a nonuniform FDTD," *Proc. 1997 North American Radio Science Meeting*, Montreal, Canada, p. 666.

[16] Tinniswood, A. D., G. Lazzi, and O. P. Gandhi, "Use of the expanding-grid FDTD technique for simulation of CAD-derived personal wireless telephones," *IEEE Trans. Microwave Theory and Techniques*, submitted.

[17] Okoniewski, M., E. Okoniewska, and M. A. Stuchly, "Three-dimensional subgridding algorithm for FDTD," *IEEE Trans. Antennas and Propagation*, Vol. 45, 1997, pp. 422–429.

[18] Luebbers, R. J., F. P. Hunsberger, K. S. Kunz, R. B. Standler, and M. Schneider, "A frequency-dependent finite-difference time-domain formulation for dispersive materials," *IEEE Trans. Electromagnetic Compatibility*, Vol. 32, 1990, pp. 222–227.

[19] Bui, M. D., S. S. Stuchly, and G. I. Costache, "Propagation of transients in dispersive dielectric media," *IEEE Trans. Microwave Theory and Techniques*, Vol. 39, 1991, pp. 1165–1172.

[20] Sullivan, D. M., "A frequency-dependent FDTD method for biological applications," *IEEE Trans. Microwave Theory and Techniques*, Vol. 40, 1992, pp. 532–539.

[21] Luebbers, R. J., F. Hunsberger, and K. S. Kunz, "FDTD for Nth-order dispersive media," *IEEE Trans. Antennas and Propagation*, Vol. 40, 1992, pp. 1297–1301.

[22] Luebbers, R. J., F. P. Hunsberger, and K. S. Kunz, "A frequency-dependent finite-difference time-domain formulation for transient propagation in plasma," *IEEE Trans. Antennas and Propagation*, Vol. 39, 1991, pp. 29–34.

[23] Joseph, R. M., S. C. Hagness, and A. Taflove, "Direct time integration of Maxwell's equations in linear dispersive media with absorption for scattering and propagation of femtosecond electromagnetic pulses," *Optics Letters*, Vol. 16, 1991, pp. 1412–1414.

[24] Gandhi, O. P., B. Q. Gao, and J. Y. Chen, "A frequency-dependent finite-difference time-domain formulation for induced current calculations in human beings," *Bioelectromagnetics*, Vol. 13, 1992, pp. 543–555.

[25] Gandhi, O. P., B. Q. Gao, and J. Y. Chen, "A frequency-dependent finite-difference time-domain formulation for general dispersive media," *IEEE Trans. Microwave Theory and Techniques*, Vol. 41, 1993, pp. 658–665.

[26] Furse, C. M., J. Y. Chen, and O. P. Gandhi, "A frequency-dependent finite-difference time-domain method for induced current and SAR calculations for a heterogeneous model of the human body," *IEEE Trans. Electromagnetic Compatibility*, Vol. 36, 1994, pp. 128–133.

[27] Gandhi, O. P., and C. M. Furse, "Currents induced in the human body for exposure to ultra-wideband electromagnetic pulses," *IEEE Trans. Electromagnetic Compatibility*, Vol. 39, 1997, pp. 174–180.

[28] U. S. Federal Communications Commission, *Guidelines for Evaluating the Environmental Effects of Radiofrequency Radiation*, FCC 96–326, Aug. 1, 1996.

[29] Furse, C. M., Q. S. Yu, and O. P. Gandhi, "Validation of the finite-difference time-domain method for near-field bioelectromagnetic simulations," *Microwave and Optical Technology Lett.*, Vol. 16, 1997, pp. 341–345.

[30] King, R. W. P., "The electromagnetic field of a horizontal electric dipole in the presence of a three-layered region: supplement," *J. Applied Physics*, Vol. 74, 1993, pp. 4845–4848.

[31] Kuster, N., and Q. Balzano, "Energy absorption mechanism by biological bodies in the near field of dipole antennas above 300 MHz," *IEEE Trans. Vehicular Technology*, Vol. 41, 1992, pp. 17–23.

[32] Gandhi, O. P., G. Lazzi, and C. M. Furse, "Electromagnetic absorption in the human head and neck for mobile telephones at 835 and 1900 MHz," *IEEE Trans. Microwave Theory and Techniques*, Vol. 44, 1996, pp. 1884–1897.

[33] Dhondt, G., and L. Martens, "A canonical case with an analytical solution for the comparison of electromagnetic field solvers," *Proc. COST 244 Meeting on Reference Models for Bioelectromagnetic Test of Mobile Communication Systems*, Simunic, D., ed., Rome, Italy, 1995, pp. 98–104.

[34] Schwan, H. P., and K. Li, "Hazards due to total body irradiation," *Proc. IRE,* Vol. 44, 1956, pp. 1572–1581.

[35] Johnson, C. C., C. H. Durney, and H. Massoudi, "Electromagnetic power absorption in anisotropic tissue media," *IEEE Trans. Microwave Theory and Techniques*, Vol. MTT-23, 1975, pp. 529–532.

[36] Kritikos, H. N., and H. P. Schwan, "Hot spots generated in conducting spheres by electromagnetic waves and biological implications," *IEEE Trans. Biomedical Engineering*, Vol. BME-19, 1972, pp. 53–58.

[37] Lin, J. C., A. W. Guy, and C. C. Johnson, "Power deposition in a spherical model of man exposed to 1-20 MHz electromagnetic fields," *IEEE Trans. Microwave Theory and Techniques*, Vol. MTT-21, 1973, pp. 791–797.

[38] Joines, W. T., and R. J. Spiegel, "Resonance absorption of microwaves by the human skull," *IEEE Trans. Biomedical Engineering*, Vol. BME-21, 1974, pp. 46–48.

[39] Weil, C. M., "Absorption characteristics of multi-layered sphere models exposed to UHF/ microwave radiation," *IEEE Trans. Biomedical Engineering*, Vol. BME-22, 1975, pp. 468–476.

[40] Stratton, J. A., *Electromagnetic Theory,* New York: McGraw-Hill, 1941, Section 9.25.

[41] Durney, C. H., C. C. Johnson, and H. Massoudi, "Long-wavelength analysis of plane-wave irradiation of a prolate spheroid model of man," *IEEE Trans. Microwave Theory and Techniques*, Vol. MTT-23, 1975, pp. 246–254.

[42] Johnson, C. C., C. H. Durney, and H. Massoudi, "Long-wavelength electromagnetic power absorption in prolate spheroidal models of man and animals," *IEEE Trans. Microwave Theory and Techniques*, Vol. MTT-23, 1975, pp. 739–747.

[43] Massoudi, H., C. H. Durney, and C. C. Johnson, "Long-wavelength analysis of plane-wave irradiation of an ellipsoidal model of man," *IEEE Trans. Microwave Theory and Techniques*, Vol. MTT-25, 1977, pp. 41–46.

[44] Massoudi, H., C. H. Durney, P. W. Barber, and M. F. Iskander, "Electromagnetic absorption in multi-layered cylindrical models of man," *IEEE Trans. Microwave Theory and Techniques*, Vol. MTT-27, 1979, pp. 825–830.

[45] Durney, C. H., "Electromagnetic dosimetry for models of humans and animals: a review of theoretical and numerical techniques," *Proc. IEEE*, Vol. 68, 1980, pp. 33–40.

[46] Livesay, D. E., and K. Chen, "Electromagnetic fields induced inside arbitrary shaped biological bodies," *IEEE Trans. Microwave Theory and Techniques*, Vol. MTT-22, 1974, pp. 1273–1280.

[47] Chen, K. M., and B. S. Guru, "Internal EM field and absorbed power density in human torsos induced by 1-500-MHz EM waves," *IEEE Trans. Microwave Theory and Techniques*, Vol. MTT-25, 1977, pp. 746–755.

[48] Chen, K. M., and B. S. Guru, "Induced EM fields inside human bodies irradiated by EM waves of up to 500 MHz," *J. Microwave Power*, Vol. 12, 1977, pp. 173–183.

[49] Hagmann, M. J., O. P. Gandhi, and C. H. Durney, "Numerical calculation of electromagnetic energy deposition for a realistic model of man," *IEEE Trans. Microwave Theory and Techniques*, Vol. MTT-27, 1979, pp. 804–809.

[50] Hagmann, M. J., and O. P. Gandhi, "Numerical calculation of electromagnetic energy deposition in man with ground and reflector effects," *Radio Science*, Vol. 14(6S), 1979, pp. 23–29.

[51] DeFord, J. F., O. P. Gandhi, and M. J. Hagmann, "Moment-method solutions and SAR calculations for inhomogeneous models of man with a large number of cells," *IEEE Trans. Microwave Theory and Techniques*, Vol. MTT-31, 1983, pp. 848–851.

[52] Gandhi, O. P., J. F. DeFord, and H. Kanai, "Impedance method for calculation of power deposition patterns in magnetically induced hyperthermia," *IEEE Trans. Biomedical Engineering*, Vol. BME-31, 1984, pp. 644–651.

[53] Gandhi, O. P., and J. F. DeFord, "Calculation of EM power deposition for operator exposure to RF induction heaters," *IEEE Trans. Electromagnetic Compatibility*, Vol. EMC-30, 1988, pp. 63–68.

[54] Eycleshymer, A. C., and D. M. Schoemaker, *A Cross-Section Anatomy*, New York: Appleton-Century-Crofts, 1970.

[55] Chen, J. Y., and O. P. Gandhi, "RF currents induced in an anatomically based model of a human for plane-wave exposures (20-100 MHz)," *Health Physics*, Vol. 57, 1989, pp. 89–98.

[56] Gandhi, O. P., "Numerical methods for specific absorption rate calculations," Chap. 6 in *Biological Effects and Medical Applications of Electromagnetic Energy*, Gandhi, O. P., ed., Englewood Cliffs, NJ: Prentice Hall, 1990.

[57] Tinniswood, A. D., C. M. Furse, and O. P. Gandhi, "Computations of SAR distributions for two anatomically based models of the human head using CAD files of commercial telephones and the parallelized FDTD code," *IEEE Trans. Antennas and Propagation*, in press.

[58] Gandhi, O. P., and J. Y. Chen, "Electromagnetic absorption in the human head from experimental 6-GHz transceivers," *IEEE Trans. Electromagnetic Compatibility*, Vol. 37, 1995, pp. 547–558.

[59] Jensen, M. A., and Y. Rahmat-Samii, "EM interaction in handset antennas and a human in personal communications," *Proc. IEEE*, Vol. 83, 1995, pp. 7–17.

[60] Okoniewski, M., and M. A. Stuchly, "A study of handset antenna and human body interaction," *IEEE Trans. Microwave Theory and Techniques*, Vol. 44, 1996, pp. 1855–1864.

[61] Dimbylow, P. J., and S. M. Mann, "SAR calculations in an anatomically based realistic model of the head for mobile communication transceivers at 900 MHz and 1.8 GHz," *Physics in Medicine and Biology*, Vol. 39, 1994, pp. 1537–1553.

[62] Hombach, V., K. Meier, M. Burkhardt, E. Kuhn, and N. Kuster, "The dependence of EM energy absorption upon human head modeling at 900 MHz," *IEEE Trans. Microwave Theory and Techniques*, Vol. 44, 1996, pp. 1865–1873.

[63] Tinniswood, A. D., C. M. Furse, and O. P. Gandhi, "Power deposition in the head and neck of an anatomically based model for plane-wave exposures," *Physics in Medicine and Biology,* in press.

[64] Furse, C. M., and O. P. Gandhi, "Calculation of electric fields and currents induced in a millimeter-resolution human model at 60 Hz using the FDTD method with a novel time-to-frequency-domain conversion," *Bioelectromagnetics*, in press.

[65] DeMoerloose, J., T. W. Dawson, and M. A. Stuchly, "Application of FDTD to quasi-static field analysis," *Radio Science*, in press.

[66] Kaune, W. T., and M. F. Gillis, "General properties of the interaction between animals and ELF electric fields," *Bioelectromagnetics*, Vol. 2, 1981, pp. 1–11.

[67] Guy, A. W., S. Davidow, G. Y. Yang, and C. K. Chou, "Determination of electric current distributions in animals and humans exposed to a uniform 60-Hz high-intensity electric field," *Bioelectromagnetics,* Vol. 3, 1982, pp. 47–71.

[68] Gandhi, O. P., Y. G. Gu, J. Y. Chen, and H. I. Bassen, "Specific absorption rates and induced current distributions in an anatomically based human model for plane-wave exposures," *Health Physics*, Vol. 63, 1992, pp. 281–290.

[69] Mautz, J. R., "Mie series solution for a sphere," *IEEE Trans. Microwave Theory and Techniques*, Vol. MTT-26, 1978, p. 375.

[70] Deno, D. W., "Currents induced in the human body by high-voltage transmission line electric field: measurement and calculation of distribution and dose," *IEEE Trans. Power Apparatus and Systems*, Vol. PAS-96, 1977, pp. 1517–1527.

[71] DiPlacido, J., C. H. Shih, B. J. Ware, "Analysis of the proximity effects in electric field measurements," *IEEE Trans. Power Apparatus and Systems*, Vol. PAS-97, 1978, pp. 2167–2177.

[72] Gandhi, O. P., "Electromagnetic energy absorption in humans and animals," Chap. 8 in *Biological Effects and Medical Applications of Electromagnetic Energy*, Gandhi, O. P., ed., Englewood Cliffs, NJ: Prentice Hall Inc., 1990.

[73] Durney, C. H., et al., *Radiofrequency Radiation Dosimetry Handbook*, 2nd ed., Report SAM-TR–78–22, USAF School of Aerospace Medicine, Brooks AFB, TX, 1978.

[74] Lazzi, G., and O. P. Gandhi, "Realistically tilted and truncated anatomically based models of the human head for dosimetry of mobile phones," *IEEE Trans. Electromagnetic Compatibility*, Vol. 39, 1997, pp. 55–61.

[75] Luebbers, R., L. Chen, T. Uno, and S. Adachi, "FDTD calculations of radiation patterns, impedance, and gain for a monopole antenna on a conducting box," *IEEE Trans. Antennas and Propagation*, Vol. 40, 1992, pp. 1577–1582.

[76] Lazzi, G., S. S. Pattnaik, C. M. Furse, and O. P. Gandhi, "Comparison of FDTD-computed and measured radiation patterns of commercial mobile telephones in presence of the human head," *IEEE Trans. Antennas and Propagation*, in press.

[77] Meyer, J. L., D. S. Kapp, P. Fessenden, and G. H. Hahn, "Hyperthermia oncology: Current biology, physics, and clinical results," *Pharmac. Ther.,* Vol. 42, 1989, pp. 251–288.

[78] Samulski, T. V., D. S. Kapp, P. Fessenden, and A. Lohrbach, "Heating deep-seated eccentrically located tumors with an annular phased array system: A comparative clinical study using two annular array operating configurations," *Int. J. Radiation Oncology, Biology and Physics*, Vol.. 13, 1987, pp. 83–94.

[79] Chen, J. Y., and O. P. Gandhi, "Numerical simulation of annular phased arrays of dipoles for hyperthermia of deep-seated tumors," *IEEE Trans. Biomedical Engineering*, Vol. 39, 1992, pp. 209–216.

[80] James, B. J., and D. M. Sullivan, "Direct use of CT scans for hyperthermia treatment planning," *IEEE Trans. Biomedical Engineering*, Vol. 39, 1992, pp. 845–851.

[81] Sullivan, D. M., R. Ben-Yosef, and D. S. Kapp, "Stanford 3D hyperthermia treatment planning system: technical review and clinical summary," *Int. J. Hyperthermia*, Vol. 9, 1993, pp. 627–643.

[82] Magin, R. L., R. P. Liburdy, and B. Persson, eds., *Biological Effects and Safety Aspects of Nuclear Magnetic Resonance Imaging and Spectroscopy*, Annals New York Acad. Sciences, Vol. 649, 1992.

[83] Vaughan, J. T., H. P. Hetherington, J. O. Otu, J. W. Pan, and G. M. Pohost, "High-frequency volume coils for clinical NMR imaging and spectroscopy," *Magnetic Resonance in Medicine*, Vol. 32, 1994, pp. 206–218.

[84] Zhang, N., M. S. Roos, S. T. Wang, and J. T. Vaughan, "An experimental study of a head coil for proton imaging and spectroscopy," presented at 2nd Meeting of Society of Magnetic Resonance, San Francisco, CA, Aug. 1994.

[85] Hayes, C. E., W. A. Edelstein, J. F. Schenck, O. M. Mueller, and M. Eash, "An efficient, highly homogeneous radio frequency coil for whole-body NMR imaging at 1.5 Tesla," *J. Magnetic Resonance*, Vol. 63, 1985, pp. 622–628.

[86] *Magnetic Resonance Diagnostic Devices: Panel Recommendation and Report on Petitions for MR Reclassification*, U.S. Food and Drug Administration, Fed. Reg. 53:7575, 1988.

[87] *Revised Advice on Acceptable Restrictions of Patient and Volunteer Exposure During Clinical Magnetic Resonance Diagnostic Procedures*, National Radiological Protection Board, U.K., 1990.

[88] Gandhi, O. P., X. B. Chen, X. J. Yuan, and J. Y. Chen, "Dosimetry for time-varying magnetic fields in MR imaging," *Proc. 16th Ann. Meeting Bioelectromagnetics Society*, Copenhagen, Denmark, June 1994, p. 18.

[89] Gandhi, O. P., X. B. Chen, and D. Wu, "Electromagnetic coupling to an anatomically based model of the human body for exposure to time-varying magnetic fields of MRI," *Magnetic Resonance in Medicine*, submitted.

[90] Jin, J. M., G. Shen, and T. Perkins, "On the field inhomogeneity of birdcage coils," *Magnetic Resonance in Medicine,* Vol. 32, 1994, pp. 418–422.

[91] Chen, J., and J. M. Jin, "Calculation of SAR and B_1-field within human head excited by MRI birdcage coils," *Proc. IEEE Antennas and Propagation Soc. Int. Symp.*, Montreal, Canada, July 1997, pp. 1210–1213.

Chapter 12

Imaging and Inverse Problems in Electromagnetics

Weng Cho Chew

12.1 INTRODUCTION

The attempt to image an object that invisible to our naked eyes is an important quest of scientists in many disciplines. The advent of X-rays brought much enthusiasm to the scientific community even though, to the detriment of some, it was discovered to have an adverse radiation effect. An X-ray picture is just a shadow cast on a photographic film by denser media that are more opaque to X-rays compared with less dense media. Due to the blockage effect, this simple picture is undesirable in many aspects, and very soon, scientists and engineers discovered how to make three-dimensional pictures out of a series of shadows cast by X-rays. This is known as X-ray tomography, which later developed to become computer-aided tomography (CAT) [1,2].

Another very significant modality is the use of nuclear magnetic resonance imaging, also known as MRI. MRI makes use of the fact that spin echoes of protons have different resonance frequencies depending on the ambient static magnetic field. Just as for a guitar string under variable tension, the stronger the magnetic field, the higher the resonance frequency of the proton spin. Hence, by immersing a body in a static magnetic field that has a linear gradient in a three-dimensional space, the spins echo with different frequencies when excited by a time-varying electromagnetic signal. By performing the same experiment again by varying the linear gradient over a number of directions, enough data diversity can be generated so that a two- or three-dimensional image of an object can be reconstructed [3].

An often-used modality is ultrasonic imaging [4]. This works almost like X-ray tomography. However, due to the lower frequencies and longer wavelengths of ultrasonic waves, they do not travel exactly in a straight line but diffract around objects. Also, there may be multiple-scattering effects within an object when ultrasonic waves enter it. Hence, when ultrasonic images are reconstructed based upon assuming straight rays and no multiple scattering, image distortion arises. A new imaging scheme called diffraction tomography (DT) attempts to correct for the diffraction effect but not for the multiple-scattering effect [5].

To date, electromagnetic (microwave and radio wave) imaging has not found wide medical applications due to its longer wavelength and hence, lower resolution. The wavelength can be reduced by using higher frequencies, but the ensuing material loss is also higher, greatly reducing the penetration depth. However, electromagnetic imaging has found a niche in nondestructive evaluation, ground-penetrating radar [6], synthetic-aperture radar (SAR) [7], target identification, and electrical well logging [8]. Many rugged and constrained environments are not amenable to having X-ray or nuclear magnetic resonance equipment on site. Also, it is often much easier to couple electromagnetic signals into material media at an air-solid interface compared to ultrasonic signals. At such interfaces, the acoustic impedance contrast between the air and the solid is often huge while the electromagnetic impedance contrast is generally lower. The high impedance contrast gives rise to a small transmission coefficient for acoustic signals, but a reasonable one for electromagnetic signals. Electromagnetic waves can propagate long distances with little attenuation in air or vacuum. Consequently, electromagnetic waves are extremely useful for remote sensing and imaging using SAR techniques where the sensor-target separation is often enormous [7].

In this chapter, we are concerned primarily with techniques for solving inverse problems using electromagnetic waves and how such techniques can be applied to imaging. We address specific key issues related to electromagnetic imaging and inverse problems.

12.2 INVERSE SCATTERING

In inverse scattering, one attempts to infer the profile of a scattering object from measurement data collected at a distance away from the scatterer (see Fig. 12.1). In addition to obtaining the shape of an object, a quantitative description of its permittivity, velocity, or conductivity profile is also obtainable from an inverse-scattering experiment. This can be extremely valuable diagnostic information [9].

The inverse-scattering solution is sought from the field perturbation, or the scattered field induced by the presence of the object in the transmitted fields. This solution is often difficult to obtain, especially when multiple-scattering effects are present within the object. Such effects can cause the scattered field to be nonlinearly related to the *object function* to be reconstructed. Here, the term, object function, is defined to be a function that describes the velocity, permittivity, or conductivity distribution of an object.

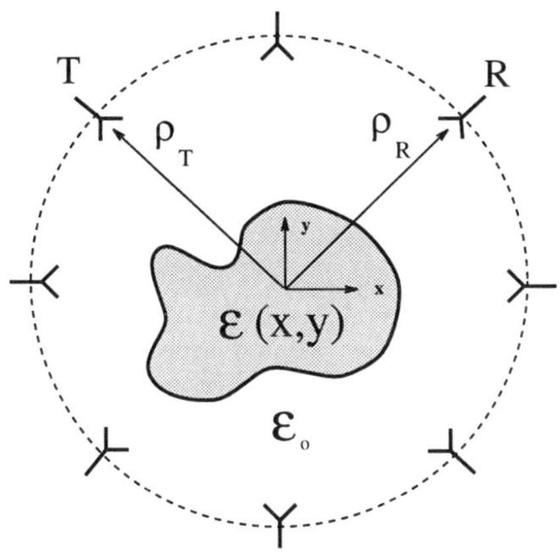

Fig. 12.1 An inverse-scattering experiment where an object is illuminated by the transmitted field and the scattered field is measured at the receivers.

The solution to the inverse problem is often nonunique because high-spatial-frequency portions of the object give rise to evanescent waves. These waves are exponentially small at the receiver locations, and in practice are not measurable unless the receivers are very close to the scatterer. Hence, the scattering operator is a low-pass filter that removes high-spatial-frequency information about the object function. As is well known, all low-pass filtering operations do not have a unique inverse operation unless some assumptions are made about the solution.

In the past, many methods have been proposed to solve particular classes of inverse-scattering problems by making certain assumptions about the object or the scattering process. For example, in CAT, a ray picture is used wherein waves are assumed to propagate in straight lines but all diffraction and multiple-scattering effects are ignored. DT takes into account wave diffraction but ignores multiple scattering and assumes that the object contrast is small. Methods have been proposed to solve the inverse-scattering problem "exactly" for one-dimensional objects such as planar layered media [10]. However, to date none of these "exact" methods have been shown to be computationally stable for objects having multidimensional spatial variations.

In general, the relationship between the scattered field and the scattering object is a nonlinear one. This nonlinearity arises from multiple-scattering effects within the object [11], as seen by examining Fig. 12.2. Assuming that when only scatterer 1 is present, the scattered field is $\vec{E}_{1S}$, and when only scatterer 2 is present, the scattered field is $\vec{E}_{2S}$. When both scatterers are simultaneously present, the scattered field is not just $\vec{E}_{1S} + \vec{E}_{2S}$, but $\vec{E}_{1S} + \vec{E}_{2S} + \vec{E}_{MS}$ where $\vec{E}_{MS}$ is a result of multiple scattering between the two scatterers.

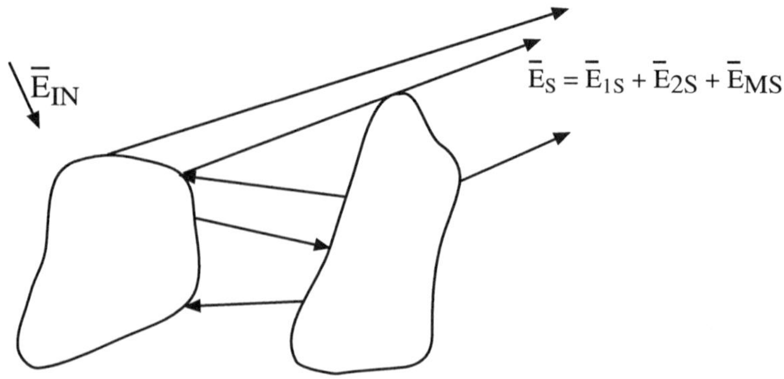

Fig. 12.2 Multiple scattering gives rise to nonlinearity in the inverse problem that precludes the use of linear superposition.

This nonlinear relationship prevents a simple solution to the inverse-scattering problem. One way to solve this nonlinear problem has been to cast it into an optimization format. Here, a cost functional is minimized by applying a gradient-search technique. With such an approach, a forward-scattering solution is sought whenever the gradient is required. Since scattering by an arbitrary object has to be obtained by a numerical method, the nonlinear inverse scattering method is often computationally intensive. However, this method can have an important technological impact if the computational burden is reduced to an acceptable level. Overall, some of the issues that nonlinear inverse-scattering theories should address are [12]:

- Resolution of the reconstructed object;
- Maximum object contrast that can be accurately inverted;
- Maximum object size that can be inverted;
- Algorithm speed;
- Validity for objects with two- and three-dimensional spatial variation;
- Robustness and numerical stability relative to noisy or imperfect measured data;
- Validity for practical scattering measurement geometries;
- Generality allowing application to other wave equations and scattering phenomena (for example, ultrasound as well as electromagnetic waves);
- Minimizing the amount of a-priori information about the object.

One of the most pressing problems for nonlinear inverse-scattering theory is the speed of solution. As mentioned previously, solving the inverse problem numerically often involves solving the forward-scattering problem numerically. However, the latter solution is by itself computationally intensive. Hence, there is considerable need to develop algorithms that can either rapidly solve forward scattering or completely avoid the need for such data. For inhomogeneous volumetric scatterers, the FDTD method [13, 14] is an efficient means to solve the forward-scattering problem in the time domain, providing multifrequency data as well.

A linear relationship between the scattered field and the object function can sometimes be found for certain cases. Hence, inverse-scattering theories are categorized into linear inverse-scattering theories and nonlinear inverse-scattering theories.

12.3 LINEAR INVERSE PROBLEMS

In the linear inverse-scattering theories, approximations are made such that a linear relationship exists between the measured data and the object. Then, the information about the object is obtained by solving the associated set of linear equations.

Such a linear relationship can be established by a number of physical approximations. For instance, in X-ray tomography or CAT, an X-ray beam is assumed to travel in a straight-line path and traverse an object only once. In fact, this is a very good description of how such a beam interacts with a material structure. Here, the total attenuation of the beam is linearly related to the integral sum of the incremental attenuation experienced by the beam as it traverses the path through the structure [11]. Consequently, back-projection tomography can be used to combine many shadows cast by an object on different X-ray films to obtain a composite three-dimensional image of the object.

In MRI, a material object is immersed in a static magnetic field having a linear gradient. The geometry is such that there is a plane in space (the iso-frequency plane) where all proton spins experience the same static magnetic field and hence echo with the same frequency. These spins can be set to precess by a narrowband RF pulse. Once set in motion, the precession frequency (the Larmor frequency) can be altered by a linear gradient in the iso-frequency plane. Therefore, protons at different parallel straight lines in this plane can have different precession frequencies. The received RF signal at a particular frequency is proportional to the integral sum of the resonating spin densities along a straight line [15]. The summation of an object function along a straight line is also known as a projection. As discussed later. a back-projection algorithm using the projection-slice theorem can be used to reconstruct the object function efficiently [1, 16].

When either the Born or Rytov approximation [17] is used to calculate the scattered field, a linear relationship exists between the scattered field and the object. These approximations are valid when the contrast of the scatterer is small so that only single-scattering is important. These approximations account for diffraction effects but not multiple scattering. When the wavelength of the probing wave is not very short relative to the key dimensions of the object and its internal structure, the ray picture of a wave is approximate and furthermore the wave diffracts around inhomogeneities. Moreover, if the contrast of the scatterer is weak, then the Born or Rytov approximation can be a superior approximation compared to the ray picture, which is a geometrical optics approximation. This is the case where diffraction tomography can be used.

In diffraction tomography, the Born or Rytov approximation is used to calculate the scattered field. As shown later, a Fourier transform relationship exists between the scattered field and the object function. The Fourier spectrum of the object function can be recovered using a filtered back-propagation algorithm as in [5] and the object therefore determined. Diffraction tomography is often used in ultrasonic imaging [4] to achieve good reconstruction when multiple-scattering effects are small. However, when multiple scattering is important, as in the case for objects having strong contrasts, the technique breaks down [18].

12.3.1 Back-Projection Tomography

The back-projection algorithm is particularly useful when the measured phase or attenuation is a linear functional of the object, for instance as in X-ray imaging. The WKB approximation [11] says that at high frequencies, a wave propagates mainly in the forward direction with very little reflection; it is essentially a theory of small reflections. Moreover, the phase shift or attenuation of the wave as it propagates through an inhomogeneous medium at high frequencies is, to a very good approximation, governed by the factor

$$\exp\left[j\omega \int_a^b s(z')\, dz' \right] \tag{12.1}$$

The slowness $s(z)$ is complex if the medium is lossy. In X-ray tomography, only the attenuation is measured, as the wavelength is too short for phase measurement. On the other hand, in ultrasound tomography, when the wave physics is approximated by a ray picture, one measures the delay of a pulse through a body which is related to the phase shift and the phase velocity of the object medium. This delay, using the physical picture of the WKB approximation, is given by

$$\tau = \int_a^b s(z')\, dz' \tag{12.2}$$

where $s(z)$ is the slowness of the wave. Moreover, in the above physical picture, the path of the ray through the medium is assumed to be a straight line. This is only true at high frequencies as in X-rays or if the contrast of the inhomogeneity is small. At ultrasonic frequencies, this physical picture is less accurate. Unlike X-ray tomography, ultrasonic tomography with the back-projection algorithm is fraught with potential distortions.

If the object is described by its slowness or its attenuation profile, $s(x, y)$, a single experiment (depicted in Fig. 12.3) then yields

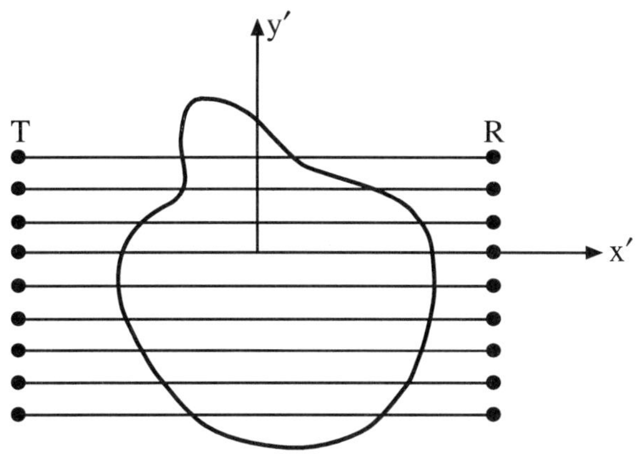

Fig. 12.3 A projection tomography measurement scheme.

$$P(y') = \int_{-\infty}^{\infty} s(x', y')\, dx' \tag{12.3}$$

where (x', y') are the coordinates of the experiment. Hence, $P(y')$ is the projection of the function $s(x', y')$. It is regarded as the shadow cast by $s(x', y')$ on an X-ray film when $s(x', y')$ is illuminated by rays pointing in the x' direction. Moreover, it is related to the Fourier transform of $s(x, y)$. To see this, we write

$$s(x', y') = \frac{1}{(2\pi)^2} \iint_{-\infty}^{\infty} dk'_x\, dk'_y\, e^{jk'_x x' + jk'_y y'}\, S(k'_x, k'_y) \tag{12.4}$$

Then,

$$P(y') = \frac{1}{2\pi} \int_{-\infty}^{\infty} dk'_y\, e^{jk'_y y'}\, S(0, k'_y) \tag{12.5}$$

Hence, from a single projection, a slice of the Fourier transform of $s(x, y)$ ($S(k'_x, k'_y)$ at $k'_x = 0$) is derivable by inverse Fourier transformation. This is also known as the projection-slice theorem [16]. Consequently, we have

$$S(0, k'_y) = \int_{-\infty}^{\infty} dy'\, e^{-jk'_y y'}\, P(y') \tag{12.6}$$

To get a different slice of the Fourier transform, we need only perform the experiment at a different angle. Therefore, by performing the experiment with angles ranging from 0° to 180°, $S(k_x, k_y)$ is filled out in the whole Fourier space. Knowing $S(k_x, k_y)$, we then reconstruct the object, $s(x, y)$, as illustrated in Fig. 12.4.

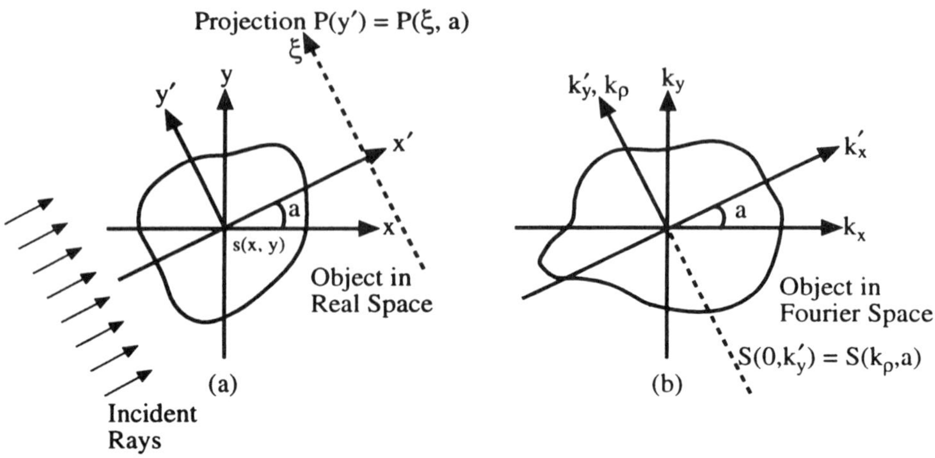

Fig. 12.4 The object in real space and in Fourier space. A projection $P(y') = \int s(x', y')dx'$ is related to $S(0, k'_y)$, a slice in Fourier space.

As mentioned previously, the straight-line approximation for an X-ray beam is an extremely good one because X-ray photons have high frequencies and momenta and are not significantly diffracted by inhomogeneities. For ultrasound, however, some diffraction causing nonstraight-line projection can occur, giving rise to distortions of the reconstructed object. We discuss how the diffraction effect is accounted for in a later section on diffraction tomography. Next, we discuss Radon transforms, which form the mathematical basis of back-projection tomography.

12.3.2 Radon Transforms

The Radon transform [19] was first used for tomographic reconstruction by Bracewell [20]. In fact, it can be shown that back-projection tomographic reconstruction is expressible as a Radon transform. We shall consider here the two-dimensional and three-dimensional Radon transforms [21].

Two Dimensions

As shown in (12.6), a slice of the Fourier transform of $s(x', y')$ is obtainable from the projection $P(y')$. Then, if all the slices of the Fourier transforms are known at different angles, the two-dimensional Fourier transform, $S(k_x, k_y)$, is retrievable to reconstruct $s(x, y)$ via

$$s(x, y) = \frac{1}{(2\pi)^2} \iint_{-\infty}^{\infty} dk_x\, dk_y\, e^{jk_x x + jk_y y}\, S(k_x, k_y) \tag{12.7}$$

Because the slices have symmetry in cylindrical coordinates, (12.7) is more conveniently written in cylindrical coordinates to give

$$s(\rho, \phi) = \frac{1}{(2\pi)^2} \int_0^{\pi} d\alpha \int_{-\infty}^{\infty} dk_\rho \left|k_\rho\right| e^{jk_\rho \rho \cos(\alpha - \phi)}\, S(k_\rho, \alpha) \tag{12.8}$$

Then, for a fixed α, $S(k_\rho, \alpha)$ is just a slice of the Fourier transform retrievable from the projection via

$$S(k_\rho, \alpha) = \int_{-\infty}^{\infty} d\xi\, e^{-jk_\rho \xi}\, P(\xi, \alpha) \tag{12.9}$$

where $P(\xi, \alpha) = P(y')$, the projection obtained for a given α.

Now, after substituting (12.9) into (12.8), performing the k_ρ integration first, and using integration by parts, we arrive at

$$s(\rho, \phi) = \frac{1}{2\pi^2} \int_0^{\pi} d\alpha \;\; \text{P.V.} \int_{-\infty}^{\infty} \frac{\partial P(\xi, \alpha)/\partial \xi}{\rho \cos(\alpha - \phi) - \xi}\, d\xi \tag{12.10}$$

To obtain the above, we have made use of the fact that

$$\text{P.V.} \int_{-\infty}^{\infty} dy\, \frac{e^{-jxy}}{y} = -j\pi\, \text{sgn}(x) \tag{12.11}$$

Moreover, from Fourier transforms, it then follows that

$$\text{P.V.} \frac{1}{y} = -\frac{j}{2} \int_{-\infty}^{\infty} dx\, \text{sgn}(x)\, e^{jxy} \tag{12.12}$$

and

$$\frac{\partial}{\partial y}\left(\text{P.V.}\frac{1}{y}\right) = \frac{1}{2}\int_{-\infty}^{\infty} dx|x|\, e^{jxy} \tag{12.13}$$

which is used to obtain (12.10).

The projection, $P(\xi, \alpha)$, is written as a two-dimensional integral:

$$P(\xi, \alpha) = \iint_{-\infty}^{\infty} dx\, dy\, \delta(\hat{\xi}\cdot\overline{\rho} - \xi)\, s(x, y) \tag{12.14}$$

where $\overline{\rho} = \hat{x}\,x + \hat{y}\,y$ and $\hat{\xi}$ is a unit vector in the ξ direction. Further, in cylindrical coordinates, (12.14) becomes

$$P(\xi, \alpha) = \int_0^{2\pi} d\phi \int_0^{\infty} \rho\, d\rho\, \delta\left[\rho\sin(\phi - \alpha) - \xi\right] s(\rho, \phi) \tag{12.15}$$

Equations (12.10) and (12.15) form the Radon-transform pair in two dimensions. These equations are essential in the two-dimensional back-projection algorithm. Note that (12.10) implies that the object, $s(\rho, \phi)$, can be reconstructed from the measurement, $P(\xi,\alpha)$. Moreover, (12.15) is equivalent to the experiment to obtain the projections of the object, $s(\rho, \phi)$.

Three Dimensions

In three dimensions, it is easy to prove in a similar manner that a two-dimensional projection of $s(x', y', z')$ is related to its three-dimensional transform as

$$P(z') = \iint_{-\infty}^{\infty} dx'\, dy'\, s(x', y', z') = \frac{1}{2\pi}\int_{-\infty}^{\infty} dk_z'\, e^{jk_z'z'}\, S(0, 0, k_z') \tag{12.16}$$

Hence,

$$S(0, 0, k_z') = \int_{-\infty}^{\infty} dz'\, e^{-jk_z'z'}\, P(z') \tag{12.17}$$

Moreover, in spherical coordinates, we write $s(r, \theta, \phi)$ as

$$s(r, \theta, \phi) = \frac{1}{(2\pi)^3}\int_0^{\pi} d\beta \int_0^{\pi} d\alpha\, \sin\alpha \int_{-\infty}^{\infty} dk\, k^2\, S(k, \beta, \alpha) e^{j\vec{k}\cdot\vec{r}} \tag{12.18}$$

In addition, analogous to the two-dimensional Radon transform, (12.17) is written as

$$S(k, \beta, \alpha) = \int_{-\infty}^{\infty} d\xi \, e^{-jk\xi} P(\xi, \beta, \alpha) \tag{12.19}$$

On substitution into (12.18), we now have

$$s(r, \theta, \phi) = \frac{1}{(2\pi)^3} \int_0^{\pi} d\beta \int_0^{\pi} d\alpha \, \sin\alpha \int_{-\infty}^{\infty} d\xi \, P(\xi, \beta, \alpha) \int_{-\infty}^{\infty} dk \, k^2 \, e^{jk(\hat{k}\cdot\bar{r} - \xi)} \tag{12.20}$$

Note that in the above, $\hat{k}$ is a unit vector pointing in the direction of $\vec{k}$, and hence, is a function of α and β. Further, the innermost integral is integrable as derivatives of a Dirac delta function. Consequently, the $d\xi$ integral can be performed in a closed form. Finally, we have

$$s(r, \theta, \phi) = \frac{-1}{(2\pi)^2} \int_0^{\pi} d\beta \int_0^{\pi} d\alpha \, \sin\alpha \, \frac{\partial^2}{\partial^2(\hat{k}\cdot\bar{r})} P(\hat{k}\cdot\bar{r}, \beta, \alpha) \tag{12.21}$$

Since

$$P(\xi, \beta, \alpha) = \int d\bar{r}' \, \delta\left(\hat{\xi}\cdot\bar{r}' - \xi\right) s(\bar{r}') \tag{12.22}$$

(12.21) and (12.22) form a Radon-transform pair in three dimensions.

As a final note, a general formula can be derived for inverse Radon transforms in even dimensions, while another formula can be derived for inverse Radon transforms in odd dimensions [21].

12.3.3 Diffraction Tomography

In projection tomography, one assumes that waves propagate as straight-line rays with no diffraction. This is not a correct assumption at longer wavelengths where the diffraction phenomenon is important. Here, reconstruction using straight-line rays yields distorted images. However, diffraction can be incorporated in the context of the Born and Rytov approximations and a reconstruction algorithm developed accordingly. This technique is termed diffraction tomography [5].

Consider a transmitter-receiver configuration as shown in Fig. 12.5. The scattered field can be obtained using the first-order Born approximation (see Appendix 12A.2):

$$\phi_{\text{sca}}(\bar{\rho}_{\text{R}}) = \int d\bar{\rho}' \, g(\bar{\rho}_{\text{R}}, \bar{\rho}') \, O(\bar{\rho}') \, \phi_{\text{inc}}(\bar{\rho}') \tag{12.23}$$

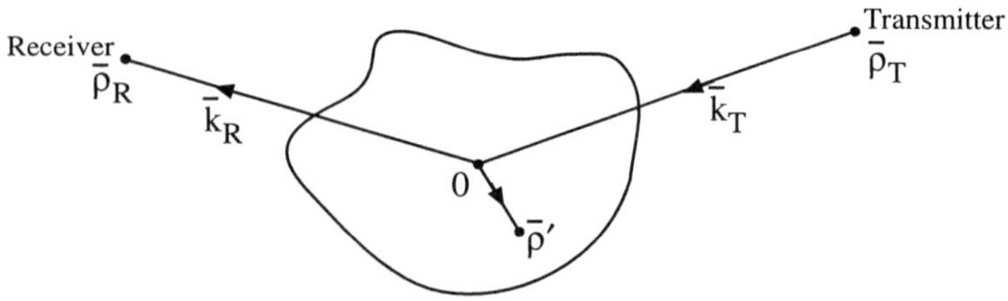

Fig. 12.5 A transmitter-receiver measurement in diffraction tomography.

where two-dimensional scattering is assumed and $O(\overline{\rho}') = k^2 - k_0^2$ is the object function to be reconstructed. (For the Rytov approximation, (12.23) is proportional to the phase perturbation. See Appendix 12A.3.) In two dimensions, the Green's function is:

$$g(\overline{\rho}_R, \overline{\rho}') = \frac{j}{4} H_0^{(1)}\left(k_0 \left|\overline{\rho}_R - \overline{\rho}'\right|\right) \tag{12.24}$$

If the receiver is in the far field of the scatterer, then we have, approximately

$$g(\overline{\rho}_R, \overline{\rho}') \cong \frac{j}{4}\sqrt{\frac{2}{j\pi\, k_0\, \rho_R}}\, e^{jk_0\rho_R - jk_0\hat{\rho}_R\cdot\overline{\rho}'} \tag{12.25}$$

Also, an incident field generated by a uniform line source is

$$\phi_{inc}(\overline{\rho}') = \frac{j}{4} H_0^{(1)}\left(k_0 \left|\overline{\rho}' - \overline{\rho}_T\right|\right) \tag{12.26}$$

And, if the transmitter is also in the far field of the object, we have

$$\phi_{inc}(\overline{\rho}') \cong \frac{j}{4}\sqrt{\frac{2}{j\pi\, k_0\, \rho_T}}\, e^{jk_0\rho_T - jk_0\hat{\rho}_T\cdot\overline{\rho}'} \tag{12.27}$$

Finally, after defining $\vec{k}_R = k_0\,\hat{\rho}_R$ and $\vec{k}_T = -\,k_0\,\hat{\rho}_T$ and substituting the above into (12.23), we have

$$\phi_{sca}(\overline{\rho}_R) \cong \frac{j}{8\pi\, k_0\sqrt{\rho_T\,\rho_R}}\, e^{jk_0(\rho_T+\rho_R)} \int d\overline{\rho}'\, e^{-j(\vec{k}_R - \vec{k}_T)\cdot\overline{\rho}'}\, O(\overline{\rho}') \tag{12.28}$$

Note that (12.28) is a Fourier-transform integral. Consequently,

$$\phi_{sca}(\bar{\rho}_R) \cong \frac{j}{8\pi k_0 \sqrt{\rho_T \rho_R}} e^{jk_0(\rho_T + \rho_R)} \tilde{O}(\vec{k}_R - \vec{k}_T) \tag{12.29}$$

where $\tilde{O}(\vec{k})$ is the Fourier transform of $O(\bar{\rho})$. Therefore, the scattered field under the Born approximation is related to the Fourier transform of the object.

Observe that the lengths of the vectors $\vec{k}_T$ and $\vec{k}_R$ are equal to k_0. Hence, $\vec{k}_R - \vec{k}_T$, the argument of $\tilde{O}$, can span only a finite region in the Fourier space. For instance, if $\vec{k}_T$ is fixed and the receiver is moved so that $\vec{k}_R$ changes direction, then the locus swept out by $\vec{k}_R - \vec{k}_T$ is as shown in Fig. 12.6.

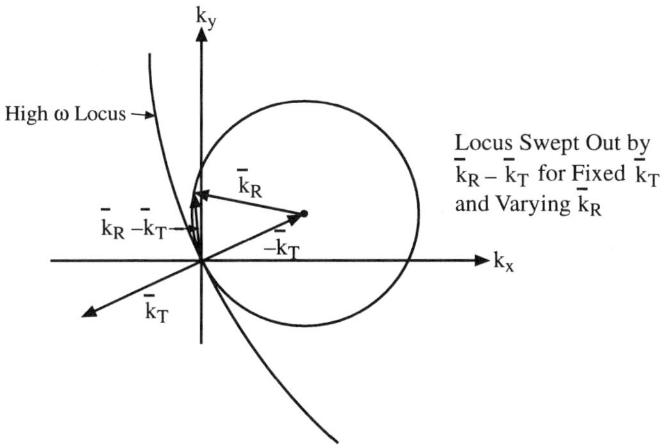

Fig. 12.6 Locus swept out by $\vec{k}_R - \vec{k}_T$ for a fixed $\vec{k}_T$ and varying $\vec{k}_R$.

Further, if the transmitter is moved so that $\vec{k}_T$ changes direction as well as $\vec{k}_R$, then the combination of varying the directions of $\vec{k}_T$ and $\vec{k}_R$ sweeps out a larger circle of radius $2k_0$ with area $4\pi k_0^2$. Consequently, $\tilde{O}(\vec{k})$ is known over the area of this circle, which is larger for larger k_0. Therefore, $O(\bar{\rho})$ can be reconstructed from $\tilde{O}(\vec{k})$ by performing an inverse Fourier transform. However, since $\tilde{O}(\vec{k})$ is known only for $|\vec{k}| \leq k_0$, only a low-pass-filtered version of $O(\bar{\rho})$ is retrieved in this reconstruction. From this discussion, it is also clear that another way of filling out the object in the Fourier space is to interrogate it with a broadband signal.

It is interesting to note that when the frequency becomes very high, the locus swept out by $\vec{k}_R - \vec{k}_T$ (for $\vec{k}_R$ and $\vec{k}_T$ to be approximately in the same direction) passes through the origin almost like a straight line. This is reminiscent of the projection-slice theorem, indicating that projection tomography is a special case of diffraction tomography. Hence, in the high-frequency limit, we need only perform forward scattering measurements with the receivers sweeping through small angles, and a straight-line slice in the Fourier space is recovered. This experiment can be performed as shown in Fig. 12.7. Here, the transmitter is far from the object, so the receivers need only sweep out a small angle in the forward-scattering direction. Note that the rays are almost parallel as in the case of projection tomography.

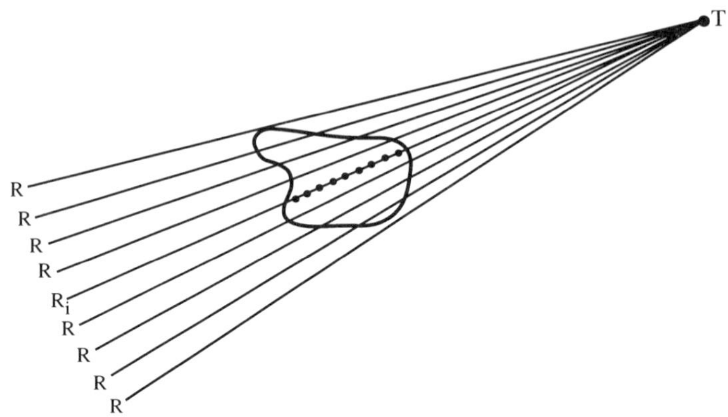

Fig. 12.7 The limit when diffraction tomography becomes projection tomography.

Under the Born approximation in the high-frequency limit, all of the scattering centers on the object (indicated by dots in Fig. 12.7) scatter in phase (coherently) at receiver R_i. But, all of the other scattering centers scatter incoherently to R_i. Hence, R_i measures the cumulative amplitudes of all of the scattered waves from all of the scattering centers aligned between T and R_i, or R_i measures a projection of the object.

Under the Rytov approximation, all of the scattering centers are aligned to contribute cumulatively to the phase perturbation of the wave. But at lower frequencies, this physical picture does not hold since the diffraction effect is more important.

12.4 NONLINEAR INVERSE-SCATTERING METHODS

Nonlinear inverse-scattering theories have been proposed to deal with problems involving substantial multiple-scattering effects [22–28]. Like the Gel'fand-Levitan-Marchenko method [11], many of these nonlinear theories have been developed for one-dimensional objects but have not been verified as being useful for multidimensional inverse scattering. Some of these methods require an impractical measurement apparatus and have never been successfully numerically implemented.

The only general nonlinear inverse scattering theory valid for higher dimensions that has been verified to date involves numerical methods. These numerical methods are usually iterative, and a forward-scattering solver is needed in each iteration [29 – 39]. The Born and the distorted-Born iterative methods have been proposed and verified as methods of solving the nonlinear inverse-scattering problem for dielectric bodies [40 – 60].

12.4.1 Born Iterative Method

In the Born iterative method (BIM) [42], the background medium is assumed to be homogeneous so that the Green's function is known in closed form. The field inside the scatterer is found numerically for an initial guess of its permittivity profile. Then, an improved profile is obtained by comparing the difference between the estimated and measured scattered field. This process is repeated until a convergent solution is obtained.

The scattering by an inhomogeneous body can be formulated in terms of the volume integral equation

$$\phi(\bar{r}) = \phi_{\text{inc}}(\bar{r}) + \omega^2\mu \int_V d^3\bar{r}'\, g_0(\bar{r}-\bar{r}')\, \delta\varepsilon_0(\bar{r}')\, \phi(\bar{r}') \tag{12.30}$$

where $\delta\varepsilon_0(\bar{r}') = \varepsilon(\bar{r}') - \varepsilon_0$. The second term on the right-hand side of (12.30) is the scattered field, which is clearly a nonlinear functional of $\delta\varepsilon_0(\bar{r}')$ because $\phi(\bar{r}')$ in the integral is also a function of $\delta\varepsilon_0(\bar{r}')$. In the inverse-scattering problem, we are interested in solving for $\delta\varepsilon_0(\bar{r}')$ from the measurement of the scattered field outside the target body. Hence, this constitutes a nonlinear problem, or a nonlinear inverse-scattering problem.

The problem could be linearized using the Born approximation by replacing the total internal field, $\phi(\bar{r}')$, inside the scatterer by $\phi_{\text{inc}}(\bar{r}')$. This yields

$$\phi_{\text{sca}}(\bar{r}) \cong \omega^2\mu \int_V d^3\bar{r}'\, g_0(\bar{r}-\bar{r}')\, \delta\varepsilon_0(\bar{r}')\, \phi_{\text{inc}}(\bar{r}') \tag{12.31}$$

Since ϕ_{sca} is known outside the scatterer, (12.31) is a linear integral equation in the unknown $\delta\varepsilon_0(\bar{r})$.

We note that the Born approximation is poor when the scattered field is not weak compared with the incident field. This occurs when the contrast of the scatterer, $\varepsilon - \varepsilon_0$, is substantial or when the size of the scatterer is large compared with the wavelength.

One way to overcome this problem is to seek an increasingly better approximation of $\phi(\bar{r}')$. First, the zeroth-order estimate to $\delta\varepsilon_0(\bar{r}')$ is found by the Born approximation. This estimate, $\delta\varepsilon_0^{(0)}(\bar{r}')$, is then used to calculate the zeroth-order estimate of the scattered field, $\phi_{\text{sca}}^{(0)}(\bar{r})$, everywhere including inside the scatterer. Then, this is used together with $\phi_{\text{inc}}(\bar{r})$ to obtain a better estimate of $\phi(\bar{r})$ within the scatterer. We denote this by $\phi^{(1)}(\bar{r}) = \phi_{\text{inc}}(\bar{r}) + \phi_{\text{sca}}^{(0)}(\bar{r})$. By so doing, we obtain a new approximation for the scattered field:

$$\phi_{\text{sca}}(\bar{r}) \;\cong\; \omega^2\mu\int_V d^3\bar{r}'\; g_0(\bar{r}-\bar{r}')\;\delta\varepsilon_0(\bar{r}')\;\phi^{(1)}(\bar{r}') \tag{12.32}$$

With $\phi^{(1)}$ known, (12.32) is a linear integral equation in $\delta\varepsilon_0(\bar{r}')$ and hence, can be solved as a linear inverse problem. This method yields a new $\delta\varepsilon_0(\bar{r}')$, denoted by $\delta\varepsilon_0^{(1)}(\bar{r}')$, which is then used to obtain a better estimate of $\phi(\bar{r}')$ within the scatterer. This procedure is repeated until convergence is reached. We call this the Born iterative method for solving the inverse problem. This method improves over the simple Born inversion and allows scatterers of relatively large size or high contrast to be reconstructed. However, the inverse problem can no longer be solved in the framework of diffraction tomography. A numerical method to solve the linearized inverse problem in (12.31) is needed. This is usually cast as an optimization problem, as described later.

12.4.2 Distorted Born Iterative Method

While the Born iterative method is simple to implement, it may not be the best method to solve the nonlinear inverse-scattering problem because it does not offer second-order convergence. An improved method with second-order convergence is the distorted Born iterative method (DBIM) [40, 43].

With DBIM, the background medium is not constrained to be homogeneous and is updated at each iteration. The provides the added difficulty that the Green's function is not available in closed form and must be sought via a numerical method. However, DBIM usually converges faster than BIM and works better for objects with large contrast. This advantage is tempered by the more robust performance of BIM in the presence of noise [43]. Nevertheless, a convergent solution can be achieved with DBIM.

In formulating DBIM, we note that an integral equation describing scattering can be derived if we know the Green's function for the following partial differential equation:

$$\left[\nabla^2 + k_b^2(\bar{r})\right] g_b(\bar{r},\bar{r}') \;=\; -\,\delta(\bar{r}-\bar{r}') \tag{12.33}$$

where $k_b(\bar{r})$ is a wavenumber of the known inhomogeneous background. Then, given an equation involving an arbitrary inhomogeneity described by $k^2(\bar{r})$,

$$\left[\nabla^2 + k^2(\bar{r})\right] \phi(\bar{r}) \;=\; S(\bar{r}) \tag{12.34}$$

where $S(\bar{r})$ is a source, we rewrite (12.34) as

$$\left[\nabla^2 + k_b^2(\bar{r})\right] \phi(\bar{r}) \;=\; \delta(\bar{r}) \;-\; (k^2 - k_b^2)\,\phi(\bar{r}) \tag{12.35}$$

Thinking of the extra term on the right-hand side of (12.35) as an induced-source distribution, we use the principle of linear superposition to obtain the solution:

$$\phi(\bar{r}) = -\int_{V_s} d^3\bar{r}'\, g_b(\bar{r},\bar{r}')\, S(\bar{r}') + \int_V d^3\bar{r}'\, g_b(\bar{r},\bar{r}') \left[k^2(\bar{r}') - k_b^2(\bar{r})\right] \phi(\bar{r}') \tag{12.36}$$

where V_s is the volume supporting $S(\bar{r})$, and V is the volume supporting $k^2(\bar{r}) - k_b^2(\bar{r})$. The first term in (12.36) is the field when only the inhomogeneity, $k_b^2(\bar{r})$, is present. Hence, we call it $\phi_{\text{inc}, b}(\bar{r})$, the incident field in the presence of the background inhomogeneity. The second term is just the scattered field due to the difference between $k^2(\bar{r})$ and $k_b^2(\bar{r})$. We call it $\phi_{\text{scat}, b}(\bar{r})$, which is small if $k^2(\bar{r})$ is close to $k_b^2(\bar{r})$. Hence, (12.36) is rewritten as

$$\phi(\bar{r}) = \phi_{\text{inc}, b}(\bar{r}) + \omega^2\mu \int_V d^3\bar{r}'\, g_b(\bar{r}, \bar{r}')\, \delta\varepsilon(\bar{r}')\, \phi(\bar{r}') \tag{12.37}$$

where $\delta\varepsilon(\bar{r}') = \varepsilon(\bar{r}') - \varepsilon_b(\bar{r}')$. Note that $\phi_{\text{inc}, b}(\bar{r})$ satisfies

$$\phi_{\text{inc}, b}(\bar{r}) = \phi_{\text{inc}}(\bar{r}) + \omega^2\mu \int_V d^3\bar{r}'\, g_0(\bar{r}, \bar{r}')\, \varepsilon_b(\bar{r}')\, \phi_{\text{inc}, b}(\bar{r}') \tag{12.38}$$

From (12.37), we deduce that

$$\phi_{\text{scat}, b}(\bar{r}) = \omega^2\mu \int_V d^3\bar{r}'\, g_b(\bar{r}, \bar{r}')\, \delta\varepsilon(\bar{r}')\, \phi(\bar{r}') \tag{12.39}$$

where

$$\phi_{\text{scat}, b}(\bar{r}) = \phi(\bar{r}) - \phi_{\text{inc}, b}(\bar{r}) \tag{12.40}$$

Once $\varepsilon_b(\bar{r})$ is assumed known, $\phi_{\text{inc}, b}(\bar{r})$ is calculated by solving (12.38). If the total field $\phi(\bar{r})$ is measured at the receivers, then $\phi_{\text{scat}, b}(\bar{r})$ is obtained at the receivers via (12.40). Essentially, the left-hand side of (12.39) is known at the receivers via measurement and solving (12.38).

The goal of the inverse problem is to solve (12.39) for $\delta\varepsilon(\bar{r})$ from which we find $\varepsilon(\bar{r})$. However, from (12.39), it is clear that this is a nonlinear inverse-scattering problem because $\phi(\bar{r})$ is also a functional of $\delta\varepsilon$, as is clear from (12.37). But if $\varepsilon(\bar{r})$ is close to $\varepsilon_b(\bar{r})$, then $\phi_{\text{scat}, b}(\bar{r}) << \phi_{\text{inc}, b}(\bar{r})$, as is evident from (12.37). Therefore, we approximate $\phi(\bar{r})$ with $\phi_{\text{inc}, b}(\bar{r})$ in (12.39) and linearize this equation as

$$\phi_{\text{scat}, b}(\bar{r}) \cong \omega^2\mu \int_V d^3\bar{r}'\, g_b(\bar{r}, \bar{r}')\, \delta\varepsilon(\bar{r}')\, \phi_{\text{inc}, b}(\bar{r}') \tag{12.41}$$

The approximation of $\phi(\bar{r})$ with $\phi_{\text{inc}, b}(\bar{r})$ in (12.39) is called the distorted Born approximation, or the distorted-wave Born approximation [11]. If $\delta\varepsilon$ is small, it can be easily seen that the error in this approximation is of order $(\delta\varepsilon)^2$ (i.e., second order).

As described later, the linearized integral equation (12.41) can be numerically solved to obtain $\delta\varepsilon$. However, $\delta\varepsilon$ obtained in this manner is not strictly correct because of the distorted Born approximation. Fortunately, this approximate $\delta\varepsilon$ can be used to improve on the approximated $\varepsilon(\bar{r})$. Then $\varepsilon(\bar{r})$ can be used as the new $\varepsilon_b(\bar{r})$. With this new $\varepsilon_b(\bar{r})$, the $\phi_{\text{scat}, b}(\bar{r})$, $g_b(\bar{r}, \bar{r}')$, and $\phi_{\text{inc}, b}(\bar{r})$ in (12.41) are recalculated and updated. Next, a new $\delta\varepsilon$ is obtained and the process is repeated until convergence is attained. Since the error in (12.41) is of second order, this method has second-order convergence. We call this method of solving the inverse-scattering problem the distorted Born iterative method. Because it is similar to the Newton-Raphson method in solving a one-dimensional nonlinear equation, this technique is sometimes called the Newton's method [36] or the Newton-Kantorovich method [35]. However, the term DBIM may be preferred since it refers to the physical basis of (12.41) (i.e., the distorted-wave Born approximation).

12.4.3 Fréchet Derivative

Let us consider a vector, $\vec{f}$, that is nonlinearly related to another vector, $\vec{g}$, via the function, $\mathbf{G}$. The relationship is written as

$$\vec{f} = \mathbf{G}(\vec{g}) \tag{12.42}$$

or in explicit notation

$$f_i = G_i(g_1, \ldots, g_N), \qquad i = j, \ldots, M. \tag{12.43}$$

We find the first variation in f_i induced by first variation in g_j as

$$\delta f_i = \sum_{j=1}^{N} \frac{\partial G_i}{\partial g_j} \delta g_j \tag{12.44}$$

The above is written in matrix notation as

$$\delta\vec{f} = \overline{\mathbf{F}} \cdot \delta\vec{g} \tag{12.45}$$

where

$$[\overline{\mathbf{F}}]_{ij} = \frac{\partial G_i}{\partial g_j} \tag{12.46}$$

$\overline{\mathbf{F}}$ is the gradient of the multidimensional function also known as its Fréchet derivative.

The analog of a Fréchet derivative exists for a continuum, infinite-dimensional space. Consider the continuous analog of (12.42) or (12.43), which is

$$f(x) = G[x, g(y)], \qquad 0 < x < a, \qquad 0 < y < b \tag{12.47}$$

where G is a nonlinear functional of g. The analog of (12.44) is

$$\delta f(x) = \int_0^b dy\, \frac{\delta G(x, g)}{\delta g(y)}\, \delta g(y) \tag{12.48}$$

where $\delta G / \delta g$ is a functional derivative analogous to $\partial G_i / \partial g_j$. The integral operator $\int_0^b dy\, (\delta G / \delta g) \cdot$ is often called the Fréchet derivative operator [37].

12.4.4 Relationship to Distorted Born Approximation

As discussed previously, the total field in a scattering problem can be written as

$$\phi(\bar{r}) = \phi_{\text{inc},\, b}(\bar{r}) + \omega^2 \mu \int d^3\bar{r}'\, g_b(\bar{r}, \bar{r}')\, \delta\varepsilon(\bar{r}')\, \phi(\bar{r}') \tag{12.49}$$

where $\delta\varepsilon(\bar{r}') = \varepsilon(\bar{r}') - \varepsilon_b(\bar{r}')$. When $\phi(\bar{r}')$ is given by the distorted Born approximation such that

$$\phi(\bar{r}') \cong \phi_{\text{inc},\, b}(\bar{r}') \tag{12.50}$$

the error in (12.50) is of $O(\delta\varepsilon)$ as is clear from (12.49). Hence, in the approximate integral such that

$$\phi(\bar{r}) \approx \phi_{\text{inc},\, b}(\bar{r}) + \omega^2 \mu \int d^3\bar{r}'\, g_b(\bar{r}, \bar{r}')\, \delta\varepsilon(\bar{r}')\, \phi_{\text{inc},\, b}(\bar{r}') + O(\delta\varepsilon)^2 \tag{12.51}$$

the error is clearly of second order. Therefore, the first-order variation in $\phi(\bar{r})$ with respect to the first-order variation in ε about a background ε_b is

$$\delta\phi(\bar{r}) = \omega^2 \mu \int d^3\bar{r}'\, g_b(\bar{r}, \bar{r}')\, \phi_{\text{inc},\, b}(\bar{r}')\, \delta\varepsilon(\bar{r}') \tag{12.52}$$

Comparing (12.52) and (12.48), we notice that the integral operator

$$\omega^2 \mu \int d^3\bar{r}'\, g_b(\bar{r}, \bar{r}')\, \phi_{\text{inc},\, b}(\bar{r}') \tag{12.53}$$

is the Fréchet derivative operator. Hence, the distorted Born approximation yields the Fréchet derivative operator directly [47].

12.5 NUMERICAL SOLUTION TO THE INVERSE PROBLEM

The above discussion sets the stage for introducing the numerical methods used to solve the nonlinear inverse problem. This problem is first linearized locally around an estimated solution, and the linearized equation is then solved. Then, the new solution is used to obtain an improvement to the estimated solution. The procedure is repeated until convergence is reached.

12.5.1 Discretization of the Integral Equation

In order to solve the integral equation (12.41) for the unknown $\delta\varepsilon(\bar{r}')$, we notice that $\phi_{\text{scat},\,b}$ is just a function of frequency, transmitter, and receiver locations. Hence, we rewrite (12.41) as

$$\phi_{\text{scat},\,b}(\bar{r}_{\text{R}}, \bar{r}_{\text{T}}, \omega) = \omega^2 \mu \int d^3\bar{r}' \; g_b(\bar{r}_{\text{R}}, \bar{r}', \omega)\, \phi_{\text{inc},\,b}(\bar{r}', \bar{r}_{\text{T}}, \omega)\, \delta\varepsilon(\bar{r}') \tag{12.54}$$

We expressed $\phi_{\text{inc},\,b}(\bar{r}', \bar{r}_{\text{T}}, \omega)$ explicitly as a function of frequency and transmitter location $\bar{r}_{\text{T}}$. We also expressed the explicit frequency dependence of other pertinent quantities, and assumed that $\delta\varepsilon(\bar{r}')$ is frequency independent. Equation (12.54) is discretized as

$$\phi_{\text{scat},\,b}(\bar{r}_{\text{R}_i}, \bar{r}_{\text{T}_j}, \omega_k) = \sum_{n=1}^{N} \Delta v_n\, \omega_k^2\, \mu\, g_b(\bar{r}_{\text{R}_i}, \bar{r}_n, \omega_k)\, \phi_{\text{inc},\,b}(\bar{r}_n, \bar{r}_{\text{T}_j}, \omega_k)\, \delta\varepsilon(\bar{r}_n) \tag{12.55}$$

where $\bar{r}_n$ is a location within the object to be reconstructed, and volume element, Δv_n, depends on the integration rule used. We write (12.55) as

$$\vec{\phi}_{\text{scat},b} = \overline{\mathbf{M}} \cdot \delta\vec{\varepsilon} \tag{12.56}$$

where $\left[\vec{\phi}_{\text{scat},b}\right]_m$ represents the data measured with different transmitter-receiver and frequency combinations, that is:

$$\left[\vec{\phi}_{\text{scat},\,b}\right]_m = \phi_{\text{scat},\,b}(\bar{r}_{\text{R}_i}, \bar{r}_{\text{T}_j}, \omega_k) \tag{12.57}$$

$$[\overline{\mathbf{M}}]_{mn} = \Delta v_n\, \omega_k^2\, \mu\, g_b(\bar{r}_{\text{R}_i}, \bar{r}_n, \omega_k)\, \phi_{\text{inc},\,b}(\bar{r}_n, \bar{r}_{\text{T}_j}, \omega_k) \tag{12.58}$$

$$[\delta\vec{\varepsilon}]_n = \delta\varepsilon(\bar{r}_n) \tag{12.59}$$

and the index, m, represents all combinations of (i, j, k). Comparing (12.56) and (12.45), we note that $\overline{\mathbf{M}}$ is just the gradient matrix. Hence, the distorted Born approximation also yields the gradient matrix directly.

12.5.2 Minimization of a Cost Functional

We described how to convert a linearized inverse problem as indicated by (12.31) or (12.41) to a matrix equation as given by (12.56). However, solution of (12.56) for $\delta\vec{\varepsilon}$ is not advised because matrix $\overline{\mathbf{M}}$ is ill-conditioned and might not have a unique inverse. There are two main reasons for this ill-conditioning:

1. $\overline{\mathbf{M}}$ is an $M \times N$ matrix. It is possible for $M < N$ if the number of measurement data points is less than the number of pixels used to approximate the object function. Then, the range space of $\overline{\mathbf{M}}$ would have a smaller dimension than the domain, implying that the inverse does not exist.
2. The inverse problem is faced with fundamental scattering physics that contributes to its nonuniqueness. $\overline{\mathbf{M}}$ maps the object function, $\delta\vec{\varepsilon}$, to the measurement data, $\phi_{\text{scat},\,b}$. Many high-spatial-frequency components of ε are not mapped to ϕ_{scat} because these components generate evanescent waves that do not reach the receiver locations. Therefore, these fine-grained details are lost in the inversion process.

Consider point 2 in more detail. The induced current in the scatterer is produced via the term, $\delta\varepsilon(\bar{r}')\,\phi_{\text{inc},\,b}(\bar{r}')$. Therefore, if $\delta\varepsilon(\bar{r}')$ has high-spatial-frequency components, so does the induced current. From (12.54), this induced current radiates via the Green's function, $g_b(\bar{r}, \bar{r}')$, which accounts for multiple scattering within the inhomogeneous background, ε_b. For simplicity, assume that the induced current radiates via the free-space Green's function, $g_0(\bar{r} - \bar{r}')$, so that only single-scattering physics occurs as the radiated field leaves the scatterer. As is clear from the diffraction tomography analysis, if the observation point is in the far field of the scatterer, the operator $g_0(\bar{r} - \bar{r}')$ only furnishes Fourier components of the induced current on the sphere where $|\vec{k}| = k_0$. However, the induced current manifestly contains Fourier components outside this sphere. These components correspond to the evanescent waves, whose information therein is lost if one measures the data in the far field.

Therefore, to solve (12.56) uniquely, we resort to an optimization technique where the error is minimized as well as the norm of the solution, $\delta\vec{\varepsilon}$. Hence, we seek to minimize a cost functional

$$I = \left\| \vec{\phi}_{\text{scat},b} - \overline{\mathbf{M}} \cdot \delta\vec{\varepsilon} \right\|_{\overline{\mathbf{C}}}^2 + \delta_t \left\| \delta\vec{\varepsilon} \right\|_{\overline{\mathbf{D}}}^2 \tag{12.60}$$

where the norm is defined to be

$$\|\mathbf{V}\|_{\overline{\mathbf{A}}}^2 = \mathbf{V}^\dagger \cdot \overline{\mathbf{A}} \cdot \mathbf{V} \tag{12.61}$$

and $\overline{\mathbf{A}}$ is a positive definite matrix.

In (12.60), $\overline{\mathbf{C}}$ is chosen so that different weights are given to different measurement data. For example, the transmitters and receivers may be at different distances from the target or the scatterer, and the measurement data may be of disparate amplitudes.

Then, $\overline{\mathbf{C}}$ may be chosen to scale the data so that equal weights are given to the data irrespective of their measured amplitudes. In another case, the measurement data of the far transmitters and receivers may be boosted to yield about the same amplitude as those from the near receivers. Statistical analysis may be used to derive $\overline{\mathbf{C}}^{-1}$, which in this case is related to the covariance matrix [37].

In (12.60), $\overline{\mathbf{D}}$ is chosen to enforce some regularity in $\delta\vec{\varepsilon}$. For instance, if $\overline{\mathbf{D}}$ is an identity matrix, then the second term in (12.60) corresponds to the energy norm or the L_2 norm of $\delta\vec{\varepsilon}$. If $\overline{\mathbf{D}}$ is a finite-difference operator, then when we minimize $\| \delta\vec{\varepsilon} \|_{\overline{\mathbf{D}}}^2$, $\delta\vec{\varepsilon}$ has to be sufficiently smooth so that its norm as defined would not be overly large. Such a method of determining the solution to (12.56) is known as the Tikhonov regularization method [61]. Cross-validation analysis based on the noise property of the experimental data can be used to derive an optimal $\overline{\mathbf{D}}$ and δ_t [62].

Further, in (12.60), δ_t has to be chosen small enough so that the goal of solving (12.56) is not overwhelmed by the need to minimize the norm of $\delta\vec{\varepsilon}$. However, if δ_t is chosen to be too small, the resultant equation may be ill-conditioned again. The optimal choice of δ_t is related to the noise characteristic of the measurement data.

By writing I as

$$I = \left(\vec{\phi}_{\text{scat}}^{\dagger} - \delta\vec{\varepsilon}^{\dagger} \cdot \overline{\mathbf{M}}^{\dagger}\right) \cdot \overline{\mathbf{C}} \cdot \left(\vec{\phi}_{\text{scat}} - \overline{\mathbf{M}} \cdot \delta\vec{\varepsilon}\right) + \delta_t\, \delta\vec{\varepsilon}^{\dagger} \cdot \overline{\mathbf{D}} \cdot \delta\vec{\varepsilon} \tag{12.62}$$

and then taking the first variation of I with respect to the first variation in $\delta\vec{\varepsilon}$, and then setting the first variation to zero to obtain the minimums of I, we obtain

$$\left[\overline{\mathbf{M}}^{\dagger} \cdot \overline{\mathbf{C}} \cdot \overline{\mathbf{M}} + \delta_t \overline{\mathbf{D}}\right] \cdot \delta\vec{\varepsilon} = \overline{\mathbf{M}}^{\dagger} \cdot \overline{\mathbf{C}} \cdot \vec{\phi}_{\text{scat}} \tag{12.63}$$

This equation can be solved iteratively by the conjugate gradient approach.

To understand the role of δ_t, we note that $\overline{\mathbf{M}}^{\dagger} \cdot \overline{\mathbf{C}} \cdot \overline{\mathbf{M}}$ is a semi-positive-definite matrix and factorize it as

$$\overline{\mathbf{M}}^{\dagger} \cdot \overline{\mathbf{C}} \cdot \overline{\mathbf{M}} = \overline{\mathbf{S}}^{\dagger} \cdot \overline{\boldsymbol{\lambda}} \cdot \overline{\mathbf{S}} \tag{12.64}$$

where $\overline{\boldsymbol{\lambda}}$ is a diagonal matrix containing the eigenvalues of $\overline{\mathbf{M}}^{\dagger} \cdot \overline{\mathbf{C}} \cdot \overline{\mathbf{M}}$ and $\overline{\mathbf{S}}$ is a unitary matrix (hence, $\overline{\mathbf{S}}^{\dagger} = \overline{\mathbf{S}}^{-1}$). If $\overline{\mathbf{D}}$ is an identity matrix $\overline{\mathbf{I}}$, then (12.63) is rewritten as

$$\overline{\mathbf{S}}^{\dagger} \cdot \left[\overline{\boldsymbol{\lambda}} + \delta_t \overline{\mathbf{I}}\right] \cdot \overline{\mathbf{S}} \cdot \delta\vec{\varepsilon} = \overline{\mathbf{M}}^{\dagger} \cdot \overline{\mathbf{C}} \cdot \vec{\phi}_{\text{scat}} \tag{12.65}$$

Because $\overline{\mathbf{M}}$ has a null space, some of the eigenvalues of $\overline{\mathbf{M}}^{\dagger} \cdot \overline{\mathbf{C}} \cdot \overline{\mathbf{M}}$ are zero, and hence, $\overline{\mathbf{M}}^{\dagger} \cdot \overline{\mathbf{C}} \cdot \overline{\mathbf{M}}$ does not have an inverse. Therefore, if $\delta_t = 0$, (12.63) does not have a unique solution. Even then, we can define a pseudo-inverse of (12.63) such that

$$\left(\overline{\mathbf{M}}^{\dagger} \cdot \overline{\mathbf{C}} \cdot \overline{\mathbf{M}}\right)^{I} = \overline{\mathbf{S}}^{\dagger} \cdot \overline{\boldsymbol{\lambda}}^{a} \cdot \overline{\mathbf{S}} \tag{12.66}$$

where $\overline{\boldsymbol{\lambda}}^a$ is a diagonal matrix whose entries are the reciprocal of the eigenvalues when they are not zero, and whose entries are zeros if the corresponding eigenvalues are zero. Such a solution to (12.63) when $\delta_t = 0$ is also called the least-norm or minimum-energy solution. The solution found in this manner also satisfies (12.56).

However, such a manner of finding the solution has two major drawbacks. The factorization as expressed by (12.64) is a procedure requiring $O(N^3)$ operation count where N is the dimension of $\overline{\mathbf{M}}^\dagger \cdot \overline{\mathbf{C}} \cdot \overline{\mathbf{M}}$. This operation count becomes inordinately large when N is large. The second problem is that some of the eigenvalues in (12.64) are not identically zero but are exponentially small. These correspond to information contained in the evanescent spectrum that is exponentially small at the receivers. Hence, there is no clear distinction between the zero eigenvalues and the nonzero eigenvalues. When the measurement data are noisy, the small eigenvalues could potentially amplify the noise in the measurement data. Therefore, in finding the pseudo-inverse given by (12.66), a threshold is chosen so that eigenvalues smaller than the threshold are set to zero and do not participate in the pseudo-inverse of $\overline{\boldsymbol{\lambda}}$.

The dimensionally large operations count can become overwhelming when N is large. Therefore, the use of an iterative solution to (12.64) is often the best approach. By looking at (12.64), we note that the role of $\delta_t \mathbf{I}$ is to pad the zero eigenvalues or the small eigenvalues of $\overline{\boldsymbol{\lambda}}$ with a nonzero value so that the solution to (12.63) can be uniquely obtained. If $\overline{\mathbf{D}}$ is positive definite or if the null-space of $\overline{\mathbf{D}}$ and $\overline{\mathbf{M}}^\dagger \cdot \overline{\mathbf{C}} \cdot \overline{\mathbf{M}}$ are non-overlapping, then the solution to (12.63) is guaranteed to be unique.

12.5.3 Back-Propagation and Correlation

In solving inverse scattering as an optimization problem, we end up with an equation of the form

$$\left[\overline{\mathbf{M}}^\dagger \cdot \overline{\mathbf{C}} \cdot \overline{\mathbf{M}} + \delta_t \overline{\mathbf{D}}\right] \cdot \delta\vec{\varepsilon} = \overline{\mathbf{M}}^\dagger \cdot \vec{\phi}_{\text{scat},\, b} \tag{12.67}$$

The above equation is most efficiently solved by iterative methods such as the conjugate gradient method. In an iterative method, the dominant computational cost at each iteration is the cost of the matrix vector multiplication. Since $\overline{\mathbf{D}}$ is usually a diagonal or a banded-diagonal matrix, the dominant cost arises from

$$\overline{\mathbf{M}}^\dagger \cdot \overline{\mathbf{C}} \cdot \left(\overline{\mathbf{M}} \cdot \delta\vec{\varepsilon}\right) \tag{12.68}$$

The operator $\overline{\mathbf{M}}$ maps the object function $\delta\vec{\varepsilon}$ to the measurement data $\vec{\phi}_{\text{scat},b}$ as indicated by (12.56). $\overline{\mathbf{C}}$, sometimes called the covariant matrix, is usually diagonal to reduce computational cost. Thus, it maps measurement data to measurement data. On the other hand, the operator $\overline{\mathbf{M}}^\dagger$ maps a function defined in the space of measurement data $\vec{\phi}_{\text{scat},b}$ to a function defined in the space of object functions, $\delta\vec{\varepsilon}$. If $\overline{\mathbf{M}}$ is an $N \times N$ matrix, the cost of performing (12.68) is of order N^2. But if an efficient forward solver is available, the operations described by (12.68) can be performed more efficiently.

Writing $\overline{\mathbf{M}} \cdot \delta\vec{\varepsilon}$ explicitly, we have for $m = (i, j, k)$:

$$\left[\overline{\mathbf{M}} \cdot \delta\vec{\varepsilon}\right]_m = \omega_k^2 \mu \int_V d^3\bar{r}' \, g_b(\bar{r}_{R_i}, \bar{r}') \, \phi_{\text{inc}, b}(\bar{r}', \bar{r}_{T_j}) \, \delta\varepsilon(\bar{r}') \tag{12.69}$$

$\phi_{\text{inc}, b}(\bar{r}', \bar{r}_{T_j})$ can be updated with one call to the forward solver. With $\phi_{\text{inc}, b}$ known, the induced current

$$j_b(\bar{r}', \bar{r}_{T_j}) = \omega_k^2 \, \mu \, \phi_{\text{inc}, b}(\bar{r}', \bar{r}_{T_j}) \, \delta\varepsilon(\bar{r}') \tag{12.70}$$

can be found in N operations where N is the number of pixels used to describe $\delta\varepsilon(\bar{r}')$. Next, we write (12.67) as

$$\left[\overline{\mathbf{M}} \cdot \delta\vec{\varepsilon}\right]_m = \int_V d^3\bar{r}' \, g_b(\bar{r}_{R_i}, \bar{r}') \, j_b(\bar{r}', \bar{r}_{T_j}) = \tilde{\phi}_{\text{scat}, b}(\bar{r}_{R_i}, \bar{r}_{T_j}, \omega_k) \tag{12.71}$$

The above integral generates the field due to the source $j_b(\bar{r}', \bar{r}_{T_j})$ in the inhomogeneous background ε_b. The above integration need not be done explicitly, but can be effected by one call to the forward solver. We note that $\tilde{\phi}_{\text{scat}, b}(\bar{r})$ as given by (12.71) satisfies the following equation:

$$\left[\nabla^2 + k_b^2(\bar{r})\right] \tilde{\phi}_{\text{scat}, b}(\bar{r}) = -\, j_b(\bar{r}, \bar{r}_{T_j}) \tag{12.72}$$

Hence, $\tilde{\phi}_{\text{scat}, b}(\bar{r})$ is a solution to the following integral equation:

$$\tilde{\phi}_{\text{scat}, b}(\bar{r}) = \phi_{inc}(\bar{r}) + \omega^2 \mu \int d^3\bar{r}' \, g_0(\bar{r} - \bar{r}') \, [\varepsilon_b(\bar{r}') - \varepsilon_0] \, \tilde{\phi}_{\text{scat}, b}(\bar{r}') \tag{12.73}$$

where $\phi_{inc}(\bar{r})$ is the field produced by $j_b(\bar{r}, \bar{r}_{T_j})$ via the homogeneous medium Green's function.

Therefore, (12.71) is implemented by first calling a forward solver to generate $j_b(\bar{r}, \bar{r}_{T_j})$ via (12.70), and then solving either (12.72) or (12.73). If the forward problem can be solved efficiently in less than $O(N^2)$ operations, then the matrix-vector multiply $\overline{\mathbf{M}} \cdot \delta\vec{\varepsilon}$ can be implemented in less than $O(N^2)$ operations.

The second step in (12.67) is to perform the operation $\overline{\mathbf{M}}^{\dagger} \cdot \vec{\tilde{\phi}}_{\text{scat}, b}$. Written explicitly, this is

$$\sum_{j,k} \Delta\bar{r} \, \phi^*_{\text{inc}, b}(\bar{r}_n, \bar{r}_{T_j}) \sum_i \omega_k^2 \, \mu \, g_b^*(\bar{r}_{R_i}, \bar{r}_n) \, \tilde{\phi}_{\text{scat}, b}(\bar{r}_{R_i}, \bar{r}_{T_j}) = \delta\tilde{O}(\bar{r}_n) \tag{12.74}$$

Equation (12.74) is equivalent to taking the field $\tilde{\phi}_{\text{scat}}(\bar{r}_{R_i}, \bar{r}_{T_j})$ as sources located at $\bar{r}_{R_i}$ for each $\bar{r}_{T_j}$, and propagating the field generated from $\bar{r}_{R_i}$ to $\bar{r}_n$ with the Green's function

$g_b^*(\bar{r}_{R_i}, \bar{r}_n)$. Since a complex conjugation in the frequency domain is equivalent to a time reversal in the time domain, this operation is also known as back-propagation. After back-propagating the field to locations $\bar{r}_n$ from $\bar{r}_{R_i}$, then the field at $\bar{r}_n$ is multiplied by $\phi_{inc}^*(\bar{r}_n, \bar{r}_{T_j})$ generated by the corresponding transmitter located at $\bar{r}_{T_j}$. This is repeated for different transmitters and the results are added or stacked together. The multiplication by $\phi_{inc}^*(\bar{r}_n, \bar{r}_{T_j})$ in the frequency domain corresponds to a correlation in the time domain. Therefore, the operations needed to implement (12.74) are also known as back-propagation and correlation [37].

12.5.4 FDTD Implementation of the Inversion Algorithm

The above algorithm has been described in the frequency domain for multifrequency data. However, it is sometimes more efficient to implement the algorithm directly in the time domain especially if a robust forward time-domain solver is available. This is the case for FDTD. In the following discussion, we derive the time-domain equivalence of the inversion algorithm [45, 12].

One important difference is the definition of the cost function to be minimized and the norm involved. In (12.60), two norms are involved. The first norm is with respect to functions defined in the measurement data space, while the second norm is defined with respect to functions defined in the object space.

For instance, in the data space, we choose the following norm:

$$\left\| \phi_{\text{scat}, b} \right\|^2 = \sum_{i,j,k} \left| \phi_{\text{scat}, b}(\bar{r}_{R_i}, \bar{r}_{T_j}, \omega_k) \right|^2 \approx \sum_{i,j} \int d\omega \left| \phi_{\text{scat}, b}(\bar{r}_{R_i}, \bar{r}_{T_j}, \omega) \right|^2 \tag{12.75}$$

By invoking Parseval's theorem, the above norm is expressed in the time domain as

$$\left\| \phi_{\text{scat}, b} \right\|^2 \approx 2\pi \sum_{i,j} \int dt \left| \phi_{\text{scat}, b}(\bar{r}_{R_i}, \bar{r}_{T_j}, t) \right|^2 \tag{12.76}$$

Consequently, the norm in (12.60) is defined directly with the measurement data in the time domain. The matrix operator is then a linear operator that maps an object function (defined in the support of the object) to a field defined at the transmitter-receiver locations in the time domain. As such, the concept of propagation, back-propagation, and correlation can be expressed so that an FDTD solver is used directly.

Equation (12.69) is expressed directly in the time domain for $m = (i, j, t)$ as

$$\left[\overline{\mathbf{M}} \cdot \delta\bar{\varepsilon} \right]_m \approx \mu \int_V d^3\bar{r}' g_b(\bar{r}_{R_i}, \bar{r}', t) * \frac{\partial^2}{\partial t^2} \phi_{\text{inc}, b}(\bar{r}', \bar{r}_{T_j}, t)\, \delta\varepsilon(\bar{r}') \tag{12.77}$$

where the asterisk denotes a convolution in time. The above operation can be implemented by calling a forward FDTD solver to process the wave equation:

$$\left[\nabla^2 - \frac{1}{\mu\varepsilon_b(\bar{r})}\frac{\partial^2}{\partial t^2}\right]\tilde{\phi}_{\text{scat},\,b}(\bar{r},t) = -j_b(\bar{r},\bar{r}_{\text{T}_j},t) \tag{12.78}$$

where the source is given by

$$j_b(\bar{r}',\bar{r}_{\text{T}_j},t) = \mu\,\delta\varepsilon(\bar{r}')\frac{\partial^2}{\partial t^2}\phi_{\text{inc},\,b}(\bar{r}',\bar{r}_{\text{T}_j},t) \tag{12.79}$$

To perform the operation $\overline{\mathbf{M}}^{\dagger}\cdot\vec{\tilde{\phi}}_{\text{scat},\,b}$ in the time domain, we rewrite it explicitly as

$$\sum_j \Delta\bar{r}\,\phi_{\text{inc},\,b}(\bar{r}_n,\bar{r}_{\text{T}_j},-t) * \sum_i \mu\, g_b(\bar{r}_{\text{R}_i},\bar{r}_n,-t) * \frac{\partial^2}{\partial t^2}\tilde{\phi}_{\text{scat},\,b}(\bar{r}_{\text{R}_i},\bar{r}_{\text{T}_j},t) = \delta\tilde{O}(\bar{r}_n) \tag{12.80}$$

By a simple manipulation, (12.80) is written as

$$\sum_j \Delta\bar{r}\,\phi_{\text{inc},\,b}(\bar{r}_n,\bar{r}_{\text{T}_j},t) * \sum_i \mu\, g_b(\bar{r}_{\text{R}_i},\bar{r}_n,t) * \frac{\partial^2}{\partial t^2}\tilde{\phi}_{\text{scat},\,b}(\bar{r}_{\text{R}_i},\bar{r}_{\text{T}_j},-t) = \delta\tilde{O}(\bar{r}_n) \tag{12.81}$$

To implement the above operation, we first time-reverse the field $\tilde{\phi}_{\text{scat},\,b}(\bar{r}_{\text{R}_i},\bar{r}_{\text{T}_j},t)$. This field is then treated as a source, and operated upon by $\mu\, g_b(\bar{r}_{\text{R}_i},\bar{r}_n,t)$ to generate fields within the scatterer. This latter step can be achieved by calling an FDTD solver with the aforementioned source. This part is known as back-propagation, since the received fields are back-propagated into the object or the scatterer. The last step is the correlation with $\phi_{\text{inc},\,b}(\bar{r}_n,\bar{r}_{\text{T}_j},t)$.

Therefore, the complete matrix-vector multiplication can be implemented by three calls to the forward FDTD solver. The first call generates $\phi_{\text{inc},\,b}(\bar{r}_n,\bar{r}_{\text{T}_j},t)$ for all points within the scatterer. The second call propagates the field generated by the induced sources within the scatterer to the receiver location. The third call back-propagates the received fields at the receivers into the object, and correlates with the $\phi_{\text{inc},\,b}(\bar{r}_n,\bar{r}_{\text{T}_j},t)$ generated from the first call. Note that this procedure has to be repeated for every transmitter location. Hence, the CPU cost per iteration in the DBIM is proportional to $N_{\text{T}}T_{\text{FDTD}}$, where N_{T} is the number of transmitters and T_{FDTD} is the CPU time needed to solve the forward-scattering problem using FDTD. For non-resonant problems, $T_{\text{FDTD}} \propto N^{1.5}$ in two dimensions where N is the number of unknowns in the scattering problem [11, p. 241].

The storage requirement of the above algorithm is proportional to N since the storage requirement of FDTD is $O(N)$. The $\phi_{\text{inc},\,b}(\bar{r}_n,\bar{r}_{\text{T}_j},t)$ can be divided into different time bins, stored for each transmitter and time bin, and discarded after the correlation operation is performed. This requires no more than $O(N)$ storage.

12.5.5 The Conjugate Gradient Minimization Method

In the previous discussion, we first linearized the inverse-scattering problem and then solve the linearized system as an optimization problem. An alternative is to cast the entire problem initially in terms of nonlinear optimization by defining a cost functional, and then use the conjugate gradient method to perform the minimization of the cost functional.

In this case, we define the cost functional as

$$I = \left\| \vec{\phi}_{\text{scat}}(\vec{\varepsilon}) - \vec{\phi}_{\text{scat}}^{\text{meas}} \right\|_{\overline{\mathbf{C}}}^2 + \delta_t \left\| \delta\vec{\varepsilon} \right\|_{\overline{\mathbf{D}}}^2 \tag{12.82}$$

where $\phi_{\text{scat}}^{\text{meas}}$ is the measured field and $\phi_{\text{scat}}(\vec{\varepsilon})$ is a nonlinear functional of $\varepsilon(\bar{r})$, which is equal to $\varepsilon_b(\bar{r}) + \delta\vec{\varepsilon}$. Once the Fréchet derivative of the nonlinear functional, $\phi_{\text{scat}}(\vec{\varepsilon})$, is known, the Hessian of the cost function can be found. Both the steepest-descent method and the conjugate gradient method can be used to minimize the cost functional [47].

12.6 NOVEL INVERSE-SCATTERING METHODS

The inverse-scattering problem is highly nonlinear. Both BIM and DBIM might not lead to convergent solutions because the cost functional may have a number of local minima. Various methods have been proposed to extricate a solution that is stagnated in a local minimum (simulated annealing, genetic algorithms, etc.). These methods generally require a large number of forward solutions to the scattering problem, and hence, are hard to apply unless the scattering problem is small. In the following, we describe some novel techniques that may help to extricate a solution from a local minimum.

12.6.1 The Local-Shape-Function T-Matrix Method

A new inverse-scattering method known as the local-shape-function (LSF) method [48, 52] has been developed to invert strong metallic scatterers. This technique maps a scatterer with infinite conductivity into a problem with a scatterer described by a binary function that ranges between 0 and 1. By so doing, the extremely nonlinear problem of inverse scattering for a metallic object is mapped into a space where the nonlinearity is significantly reduced. Eventually, this method allows one to iteratively reconstruct metallic scatterers where the application of BIM or DBIM would converge extremely slowly or not at all. Later, this method was extended to the use of T-matrices to describe each individual scatterer. It was found that the LSF method can in general invert scatterers with a larger dielectric contrast [52, 57] than BIM or DBIM. This is especially true for TE-polarized waves or variable-density acoustic waves.

Although LSF theory has been applied to both CW and transient excitation, the LSF algorithm is more simply derived in the frequency domain. First, the scattering volume, V, is discretized by dividing it into N subvolumes, V_i, $i = 1, \ldots, N$. Each subvolume is viewed as an individual subscatterer.

The total field may be expanded in cylindrical harmonics and written as [11, p. 465]:

$$\phi(\bar{r}) = \mathcal{R}g\vec{\psi}^t(k_0,\bar{r})\cdot\vec{\mathrm{e}}_0 + \sum_{i=1}^{N}\vec{\psi}^t(k_0,\bar{r}_i)\cdot\vec{\mathrm{a}}_i \tag{12.83}$$

In the above, $\mathcal{R}g\vec{\psi}^t(k_0,\bar{r})$ is a vector that contains Bessel function cylindrical harmonics, and $\vec{\psi}^t(k_0,\bar{r}_i)$ is one that contains Hankel function cylindrical harmonics.

Enforcing multiple-scattering boundary conditions on each subscatterer, i, we have

$$\vec{\mathrm{a}}_i = \overline{\mathbf{T}}_i\cdot\left(\overline{\boldsymbol{\beta}}_{i0}\cdot\vec{\mathrm{e}}_0 + \sum_{\substack{j=1\\ j\neq i}}^{N}\overline{\boldsymbol{\alpha}}_{ij}\cdot\vec{\mathrm{a}}_j\right), \qquad i = 1, 2, \ldots, N \tag{12.84}$$

where $\overline{\mathbf{T}}_i$ is the T-matrix for the isolated subscatterer, i, and $\overline{\boldsymbol{\alpha}}_{ij}$ and $\overline{\boldsymbol{\beta}}_{i0}$ are translation matrices. Equation (12.84) is written as

$$\vec{\bar{\mathrm{a}}} = \left[\overline{\mathbf{D}}(\tilde{\gamma})\cdot\overline{\mathbf{A}} + \overline{\mathbf{I}}\right]^{-1}\cdot\overline{\mathbf{D}}(\tilde{\gamma})\cdot\overline{\mathbf{B}}\cdot\vec{\mathrm{e}}_0 \tag{12.85}$$

where $\overline{\mathbf{A}}$ and $\overline{\mathbf{B}}$ are known matrices and $\overline{\mathbf{D}}(\tilde{\gamma})$ converts the vector, $\tilde{\gamma}$, into a diagonal matrix. Here $\tilde{\gamma}$ is a vector containing the T-matrices of each individual subscatterer. These T-matrices can in turn be related to the physical parameters of the scatterers.

Equation (12.84) is important because a_i expresses the amplitude of the field emitted from the ith subvolume including multiple scattering from the other subvolumes, $V_j,\ j\neq i$. For inverse scattering, we linearize the problem such that

$$\phi_{\text{scat}}(\bar{r}) \approx \overline{\psi}^t\cdot\vec{\bar{\mathrm{a}}}_b + \overline{\mathbf{F}_{\mathbf{b}}}\cdot(\delta\vec{\tilde{\gamma}}) \tag{12.86}$$

where $\vec{\tilde{\gamma}} = \vec{\tilde{\gamma}}_b + \delta\vec{\tilde{\gamma}}$ and $\overline{\mathbf{F}_{\mathbf{b}}}$ is the Fréchet derivative or the gradient matrix of the scattered field at $\vec{\tilde{\gamma}} = \vec{\tilde{\gamma}}_b$. Using a perturbation expansion of (12.85), the Fréchet derivative is expressed as

$$\overline{\mathbf{F}_{\mathbf{b}}} = \overline{\psi}^t\cdot\left[\overline{\mathbf{D}}(\vec{\tilde{\gamma}}_b)\cdot\overline{\mathbf{A}} + \overline{\mathbf{I}}\right]^{-1}\cdot\overline{\mathbf{D}}\left\{\left[\overline{\mathbf{A}}\cdot\overline{\mathbf{D}}(\vec{\tilde{\gamma}}_b) + \overline{\mathbf{I}}\right]^{-1}\cdot\overline{\mathbf{B}}\cdot\vec{\mathrm{e}}_0\right\} \tag{12.87}$$

The above represents the gradient matrix as a cascade of one propagator that maps the incident field into the object space and a second propagator that maps the object field back out to the receivers. Hence, the N-dimensional gradient can be implemented with only two calls to a forward-scattering solver.

The LSF method has been applied to process ultrasonic data provided by Thermotrex [63]. For such data, the method converges better than BIM and DBIM [56, 57]. Fig. 12.8 shows the reconstruction of a typical ultrasonic scatterer using the LSF method on simulated data. Some experimental results using this technique for real data are described in Section 12.8.

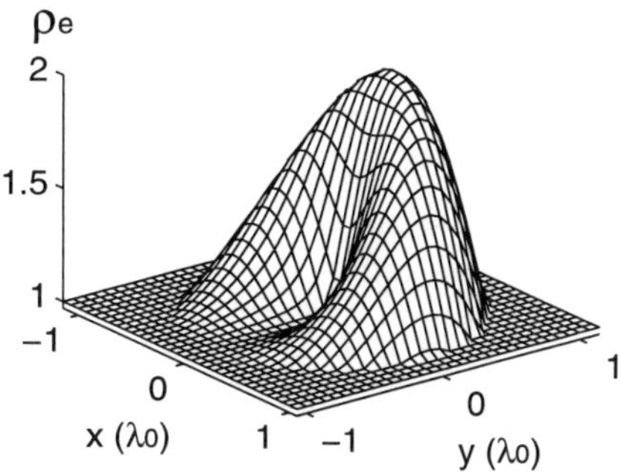

(a) original profile

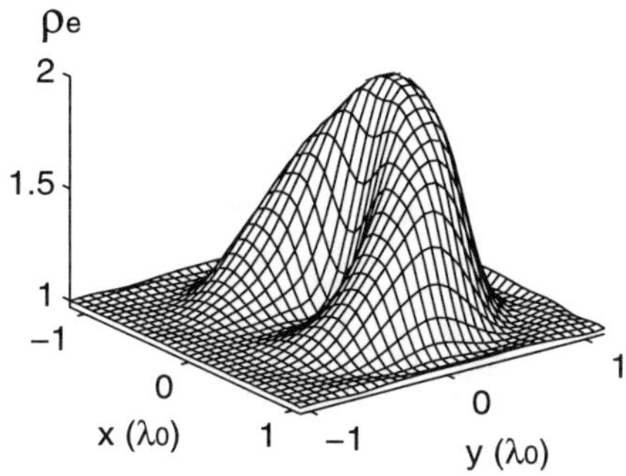

(b) reconstructed profile

Fig. 12.8 True and reconstructed density profiles obtained using the LSF method for an ultrasonic scatterer (no compressibility variation). Sixteen transmitters and 32 receivers are employed. The reconstructed area is $2.26\lambda_0 \times 2.26\lambda_0$ and is divided into a 32×32 mesh.

12.6.2 Frequency-Hopping Approach

Nonlinearity in inverse scattering is primarily due to multiple scattering of waves in the object to be reconstructed. The nonlinearity is exacerbated at higher frequencies for objects that are large compared to the wavelength. Due to this nonlinearity, inverse-scattering algorithms tend to stagnate in local minima for large problems. Recently, the use of frequency hopping has been reported as a means to mitigate such nonlinear effects [54, 55]. The idea is to collect measurement data over a number of frequencies starting with a frequency for which the wavelength is about the size of the scatterer. Then, the inverse-scattering problem is first solved at a lower frequency where the problem is more linear. The solution found at this lower frequency is then used as an initial guess to the solution at a higher frequency. By stepping over a range of gradually increasing frequencies, we can reconstruct an object spanning over 10λ. This is generally not possible if only high-frequency data are available. Figs. 12.9 and 12.10 show sample reconstructions using the frequency-hopping approach where an object as large as 10λ can be reconstructed.

12.7 OBSERVATION OF SUPERRESOLUTION

Superresolution, that is resolution beyond the Rayleigh criterion of 0.5λ, has been observed for the BIM and DBIM algorithms. When either the BIM or DBIM method is used to process the scattering data from two closely spaced point objects, the objects

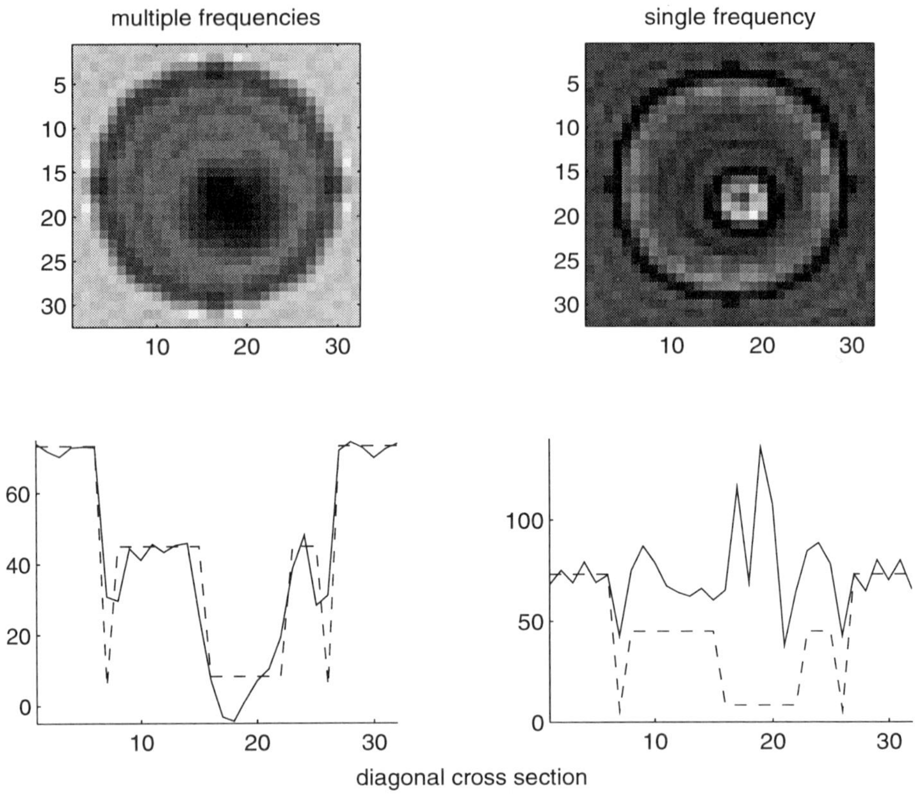

Fig. 12.9 Comparison of frequency-hopping and monochromatic reconstructions (image and profile of the real part of the permittivity). Left: frequency-hopping using data at 1, 2, and 3 GHz; Right: monochromatic using data at 3 GHz. The reconstructed area is $5.56\lambda_w \times 5.56\lambda_w$, where λ_w is the wavelength in water.

can be resolved when they are spaced by as little as 0.1λ apart [45, 47]. This is because these nonlinear reconstruction algorithms account for the multiple-scattering physics between the objects. The multiple-scattering also converts evanescent waves (where high-resolution information is hidden) into propagating waves that can reach the remotely located receivers. Hence, the field at the receivers contains superresolution information if an inversion algorithm accounts for multiple scattering. Fig. 12.11 depicts sample results for the horizontal resolution of two point objects placed 0.1λ apart.

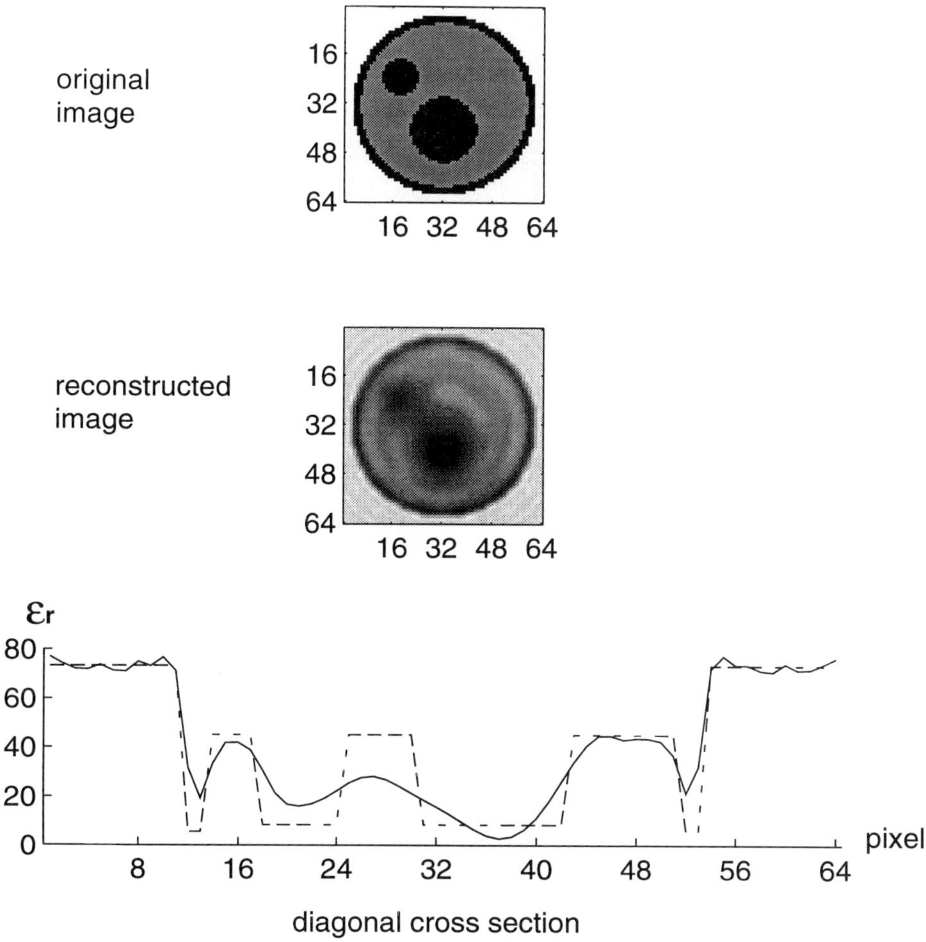

Fig. 12.10 Reconstruction of the real part of the permittivity for an object resembling a human arm using the frequency-hopping approach for data at 0.5, 0.9, 1.65, and 3 GHz. The reconstructed area is $10.2\,\lambda_w \times 10.2\,\lambda_w$.

12.8 WORK WITH EXPERIMENTAL DATA

All of the inverse-scattering methods described thus far (BIM, DBIM, and LSF) have been used to process real data. The CW implementation of DBIM was used to process microwave scattering data collected by a group in Barcelona, Spain [64]. Sample results

(a)

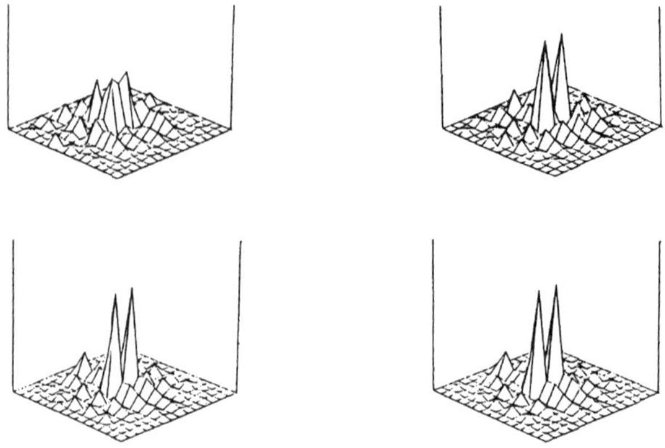

(b)

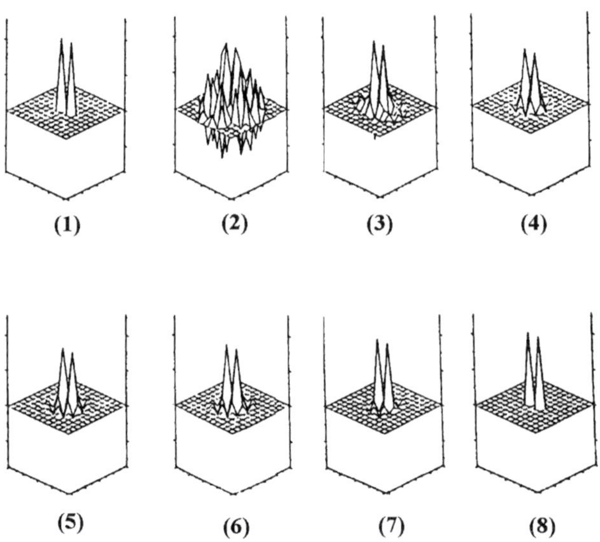

Fig. 12.11 (a) Horizontal resolution of two point objects placed 0.1 λ apart with contrast of 2:1. Results of the second, fourth, sixth, and eighth iterations using time-domain BIM. (b) Reconstruction of two single-pixel pulses separated by one pixel with the same contrast using the CW distorted BIM at 100 MHz where the diameter of the reconstructed area is 3.0 meters: (1) original distribution; (2) first-order approximation; (3) – (7) results of iterations 2 through 6; (8) final convergent solution after 20 iterations.

for the reconstructed microwave image of a human arm at 2.5 GHz are shown in Fig. 12.12. Previously, DT was used to process the data yielding results similar to Iteration 1 of Fig. 12.12(b). Due to high contrasts present in the human body, the reconstruction is not very good. Using DBIM, the processed image was significantly improved.

The DBIM was also used to process data from a subsurface impulse radar at the Advanced Construction Technology Center of the University of Illinois at Urbana-Champaign (in collaboration with the Electromagnetics Laboratory. This radar is used for nondestructive evaluation purposes to locate and identify cracks and voids in concrete structures, as well as metallic reinforcement bars (rebars) and utility pipes. The Electromagnetics Laboratory has also developed a step-frequency radar involving an array of Vivaldi antennas [50]. This apparatus is shown in Fig. 12.13.

Using DBIM, reconstructions of the step-frequency radar data were of a higher quality than that attainable previously using a method similar to DT involving straight back-propagation (see Fig. 12.14). It is currently believed that a new experimental system needs to be designed and built to achieve more sensitive measurements with more data diversity in order to realize the capability demonstrated using simulated data. A particular goal of this experimental system is the demonstration of the superresolution phenomenon at microwave frequencies.

The LSF T-matrix method was also demonstrated experimentally by processing ultrasonic experimental data from Thermotrex [63]. Fig. 12.15 shows the processing of ultrasonic data [56, 57]. Fig. 12.16 shows the processing of the Ipswich experimental data [58] using the LSF method.

12.9 THREE-DIMENSIONAL RECONSTRUCTION

The Electromagnetics Laboratory of the University of Illinois at Urbana-Champaign has obtained preliminary results of three-dimensional reconstruction using DBIM and the LSF T-matrix method. Fig. 12.17 shows a sample three-dimensional reconstruction using LSF. The results to date indicate that LSF performs better than DBIM [59, 60]. The reason is that LSF accounts for the effect of the dyadic Green's function singularity, whereas DBIM does not. The three-dimensional dyadic Green's function is inherently more singular than in two dimensions, and hence, makes the inverse problem more nonlinear. However, both of these inverse-scattering algorithms are computationally intensive, mandating that only simple objects can be reconstructed at this point. A key goal is to reduce the computational burden.

12.10 FAST FORWARD SOLVERS

A forward-scattering solver is required in all of the multidimensional, nonlinear inverse-scattering theories that can be applied to practical problems. Often, the forward solver is the computational bottleneck. Hence, if a high-speed forward solver is used, the inverse-scattering problem can be expedited. Recently, significant advances have been made in

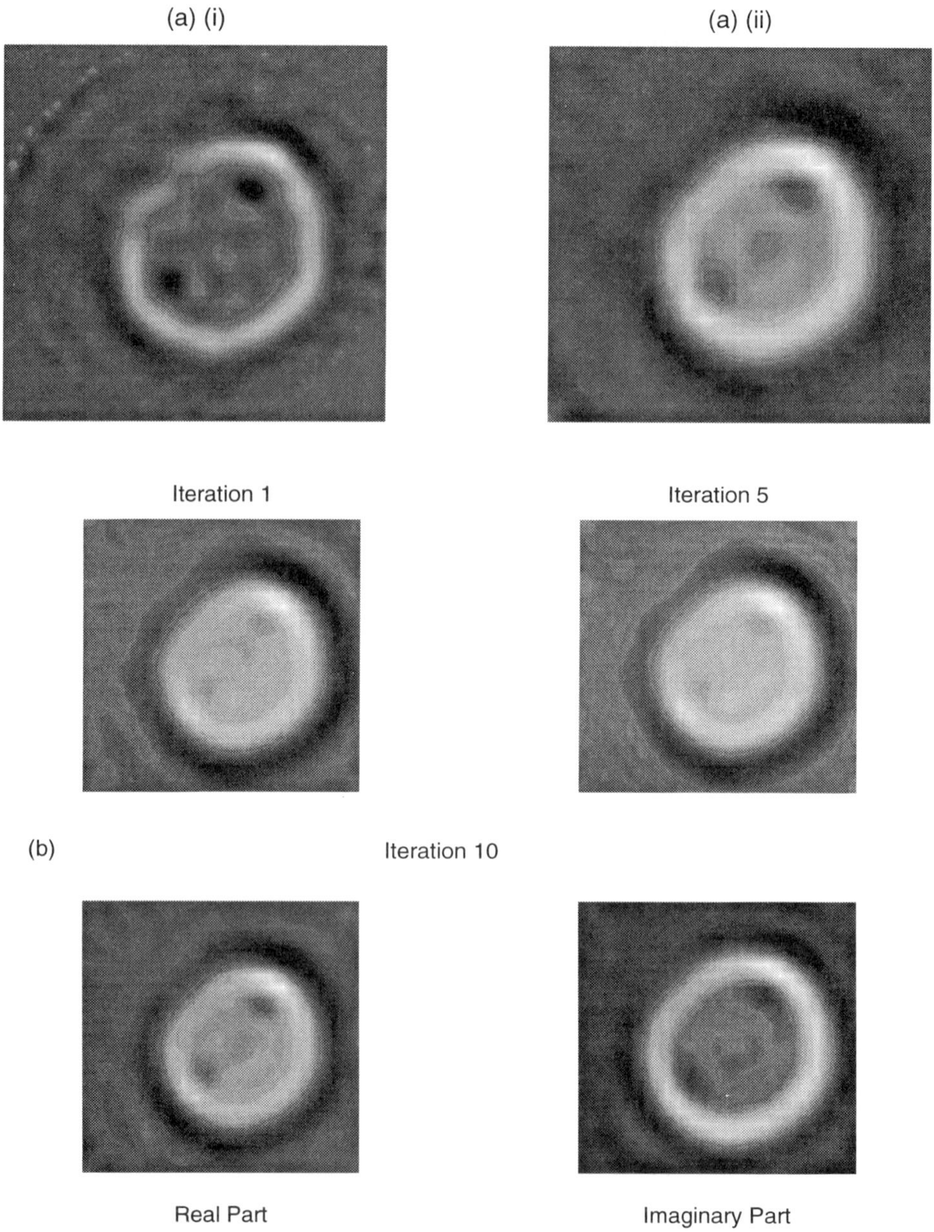

Fig. 12.12 (a) Reconstructed 2.5-GHz microwave image of human arm immersed in water. Sixty-four transmitters and 33 receivers are used with two forward solvers: (i) the fast recursive algorithm; (ii) CG-FFT. (b) Reconstruction of the image at different iterations of DBIM. The first iteration corresponds to a linear theory and resembles the solution from DT.

Fig. 12.13 Photograph of a broadband switched antenna array containing 11 identical Vivaldi antennas and two microwave switches enclosed in a polystyrene housing [50].

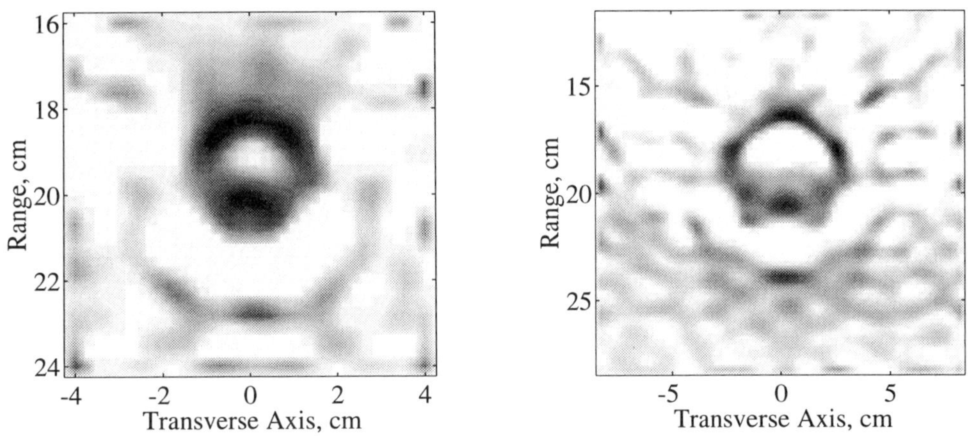

Fig. 12.14 DBIM permittivity reconstruction from microwave data for the case of a hollow PVC pipe in air. Left: 2.7-cm pipe diameter; right: 4.8-cm pipe diameter [50].

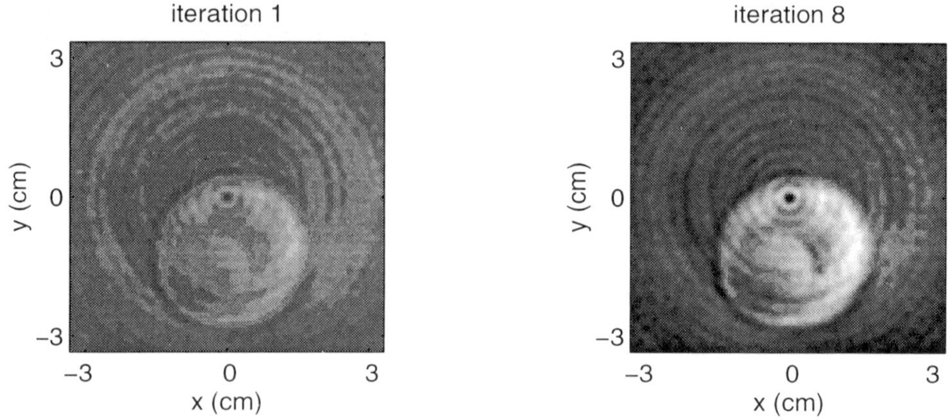

(a) Images of the real part of the modulus.

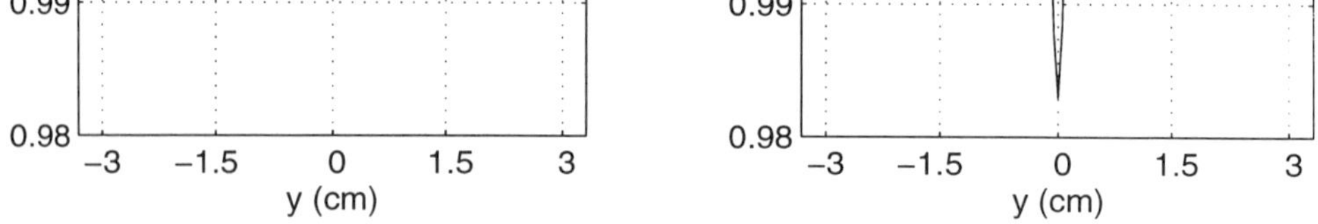

(b) y-axis cross sections of the modulus.

Fig. 12.15 Comparison of the dual-balloon reconstructions for iteration 1 and iteration 8. The frequency is 312.5 kHz and $\lambda_0 = 4.8$ mm. The reconstruction area is $14\lambda_0 \times 14\lambda_0$ and is divided into 127×127 cells. 128×102 transmitter-receiver paired data are used.

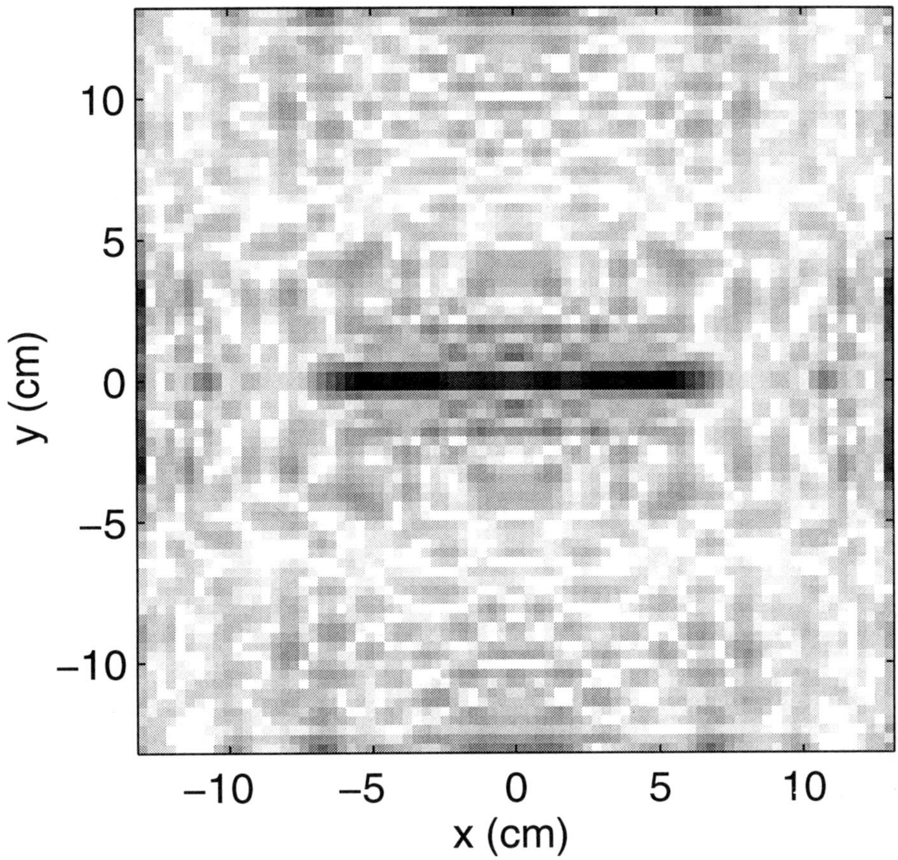

Fig. 12.16 Reconstruction of a PEC strip. The approximate width measured from the image is between 11.5 and 12.0 cm. (The exact size of the strip is not known.) The image region is a square that consists of 80×80 cells.

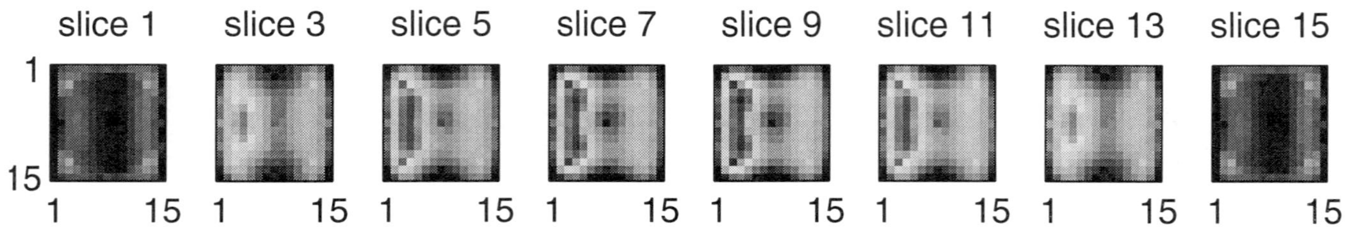

(a)

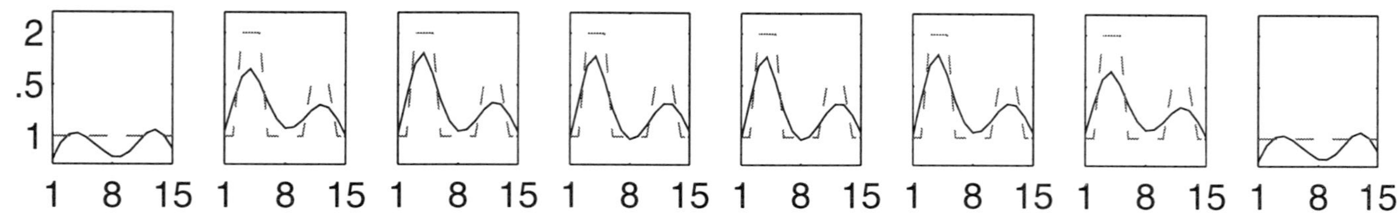

(b)

Fig. 12.17 Reconstruction of a 0.6λ cube divided into $15 \times 15 \times 15$ cells with 6 plane-wave illuminations and 258 receivers: (a) slice images along the z axis; (b) cross-section relative-permittivity profiles of (a) where the solid line is the reconstructed profile and the dashed line is the true profile.

the development of fast algorithms for forward scattering [65–82]. One such development is an electrodynamic version of a multilevel, fast multipole algorithm that has been shown to be efficient relative to both computer time and memory. Problems having over two million unknowns arising from a surface integral equation (with an equivalent dense matrix system) have been solved using the SGI Power Challenge Array.

In the FDTD area, much progress has been made in the area of PML ABCs [83]. (See Chapter 5.) Such absorbing boundaries allow the easier parallelization of the FDTD code involved [84], and may be helpful to the efficient solution of the inverse-scattering problem.

12.11 SUMMARY AND CONCLUSIONS

This chapter reviewed various methods to obtain images of objects, some of which are intimately related to the use of electromagnetic waves. Topics included:

- Issues arising in inverse-scattering theories;
- Linear inverse-problem techniques, including back-projection tomography, Radon transforms, and diffraction tomography;
- Nonlinear inverse-scattering methods, including the Born iterative method, the distorted Born iterative method, and the Fréchet derivative;
- Numerical solution to the inverse problem, including discretization of the integral equation, minimization of a cost functional, back-propagation and correlation, FDTD implementation of the inversion algorithm, and the conjugate gradient minimization method;
- Novel inverse-scattering methods, including the local-shape-function T-matrix method and the frequency-hopping approach;
- Observation of superresolution;
- Work with experimental data;
- Three-dimensional reconstruction example;
- Impact of fast forward solvers.

Linear inverse problems can be solved in $O(N \log N)$ operations, allowing large reconstructions. However, nonlinear inverse problems have to be solved iteratively. These require $O(N_t N_{iter} N^{\alpha})$ operations, where N_t is the number of transmitters involved, N_{iter} is the number of iterations needed, and N^{α} is the number of operations needed to solve a forward-scattering problem. For two-dimensional FDTD problems, $\alpha = 1.5$, while for three-dimensional problems, $\alpha = 1.33$. New technologies to rapidly solve the forward-scattering problem are desirable to expedite the solution to the inverse problem.

REFERENCES

[1] Raviv, J, J. F. Greenleaf, and G. T. Herman, eds., *Computer Aided Tomography and Ultrasonics in Medicine*, Elsevier North-Holland, 1979.

[2] Kak, A. C., "Computerized tomography with X-ray, emission, and ultrasound sources," *Proc. IEEE*, Vol. 67, 1979, pp. 1245–1272.

[3] Krestel, E., ed., *Imaging Systems for Medical Diagnosis: Fundamentals and Technical Solutions, X-Ray Diagnostics, Computed Tomography, Nuclear Medical Diagnostics, Magnetic Resonance Imaging, and Ultrasound Technology*, New York: Wiley, 1990.

[4] Lee, H. and G. Wade, *Modern Acoustic Imaging*, New York: IEEE Press, 1986.

[5] Devaney, A. J., "A filtered back-propagation algorithm for diffraction tomography," *Ultrasonic Imaging*, Vol. 4, 1982, pp. 336–360.

[6] Maser, K. R., *Ground Penetrating Radar Surveys to Characterize Pavement Layer Thickness Variations at GPR Sites*, Strategic Highway Research Program, 1994.

[7] Curlander, J. C., R. N. McDonough, *Synthetic Aperture Radar: Systems and Signal Processing*, New York: Wiley, 1991.

[8] Chew, W. C. and Q. H. Liu, "Inversion of induction tool measurements using the distorted Born iterative method and CG-FFT," *IEEE Trans. Geosci. Remote Sensing*, Vol. 32, 1994, pp. 878–884.

[9] Chew, W. C., G. P. Otto, W. H. Weedon, J. H. Lin, C. C. Lu, Y. M. Wang, and M. Moghaddam, "Nonlinear diffraction tomography- the use of inverse scattering for imaging," *Int. J. Imaging Sys. Tech.*, Vol. 7, 1996, pp. 16–24.

[10] Bube, K. P. and R. Burridge, "The one-dimensional problem of reflection seismology," *SIAM Review*, Vol. 25, 1983, pp. 497–559.

[11] Chew, W. C., *Waves and Fields in Inhomogeneous Media*, New York: Van Nostrand Reinhold, 1990.

[12] Weedon, W. H. and W. C. Chew, "Broadband microwave inverse scattering for nondestructive evaluation (NDE)," *Proc. Quantitative Nondestructive Evaluation Conf.*, Bowdoin College, Maine, 1993.

[13] Taflove, A. and M. E. Brodwin, "Numerical solution of steady-state electromagnetic scattering problems using the time-dependent Maxwell's equations," *IEEE Trans. Microwave Theory Tech.*, Vol. MTT-23, 1975, pp. 623–630.

[14] Taflove, A., *Computational Electrodynamics: The Finite-Difference Time-Domain Method*, Norwood, MA: Artech House, 1995.

[15] Lauterbur, P. C., "Medical imaging by nuclear magnetic resonance zeugmatography," *IEEE Trans. Nuclear Sci.*, Vol. 27, 1973, pp. 2808–2811.

[16] Herman, G. T., *Image Reconstruction from Projections*, New York: Academic Press, 1980.

[17] Born, M. and E. Wolf, *Principles of Optics*, New York: Pergamon Press, 1st ed., 1959, 6th ed., 1980.

[18] Slaney, M., A. C. Kak, and L. E. Larsen, "Limitations of imaging with first-order diffraction tomography," *IEEE Trans. Microwave Theory Tech.*, Vol. MTT-32, 1984, pp. 860–874.

[19] Radon, J., "Uber die bestimmung von funktionen durch ihre intergralwerte langs gewisser mannigfaltigkaiten," *Berichte Saechsische Akademie der Wissenschaften*, Vol. 69, 1917, pp. 262–277.

[20] Bracewell, R. N., "Strip integration in radio astronomy," *Aust. J. Phys.*, Vol. 9, 1956, pp. 198–217.

[21] Deans, S. R., *The Radon Transform and Some of Its Applications*, New York: Wiley, 1983.

[22] Lin, F. C. and M. A. Fiddy, "Image estimation from scattered-field data," *Int. J. Imag. Sys. Tech.*, Vol. 2, 1990, pp. 76–95.

[23] Baysal, E., D. D. Kosloff, and J. W. C. Sherwood, "Reverse time migration," *Geophysics*, Vol. 48, 1983, pp. 1514–1524.

[24] Berryman, J. G., "Weighted least-square criteria for seismic travel time tomography," *IEEE Trans. Geosci. Remote Sensing*, Vol. GE-27, 1989, pp. 302–309.

[25] Rose, J. H., M. Cheney, and B. DeFacio, "The connection between time- and frequency-domain three-dimensional inverse-scattering methods," *J. Math. Phys.*, Vol. 25, 1984, pp. 2995–3000.

[26] Colton, D. and P. Monk, "The inverse-scattering problem for time-harmonic acoustic waves in an inhomogeneous medium," *Quart. J. Mech. Appl. Math.*, Vol. 41, 1988, pp. 97–125.

[27] Angell, T. S., R. E. Kleinman, and G. F. Roach, "An inverse-transmission problem for the Helmholtz equation," *Inverse Problems*, Vol. 3, 1987, pp. 149–180.

[28] Habashy, T. M. and R. Mittra, "On some inverse methods in electromagnetics," *J. Electromag. Waves Appl.*, Vol. 1, 1987, pp. 25–28.

[29] Bolomey, J.-C., C. Durix, and D. Lesselier, "Determination of conductivity profiles by time-domain reflectometry," *IEEE Trans. Antennas Propagat.*, Vol. AP-27, 1979, pp. 244–248.

[30] Johnson, S. A. and M. L. Tracy, "Inverse-scattering solutions by a sinc-basis, multiple-source, moment method - Part I: Theory," *Ultrasonic Imaging*, Vol. 5, 1983, pp. 361–375.

[31] Johnson, S. A. and M. L. Tracy, "Inverse-scattering solutions by a sinc-basis, multiple-source, moment method - Part II: Numerical evaluations," *Ultrasonic Imaging*, Vol. 5, 1983, pp. 376–392.

[32] Cavicchi, T. J. and W. D. O'Brien, "Numerical study of high-order diffraction tomography via the sinc-basis moment method," *Ultrasonic Imaging*, Vol. 11, 1989, pp. 42–74.

[33] Kleinman, R. E. and P. M. van den Berg, "Nonlinearized approach to profile inversion," *Int. J. Imag. Sys. Tech.*, Vol. 2, 1990, pp. 119–126.

[34] Bolomey, J.-C. and C. Pichot, "Microwave tomography: from theory to practical imaging systems," *Int. J. Imag. Sys. Tech.*, Vol. 2, 1990, pp. 144–156.

[35] Joachimowicz, N., C. Pichot, and J.-P. Hugonin, "Inverse scattering: An iterative numerical method for electromagnetic imaging," *IEEE Trans. Antennas Propagat.*, Vol. AP-39, 1991, pp. 1742–1752.

[36] Tarantola, A., "The seismic reflection inverse problem," in *Inverse Problems of Acoustic and Elastic Waves*, F. Santosa, Y. H. Pao, W. Symes, and C. Holland, eds., New York: Soc. Industr. Appl. Math., 1984.

[37] Tarantola, A., *Inverse Problem Theory*, New York: Elsevier, 1987.

[38] Crase, E., A. Pica, M. Noble, J. McDonald, and A. Tarantola, "Robust nonlinear waveform inversion: application to real data," *Geophysics*, Vol. 55, 1990, pp. 527–538.

[39] Mora, P., "Nonlinear two-dimensional elastic inversion of multi-offset seismic data," *Geophysics*, Vol. 52, 1987, pp. 1211–1228.

[40] Chew, W. C. and S. L. Chuang, "Profile inversion of a planar medium with a line source or a point source," presented at *IEEE Int. Geosci. Remote Sensing Soc. Symp.*, Strasbourg, France, 1984.

[41] Habashy, T. M., W. C. Chew, and E. Y. Chow, "Simultaneous reconstruction of permittivity and conductivity profiles in a radially inhomogeneous slab," *Radio Sci.*, Vol. 21, 1986, pp. 635–645.

[42] Wang, Y. M. and W. C. Chew, "An iterative solution of the two-dimensional electromagnetic inverse-scattering problem," *Int. J. Imag. Sys. Tech.*, Vol. 1, 1989, pp. 100–108.

[43] Chew, W. C. and Y. M. Wang, "Reconstruction of two-dimensional permittivity distribution using the distorted Born iterative method," *IEEE Trans. Medical Imag.*, Vol. 9, 1990, pp. 218–225.

[44] Wang, Y. M. and W. C. Chew, "Limited-angle inverse-scattering problems and their applications for geophysical explorations," *Int. J. Imag. Sys. Tech.*, Vol. 2, 1990, pp. 96–111.

[45] Moghaddam, M. and W. C. Chew, "Nonlinear two-dimensional velocity profile inversion in the time domain," *IEEE Trans. Geosci. Remote Sensing*, Vol. GE-30, 1992, pp. 147–156.

[46] Wang, Y. M. and W. C. Chew, "Accelerating the iterative inverse-scattering algorithms by using the fast recursive operator algorithm," *Radio Sci.*, Vol. 27, 1992.

[47] Moghaddam, M., W. C. Chew, and M. Oristaglio, "Comparison of the Born iterative method and Tarantola's method for an electromagnetic time-domain inverse problem," *Int. J. Imag. Sys. Tech.*, Vol. 3, 1991, pp. 318–333.

[48] Chew, W. C. and G. P. Otto, "Microwave imaging of multiple metallic cylinders using shape functions," *IEEE Microwave Guided Wave Lett.*, Vol. 2, 1992, pp. 284–286.

[49] Otto, G. P. and W. C. Chew, "Inverse scattering of H_z waves using local shape function imaging: A T-matrix formulation," *Int. J. Imag. Sys. Tech.*, Vol. 5, 1994, pp. 22–27.

[50] Weedon, W. H. and W. C. Chew, "Time-domain inverse scattering using the local shape function (LSF) method," *J. Inverse Problems*, Vol. 9, 1993, pp. 551–564.

[51] Moghaddam, M. and W. C. Chew, "Variable-density linear acoustic inverse problem," *Ultrasonic Imaging*, Vol. 15, 1993, pp. 255–266.

[52] Otto, G. P. and W. C. Chew, "Time-harmonic impedance tomography using the T-matrix method," *IEEE Trans. Med. Imag.*, Vol. 13, 1994, pp. 508–516.

[53] Chew, W. C., Y. M. Wang, G. Otto, D. Lesselier, and J.-C. Bolomey, "On inverse-source method of solving inverse-scattering problems," *J. Inverse Problems*, Vol. 10, 1994, pp. 547–553.

[54] Borup, D. T., S. A. Johnson, W. W. Kim, and M. J. Berggren, "Nonperturbative diffraction tomography via Gauss-Newton iteration applied to the scattering integral equation," *Ultrasonic Imaging*, Vol. 14, 1992, pp. 69–85.

[55] Chew, W. C. and J. H. Lin, "A frequency-hopping approach for microwave imaging of large inhomogeneous bodies," *IEEE Microwave Guided Wave Lett.*, Vol. 5, 1995, pp. 439–441.

[56] Lu, C. C., J. H. Lin, W. C. Chew and G. P. Otto, "Image reconstruction with acoustic measurement using distorted Born iterative method," *Ultrasonic Imaging*, Vol. 18, 1996. pp. 140–156.

[57] Lin, J. H. and W. C. Chew, "Ultrasonic imaging by local shape function method with CGFFT," *IEEE Trans. Ultrasonics, Ferroelectrics, and Frequency Control*, Vol. 43, 1996, pp. 956–969.

[58] Lu, C. C. and W. C. Chew, "Processing Ipswich data with the local shape function method," *IEEE Antennas and Propagation Magazine*, Vol. 38, no. 3, 1996, pp. 51–53.

[59] Gan, H. and W. C. Chew, "3-D inhomogeneous inversion for microwave imaging using distorted born iterative method and BCG-FFT," *Proc. Prog. In Electromag. Research Symp.*, July 1995, University of Washington, Seattle, p. 611.

[60] Lin, J. H. and W. C. Chew, "Three-dimensional electromagnetic inverse problem by local shape function method with CGFFT," *J. Opt. Soc. Am.*, accepted for publication.

[61] Tikhonov, A. N., "On the problems with approximately specified information," in *Ill-Posed Problems in the Natural Sciences*, A. N. Tikhonov and A. V. Goncharsky, eds., Moscow: MIR Publishers, 1987.

[62] Franchois, A. and C. Pichot, "Microwave imaging - complex permittivity reconstruction with a Levenberg-Marquardt method," *IEEE Trans. Antennas Propagat.*, Vol. AP-45, 1997, pp. 203–215.

[63] Andre, M. P., P. J. Martin, G. P. Otto, L. K. Olson, T. K. Barrett, B. A. Spivey, and D. A. Palmer, "A new consideration of diffraction computed tomography for breast imaging: Studies in phantoms and patients," *Acoust. Imag.*, Vol. 21, 1995, pp. 379–390.

[64] Broquetas, A., J. Romeu, J. M. Rius, A. R. Elias-Fuste, A. Cardama, and L. Jofre, "Cylindrical geometry: a further step in active microwave tomography," *IEEE Trans. Microwave Theory Tech.*, Vol. MTT-39, 1991, pp. 836–844.

[65] Chew, W. C., "An N^2 algorithm for the multiple-scattering solution of N scatterers," *Microwave and Optical Tech. Lett.*, Vol. 2, 1989, pp. 380–383.

[66] Wang, Y. M. and W. C. Chew, "An efficient algorithm for solution of a scattering problem," *Microwave and Optical Tech. Lett.*, Vol. 3, 1990, pp. 102–106.

[67] Chew, W. C. and Y. M. Wang, "A fast algorithm for solution of a scattering problem using a recursive aggregate tau matrix method," *Microwave and Optical Tech. Lett.*, Vol. 3, 1990, pp. 164–169.

[68] Wang, Y. M. and W. C. Chew, "Solution of large 2-D inhomogeneous scatterers for TM waves by fast recursive algorithm," *Microwave and Optical Tech. Lett.*, Vol. 4, no. 3, 1991.

[69] Chew, W. C., L. Gurel, Y. M. Wang, G. Otto, R. Wagner, and Q. H. Liu, "A generalized recursive algorithm for wave-scattering solutions in two dimensions," *IEEE Trans. Microwave Theory Tech.*, Vol. MTT- 40, 1992, pp. 716–723.

[70] Chew, W. C. and C. C. Lu, "The use of Huygen's equivalence principle for solving the volume integral equation of scattering," *IEEE Trans. Antennas Propagat.*, Vol. AP-41, 1993, pp. 897–904.

[71] Lu, C. C. and W. C. Chew, "A fast algorithm for solving hybrid integral equation," *IEE Proc. – H*, Vol. 140, 1993, pp. 455–460.

[72] Chew, W. C., "Fast algorithms for wave scattering developed at the Electromagnetics Laboratory, University of Illinois," *IEEE Antennas and Propagat. Magazine*, Vol. 35, no. 4, 1993, pp. 22–32.

[73] Liu, Q. H. and W. C. Chew, "Applications of the CG-FFHT method with an improved FHT algorithm," *Radio Science*, Vol. 29, 1994, pp. 1009–1022.

[74] Chew, W. C., C. C. Lu and Y. M. Wang, "Efficient computation of three-dimensional scattering of vector electromagnetic waves," *J. Opt. Soc. Am. A*, Vol. 11, 1994, pp. 1528–1537.

[75] Lu, C. C. and W. C. Chew, "A multilevel algorithm for solving boundary integral equation of scattering," *Microwave and Optical Tech. Lett.*, Vol. 7, 1994, pp. 466–470.

[76] Michielssen, E. and A. Boag, "Multilevel evaluation of electromagnetic fields for the rapid solution of scattering problems," *Microwave and Optical Tech. Lett.*, Vol. 7, 1994, pp. 790–795.

[77] Wagner, R. L. and W. C. Chew, "A ray-propagation fast multipole algorithm," *Microwave and Optical Tech. Lett.*, Vol. 7, 1994, pp. 435–438.

[78] Song, J. M. and W. C. Chew, "Fast multipole method solution using parametric geometry," *Microwave and Optical Tech. Lett.*, Vol. 7, 1994, pp. 760–765.

[79] Lu, C. C. and W. C. Chew, "The use of Huygen's equivalence principle for solving 3-D volume integral equation of scattering," *IEEE Trans. Antennas Propagat.*, Vol. AP-43, 1995, pp. 500–507.

[80] Lu, C. C. and W. C. Chew, "Fast far-field approximation for calculating the RCS of large objects," *Microwave and Optical Tech. Lett.*, Vol. 8, 1995, pp. 238–241.

[81] Song, J. M. and W. C. Chew, "Multilevel fast multipole algorithm for solving combined-field integral equation of electromagnetic scattering," *Microwave and Optical Tech. Lett.*, Vol. 10, 1995, pp. 14–19.

[82] Chew, W. C., J. M. Jin, C. C. Lu, E. Michielssen, and J. M. Song, "Fast solution methods in electromagnetics," *IEEE Trans. Antennas Propagat.*, Vol. 45, 1997, pp. 533–543.

[83] Berenger, J., "A perfectly matched layer for the absorption of electromagnetic waves," *J. Comput. Phys.*, Vol. 114, 1994, pp. 185–200.

[84] Chew, W. C. and W. Weedon, "A 3-D perfectly matched medium from modified Maxwell's equations with stretched coordinates," *Microwave and Optical Tech. Lett.*, Vol. 7, 1994, pp. 599–604.

[85] Tatarski, V. I., *Wave Propagation in a Turbulent Medium*, New York: McGraw-Hill, 1961.

[86] Keller, J. B., "Accuracy and validity of the Born and Rytov approximations," *J. Opt. Soc. Am.*, Vol. 59, 1969, pp. 1003–1004.

[87] Crane, R. K., "Spectra of ionospheric scintillation," *J. Geophys. Res.*, Vol. 81, 1976, pp. 2041–2050.

APPENDIX 12A: THE BORN AND RYTOV APPROXIMATIONS

For many inverse-scattering problems, especially when the scattering from the inhomogeneity is weak, it suffices to derive approximate solutions. We discuss two such approaches, the Born approximation [17, p. 453], which works better at low frequencies, and the Rytov approximation [85], which works better at higher frequencies. We first derive the volume integral equation for scattering, referring to Fig. 12A.1.

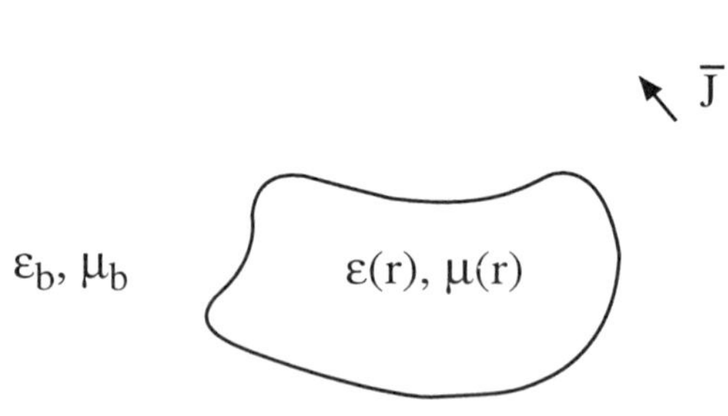

Fig. 12A.1 A current source radiating in the vicinity of a general inhomogeneity.

12A.1 Derivation of Volume Integral Equation

Consider the scalar wave equation

$$\left[\nabla^2 + k^2(\bar{r})\right]\phi(\bar{r}) = q(\bar{r}) \tag{12A.1}$$

where $k^2(\bar{r}) = \omega^2 \mu(\bar{r})\varepsilon(\bar{r})$ represents an inhomogeneous medium over a finite domain, V, and $k^2 = k_b^2 = \omega^2 \mu_b \varepsilon_b$ outside V. Next, we define a Green's function satisfying

$$\left[\nabla^2 + k_b^2\right] g(\bar{r}, \bar{r}') = -\delta(\bar{r} - \bar{r}') \tag{12A.2}$$

Then, (12A.1) is rewritten as

$$\left[\nabla^2 + k_b^2\right]\phi(\bar{r}) = q(\bar{r}) - \left[k^2(\bar{r}) - k_b^2\right]\phi(\bar{r}) \tag{12A.3}$$

Note that the right-hand side of (12A.3) can be considered an equivalent source. Since the Green's function corresponding to the differential operator on the left-hand side of (12A.3) is known, by the principle of linear superposition we write

$$\phi(\bar{r}) = -\int_{V_s} d^3\bar{r}'\, g(\bar{r}, \bar{r}')\, q(\bar{r}') + \int_{V} d^3\bar{r}'\, g(\bar{r}, \bar{r}')\left[k^2(\bar{r}') - k_b^2\right]\phi(\bar{r}') \tag{12A.4}$$

The first term on the right-hand side is just the field due to the source in the absence of the inhomogeneity, and hence, is the incident field. Therefore, (12A.4) becomes

$$\phi(\bar{r}) = \phi_{\text{inc}}(\bar{r}) + \int_{V} d^3\bar{r}'\, g(\bar{r}, \bar{r}')\left[k^2(\bar{r}') - k_b^2\right]\phi(\bar{r}') \tag{12A.5}$$

In the above equation, if the total field, $\phi(\bar{r}')$, inside V is known, then $\phi(\bar{r})$ can be calculated everywhere. However, $\phi(\bar{r})$ is unknown at this stage. To solve for $\phi(\bar{r})$, an integral equation has to be formulated. To this end, we impose (12A.5) for $\bar{r}$ in V. Then, $\phi(\bar{r})$ on the left-hand side and on the right-hand side are the same unknown defined over the same domain. Consequently, (12A.5) becomes the desired integral equation

$$\phi_{\text{inc}}(\bar{r}) = \phi(\bar{r}) - \int_{V} d^3\bar{r}'\, g(\bar{r}, \bar{r}')\left[k^2(\bar{r}') - k_b^2\right]\phi(\bar{r}'), \qquad \bar{r} \in V \tag{12A.6}$$

In the above, the unknown, $\phi(\bar{r})$, is defined over a volume, V, over which the integration is performed, and hence the name, volume integral equation.

12A.2 Born Approximation

In the cases when $k^2 - k_b^2$ is small, or where the contrast of the scatterer is weak so that the second term on the right of (12A.5) is small compared to the first term, we approximate

$$\phi(\bar{r}) \cong \phi_{\text{inc}}(\bar{r}) \tag{12A.7}$$

Then, the total field in (12A.5) is approximately calculated as

$$\phi(\bar{r}) = \phi_{\text{inc}}(\bar{r}) + \int_V d^3\bar{r}'\, g(\bar{r}, \bar{r}')\, O(\bar{r}')\, \phi_{\text{inc}}(\bar{r}') \tag{12A.8}$$

where $O(\bar{r}') = k^2 - k_b^2$. The above is known as the first-order Born approximation. It is also the first-order approximation in the Neumann series expansion of the integral equation in (12A.6), or the Taylor series approximation of $\phi(\bar{r})$ using $(k^2 - k_b^2)$ as a small parameter.

Since the Born approximation is good only when the second term is much smaller than the first term in (12A.8), we can establish its regime of validity. (The regime of validity of the Born approximation has also been discussed by Keller [86].) First, notice that for the homogeneous background case,

$$g(\bar{r}, \bar{r}') = \frac{e^{jk_b|\bar{r} - \bar{r}'|}}{4\pi|\bar{r} - \bar{r}'|} \tag{12A.9}$$

If the size of the scatterer is of the order, L, and $k_b L << 1$, then by dimensional analysis we have

$$g(\bar{r}, \bar{r}') \sim \frac{1}{L} \tag{12A.10a}$$

Then,

$$O(\bar{r}) = (k^2 - k_b^2) \sim k_b^2 \Delta\varepsilon_r \tag{12A.10b}$$

where $\Delta\varepsilon_r = \varepsilon/\varepsilon_b - 1$ and

$$\int d^3\bar{r}' \sim L^3 \tag{12A.10c}$$

Therefore, we see that the second term in (12A.8) is of the order of

$$(k_b L)^2 \Delta\varepsilon_r \phi_{\text{inc}} \tag{12A.11}$$

However, since $k_b L << 1$, in order for the scattered field to be much smaller than the incident field, the constraint is

$$k_b^2 \, L^2 \, \Delta\varepsilon_r \;<<\; 1 \tag{12A.12}$$

Moreover, in the long-wavelength limit, $k_b L << 1$, and this constraint could be met even when $\Delta\varepsilon_r > 1$. Hence, in the scalar wave case, the Born approximation becomes exceedingly good at low frequencies. This is not true for the vector electromagnetic field.

However, in the short-wavelength limit, the phase of the wave as it propagates inside the object becomes important. For example, if $\phi_{\text{inc}} \sim \exp(j\,\vec{k}_b \cdot \bar{r})$, then the total field inside a tenuous inhomogeneity consists of a linear superposition of plane waves of the type $\phi \sim \exp(j\,\vec{k} \cdot \bar{r})$ (which is motivated by the WKB-type approximation) when $k \to \infty$. (Strictly speaking, it should be $\phi \sim \exp(j \int k\,dz')$, but this distinction is not important for this order-of-magnitude argument.) Therefore, we write

$$\begin{aligned} \phi \;&\sim\; \exp\left(j\vec{k}_b \cdot \bar{r}\right) \exp\left[j(\vec{k} - \vec{k}_b) \cdot \bar{r}\right] \\ &\sim\; \phi_{\text{inc}} \exp\left[j(\vec{k} - \vec{k}_b) \cdot \bar{r}\right] \end{aligned} \tag{12A.13}$$

Hence, $\phi \cong \phi_{\text{inc}}$ only if $(k - k_b)L << 1$. Consequently, at high frequencies, the Born approximation is valid only if

$$k_b \, L \Delta\varepsilon_r \;<<\; 1, \qquad k_b \, L \to \infty \tag{12A.14}$$

Note further that the above is a much more stringent restriction than (12A.12).

The Born approximation assumes single scattering. Moreover, note that in (12A.8), the incident wave enters the scatterer with no distortion, induces a polarization current proportional to $(k^2 - k_b^2)\,\phi_{\text{inc}}$, and causes a reradiation or scattering. Since the incident field is unaffected while it gives rise to a scattered field, the Born approximation violates energy conservation. However, because of the symmetry of the dyadic Green's function, reciprocity is still preserved.

Finally, it is important to note that the above constraints are for a three-dimensional space. In a one- or two-dimensional space, they have to be rederived.

12A.3 Rytov Approximation

We have seen that the polarization-charge effect is unimportant at high frequencies when the wavelength is much smaller than the size of the inhomogeneity. If this is the case, the study of the vector electromagnetic wave equation can be reduced to the study of the scalar wave equation. Therefore, the pertinent equation is

$$\left[\nabla^2 + k^2(\bar{r})\right] \phi(\bar{r}) \;=\; 0 \tag{12A.15}$$

To derive the Rytov approximation, we first let

$$\phi(\bar{r}) = \exp[j\,\psi(\bar{r})] \tag{12A.16}$$

Then,

$$\nabla\phi(\bar{r}) = j\,\phi(\bar{r})\,\nabla\psi(\bar{r}) \tag{12A.17a}$$

$$\nabla\cdot\nabla\phi(\bar{r}) = \left\{j\nabla^2\psi(\bar{r}) - [\nabla\psi(\bar{r})]^2\right\}\phi(\bar{r}) \tag{12A.17b}$$

Using the above in (12A.15), we have

$$j\nabla^2\psi(\bar{r}) - (\nabla\psi)^2 + k^2(\bar{r}) = 0 \tag{12A.18}$$

At this point, (12A.18) is still exact but nonlinear. However, we can solve (12A.18) perturbatively by letting

$$\psi(\bar{r}) \sim \psi_0(\bar{r}) + \psi_1(\bar{r}) \tag{12A.19}$$

Here, $\psi_0(\bar{r})$ is assumed to satisfy the equation

$$j\nabla^2\psi_0(\bar{r}) - (\nabla\psi_0)^2 + k_b^2(\bar{r}) = 0 \tag{12A.20}$$

that is, it is the solution in some background medium with wave number, k_b. After substituting (12A.19) into (12A.18), we have consequently

$$j\nabla^2\psi_1(\bar{r}) - 2(\nabla\psi_0)\cdot(\nabla\psi_1) - (\nabla\psi_1)^2 + O(\bar{r}) = 0 \tag{12A.21}$$

where $O(\bar{r}) = k^2 - k_b^2$. At this point, the above equation is nonlinear, but it could be simplified by using the identity that

$$\nabla^2(\phi_0\,\psi_1) = \psi_1\,\nabla^2\phi_0 + 2(\nabla\phi_0)\cdot(\nabla\psi_1) + \phi_0\,\nabla^2\psi_1 \tag{12A.22}$$

where $\phi_0 = \exp[j\psi_0(\bar{r})]$. Since $\nabla^2\phi_0 = -k_b^2\,\phi_0$ and $\nabla\phi_0 = j(\nabla\psi_0)\phi_0$, we have

$$\nabla^2(\phi_0\,\psi_1) = -k_b^2\,\psi_1\,\phi_0 + 2j\,\phi_0(\nabla\psi_0)\cdot(\nabla\psi_1) + \phi_0\,\nabla^2\psi_1 \tag{12A.23}$$

Then, on multiplying (12A.21) by $j\phi_0$ and using (12A.23), we obtain

$$\nabla^2(\phi_0\,\psi_1) + k_b^2\,\phi_0\,\psi_1 = -j\phi_0(\nabla\psi_1)^2 + j\phi_0\,O(\bar{r}) \tag{12A.24}$$

Equation (12A.24) is still exact at this point. But if we assume that ψ_1 is small so that $(\nabla\psi_1)^2$ is even smaller, (12A.24) is approximated as

$$\left(\nabla^2 + k_b^2\right)\phi_0\,\psi_1 \;=\; j\phi_0\,O(\bar{r}) \tag{12A.25}$$

The solution to (12A.25) is then

$$\psi_1(\bar{r}) \;=\; -\frac{j}{\phi_0(\bar{r})}\int d\bar{r}'\,g(\bar{r},\bar{r}')\,\phi_0(\bar{r}')\,O(\bar{r}') \tag{12A.26}$$

The above is known as the Rytov approximation. The total solution is

$$\phi(\bar{r}) \;\cong\; \phi_0(\bar{r})\exp\left[j\,\psi_1(\bar{r})\right] \tag{12A.27}$$

The Rytov approximation is valid when the first term on the right of (12A.24) is much smaller than the second term, or

$$\left(\nabla\psi_1\right)^2 \;<<\; O(\bar{r}) \tag{12A.28}$$

Moreover, this approximation attempts to correct for the phase of the wave as it propagates through the inhomogeneous media. Hence, it shares some similarity with the WKB approximation. (The regime of validity of the Rytov approximation has also been discussed by Keller [86] and Crane [87].)

Applying dimensional analysis, it is shown from (12A.26) that

$$\psi_1(\bar{r}) \;\sim\; k_b^2\,L^2\,\Delta\varepsilon_r \quad \text{when} \quad k_b\,L \to 0 \tag{12A.29}$$

Further, assuming that $\nabla \sim 1/L$ when $k_b L \to 0$, then the use of (12A.29) in (12A.28) yields the condition that

$$\left(k_b\,L\right)^2\Delta\varepsilon_r \;<<\; 1 \tag{12A.30}$$

which is the same as (12A.12). Again, this is dependent on dimensions.

On the other hand, when the frequency tends to infinity, the field inside the inhomogeneity is of the form $\exp(j\,\vec{k}\cdot\bar{r})$. Then,

$$\begin{aligned}\phi(\bar{r}) &\sim \exp\left(j\,\vec{k}\cdot\bar{r}\right) \sim \exp\left(j\,\vec{k}_b\cdot\bar{r}\right)\exp\left[j\,(\vec{k}-\vec{k}_b)\cdot\bar{r}\right] \\ &\sim \phi_0\exp\left[j\psi_1(\bar{r})\right]\end{aligned} \tag{12A.31}$$

Therefore, $\psi_1(\bar{r}) \cong (\vec{k} - \vec{k}_b) \cdot \bar{r}$ and

$$\psi_1(\bar{r}) \sim k_b L \Delta\varepsilon_r , \qquad k_b L \rightarrow \infty \tag{12A.32}$$

Moreover, assuming that $\nabla \sim 1/L$ when $k_b L \rightarrow \infty$, the use of (12A.32) in (12A.28) yields the constraint

$$\Delta\varepsilon_r << 1 \tag{12A.33}$$

which is more relaxed than (12A.14). (Unlike (12A.8), $\nabla \sim 1/L$ here instead of k_b because ∇ operates on ψ_1, which is the phase variation.)

In the Rytov approximation, the correction, $\psi_1(\bar{r})$, occurs as a phase term. However, the magnitude of the correction to $\phi_0(\bar{r})$ is always unity even when $\psi_1(\bar{r})$ is bad. Hence, the Rytov approximation breaks down more gracefully than the Born approximation. Further, the form given by (12A.27) or (12A.31) is more suitable for the field inside the inhomogeneity. Outside the scatterer, the constraint is again given by (12A.14) for high frequencies.

Note that the Born approximation for the scalar wave equation is of the form

$$\phi_1(\bar{r}) = -\int d^3\bar{r}' \, g(\bar{r}, \bar{r}') \, \phi_0(\bar{r}') \, O(\bar{r}') \tag{12A.34}$$

where $\phi_1(\bar{r})$ is the scattered field and $\phi_0(\bar{r})$ is the incident field. However, when $\psi_1(\bar{r})$ is very small, we rewrite (12A.27) as

$$\phi(\bar{r}) \cong \phi_0(\bar{r}) + j\,\psi_1(\bar{r})\,\phi_0(\bar{r}) \tag{12A.35}$$

Therefore, $\phi_1(\bar{r}) \sim j\,\psi_1(\bar{r})\,\phi_0(\bar{r})$ if $\psi_1(\bar{r})$ is very small. Then, on multiplying (12A.26) by $j\phi_0(\bar{r})$, we recover (12A.34). Hence, the Born and Rytov approximations reduce to the same approximation when the scattered field is very weak.

Both the Born and the Rytov approximations assume that the scattered field is linearly proportional to the inhomogeneity, $O(\bar{r})$. As such, this linearized approximation makes them particularly suitable for solving inverse-scattering problems where the objects to be reconstructed are weak scatterers.

List of Acronyms

ABC	absorbing boundary condition
ADE	auxiliary differential equation
ADI	alternating direction implicit
BIM	Born iterative method
BOR	body of revolution
CAD	computer-aided design
CAT	computer-aided tomography
CEM	computational electromagnetics
CESS	combined electromagnetic and solid-state
CFD	computational fluid dynamics
CFS	complex-frequency-shifted
CP-FDTD	contour-path FDTD
CPW	coplanar waveguide
DBIM	distorted Born iterative method
DFT	discrete Fourier transform
DN-FDTD	divergenceless nonorthogonal FDTD
DSI	discrete surface integral
DT	diffraction tomography
EMCC	Electromagnetic Code Consortium
EMP	electromagnetic pulse
ENO	essentially non-oscillatory
FD-NFFF	frequency-domain near-field to far-field
FDTD	finite-difference time-domain
FET	field-effect transistor
FETD	finite-element time-domain
FFT	fast Fourier transform

FSS	frequency-selective surface
FVTD	finite-volume time-domain
GN-FDTD	globally nonorthogonal FDTD
GPR	ground-penetrating radar
GY	generalized Yee
HBM	harmonic-balance method
H-H	horizontal-horizontal
HTSA	hybrid thin-slot algorithm
IF	intermediate frequency
JJ	Josephson junction
LN-FDTD	locally nonorthogonal FDTD
LSF	local shape function
MCM	multichip module
MESFET	metal-semiconductor field-effect transistor
MMP	multiple multipole
MoM	method of moments
MRI	magnetic resonance imaging
MRTD	multiresolution time-domain
N-FDTD	nonorthogonal FDTD
NFFF	near-field to far-field
NMR	nuclear magnetic resonance
PBG	photonic bandgap
PCB	printed circuit board
PDE	partial differential equation
PEC	perfect electric conductor
PGS	periodic gain structure
PGY	planar generalized Yee
PIFA	planar inverted-*F* antenna
PMC	perfect magnetic conductor
PML	perfectly matched layer
PSTD	pseudospectral time-domain
RAM	random-access memory
RC	recursive convolution
RCS	radar cross section
RLC	resistance, inductance, capacitance
SAR	specific absorption rate
SFR	step-frequency radar
SIBC	surface-impedance boundary condition
SIR	subsurface impulse radar
TD-NFFF	time-domain near-field to far-field
TE	transverse electric
TEM	transverse electromagnetic
TLM	transmission-line matrix

TM	transverse magnetic
TSF	thin-slot formulation
UGS	uniform-gain structure
UPML	uniaxial perfectly matched layer
UTD	uniform theory of diffraction
VCSEL	vertical-cavity surface-emitting laser
VLSI	very-large-scale integration
V-V	vertical-vertical
WDM	wavelength-division multiplexing
WGM	whispering-gallery mode
WWW	World Wide Web

About the Chapter Authors

Allen Taflove (Editor) is a professor in the Department of Electrical and Computer Engineering, McCormick School of Engineering, Northwestern University, Evanston, Illinois. Since 1972, he has pioneered basic theoretical approaches and engineering applications of finite-difference time-domain computational electromagnetics; in fact, he coined the FDTD acronym in a 1980 IEEE paper. In 1990, he was named an IEEE Fellow for this work. Since 1990, he has given scores of presentations on FDTD methods and applications in the United States and abroad, helping to raise awareness of this powerful technique within the engineering community. In 1990 and 1991, he was a Distinguished National Lecturer for the IEEE Antennas and Propagation Society, and in 1992 he was Chairman of the Technical Program of the IEEE Antennas and Propagation Society International Symposium in Chicago.

Prof. Taflove originated several innovative programs in the McCormick School, including the Honors Program in Undergraduate Research and the Undergraduate Design Competition, and was named McCormick Faculty Adviser of the Year in 1991. He served as the principal investigator on more than three-dozen grants and contracts and as the thesis adviser of 11 Ph.D. recipients. He holds 11 U.S. patents. His current research interests include the electromagnetic design of nanofabricated, VLSI-density optical and optoelectronic microcircuits; the design of pulsed, confocal microwave detection and imaging systems for early-stage breast cancer; and the development of numerical electromagnetic wave diffraction coefficients for building structures in wireless microcells.

Prof. Taflove is a member of IEEE COMAR; Eta Kappa Nu; Tau Beta Pi; Sigma Xi; URSI Commissions B, D, and K; the Electromagnetics Academy; AAAS; and New York Academy of Sciences. His biographical listings include *Who's Who in Engineering*, *Who's Who in America*, *Who's Who in Science and Engineering*, and *Who's Who in American Education*. He can be contacted via the Internet at `taflove@ece.nwu.edu`.

Kurt L. Shlager (Chapter 1) is with Lockheed-Martin Missiles and Space Company, Sunnyvale, California, where he is presently a specialist involved in antenna analysis, design, and development. He received the B.S. in electrical engineering from Southeastern Massachusetts University (now University of Massachusetts at Dartmouth) in 1985; the M.S. in electrical engineering from Tufts University, Medford, Massachusetts, in 1987; and the Ph.D. in electrical engineering from the Georgia Institute of Technology, Atlanta, in 1995. From 1985 to 1986, he was with the antenna-development group of GTE Government Systems, Westborough, Massachusetts. He was with Lockheed-Martin Missiles and Space Company from 1987 to 1990, doing research and development in electromagnetic scattering. After completing his Ph.D. studies in 1995, he joined the antenna-systems laboratory of TRW Space and Electronics Group, Redondo Beach, California. He returned to Lockheed-Martin in 1996. Dr. Shlager's current research interests are computational electromagnetics, including the finite-difference time-domain method; electromagnetic scattering; and antenna analysis and design.

John B. Schneider (Chapter 1) is an assistant professor in the School of Electrical Engineering and Computer Science at Washington State University. He received the B.S. degree in electrical engineering from Tulane University and both the M.S. and Ph.D. degrees in electrical engineering from the University of Washington. His research interests include the use of computational methods to analyze acoustic, elastic, and electromagnetic wave propagation. Prof. Schneider received an Office of Naval Research Young Investigator award in 1996.

Eli Turkel (Chapter 2) is a professor in the Department of Mathematics at Tel Aviv University, Tel Aviv, Israel. He received the B.A. degree from Yeshiva University in 1965 and the Ph.D. from the Courant Institute, New York University, in 1970. He has worked in many fields including ballistics, plasma physics, CFD, seismology, acoustics, and CEM, and he is an associate fellow of the AIAA. He has consulted at the ICASE research center in NASA Langley for over 23 years, and he regularly visits the ICOMP center in NASA Lewis and the RIACS center in NASA Ames. In 1992, he received a group award from NASA as part of the ICASE numerical analysis group. His research in CEM includes high-order methods and the treatment of boundaries both in the far field and at solid bodies and interfaces. In 1980, he and Alvin Bayliss developed a series of absorbing boundary conditions based on asymptotic expansions for far-field artificial surfaces.

Linda P. B. Katehi (Chapter 3) is a professor in the Department of Electrical Engineering and Computer Science at the University of Michigan, Ann Arbor. She received the B.S.E.E. degree from the National Technical University of Athens, Greece, in 1977, and the M.S.E.E. and Ph.D. degrees from the University of California, Los Angeles, in 1981 and 1984, respectively. In September 1984, she joined the faculty of the EECS Department of the University of Michigan. Since then, her interests have included the theoretical and experimental development and characterization of microwave and millimeter-wave printed circuits; the computer-aided design of VLSI interconnects; the development and characterization of micromachined circuits for millimeter-wave and submillimeter-wave applications; and the development of low-loss lines for terahertz-frequency applications. She has also been studying theoretically and experimentally uniplanar radiating structures for hybrid-monolithic oscillator and mixer designs.

Prof. Katehi was awarded two best-paper prizes of the IEEE Antennas and Propagation Society (AP-S): the R. W. P. King Prize in 1984 and the S. A. Schelkunoff Prize in 1985. She received the NSF Presidential Young Investigator Award and the URSI Young Scientist Fellowship in 1987, the Humboldt Research Award and the University of Michigan Faculty Recognition Award in 1994, and the IEEE Microwave Theory and Techniques Society (MTT-S) Microwave Prize in 1996. She is a Fellow of IEEE and a member of IEEE AP-S, MTT-S, Sigma XI, Hybrid Microelectronics, and URSI Commission D. From 1992 to 1995, she served on the IEEE AP-S ADCOM. Prof. Katehi is also an associate editor for *IEEE Transactions on Microwave Theory and Techniques*.

James F. Harvey (Chapter 3) is a program manager at the U.S. Army Research Office (ARO), Triangle Park, North Carolina. He received the B.S. degree in engineering from the U.S. Military Academy, West Point, New York, in 1964; the M.A. degree in physics from Dartmouth College, Hanover, New Hampshire, in 1972; and the Ph.D. degree in applied science from the University of California, Davis, in 1991, with research performed at Lawrence-Livermore National Laboratory.

Dr. Harvey's primary responsibility at ARO is within the fields of electromagnetics, millimeter-wave circuit integration, low-power / minimum-power system design, and the electronic aspects of demining. His personal research interests are in the fields of quasi-optics and millimeter-wave devices. He is active in IEEE MTT-S, the IEEE Lasers and Electro-Optics Society, URSI, and the American Physical Society.

Emmanouil M. Tentzeris (Chapter 3) is an assistant professor in the Department of Electrical and Computer Engineering, Georgia Institute of Technology, Atlanta. He received the Diploma Degree summa cum laude in electrical engineering from the National Technical University of Athens, Greece, in 1992, and the M.S. and Ph.D. degrees in electrical engineering from the University of Michigan, Ann Arbor, in 1993 and 1998, respectively. He has been the author or co-author of more than 25 papers in refereed journals and conference proceedings, and was awarded the Best Paper Prize from the International Microelectronics and Packaging Society in 1997. His research interests include the development of novel numerical techniques and the application of multiresolution analysis to the simulation of microwave circuits used in wireless or satellite communication systems.

Stephen D. Gedney (Chapters 4 and 5) is an associate professor in the Department of Electrical Engineering at the University of Kentucky, Lexington. He received the B.Eng.-Honors degree from McGill University, Montreal, in 1985, and the M.S. and Ph.D. degrees in electrical engineering from the University of Illinois, Urbana-Champaign, in 1987 and 1991, respectively. While on the faculty at the University of Kentucky, he was named a NASA / ASEE Faculty Fellow at Jet Propulsion Laboratory, Pasadena, California. There, he developed techniques for full-wave analysis of printed microwave circuits and antennas. In 1996, he was a visiting professor at Hughes Research Laboratories in Malibu, California, where he engaged in microwave circuit analysis and design and developed higher-order Nystrom solution schemes for electromagnetic scattering simulations. Also in 1996, he received the NSF CAREER Award. His current research is in the field of computational electromagnetics. Specifically, his emphasis is on the development of sequential and parallel algorithms for the rigorous analysis of microwave circuits and antennas, microwave and electronic packaging, and electromagnetic scattering.

J. Alan Roden (Chapter 4) is with the Signature Technology Laboratory at the Georgia Tech Research Institute, Atlanta. He received the B.S. degree from the University of Tennessee at Chattanooga in 1984; the M.S. degree in electrical engineering from North Carolina State University, Raleigh in 1988; and the Ph.D. degree in electrical engineering from the University of Kentucky, Lexington, in 1997. His previous industrial experience includes a position as a product-development engineer with IBM, Research Triangle Park, North Carolina, where he was involved with high-speed communications adapter development. Currently, his primary research interest is the development of time-dependent solutions of Maxwell's equations for the analysis of large-scale electromagnetic compatibility and radiation problems.

Niel Madsen (Chapter 4) heads a computational electromagnetics group at the Lawrence-Livermore National Laboratory, Livermore, California. He received the Ph.D. degree in applied mathematics from the University of Maryland in 1969. His previous professional employment includes positions at the Los Alamos National Laboratory and the Bettis Atomic Power Laboratory. Currently, his research interests include the numerical solution of partial differential equations, numerical algorithms and parallel computing, computational electromagnetics, and radiation transport.

Alireza H. Mohammadian (Chapter 4) is a senior scientist at Rockwell Science Center, Thousand Oaks, California. He received the M.S. degree in electrical engineering from Michigan Technological University in 1976 and the Ph.D. degree, specializing in electromagnetics and antennas, in 1980 from the University of Michigan, Ann Arbor. His professional experience includes a postdoctoral research associate position at the University of Adelaide, South Australia, and a staff scientist position with Antech Antenna Technologies in Montreal, Canada. From 1984 to 1988, he was an assistant professor of electrical and computer engineering at the University of Michigan,

Dearborn. His technical interests are in applied and computational electromagnetics, antennas, electromagnetic compatibility, RF and microwaves, and electronic packaging. He has over 50 refereed journal and conference publications. He is a senior member of IEEE; a member of the IEEE AP-S, MTT-S, and Electromagnetic Compatibility (EMC) societies; on the editorial board of *IEEE Transactions on Microwave Theory and Techniques*; and a member of Tau Beta Pi and Sigma Xi.

William F. Hall (Chapter 4) is a principal scientist at Rockwell Science Center, Thousand Oaks, California. He received the Ph.D. degree in physics at the University of California in 1964. He has made significant scientific contributions in charged-particle scattering in crystals, properties of magnetic systems, viscoelastic effects in polymer solutions, and characteristics of compound semiconductor interfaces. In addition to his research in these areas, he has collaborated in the investigation of a number of diverse topics, including high-temperature superconductivity, distributed-feedback lasers, magnetic suspension viscosity, dielectric properties of condensed matter, and various applications of electromagnetic theory. Dr. Hall has over 60 publications.

Vijaya Shankar (Chapter 4) is Director of the Computational Sciences Function at Rockwell Science Center, Thousand Oaks, California, and adjunct professor at the University of California, Davis. He conducts and coordinates research in nonlinear mechanics involving a wide variety of problems in mathematical physics. He received the M.S. and Ph.D. degrees in aerospace engineering from Iowa State University in 1977, and he joined Rockwell in 1976.

Dr. Shankar has received numerous recognitions, including the 1984 Rockwell Engineer-of-the-Year Award, the 1985 Lawrence Sperry Award, the 1986 NASA Public Service Medal for exceptional scientific achievement, the 1990 CRAY Gigaflop Performance Award, the 1991 AIAA Dryden Research Lectureship, and the 1992 Distinguished Engineering Achievements Award from the Institute for the Advancement of Engineering (IAE). From Iowa State University, he received the 1985 Outstanding Young Alumnus Award and the 1991 Professional Achievement Citation in Engineering. He is a Fellow of AIAA and IAE, associate editor of *AIAA Journal*, and a member of the editorial board of *Journal on Computing Systems in Engineering*. He has authored or co-authored over 50 technical publications.

Christopher M. Rowell (Chapter 4) is a member of the technical staff at Rockwell Science Center, Thousand Oaks, California. He received the M.S. degree in aerospace engineering from Georgia Institute of Technology in 1990. His work at the Science Center has involved parallelizing structured and unstructured CEM codes. His experience with parallel computing platforms involves the IBM SP2, Intel Paragon, and nCUBE2 architectures, as well as distributed-workstation clusters. He has helped in developing in-house structured and unstructured grid-generation tools, and has generated grids for large complex geometries, including full aircraft and warship configurations up to 1 GHz. He has also written object-oriented graphical user interfaces for CEM solvers. He has authored and co-authored a number of technical papers in these areas.

James G. Maloney (Chapters 6 and 7) is a senior research engineer in the Signature Technology Laboratory of Georgia Tech Research Institute, Atlanta. He received the B.E.E. and Ph.D. degrees from Georgia Institute of Technology, Atlanta, in 1987 and 1992, respectively. There, he was a graduate research and teaching assistant, and from 1989 to 1992 he was a JSEP Fellow. In 1994, Dr. Maloney was awarded the R. W. P. King Best Paper Prize by IEEE AP-S, and was named honorable mention for the S. A. Schelkunoff Best Paper Prize. He has authored more than 25 journal publications and conference papers.

His current research interests are in the application of cutting-edge computational techniques to assist in understanding electromagnetic phenomenology. Specifically, his interest areas include ultrawideband, transient, conformal, and low-signature antenna and radome systems; advanced electromagnetic materials realized using photonic bandgap structures and composite random materials; and adaptive processing and near-field scanning. He was recently awarded a patent entitled "Electromagnetic Antenna and Transmission Line Utilizing Photonic Bandgap Material."

Morris P. Kesler (Chapter 6) is a senior research engineer at the Georgia Tech Research Institute, Atlanta, and an adjunct professor in the School of Electrical and Computer Engineering at Georgia Institute of Technology, Atlanta. He received the B.S. and M.S. degrees in 1984 and the Ph.D. degree in 1988, all in electrical engineering from Massachusetts Institute of Technology. His dissertation involved the development of a near-infrared femtosecond laser system and the study of subpicosecond dynamics in GaAs semiconductor lasers. From 1988 to 1990, he was a postdoctoral fellow with IBM Research in Yorktown Heights, New York and with the IBM Zurich Research Laboratory where he investigated properties of quantum well lasers. In 1990, he joined the Signature Technology Laboratory of the Georgia Tech Research Institute. Since that time, he has been involved in a variety of research efforts including near-field scattering, photoconductive switching, photonic bandgap structures, and FDTD numerical techniques. He was recently awarded a patent entitled "Electromagnetic Antenna and Transmission Line Utilizing Photonic Bandgap Material." His current research interests include electromagnetic materials and scattering, electromagnetic modeling and computation, and adaptive-processing techniques.

Glenn S. Smith (Chapter 7) is Regents' Professor of Electrical and Computer Engineering at the Georgia Institute of Technology, Atlanta. He received the B.S.E.E. degree from Tufts University, Medford, Massachusetts, in 1967, and the S.M. and Ph.D. degrees in applied physics from Harvard University, Cambridge, Massachusetts, in 1968 and 1972, respectively. From 1972 to 1975, he served as a postdoctoral fellow at Harvard University and also as a part-time research associate and instructor at Northeastern University, Boston. In 1975, he joined the faculty of Georgia Tech.

Prof. Smith is author of the book, *An Introduction to Classical Electromagnetic Radiation* (Cambridge, 1997) and co-author of the book, *Antennas in Matter: Fundamentals, Theory and Applications* (MIT Press, 1981). He also authored the chapter "Loop Antennas" in the McGraw-Hill *Antenna Engineering Handbook* (1993). He is a member of Tau Beta Pi, Eta Kappa Nu, and Sigma Xi; a Fellow of IEEE; and a member of URSI Commissions A and B. His technical interests include electromagnetic theory and measurements, antennas and wave propagation in materials, and the radiation and reception of pulses by antennas.

Bijan Houshmand (Chapter 8) joined the Jet Propulsion Laboratory of the California Institute of Technology in 1994. He was a visiting assistant professor at the University of Illinois, Urbana-Champaign, and the University of California, Los Angeles, from 1990 to 1994. He has been involved in development of FDTD algorithms for application to microwave circuits that include nonlinear and active elements. His research interests also include interferometric and polarimetric measurements and modeling for remote sensing applications. Dr. Houshmand is a member of URSI Commission D.

Tatsuo Itoh (Chapter 8) is a professor of electrical engineering and the holder of the TRW Endowed Chair in microwave and millimeter-wave electronics at UCLA. He is currently the Director of UCLA's Joint Services Electronics Program (JSEP) and Multidisciplinary University Research Initiative (MURI). He received the B.S. and M.S. degrees from Yokohama National University, Japan, in 1964 and 1968, respectively, and the Ph.D. degree in electrical engineering from the University of Illinois, Urbana-Champaign, in 1969. His past professional and academic appointments include positions with the University of Illinois; SRI International, Menlo Park, California; the University of Kentucky, Lexington; and the University of Texas, Austin, where in 1983, he was selected to hold the Hayden Head Centennial Professorship of Engineering.

Prof. Itoh is a Fellow of IEEE, a member of the Institute of Electronics and Communication Engineers of Japan, and a member of Commissions B and D of URSI. He served as Editor of *IEEE Transactions on Microwave Theory and Techniques* from 1983 through 1985, and Editor of *IEEE Microwave and Guided Wave Letters* from 1991 through 1994. He also was Vice President of IEEE MTT-S in 1989, becoming President in 1990. He was Chairman of the U.S. National Committee of URSI, Commission D, from 1988 to 1990; Vice Chairman of Commission D of the International URSI from 1991 to 1993; and Chairman of the same commission from 1993 to 1996. Currently, he is on the Long Range Planning Committee of URSI. Prof. Itoh has served on a number of scientific advisory boards and committees around the world, and as a visiting professor at a number of international university locations.

Melinda J. Piket-May (Chapter 8) is an assistant professor in the Department of Electrical and Computer Engineering, University of Colorado, Boulder. She received the B.S. degree in electrical engineering in 1988 from the University of Illinois, Urbana-Champaign, and the M.S. and Ph.D. degrees in electrical engineering from Northwestern

University, Evanston, Illinois, in 1990 and 1993, respectively. During this time period, she interned for several academic quarters at Fermi National Accelerator Laboratory, Naval Research Laboratory, and Cray Research, Inc. At the University of Colorado, she has an active industrial-based research program in computational electromagnetics with applications in high-speed analog and digital design, EMC/EMI, solar-cell design, and wireless communications. Her research includes the development of general methods to extract information from the numerical simulations, as well as means to reduce phase error via high-order FDTD schemes. She is also very active in engineering education, with a focus on moving toward an interactive environment where the student is in charge of his or her learning. She works on undergraduate engineering design issues and incorporating research into the classroom in an interactive and meaningful way.

Prof. Piket-May received a 1996 URSI Young Scientist Award, and was named a Sloan New Faculty Fellow in 1997. Also in 1997, she was the recipient of the NSF CAREER award for her research and teaching activities. She is currently mentoring an NSF postdoctoral fellow in engineering education. Recently, she was elected by the membership of IEEE AP-S to serve on the Society's ADCOM.

Samir M. El-Ghazaly (Chapter 9) is an associate professor in the Department of Electrical Engineering at Arizona State University, Tempe. He received the Ph.D. degree in electrical engineering from the University of Texas, Austin, in 1988. His past appointments include positions at Cairo University, Egypt; Université de Lille, France; University of Ottawa, Canada; University of Texas, Austin; Jet Propulsion Lab, Pasadena, California; and CST-Motorola. His research interests include microwave and millimeter-wave semiconductor devices and passive circuits, semiconductor device simulations, ultrashort pulse propagation, and wave-device interactions. He is a senior member of IEEE and a member of URSI Commissions A and D, Tau Beta Pi, Sigma Xi, and Eta Kappa Nu. Currently, he is the secretary of the U.S. National Committee of URSI, Commission A. He has been a member of the Technical Program Committee of the IEEE International Microwave Symposium since 1991, and he is a member of the editorial board of *IEEE Transactions on Microwave Theory and Techniques*. He was the Chairman of the IEEE-Waves and Devices Group, Phoenix Section, and has been Chapter Funding Coordinator for IEEE MTT-S since 1992. Currently, he is the Chairman of the Chapter Activities Committee of IEEE MTT-S, as well as General Chairman of the upcoming 2001 IEEE MTT-S International Microwave Symposium.

Susan C. Hagness (Chapter 10) is an assistant professor in the Department of Electrical and Computer Engineering, University of Wisconsin, Madison. Her undergraduate and graduate education was in electrical engineering at Northwestern University, Evanston, Illinois, where she received the B.S. degree with highest distinction in 1993 and the Ph.D. degree in 1998. In 1992, she was the recipient of the Barry M. Goldwater Foundation Scholarship, and the Eta Kappa Nu Norman R. Carson Award as the Outstanding Junior Electrical Engineering Student in the United States.

In 1993, she was selected for the AT&T Bell Labs Graduate Fellowship for Women in Science and Engineering, and the National Science Foundation Graduate Fellowship. In 1997, she was awarded the Northwestern University Cabell Dissertation-Year Fellowship. Her current research interests are in optical and biomedical applications of computational electrodynamics. She is continuing her work on FDTD algorithms for modeling optical gain media, microcavity resonators, and photonic crystals suitable for high-density photonic integrated circuits in communications and computing applications. In addition, she is actively involved in developing technology for detecting and imaging early-stage breast cancers using pulsed, focused, nonionizing electromagnetic energy. Over the course of her studies, she has authored or co-authored eight refereed journal papers, two book chapters, one book section, and 23 refereed conference papers. Prof. Hagness is a member of Tau Beta Pi and Eta Kappa Nu.

Om P. Gandhi (Chapter 11) is a professor and Chairman of the Department of Electrical Engineering at the University of Utah, Salt Lake City. He is the author or co-author of several book chapters and many journal articles on electromagnetic dosimetry, microwave tubes, and solid-state devices. He also edited the book, *Biological Effects and Medical Applications of Electromagnetic Energy* (Prentice-Hall, 1990), and co-edited the book, *Electromagnetic Biointeraction* (Plenum Press, 1989). Prof. Gandhi was elected a Fellow of IEEE in 1979, and received the Distinguished Research Award from the University of Utah for the 1979–1980 academic year. He was President of the Bioelectromagnetics Society from 1992 to 1993, Co-Chairman of IEEE SCC 28.IV Subcommittee on the RF Safety Standards from 1988 to 1997, and Chairman of the IEEE Committee on Man and Radiation (COMAR). In 1995, he received the d'Arsonval Medal of the Bioelectromagnetics Society for pioneering contributions to the field of bioelectromagnetics.

Weng Cho Chew (Chapter 12) is a professor in the Department of Electrical and Computer Engineering at the University of Illinois, Urbana-Champaign, where he is also Director of both the Center for Computational Electromagnetics and the Electromagnetics Laboratory. He received the B.S. degree in 1976, the M.S. and Engineer's degrees in 1978, and the Ph.D. degree in 1980, all in electrical engineering from the Massachusetts Institute of Technology, Cambridge. His research interests are in the areas of waves in inhomogeneous media for various sensing applications, integrated circuits, microstrip antenna applications, and fast algorithms for solving wave scattering and radiation problems. He is the originator of several fast algorithms for solving electromagnetics scattering and inverse problems. Before coming to the University of Illinois, he was a department manager and a program leader at Schlumberger-Doll Research. He is an IEEE Fellow, and was an NSF Presidential Young Investigator. He has been active with various journals and societies. He authored the book, *Waves and Fields in Inhomogeneous Media*, and authored or co-authored over 190 journal publications and over 220 conference publications.

Subject Index

The Artech House Antenna Library

Helmut E. Schrank, *Series Editor*

Advanced Technology in Satellite Communication Antennas: Electrical and Mechanical Design, Takashi Kitsuregawa

Advances in Computational Electrodynamics: The Finite-Difference Time-Domain Method, Allen Taflove, *editor*

Analysis Methods for Electromagnetic Wave Problems, Volume 2, Eikichi Yamashita, *editor*

Analysis of Wire Antennas and Scatterers: Software and User's Manual, A. R. Djordjević, M. B. Bazdar, G. M. Bazdar, G. M. Vitosevic, T. K. Sarkar, and R. F. Harrington

Antenna-Based Signal Processing Techniques for Radar Systems, Alfonso Farina

Antenna Engineering Using Physical Optics: Practical CAD Techniques and Software, Leo Diaz and Thomas Milligan

Antenna Design With Fiber Optics, A. Kumar

Broadband Patch Antennas, Jean-François Zürcher and Fred E. Gardiol

CAD for Linear and Planar Antenna Arrays of Various Radiating Elements: Software and User's Manual, Miodrag Mikavica and Aleksandar Nešić

CAD of Aperture-fed Microstrip Transmission Lines and Antennas: Software and User's Manual, Naftali Herscovici

CAD of Microstrip Antennas for Wireless Applications, Robert A. Sainati

The CG-FFT Method: Application of Signal Processing Techniques to Electromagnetics, Manuel F. Cátedra, Rafael P. Torres, José Basterrechea, and Emilio Gago

Electromagnetic Waves in Chiral and Bi-Isotropic Media, I.V. Lindell, S.A. Tretyakov, A.H. Sihvola, A. J. Viitanen

Fixed and Mobile Terminal Antennas, A. Kumar

Handbook of Antennas for EMC, Thereza Macnamara

Four-Armed Spiral Antennas, Robert G. Corzine and Joseph A. Mosko

Introduction to Electromagnetic Wave Propagation, Paul Rohan

Mobile Antenna Systems, K. Fujimoto and J. R. James

Modern Methods of Reflector Antenna Analysis and Design, Craig Scott

Moment Methods in Antennas and Scattering, Robert C. Hansen, *editor*

Monopole Elements on Circular Ground Planes, M. M. Weiner et al.

Near-Field Antenna Measurements, D. Slater

Passive Optical Components for Optical Fiber Transmission, Norio Kashima

Phased Array Antenna Handbook, Robert J. Mailloux

Polariztion in Electromagnetic Systems, Warren Stutzman

Practical Simulation of Radar Antennas and Radomes, Herbert L. Hirsch and Douglas C. Grove

Quick Finite Elements for Elctromagnetic Waves, Giuseppe Pelosi, Roberto Coccioli, and Stefano Selleri

Radiowave Propagation and Antennas for Personal Communications, Second Edition, Kazimierz Siwiak

Small-Aperture Radio Direction-Finding, Herndon Jenkins

Solid Dielectric Horn Antennas, Carlos Salema, Carlos Fernandes, and Rama Kant Jha

Spectral Domain Method in Electromagnetics, Craig Scott

Understanding Electromagnetic Scattering Using the Moment Method: A Practical Approach, Randy Bancroft

Waveguide Components for Antenna Feed Systems: Theory and CAD, J. Uher, J. Bornemann, and Uwe Rosenberg

For further information on these and other Artech House titles, including previously considered out-of-print books now available through our In-Print-Forever™ (IPF™) program, contact:

Artech House
685 Canton Street
Norwood, MA 02062
781-769-9750
Fax: 781-769-6334
Telex: 951-659
e-mail: artech@artech-house.com

Artech House
Portland House - Stag Place
London SW1E 5XA England
+44 (0) 171-973-8077
Fax: +44 (0) 171-630-0166
Telex: 951-659
e-mail: artech-uk@artech-house.com

Find us on the World Wide Web at:
www.artech-house.com